MECHANICAL BEHAVIOR OF MATERIALS

Engineering Methods for Deformation, Fracture, and Fatigue

Norman E. Dowling

College of Engineering
Virginia Polytechnic Institute and State University
Blacksburg, Virginia

Prentice Hall
Upper Saddle River, New Jersey 07458

Library of Congress Cataloging-in-Publication Data

Dowling, Norman E.
    Mechanical behavior of materials : engineering methods for
    deformation, fracture, and fatigue / Norman E. Dowling.
        p. cm.
        Includes bibliographical references and index.
    ISBN 0-13-579046-8
    1. Materials.   2. Materials—Testing.   I. Title.
    TA404.8.D68   1993
    620.1'1292—dc20   92-14544
                                                  CIP

Acquisitions editor: Doug Humphrey
Supplements editor: Alice Dworkin
Production and interior design: Fred Dahl
Prepress buyer: Linda Behrens
Manufacturing buyer: David Dickey

©1993 by Prentice-Hall, Inc.

Simon & Schuster Company / A Viacom Company

Upper Saddle River, New Jersey 07458

Printed in the United States of America

10 9 8 7 6

ISBN 0-13-579046-8

Prentice-Hall International (UK) Limited, *London*
Prentice-Hall of Australia Pty. Limited, *Sydney*
Prentice-Hall Canada Inc., *Toronto*
Prentice-Hall Hispanoamericana, S.A., *Mexico*
Prentice-Hall of India Private Limited, *New Delhi*
Prentice-Hall of Japan, Inc., *Tokyo*
Prentice-Hall of Southeast Asia Pte. Ltd., *Singapore*
Editora Prentice-Hall do Brasil, Ltda., *Rio de Janeiro*

Contents

7 Yielding and Fracture under Combined Stresses, *233*

8 Fracture of Cracked Members, *276*

9 Fatigue of Materials: Introduction and Stress-Based Approach, *339*

10 Stress-Based Approach to Fatigue: Notched Members, *397*

11 Fatigue Crack Growth, *456*

Preface

Designing machines, vehicles, and structures that are safe, reliable, and economical requires both efficient use of materials and assurance that structural failure will not occur. It is therefore appropriate for undergraduate engineering majors to study the mechanical behavior of materials, specifically such topics as deformation, fracture, and fatigue. This book provides a text for courses in this area at the junior or senior undergraduate level, and it will also serve as an introductory reference for practicing engineers. Use for a first-year graduate level course is also a possibility, especially if supplementary, more advanced literature is provided.

Up-to-date but still introductory coverage is given of traditional topics in mechanical behavior of materials, such as materials testing, yielding and plasticity, stress-based fatigue analysis, and creep. The relatively new methods of fracture mechanics and strain-based fatigue analysis are also considered, and are in fact treated in some detail.

Emphasis is placed on analytical and predictive methods that are useful to the engineering designer in avoiding structural failure. These methods are developed from an engineering mechanics viewpoint, and the resistance of materials to failure is quantified by properties such as yield strength, fracture toughness, and stress-life curves for fatigue or creep. The intelligent use of materials property data requires some understanding of how the data are obtained, so that their limitations and significance are clear. Thus, the materials tests used in various areas are generally discussed prior to considering the analytical and predictive methods.

In many of the areas covered, the existing technology is more highly developed for metals than for nonmetals. Nevertheless, data and examples for nonmetals, such as polymers and ceramics, are included where appropriate. Highly anisotropic materials, such as continuous fiber composites, are also considered, but only to a limited extent. Detailed treatment of these complex materials is not attempted here.

The remainder of this preface provides some comments on use of this book by anyone, whether student, instructor, or practicing engineer.

PREREQUISITES

Elementary mechanics of materials, also called strength of materials or mechanics of deformable bodies, provides an introduction to the subject of analyzing stresses and strains in engineering components, such as beams and shafts, for linear-elastic behavior. Completion of a standard (typically sophomore) course of this type is an essential prerequisite to the treatment provided here.

Many engineering curricula include an introductory (again, typically sophomore) course in materials science, including such subjects as crystalline and noncrystalline structure, dislocations and other imperfections, deformation mechanisms, processing of materials, and naming systems for materials. Prior exposure to this area of study is also recommended. However, as such a prerequisite may be missing, limited introductory coverage is given in Chapters 2 and 3. These two chapters can be used as a brief review, or skipped entirely when a course in materials science has been taken by the students.

Basic mathematics is also needed and should include elementary calculus. As discussed below, a personal computer can be used with this book. If this is done, a course in, or experience in programming with, a language such as Basic or Fortran is needed. Some limited background in numerical analysis is also helpful for computer applications.

REFERENCES AND BIBLIOGRAPHY

Each chapter contains a list of *References* near the end that identifies sources of additional reading and information. These lists are in some cases divided into categories such as general references, sources of material properties, and useful handbooks. Where these references are mentioned in the text, one or more authors' names and years of publication are given, allowing the reference to be quickly found in the list at the end of that chapter.

Where specific data or illustrations from other publications are used, these sources are identified by information in brackets, such as [Richards 61] or [ASM 88], where the two-digit numbers indicate the year of publication. All such *Bibliography* items are listed in a single section near the end of the book. The order of listing is alphabetical according to the first author's name or the organization's acronym, and then according to the year of publication. Any multiple items in a given year are distinguished by a letter, as in [Peterson 59a] and [Peterson 59b].

UNITS

The International System of Units (SI) is emphasized, but U.S. Customary Units are also included in most tables of data. On graphs, the scales are either SI or dual, except for a few cases of other units where illustrations from another publication are used in their original form. Only SI units are given in most exercises and where values are given in the text, as the use of dual units in these situations invites confusion.

The SI unit of force is the newton (N), and the U.S. unit is the pound (lb). It is often convenient to employ thousands of newtons (kilonewtons, kN) or thousands of pounds (kilopounds, kip). Stresses and pressures in SI units are thus given in newtons per square meter, N/m^2, which in the SI system is given the special name of pascal (Pa). Millions of pascals (megapascals, MPa) are generally appropriate for our use.

$$1 \text{ MPa} = 1\frac{\text{MN}}{\text{m}^2} = 1\frac{\text{N}}{\text{mm}^2}$$

where the latter equivalent form using millimeters (mm) is sometimes convenient. In U.S. units, stresses are generally given in kilopounds per square inch (ksi).

These units and others frequently used are listed, along with conversion factors, inside the front cover. As an illustrative use of this listing, let us convert a stress of 20 ksi to MPa. Since 1 ksi is equivalent to 6.895 MPa, we have

$$20 \text{ ksi} = 20 \text{ ksi} \left(6.895 \frac{\text{MPa}}{\text{ksi}} \right) = 137.9 \text{ MPa}$$

Conversion in the opposite direction involves dividing by the equivalence value.

$$137.9 \text{ MPa} = \frac{137.9 \text{ MPa}}{\left(6.895 \frac{\text{MPa}}{\text{ksi}} \right)} = 20 \text{ ksi}$$

It is also useful to note that strains are dimensionless quantities, so that no units are necessary. Strains are most commonly given as straightforward ratios of length change to length, but percentages are sometimes used, $\varepsilon_\% = 100\varepsilon$.

MATHEMATICAL CONVENTIONS

Standard practice is followed in most cases. The function *log* is understood to indicate logarithms to the base ten, and the function *ln* to indicate logarithms to the base $e = 2.718\ldots$, that is, natural logarithms. To indicate selection of the largest of several values, the nonstandard function MAX( ) is employed.

NOMENCLATURE

In journal articles and in other books, and in various test standards and design codes, a wide variety of different symbols are used for certain variables that are needed. This situation is handled by using a consistent set of symbols throughout, while following the most common conventions wherever possible. However, a few exceptions or modifications to common practice are necessary to avoid confusion.

For example, K is used for the stress intensity of fracture mechanics, but not for stress concentration factor, which is designated k. Also, H is used instead of K or k for the strength coefficient describing certain stress-strain curves. The symbol S is used for nominal or average stress, whereas σ is the stress at a point, and also the stress at any point in a uniformly stressed member. Dual use of symbols is avoided except where the different usages occur in separate portions of the book. A list of the more commonly used symbols is given inside the back cover. More detailed lists are given near the end of each chapter in a section on *New Terms and Symbols*.

USE AS A TEXT

The various chapters are constituted so that considerable latitude is possible in choosing topics for study. A semester-length course could include at least portions of all chapters through 11, and also portions of Chapter 15. This covers the introductory and review topics in Chapters 1 to 6, and yield and fracture criteria for uncracked material in Chapter 7. Fracture mechanics is applied to static fracture in Chapter 8, and to fatigue crack growth in Chapter 11. Also, Chapters 9 and 10 cover the stress-based approach to fatigue, and Chapter 15 covers creep. If time permits, some topics on plastic deformation could be added from Chapters 12 and 13, and from Chapter 14 on the strain-based approach to fatigue.

Particular portions of certain chapters are not strongly required as preparation for the remainder of that chapter, nor are they crucial for later chapters. Thus, although the topics involved are generally important in their own right, they may be omitted or delayed if desired without serious loss of continuity. These include the following:

Section	Topic
4.6	Anisotropic elasticity
5.5	True stress-strain for tension test
5.6 to 5.9	Compression, hardness, impact, and bending tests
7.7 and 7.8	Failure criteria for brittle materials
8.6 to 8.8	Plastic zones, toughness testing, and nonlinear fracture mechanics
11.7	Crack growth for variable amplitude loading
11.9	Plasticity and limitations of LEFM for crack growth
13.3	Residual stresses and strain for bending

After completion of Chapter 8 on fracture mechanics, it may be useful to proceed directly to Chapter 11, which extends the topic to fatigue crack growth. This can be done by passing over all of Chapters 9 and 10 except Sections 9.1 to 9.3.

Various options exist for limited but still coherent coverage of the relatively advanced topics in Chapters 12 through 15. For example, it might be useful to include some material from Chapter 14 on strain-based fatigue, but not all details. Section 14.4 on multiaxial stresses could then be omitted, as could all of Section 14.5 except 14.5.1, which limits the coverage to constant amplitude loading. If this is done, then Section 12.3 on multiaxial stresses is not needed beforehand, and the only text from Chapter 13 that is essential is the introductory Section 13.1 and Section 13.5 on notched members.

An alternative that might be especially appropriate for a graduate course is to cover Chapters 12 and 13 on plastic deformation prior to Chapters 9 and 10 on stress-based fatigue. This permits detailed discussion of some of the rough empirical estimates used in stress-based fatigue analysis, and it gives enhanced appreciation of the need for strain-based fatigue analysis, Chapter 14, which can then be covered next. In Chapter 15, Sections 15.1 to 15.4 provide a reasonable introduction to the topic of creep that does not depend heavily on any other material beyond Chapter 5. Section 15.5 and those following are relatively independent of one another, but some of these do depend on particular portions of Chapters 12 and 13.

USE WITH DIGITAL COMPUTERS

Use of a personal computer or other digital computer is not essential, but some end-of-chapter problems are included that lend themselves to solution by computer. These are identified by the words *PC Problem* in parentheses at the end of the problem statement. Some of these problems involve straightforward computations that would be tedious if not done by computer. In other cases, a particular type of numerical analysis is required.

Three classes of numerical analysis are encountered: (1) the fitting of least-squares lines to data, (2) numerical integration, and (3) solution of equations where closed-form manipulation is not possible. In the least-squares fits, only linear relationships of the variables or their logarithms are involved. The numerical integration that is needed can be done by basic methods such as Simpson's rule, or the modified version thereof described in Chapter 11. For solving equations, simple trial and error computation or Newton's method will suffice, as the equations involved are in all cases well-behaved. The methods of numerical analysis that are needed are described in any introductory text on the subject, such as Gerald and Wheatley (1989), or Nakamura (1991), both of which are listed at the end of this preface. However, all numerical analysis that is needed can be replaced by approximate graphical methods as described in the text, so that lack of background in this area does not present a difficulty.

Use of large software packages for analysis and design of mechanical systems is becoming commonplace. These include not only finite element stress analysis programs but also programs for fatigue life and fracture mechanics analysis. No attempt is made to discuss these in the text. If students have such software available to them, a constructive way to use it would be to require that some problems be solved both manually and with the software. This will promote an understanding of the principles and assumptions employed by the software.

REFERENCES

ASTM. 1989 *Standard Practice for Use of the International System of Units*, ASTM Standard No. E380, Am. Soc. for Testing and Materials, Philadelphia, Penn.

GERALD, C. F. and P. O. WHEATLEY. 1989 *Applied Numerical Analysis*, 4th ed., Addison-Wesley, Reading, Mass.

NAKAMURA, S. 1991 *Applied Numerical Methods with Software*, Prentice Hall, Englewood Cliffs, N.J.

Acknowledgments

I am indebted to numerous colleagues for aid of various kinds, such as participating in helpful discussions, contributing illustrations, reviewing portions of the manuscript, and providing comments from trial classroom use. These include: B. E. Boardman (Deere & Co.), C. Q. Bowles (Missouri University), K. H. Donaldson (MTS Systems Corp.), J. P. Gallagher (University of Dayton), G. R. Halford (NASA Lewis), B. M. Hillberry (Purdue University), K. L. Jerina (Washington University), R. W. Landgraf (Viginia Tech), A. Madeyski (Westinghouse Electric), D. L. McDowell (Georgia Tech), M. R. Mitchell (Rockwell International), H. Nisitani (Kyushu University, Japan), H. S. Pearson (Pearson Testing Labs), K. Rahka (Technical Research Centre of Finland), C. G. Rhodes (Rockwell International), B. I. Sandor (University of Wisconsin), W. N. Sharpe (Johns Hopkins University), E. E. Underwood (Georgia Tech), and J. H. Underwood (U.S. Army Armament RD & E Center). However, there are numerous others, not just named, to whom I extend my thanks as well.

It is appropriate to acknowledge my teachers and mentors who had important influences in shaping the career that is now punctuated by publication of this book. These include especially J. C. McCormac and the late R. W. Moorman at Clemson University, JoDean Morrow and the late J. O. Smith at the University of Illinois, and W. G. Clark, Jr. and E. T. Wessel at Westinghouse Research Laboratories.

Encouragement and support in several forms was provided in the Engineering Science and Mechanics Department at Virginia Tech. Department heads Daniel Frederick and then E. G. Henneke were most helpful. Photographs of test specimens and fractures were taken by R. A. Simonds. The initial form of the manuscript and numerous revisions were typed by Paula W. Lee.

The Media Services Department at Virginia Tech performed additional photography and the computer drafting of illustrations. The latter were prepared by Catherine E. Gorman and Beulah M. Prestrude. Coding of the manuscript for typesetting was done by Terry J. Abrams, of Blacksburg, VA, using LATEX.

Most importantly, I wish to thank my family for their support and patience during this work, especially my wife Nancy and children Emily and Stuart. My late parents, Virginia and Havelock Dowling, did not see the completion of this effort, but their contribution was nevertheless significant.

1

Introduction

1.1 INTRODUCTION

Designers of machines, vehicles, and structures must achieve acceptable levels of performance and economy, while at the same time striving to guarantee that the item is both safe and durable. To assure performance, safety, and durability, it is necessary to avoid excess deformation, that is, bending, twisting, or stretching, of the component parts of the engineered item. In addition, cracking in components must be avoided entirely, or strictly limited, so that it does not progress to the point of complete fracture.

The study of deformation and cracking in materials is called *mechanical behavior of materials*. Knowledge of this area provides the basis for avoiding these types of failure in engineering applications. One aspect of the subject is the physical testing of samples of materials by applying forces and deformations. Once the behavior of a given material is quantitatively known from testing, or from published test data, its chances of success in a particular engineering design can be evaluated.

The most basic concern in design to avoid structural failure is that the *stress* in a component must not exceed the *strength* of the material, where the strength is simply the stress that causes a deformation or cracking failure. Additional complexities or particular causes of failure often require further analysis, such as the following:

1. Stresses are often present that act in more than one direction, that is, the state of stress is biaxial or triaxial.
2. Real components may contain flaws or even cracks that must be specifically considered.
3. Stresses may be applied for long periods of time.
4. Stresses may be repeatedly applied and removed, or the direction of stress repeatedly reversed.

In the remainder of this introductory chapter, we will define and briefly discuss various types of material failure, and we will consider the relationships of mechanical behavior of materials to engineering design, to new technology, and to the economy.

1.2 TYPES OF MATERIAL FAILURE

A *deformation failure* is a change in shape or size of a component that is sufficient for its function to be lost or impaired. Cracking to the extent that a component is separated into two or more pieces is termed *fracture*. *Corrosion* is the loss of material due to chemical action, and *wear* is surface removal due to abrasion or sticking between solid surfaces that touch. If wear is caused by a fluid (gas or liquid), it is called *erosion*, which is especially likely if the fluid contains hard particles. Although corrosion and wear are also of great importance, this book primarily considers deformation and fracture.

The basic types of material failure that are classified as either deformation or fracture are indicated in Fig. 1.1. Since several different causes exist, it is important to correctly identify the ones that may apply to a given design, so that the appropriate analysis methods can be chosen to predict the behavior. With such a need for classification in mind, the various types of deformation and fracture are defined and briefly described below.

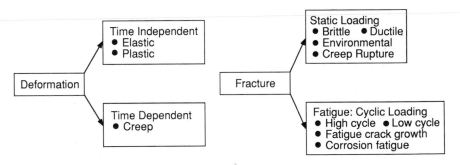

Figure 1.1 Basic types of deformation and fracture.

1.2.1 Elastic and Plastic Deformation

Deformations are quantified in terms of normal and shear strain in elementary mechanics of materials. The cumulative effect of the strains in a component is a deformation, such as a bend, twist, or stretch. Deformations are sometimes essential for function, as in a spring. Excessive deformation, especially if permanent, is often harmful.

Deformation that appears quickly upon loading can be classed as either elastic deformation or plastic deformation, as illustrated in Fig. 1.2. *Elastic deformation* is recovered immediately upon unloading. Where this is the only deformation present, stress and strain are usually proportional. For axial loading, the constant of proportionality is the *modulus of elasticity, E,* as defined in Fig. 1.2(b). An example of failure by elastic deformation is a tall building that sways in the wind and causes discomfort to the occupants, although there may be only remote chance of collapse. Elastic deformations are analyzed by the methods of elementary mechanics of materials and extensions of this general approach, as in books on theory of elasticity and structural analysis.

Plastic deformation is not recovered upon unloading and is therefore permanent. The difference between elastic and plastic deformation is illustrated in Fig. 1.2(c). Once plastic deformation begins, only a small increase in stress usually causes a relatively large additional deformation. This process of relatively easy further deformation is called *yielding,* and the value of stress where this behavior begins to be important for a given material is called the *yield strength, σ_o.*

Materials capable of sustaining large amounts of plastic deformation are said to behave in a *ductile* manner, and those that fracture without very much plastic deformation behave in a *brittle* manner. Ductile behavior occurs for many metals, such as low-strength steels, copper, and lead, and for some plastics, such as polyethylene. Brittle behavior

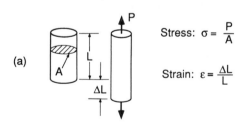

Stress: $\sigma = \dfrac{P}{A}$

Strain: $\varepsilon = \dfrac{\Delta L}{L}$

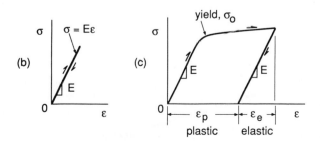

Figure 1.2   Axial member (a), subject to loading and unloading showing elastic deformation (b), and both elastic and plastic deformation (c).

occurs for glass, stone, acrylic plastic, and some metals, such as the high-strength steel used to make a file. (Note that the word *plastic* is used both as the common name for polymeric materials and in identifying plastic deformation, which can occur in any type of material.)

Tension tests are often employed to assess the strength and ductility of materials as illustrated in Fig. 1.3. Such a test is done by slowly stretching a bar of the material in tension until it breaks (fractures). The *ultimate tensile strength*, σ_u, which is the highest stress reached before fracture, is obtained along with the yield strength and the strain at fracture, ε_f. The latter is a measure of ductility and is usually expressed as a percentage, being called the *percent elongation*. Materials having high values of both σ_u and ε_f are said to be *tough*, and tough materials are generally desirable for use in design.

Large plastic deformations virtually always constitute failure. For example, collapse of a steel bridge or building during an earthquake could occur due to plastic deformation. However, plastic deformation can be relatively small but still cause malfunction of a component. For example, in a rotating shaft, a slight permanent bend results in unbalanced rotation, which in turn may cause vibration and perhaps early failure of the bearings supporting the shaft.

Buckling is deformation due to compressive stress that causes large changes in alignment of columns or plates, perhaps to the extent of folding or collapse. Either elastic or plastic deformation, or a combination of both, can dominate the behavior. Buckling is generally considered in books on elementary mechanics of materials and structural analysis.

1.2.2 Creep Deformation

Creep is deformation that accumulates with time. Depending on the magnitude of the applied stress and its time of application, the deformation may become so large that

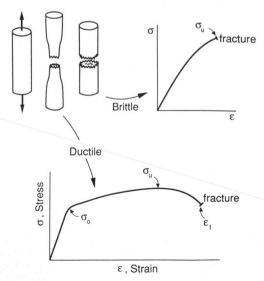

Figure 1.3   Tension test showing brittle and ductile behavior. There is little plastic deformation for brittle behavior, but a considerable amount for ductile behavior.

a component can no longer perform its function. Plastics and low-melting-temperature metals may creep at room temperature, and virtually any material will creep upon approaching its melting temperature. Creep is thus often an important problem where high temperature is encountered, as in gas-turbine aircraft engines. Buckling can occur in a time-dependent manner due to creep deformation.

An example of an application involving creep deformation is the design of tungsten lightbulb filaments. The situation is illustrated in Fig. 1.4. Sagging of the filament coil between its supports increases with time due to creep deformation caused by the weight of the filament itself. If too much deformation occurs, the adjacent turns of the coil touch one another, causing an electrical short and local overheating, which quickly leads to failure of the filament. The coil geometry and supports are therefore designed to limit the stresses caused by the weight of the filament, and a special tungsten alloy that creeps less than pure tungsten is used.

1.2.3 Fracture Under Static and Impact Loading

Rapid fracture can occur under loading that does not vary with time, or which changes only slowly, called *static loading*. If such a fracture is accompanied by little plastic deformation, it is called a *brittle fracture*. This is the normal mode of failure of glass and other materials that are resistant to plastic deformation. If the loading is applied very rapidly, called *impact loading*, brittle fracture is more likely to occur than for static loading.

If a crack or other sharp flaw is present, brittle fracture can occur even in ductile steels or aluminum alloys, or in other materials that are normally capable of deforming plastically by large amounts. Such situations are analyzed by the special technology

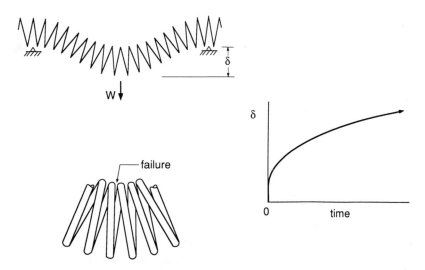

Figure 1.4   A tungsten lightbulb filament sagging under its own weight. The deflection increases with time due to creep and can lead to touching of adjacent coils, which causes bulb failure.

called *fracture mechanics*, which is the study of cracks in solids. Resistance to brittle fracture in the presence of a crack is measured by a material property called the *fracture toughness*, K_{Ic}, as illustrated in Fig. 1.5. Materials with high strength generally have low fracture toughness, and vice versa. This trend is illustrated for several classes of high-strength steel in Fig. 1.6.

Ductile fracture can also occur. This type of fracture is accompanied by significant plastic deformation and is sometimes a gradual tearing process. Fracture mechanics and brittle or ductile fracture are especially important in the design of pressure vessels and large welded structures, such as bridges and ships. Fracture may occur as a result of a combination of stress and chemical effects, and this is called *environmental cracking*. Problems of this type are a particular concern in the chemical industry, but also occur widely elsewhere. For example, some low-strength steels are susceptible to cracking in

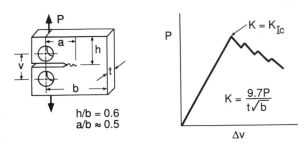

Figure 1.5   Fracture toughness test. K is a measure of the severity of the combination of crack size, geometry, and load. K_{Ic} is the particular value, called the *fracture toughness*, where the material fails.

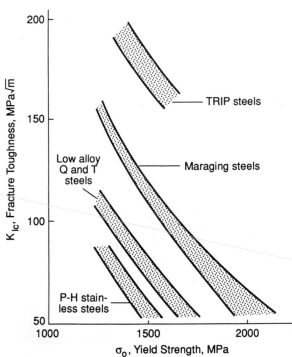

Figure 1.6   Decreased fracture toughness, as yield strength is increased by heat treatment, for various classes of high-strength steel. (Adapted from [Knott 79]; used with permission.)

caustic (basic or high pH) chemicals such as NaOH, and high-strength steels may crack in the presence of hydrogen or hydrogen sulfide gas. The term *stress-corrosion cracking* is also used to describe such behavior. This latter term is especially appropriate where material removal by corrosive action is also involved, which is not the case for all types of environmental cracking. Photographs of cracking caused by a hostile environment are shown in Fig. 1.7. Creep deformation may proceed to the point that separation into two pieces occurs. This is called *creep rupture* and is similar to ductile fracture except that the process is time dependent.

1.2.4 Fatigue Under Cyclic Loading

A common cause of fracture is *fatigue*, which is failure due to repeated loading. In general, one or more tiny cracks start in the material, and these grow until complete failure occurs. A simple example is breaking a piece of wire by bending it back and forth a number of times. Crack growth during fatigue is illustrated in Fig. 1.8, and a fatigue fracture is shown in Fig. 1.9.

Prevention of fatigue fracture is a vital aspect of design for machines, vehicles, and structures that are subjected to repeated loading or vibration. For example, trucks

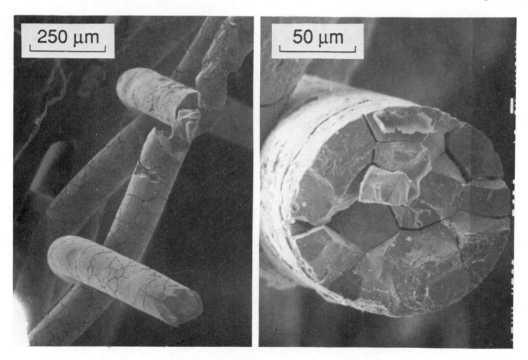

Figure 1.7 Stainless steel wires broken as a result of environmental attack. These were employed in a filter exposed at 300°C to a complex organic environment that included molten nylon. Cracking occurred along the boundaries of the crystal grains of the material. (Photos by W. G. Halley; courtesy of R. E. Swanson.)

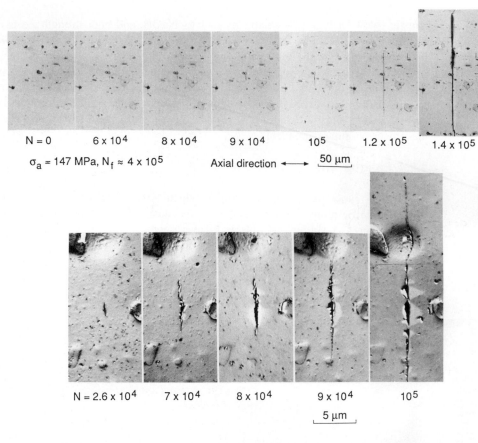

Figure 1.8 Development of a fatigue crack during rotating bending of a precipitation-hardened aluminum alloy. Photographs at various numbers of cycles are shown for a test requiring 400,000 cycles for failure. The sequence below shows more detail of the middle portion of the sequence above. (Photos courtesy of Prof. H. Nisitani, Kyushu University, Fukuoka, Japan. Published in [Nisitani 81]; reprinted with permission from *Engineering Fracture Mechanics*, Pergamon Press, Oxford, UK.)

passing over bridges cause fatigue in the bridge, and sailboat rudders and bicycle pedals can fail in fatigue. Vehicles of all types, including automobiles, tractors, helicopters, and airplanes, are subject to this problem and must be extensively analyzed and tested to avoid it. For example, some of the parts of a helicopter that require careful design to avoid fatigue problems are shown in Fig. 1.10.

If the number of repetitions (cycles) of the load is large, say millions, then the situation is termed *high-cycle fatigue*. Conversely, *low-cycle fatigue* is caused by a relatively small number of cycles, say tens, hundreds, or thousands. Low-cycle fatigue is generally accompanied by significant amounts of plastic deformation, whereas high-cycle fatigue is associated with relatively small deformations that are primarily elastic. Re-

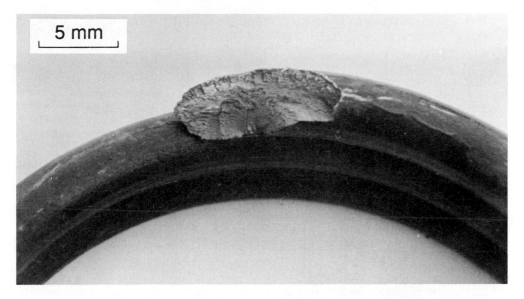

Figure 1.9   Fatigue failure of a garage door spring that occurred after 15 years of service. (Photo by R. A. Simonds; sample contributed by R. S. Alvarez, Blacksburg, Va.)

peated heating and cooling can cause a cyclic stress due to differential thermal expansion and contraction, resulting in *thermal fatigue.*

Cracks may be initially present in a component from manufacture, or they may start early in the service life. Emphasis must then be placed on the possible growth of these cracks by fatigue, as this can lead to a brittle or ductile fracture once the cracks are sufficiently large. Such situations are identified by the term *fatigue crack growth* and may also be analyzed by the previously mentioned technology of fracture mechanics. For example, analysis of fatigue crack growth is used to schedule inspection and repair of large aircraft, in which cracks are commonly present.

Such analysis is useful in preventing problems similar to the fuselage (main body) failure in 1988 of a passenger jet as shown in Fig. 1.11. The problem in this case started with fatigue cracks at rivet holes in the aluminum structure. These cracks gradually grew during use of the airplane, joining together and forming a large crack that caused a major fracture, resulting in separation of a large section of the structure. The failure could have been avoided by more frequent inspection and repair of cracks before they grew to a dangerous extent.

1.2.5 Combined Effects

Two or more of the above described types of failure may act together to cause effects greater than would be expected by their separate action, that is, there is a *synergistic effect.* Creep and fatigue may produce such an enhanced effect where there is cyclic loading at high temperature. This occurs in steam turbines in electric power plants and in gas-turbine aircraft engines.

Figure 1.10 Main mast region of a helicopter showing inboard ends of blades, their attachment, and the linkages and mechanism that controls the pitch angles of the rotating blades. The cylinder above the rotors is not ordinarily present, but is part of instrumentation used to monitor strains in the rotor blades for experimental purposes. (Photo courtesy of Bell Helicopter Textron, Inc., Ft. Worth, Tx.)

Wear due to small motions between fitted parts may combine with cyclic loading to produce surface damage followed by cracking, which is called *fretting fatigue*. This may cause failure at surprisingly low stress levels for certain combinations of materials. For example, fretting fatigue could occur where a gear is fastened on a shaft by shrink fitting or press fitting. Similarly, *corrosion fatigue* is the combination of cyclic loading and corrosion. It is often a problem in cyclically loaded components of steel that must operate in seawater, such as the structural members of offshore oil well platforms.

Material properties may degrade with time due to various environmental effects. For example, the ultraviolet content of sunlight causes some plastics to become brittle, and wood decreases in strength with time, especially if exposed to moisture. As a further example, steels become brittle if exposed to neutron radiation over long periods of time, and this affects the retirement life of nuclear reactors.

1.3 DESIGN AND MATERIALS SELECTION

Design is the process of choosing the geometric shape, materials, manufacturing method, and other details needed to completely describe a machine, vehicle, structure, or other

Figure 1.11   Fuselage failure in a passenger jet that occurred in 1988. (Photo courtesy of J. F. Wildey II, National Transportation Safety Board, Washington, DC; see [NTSB 89] for more detail.)

engineered item. This process involves a wide range of activities and objectives. It is first necessary to assure that the item is capable of performing its intended function. For example, an automobile should be capable of the necessary speeds and maneuvers while carrying up to a certain number of passengers and additional weight, and the refueling and maintenance requirements should be reasonable as to frequency and cost.

However, any engineered item must meet additional requirements: The design must be such that it is physically possible and economical to manufacture the item. Certain standards must be met as to esthetics and convenience of use. Environmental pollution needs to be minimized, and hopefully the materials and type of construction are chosen so that recycling of the materials used is later possible. Finally, the item must be safe and durable.

Safety is affected not only by design features such as seat belts in automobiles but also by avoiding structural failure. For example, excessive deformation or fracture of an automobile axle or steering component can cause a serious accident. *Durability* is the capacity of an item to survive its intended use for a suitably long period of time, so that good durability minimizes the cost of maintaining and replacing the item. For example, more durable automobiles cost less to drive than otherwise similar ones that experience more repairs and shorter life due to such gradually occurring processes as fatigue, creep, wear, and corrosion. In addition, durability is important to safety, as poor durability can lead to a structural failure or malfunction that can cause an accident.

1.3.1 Iterative and Stepwise Nature of Design

A flow chart showing some of the steps necessary to complete a mechanical design is shown in Fig. 1.12. The logic loops shown by arrows indicate that the design process is fundamentally iterative in nature. In other words, there is a strong element of trial and error where an initial design is done and then analyzed, tested, and subjected to trial production. Changes may be made at any stage of the process to satisfy requirements not

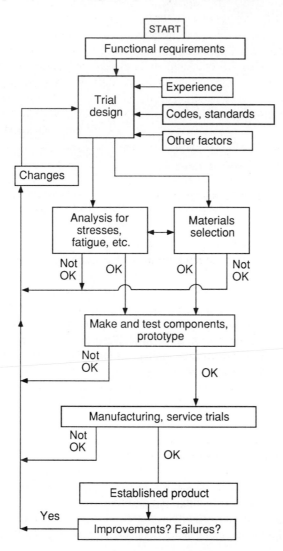

Figure 1.12 Steps in the design process related to avoiding structural failure. (Adapted from [Dowling 87b]; used with permission; © Society of Automotive Engineers.)

previously considered or problems just discovered. Changes may in turn require further analysis or testing. All of this must be done while observing constraints on time and cost.

Each step involves a *synthesis* process in which all of the various concerns and requirements are considered together. Compromises between conflicting requirements are usually necessary, and continual effort is needed to maintain simplicity, practicability, and economy. For example, the cargo weight limit of an aircraft cannot be made too large without causing unacceptable limits on the weight of fuel that can be carried, and therefore also on flight distance. Prior individual or organizational experience may have important influences on the design. Also, certain design codes and standards may be used as an aid, and sometimes they are required by law. These are generally developed and published by either professional societies or governmental units, and one of their main purposes is often to assure safety and durability. An example is the *Manual of Steel Construction* published by the American Institute of Steel Construction.

One difficult and sometimes tricky step in design is estimation of the applied loads (forces or combinations of forces). Even rough estimates are often difficult to make, especially for vibratory loads resulting from such sources as road roughness or air turbulence. It is sometimes possible to use measurements from a similar item that is already in service, but this is clearly impossible if the item being designed is unique. Once at least rough estimates (or assumptions) are made of the loads, then stresses in components can be calculated.

The initial design is often made based on avoiding stresses that exceed the yield strength of the material. Then the design is checked by more refined analysis, and changes are made as necessary to avoid more subtle modes of material failure, such as fatigue, brittle fracture, and creep. The geometric shape or size may be changed to lower the magnitude or alter the distribution of stresses and strains to avoid one of these problems, or the material may be changed to one more suitable to resist a particular failure mode.

1.3.2 Safety Factors

In making design decisions that involve safety and durability, the concept of a *safety factor* is often used. The safety factor in load is the ratio of the load necessary to cause structural failure to the expected service load.

$$X_1 = \frac{\text{failure load}}{\text{service load}} \tag{1.1}$$

For example, if $X_1 = 2$, the load necessary to cause failure is twice as large as the highest load expected in service. Safety factors provide a degree of assurance that unexpected events in service do not cause failure. They also allow some latitude for the usual lack of complete input information for the design process and for the approximations and assumptions that are often necessary. Safety factors must be larger where there are greater uncertainties or where the consequences of failure are severe. Values in the range $X_1 = 1.5$ to 3.0 are common.

Safety factors in load are sometimes supplemented by, or replaced by, safety factors in life. This safety factor is the ratio of the expected life to failure to the desired service life. Life is measured by time or by events such as the number of flights of an aircraft.

$$X_2 = \frac{\text{failure life}}{\text{desired service life}} \qquad (1.2)$$

For example, if a helicopter part is expected to fail after ten years of service, and if it is to be replaced after two years, there is a safety factor of five on life. Safety factors in life are used where deformation or cracking progresses gradually with time, as for creep or fatigue. As the life is generally quite sensitive to small changes in load, values of this factor are relatively large, typically in the range $X_2 = 5$ to 20.

1.3.3 Prototype and Component Testing

Even though mechanical behavior of materials considerations may be involved in the design process from its early stages, testing is still often necessary to verify safety and durability. This arises because of the assumptions and imperfect knowledge reflected in many engineering estimates of strength or life.

A prototype, or trial model, is often made and subjected to *simulated service testing* to demonstrate whether or not a machine or vehicle functions properly. For example, a prototype automobile is generally run on a test course that includes rough roads, bumps, quick turns, etc. Loads may be measured during simulated service testing, and these are used to improve the initial design, as the early estimates of load may have been quite uncertain. A prototype may also be subjected to simulated service testing until either a mechanical failure occurs, perhaps by fatigue, creep, wear, or corrosion, or the design is proven to be reliable. This is called *durability testing* and is commonly done for new models of automobiles, tractors, etc. A photograph of an automobile set up for such a test is shown in Fig. 1.13.

For very large items, and especially for one-of-a-kind items, it may be impractical or uneconomical to test a prototype of the entire item. Components of the item may then be tested. For example, wings, tail sections, and fuselages of large aircraft are separately tested to destruction under repeated loads that cause fatigue cracking in a manner similar to actual service. Individual joints and members of offshore oil well structures are similarly tested. Component testing may also be done as a prelude to testing of a full prototype. An example of this is the testing of a new design of an automobile axle prior to manufacture and testing of the first prototype of the entire automobile.

Various sources of loading and vibration in machines, vehicles, and structures can be simulated using digital computers, as can the resulting deformation and fracture of the material. This technology is now often used to reduce the need for prototype and component testing, thus accelerating the design process. However, computer simulations are only as good as the simplifying assumptions used in analysis, and the limitations on input data, which are always present. Thus, physical testing will always be needed at least as a final check on the design process.

Figure 1.13 Road simulation test of an automobile using loads applied at all four wheels. (Photo courtesy of MTS Systems Corp., Minneapolis, Minn.)

1.3.4 Service Experience

Design changes may also be made as a result of experience with a limited production run of a new product. Purchasers of the product may use it in a way not anticipated by the designer, resulting in failures and necessitating design changes. For example, early models of surgical implants, such as hip joints and pin supports for broken bones, have experienced failure problems that necessitated changes in both geometry and material.

The design process often continues even after a product is established and widely distributed. Long-term usage may uncover additional problems that need to be corrected in new items. If the problem is severe, perhaps safety related, changes may be needed in items already in service. Recalls of automobiles are an example of this, and a portion of these involve problems of deformation or fracture.

1.4 TECHNOLOGICAL CHALLENGE

In recent history, technology has advanced and changed at a rapid rate to meet human needs. Some of the advances from 1500 A.D. to the present are charted in the first column of Table 1.1. The second column shows the improved materials, and the third

TABLE 1.1 SOME MAJOR TECHNOLOGICAL ADVANCES FROM 1500 A.D., THE PARALLEL DEVELOPMENTS IN MATERIALS AND MATERIALS TESTING, AND FAILURES RELATED TO BEHAVIOR OF MATERIALS

Year	Technological Advance	New Materials Introduced	Materials Testing Advances	Failures
1500	Dikes Canals Pumps Telescope	(Stone, brick, wood, copper, bronze, and cast and wrought iron in use)	Tension (L. da Vinci) Tension, bending (Galileo) Pressure burst (Mariotte) Elasticity (Hooke)	
1700	Steam engine Cast iron bridge	Malleable cast iron	Shear, torsion (Coulomb)	
1800	Railroad industry Suspension bridge Internal combustion engine	Portland cement Vulcanized rubber Bessemer steel	Fatigue (Wöhler) Plasticity (Tresca) Universal testing machines	Steam boilers Railroad axles Iron bridges
1900	Electric power Powered flight Vacuum tube	Alloy steels Aluminum alloys Synthetic plastics	Hardness (Brinell) Impact (Izod, Charpy) Creep (Andrade)	Quebec bridge Boston molasses tank
1920	Gas-turbine engine Strain gage	Stainless steel Tungsten carbide	Fracture (Griffith)	Railroad wheels, rails Automotive components
1940	Controlled fission Jet aircraft Transistor; computer Sputnik	Ni-base alloys Ti-base alloys Fiberglass	Electronic testing machine Low-cycle fatigue (Coffin, Manson) Fracture mechanics (Irwin)	Liberty ships Comet airliner Turbine generators
1960	Laser Microprocessor Moon landing	HSLA steels High-performance composites	Closed-loop testing machine Fatigue crack growth (Paris) Computer control testing	F-111 aircraft DC-10 aircraft Highway bridges
1980	Space station Magnetic levitation	Tough ceramics	Multiaxial testing Direct digital control	Alex. Kielland rig Surgical implants

Source: [Herring 89], [Landgraf 80], [Timoshenko 83], [Whyte 75], *Encyclopedia Britannica.*

the materials testing capabilities that were necessary to support these advances. Representative technological failures involving deformation or fracture are also shown. These and other types of failures further stimulated improvements in materials, and in testing and analysis capability, by having a feedback type of effect. Such interactions among technological advances, materials, testing, and failures are still underway today and will continue into the foreseeable future.

As a particular example, consider improvements in engines. At the turn of the century, steam engines were in wide use, and the internal combustion engine had been invented and was being improved. Gas-turbine engines became practical for propulsion during World War II when they were used in the first jet aircraft. Higher operating temperatures in engines provide greater efficiency, with temperatures increasing over the years as shown in Fig. 1.14. To resist the higher temperatures, improved low-alloy steels and then stainless steels were developed, followed by increasingly sophisticated metal alloys based on nickel and cobalt. However, failures due to such causes as creep, fatigue, and corrosion still occurred and had major influences on engine development.

At present, the most likely route to further improvement in engines appears to lie in developing new ceramic materials, as this class of materials is inherently lighter in weight and more temperature resistant than metals. However, the usual brittleness of ceramics must be avoided in the new materials developed. Note that improved gas-

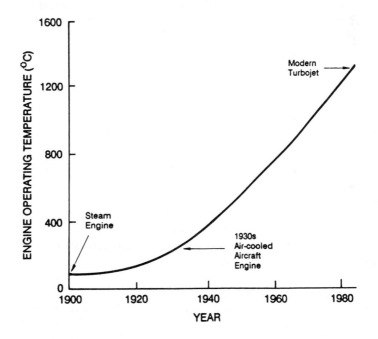

Figure 1.14   The steep climb in operating temperatures of engines during this century as made possible by improved materials. (From [NRC 89]; reprinted with permission; copyright © 1989 by the National Academy of Sciences.)

turbine engines could be used in aircraft, in small power generation stations, and perhaps even in automobiles.

Technology currently faces major challenges. Some of these arise from the need for greater energy efficiency, while at the same time maintaining environmental quality. For example, in transportation, the challenges include lighter weight vehicle structures, improved engines, and new types of engines. Another example is nuclear fusion, which has great promise, but involves severe radiation damage to materials and severe thermal fatigue associated with stresses caused by temperature changes.

In general, the challenges of advancing technology require not only improved materials but also more careful analysis in design, and more detailed information on materials behavior, than before. Furthermore, there has recently been a desirable increased awareness of safety and warranty issues. Manufacturers of machines, vehicles, and structures now find it appropriate not just to maintain current levels of safety and durability, but to improve these at the same time that the other technological challenges are being met.

1.5 ECONOMIC IMPORTANCE OF FRACTURE

The U.S. Department of Commerce, National Institute of Standards and Technology, formerly the National Bureau of Standards, completed a study in 1983 of the economic effects of fracture of materials in the United States. The total costs per year are large, specifically $119 billion in 1982 dollars. This is 4% of the gross national product (GNP), therefore representing a significant use of resources and manpower. The definition of fracture used for the study was quite broad, including not only fracture in the sense of cracking, but also deformation and related problems such as delamination. Wear and corrosion are not included. Separate studies indicate that adding these to obtain the total cost for materials durability would increase the total to roughly 10% of the GNP.

In the fracture study, the costs were considered to include the extra costs of designing machines, vehicles, and structures beyond the requirements of resisting simple yielding failure of the material. Note that resistance to fracture necessitates the use of more raw materials, or of more expensive materials or processing, to give components the extra strength necessary. Also, additional analysis and testing is needed in the design process, as discussed above in connection with Fig. 1.12, to avoid such problems. The extra use of materials and other activities all involve extra costs for manpower and facilities. There are also significant expenses associated with fracture for repair, maintenance, and replacement parts. Inspection of newly manufactured parts for flaws, and of parts in service for developing cracks, involves considerable cost. There are also costs such as recalls, litigation, insurance, etc., collectively called *product liability costs*, that add to the total.

Annual costs for the individual economic sectors that are most affected by fracture are shown in Fig. 1.15. About 10% of the total is associated with motor vehicles, and about 5% with aircraft. Note that fatigue cracking is the major cause of fracture problems in these two sectors, which are the two most affected. However, brittle and ductile fracture, environmental cracking, and creep are also important for these and other sectors. Several other economic sectors experience annual costs of $2 billion or more as indicated.

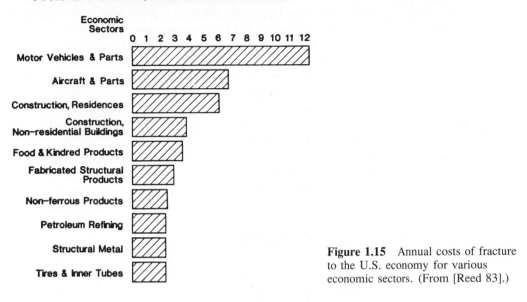

Figure 1.15   Annual costs of fracture to the U.S. economy for various economic sectors. (From [Reed 83].)

The study found that roughly one-third of this $119 billion annual cost could be eliminated through better use of current technology. Another third could perhaps be eliminated over a longer time period through research and development, that is, by obtaining new knowledge and developing ways to put this knowledge to work. And the final roughly one-third would be difficult to eliminate without major research breakthroughs. This situation is summarized by Fig. 1.16.

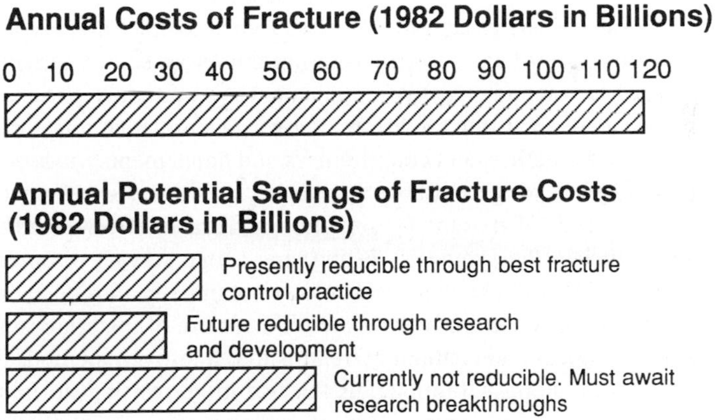

Annual Costs of Fracture (1982 Dollars in Billions)

0   10   20   30   40   50   60   70   80   90   100   110   120

Annual Potential Savings of Fracture Costs (1982 Dollars in Billions)

Presently reducible through best fracture control practice

Future reducible through research and development

Currently not reducible. Must await research breakthroughs

Figure 1.16   Potential for savings of the costs of fracture. (Adapted from [Reed 83].)

Hence, noting that two-thirds of these costs could be eliminated by improved use of currently available technology, or by technology that could be developed in a reasonable time, there is a definite economic incentive for learning about deformation and fracture. Engineers with knowledge in this area can help companies they work for avoid costs due to structural failures, and they can help make the design process more efficient, hence more economical and faster, by early attention to such potential problems. Benefits to society result, such as lower costs to the consumer and improved safety.

1.6 SUMMARY

Mechanical behavior of materials is the study of the deformation and fracture of materials. Materials tests are used to evaluate the behavior of a material, such as its resistance to failure in terms of the yield strength or fracture toughness. The material's strength is compared with the stresses expected for a component in service to assure that the design is adequate.

Different methods of testing materials and of analyzing trial engineering designs are needed for different types of material failure. These failure types include elastic, plastic, and creep deformation. Elastic deformation is recovered immediately upon unloading, whereas plastic deformation is permanent. Creep is deformation that accumulates with time. Other types of material failure involve cracking, such as brittle or ductile fracture, environmental cracking, creep rupture, and fatigue. Brittle fracture can occur due to static loads and involves little deformation, whereas ductile fracture involves considerable deformation. Environmental cracking is caused by a hostile chemical environment, and creep rupture is a time-dependent and usually ductile fracture. Fatigue is failure due to repeated loading and involves the gradual development and growth of cracks. A special method called fracture mechanics is used to specifically analyze cracks in engineering components.

Engineering design is the process of choosing all details necessary to describe a machine, vehicle, or structure. A variety of activities are involved, one of which is to assure that safety and durability are not compromised by structural failure caused by deformation or fracture. Design is fundamentally an iterative (trial and error) process, and it is necessary at each step to perform a synthesis in which all concerns and requirements are considered together, and compromises and adjustments made as necessary. Prototype and component testing, and also monitoring of service experience, are often important in the later stages of design. Deformation and fracture may need to be analyzed in one or more stages of the synthesis, testing, and actual service of an engineered item.

Advancing and changing technology continually introduces new challenges to the engineering designer, demanding more efficient use of materials and improved materials. Thus, the historical and continuing trend is that improved methods of testing and analysis have developed along with materials that are more resistant to failure.

Deformation and fracture are of major economic importance, especially in the motor vehicle and aircraft sectors. The costs involved in avoiding fracture and in paying for its consequences in all sectors of the economy are on the order of 4% of the GNP.

NEW TERMS AND SYMBOLS

brittle fracture

creep

ductile fracture

durability testing

elastic deformation

environmental cracking

fatigue

fatigue crack growth

fracture mechanics

fracture toughness, K_{Ic}

high-cycle fatigue

low-cycle fatigue

modulus of elasticity, E

percent elongation, $100\varepsilon_f$

plastic deformation

safety factor, X

simulated service testing

static loading

synergistic effect

synthesis

thermal fatigue

ultimate tensile strength, σ_u

yield strength, σ_o

REFERENCES

DIETER, G. E. 1983 *Engineering Design: A Materials and Processing Approach*, McGraw-Hill, New York.

HERRING, S. D. 1989 *From the Titanic to the Challenger: An Annotated Bibliography on Technological Failures of the Twentieth Century*, Garland Publishing, Inc., New York.

NRC. 1989 *Materials Science and Engineering for the 1990's: Maintaining Competitiveness in the Age of Materials*, Report of the National Research Council, Committee on Materials Science and Engineering; National Academy Press, Washington, DC.

REED, R. P., J. H. SMITH, and B. W. CHRIST. 1983 "The Economic Effects of Fracture in the United States: Part 1," Special Pub. No. 647-1, U.S. Dept. of Commerce, National Bureau of Standards, U.S. Government Printing Office, Washington, DC.

WHYTE, R. R., ed. 1975 *Engineering Progress Through Trouble*, The Institution of Mechanical Engineers, London.

WULPI, D. J. 1985 *Understanding How Components Fail*, Am. Soc. for Metals, Metals Park, Oh.

PROBLEMS AND QUESTIONS

Section 1.2

1.1. Classify each of the following failures by identifying its category in Fig. 1.1. Explain the reasons for each choice in one or two sentences.

(a) The plastic frames on eyeglasses gradually spread and become loose.

(b) A glass bowl with a small crack breaks into two pieces when it is immersed, while still hot, into cold water.

(c) The bottom of an aluminum frying pan buckles when it is immersed, while still hot, into cold water.

(d) A copper water pipe freezes and develops a lengthwise split that causes a leak.

(e) The steel radiator fan blades in an automobile develop small cracks near the base of the blades.

1.2. Repeat Prob.1.1 for the following:

(a) A child's plastic tricycle, used in rough play to make skidding turns, develops cracks where the handlebars join the frame.

(b) An aluminum baseball bat develops a crack.

(c) A large steel artillery tube (barrel), which previously had cracks emanating from the rifling, suddenly bursts into pieces. Classify both the cracks and the final fracture.

(d) The fuselage (body) of a passenger airliner breaks into two pieces, with the fracture starting from cracks that had previously initiated at the corners of window cutouts in the aluminum-alloy material. Classify both the cracks and the final fracture.

(e) The nickel-alloy blades in an aircraft turbine engine lengthen during service and rub the casing.

1.3. Think of four deformation or fracture failures that have actually occurred, either from your personal experience or from items that you have read about in newspapers, magazines, or books. Classify each according to a category in Fig. 1.1, and briefly explain the reason for your classification.

1.4. In private residences, heavy garage doors that lift to open often have a pair of large steel springs, which can be either linear springs or torsional springs. These are preloaded, that is, they are always under tension (or torsion), to counterbalance the weight of the door. The steel spring material is not thought to be subject to creep at ordinary temperatures under steady load.

(a) Why do the springs sometimes gradually lose their preload and need to be tightened?

(b) Why do the springs sometimes break?

Section 1.3

1.5 As an engineer, you work for a company that makes bicycles. Some bicycles that have been in use for several years have begun to develop failures starting from cracks in the spindle that holds the pedal, in particular where this spindle threads into the crank. What is the most likely cause of these cracks? Describe some of the steps that you might take to redesign this part and to verify that your new design will solve this problem.

1.6 Repeat Prob.1.5 for failures in the cast aluminum bracket used to attach the rudder of a small recreational sailboat.

1.7 Discuss in your own words, in two or three paragraphs, the moral implications and importance of good engineering judgment, as related to compromises in the design process, which are necessitated by limitations of time and cost.

1.8 Briefly describe the legal implications that might need to be considered in the design of small one-person nonenclosed recreational aircraft.

2

Structure and Deformation in Materials

2.1 INTRODUCTION

A wide variety of materials are used in applications where resistance to mechanical loading is necessary. These are collectively called *engineering materials* and can be broadly classified as metals and alloys, polymers, ceramics and glasses, and composites. Some typical members of each class are given in Table 2.1.

Differences among the classes of materials as to chemical bonding and microstructure affect mechanical behavior, giving rise to relative advantages and disadvantages among the classes. The situation is summarized by Fig. 2.1. For example, the strong chemical bonding in ceramics and glasses imparts mechanical strength and stiffness (high E), and also temperature and corrosion resistance, but causes brittle behavior. In contrast, many polymers are relatively weakly bonded between the chain molecules, in which case the material has low strength and stiffness and is susceptible to creep deformation.

Starting from the size scale of primary interest in engineering, roughly one meter, there is a span of ten orders of magnitude in size down to the scale of the atom, which is around 10^{-10} m. This situation and various intermediate size scales of interest are

TABLE 2.1   CLASSES AND EXAMPLES OF ENGINEERING MATERIALS

Metals and Alloys	Ceramics and Glasses
Irons and steels	Clay products
Aluminum alloys	Concrete
Titanium alloys	Alumina (Al_2O_3)
Copper alloys; brasses, bronzes	Tungsten carbide (WC)
Magnesium alloys	Titanium aluminide (Ti_3Al)
Nickel-base superalloys	Silica (SiO_2) glasses

Polymers	Composites
Polyethylene (PE)	Plywood
Polyvinyl chloride (PVC)	Cemented carbides
Polystyrene (PS)	Fiberglass
Nylons	Graphite-epoxy
Epoxies	SiC-aluminum
Rubbers	Aramid-aluminum laminate (ARALL)

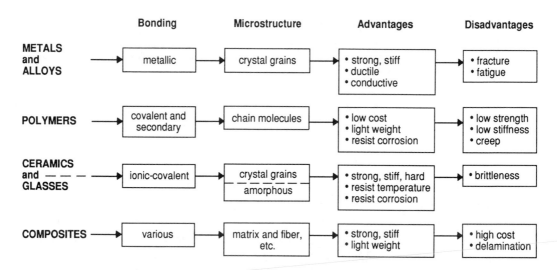

Figure 2.1   General characteristics of the major classes of engineering materials.

indicated in Fig. 2.2. At any given size scale, an understanding of the behavior can be sought by looking at what happens at a smaller scale: The behavior of a machine, vehicle, or structure is explained by the behavior of its component parts, and the behavior of these can in turn be explained using small (10^{-1} to 10^{-2} m) test specimens of the material. Similarly, the macroscopic behavior of the material is explained by the behavior of crystal grains, defects in crystals, polymer chains, and other microstructural features that exist in the size range of 10^{-3} to 10^{-9} m. Thus, knowledge of behavior over the entire range of sizes from 1 m down to 10^{-10} m contributes to understanding and predicting the performance of machines, vehicles, and structures.

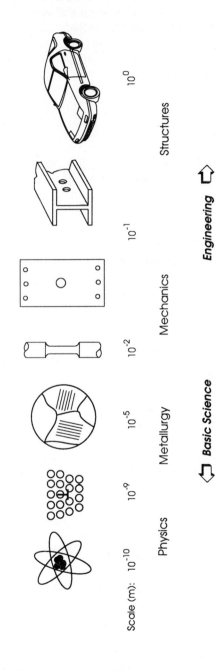

Figure 2.2 Size scales and disciplines involved in the study and use of engineering materials. (Illustration courtesy of R. W. Landgraf, Virginia Tech, Blacksburg, Va.)

This chapter reviews some of the fundamentals needed to understand mechanical behavior of materials. We will start at the lower end of the size scale in Fig. 2.2 and progress upward. The individual topics include chemical bonding, crystal structures, defects in crystals, and the physical causes of elastic, plastic, and creep deformation. The next chapter will then apply these concepts in discussing each of the classes of engineering materials in more detail.

2.2 BONDING IN SOLIDS

There are several types of chemical bonds that hold atoms and molecules together in solids. Three types—*ionic*, *covalent*, and *metallic*—bonds, are collectively termed *primary* bonds. Primary bonds are strong and stiff and do not easily melt with increasing temperature, being responsible for the bonding of metals and ceramics, and providing the relatively high elastic modulus (E) in these materials. *Van der Waals* and *hydrogen* bonds are relatively weak and so are called *secondary* bonds. These are important in determining the behavior of liquids and in linking the carbon-chain molecules in polymers, so that these materials can exist as solids.

2.2.1 Primary Chemical Bonds

The three types of primary bonds are illustrated in Fig. 2.3. Ionic bonding involves the transfer of one or more electrons between atoms of different types. Note that the outer shell of electrons surrounding an atom is stable if it contains eight electrons. (Except that the stable number is two for the single shell of hydrogen or helium.) Hence, a metal atom such as sodium with only one electron in its outer shell can donate it to an atom such as chlorine, which has seven. Both then have eight in the outer shell, since the next shell of sodium has eight. The atoms become charged ions, such as Na^+ and Cl^-, which attract one another and form a chemical bond due to their opposite electrostatic charges. A collection of such charged ions, equal numbers of each in this case, form

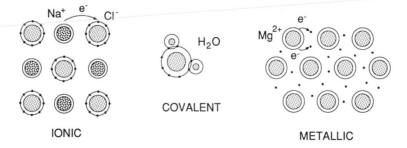

Figure 2.3 The three types of primary chemical bond. Electrons are transferred in ionic bonding, as in NaCl; shared in covalent bonding, as in water; and given up to a common "cloud" in metallic bonding, as in magnesium metal.

an electrically neutral solid by arranging themselves into a regular crystalline array, as shown in Fig. 2.4.

The number of electrons transferred may differ from one. For example, in the salt $MgCl_2$, and in the oxide MgO, two electrons are transferred to form an Mg^{2+} ion. Many common salts, oxides, and other solids have bonds that are mostly or partially ionic. These materials tend to be hard and brittle.

Covalent bonding involves the sharing of electrons and occurs where the outer shells are half full or more than half full. For example, two hydrogen atoms each share an electron with an oxygen atom to make water, H_2O, or two chlorine atoms share one electron to form the diatomic molecule Cl_2. The tight covalent bonds make such simple molecules relatively independent of one another, so that collections of them tend to form liquids or gases at ambient temperatures.

Metallic bonding is responsible for the usually solid form of metals and alloys. Noting that for metals the outer shell of electrons is in most cases less than half full, each atom donates one or more electrons to a "cloud" of electrons. These electrons are shared in common by all of the metal atoms, which have become positively charged ions as a result of giving up electrons. The metal ions are thus held together by their mutual attraction to the electron cloud.

2.2.2 Discussion of Primary Bonds

Covalent bonds have the property not shared by the other two types of primary bonds of being strongly directional. This arises from covalent bonds being dependent on the sharing of electrons with specific neighboring atoms, whereas ionic and metallic solids are held together by electrostatic attraction involving all neighboring ions.

A continuous arrangement of covalent bonds can form a three-dimensional network to make a solid. An example is carbon in the form of diamond, in which each carbon atom shares an electron with four adjacent ones. These atoms are arranged at equal angles to one another in three-dimensional space, as illustrated in Fig. 2.5. As a result of the strong and directional bonds, the crystal is very hard and stiff. Another important continuous arrangement of covalent bonds is the carbon chain. For example, in the

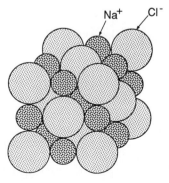

Figure 2.4 Three-dimensional crystal structure of NaCl, consisting of two interpenetrating FCC structures.

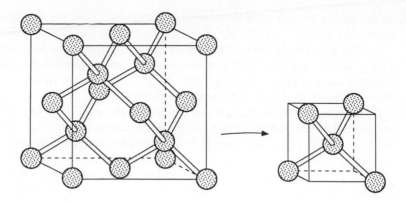

Figure 2.5 Diamond cubic crystal structure of carbon. As a result of the strong and directional covalent bonds, diamond has the highest melting temperature, the highest hardness, and the highest elastic modulus E, of all known solids.

gas *ethylene*, C_2H_4, each molecule is formed by covalent bonds as shown in Fig. 2.6. However, if the double bond between the carbon atoms is replaced by a single bond to each of two adjacent carbon atoms, then a long chain molecule can form. The result is the polymer called *polyethylene*.

Many solids, such as SiO_2 and other ceramics, have chemical bonds that have a mixed ionic-covalent character. The examples given above of NaCl for ionic bonding and of a diamond for covalent bonding do represent cases of nearly pure bonding of these types, but mixed bonding is more common.

2.2.3 Secondary Bonds

Secondary bonds occur due to the presence of an electrostatic dipole, which can be induced by a primary bond. For example, in water, the side of a hydrogen atom away

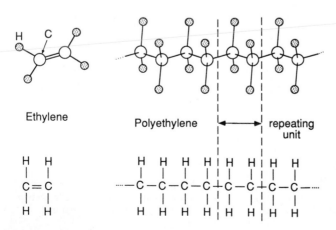

Figure 2.6 Molecular structures of ethylene gas (C_2H_4) and polyethylene polymer. The double bond in ethylene is replaced by two single bonds in polyethylene, permitting formation of the chain molecule.

from the covalent bond to the oxygen atom has a positive charge, due to the sole electron being predominantly on the side toward the oxygen atom. Conservation of charge over the entire molecule then requires a negative charge on the exposed portion of the oxygen atom. The dipoles formed cause an attraction between adjacent molecules, as illustrated in Fig. 2.7.

Such bonds, termed *permanent dipole bonds*, occur between various molecules. They are relatively weak but are nevertheless sometimes sufficient to bind materials into solids, water ice being an example. Where the secondary bond involves hydrogen, as in the case of water, it is stronger than other dipole bonds and is called a *hydrogen bond*.

Van der Waals bonds arise from the fluctuating positions of electrons relative to an atom's nucleus. The uneven distribution of electric charge which thus occurs causes a weak attraction between atoms or molecules. This type of bond can also be called a *fluctuating dipole bond*, being distinguished from a permanent dipole bond because the dipole is not fixed in direction as it is in a water molecule.

In polymers, covalent bonds form the chain molecules and attach hydrogen and other atoms to the carbon backbone. Hydrogen bonds and other secondary bonds occur between the chain molecules and tend to prevent them from sliding past one another. This is illustrated in Fig. 2.8 for polyvinyl chloride. The relative weakness of the secondary bonds accounts for the low melting temperatures and the low strengths and stiffnesses of these materials.

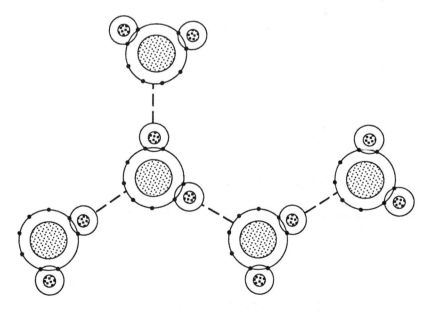

Figure 2.7 Oxygen-to-hydrogen secondary bonds between water (H_2O) molecules.

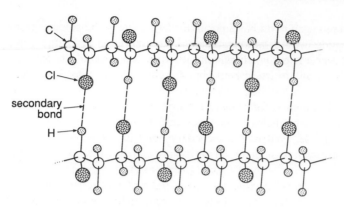

Figure 2.8 Hydrogen-to-chlorine secondary bonds between chain molecules in polyvinyl chloride.

2.3 STRUCTURE IN CRYSTALLINE MATERIALS

Metals and ceramics are composed of aggregations of small grains, each of which is an individual crystal. In contrast, glasses have an amorphous or noncrystalline structure. Polymers are composed of chain-like molecules, which are sometimes arranged in regular arrays in a crystalline manner.

2.3.1 Basic Crystal Structures

The arrangement of atoms (or ions) in crystals can be described in terms of the smallest grouping that can be considered to be a building block for a perfect crystal. Such a grouping, called a *unit cell*, can be classified according to the lengths and angles involved. There are seven basic types of unit cell, three of which are shown in Fig. 2.9. If all three angles are 90° and all distances are the same, the crystal is classed as *cubic*. But if one distance is not equal to the other two, the crystal is *tetragonal*. If in addition one angle is 120°, while the other two remain at 90°, the crystal is *hexagonal*. The four additional types are orthorombic, rhombohedral, monoclinic, and triclinic.

For a given type of unit cell, various arrangements of atoms are possible, with each such arrangement being called a *crystal structure*. Three crystal structures having a cubic unit cell are the *primitive cubic* (PC), *body-centered cubic* (BCC), and *face-centered*

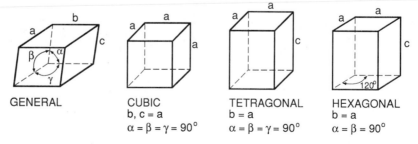

Figure 2.9 The general case of a unit cell in a crystal and three of the seven basic types.

cubic (FCC) structures. These are illustrated in Fig. 2.10. Note that the PC structure has atoms only at the corners of the cube, whereas the BCC structure also has one in the center of the cube. The FCC structure has atoms at the cube corners and in the center of each surface. The PC structure occurs only rarely, but the BCC structure is found in a number of common metals, such as chromium, iron, molybdenum, sodium, and tungsten. Similarly, the FCC structure is common for metals, as in silver, aluminum, lead, copper, and nickel.

The *hexagonal close-packed* (HCP) crystal structure is also common in metals. Although the unit cell is the hexagonal one, as shown in Fig. 2.9, it is useful to illustrate this structure using a larger grouping that forms a hexagonal prism as shown in Fig. 2.10. Two parallel planes have atoms at the corners and center of a hexagon, and there are three additional atoms halfway between these planes as shown. Some common metals having this structure are beryllium, magnesium, titanium, and zinc.

A given metal or other material may change its crystal structure with temperature or pressure, or with the addition of alloying elements. For example, the BCC structure of iron changes to FCC above 910°C, and back to BCC above 1390°C. These phases are often called, respectively, alpha iron, gamma iron, and delta iron, denoted α-Fe, γ-Fe, and δ-Fe. Also, the addition of about 10% nickel or manganese changes the crystal structure to FCC even at room temperature. Similarly, HCP titanium is called α-Ti, whereas β-Ti has a BCC structure and occurs above 885°C, although it can also exist at room temperature as a result of alloying and processing.

2.3.2 More Complex Crystal Structures

Compounds formed by ionic or covalent bonding, such as ionic salts and ceramics, have more complex crystal structures than elemental materials. This is due to the necessity of accommodating more than one type of atom and to the directional aspect of even partially covalent bonds. However, the structure can often be thought of as an elaboration of one of the basic crystal structures. For example, NaCl is an FCC arrangement of Cl^- ions with Na^+ ions at intermediate positions, so that these also form an FCC structure that is merged with the one for the Cl^- ions. See Fig. 2.4. Many important ionic salts and ceramics have this structure, including oxides such as MgO and FeO, and carbides such as TiC and ZrC.

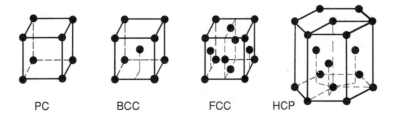

Figure 2.10   Four crystal structures: primitive cubic (PC), body-centered cubic (BCC), face-centered cubic (FCC), and hexagonal close-packed (HCP) structures.

In the *diamond cubic* structure of carbon, half of the atoms form an FCC structure, and the other half lie at intermediate positions as required by the tetragonal bonding geometry and also form an FCC structure. See Fig. 2.5. Another solid with a diamond cubic structure is SiC, in which Si and C atoms occupy alternate sites in the same structure as in Fig. 2.5. The ceramic Al_2O_3 has a crystal structure with a hexagonal unit cell, aluminum atoms occurring in two-thirds of the spaces available between the oxygen atoms. As a final example of a ceramic structure, consider the ceramic compound $BaTiO_3$. This is a combined FCC and BCC structure, with Ba^{2+} ions at the cube corners, O^{2-} at the face centers, and Ti^{4+} at the body center. See Fig. 2.11. The mineral perovskite, $CaTiO_3$, has this same structure. Many compounds have even more complex crystal structures than these examples.

Polymers may be amorphous in that the structure is an irregular tangle of chain molecules. Alternatively, portions or even most of the material may have the chains arranged in a regular manner under the influence of the secondary bonds between the chains. This is illustrated in Fig. 2.12.

2.3.3 Defects in Crystals

Ceramics and metals in the form used for engineering applications are composed of crystalline *grains* that are separated by *grain boundaries*. This is shown for a metal in Fig. 2.13. Grain sizes vary widely from as small as 1 μm to as large as 10 mm,

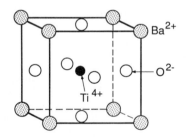

Figure 2.11 The crystal structure of the ceramic material $BaTiO_3$.

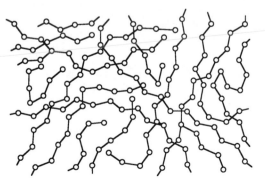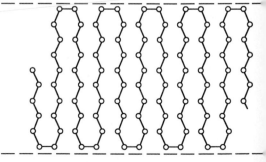

Figure 2.12 Two-dimensional schematics of amorphous structure (left) and crystalline structure (right) in a polymer.

Figure 2.13 Crystal grain structure in the alpha form of the alloy Ti-6Al, which is titanium with 6% aluminum. (Photo courtesy of C. G Rhodes, Rockwell International Science Center, Thousand Oaks, Ca.)

depending on the material and its processing. Even within grains the crystals are not perfect, with defects occurring that can be classed as *point defects*, *line defects*, or *surface defects*. Grain boundaries and crystal defects within grains can both have large effects on mechanical behavior. In discussing these, it is useful to use the term *lattice plane* to describe the regular parallel planes of atoms in a perfect crystal, and the term *lattice site* to describe the position of one atom.

Some types of point defects are illustrated in Fig. 2.14. A *substitutional impurity* occupies a normal lattice site but is an atom of a different element than the bulk material. A *vacancy* is the absence of an atom at a normally occupied lattice site, and an *interstitial* is an atom occupying a position between normal lattice sites. If the interstitial is of the same type as the bulk material, it is called a *self interstitial*, and if it is of another kind it is called an *impurity interstitial*.

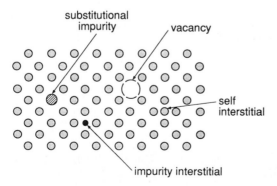

Figure 2.14 Four types of point defect in a crystalline solid.

Relatively small impurity atoms often occupy interstitial sites in materials with larger atoms. An example is carbon in solid solution in iron. If the impurity atoms are of similar size to those of the bulk material, they are more likely to appear as substitutional impurities. This is the normal situation where two metals are alloyed, that is, melted together. An example is the addition of 10 to 20% chromium to iron, and in some cases also of 10 to 20% nickel, to make a stainless steel.

Line defects are called *dislocations* and are the edges of surfaces where there is a relative displacement of lattice planes. One type is an *edge dislocation* and the other is a *screw dislocation*, both of which are illustrated in Fig. 2.15. The edge dislocation can be thought of as the border of an extra plane of atoms as shown in (a). The *dislocation line* shown identifies the edge of the extra plane, and the special symbol indicated is sometimes used.

The screw dislocation can be explained by assuming that a perfect crystal is cut as shown in Fig. 2.15(b). The crystal is then displaced parallel to the cut, and finally reconnected into the configuration shown. The dislocation line is the edge of the cut and hence also the border of the displaced region. Dislocations in solids generally have a combined edge and screw character and form curves and loops. Where many are present, complex tangles of dislocation lines may form.

Grain boundaries can be thought of as a class of surface defect where the lattice planes change orientation by a large angle. Within a grain, there may also be *low-angle boundaries*. An array of edge dislocations can form such a boundary as shown in Fig. 2.16. Several low-angle boundaries may exist within a grain, separating regions of slightly different lattice orientation, which are called *subgrains*.

There are additional types of surface defects. A *twin boundary* separates two regions of a crystal where the lattice planes are a mirror image of one another. If the lattice planes are not in the proper sequence for a perfect crystal, a *stacking fault* is said to exist.

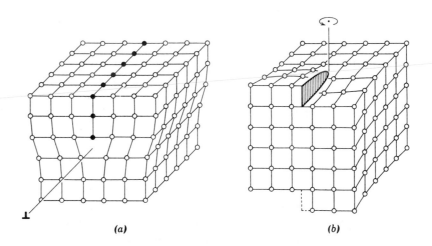

(a)   *(b)*

Figure 2.15 The two basic types of dislocations: (a) edge dislocation, and (b) screw dislocation. (From [Hayden 65] p. 63; used with permission.)

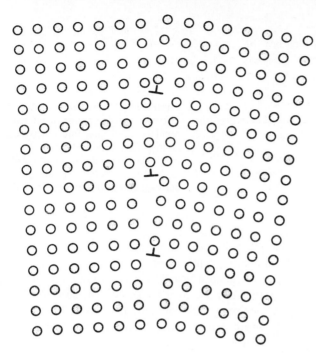

Figure 2.16   Low-angle boundary in a crystal formed by an array of edge dislocations. (From [Boyer 85] p. 2.15; used with permission.)

2.4 ELASTIC DEFORMATION AND THEORETICAL STRENGTH

The discussion of bonding and structure in solids can be extended to consider the physical mechanisms of deformation as viewed at the size scales of atoms, dislocations, and grains. Recall from Chapter 1 that there are three basic types of deformation: elastic, plastic, and creep deformation. Elastic deformation is discussed first below, and this leads to some rough theoretical estimates of strength for solids.

2.4.1 Elastic Deformation

Elastic deformation is associated with stretching but not breaking the chemical bonds between the atoms in a solid. If an external stress is applied to a material, the distance between the atoms changes by a small amount that depends on the material and the details of its structure and bonding. These distance changes, when accumulated over a piece of material of macroscopic size, are called elastic deformations.

   If the atoms in a solid were very far apart, there would be no forces between them. As the distance x between atoms is decreased, they begin to attract one another according to the type of bonding that applies to the particular case. This is illustrated by the upper curve in Fig. 2.17. A repulsive force also acts that is associated with resistance to overlapping of the electron shells of the two atoms. This repulsive force is smaller than the attractive force at relatively large distances, but it increases more rapidly, becoming larger at short distances. The total force is thus attractive at large distances, repulsive at

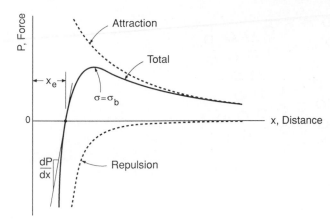

Figure 2.17   Variation with the distance between atoms of attractive, repulsive, and total forces between atoms. The slope dP/dx at the equilibrium spacing x_e is proportional to the elastic modulus E; the stress σ_b, corresponding to the peak in total force, is the theoretical cohesive strength.

short distances, and zero at one particular distance x_e, which is the equilibrium atomic spacing. This is also the point of minimum potential energy.

Elastic deformations of engineering interest usually represent only a small perturbation about the equilibrium spacing, typically less than 1% strain. The slope of the total force curve over this small region is approximately constant. Let us express force on a unit area basis as stress, $\sigma = P/A$, where A is the cross-sectional area of material per atom. Also, note that strain is the ratio of the change in x to the equilibrium distance x_e.

$$\sigma = \frac{P}{A}, \qquad \varepsilon = \frac{x - x_e}{x_e} \tag{2.1}$$

Since the elastic modulus E is the slope of the stress-strain relationship, we have

$$E = \frac{d\sigma}{d\varepsilon}\bigg|_{x=x_e} = \frac{x_e}{A}\frac{dP}{dx}\bigg|_{x=x_e} \tag{2.2}$$

Hence, E is fixed by the slope of the total force curve at $x = x_e$, which is illustrated in Fig. 2.17.

2.4.2 Trends in Elastic Modulus Values

Strong primary chemical bonds are resistant to stretching and so result in a high value of E. For example, the strong covalent bonds in diamond yield a value around $E = 1000$ GPa, whereas the weaker metallic bonds in metals give values generally within a factor of three of $E = 100$ GPa. In polymers, E is determined by the combination of covalent bonding along the carbon chains and the much weaker secondary bonding between chains. At relatively low temperatures, many polymers exist in a glassy or crystalline state. The modulus is then on the order of $E = 3$ GPa, but it varies considerably above and below this level depending on the chain-molecule structure and other details. If the temperature is increased, the secondary bonds melt, causing the elastic modulus to decrease, often dramatically. This trend is shown for polystyrene in Fig. 2.18.

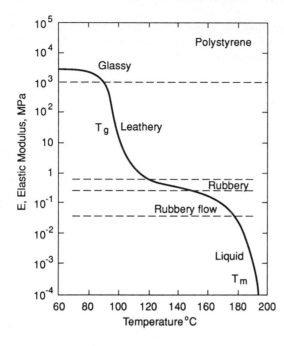

Figure 2.18 Variation of elastic modulus with temperature for polystyrene. (Data from [Tobolsky 65].)

The temperature where the rapid decrease in E occurs varies for different polymers and is called the *glass transition temperature*, T_g. Melting does not occur until a somewhat higher temperature, T_m, provided that chemical decomposition does not occur first. Above T_g, the elastic modulus is typically within a factor of ten of the very low value $E = 10$ MPa. A degree of solidity of the material is still possible only due to such mechanisms as tangling of the long-chain molecules and occasional covalent bonds between chains, called *cross-linking*. A polymer has a leathery or rubbery character above its T_g; vulcanized natural rubber and synthetic rubbers at room temperature are examples.

For single crystals, E varies with the direction relative to the crystal structure; that is, crystals are more resistant to elastic deformation in some directions than in others. But in a polycrystalline aggregate of randomly oriented grains, an averaging effect occurs, so that E is the same in all directions. This latter situation is at least approximated for most engineering metals and ceramics.

2.4.3 Theoretical Strength

A value for the *theoretical cohesive strength* of a solid can be obtained by using solid-state physics to estimate the tensile stress necessary to break primary chemical bonds, which is the stress σ_b corresponding to the peak value of force in Fig. 2.17. Values on the order of $\sigma_b = E/10$ are generally obtained for various materials. Hence, for diamond, $\sigma_b \approx 100$ GPa, and for a typical metal, $\sigma_b \approx 10$ GPa.

Rather than the bonds being simply pulled apart in tension, another possibility is shear failure. A simple calculation can be done to obtain an estimate of the *theoretical shear strength.* Consider two planes of atoms being forced to move slowly past one another as in Fig. 2.19. The shear stress τ required first increases rapidly with displacement x, then decreases and passes through zero as the atoms pass opposite one another at the unstable equilibrium position $x = b/2$. The stress changes direction beyond this as the atoms try to snap into a second stable configuration at $x = b$. A reasonable estimate is a sinusoidal variation

$$\tau = \tau_b \sin \frac{2\pi x}{b} \tag{2.3}$$

where τ_b is the maximum value as τ varies with x; hence, it is the theoretical shear strength.

The initial slope of the stress-strain relationship must be the shear modulus, G, in a manner analogous to E for the tension case discussed above. Noting that the shear strain for small values of displacement is $\gamma = x/h$, we have

$$G = \frac{d\tau}{d\gamma}\bigg|_{x=0} = h\frac{d\tau}{dx}\bigg|_{x=0} \tag{2.4}$$

Obtaining $d\tau/dx$ from Eq. 2.3 and substituting its value at $x = 0$ gives τ_b.

$$\tau_b = \frac{Gb}{2\pi h} \tag{2.5}$$

The ratio b/h varies with the crystal structure and is generally around 0.5 to 1, so this estimate is on the order of $G/10$.

In a tension test, the maximum shear stress occurs on a plane $45°$ to the direction of uniaxial stress and is half as large. Thus, a theoretical estimate of shear failure in a tension test is

$$\sigma_b = 2\tau_b = \frac{Gb}{\pi h} \tag{2.6}$$

Since G is in the range $E/2$ to $E/3$, this estimate gives a value similar to the $\sigma_b = E/10$ estimate previously mentioned based on tensile breaking of bonds. Estimates

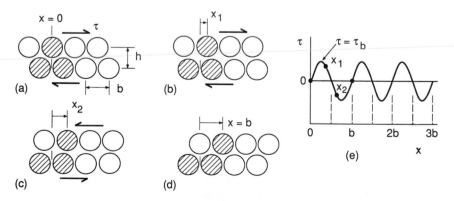

Figure 2.19   Basis of estimates of theoretical shear strength, where it is assumed that entire planes of atoms shift simultaneously relative to one another.

of theoretical strength are discussed in more detail in the first chapter of Kelly and Macmillan (1986).

Theoretical tensile strengths around $\sigma_b = E/10$ are larger than the typical strengths of solids by a large amount, typically by a factor of 10 to 100. This discrepancy is thought to be due mainly to the imperfections present in most crystals, which decrease the strength. However, small whiskers can be made that are nearly perfect single crystals. Also, thin fibers and wires may have a crystal structure such that strong chemical bonds are aligned with the length direction. Tensile strengths in such cases are indeed much higher than for larger and more imperfect samples of material. Strengths in the range $E/100$ to $E/20$, that is, one-tenth to one-half of the theoretical strength, have been achieved in this way, lending credence to the estimates. Some representative data are given in Table 2.2.

TABLE 2.2   ULTIMATE STRENGTHS IN TENSION OF SINGLE CRYSTAL WHISKERS AND STRONG FIBERS AND WIRES.

Material	Elastic Modulus E, GPa (10^3 ksi)		Tensile Strength σ_u, GPa (ksi)		Ratio E/σ_u
(a) Whiskers					
SiC	700	(102)	21.0	(3050)	33
Graphite	686	(99.5)	19.6	(2840)	35
Al_2O_3	420	(60.9)	22.3	(3230)	19
α-Fe	196	(28.4)	12.6	(1830)	16
Si	163	(23.6)	7.6	(1100)	21
NaCl	42	(6.09)	1.1	(160)	38
(b) Fibers and wires					
SiC	616	(89.3)	8.3	(1200)	74
Tungsten (0.26 μm diameter)	405	(58.7)	24.0	(3500)	17
Tungsten (25 μm diameter)	405	(58.7)	3.9	(570)	104
Al_2O_3	379	(55.0)	2.1	(300)	180
Graphite	256	(37.1)	5.5	(800)	47
Iron	220	(31.9)	9.7	(1400)	23
Linear polyethylene	160	(23.2)	4.6	(670)	35
Drawn silica glass	73.5	(10.7)	10.0	(1450)	7.4

Source: Data in [Kelly 86].

2.5 INELASTIC DEFORMATION

As discussed above, elastic deformation involves the stretching of chemical bonds. When the stress is removed, the deformation disappears. More drastic events can occur that have the effect of rearranging the atoms so that they have new neighbors after the deformation is complete. This causes a permanent or inelastic deformation that does not disappear when the stress is removed.

2.5.1 Plastic Deformation by Dislocation Motion

Single crystals of pure metals that are macroscopic in size and which contain a few dislocations are observed to yield in shear at very low stresses. For example, for iron and other BCC metals, this occurs around $\tau_o = G/3000$, that is about $\tau_o = 30$ MPa. For FCC and HCP metals, even lower values are obtained around $\tau_o = G/100,000$, or typically $\tau_o = 0.5$ MPa. Thus, shear strengths for imperfect crystals of pure metals can be lower than the theoretical value for a perfect crystal of $\tau_b = G/10$ by at least a factor of 300 and sometimes by as much as a factor of 10,000.

This large discrepancy can be explained by the fact that plastic deformation occurs by motion of dislocations under the influence of a shear stress as illustrated in Fig. 2.20. As a dislocation moves through the crystal, plastic deformation is in effect proceeding one atom at a time, rather than occurring simultaneously over an entire plane as implied by Fig. 2.19. This incremental process can occur much more easily than simultaneous breaking of all the bonds as assumed in the theoretical shear strength calculation for a perfect crystal.

The deformation resulting from dislocation motion proceeds for edge and screw dislocations as illustrated in Figs. 2.21 and 2.22, respectively. The plane in which the dislocation line moves is called the *slip plane*, and where the slip plane intersects a free surface, a *slip step* is formed. Since dislocations in real crystals are usually curved and thus have both edge and screw character, plastic deformation actually occurs by a combination of the two types of dislocation motion.

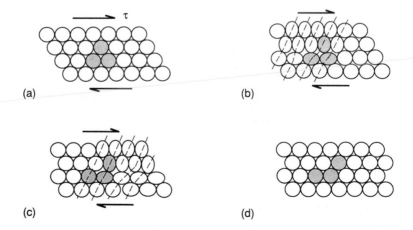

Figure 2.20   Shear deformation occurring in an incremental manner due to dislocation motion. (Adapted from [Van Vlack 89] p. 265; ©1989 by Addison-Wesley Publishing Company, Inc.; reprinted with permission of the publisher.)

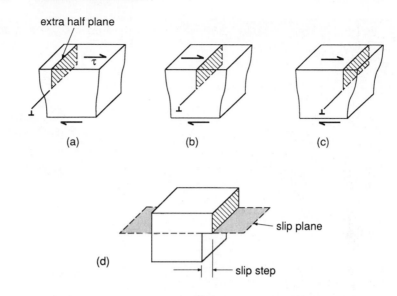

Figure 2.21   Slip caused by the motion of an edge dislocation.

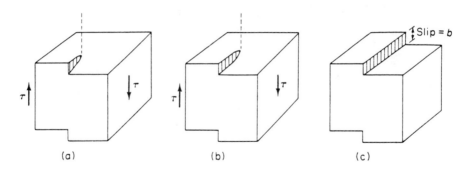

Figure 2.22   Slip caused by the motion of a screw dislocation. (From [Felbeck 84] p. 118; ©1984 by Prentice Hall, Englewood Cliffs, N.J.; reprinted with permission.)

Plastic deformation is often concentrated in bands called *slip bands*. These are regions where the slip planes of numerous dislocations are concentrated, hence they are regions of intense plastic shear deformation separated by regions of little shear. Where slip bands intersect a free surface, steps are formed as a result of the combined slip steps of numerous dislocations. See Fig. 2.23.

For a given crystal structure, such as BCC, FCC, or HCP, slip is easier on certain planes, and within these planes in certain directions. For metals, the most common planes and directions are shown in Fig. 2.24. The preferred planes are those on which the atoms are relatively close together, called *close-packed planes*, such as the base plane for the HCP crystal. Similarly, the preferred slip directions within a given plane

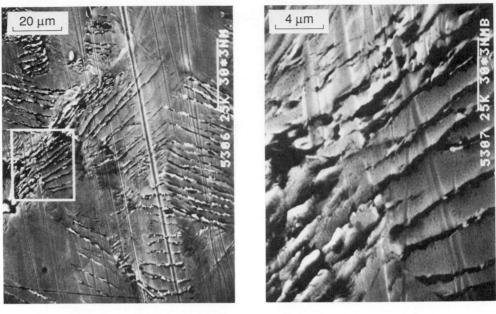

Figure 2.23   Slip bands and slip steps caused by the motion of many dislocations resulting from cyclic loading of AISI 1010 steel. (Photos courtesy of R. W. Landgraf, Virginia Tech, Blacksburg, Va.)

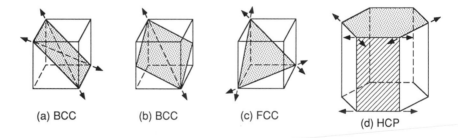

(a) BCC   (b) BCC   (c) FCC   (d) HCP

Figure 2.24   Some slip planes and directions frequently observed for BCC, FCC, and HCP crystal structures. Considering symmetry, there are additional combinations of slip plane and direction similar to each of these, giving a total of twelve slip systems similar to each of (a), (b), and (c), and three for each of the two cases in (d). (Adapted from [Hayden 65] p. 100; used with permission.)

are the *close-packed directions* in which the distances between atoms is smallest. This is the case because a dislocation can more easily move if the distance to the next atom is smaller. Also, atoms in adjacent planes project less into the spaces between atoms in the close-packed planes than for other planes, so there is less interference with slip displacement.

2.5.2 Discussion of Plastic Deformation

The result of plastic deformation (yielding) is that atoms change neighbors, returning to a stable configuration with new neighbors after the dislocation has passed. Note that this is a fundamentally different process than elastic deformation, which is merely the stretching of chemical bonds. Elastic deformation occurs as an essentially independent process along with plastic deformation. When a stress causing yielding is removed, the elastic strain is recovered just as if there had been no yielding. See Fig. 1.2.

Metals used in load-resisting applications have strengths considerably above the very low values observed in crystals of pure metals with defects, but not nearly as high as the very high theoretical value for a perfect crystal. This is illustrated for irons and steels, which are composed mostly of iron, in Fig. 2.25. If there are obstacles that impede dislocation motion, the strength may be increased by a factor of ten or more above the low value for a pure metal crystal. Grain boundaries have this effect, as does a second phase of hard particles dispersed in the metal. Alloying also increases strength as the different-sized atoms make dislocation motion more difficult. If a large number of dislocations are present, these interfere with one another, forming dense tangles and blocking free movement.

In nonmetals and compounds where the chemical bonding is covalent or partially covalent, the directional nature of the bonds makes dislocation motion difficult. Materials in this class include the crystals of carbon, boron, and silicon, and also compounds formed between metals and nonmetals, such as metal carbides, borides, nitrides, oxides, and other ceramics. At ambient temperatures, these materials are hard and brittle and do

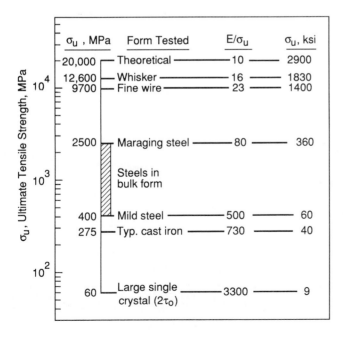

Figure 2.25 Ultimate tensile strengths for irons and steels in various forms. Note that steels are mostly composed of iron and contain small to moderate amounts of other elements. (Data from [Boyer 85], [Hayden 65], and [Kelly 86].)

not generally fail by yielding due to dislocation motion. Instead, the strength falls below the high theoretical value for a perfect crystal mainly because of the weakening effect of small cracks and fissures that are present in the material. However, some dislocation motion does occur, especially for temperatures above about half of the (usually high) melting temperature, where T_m is measured relative to absolute zero.

2.5.3 Creep Deformation

In addition to elastic and plastic deformation as already described, materials deform by mechanisms that result in markedly time-dependent behavior, called creep. Under constant stress, the strain varies with time as shown in Fig. 2.26. There is an initial elastic deformation, and following this the strain slowly increases as long as the stress is maintained. If the stress is removed, the elastic strain ε_e is quickly recovered, and a portion of the creep strain may be recovered slowly with time; the rest remains as permanent deformation.

In crystalline materials, that is, in metals and ceramics, one important mechanism of creep is *diffusional flow* of vacancies. Vacancies are more easily accommodated near grain boundaries under compressive strain, and less easily accommodated near those under tensile strain. Thus, they tend to move in a slow time-dependent manner, that is, to diffuse, from some regions within a grain to others as illustrated in Fig. 2.27. As indicated, movement of a vacancy in one direction is equivalent to movement of an atom in the opposite direction. The overall effect is a change in the shape of the grain, contributing to a macroscopic creep strain.

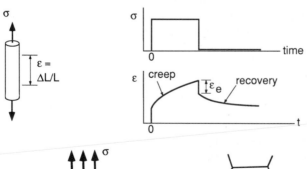

Figure 2.26 Accumulation of creep strain with time under constant stress, and partial recovery after removal of the stress.

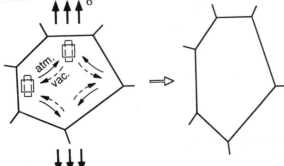

Figure 2.27 Mechanism of creep by diffusion of vacancies within a crystal grain.

Some other creep mechanisms that operate in crystalline materials include special dislocation motions that can circumvent obstacles in a time-dependent manner. There may also be sliding of grain boundaries and the formation of cavities along grain boundaries. Creep behavior in crystalline materials is strongly temperature dependent, typically becoming an important engineering consideration around 0.3 to $0.6T_m$, where T_m is the absolute melting temperature.

Different creep mechanisms operate in amorphous (noncrystalline) glasses and in polymers. One of these is viscous flow in the manner of a very thick liquid. This occurs in polymers at temperatures substantially above the glass transition temperature T_g and approaching T_m. The chain-like molecules simply slide past one another in a time-dependent manner. Around and below T_g, more complex behavior involving segments of chains and obstacles to chain sliding become important. In this case, much of the creep deformation may disappear slowly with time after removal of an applied stress, as illustrated in Fig. 2.26. Creep is a major limitation on the engineering application of any polymer above its T_g, which is generally in the range -100 to $200°C$ for common polymers.

Additional discussion on mechanisms of creep deformation is given later in Chapter 15.

2.6 SUMMARY

Atoms and molecules in solids are held together by primary chemical bonds of three kinds: ionic, covalent, and metallic bonds. Secondary bonds, especially hydrogen bonds, also influence the behavior. Covalent bonds are strong and directional and therefore resist deformation. This contributes to the high strength and brittleness of ceramics and glasses, as these materials are bound by covalent or mixed ionic-covalent bonds. Metallic bonds in metals do not have such a directionality and therefore deform more easily. Polymers are composed of carbon-chain molecules formed by covalent bonds. However, they may deform easily by relative sliding between the chain molecules where this is prevented only by secondary bonds.

A variety of crystal structures exist in solid materials. Three of particular importance for metals are the body-centered cubic (BCC), face-centered cubic (FCC), and hexagonal close-packed (HCP) structures. The crystal structures of ceramics are often elaborations of these simpler structures, but greater complexity exists due to the necessity in these compounds of accommodating more than one type of atom. Crystalline materials (metals and ceramics) are composed of aggregations of small crystal grains. Numerous defects, such as vacancies, interstitials, and dislocations are usually present in these grains.

Elastic deformation, being caused by the stretching of chemical bonds, disappears if the stress is removed. The elastic modulus E is therefore higher if the bonding is stronger, being highest in covalent solids such as diamond. Metals have a value of E about ten times lower than that for highly covalent solids, and polymers have a value of E that is generally lower by an additional factor of ten or more, due to the influence of the chain-molecule structure and secondary bonds. Above the glass-transition temperature for a given polymer, the secondary bonds melt, and E is further lowered by a large amount, then being typically smaller than for diamond by a factor on the order of 10^5.

Estimates of the theoretical tensile strength to break chemical bonds in perfect crystals give values on the order of $E/10$. However, strengths approaching such a high value are realized only in tiny perfect single crystals and in fine wires with an aligned structure. Strengths in large samples of material are much lower as these are weakened by defects. In ceramics, the defects of importance are small cracks and fissures that contribute to brittle behavior.

In metals, the defects that lower the strength are primarily dislocations. These move under the influence of applied stresses and cause yielding behavior. In large single crystals containing a few dislocations, yielding occurs at very low stresses that are lower than the theoretical value by a factor of 300 or more. Strengths are increased above this value if there are obstacles to dislocation motion, such as grain boundaries, hard second-phase particles, alloying elements, and dislocation entanglements. The resulting strength for engineering metals in bulk form may be as high as one tenth of the theoretical value of $E/10$, that is, around $E/100$.

Materials are also subject to time-dependent deformation called creep. Such deformation is especially likely at temperatures approaching melting. Physical mechanisms vary with material and temperature. Examples include diffusion of vacancies in metals and ceramics and sliding of chain molecules in polymers.

The necessarily brief treatment given in this chapter on structure and deformation in materials represents only a minimal introduction to the topic. More detail is given in a number of excellent books, a few of which are listed as References at the end of this chapter.

NEW TERMS AND SYMBOLS

body-centered cubic (BCC) structure
close-packed planes, directions
covalent bond
diamond cubic structure
edge dislocation
face-centered cubic (FCC) structure
glass transition temperature, T_g
grain boundary
hexagonal close-packed (HCP) structure
interstitial
ionic bond
lattice plane; lattice site

melting temperature, T_m
metallic bond
screw dislocation
secondary (hydrogen) bond
slip plane
slip step
substitutional impurity
theoretical cohesive strength, $\sigma_b \approx E/10$
theoretical shear strength, $\tau_b \approx G/10$
unit cell
vacancy

REFERENCES

ASHBY, M. F. and D. R. H. JONES. 1980 *Engineering Materials 1: An Introduction to Their Properties and Applications*, Pergamon Press, Oxford, UK.

BOYER, H. E. and T. L. GALL, eds. 1985 *Metals Handbook: Desk Edition*, American Society for Metals, Metals Park, Oh.

FLINN, R. A. and P. K. TROJAN. 1990 *Engineering Materials and Their Applications*, Houghton Mifflin Co., Boston, Ma.

HAYDEN, H. W., W. G. MOFFATT and J. WULFF. 1965 *The Structure and Properties of Materials, Vol III: Mechanical Behavior*, John Wiley, New York.

KELLY, A. and N. H. MACMILLAN. 1986 *Strong Solids*, 3rd ed., Clarendon Press, Oxford, UK.

MOFFATT, W. G., G. W. PEARSALL and J. WULFF. 1964 *The Structure and Properties of Materials, Vol. I: Structure*, John Wiley, New York.

PROBLEMS AND QUESTIONS

Section 2.4

2.1 Based on your knowledge of chemistry, qualitatively explain the values of the elastic moduli given in Table 2.2(a) for SiC and NaCl. (You may consult a chemistry text if desired.)

2.2 Table 2.2(b) gives a value of $E = 160$ GPa for a fiber of linear polyethylene, in which the polymer chains are aligned with the fiber axis. Why is this value so much higher than the typical $E = 3$ GPa mentioned for polymers, being in fact almost as high as the value for iron and steel?

2.3 Consider Fig. 2.17 and assume that it is possible to make accurate measurements of the elastic modulus E for high stresses in both tension and compression. Describe the expected variation of E with stress.

2.4 Consider Fig. 2.17 and two atoms that are initially an infinite distance apart, $x = \infty$, at which point the potential energy of the system is $U = 0$. If they are brought together to $x = x_1$, the potential energy is related to the total force P by

$$\left. \frac{dU}{dx} \right|_{x=x_1} = P$$

Based on this, qualitatively sketch the variation of U with x. What happens at $x = x_e$? What is the significance of $x = x_e$ in terms of the potential energy?

2.5 In Table 2.2, compare the strengths of Al_2O_3 whiskers versus Al_2O_3 fibers, and also compare the two diameters of tungsten wire to each other. Can you explain the large differences observed?

2.6 Consult Flinn and Trojan (1990) or Moffatt et al. (1964) in the References, or other materials science or chemistry text, and study the crystal structure of carbon in the form of graphite. How does the structure differ from that of diamond? Why is graphite in bulk form usually soft and weak? And how could a whisker of such a material have the high strength and elastic modulus indicated in Table 2.2(a)?

Section 2.5

2.7 Identify the slip plane and the slip step in Fig. 2.22(c) for plastic deformation by motion of a screw dislocation.

2.8 Cold working a metal by rolling it to a lesser thickness or hammering it introduces a large number of dislocations into the crystal structure. Would you expect the yield strength to be affected by this, and if so should it increase or decrease, and why? Also, answer the same question for the elastic modulus.

2.9 In metals, grain size d is observed to be related to yield strength by

$$\sigma_o = A + Bd^{-1/2}$$

where A and B are constants for a given material. Does this trend make physical sense? Can you explain qualitatively why this equation is reasonable? What physical interpretation can you make of the constant A?

2.10 An important group of polymers called thermosetting plastics forms a network structure by means of covalent bonds between the chain molecules. How would you expect these to differ from other polymers as to the value of the elastic modulus and the resistance to creep deformation, and why?

2.11 In ionic solids such as NaCl, plastic deformation is generally limited to those planes where the relative slip does not result in bringing ions of like charge in close proximity to one another. With this in mind, identify two likely slip planes, and two unlikely ones, based on a two-dimensional sketch similar to the one in Fig. 2.3. Can you extend your thinking to three dimensions using Fig. 2.4 to identify at least one likely and one unlikely slip plane?

3

A Survey of Engineering Materials

3.1 INTRODUCTION

Materials used for resistance to mechanical loading, which are here termed *engineering materials*, can belong to any of four major classes: metals and alloys, polymers, ceramics and glasses, and composites. The first three of these categories have already been discussed to an extent in the previous chapter from the viewpoint of structure and deformation mechanisms. Examples of members of each class are given in Table 2.1, and their general characteristics are in Fig. 2.1.

In this chapter, each major class of materials is considered in more detail. Groups of related materials within each major class are identified, the effects of processing variables are summarized, and the systems used for naming various materials are described. Metals and alloys are the dominant engineering materials in current use in many applications, and so more space is devoted to these than to the others. However, polymers, ceramics and glasses, and composites are also of major importance. Recent improvements in nonmetallic and composite materials have resulted in a trend where they are replacing metals in some applications.

Design engineers need a general knowledge of the composition, structure, and characteristics of materials. Such knowledge serves as a guide in selecting candidate materials for use in particular situations. More specific quantitative prediction of the mechanical behavior, using methods described in later chapters, can then be pursued in finalizing the engineering design.

3.2 ALLOYING AND PROCESSING OF METALS

Approximately 80% of the one-hundred-plus elements in the periodic table can be classed as metals. A number of these possess combinations of availability and properties that lead to their use as *engineering metals* where mechanical strength is needed. The most widely used engineering metal is iron, which is the main constituent of the iron-based alloys termed steels. Some other structural metals that are widely used are aluminum, copper, titanium, magnesium, nickel, and cobalt. Additional common metals, such as zinc, lead, tin, and silver, are used where the stresses are quite low, as in various low-strength cast parts and solder joints. The *refractory metals*, notably molybdenum, niobium, tantalum, tungsten, and zirconium, have melting temperatures somewhat or even substantially above that of iron (1536°C). Relatively small quantities of these are used as engineering metals for specialized applications, particularly where high strength is needed at a very high temperature. Some properties and uses for selected engineering metals are given in Table 3.1.

A metal alloy is a usually melted-together combination of two or more chemical elements, where the bulk of the material consists of one or more metals. A wide variety of metallic and nonmetallic chemical elements are used in alloying the principal engineering metals. Some of the more common ones are boron, carbon, magnesium, silicon, vanadium, chromium, manganese, nickel, copper, zinc, molybdenum, and tin. The amounts and combinations of alloying elements used with various metals have major effects on their strength, ductility, temperature resistance, corrosion resistance, and other properties.

For a given alloy composition, the properties are further affected by the particular processing used. Processing includes *heat treatment, deformation*, and *casting*. In heat treatment, a metal or alloy is subjected to a particular schedule of heating, holding at temperature, and cooling that causes desirable physical or chemical changes. Deformation is the process of forcing a piece of material to change its thickness or shape. Some of the means of doing so are *forging, rolling, extruding*, and *drawing* as illustrated in Fig. 3.1. Casting is simply the pouring of melted metal into a mold so that it conforms to the shape of the mold when it solidifies. Heat treatment and deformation or casting may be used in combination, and particular alloying elements are often added because they influence such processing in a desirable way. Metals that are subjected to deformation as the final processing step are termed *wrought metals* to distinguish them as a group from cast metals.

The details of alloying and processing are chosen so that the material has appropriate temperature resistance, corrosion resistance, strength, ductility, and other required

TABLE 3.1   PROPERTIES AND USES FOR SELECTED ENGINEERING METALS AND THEIR ALLOYS

Metal	Melting Temp.	Density	Elastic Modulus	Typical Strength	Uses; Comments
	T_m °C	ρ g/cm^3	E GPa (10^3 ksi)	σ_u MPa (ksi)	
Iron (Fe) and steel	1536	7.87	212 (30.7)	200 to 2500 (30 to 360)	Diverse: structures, machine and vehicle parts, tools. Most widely used engineering metal.
Aluminum (Al)	660	2.70	70 (10.2)	140 to 550 (20 to 80)	Aircraft and other lightweight structure and parts.
Titanium (Ti)	1668	4.51	120 (17.4)	340 to 1200 (50 to 170)	Aircraft structure and engines; industrial machine parts.
Copper (Cu)	1083	8.96	130 (18.8)	170 to 1000 (25 to 150)	Electrical conductors; corrosion-resistant parts, valves, pipes. Alloyed to make bronze and brass.
Magnesium (Mg)	650	1.74	45 (6.5)	170 to 340 (25 to 50)	Parts for high-speed machinery; aerospace parts.
Nickel (Ni)	1453	8.9	210 (30.5)	340 to 1400 (50 to 200)	Jet engine parts; alloying addition for steels.
Cobalt (Co)	1495	8.85	211 (30.6)	230 to 1000 (33 to 150)	Jet engine parts; wear resistant coatings.
Tungsten (W)	3410	19.3	411 (59.6)	860 (125)	Electrodes, light bulb filaments, flywheels, gyroscopes.
Lead (Pb)	327	11.3	16 (2.3)	14 to 70 (2 to 10)	Corrosion resistant piping; weights, shot. Alloyed with tin in solders.

Notes: The values of T_m, ρ, and E are only moderately sensitive to alloying. Ranges for σ_u and uses include alloys based on these metals.

Source: Data in [Boyer 85].

characteristics for its intended use. Recalling that plastic deformation is due to the motion of dislocations, the yield strength of a metal or alloy can usually be increased by intro-ducing obstacles to dislocation motion. Such obstacles can be tangles of dislocations, grain boundaries, distorted crystal structure due to impurity atoms, or small particles dispersed in the crystal structure. Some of the principal processing methods used for strengthening metals are listed along with the type of obstacle in Table 3.2. We will now discuss each of these methods.

3.2.1 Cold Work and Annealing

Cold work is the severe deforming of a metal at ambient temperature, often by rolling or drawing. This causes a dense array of dislocations and in general disorders the crystal structure, resulting in an increase in yield strength and a decrease in ductility.

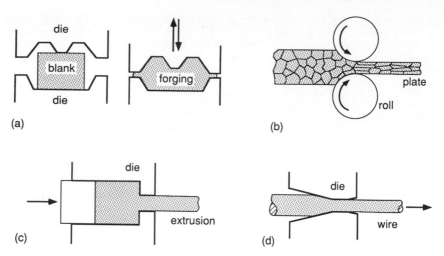

Figure 3.1 Some methods of forming metals into useful shapes: (a) forging employs compression or hammering; other methods include (b) rolling, (c) extrusion, and (d) drawing.

TABLE 3.2 STRENGTHENING METHODS FOR METALS AND ALLOYS

Method	Features That Impede Dislocation Motion
Cold work	High dislocation density causing tangles
Grain refinement	Changes in crystal orientation and other irregularities at grain boundaries
Solid solution strengthening	Interstitial or substitutional impurities distorting the crystal lattice
Precipitation hardening	Fine particles of a hard material precipitating out of solution upon cooling
Multiple phases	Discontinuities in crystal structure at phase boundaries
Quenching and tempering	Multiphase structure of martensite and Fe_3C precipitates in BCC iron

Strengthening occurs because the large number of dislocations form dense tangles that act as obstacles to further deformation. Hence, controlled amounts of cold work can be used to vary the properties. For example, this is done for copper and its alloys.

The effects of cold work can be partially or completely reversed by heating the metal to such a high temperature that new crystals form within the solid material, a process called *annealing*. If this is done following severe cold work, the recrystallized grains are at first quite small. Cooling the material at this stage creates a situation where strengthening is said to be due to *grain refinement*, because the grain boundaries impede dislocation motion. A long annealing time, or annealing at a higher temperature, causes the grains to coalesce into larger sizes, resulting in a loss of strength but a gain in ductility. The microstructural changes involved in cold working and annealing are illustrated in Fig. 3.2.

3.2.2 Solid Solution Strengthening

Solid solution strengthening occurs as a result of impurity atoms distorting the crystal lattice and thus making dislocation motion more difficult. Note that alloying elements are said to form a solid solution with the major constituent if their atoms are incorporated into the crystal structure in an orderly manner. The atoms providing the strengthening may be located at either interstitial or substitutional lattice positions. Atoms of much smaller size than those of the major constituent usually form interstitial alloys, as for hydrogen, boron, carbon, nitrogen, and oxygen in metals. Substitutional alloys may be formed by combinations of two or more metals, especially if the atomic sizes are similar and the preferred crystal structures are the same.

As might be expected, the effect of a substitutional impurity is greater if the atomic size differs more from that of the major constituent. This is illustrated by the effects of various percentages of alloying elements in copper in Fig. 3.3. Zinc and nickel have

Figure 3.2   Microstructures of 70% Cu, 30% Zn brass in three conditions: cold worked (above); annealed 1 hour at 375°C (left); and annealed 1 h at 500°C (right). (Photos courtesy of Olin Corp., New Haven, Ct.)

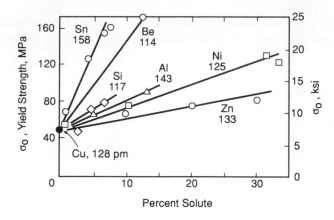

Figure 3.3 Effect of alloying on the yield strength of copper. Atomic sizes are given in picometers, that is, 10^{-12} m, and yield strengths correspond to 1% strain. (Adapted from [French 50]; used with permission.)

atomic sizes that do not differ very much from that of copper, so that the strengthening effect is small. But the small atoms of beryllium and the large ones of tin have a dramatic effect.

3.2.3 Precipitation Hardening and Other Multiple Phase Effects

The solubility of a particular impurity species in a given metal may be quite limited if the two elements have dissimilar chemical and physical properties, but this limited solubility usually increases with temperature. Such a situation may provide an opportunity for strengthening due to *precipitation hardening*. Consider an impurity that exists as a solid solution while the metal is held at a relatively high temperature, but also assume that the amount present exceeds the solubility limit for room temperature. Upon cooling, the impurity tends to precipitate out of solution, sometimes forming a chemical compound in the process. The precipitate is said to constitute a *second phase* as the chemical composition differs from that of the surrounding material. The yield strength may be increased substantially if the second phase has a hard crystal structure that resists deformation, and particularly if it exists as very small particles that are distributed fairly uniformly.

For example, aluminum with around 4% copper forms strengthening precipitates of the chemical compound $CuAl_2$. The means of achieving this is illustrated in Fig. 3.4. Slow cooling allows the impurity atoms to move relatively long distances, and the precipitate forms along the grain boundaries where it has little benefit. However, substantial benefit can be achieved by rapid cooling to form a supersaturated solution and then reheating to an intermediate temperature for a limited time. The reduced movement of the impurities at the intermediate temperature causes the precipitate to form as fairly uniformly scattered small particles. However, if the intermediate temperature is too high, or the holding time too long, the particles coalesce into larger ones and some of the benefit is lost, the particles being too far apart to effectively impede dislocation motion. Thus, for a given temperature, there is a precipitation (aging) time that gives the maximum

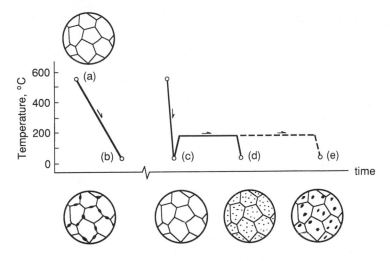

Figure 3.4   Precipitation hardening of aluminum alloyed with 4% Cu. Slow cooling from a solid solution (a) produces grain boundary precipitates (b). Rapid cooling to obtain a supersaturated solution (c) can be followed by aging at a moderate temperature to obtain fine precipitates within grains (d), but overaging gives coarse precipitates (e).

effect. The resulting trends in strength are illustrated for a commercial aluminum alloy, where similar precipitation hardening occurs, in Fig. 3.5.

If an alloy contains interspersed regions of more than one chemical composition, as for precipitate particles as just discussed, a *multiple phase* situation is said to exist. Other multiple phase situations involve needle-like or layered microstructural features, or crystal grains of more than one type. For example, some titanium alloys have a two-phase structure involving grains of both alpha (HCP) and beta (BCC) crystal structures. Multiple phases increase strength because the discontinuities in the crystal structure at the phase boundaries make dislocation motion more difficult, and also because one phase

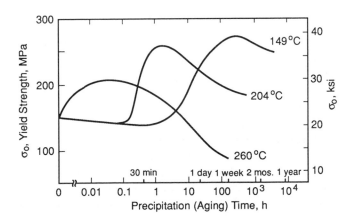

Figure 3.5   Effects of precipitation (aging) time and temperature on the resulting yield strength in aluminum alloy 6061. (Adapted from [Boyer 85] p. 6.7; used with permission.)

may be resistant to deformation. The two-phase (alpha-beta) structure just mentioned for titanium provides a portion of the strengthening for the highest strength titanium alloys. Also, the processing of steels by *quenching and tempering*, which will be discussed below, owes its benefits to multiple phase effects.

3.3 IRONS AND STEELS

Iron-based alloys, also called ferrous alloys, include cast irons and steels and are the most widely used structural metals. *Steels* consist primarily of iron and contain some carbon and manganese, and often additional alloying elements. They are distinguished from nearly pure iron, which is called *ingot iron*, and also from *cast irons*, which contain carbon in excess of its solubility limit and from 1 to 3% silicon. Irons and steels can be divided into various classes depending on their alloy compositions and other characteristics as indicated in Table 3.3. Some examples of particular irons or steels and their alloy compositions are given in Table 3.4.

A wide variation in properties exists for various steels as illustrated in Fig. 3.6. Pure iron is quite weak but is strengthened considerably by the addition of small amounts of carbon. Additional alloying with small amounts of niobium, vanadium, copper, or other elements permits strengthening by grain refinement, precipitation, or solid solution effects. If sufficient carbon is added for quenching and tempering to be effective, a major increase in strength is possible. Additional alloying and special processing can be combined with quenching and tempering and/or precipitation hardening to achieve even higher strengths.

TABLE 3.3 COMMONLY ENCOUNTERED CLASSES OF IRONS AND STEELS

Class	Distinguishing Features	Typical Uses	Source of Strengthening
Cast iron	More than 2% C and 1 to 3% Si	Pipes, valves, gears, engine blocks	Ferrite-pearlite structure as affected by free graphite
Plain-carbon steel	Principal alloying element is carbon up to 1%	Structural and machine parts	Ferrite-pearlite structure if low carbon; quenching and tempering if medium to high carbon
Low-alloy steel	Metallic elements totaling up to 5%	High-strength structural and machine parts	Grain refinement, precipitation, and solid solution if low carbon; otherwise quenching and tempering
Stainless steel	At least 10% Cr; does not rust	Corrosion resistant piping and nuts and bolts; turbine blades	Quenching and tempering if < 15% Cr and low Ni; otherwise cold work or precipitation
Tool steel	Heat treatable to high hardness and wear resistance	Cutters, drill bits, dies	Quenching and tempering, etc.

TABLE 3.4   SOME TYPICAL IRONS AND STEELS

Description	Identification	UNS No.	Principal Alloying Elements, Typical % by Weight							
			C	Cr	Mn	Mo	Ni	Si	V	Other
Ductile cast iron	ASTM A395	F32800	3.5	—	—	—	—	2	—	—
Low-carbon steel	AISI 1020	G10200	0.2	—	0.45	—	—	0.2	—	—
Medium-carbon steel	AISI 1045	G10450	0.45	—	0.75	—	—	0.2	—	—
High-carbon steel	AISI 1095	G10950	0.95	—	0.4	—	—	0.2	—	—
Low-alloy steel	AISI 4340	G43400	0.40	0.8	0.7	0.25	1.8	0.2	—	—
HSLA steel	ASTM A588-A	K11430	0.15	0.5	1.1	—	—	0.2	0.05	0.3 Cu
Martensitic stainless steel	AISI 403	S40300	0.15	12	1.0	—	0.6	0.5	—	—
Austenitic stainless steel	AISI 310	S31000	0.25	25	2.0	—	20	1.5	—	—
Precipitation hardening stainless steel	17-4 PH	S17400	0.07	17	1.0	—	4	1.0	—	4 Cu 0.3 (Nb+Ta)
Tungsten high-speed tool steel	AISI T1	T12001	0.75	3.8	0.25	—		0.2	0.3	1.1	18 W
18 Ni maraging steel	ASTM A538-C	K93120	0.01	—	—		5	18	—	—	9 Co, 0.7 Ti

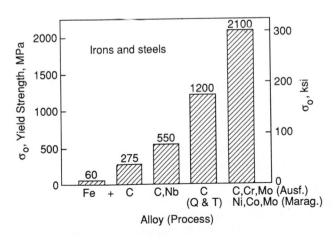

Figure 3.6 Effects of alloying additions and processing (x-axis) on the yield strength of steel. Alloying iron with carbon and other elements provides substantial strengthening, but even higher strengths can be achieved by heat treating using quenching and tempering. The highest strengths are obtained by combining alloying with special processing, such as ausforming or maraging. (Adapted from an illustration courtesy of R. W. Landgraf, Virginia Tech, Blacksburg, Va.)

3.3.1 Naming Systems for Irons and Steels

A number of different organizations have developed naming systems and specifications for various irons and steels that give the required alloy composition and sometimes required mechanical properties. These include the American Iron and Steel Institute (AISI), the Society of Automotive Engineers (SAE), the American Society for Testing and Materials (ASTM), and the American Society of Mechanical Engineers (ASME). In addition, SAE and ASTM have recently cooperated to develop a new Unified Numbering System (UNS) that gives designations not only for irons and steels but for all other metal alloys. See the *Metals Handbook* (Boyer and Gall, 1985) for an introduction to various naming systems, and the current publication on the UNS System (SAE, 1989), for a description of those designations and their equivalence with other specifications.

The AISI and SAE designations for various steels are coordinated between the two organizations and are nearly identical. Details for common carbon and low-alloy steels are given in Table 3.5. Note that in this case there is usually a four-digit number. The first two digits specify the alloy content other than carbon, and the second two give the carbon content in hundredths of a percent. For example, AISI 1340 (or SAE 1340) contains 0.40% carbon with 1.75% manganese as the only other alloying element. (Percentages of alloys are always given on the basis of weight.)

The UNS system has a letter followed by a five-digit number. The letter indicates the category of alloy, such as *F* for cast irons, *G* for carbon and low-alloy steels in the AISI–SAE naming system, *K* for various special-purpose steels, *S* for stainless steels, and *T* for tool steels. For carbon and low-alloy steels, the number is in most cases the same as that used by AISI and SAE except that a zero is added at the end. Thus, AISI 1340 is the same steel as UNS G13400.

Some particular classes of irons and steels will now be considered.

3.3.2 Cast Irons

Cast irons in various forms have been used for hundreds of years and continue to be relatively inexpensive and useful materials. The iron is not highly refined subsequent to extraction from ore or scrap, and it is formed into useful shapes by melting and pouring into molds. Prior to the fourteenth century, the temperature that could be obtained in a furnace did not allow iron to be melted. This resulted in the use of *wrought iron*, which is heated and forged into useful shapes but never melted in processing. Several different types of cast iron exist. All contain large amounts of carbon, typically 2 to 4% by weight, and also 1 to 3% silicon. The large amount of carbon present exceeds the 2% that can be held in solid solution, and in most cast irons the excess is present in the form of graphite.

Gray iron contains graphite in the form of flakes as seen in Fig. 3.7(left). These flakes easily develop into cracks under tensile stress, so that gray iron is relatively weak and brittle in tension. In compression, the strength and ductility are both considerably higher than for tension. *Ductile iron*, also called *nodular iron*, contains graphite in the more nearly spherical form of nodules as seen in Fig. 3.7(right). This is achieved by careful control of impurities and by adding small amounts of magnesium or other

TABLE 3.5 SUMMARY OF THE AISI-SAE DESIGNATIONS FOR COMMON CARBON AND LOW-ALLOY STEELS

Designation[1]	Approx. Alloy Content, %	Designation	Approx. Alloy Content, %
Carbon steels		*Nickel-molybdenum steels*	
10XX	Plain carbon	46XX	Ni 0.85 or 1.82; Mo 0.25
11XX	Resulfurized	48XX	Ni 3.50; Mo 0.25
12XX	Resulfurized and rephosphorized		
15XX	Mn 1.00 to 1.65		
Manganese steels		*Chromium steels*	
13XX	Mn 1.75	50XX(X)	Cr 0.27 to 0.65
		51XX(X)	Cr 0.80 to 1.05
		52XXX	Cr 1.45
Molybdenum steels		*Chromium-vanadium steels*	
40XX	Mo 0.25	61XX	Cr 0.6 to 0.95; V 0.15
44XX	Mo 0.40 or 0.52		
Chromium-molybdenum steels		*Silicon-manganese steels*	
41XX	Cr 0.50 to 0.95; Mo 0.12 to 0.30	92XX	Si 1.40 or 2.00; Mn 0.70 to 0.87; Cr 0 or 0.70
Nickel-chromium-molybdenum steels		*Boron steels[2]*	
43XX	Ni 1.82; Cr 0.50 or 0.80; Mo 0.25	YYBXX	B 0.0005 to 0.003
47XX	Ni 1.45; Cr 0.45; Mo 0.20 or 0.35		
81XX	Ni 0.30; Cr 0.40; Mo 0.12		
86XX	Ni 0.55; Cr 0.50; Mo 0.20		
87XX	Ni 0.55; Cr 0.50; Mo 0.25		
94XX	Ni 0.45; Cr 0.40; Mo 0.12		

Notes: [1]Replace "XX" or "XXX" with carbon content in hundredths of a percent, such as AISI 1045 having 0.45% C, or 52100 having 1.00% C. [2]Replace "YY" with any two digits from earlier in table to indicate the additional alloy content.

elements that aid in nodule formation. As a result of the different form of the graphite, ductile iron has considerably greater strength and ductility in tension than gray iron.

In *malleable iron*, special heat treatment is used to obtain a result similar to that for ductile iron. *White iron* is formed by rapid cooling of a melt that would otherwise form gray iron. The excess carbon is in the form of a multiphase network involving large amounts of iron carbide, Fe_3C, also called *cementite*. This very hard and brittle phase results in the bulk material also being hard and brittle. In addition, various alloying elements are used in making special purpose cast irons that have improved response to processing or desirable properties such as resistance to heat or corrosion.

3.3.3 Carbon Steels

Plain-carbon steels contain carbon, in amounts usually less than 1%, as the alloying element that controls the properties. They also contain limited amounts of manganese

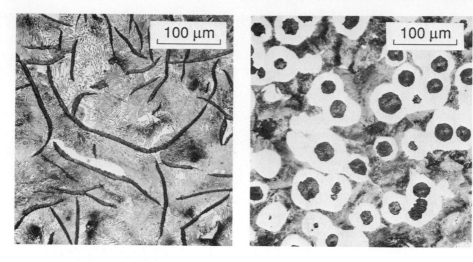

Figure 3.7 Microstructures of gray cast iron (left) and ductile (nodular) cast iron (right). The graphite flakes on the left are the heavy dark bands, and the graphite nodules on the right are the dark shapes. In gray iron (left), the fine lines are a pearlitic structure similar to that in mild steel. (Photos courtesy of Deere and Co., Moline, Il.)

and (generally undesirable) impurities, such as sulfur and phosphorus. The more specific terms *low-carbon steel* and *mild steel* are often used to indicate a carbon content of less than 0.25%, such as AISI 1020 steel. These steels have relatively low strength but excellent ductility. The structure is a mixture of BCC iron, also called α-iron or *ferrite*, and *pearlite*. Pearlite is a layered two-phase structure of ferrite and cementite (Fe_3C) as seen in Fig. 3.8(left). Low-carbon steels can be strengthened somewhat by cold working, but only minor strengthening is possible by heat treatment. Uses include structural steel for buildings and bridges, and sheet metal applications, such as automobile bodies.

Medium-carbon steels, with carbon content around 0.3 to 0.6%, and *high-carbon steels*, with carbon content around 0.7 to 1% and above, have higher strengths than low-carbon steels as a result of the presence of more carbon. In addition, the strength can be increased significantly by heat treatment using the quenching and tempering process, increasingly so for higher carbon contents. However, high strengths are accompanied by loss of ductility, that is, by more brittle behavior. Medium-carbon steels have a wide range of uses as shafts and other components of machines and vehicles. High-carbon steels are limited to uses where their high hardness is beneficial and the low ductility is not a serious disadvantage, as in cutting tools and springs.

In *quenching and tempering*, the steel is first heated to about 850°C so that the iron changes to the FCC phase known as γ-iron or *austenite*, with carbon being in solid solution. A supersaturated solution of carbon in iron is then formed by rapid cooling, called *quenching*, which can be accomplished by immersing the hot metal into water or oil. After quenching, a structure called *martensite* is present, which has a BCC lattice distorted by interstitial carbon atoms. The martensite exists as either groupings of parallel

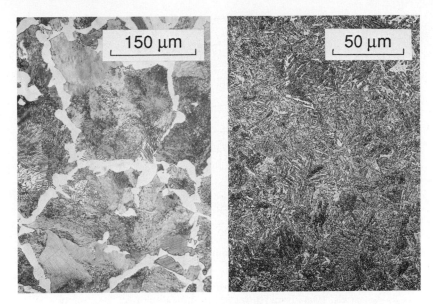

Figure 3.8   Steel microstructures: ferrite-pearlite structure in normalized AISI 1045 steel (left), with ferrite being the light-colored areas and pearlite the striated regions; quenched and tempered structure in AISI 4340 steel (right). (Left photo courtesy of Deere and Co., Moline, Il.)

thin crystals (laths), or as more randomly oriented thin plates, surrounded by regions of austenite.

As-quenched steel is very hard and brittle due to the two phases present, the distorted crystal structure, and a high dislocation density. To obtain a useful material, it must be subjected to a second stage of heat treatment at a lower temperature, called *tempering*. This causes removal of some of the carbon from the martensite and the formation of dispersed particles of Fe_3C. Tempering lowers the strength but increases the ductility. The effect is greater for higher tempering temperatures and varies with carbon content and alloying as illustrated in Fig. 3.9. The microstructure of a quenched and tempered steel is shown in Fig. 3.8(right).

3.3.4 Low-Alloy Steels

In *low-alloy steels*, also often called simply *alloy steels*, small amounts of alloying elements totaling no more than about 5% are added to improve various properties or the response to processing. Percentages of the principal alloying elements are given for some of these in Table 3.5. As examples of the effects of alloying, sulfur improves machineability, and molybdenum and vanadium promote grain refinement. The combination of alloys used in the steel AISI 4340 gives improved strength and toughness, that is, resistance to failure due to a crack or sharp flaw. In this steel, the metallurgical changes during quenching proceed at a relatively slow rate so that quenching and tempering is

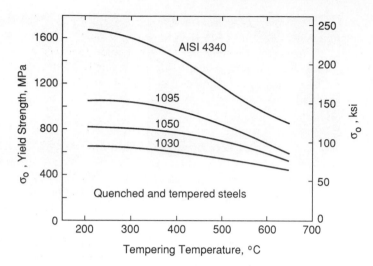

Figure 3.9   Effect of tempering temperature on the yield strength for several steels. (Data from [Boyer 85] p. 4.21.)

effective in components as thick as 100 mm. Note that the corresponding plain-carbon steel, AISI 1040, requires very rapid quenching that cannot be achieved except within about 5 mm of the surface.

Various special-purpose low-alloy steels are used that may not fit any of the standard AISI-SAE designations. Many of these are described in the *ASTM Standards* (ASTM, 1990), where requirements are placed on mechanical properties in addition to alloy content. Some of these are classified as *high-strength low-alloy* (HSLA) steels, which have a low carbon content and a ferritic-pearlitic structure, with small amounts of alloying resulting in higher strengths than other low-carbon steels. Examples include structural steels as used in buildings and bridges, such as ASTM A242, A441, A572, and A588. Note that use of the words "high-strength" here can be somewhat misleading, as the strengths are high for a low-carbon steel, but not nearly as high as for many quenched and tempered steels. The low-alloy steels used for pressure vessels, such as ASTM A302, A517, and A533, constitute an additional group of special purpose steels.

3.3.5 Stainless Steels

Steels containing at least 10% chromium are called *stainless steels* because they have good corrosion resistance, that is, they do not rust. These alloys also often have improved resistance to high temperature. A separate system of AISI designations is used which employs a three-digit number, such as AISI 316 and AISI 403, with the first digit indicating a particular class of stainless steels. The corresponding UNS designations often use the same digits, such as S31600 and S40300 for the two just listed.

The 400-series stainless steels have carbon in various percentages and also small amounts of metallic alloying elements in addition to the chromium. If the chromium content is less than about 15%, as in types 403, 410, and 422, the steel can in most cases be heat treated using quenching and tempering to have a martensitic structure, so that it is called a *martensitic stainless steel*. Uses include tools and blades in steam turbines. However, if the chromium content is higher, typically 17 to 25%, the result is a *ferritic stainless steel* that can be strengthened only by cold work, and then only modestly. These are used where high strength is not as essential as high corrosion resistance, as in architectural use.

The 300-series stainless steels, such as types 304, 310, 316, and 347, contain around 10 to 20% nickel in addition to 17 to 25% chromium. The nickel further enhances corrosion resistance and results in the FCC crystal structure being stable even at low temperatures, so that these are termed *austenitic stainless steels*. These materials either are used in the annealed condition or are strengthened by cold work, and they have excellent ductility and toughness. Uses include nuts and bolts, and pressure vessels and piping.

Another group is the *precipitation-hardening stainless steels*. These are strengthened as the name implies and are used in various high-stress applications where resistance to corrosion and high temperature are required, as in heat-exchanger tubes and turbine blades. An example is 17-4 PH stainless steel (UNS S17400), which contains 17% chromium and 4% nickel, hence its name, and also 4% copper and smaller amounts of other elements.

3.3.6 Tool Steels and Other Special Steels

Tool steels are specially alloyed and processed to have high hardness and wear resistance for use in cutting tools and special components of machinery. Most contain several percent chromium, some have quite high carbon contents in the 1 to 2% range, and some contain fairly high percentages of molybdenum and/or tungsten. Strengthening generally involves quenching and tempering or related heat treatments. The AISI designations are in the form of a letter followed by a one or two-digit number. For example, tool steels M1, M2, etc., contain 5 to 10% molybdenum and smaller amounts of tungsten and vanadium, and tool steels T1, T2, etc. contain substantial amounts of tungsten, typically 18%.

The tool steel H11, containing 0.4% carbon, 5% chromium, and modest amounts of other elements, is used in various high-stress applications. It can be fully strengthened in thick sections up to 150 mm and retains moderate ductility and toughness even at very high yield strengths around 2100 MPa and above. This is achieved by the *ausforming* process, which involves deforming the steel at a high temperature within the range where the austenite (FCC) crystal structure exists. An extremely high dislocation density and a very fine precipitate are introduced, which combine to provide additional strengthening that is added to the usual martensite strengthening due to quenching and tempering. Ausformed H11 is one of the strongest steels that has reasonable ductility and toughness.

Various additional specialized high-strength steels have names that are nonstandard trade names. Examples include 300M, which is AISI 4340 modified with 1.6% silicon and some vanadium, and D-6a steel used in aerospace applications. Maraging steels contain 18% nickel and other alloying elements and have high strength and toughness due to a combination of a martensitic structure and precipitation hardening.

3.4 NONFERROUS METALS

Quenching and tempering to produce a martensitic structure is the most effective means of strengthening steels. In the common nonferrous metals, martensite may not occur, and where it does occur the effect is not as large as in steels. Hence, the other methods of strengthening, which are generally less effective, must be used. The higher strength nonferrous metals often employ precipitation hardening.

For example, consider the strength levels achievable in aluminum alloys as illustrated by Fig. 3.10. Annealed pure aluminum is very weak and can be strengthened only by cold work. Adding magnesium provides solid-solution strengthening, and the resulting alloy can be cold worked. Further strengthening is possible by precipitation hardening, which is achieved by various combinations of alloying elements and aging treatments. However, the highest strength available is only about 25% of that for the highest strength steel. Aluminum is nevertheless widely used, as in aerospace applications, where its light weight and corrosion resistance are major advantages that offset the disadvantage of lower strength than some steels.

We will now proceed to discuss the nonferrous metals that are commonly used in structural applications.

3.4.1 Aluminum Alloys

For aluminum alloys produced in wrought form, as by rolling or extruding, the naming system used involves a four-digit number. The first digit specifies the major alloying elements as listed in Table 3.6. Subsequent digits are then assigned to indicate specific

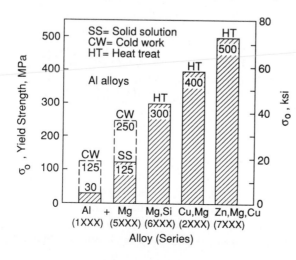

Figure 3.10 Effects of alloying additions and processing (x-axis) on the yield strength of aluminum alloys. Pure aluminum can be strengthened by cold work, and alloying increases strength due to solid-solution hardening. The higher strength alloys are heat treated to produce precipitation hardening. (Adapted from an illustration courtesy of R. W. Landgraf, Virginia Tech, Blacksburg, Va.)

TABLE 3.6   NAMING SYSTEM FOR COMMON WROUGHT ALUMINUM ALLOYS

Series	Major Additions	Other Frequent Additions	Heat Treatable
1XXX	None	None	No
2XXX	Cu	Mg, Mn, Si	Yes
3XXX	Mn	Mg, Cu	No
4XXX	Si	None	Most no
5XXX	Mg	Mn, Cr	No
6XXX	Mg, Si	Cu, Mn	Yes
7XXX	Zn	Mg, Cu, Cr	Yes

Processing Designations		Common TX Treatments	
-F	As fabricated	-T3	Cold worked, then naturally aged
-O	Annealed	-T4	Naturally aged
-H1X	Cold worked	-T6	Artificially aged
-H2X	Cold worked, then partially annealed	-T8	Cold worked, then artificially aged
-H3X	Cold worked, then stabilized	-TX51	Stress relieved by stretching
-TX	Solution heat treated, then aged		

alloys, with some examples being given in Table 3.7. The UNS numbers for wrought alloys are similar except that A9 precedes the four-digit number. Following the four-digit number, a processing code is used, as in 2024-T4, as detailed in Table 3.6.

For codes involving cold work, HXX, the first number indicates whether only cold work is used (H1X), or whether cold work is followed by partial annealing (H2X), or by a stabilizing heat treatment (H3X). The latter is a low temperature heat treatment that prevents subsequent gradual changes in the properties. The second digit indicates the degree of cold work, HX8 for the maximum effect of cold work on strength, and HX2, HX4, and HX6 for one-fourth, one-half, and three-fourths as much effect, respectively.

Processing codes of the form TX all involve a *solution heat treatment* at a high temperature to create a solid solution of alloying elements. This may or may not be followed by cold work, but the material is always subsequently *aged*, during which precipitation hardening occurs. *Natural aging* occurs at room temperature, whereas *artificial aging* involves a second stage of heat treatment as in Fig. 3.4. Additional digits following HXX or TX describe additional variations in processing, such as T651 for a T6 treatment in which the material is also stretched up to 3% in length to relieve residual (locked-in) stresses.

TABLE 3.7　SOME TYPICAL WROUGHT ALUMINUM ALLOYS

Identification	UNS No.	Principal Alloying Elements, Typical % by Weight					
		Cu	Cr	Mg	Mn	Si	Other
1100-O	A91100	0.12	—	—	—	—	—
2014-T6	A92014	4.4	—	0.5	0.8	0.8	—
2024-T4	A92024	4.4	—	1.5	0.6	—	—
2219-T851	A92219	6.3	—	—	0.3	—	0.1 V, 0.18 Zr
3003-H14	A93003	0.12	—	—	1.2	—	—
4032-T6	A94032	0.9	—	1.0	—	12.2	0.9 Ni
5052-H38	A95052	—	0.25	2.5	—	—	—
6061-T6	A96061	0.28	0.2	1.0	—	0.6	—
7075-T651	A97075	1.6	0.23	2.5	—	—	5.6 Zn

The alloy content determines the response to processing. Alloys in the 1XXX, 3XXX, and 5XXX series, and most of those in the 4XXX series, do not respond to precipitation-hardening heat treatment. These alloys achieve some of their strength from solid solution effects, and all can be strengthened beyond the annealed condition by cold work. The alloys capable of the highest strengths are those that do respond to precipitation hardening, namely the 2XXX, 6XXX, and 7XXX series. Additional details of the response to processing are also affected by the alloy content. For example, 2024 can be precipitation hardened by natural aging, but 7075 and similar alloys require artificial aging.

Aluminum alloys produced in cast form have a similar but separate naming system. A four-digit number with a decimal point is used, such as 356.0-T6. Corresponding UNS numbers have A0 preceding the four-digit number and no decimal point, such as A03560.

3.4.2 Titanium Alloys

The density of titanium is considerably greater than that of aluminum, but still only about 60% of that of steel. In addition, the melting temperature is somewhat greater than for steel and far greater than for aluminum. In aerospace applications, the strength-to-weight ratio is important, and in this respect the highest strength titanium alloys are comparable to the highest strength steels. These characteristics and good corrosion resistance have led to increasing application of titanium alloys since commercial development of the material began in the 1940s.

Since only about thirty different titanium alloys are in common use, it is sufficient to identify these by simply giving the weight percentages of alloying elements, such as Ti-6Al-4V, or Ti-10V-2Fe-3Al. Three categories exist, these being the alpha and near alpha alloys, the beta alloys, and the alpha-beta alloys. Although the alpha (HCP) crystal structure is stable at room temperature in pure titanium, certain combinations of alloying elements, such as chromium along with vanadium, cause the beta (BCC) structure to be be stable, or they result in a mixed structure. Small percentages of molybdenum or nickel improve corrosion resistance, and aluminum, tin, and zirconium improve creep resistance of the alpha phase.

Alpha alloys are strengthened mainly by solid solution effects and do not respond to heat treatment. The other alloys can be strengthened by heat treatment. As in steels, a martensitic transformation occurs upon quenching, but the effect is less. Precipitation hardening and the effects of complex multiple phases are the principal means of strengthening alpha-beta and beta alloys.

3.4.3 Other Nonferrous Metals

A wide range of copper alloys are employed in diverse applications as a result of their electrical conductivity, corrosion resistance, and attractiveness. Copper is easily alloyed with various other metals, and copper alloys are generally easy to deform or to cast into useful shapes. Strengths are typically lower than for the metals already discussed, but still sufficiently high that copper alloys are often useful as engineering metals.

Percentages of alloying elements range from relatively small to quite substantial, as for 35% zinc in common yellow brass. Copper with approximately 10% tin is called bronze, although this term is also used to describe various alloys with aluminum, silicon, zinc, and other elements. Copper alloys with zinc, aluminum, or nickel are strengthened by solid-solution effects. Beryllium additions permit precipitation hardening and produce the highest strength copper alloys. Cold work is also frequently used for strengthening, often in combination with the other methods. A variety of common names are in use for various copper alloys, such as *beryllium copper*, *naval brass*, and *aluminum bronze*. The UNS numbering system with a prefix letter *C* is used for copper alloys.

Magnesium has a melting temperature near that of aluminum but a density only 65% as great, making it only 22% as dense as steel and the lightest engineering metal. This silvery-white metal is most commonly produced in cast form but is also extruded, forged, and rolled. Alloying elements do not generally exceed 10% total for all additions, the most common being aluminum, manganese, zinc, and zirconium. Strengthening methods are roughly similar to those for aluminum alloys. The highest strengths are about 60% as large, resulting in comparable strength-to-weight ratios. The naming system in common use is generally similar to that for aluminum alloys but differs as to the details. A combination of letters and numbers that identifies the specific alloy is followed by a processing designation, such as AZ91C-T6.

Superalloys are special heat-resisting alloys that are used primarily above 550°C. The major constituent is either nickel or cobalt, or a combination of iron and nickel, and percentages of alloying elements are often quite large. For example, the Ni-base alloy Udimet 500 contains 48% Ni, 19% Cr, and 19% Co, and the Co-base alloy Haynes 188 has 37% Co, 22% Cr, 22% Ni, and 14% W, with both also containing small percentages of other elements. Nonstandard combinations of trade names and letters and numerals are commonly used to identify the relatively small number of superalloys that are in common use. Some examples in addition to the two mentioned above are Waspaloy, MAR-M302, A286, and Inconel 718.

Although nickel and cobalt have melting temperatures just below that of iron, superalloys have superior resistance to corrosion, oxidation, and creep compared to steels. Many have substantial strengths even above 750°C, which is beyond the useful range for low-alloy and stainless steels. This accounts for their use in high-temperature applications, despite high cost due to the relative scarcity of nickel, chromium, and cobalt. Superalloys are often produced in wrought form, and Ni-base and Co-base alloys are also often cast. Strengthening is primarily by solid-solution effects and by various heat treatments resulting in precipitation of intermetallic compounds or metal carbides.

3.5 POLYMERS

Polymers are materials made up of long-chain molecules formed primarily by carbon-to-carbon bonds. Examples include all materials commonly referred to as plastics, most familiar natural and synthetic fibers, rubbers, and cellulose and lignin in wood. Polymers that are produced or modified by man for use as engineering materials can be classified into three groups: thermoplastics, thermosetting plastics, and elastomers.

When heated, a *thermoplastic* softens and usually melts, and then if cooled returns to its original solid condition. The process can be repeated a number of times. However, a *thermosetting plastic* changes chemically during processing at elevated temperature. It will not melt upon reheating but will instead decompose, as by charring or burning. *Elastomers* are distinguished from plastics by being capable of rubbery behavior. In particular, they can be deformed by large amounts, say 100% to 200% strain or more, with most of this deformation being recovered after removal of the stress. Examples of polymers in each of these groups are listed along with typical uses in Table 3.8.

After chemical synthesis based primarily on petroleum products, polymers are made into useful shapes by various molding and extrusion processes, two of which are illustrated in Fig. 3.11. For thermosetting plastics and elastomers that behave in a similar manner, the final stage of chemical reaction is often accomplished by the application of temperature and/or pressure, and this must occur while the material is being molded into its final shape.

Polymers are named according to the conventions of organic chemistry. These sometimes lengthy names are often abbreviated by acronyms, such as PMMA for poly-methyl methacrylate. In addition, various trade names and popular names, such as Plexiglas, Teflon, and nylon, are often used in addition to, or in place of, the chemical names.

An important characteristic of polymers is their light weight. Most have a mass density similar to that of water, that is, around $\rho = 1$ g/cm^3, and few exceed $\rho = 2$ g/cm^3. Hence, polymers are typically half as heavy as aluminum ($\rho = 2.7$ g/cm^3), and much lighter than steel ($\rho \approx 7.9$ g/cm^3). Most polymers in unmodified form

TABLE 3.8   CLASSES, EXAMPLES, AND USES OF REPRESENTATIVE POLYMERS

Polymer	Typical Uses
(a) Thermoplastics: ethylene structure	
Polyethylene (PE)	Packaging, bottles, piping
Polyvinyl chloride (PVC)	Upholstery, tubing, electrical insulation
Polypropylene (PP)	Hinges, boxes, ropes
Polystyrene (PS)	Toys, appliance housings, foams
Polymethyl methacrylate (PMMA, Plexiglas, acrylic)	Windows, lenses, clear shields
Polytetrafluoroethylene (PTFE, Teflon)	Tubing, bottles, seals
Acrylonitrile butadiene styrene (ABS)	Telephone and appliance housings, toys
(b) Thermoplastics: others	
Nylon	Gears, tire cords, tool housings
Aramids (Kevlar, Nomex)	High strength fibers
Polyoxymethylene (POM, acetal)	Gears, fan blades, pipe fittings
Polyetheretherketone (PEEK)	Coatings, fans, impellers
Polycarbonate (PC)	Safety helmets and lenses
(c) Thermosetting plastics	
Phenol formaldehyde (phenolic, Bakelite)	Electrical plugs and switches, pot handles
Melamine formaldehyde	Plastic dishes, tabletops
Urea formaldehyde	Buttons, bottle caps, toilet seats
Epoxies	Matrix for composites
Unsaturated polyesters	Fiberglass resin
(d) Elastomers	
Natural rubber; cis-polyisoprene	Shock absorbers, tires
Styrene-butadiene rubber (SBR)	Tires, hoses, belts
Polyurethane elastomers	Shoe soles, electrical insulation
Nitrile rubber	O-rings, oil seals, hoses
Polychloroprene (Neoprene)	Wet suits, gaskets

are relatively weak, with ultimate tensile strengths being typically in the range 10 to 200 MPa.

In the more detailed discussion that follows, we first consider the basic molecular structure of typical polymers in each group. This provides the background for later discussion of how the details of molecular structure affect the mechanical properties.

3.5.1 Molecular Structure of Thermoplastics

Many thermoplastics have a molecular structure related to that of the hydrocarbon gas ethylene, C_2H_4. In particular, the *repeating unit* in the chain molecule is similar to

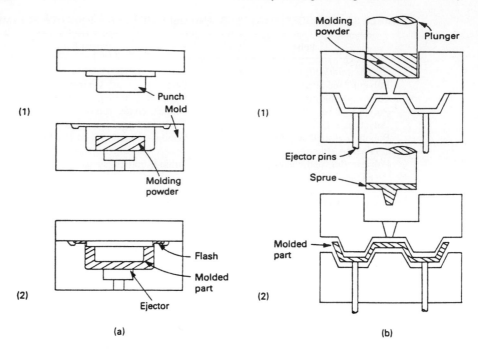

Figure 3.11   Forming of plastics by (a) compression molding and (b) transfer molding. (From [Farag 89] p. 91; used with permission.)

an ethylene molecule except that the carbon-to-carbon bond is rearranged as illustrated previously in Fig. 2.6. The molecular structures of some of the simpler polymers of this type are illustrated by giving their repeating unit structures in Fig. 3.12.

Polyethylene (PE) is the simplest case in that the only modification to the ethylene molecule is the rearranged carbon-to-carbon bond. In polyvinyl chloride (PVC), one of the hydrogen atoms is replaced by a chlorine atom, whereas polypropylene (PP) has a similar substitution of a methyl (CH_3) group. Polystyrene (PS) has a substitution of an entire benzene ring, and PMMA is based on two substitutions as shown. Polytetrafluoroethylene (PTFE), also known as Teflon, has four fluorine substitutions. Ethylene-based thermoplastics are by far the most widely used plastics, with PE, PVC, PP, and PS, accounting for more than half the weight of plastics usage.

However, other classes of thermoplastics that are used in smaller quantities are more suitable for engineering applications where high strength is needed. The *engineering plastics* include the nylons, the aramids such as Kevlar, polyoxymethylene (POM), polyethylene terephthalate (PET), polyphenylene oxide (PPO), and polycarbonate (PC). Their molecular structures are generally more complex than for the ethylene-based thermoplastics. The repeating unit structures of two of these, nylon 6 and polycarbonate, are shown as examples in Fig. 3.12. Other nylons, such as nylon 66 and nylon 12, have more complex structures than nylon 6. Kevlar belongs to the polyamide group along

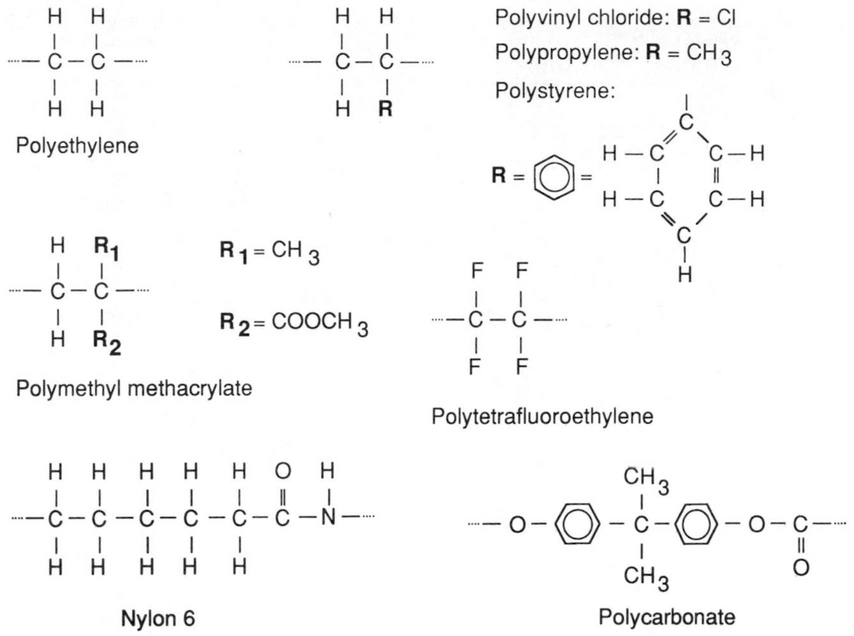

Figure 3.12   Molecular structure of several linear polymers. All but the last two are related to the polyethylene structure by simple substitutions, R, R_1 and R_2, or F.

with the nylons and has a related structure involving benzene rings, so that it is classed as an aromatic polyamide, that is, as an *aramid*.

3.5.2 Crystalline Versus Amorphous Thermoplastics

Some thermoplastics are composed partially or mostly of material where the polymer chains are arranged into an orderly crystalline structure. Examples of such *crystalline polymers* include PE, PP, PTFE, nylon, Kevlar, POM, and PEEK. A photograph of crystal structure in polyethylene is shown as Fig. 3.13.

If the chain molecules are instead arranged in a random manner, the polymer is said to be *amorphous*. Examples of amorphous polymers include PVC, PMMA, and PC. Polystyrene (PS) is amorphous in its *atactic* form where the benzene ring substitution is randomly located within each repeating unit of the molecule, but is crystalline in the *isotactic* from where the substitution occurs at the same location in each repeating unit. This same situation occurs for other polymers as well due to the regular structure of the isotactic form promoting crystallinity. If the side groups alternate their positions in a regular manner, the polymer is said to be *syndiotactic*, with a crystalline structure being likely in this case also.

Amorphous polymers are generally used around and below their respective glass transition temperatures T_g, some values of which are listed in Table 3.9. Above T_g,

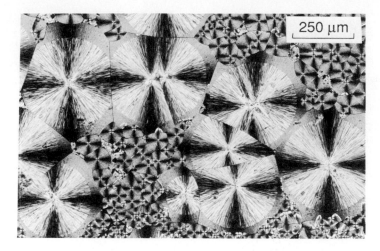
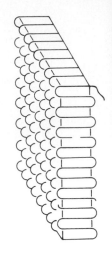

Figure 3.13 Crystal structure of polyethylene. Layers similar to that shown in the diagram are seen edge-on in the photo, being arranged in a radiating pattern to form the prominent crystalline features called *spherulites*. (Photo courtesy of A. S. Holik, General Electric Co., Schenectady, N.Y. Diagram reprinted with permission from [Geil 65]; copyright ©1965 American Chemical Society.)

TABLE 3.9 TYPICAL VALUES OF GLASS TRANSITION AND MELTING TEMPERATURES FOR VARIOUS THERMOPLASTICS AND ELASTOMERS

Polymer	Transition T_g, °C	Melting T_m, °C
(a) Amorphous thermoplastics		
Polyvinyl chloride (PVC)	87	212
Polystyrene (atactic)	100	≈ 180
Polycarbonate (PC)	150	265
(b) Primarily crystalline thermoplastics		
Low-density polyethylene (LDPE)	−110	115
High-density polyethylene (HDPE)	−90	137
Polyoxymethylene (POM)	−85	175
Polypropylene (PP)	−10	176
Nylon 6	50	215
Polystyrene (isotactic)	100	240
Polyetheretherketone (PEEK)	143	334
Aramid	375	640
(c) Elastomers		
Silicone rubber	−123	−54
Cis-polyisoprene	− 73	28
Polychloroprene	− 50	80

Source: Data in [ASM 88] pp. 50–54.

the elastic modulus decreases rapidly, and time-dependent deformation (creep) effects become pronounced, limiting the usefulness of these materials in load-resisting applications. Their behavior below T_g tends to be glassy and brittle, with the elastic modulus being on the order of $E = 3$ GPa. Amorphous polymers composed of single strand molecules are said to be *linear polymers*. Another possibility is that there is some degree of *branching* as shown in Fig. 3.14.

Crystalline polymers tend to be less brittle than amorphous polymers, and the stiffness and strength do not drop as dramatically beyond T_g. For example, such differences occur between the amorphous and crystalline forms of PS as illustrated in Fig. 3.15. As a result of this behavior, many crystalline polymers can be used above their T_g values. Crystalline polymers tend to be opaque to light, whereas amorphous polymers are transparent.

3.5.3 Thermosetting Plastics

The molecular structure of a thermosetting plastic consists of a three-dimensional network formed by frequent strong covalent bonds between chains as illustrated in Fig. 3.14(c).

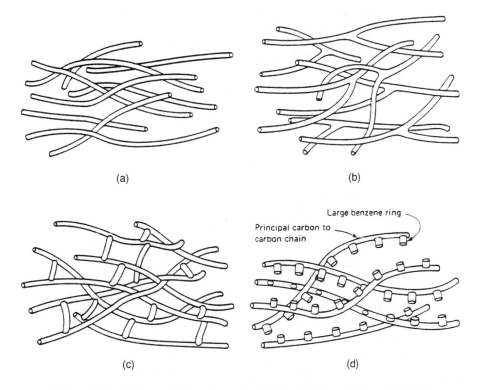

Figure 3.14   Polymer chain structures which are linear (a), branched (b), crosslinked (c), or stiffened by side groups (d). (From [Budinski 92] pp. 68–70; ©1992 by Prentice Hall, Englewood Cliffs, NJ; reprinted with permission.)

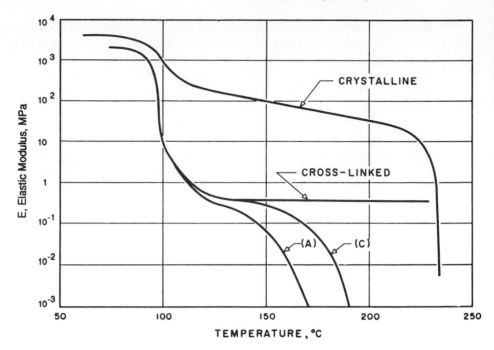

Figure 3.15 Elastic modulus versus temperature for amorphous, lightly cross-linked, and crystalline polystyrene. For amorphous samples (A) and (C), the chain lengths correspond to average molecular weights of 2.1×10^5 and 3.3×10^5, respectively. (Adapted from [Tobolsky 65] p. 75; reprinted by permission of John Wiley & Sons, Inc.; copyright ©1965 by John Wiley & Sons, Inc.)

These chain-to-chain bonds, called *cross-links*, result in a rigid and strong (but sometimes brittle) solid. Examples include the epoxies, and the phenolics such as Bakelite.

As an example, the repeating unit structure for phenol formaldehyde (phenolic) is shown in Fig. 3.16. Each unit has not only the usual carbon-to-carbon bonds to the next units in either direction along the chain, but also an additional branch involving a carbon atom that can form a cross-link to another chain. This cross-linking chemical reaction occurs during the final stage of processing, which is typically compression molding at elevated temperature. The resulting solid will thereafter not melt upon heating, generally decomposing or burning instead.

Recall that the melting of weak secondary bonds produces the glass transition temperature (T_g) effect in thermoplastics, above which deformation may occur by relative sliding between chain molecules. This situation contrasts with that for a thermosetting plastic where the relative motion between the molecules is prevented by strong and temperature resistant covalent bonds. As a result, there is no distinct T_g effect in highly cross-linked thermosetting plastics. They tend to be stiff, but rather brittle, and their temperature resistance is limited by decomposition rather than softening.

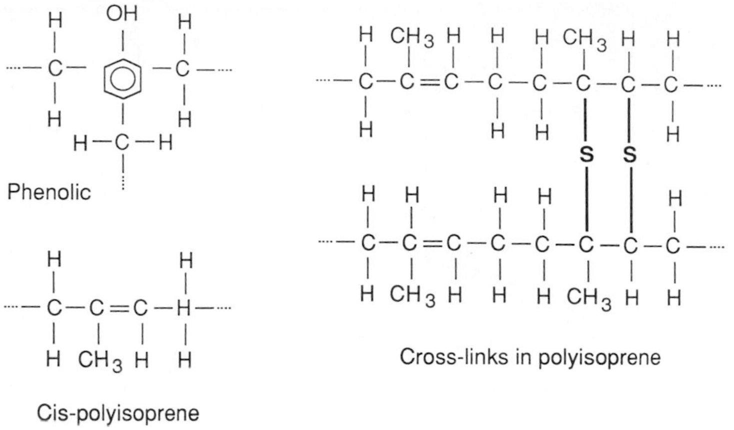

Figure 3.16 Molecular structures of a phenolic thermosetting plastic and of a synthetic similar to natural rubber, cis-polyisoprene. In the phenolics, carbon-hydrogen bonds form crosslinks, whereas in polyisoprene cross-links are formed by sulfur atoms.

3.5.4 Elastomers

Elastomers are typified by natural rubber but also include a variety of synthetic polymers with similar mechanical behavior. Some elastomers, such as the polyurethane elastomers, behave in a thermoplastic manner, but others are thermosetting materials. For example, polyisoprene is a synthetic rubber with the same basic structure as natural rubber, but lacking various impurities found in natural rubber. This structure is shown in Fig. 3.16. Adding sulfur and subjecting the rubber to pressure and a temperature around 160°C causes sulfur cross-links to form as shown. Greater degrees of cross-linking result in a harder rubber. This particular thermosetting process is called *vulcanization*.

Although cross-linking results in the rigid network structure of thermosetting plastics, typical elastomers behave in a very different manner because the cross-links occur much less frequently along the chains, specifically at intervals on the order of hundreds of carbon atoms. Also, the cross-links and the main chains themselves are flexible in elastomers rather than stiff as in the thermosetting plastics. This flexibility exists because the geometry at the carbon-to-carbon double bond causes a bend in the chain, which has a cumulative effect over long lengths of chain, such that the chain is coiled between cross-link points. Upon loading, these coils unwind between the cross-link attachment points, and after removal of the stress, the coils recover, resulting in the macroscopic effect of recovery of most of the deformation. Typical deformation response is shown in Fig. 3.17.

The initial elastic modulus is very low as it is associated only with uncoiling the chains, resulting in a value on the order of $E = 1$ MPa. Some stiffening occurs as the chains straighten. This low value of E contrasts with that for a glassy polymer below its T_g, where elastic deformation is associated with stretching the combination of covalent

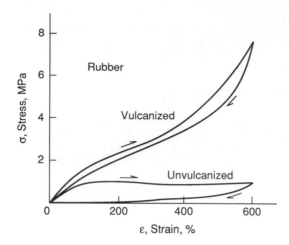

Figure 3.17 Stress-strain curves for unvulcanized and vulcanized natural rubber. (Data from [Hock 26].)

and secondary chemical bonds involved, resulting in a value of E on the order of 1000 times higher.

3.5.5 Strengthening Effects

The molecular structures of polymers are affected by the details of their chemical synthesis, such as the pressure, temperature, reaction time, presence and amount of catalysts, and cooling rate. These are often varied to produce a wide range of properties for a given polymer. Any molecular structure that tends to retard relative sliding between the chain-like molecules increases the stiffness and strength. Longer chain molecules, that is, greater molecular weight, has this effect as longer chains are more prone to becoming entangled with one another. Stiffness and strength are similarly increased by more branching in an amorphous polymer, by greater crystallinity, and by causing some cross-linking to occur in normally thermoplastic polymers. All of these effects are most pronounced above T_g, where the secondary bonds between chains have melted, and sliding can be inhibited only by the chains physically interfering with one another's motion.

For example, one variant of polyethylene, called low-density polyethylene (LDPE), has a significant degree of chain branching. These irregular branches interfere with the formation of an orderly crystalline structure, so that the degree of crystallinity is limited to about 65%. In contrast, the high-density variant HDPE has less branching, and the degree of crystallinity can reach 90%. As a result of the structural differences, LDPE is quite flexible, whereas HDPE is stronger and stiffer.

An extreme variation in the properties of rubber is possible by varying the amount of vulcanization, resulting in different degrees of cross-linking. The effect on the elastic modulus (stiffness) of the synthetic rubber polyisoprene is shown in Fig. 3.18. Unvulcanized rubber is soft and flows in a viscous manner. Cross-linking by sulfur at about 5% of the possible sites yields a rubber that is useful in a variety of applications, such as

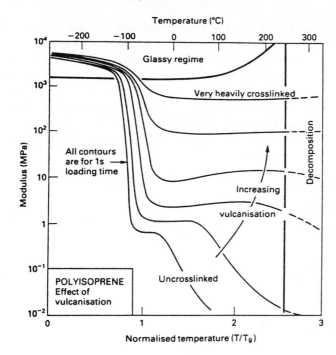

Figure 3.18 Effect of degree of cross-linking on the elastic modulus E of a synthetic similar to natural rubber. (From [Ashby 86] p. 227; reprinted with permission of Pergamon Press, Oxford, UK.)

automobile tires. A high degree of cross-linking yields a hard and tough material called *ebonite* that is capable of only limited deformation.

Another effect observed is that bulky side groups attached to the main chain interfere with molecular motion, thus also increasing strength and stiffness as suggested by Fig. 3.14(d). This is a factor in polystyrene (PS), as the benzene ring side group has this effect, causing the amorphous form of this material to be hard and brittle at room temperature.

3.5.6 Combining and Modifying Polymers

Polymers are seldom used in pure form, often being combined with one another or with other substances in various ways. *Alloying*, also called *blending*, involves melting two or more polymers together so that the resulting material contains a mixture of two or more chain types. This mixture may be fairly uniform, or the components may separate themselves into a multiphase structure. For example, PVC and PMMA are blended to make a tough plastic with good flame and chemical resistance.

Copolymerization is another means of combining two polymers, in which the ingredients and other details of the chemical synthesis are chosen so that the individual chains are composed of two types of repeating units. For example, styrene-butadiene rubber is a copolymer of three parts butadiene and one part styrene, both of which occur in most individual chain molecules. ABS plastic is a combination of three polymers, called a *terpolymer.* In particular, the acrylonitrile-styrene copolymer chain has side branches of butadiene polymer.

Among the nonpolymer substances added to modify the properties of polymers are *plasticizers.* These generally have the objective of increasing toughness and flexibility, while often decreasing strength in the process. Plasticizers are usually high-boiling-point organic liquids, the molecules of which distribute themselves through the polymer structure. The plasticizer molecules tend to separate the polymer chains and allow easier relative motion between them, that is, easier deformation. For example, plasticizers are added to PVC to make flexible vinyl, which is used as imitation leather.

Polymers are often *modified* or *filled* by adding other materials in the form of particles of fibers. For example, carbon black, which is similar to soot, is usually added to rubber, increasing its stiffness and strength in addition to the effects of vulcanization. Also, rubber particles are added to polystyrene to reduce its brittleness, with the resulting material being called high-impact polystyrene (HIPS). The microstructure of a HIPS material is shown in Fig. 3.19. If the added substance has the specific purpose of increasing strength, it is called a *reinforcement.* For example, chopped glass

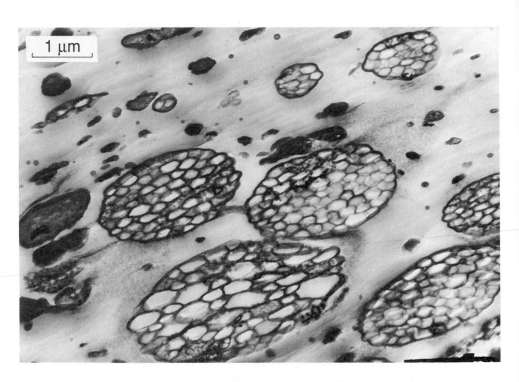

Figure 3.19 Microstructure of rubber modified polystyrene, in which the dark colored particles and networks are rubber, and all light-colored areas inside and outside of particles are polystyrene. The originally equiaxial particles were elongated somewhat when cut to prepare the surface. (Photo courtesy of R. P. Kambour, General Electric Co., Schenectady, N.Y.)

fibers are added as reinforcement to various thermoplastics to increase strength and stiffness.

Reinforcement may also take the form of long fibers or woven cloth made of high strength fibers, such as glass, carbon in the form of graphite, or Kevlar. These are often used in a matrix of a thermosetting plastic. For example, fiberglass contains glass fibers in the form of mats or woven cloth, and these are embedded in a matrix of unsaturated polyester. Such a combination is a *composite material*, which topic is considered further in a separate section near the end of this chapter.

3.6 CERAMICS AND GLASSES

Ceramics and glasses are solids that are neither metallic nor organic (carbon-chain based) materials. Ceramics thus include clay products, such as porcelain, china, and brick, and also natural stone and concrete. Ceramics used in high stress applications, called *engineering ceramics*, are often relatively simple compounds of metals, or the metalloids silicon or boron, with nonmetals such as oxygen, carbon, or nitrogen. Other ceramics include intermetallic compounds, such as nickel aluminide (NiAl), and also carbon in the form of graphite. Ceramics are predominantly crystalline, whereas glasses are amorphous. Most glass is produced by melting silica (SiO_2), which is ordinary sand, along with other metal oxides, such as CaO, Na_2O, B_2O_3, and PbO. In contrast, ceramics are usually processed not by melting but by some other means of binding the particles of a fine powder into a solid. Specific examples of ceramics and glasses and some of their properties are given in Table 3.10. The microstructure of a crystalline ceramic is shown in Fig. 3.20.

Engineering ceramics have a number of important advantages compared to metals. They are highly resistant to corrosion and wear, and melting temperatures are typically quite high. These characteristics all arise from the strong covalent or ionic-covalent chemical bonding of these compounds. Ceramics are also relatively stiff (high E) and light in weight. In addition, they are often inexpensive, as the ingredients for their manufacture are typically abundant in nature.

As discussed in the previous chapter in connection with plastic deformation, slip of crystal planes does not occur readily in ceramics as a result of the strength and directional nature of covalent bonding and the relatively complex crystal structures. This results in the mechanical behavior of ceramics being inherently brittle, and glasses are similarly affected by covalent bonding. In ceramics, the brittleness is further enhanced by the fact that grain boundaries in these crystalline compounds are relatively weaker than in metals. This arises from disrupted chemical bonds where the lattice planes are discontinuous at grain boundaries, and also from the existence of regions where ions of the same charge are in proximity. In addition, there is often an appreciable degree of porosity in ceramics, and both ceramics and glasses usually contain microscopic cracks. These discontinuities promote macroscopic cracking and thus also contribute to brittle behavior.

The processing and uses of ceramics are strongly influenced by their brittleness. As a consequence, recent efforts aimed at developing improved ceramics for engineering use

TABLE 3.10   PROPERTIES AND USES FOR SELECTED ENGINEERING AND OTHER CERAMICS

Ceramic	Melting Temp.	Density	Elastic Modulus	Typical Strength		Uses
	T_m	ρ	E	σ_u, MPa (ksi)		
	°C	g/cm³	GPa (10^3 ksi)	Tension	Compression	
Soda-lime glass	730 (soften)	2.48	74 (10.7)	≈ 50 (7)	1000 (145)	Windows, containers
Type S glass (fibers)	970 (soften)	2.49	85.5 (12.4)	4480 (650)	—	Fibers in aerospace composites
Zircon porcelain	1567	3.60	147 (21.3)	56 (8.1)	560 (81)	High-voltage electrical insulators
Magnesia, MgO	2850	3.60	280 (40.6)	140 (20.3)	840 (122)	Refractory brick, wear parts
Alumina, Al_2O_3 (99.5% dense)	2050	3.89	372 (54)	262 (38)	2620 (380)	Spark plug insulators, cutting tool inserts, fibers for composites
Zirconia, ZrO_2	2570	5.80	210 (30.4)	147 (21.3)	2100 (304)	High temperature crucibles, refractory brick, engine parts
Silicon carbide, SiC	2837	3.10	410 (60)	299 (42)	≈ 2000 (300)	Engine parts, abrasives, fibers for composites
Boron carbide, B_4C	2350	2.51	308 (44.6)	—	2900 (420)	Bearings, armour, abrasives
Silicon nitride, Si_3N_4	1900	3.2	310 (45)	≈ 580 (80)	> 3500 (500)	Turbine blades, fibers for composites cutting tool inserts

Note: Data are for materials in bulk form except for type S glass.

Source: Data in [Farag 89] p. 510, [Budinski 92] , [Ashby 86] p. 150, and [Schwartz 84] p. 2.44.

involve various means of reducing brittleness. Noting the advantages of ceramics listed above, success in this area would be of major importance as it would allow increased use of ceramics in applications such as automobile and jet engines, where lighter weights and operation at higher temperatures both result in greater fuel efficiency.

Various classes of ceramics will now be discussed separately as to their processing and uses.

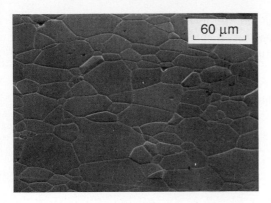 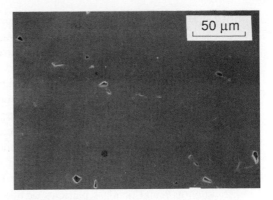

Figure 3.20 Surface (left) of near-maximum-density Al_2O_3, with grain boundaries visible. In a polished section (right), grain boundaries cannot be seen, but pores are visible as black areas. (Left photo same as in [Venkateswaran 88]; reprinted by permission of the American Ceramic Society. Right photo courtesy of D. P. H. Hasselman, Virginia Tech, Blacksburg, Va.)

3.6.1 Clay Products, Natural Stone, and Concrete

Clays consist of various silicate minerals that have a sheet-like crystal structure, an important example being kaolin, Al_2O_3-$2SiO_2$-$2H_2O$. In processing, the clay is first mixed with water to the consistency of a thick paste and then formed into a cup, dish, brick, or other useful shape. Firing at a temperature in the range 800 to 1200°C then drives off the water and melts some of the SiO_2 to form a glass that binds the Al_2O_3 and the remaining SiO_2 into a solid. The presence or addition of small amounts of minerals containing sodium or potassium enhances formation of the glass by permitting a lower firing temperature.

Natural stone is of course used without processing other than cutting it into useful shapes. The prior processing done by nature varies greatly. For example, limestone is principally crystalline calcium carbonate ($CaCO_3$) that has precipitated out of ocean water, and marble is the same mineral that has been recrystallized (metamorphosed) under the influence of temperature and pressure. Sandstone consists of particles of silica sand (SiO_2) bound together by additional SiO_2, or by $CaCO_3$, that is present due to precipitation from water solution. In contrast, igneous rocks such as granite have been melted and are multiphase alloys of various crystalline minerals.

Concrete is a combination of crushed stone, sand, and a cement paste that binds the other components into a solid. The modern cement paste, called Portland cement, is made by firing a mixture of limestone and clay at 1500°C. This forms a mixture of fine particles involving primarily lime (CaO), silica (SiO_2), and alumina (Al_2O_3), where these are in the form of tricalcium silicate ($3CaO$-SiO_2), dicalcium silicate ($2CaO$-SiO_2), and tricalcium aluminate ($3CaO$-Al_2O_3). When water is added, a *hydration* reaction starts during which water is chemically bound to these minerals by being incorporated into their crystal structures. During hydration, interlocking needle-like crystals form that bind the cement particles to each other and to the stone and sand. The reaction is rapid at first and slows with time. Even after long times, some residual water remains in small pores, between layers of the crystal structure, and chemically adsorbed to the surface of hydrated paste.

Clay products, natural stone, and concrete are used in large quantities for their familiar uses, including major use in buildings, bridges, and other large stationary structures. All are quite brittle and have poor strength in tension, but reasonable strength in compression. Concrete is very economical to use in construction and has the important advantage that it can be poured as a slurry into forms and hardened in place into complex shapes. Improved concretes continue to be developed, including some exotic varieties with quite high strength achieved by minimizing the porosity or by adding substances such as metal or glass particles or fibers.

3.6.2 Engineering Ceramics

The processing of engineering ceramics composed of simple chemical compounds involves first obtaining the compound. For example, alumina (Al_2O_3) is made from the mineral bauxite ($Al_2O_3 \cdot 2H_2O$) by heating to remove the hydrated water. Other engineering ceramics, such as ZrO_2, are also obtainable directly from naturally available minerals. But some, such as WC, SiC, and Si_3N_4, must be produced by appropriate chemical reactions starting from constituents that are available in nature. After the compound is obtained, it is ground to a fine powder if not already in this form. The powder is then compacted into a useful shape, typically by cold or hot pressing. A binding agent, such as a plastic, may be used to prevent the consolidated powder from crumbling. The ceramic at this stage is said to be in a *green* state and has little strength. Green ceramics are sometimes machined to obtain flat surfaces, holes, threads, etc., that would otherwise be difficult to achieve.

The next and final step in processing is *sintering*, which involves heating the green ceramic, typically to around 70% of its absolute melting temperature. This causes the particles to fuse and form a solid that contains some degree of porosity. Improved properties result from minimizing the porosity, that is, the volume percentage of voids. This can be done by using a gradation of particle sizes or by applying pressure during sintering. Small percentages of other ceramics may be added to the powder to improve response to processing. Also, small to medium percentages of other ceramics may be mixed with a given compound to tailor the properties of the final product.

One variation on the sintering process that aids in minimizing voids is *hot isostatic pressing*. This involves enclosing the ceramic in a sheet metal enclosure and placing this in a vessel that is pressurized with a hot gas. Some additional methods of processing that are sometimes used are *chemical vapor deposition* and *reaction bonding*. The former process involves chemical reactions among hot gases that result in a solid deposit of ceramic material onto the surface of another material. Reaction bonding combines the chemical reaction that forms the ceramic compound with the sintering process.

Engineering ceramics typically have high stiffness, light weight, and very high strength in compression. Although all are relatively brittle, their strength in tension and their fracture toughness may be sufficiently high that use in high-stress structural applications is not precluded if the details of the component design consider the limitations of the material. Increased use of ceramics in the future is likely due to their high-temperature capability.

3.6.3 Cermets; Cemented Carbides

A *cermet* is made from powders of a ceramic and a metal by sintering them together. The metal surrounds the ceramic particles and binds them together, with the ceramic constituent providing high hardness and wear resistance. *Cemented carbides* as made into cutting tools are the most important cermets. In this case, tungsten carbide (WC) is sintered with cobalt metal in amounts ranging from 3 to 25%. Other carbides are also used in the same manner, namely TiC, TaC, and Cr_3C_2, typically in combination with WC. The most frequent binder metal is cobalt, but nickel and steel are also employed.

The metal matrix of cemented carbides provides useful toughness but limits resistance to temperature and oxidation. Ordinary ceramics, such as alumina (Al_2O_3) and boron nitride (BN), are also used for cutting tools and have advantages compared to cemented carbides of greater hardness, lighter weight, and greater resistance to temperature and oxidation. But the extra care needed in working with brittle ceramics causes cemented carbides to be used except where ceramics cannot be avoided. Some of the advantages of ceramics can be obtained by chemical vapor deposition of a coating of a ceramic onto a cemented carbide tool. Ceramics used in this manner include TiC, Al_2O_3, and TiN.

3.6.4 Glasses

Pure silica (SiO_2) in crystalline form is a quartz mineral, the crystal structure of one of which is illustrated in Fig. 3.21. However, when silica is solidified from a molten state, an amorphous solid results. This occurs because the molten glass has a high viscosity due to a chain-like molecular structure, which limits the molecular mobility to the extent that perfect crystals do not form upon solidification. The three-dimensional crystal structure in Fig. 3.21 is depicted in a simplified two-dimensional form in Fig. 3.22. A perfect crystal as formed from solution in nature is represented by (a). Glass formed from molten silica has a network structure that is similar but highly imperfect, as in (b).

In processing, glasses are sometimes heated until they melt and are then poured into molds and cast into useful shapes. Alternatively, they may be heated only until soft and then formed by rolling, as for plate glass, or by blowing, as for bottles. Forming

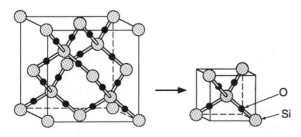

Figure 3.21   Diamond cubic crystal structure of silica, SiO_2, in its high-temperature cristobalite form. The crystal structure at ambient temperatures is a more complex arrangement of the basic tetrahedral unit shown on the right.

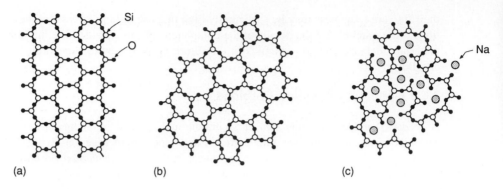

Figure 3.22 Simplified two-dimensional diagram of the structure of silica in the form of (a) quartz crystal, (b) glass, and (c) glass with a network modifier. (Part (b) adapted from [Zachariasen 32]; published 1932 by the American Chemical Society. Part (c) adapted from [Warren 38]; reprinted by permission of the American Ceramic Society.)

is made easier by the fact that the viscosity of glass varies gradually with temperature, so that the temperature can be adjusted to obtain a consistency that is appropriate to the particular method of forming. However, for pure silica, the temperatures involved are around 1800°C, which is inconveniently high. The temperature for forming can be lowered to around 800 to 1000°C by adding Na_2O, K_2O, or CaO. These oxides are called *network modifiers* because the metal ions involved tend to form nondirectional ionic bonds with oxygen atoms, resulting in terminal ends in the structure as illustrated by Fig. 3.22(c). This change in the molecular structure also causes the glass to be less brittle than pure silica glass. Commercial glasses contain varying amounts of the network modifiers as indicated by typical compositions in Table 3.11.

TABLE 3.11 TYPICAL COMPOSITIONS AND USES OF REPRESENTATIVE SILICA GLASSES

Glass	Major Components, % by Weight							Uses; Comment
	SiO_2	Al_2O_3	CaO	Na_2O	B_2O_3	MgO	PbO	
Fused silica	99	—	—	—	—	—	—	Furnace windows
Borosilicate (Pyrex)	81	2	—	4	12	—	—	Cookware, laboratory ware
Soda-lime	72	1	9	14	—	3	—	Windows, containers
Leaded	66	1	1	6	1	—	15	Tableware; also contains 9% K_2O
Type E	54	14	16	1	10	4	—	Fibers in fiberglass
Type S	65	25	—	—	—	10	—	Fibers for aerospace composites

Source: Data in [Lyle 74] and [Schwartz 84] p. 2.25.

Other oxides are added to modify the optical or electrical properties, color, or other characteristics of glass. Some oxides, such as B_2O_3, can form a glass themselves and may result in a two-phase structure. Leaded glass contains PbO, in which the lead participates in the chain structure. This modifies the glass to increase its resistivity, and also gives a high index of refraction, which contributes to the brilliance of fine crystal. The addition of Al_2O_3 increases the strength and stiffness of the glass fibers used in fiberglass and other composite materials.

3.7 COMPOSITE MATERIALS

A *composite material* is made by combining two or more materials that are mutually insoluble by mixing or bonding them in such a way that each maintains its integrity. Some composites have already been discussed, namely plastics modified by adding rubber particles, plastics reinforced by chopped glass fibers, cemented carbides, and concrete. These and many other composite materials consist of a matrix of one material that surrounds particles or fibers of a second material as shown in Fig. 3.23. Some composites involve layers of different materials, and the individual layers may themselves be composites. Materials that are melted (alloyed) together are not considered composites, even if a two-phase structure results, nor are solid solutions or precipitate structures arising from solid solutions. Some representative types and examples of composite materials and their uses are listed in Table 3.12.

Materials of biological origin are usually composites. Wood contains *cellulose* fibers surrounded by *lignin* and *hemicellulose*, all of which are polymers. Bone is composed of the fibrous protein *collagen* in a ceramic-like matrix of the crystalline mineral *hydroxylapatite*, $Ca_5(PO_4)_3OH$.

Composite materials have a wide range of uses, and their use is rapidly increasing. Man-made composites can be tailored to meet special needs such as high strength and stiffness combined with light weight. The resulting high performance (and expensive) materials are being increasingly used in aircraft, space, and defense applications. More economical composites, such as glass reinforced plastics, are continually finding new uses in a wide range of products, such as automotive components, boat hulls, sports equipment,

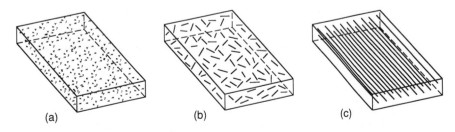

(a)   (b)   (c)

Figure 3.23 Composites reinforced by (a) particles, (b) chopped fibers or whiskers, and (c) continuous fibers. (Adapted from [Budinski 92] p. 137; ©1992 by Prentice Hall, Englewood Cliffs, NJ; reprinted with permission.)

TABLE 3.12 REPRESENTATIVE TYPES AND EXAMPLES OF COMPOSITE MATERIALS

Reinforcing Type	Matrix Type	Example	Typical Use
(a) Particulate composites			
Ductile polymer or elastomer	Brittle polymer	Rubber in polystyrene	Toys, cameras
Ceramic	Ductile metal	WC with Co metal binder	Cutting tools
Ceramic	Ceramic	Granite stone and silica sand in Portland cement	Bridges, buildings
(b) Short fiber, whisker composites			
Strong fiber	Thermosetting plastic	Chopped glass in polyester resin	Auto body panels
Ceramic	Ductile metal	SiC whiskers in Al alloy	Aircraft structural panels
(c) Continuous fiber composites			
Ceramic	Thermosetting plastic	Graphite in epoxy	Aircraft wing flaps
Ceramic	Ductile metal	Boron in Al alloy	Aircraft structure
Ceramic	Ceramic	SiC in Si_3N_4	Engine parts
(d) Laminated composites			
Stiff sheet	Foamed polymer	PVC and ABS sheets over ABS foam core	Canoes
Composite	Metal	Kevlar in epoxy between Al alloy layers (ARALL)	Aircraft structure

and furniture. Wood and concrete of course continue to be major construction materials, and new composites involving these and other materials have also come into recent use in the construction industry.

Various classes of composite materials will now be discussed.

3.7.1 Particulate Composites

Particles can have various effects on a matrix material depending on the properties of the two constituents. Ductile particles added to a brittle matrix increase the toughness as cracks have difficulty passing through the particles. An example is rubber-modified polystyrene, the microstructure of which has already been illustrated in Fig. 3.19. Another ductile particle composite made from two polymers is shown broken open in Fig. 3.24.

Particles of a hard and stiff (high E) material added to a ductile matrix increase its strength and stiffness. An example is carbon black added to rubber. As might be expected, hard particles generally decrease the fracture toughness of a ductile matrix, and this limits the usefulness of some composites of this type. However, the composite

Figure 3.24 Fracture surface of polyphenylene oxide (PPO) modified with high impact polystyrene particles. (Photo courtesy of General Electric Co., Pittsfield, Ma.)

may still be useful if it has other desirable properties that outweigh the disadvantages of limited toughness, such as the high hardness and wear resistance of cemented carbides and other cermets.

   If the hard particles in a ductile matrix are quite small and limited in quantity, the reduction in toughness is modest. In a metal matrix, a desirable strengthening effect similar to that of precipitation hardening can be achieved by sintering the metal in powder form with ceramic particles of size on the order of 0.1 μm. This is called *dispersion hardening*. The volume fraction of particles seldom exceeds 15%, and the amount may be as small as 1%. Aluminum reinforced in this manner with Al_2O_3 has improved creep resistance. Tungsten is similarly dispersion hardened with small amounts of oxide ceramics, such as ThO_2, Al_2O_3, SiO_2, and K_2O, so that it has sufficient creep resistance for use in light-bulb filaments.

3.7.2 Fibrous Composites

Strong and stiff fibers can be made from ceramic materials that are difficult to use as structural materials in bulk form, such as glass, graphite (carbon), boron, and silicon carbide (SiC). When these are embedded in a matrix of a ductile material, such as a polymer or a metal, the resulting composite can be strong, stiff, and tough. The fibers carry most of the stress, whereas the matrix holds them in place. Fibers and matrix can be seen in the photograph of a broken open composite material in Fig. 3.25. Good adhesion between fibers and matrix is important as this allows the matrix to carry the stress from one fiber to another where a fiber breaks or where one simply ends because of its limited length. Fiber diameters are typically in the size range 1 to 100 μm.

   Fibers are used in composites in a variety of different configurations, two of which are shown in Fig. 3.23. Short, randomly oriented fibers result in a composite that has similar properties in all directions. Chopped glass fibers used to reinforce thermoplastics are of this type. *Whiskers* are a special class of short fiber that consist of tiny elongated single crystals that are very strong because they are dislocation free. Diameters are 1 to 10 μm or smaller, and lengths are 10 to 100 times larger than the diameter. For example, randomly oriented SiC whiskers can be used to strengthen and stiffen aluminum alloys.

150 μm

Figure 3.25 Fracture surface showing broken fibers for a composite of Nicalon type SiC fibers in a CAS glass-ceramic matrix. (Photo by S. S. Lee; material manufactured by Corning.)

Long fibers can be woven into a cloth or made into a mat of intertwined strands. Glass fibers in both of these configurations are used with polyester resins to make common fiberglass. High performance composites are often made using long straight continuous fibers. Continuous fibers all oriented in a single direction provide maximum strength and stiffness parallel to the fibers. Since such a material is weak if stressed in the transverse direction, several thin layers with different fiber orientations are usually stacked into a laminate as shown in Fig. 3.26. For example, composites with an thermosetting plastic matrix, often epoxy, are assembled in this manner using partially cured sheets, which are called *prepregs* because they have been previously impregnated with the epoxy resin. Appropriate heat and pressure is applied to complete the cross-linking reaction while at the same time bonding the layers into a solid laminate. Fibers commonly used in this manner with an epoxy matrix include glass, graphite, boron, and the aramid polymer Kevlar.

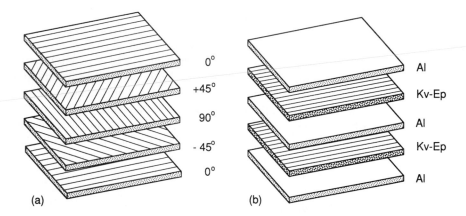

Figure 3.26 Laminated composites. Sheets having various fiber directions, as shown in (a), may be bonded together. The ARALL laminate (b) is constructed with aluminum sheets bonded to sheets of composite, with the latter being made of Kevlar fibers in an epoxy matrix.

Strengths comparable to those of structural metals are achievable with a polymer matrix as shown in Fig. 3.27. Values of stiffness (E) may also be comparable to those for structural metals but are often somewhat lower. However, in considering materials for weight-critical applications such as aircraft structure, it is more relevant to consider the strength-to-weight ratio and the stiffness-to-weight ratio. On this basis, fibrous composites are typically superior to structural metals in both strength and stiffness. Compare Figs. 3.27 and 3.28.

Due to the limitations of the matrix, polymer matrix composites have limited resistance to high temperature. Composites with an aluminum or titanium matrix have

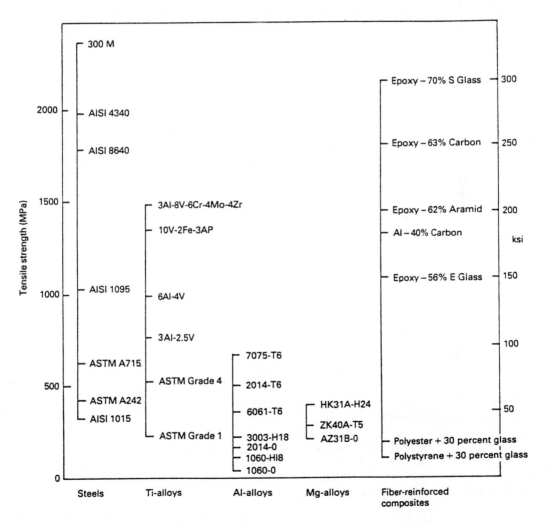

Figure 3.27 Comparison of some engineering metals and composites on the basis of tensile strength. (From [Farag 89] p. 174; used with permission.)

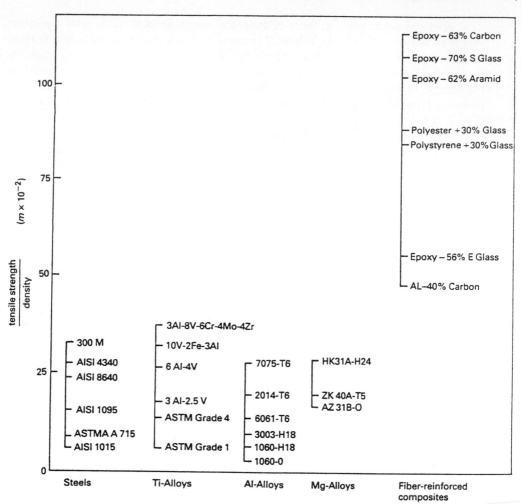

Figure 3.28 Comparison of some engineering metals and composites on the basis of specific tensile strength, that is, the strength-to-density ratio. (From [Farag 89] p. 176; used with permission.)

reasonable temperature resistance. These metals are sometimes used with continuous straight fibers of silicon carbide of fairly large diameter, around 140 μm. Other fiber types and configurations are also used. For high temperature applications, ceramic matrix composites are being developed. These materials have a matrix that is already strong and stiff, but which is brittle and has a low fracture toughness. Whiskers or fibers of another ceramic can act to retard cracking by bridging across small cracks that exist and holding them closed so that their growth is retarded. For example, whiskers of SiC in a matrix of Al_2O_3 are used in this manner. Continuous fibers can also be used, such as SiC fibers in a matrix of Si_3N_4. A cross section of a ceramic matrix composite is shown in Fig. 3.29.

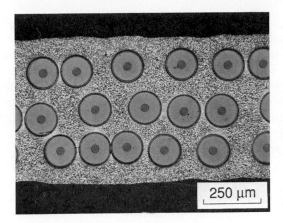

Figure 3.29 Cross section of a ceramic-ceramic composite having SiC fibers in a Ti_3Al matrix. The fibers have a carbon core, used in their manufacture as a base for chemical vapor deposition of SiC, and they also have a carbon-rich coating. (Photo courtesy of C. G. Rhodes, Rockwell International Science Center, Thousand Oaks, Ca.)

3.7.3 Laminated Composites

A material made by combining layers is called a *laminate*. The layers may differ as to the fiber orientation, or they may consist of different materials. Plywood is a familiar example of a laminate, the layers differing as to grain direction and perhaps also as to type of wood. As already noted, unidirectional composite sheets are frequently laminated as in Fig. 3.26(a). The recently developed aramid-aluminum laminate (ARALL) has layers of an aluminum alloy and a composite with unidirectional Kevlar fibers in an epoxy matrix. See Fig. 3.26(b).

Where stiffness in bending is needed along with light weight, layers of a strong and stiff material may be placed on either side of a lightweight core. Such *sandwich* materials include aluminum or fibrous composite sheets bonded on each side of a core that is made of a stiff foam. Another possibility is a core made of a honeycomb of aluminum or other material.

3.8 SUMMARY

A number of metals have combinations of properties and availability that result in their use as load-resisting engineering materials. These include irons and steels, and aluminum, titanium, copper, and magnesium. Pure metals in bulk form yield at quite low stresses, but useful levels of strength can be obtained by introducing obstacles to dislocation motion through such means as cold work, solid solution strengthening, precipitation hardening, and the introduction of multiple phases. Alloying with various amounts of one or more additional metals or nonmetals is usually needed to achieve this strengthening and to otherwise tailor the properties to obtain a useful engineering metal.

In steels, small amounts of carbon and other elements in solid solution provide some strengthening without heat treatment. For carbon contents above about 0.3%, substantially greater strengthening can be obtained by heat treating using the quenching and tempering process. Small percentages of alloying elements, such as Ni, Cr, and Mo, enhance the strengthening effect. Special steels such as stainless steels and tool steels typically include fairly substantial percentages of various alloying elements.

Considering aluminum alloys, the highest strengths in this lightweight metal are obtained by heat treatment and alloying that causes precipitation hardening (aging) to be effective. Magnesium is strengthened in a similar manner and is noteworthy as being the lightest engineering metal. Titanium alloys are somewhat heavier than aluminum but have greater temperature resistance. They are strengthened by a combination of the various methods, including multiple phase effects in the alpha-beta alloys. Superalloys are corrosion and temperature resistant metals that have large percentages of two or more of the metals nickel, cobalt, and iron.

Polymers have long chain-like molecules based on carbon. Compared to metals, they lack strength, stiffness, and temperature resistance. However, these disadvantages are offset to an extent by light weight and corrosion resistance, leading to their use in numerous low-stress applications. Polymers are classified as thermoplastics if they can be repeatedly melted and solidified. Some examples of thermoplastics are polyethylene, polymethyl methacrylate, and nylon. Contrasting behavior occurs for thermosetting plastics, which change chemically during processing and thereafter cannot be melted. Examples include phenolics and epoxies. Elastomers, such as natural and synthetic rubbers, are distinguished by being capable of deformations of at least 100% to 200%, and of then recovering most of this deformation after removal of the stress. Most elastomers behave in a thermosetting manner, but some are thermoplastic.

A given thermoplastic is usually glassy and brittle below its glass transition temperature, T_g. Above the T_g of a given polymer, the stiffness (E) is likely to be very low unless the material has a substantially crystalline structure. Stiffness and strength in polymers is also enhanced by longer lengths of the chain molecules, chain branching, bulky side groups on the chain, and cross-linking between chains. Thermosetting plastics have a molecular structure that causes a large number of cross-links to form during processing. Once these covalent bonds are formed, the material cannot later be melted, explaining the thermosetting behavior. Vulcanizing of rubber is also a thermosetting process, in which case sulfur atoms form bonds that link chain molecules.

Ceramics are nonmetallic and inorganic crystalline solids that are generally chemical compounds. Clay products, porcelain, natural stone, and concrete are fairly complex combinations of crystalline phases, primarily silica (SiO_2) and metal oxides, and $CaCO_3$ in the case of some natural stones, bound together by various means. High-strength engineering ceramics tend to be fairly simple chemical compounds, such as metal oxides, carbides, or nitrides, or intermetallic compounds. Cermets, such as cemented carbides, are ceramic materials sintered with a metal phase that acts as a binder. Glasses are amorphous (noncrystalline) materials consisting of SiO_2 combined with varying amounts of metal oxides.

All ceramics and glasses tend to be brittle compared to metals. However, many have advantages, such as light weight, high stiffness, high compressive strength, and temperature resistance, that cause them to be the most suitable materials in certain situations. An example is the use of engineering ceramics for high temperature engine parts.

Composites are combinations of two or more materials, with one generally acting as a matrix and the other as reinforcement. The reinforcement may be in the form of particles, short fibers, or continuous fibers. Composites include many common man-made materials, such as concrete, cemented carbides, and fiberglass, and other reinforced plastics, as well as biological materials, notably wood and bone. High-performance composites as used in aerospace applications generally employ high-strength fibers in a ductile matrix. The fibers are often ceramics or glass, and the matrix is typically a polymer or a lightweight metal. However, even a ceramic matrix is made stronger and less brittle by the presence of reinforcing fibers.

It is often useful to combine layers to make a laminated composite. The layers may differ as to fiber direction, or they may consist of more than one type of material, or both. High-performance composite laminates may be advantageous for use in situations such as aerospace structure, as their strength and stiffness are both quite high if compared to those of metals on a unit weight basis.

The survey of engineering materials given in this chapter should be considered to be only a summary. Numerous sources of more detailed information exist, a few of which are given in the References section of this chapter. Companies that supply materials are also often a useful source of information on their particular products.

NEW TERMS AND SYMBOLS

alpha-beta titanium alloy	laminate
annealing	low-alloy steel
austenite	martensite
casting	network modifier
cast iron	particulate composite
cemented carbide	pearlite
cementite, Fe_3C	plain-carbon steel
ceramic	plasticizer
cermet	precipitation hardening
cold work	quenching and tempering
composite material	reinforced plastic
copolymer	second phase
cross-linking	sintering
deformation processing	solid solution strengthening
dispersion hardening	stainless steel
ductile (nodular) iron	steel
elastomer	superalloy
ferrite	thermoplastic
fibrous composite	thermosetting plastic
glass	tool steel
grain refinement	vulcanization
gray iron	whisker
heat treatment	wrought metals

REFERENCES

Ashby, M. F. and D. R. H. Jones. 1986 *Engineering Materials 2: An Introduction to Microstructures, Processing and Design*, Pergamon Press, Oxford, UK.

ASM. 1987 *Engineered Materials Handbook, Vol. 1: Composites*, ASM International, Metals Park, Oh.

ASM. 1988 *Engineered Materials Handbook, Vol. 2: Engineering Plastics*, ASM International, Metals Park, Oh.

ASTM. 1990 *Annual Book of ASTM Standards, Section 1: Iron and Steel Products*, Vols. 01.01 to 01.07, Am. Soc. for Testing and Materials, Philadelphia, Pa.

Boyer, H. E. and T. L. Gall, eds. 1985 *Metals Handbook: Desk Edition*, American Society for Metals, Metals Park, Oh.

Budinski, K. G. 1992 *Engineering Materials: Properties and Selection*, 4th ed., Prentice-Hall, Englewood Cliffs, N.J.

Creyke, W. E. C., I. E. J. Sainsbury and R. Morrell. 1982 *Design with Non-Ductile Materials*, Applied Science Pubs., London, UK.

Kelly, A., ed. 1989 *Concise Encyclopedia of Composite Materials*, Pergamon Press, Oxford, UK.

Kingery, W. D., H. K. Bowen and D. R. Uhlmann. 1976 *Introduction to Ceramics*, 2nd ed., John Wiley, New York.

Morrell, R. 1985 *Handbook of Properties of Technical and Engineering Ceramics: Part 1, An Introduction for the Engineer and Designer; Part 2, Data Reviews*, Her Majesty's Stationary Office, London, UK, 1985 and 1987.

SAE. 1989 *Metals and Alloys in the Unified Numbering System*, 5th ed., Pub. No. SAE HS 1086, Soc. of Automotive Engineers, Warrendale, Pa.; also Pub. No. ASTM DS-56D, Am. Soc. for Testing and Materials, Philadelphia, Pa.

Sperling, L. H. 1986 *Introduction to Physical Polymer Science*, John Wiley, New York.

PROBLEMS AND QUESTIONS

Sections 3.2 to 3.4

3.1 For the metals in Table 3.1, compute strength-to-density and stiffness-to-density ratios, σ_u/ρ and E/ρ. Plot σ_u/ρ versus E/ρ and comment on the trends. Use both the lower and upper ranges on σ_u and connect the resulting two points with a line. How are these trends related to the uses of these metals?

3.2 Nickel and copper are mutually soluble in all percentages as substitutional alloys with an FCC crystal structure. The effect of up to 30% nickel on the yield strength of copper is shown in Fig. 3.3. Draw a qualitative graph showing how you expect the yield strength of otherwise pure Cu-Ni alloys to vary as the nickel content is varied from zero to 100%.

3.3 With reference to Fig. 3.5, note that higher aging temperatures have their maximum effect on yield strength after shorter time than lower temperatures, but the highest yield strength obtained is not as high. Why do you think these trends occur?

Section 3.5

3.4 In your own words, explain why thermosetting plastics do not have a pronounced decrease in the elastic modulus, E, at a glass transition temperature, T_g.

3.5 In *cis*-polyisoprene, which is similar to natural rubber, the H and CH_3 side units at either end of the carbon-to-carbon double bond are on the same side of the chain as shown in Fig. 3.16. These interfere with one another, causing the chain to bend slightly, and resulting in a coil over long lengths of chain. But in *trans*-polyisoprene, also called *gutta percha*, the H and CH_3 side units are on opposite sides, resulting in relatively straight rather than coiled chains. How would you expect the behavior of trans-polyisoprene to differ from that of cis-polyisoprene?

3.6 For the polymers in Table 3.9, plot T_g versus T_m after converting both to absolute temperature. Use a different plotting symbol for each class of polymers. Does there appear to be a correlation between T_g and T_m? Are there different trends for the different classes of polymers?

Section 3.6

3.7 For S-glass in Table 3.11, explain why some oxides commonly used in glass are not included and why the percentage of Al_2O_3 is high. How would you expect the strength of S-glass fibers to compare with those of E-glass?

3.8 Consider the data for ultimate strength of Al_2O_3, SiC, and glass in both bulk and fiber form in Tables 3.10 and 2.2(b). Explain the large differences between the strengths in tension and compression for these materials in bulk form, and also explain why the strengths of fibers in tension are so much greater than for bulk material.

Section 3.7

3.9 Compute strength-to-density and stiffness-to-density ratios, σ_u/ρ and E/ρ, for the first five metals in Table 3.1, for the S-glass fibers in Table 3.10, and for for the SiC and Al_2O_3 whiskers and fibers in Table 2.2(a) and (b). Use the upper limits of strength for the metals, and for SiC and Al_2O_3 use densities from Table 3.10 as approximate values. Plot σ_u/ρ versus E/ρ using different plotting symbols for metals, fibers, and whiskers. What trends do you observe? Discuss the significance of these trends in view of the possibility of making polymer or metal matrix composites containing say 50% fibers or whiskers by volume.

3.10 In December of 1986, Rutan and Yeager flew around the world without refueling in an aircraft named the *Voyager*. Consult news magazines or other sources and obtain information on the materials used to construct the aircraft. Why was the choice of materials critical? Could the flight have been accomplished with a metal aircraft?

Section 3.8

3.11 Examine a bicycle and attempt to identify the materials used for various parts. How is each material appropriate for use in the part where it is found? (You may wish to visit a bicycle shop or consult a book on bicycling.)

3.12 Look at one or more catalogs that include toys, sports equipment, tools, or appliances. Make a list of the materials named and guess at a more precise identification where trade names or abbreviated names are used. Are any composite materials identified?

4

Stress-Strain Equations and Models

4.1 INTRODUCTION

The three major types of deformation that occur in engineering materials are elastic, plastic, and creep deformation. These have already been discussed in earlier chapters from the viewpoint of physical mechanisms and general trends in behavior for metals, polymers, and ceramics. Recall that *elastic* deformation is associated with the stretching but not breaking of chemical bonds. In contrast, the two types of inelastic deformation involve processes where atoms change their relative positions, such as slip of crystal planes or sliding of chain molecules. If the inelastic deformation is time dependent, it is classed as *creep*, as distinguished from *plastic* deformation, which is not time dependent.

In engineering design and analysis, equations describing stress-strain behavior, called stress-strain relationships or *constitutive equations*, are frequently needed. For example, in elementary mechanics of materials, elastic behavior with a linear stress-strain relationship is assumed and used in calculating stresses and deflections in simple components such as beams and shafts. More complex situations of geometry and loading can be analyzed using the same basic assumptions in the form of *theory of elasticity*. This is now often accomplished by using the numerical technique called *finite element analysis* with a digital computer.

Stress-strain relationships need to consider behavior in three dimensions. In addition to elastic strains, the equations may also need to include plastic strains and creep strains. Treatment of creep strain requires the introduction of time as an additional variable. Regardless of the method used, analysis to determine stresses and deflections always requires employing appropriate stress-strain relationships for the particular material involved.

In this chapter, we will review elastic stress-strain behavior in three dimensions, also extending the topic to simple cases of anisotropy, where the elastic behavior varies with direction, as in composite materials. Plasticity and creep are discussed, and it is noted that more detailed treatments of these topics are given in later chapters.

4.2 RHEOLOGICAL MODELS

Simple mechanical devices, such as linear springs, frictional sliders, and viscous dashpots, can be used as an aid to understanding the various types of deformation. These are illustrated in Fig. 4.1. Such devices and combinations of them are called *rheological models*.

Elastic deformation is similar to the behavior of a simple linear spring characterized by its constant k. The deformation is always proportional to the force, $x = P/k$, and it is recovered instantly upon unloading. Plastic deformation is similar to the movement (without inertial effects) of a block of weight W on a horizontal plane. The static and kinetic coefficients of friction μ are assumed to be equal, so that there is a critical force for motion P_o. Note that the deformation remains if the force is removed. To return the block to its original position, it is necessary to apply a force of magnitude P_o but in the opposite direction.

Creep deformation can be subdivided into two types. *Steady-state creep* proceeds at a constant rate under constant stress. Such behavior occurs in a linear dashpot, which is an element where the velocity, $\dot{x} = dx/dt$, is proportional to the force. The constant of proportionality is the dashpot constant c, so that a constant value of force P' gives a constant velocity, $\dot{x} = P'/c$, resulting in a linear displacement versus time behavior. When the force is removed, the motion stops, so that the deformation is permanent, that is, not recovered. A dashpot could be physically constructed by placing a piston in a cylinder filled with a viscous liquid, such as a heavy oil. When a force is applied, small amounts of oil leak past the piston, allowing the piston to move. The velocity of motion will be approximately proportional to the magnitude of the force, and the displacement will remain after all force is removed.

The second type of creep, called *transient creep*, slows down as time passes. Such behavior occurs in a spring mounted parallel to a dashpot. If a constant force P' is applied, the deformation increases with time. But an increasing fraction of the applied force is needed to pull against the spring as x increases, so that less force is available

Description	Model	Input	Response

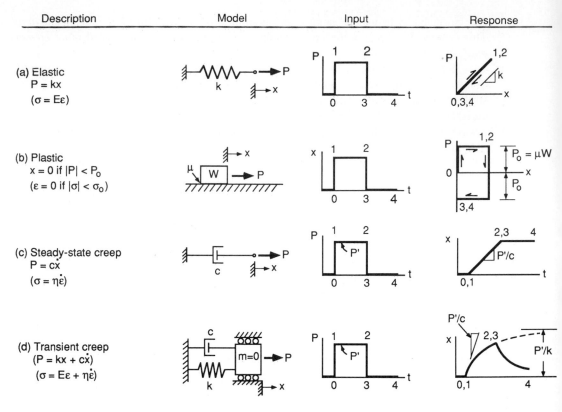

Figure 4.1 Mechanical models for four types of deformation. Responses are also shown for step inputs of force or displacement. Force P is analogous to stress, and displacement x is analogous to strain.

to the dashpot, and the rate of deformation decreases. The deformation approaches the value P'/k if the force is maintained for a long period of time. If the applied force is removed, the spring, having been extended, now pulls against the dashpot. This results in all of the deformation being recovered with time in this ideal model.

Thus, elastic and plastic deformation and two types of creep can be distinguished by considering whether or not they are time dependent, and whether or not they are recovered, according to Table 4.1. However, transient creep is best defined by its decreasing rate, as the degree of recovery of non-steady-state creep varies with material.

Rheological models may be used to represent stress and strain in a bar of material under axial loading as shown in Fig. 4.2. The model constants are related to material constants that are independent of the bar length L or area A. For elastic deformation, the constant of proportionality between stress and strain is the *elastic modulus*, also called *Young's modulus*.

$$E = \frac{\sigma}{\varepsilon} \qquad\qquad (4.1)$$

TABLE 4.1   CHARACTERISTICS OF THE VARIOUS TYPES OF DEFORMATION

Type of Deformation	Time Dependent?	Additional Distinguishing Characteristics
Elastic	No	Recovered instantly
Plastic	No	Not recovered
Steady-state creep	Yes	Constant rate; seldom recovered
Transient creep	Yes	Decreasing rate; may be recovered

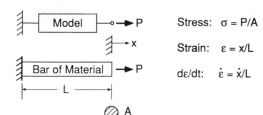

Stress: $\sigma = P/A$

Strain: $\varepsilon = x/L$

$d\varepsilon/dt$: $\dot{\varepsilon} = \dot{x}/L$

Figure 4.2   Relationship of models to stress, strain, and strain rate, in a bar of material.

Substituting the definitions of stress and strain, and also employing $P = kx$, yields the relationship between E and k.

$$E = \frac{kL}{A} \tag{4.2}$$

The material constant analogous to the dashpot constant c is called the *coefficient of tensile viscosity*.[1]

$$\eta = \frac{\sigma}{\dot{\varepsilon}} \tag{4.3}$$

where $\dot{\varepsilon} = d\varepsilon/dt$ is the strain rate. Substitutions from Fig. 4.2 and $P = c\dot{x}$ yields the relationship between η and c.

$$\eta = \frac{cL}{A} \tag{4.4}$$

For the plastic deformation model, the yield strength of the material is simply

$$\sigma_o = \frac{P_o}{A} \tag{4.5}$$

4.3 ELASTIC DEFORMATION

As discussed in Chapter 2, elastic deformation is associated with stretching the bonds between the atoms in a solid. As a result, the value of the elastic modulus, E, is quite high for strongly bound covalent solids. Metals have intermediate values, and polymers

[1] In fluid mechanics, viscosities are defined in terms of shear stresses and strains, $\eta_\tau = \tau/\dot{\gamma}$, where $\eta = 3\eta_\tau$ relates values of tensile and shear viscosity for an ideal incompressible material.

generally have low values due to the effect of secondary bonds between chain molecules. Some representative values for various materials are given in Table 4.2.

4.3.1 Elastic Constants

A material that has the same properties at all points within the solid is said to be *homogeneous*, and if the properties are the same in all directions, the material is *isotropic*. When viewed at macroscopic size scales, real materials obey these idealizations if they are composed of tiny randomly oriented crystal grains. This is at least approximately true for many metals and ceramics. Amorphous materials, such as glass and some polymers, may also be approximately isotropic and homogeneous. Hence, for the present, we will proceed with these simplifying assumptions.

TABLE 4.2 ELASTIC CONSTANTS FOR VARIOUS MATERIALS AT AMBIENT TEMPERATURE

Material	Elastic Modulus E, GPa (10^3 ksi)		Poisson's Ratio ν
(a) Metals			
Aluminum	70.3	(10.2)	0.345
Brass, 70Cu-30Zn	101	(14.6)	0.350
Copper	130	(18.8)	0.343
Iron; mild steel	212	(30.7)	0.293
Lead	16.1	(2.34)	0.44
Magnesium	44.7	(6.48)	0.291
Stainless steel, 2Ni-18Cr	215	(31.2)	0.283
Steel, hard 0.75C	201	(29.2)	0.296
Titanium	120	(17.4)	0.361
Tungsten	411	(59.6)	0.280
(b) Polymers			
ABS plastic	2.5	(0.36)	—
Epoxy	3.5	(0.51)	0.33
High impact polystyrene	1.65	(0.24)	0.34
Nylon 6	2.6	(0.38)	—
Nylon 6 with 33% glass fibers	9.3	(1.35)	—
PEEK	4.0	(0.58)	0.37
Unsaturated polyester	3.45	(0.50)	—
(c) Ceramics and glasses			
Alumina, Al_2O_3	400	(58.0)	0.22
Magnesia, MgO	300	(43.5)	0.18
Silicon carbide, SiC	396	(57.4)	0.22
Fused silica glass	70	(10.2)	0.18
Soda-lime glass	69	(10.0)	0.20
Type E glass	72.4	(10.5)	0.22

Sources: Data in [Boyer 85] p. 216, [ASM 88], [Creyke 82] p. 222, [Kelly 86] pp. 376, 392, [Kelly 89] p. 262, and [Schwartz 84] p. 2.44.

Let a bar of a homogeneous and isotropic material be subjected to an axial stress σ_x as in Fig. 4.3. The strain in the direction of the stress is

$$\varepsilon_x = \frac{L - L_i}{L_i} = \frac{\Delta L}{L_i} \tag{4.6}$$

where L is the deformed length, L_i the initial length, and ΔL the change. In a similar manner, the strain in any direction perpendicular to the stress, that is, along any diameter of the bar, is

$$\varepsilon_y = \varepsilon_z = \frac{d - d_i}{d_i} = \frac{\Delta d}{d_i} \tag{4.7}$$

This transverse strain is negative as the bar of course becomes thinner when stretched in the length direction.

The material is said to be *linear-elastic* if the stress is linearly related to these strains, and if the strains return immediately to zero after unloading in the manner of a simple linear spring. In this situation, two elastic constants are needed to characterize the material. One is the *elastic modulus*, $E = \sigma_x/\varepsilon_x$, which is the slope of the σ_x versus ε_x line in Fig. 4.3. The second constant is *Poisson's ratio*.

$$\nu = -\frac{\text{transverse strain}}{\text{longitudinal strain}} = -\frac{\varepsilon_y}{\varepsilon_x} \tag{4.8}$$

Since ε_y is negative, ν is positive. Hence, as shown in Fig. 4.3, the slope of a plot of ε_y versus ε_x is $-\nu$. Substituting ε_x from Eq. 4.8 into $E = \sigma_x/\varepsilon_x$ gives

$$\sigma_x = -\frac{E}{\nu}\varepsilon_y \tag{4.9}$$

This linear relationship is also shown in Fig. 4.3. The same situation also occurs along any other diameter, such as the z direction in Fig. 4.3.

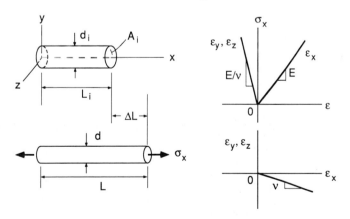

Figure 4.3   Longitudinal extension and lateral contraction used to obtain constants for a linear-elastic material that is isotropic and homogeneous.

Values of the elastic modulus vary widely for different materials. Poisson's ratio is often around 0.3 and does not vary outside the range 0 to 0.5 except under very unusual circumstances. Note that negative values of ν imply lateral expansion during axial tension, which is unlikely. As will be seen below, $\nu = 0.5$ implies constant volume, and values larger than 0.5 imply a decrease in volume for tensile loading, which is also unlikely. Values of ν for various materials are included in Table 4.2.

It should be noted that no material has perfectly linear or perfectly elastic behavior. Use of elastic constants such as E and ν should therefore be regarded as a useful approximation, or model, that often gives reasonably accurate answers. For example, most engineering metals can be modeled in this way at relatively low stresses below the yield strength, beyond which the behavior becomes nonlinear and inelastic. Also, original dimensions and cross-sectional areas are used in the above discussion to determine stresses and strains. Such an approach is appropriate for many situations of practical engineering interest, where dimensional changes are small. Except where otherwise indicated, this assumption, called *small-strain theory*, will be used.

4.3.2 Hooke's Law for Three Dimensions

Consider the general state of stress at a point as illustrated in Fig. 4.4. A complete description consists of normal stresses in three directions, σ_x, σ_y, and σ_z, and shear stresses on three planes, τ_{xy}, τ_{yz}, and τ_{zx}. Considering normal stresses first, and assuming that small strain theory applies, the strains caused by each component of stress can simply be added together. A stress in the x-direction causes a strain in the x-direction of

$$\frac{\sigma_x}{E}$$

This σ_x also causes a strain in the y-direction, from Eq. 4.9, of

$$-\frac{\nu\sigma_x}{E}$$

and the same strain the z-direction. Similarly, stresses in the y- and z-directions each cause strains in all three directions.

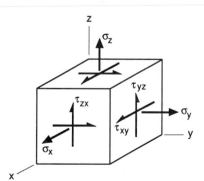

Figure 4.4 The six components needed to completely describe the state of stress at a point.

The situation can be summarized by the following table:

Resulting Strain Each Direction

Stress	x	y	z
σ_x	$\dfrac{\sigma_x}{E}$	$-\dfrac{\nu\sigma_x}{E}$	$-\dfrac{\nu\sigma_x}{E}$
σ_y	$-\dfrac{\nu\sigma_y}{E}$	$\dfrac{\sigma_y}{E}$	$-\dfrac{\nu\sigma_y}{E}$
σ_z	$-\dfrac{\nu\sigma_z}{E}$	$-\dfrac{\nu\sigma_z}{E}$	$\dfrac{\sigma_z}{E}$

Adding the columns in this table to obtain the total strain in each direction gives the equations

$$\varepsilon_x = \frac{1}{E}\left[\sigma_x - \nu\left(\sigma_y + \sigma_z\right)\right] \quad \text{(a)}$$

$$\varepsilon_y = \frac{1}{E}\left[\sigma_y - \nu\left(\sigma_x + \sigma_z\right)\right] \quad \text{(b)} \qquad (4.10)$$

$$\varepsilon_z = \frac{1}{E}\left[\sigma_z - \nu\left(\sigma_x + \sigma_y\right)\right] \quad \text{(c)}$$

The shear strains that occur on the orthogonal planes are each related to the corresponding shear stress by a constant called the *shear modulus, G*.

$$\gamma_{xy} = \frac{\tau_{xy}}{G}, \qquad \gamma_{yz} = \frac{\tau_{yz}}{G}, \qquad \gamma_{zx} = \frac{\tau_{zx}}{G} \qquad (4.11)$$

Note that the shear strain on a given plane is unaffected by the shear stresses on other planes. Hence, for shear strains, there is no effect analogous to Poisson contraction. Equations 4.10 and 4.11, taken together, are often called the *generalized Hooke's Law*.

Only two independent elastic constants are needed for an isotropic material, so that one of E, G, and ν can be considered redundant. The following equation allows any one of these to be calculated from the other two:

$$G = \frac{E}{2\left(1 + \nu\right)} \qquad (4.12)$$

This equation can be derived by considering a state of pure shear stress, as in a round bar under torsion in Fig. 4.5. Recall from elementary mechanics of materials that the state of shear stress, τ, can be equivalently represented by principal normal stresses on planes rotated 45° with respect to the directions of pure shear. Similarly, the shear strain is equivalent to normal strains as shown, also on 45° planes. Let the (x, y, z) directions for Eq. 4.10 correspond to the principal $(1, 2, 3)$ directions. The following substitutions can then be made in Eq. 4.10(a):

$$\sigma_x = \tau, \qquad \sigma_y = -\tau, \qquad \sigma_z = 0, \qquad \varepsilon_x = \frac{\gamma}{2} \qquad (4.13)$$

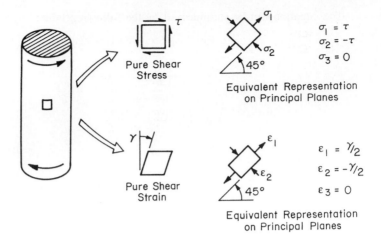

Figure 4.5 A state of pure shear stress and strain in an isotropic material.

This yields

$$\gamma = \frac{2(1+\nu)}{E}\tau \tag{4.14}$$

From the definition of G, the constant of proportionality is $G = \tau/\gamma$, so that Eq. 4.12 is confirmed.

Measured values of the three constants E, G, and ν for real materials will not generally obey Eq. 4.12 perfectly. This situation is mainly due to the material not being perfectly isotropic.

Example 4.1

A cylindrical pressure vessel 10 m long has closed ends, a wall thickness of 5 mm, and a diameter at midthickness of 3 m. If the vessel is filled with air to a pressure of 2 MPa, how much do the length, diameter, and wall thickness change, and in each case is the change an increase or a decrease? The vessel is made of a steel having elastic modulus $E = 200,000$ MPa and Poisson's ratio $\nu = 0.3$. Neglect any effects associated with the details of how the ends are attached.

Solution Attach a coordinate system to the surface of the pressure vessel as shown below, such that the z-axis is normal to the surface.

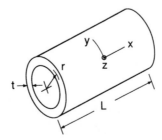

Figure E4.1

The ratio of radius to thickness, r/t, is such that it is reasonable to employ the thin-walled tube assumption, and the resulting stress equations, from any elementary text on mechanics of materials. Denoting the pressure as p, we have

$$\sigma_x = \frac{pr}{2t} = \frac{(2 \text{ MPa}) (1500 \text{ mm})}{2 (5 \text{ mm})} = 300 \text{ MPa}$$

$$\sigma_y = \frac{pr}{t} = \frac{(2 \text{ MPa}) (1500 \text{ mm})}{5 \text{ mm}} = 600 \text{ MPa}$$

The value of σ_z varies from $-p$ on the inside wall to zero on the outside, and for a thin-walled tube is everywhere sufficiently small that $\sigma_z \approx 0$ can be used. Substitute these stresses, and the known E and v into Hooke's Law, Eq. 4.10, which gives

$$\varepsilon_x = 6.00 \times 10^{-4}, \qquad \varepsilon_y = 2.55 \times 10^{-3}, \qquad \varepsilon_z = -1.35 \times 10^{-3}$$

These strains are related to the changes in length ΔL, circumference $\Delta(\pi d)$, diameter Δd, and thickness Δt, as follows:

$$\varepsilon_x = \frac{\Delta L}{L}, \qquad \varepsilon_y = \frac{\Delta (\pi d)}{\pi d} = \frac{\Delta d}{d}, \qquad \varepsilon_z = \frac{\Delta t}{t}$$

Substituting the strains from above and the known dimensions gives

$$\Delta L = 6 \text{ mm}, \qquad \Delta d = 7.65 \text{ mm}, \qquad \Delta t = -6.75 \times 10^{-3} \text{ mm} \qquad \textbf{Ans.}$$

Thus, there are small increases in length and diameter, and a tiny decrease in the wall thickness.

Example 4.2

A sample of a material subjected to a compressive stress σ_z is confined so that it cannot deform in the y-direction, but deformation is permitted in the x-direction. Assume that the material is isotropic and exhibits linear-elastic behavior. Determine the following in terms of σ_z and the elastic constants of the material:

   (a) The stress that develops in the y-direction.
   (b) The strain in the z-direction.
   (c) The strain in the x-direction.
   (d) The stiffness $E' = \sigma_z / \varepsilon_z$ in the z-direction. Is this apparent modulus equal to the elastic modulus E from a uniaxial test on the material? Why or why not?

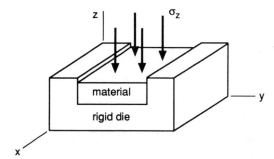

Figure E4.2

Solution Hooke's Law for the three-dimensional case, Eq. 4.10, is needed. The situation posed requires substituting $\varepsilon_y = 0$ and $\sigma_x = 0$, and also treating σ_z as a known quantity.

(a) The stress in the y-direction is obtained from Eq. 4.10(b).

$$0 = \frac{1}{E}\left[\sigma_y - v\left(0 + \sigma_z\right)\right]$$

$$\sigma_y = v\sigma_z \qquad\qquad\qquad \textbf{Ans.}$$

(b) The strain in the z-direction is given by substituting this σ_y into Eq. 4.10(c).

$$\varepsilon_z = \frac{1}{E}\left[\sigma_z - v\left(0 + v\sigma_z\right)\right]$$

$$\varepsilon_z = \frac{1 - v^2}{E}\sigma_z \qquad\qquad\qquad \textbf{Ans.}$$

(c) The strain in the x-direction is given by Eq. 4.10(a) with σ_y from above substituted.

$$\varepsilon_x = \frac{1}{E}\left[0 - v\left(v\sigma_z + \sigma_z\right)\right]$$

$$\varepsilon_x = -\frac{v\left(1 + v\right)}{E}\sigma_z \qquad\qquad\qquad \textbf{Ans.}$$

(d) The apparent stiffness in the z-direction is obtained immediately from the equation for ε_z

$$E' = \frac{\sigma_z}{\varepsilon_z} = \frac{E}{1 - v^2} \qquad\qquad\qquad \textbf{Ans.}$$

Obviously, this value differs from E, being larger, such as $E' = 1.10E$ for a typical value of $v = 0.3$. The value E is the ratio of stress to strain only for the uniaxial case, and such ratios for any other case are determined by behavior according to the three-dimensional form of Hooke's Law.

Comment Note that any value of compressive σ_z will have a negative sign, resulting in negative values for σ_y and ε_z, and a positive value for ε_x, as expected from the physical situation.

4.3.3 Volumetric Strain and Hydrostatic Stress

In stressed bodies, small volume changes occur that are associated with normal strains. Shear strains are not involved as they cause no volume change, only distortion. Consider a rectangular solid as in Fig. 4.6, where there are normal strains in three directions. The dimensions L, W, and H change by infinitesimal amounts, dL, dW, and dH, respectively, so that the normal strains are

$$\varepsilon_x = \frac{dL}{L}, \qquad \varepsilon_y = \frac{dW}{W}, \qquad \varepsilon_z = \frac{dH}{H} \qquad\qquad (4.15)$$

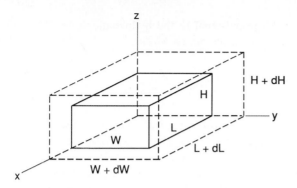

Figure 4.6   Volume change due to normal strains.

The volume, $V = LWH$, changes by an amount that can be evaluated from differential calculus, where V is considered to be a function of the three independent variables L, W, and H.

$$dV = \frac{\partial V}{\partial L}dL + \frac{\partial V}{\partial W}dW + \frac{\partial V}{\partial H}dH \tag{4.16}$$

Evaluating the partial derivatives and dividing both sides by $V = LWH$ gives

$$\frac{dV}{V} = \frac{dL}{L} + \frac{dW}{W} + \frac{dH}{H} \tag{4.17}$$

This ratio of the change in volume to the original volume is called the *volumetric strain*, or *dilitation*, ε_v. By substituting Eq. 4.15, it is seen to be simply the sum of the normal strains.

$$\varepsilon_v = \frac{dV}{V} = \varepsilon_x + \varepsilon_y + \varepsilon_z \tag{4.18}$$

For an isotropic material, the volumetric strain can be expressed in terms of stresses by substituting the generalized Hooke's Law, specifically Eq. 4.10, into Eq. 4.18. The following is obtained after collecting terms:

$$\varepsilon_v = \frac{1 - 2\nu}{E}\left(\sigma_x + \sigma_y + \sigma_z\right) \tag{4.19}$$

Note from this that $\nu = 0.5$ causes the change in volume to be zero, $\varepsilon_v = 0$, even in the presence of nonzero stresses. Also, a value of ν exceeding 0.5 would imply negative ε_v for tensile stresses, that is, a decrease in volume. This would be highly unusual, so that 0.5 appears to be an upper limit on ν that is seldom exceeded for real materials.

The average normal stress is called the *hydrostatic stress*.

$$\sigma_h = \frac{\sigma_x + \sigma_y + \sigma_z}{3} \tag{4.20}$$

Substituting this into Eq. 4.19 yields

$$\varepsilon_v = \frac{3(1 - 2\nu)}{E}\sigma_h \tag{4.21}$$

Hence, the volumetric strain is proportional to the hydrostatic stress. The constant of proportionality relating these is called the *bulk modulus*.

$$B = \frac{\sigma_h}{\varepsilon_v} = \frac{E}{3\,(1 - 2v)} \qquad (4.22)$$

Note that ε_v and σ_h are classed as *invariant* quantities. This means that they will always have the same values regardless of the choice of coordinate system. In other words, a different choice of x-y-z axes at a particular point in a material will cause the various stress and strain components to have different values, but the sum of the normal strains and the sum of the normal stresses will have the same value for any coordinate system.

4.3.4 Thermal Strains

Thermal strain is a special class of elastic strain that results from expansion with increasing temperature, or contraction with decreasing temperature. Increased temperature causes the atoms in a solid to vibrate by a larger amount. Consider a vibration over a range Δx centered on the equilibrium atomic spacing x_e of Fig. 2.17. The shape of the force versus distance curve in this region is such that the average zero-force position, x_{avg}, is greater than x_e as illustrated in Fig. 4.7. Such larger average atomic spacings accumulate over a macroscopic distance in the material to produce a dimensional increase. Similarly, decreasing the temperature causes the average spacing to decrease and approach x_e.

In isotropic materials, the effect is the same in all directions. Over a limited range of temperatures, the thermal strains at a given temperature T can be assumed to be proportional to the temperature change, ΔT.

$$\varepsilon = \alpha\,(T - T_0) = \alpha\,(\Delta T) \qquad (4.23)$$

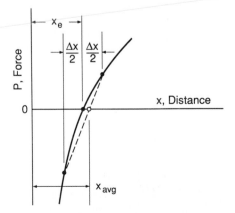

Figure 4.7   Force vs. distance between atoms. A thermal oscillation of $\pm \Delta x/2$ about the equilibrium position x_e gives an average distance x_{avg} greater than x_e.

where T_0 is a reference temperature where the strains are taken to be zero. The *coefficient of thermal expansion*, α, is seen to be in units of $1/°C$, where strain is dimensionless.

Thermal effects are generally greater at higher temperatures, that is, α increases with temperature. Hence, it may be necessary to allow for variation in α if ΔT is large. Thermal strains, and therefore values of α, are smaller where the atomic bonding is stronger. If values of α at room temperature for various materials are compared, this leads to a trend of decreasing α with increasing melting temperature, as the atomic bonding is stronger at temperatures more remote from the melting temperature. Figure 4.8 shows this trend for various materials.

In an isotropic material, since uniform thermal strains occur in all directions, Hooke's law for three dimensions from Eq. 4.10 can be generalized to include thermal effects.

$$\varepsilon_x = \frac{1}{E} \left[\sigma_x - \nu \left(\sigma_y + \sigma_z \right) \right] + \alpha \left(\Delta T \right)$$

$$\varepsilon_y = \frac{1}{E} \left[\sigma_y - \nu \left(\sigma_x + \sigma_z \right) \right] + \alpha(\Delta T) \qquad (4.24)$$

$$\varepsilon_z = \frac{1}{E} \left[\sigma_z - \nu \left(\sigma_x + \sigma_y \right) \right] + \alpha \left(\Delta T \right)$$

If free thermal expansion is prevented by geometric constraint, then a sufficient ΔT will cause large stresses that are of engineering significance to develop.

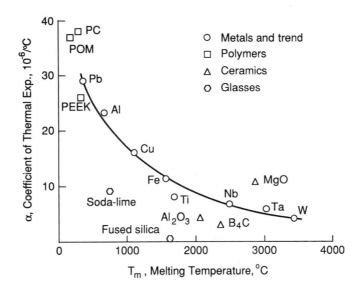

Figure 4.8 Coefficients of thermal expansion at room temperature vs. melting temperature for various materials. (Data from [Boyer 85] p. 1.44, [Creyke 82] p. 50, and [ASM 88] p. 69.)

4.4 PLASTIC DEFORMATION

Plastic deformation is time-independent deformation that is not recovered after unloading. As discussed in Chapter 2, the principal physical mechanism causing plastic deformation in metals and ceramics is slippage between planes of atoms in the crystal grains of the material, with this occurring in an incremental manner due to dislocation motion. The material's resistance to plastic deformation is roughly analogous to the friction of a block on a plane as in the rheological model of Fig. 4.1(b).

4.4.1 Stress-Strain Behavior

Consider the behavior of a bar of material under uniaxial stress as the deformation is gradually increased as in Fig. 4.9. Behavior typical of engineering metals is shown, where a distinct elastic region of slope E is followed by yielding, beyond which small increases in stress result in relatively large increases in strain. If the stress is removed, unloading will proceed along a line that is approximately straight and parallel to the original elastic slope. The portion of the total axial strain ε_x that is recovered is called the elastic strain, ε_{ex}, and the unrecovered portion is called the plastic strain, ε_{px}.

$$\varepsilon_x = \varepsilon_{ex} + \varepsilon_{px} \tag{4.25}$$

The elastic strain is still related to stress by Hooke's Law, so that $\varepsilon_{ex} = \sigma_x/E$.

$$\varepsilon_x = \frac{\sigma_x}{E} + \varepsilon_{px} \tag{4.26}$$

Now consider any direction transverse to the x-direction, such as the y- or z-direction. As shown in Fig. 4.9, the slope beyond yielding of a plot of $-\varepsilon_y$ or $-\varepsilon_z$ versus ε_x deviates from the elastic Poisson's ratio, ν, and approaches 0.5. The elastic

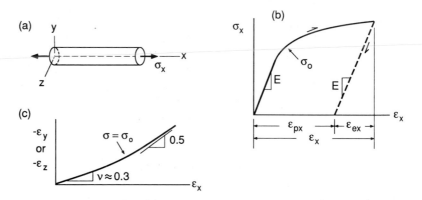

Figure 4.9 Elastic and plastic components of total strain, and the effect of plastic deformation on Poisson's ratio.

strain in any transverse direction is still related to that in the x-direction by the elastic value of Poisson's ratio, despite the fact that plastic strain is also present.

$$\varepsilon_{ey} = \varepsilon_{ez} = -\nu\varepsilon_{ex} \tag{4.27}$$

The trend of the $-\varepsilon_y$ or $-\varepsilon_z$ versus ε_x curve toward a slope of 0.5 at large strains is due to the fact that the constant analogous to Poisson's ratio for plastic strains is near this value.

$$\varepsilon_{py} = \varepsilon_{pz} = -0.5\varepsilon_{px} \tag{4.28}$$

The total strain in any transverse direction may be obtained by adding its elastic and plastic components.

$$\varepsilon_y = \varepsilon_z = -\nu\varepsilon_{ex} - 0.5\varepsilon_{px} \tag{4.29}$$

Substitution of Eqs. 4.25 and 4.29 into Eq. 4.18 gives a volumetric strain of

$$\varepsilon_v = \varepsilon_{ex}(1 - 2v) = \frac{\sigma_x}{E}(1 - 2v) \tag{4.30}$$

where the 0.5 value is seen to cause ε_v to be unaffected by the plastic strain. Note that this result is the same as Eq. 4.19 applied to the uniaxial case.

For the general case of a three-dimensional state of stress where plastic deformations occur, Hooke's Law in the form of Eq. 4.10 gives only the elastic strains ε_{ex}, ε_{ey}, and ε_{ez}. A second set of equations are needed for the plastic strains.

$$\varepsilon_{px} = \frac{1}{E_p}\left[\sigma_x - 0.5\left(\sigma_y + \sigma_z\right)\right]$$

$$\varepsilon_{py} = \frac{1}{E_p}\left[\sigma_y - 0.5\left(\sigma_x + \sigma_z\right)\right] \tag{4.31}$$

$$\varepsilon_{pz} = \frac{1}{E_p}\left[\sigma_z - 0.5\left(\sigma_x + \sigma_y\right)\right]$$

In place of the constant E, there is a variable E_p, and Poisson's ratio is replaced by 0.5. These equations and related topics will be considered in some detail later in Chapter 12. The constant of 0.5 results in plastic strain not contributing to the volume change for any state of stress. This is expected from the physical mechanism of plastic strain, as relative slip between crystal planes should not result in any volume change. [Compare (a) and (d) in Fig. 2.20, where the area occupied by the atoms in this illustration represents the volume.]

4.4.2 Rheological Models for Plastic Deformation

Recall the simple mechanical model of plastic deformation as a block on a plane of Fig. 4.1(b). There are similarities between the behavior of this model and that of a real material, as in Fig. 4.9, but also significant differences. It is therefore useful to describe certain more sophisticated rheological models that behave more like real materials, some of which are shown in Fig. 4.10.

The model of Fig. 4.1(b) appears in Fig. 4.10(a), where the block on a plane is replaced by a frictional slider, such as a spring clip, that has zero mass but identical force versus displacement response for slow loading. Also, the models are all assumed to represent bars of material of length L and cross section A, so that they give the behavior of stress and strain. On this basis, sliders move when the stress reaches the yield stress σ_o, and springs deform according to $\varepsilon = \sigma/E$.

Responses are given in Fig. 4.10 for various strain inputs to these models. The initial response of each model is either *elastic* or *rigid*, depending on whether or not there is any linear deformation prior to movement of the slider. Slider movement is analogous

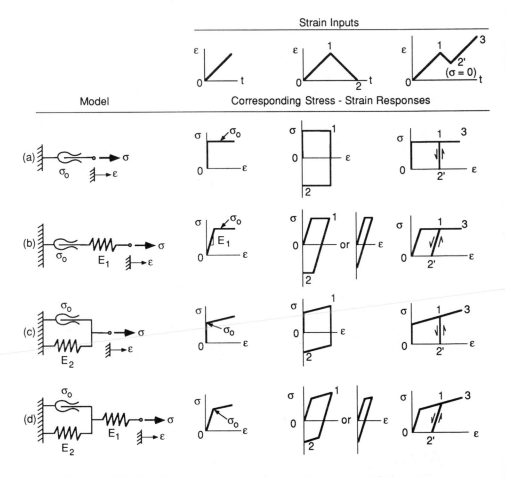

Figure 4.10 Rheological models for plastic deformation and their responses to three different strain inputs. Model (a) has behavior that is rigid, perfectly plastic, (b) elastic, perfectly plastic, (c) rigid, linear hardening, and (d) elastic, linear hardening.

to yielding of the material. After yielding, a flat σ-ε response is called *perfectly plastic* behavior, whereas a sloping line is called *linear hardening*. *Nonlinear hardening* would describe a curved stress-strain relationship after yielding as in Fig. 4.9.

Adding a spring in series with the frictional slider as in Fig. 4.10(b) results in an elastic response prior to yielding of the slider. The response after yielding is the same as for (a) except that an added elastic strain is present due to the spring.

$$\varepsilon = \varepsilon_e + \varepsilon_p = \frac{\sigma}{E_1} + \varepsilon_p \qquad \left(\varepsilon > \frac{\sigma_o}{E_1} \right) \tag{4.32}$$

where subscripts e and p refer to the individual displacements in the elastic (spring) and plastic (slider) elements.

If the spring is instead added parallel to the slider, the rigid, linear-hardening model of (c) is obtained. (Note that the vertical bar connecting the spring and the slider is assumed not to rotate, so that both the spring and the slider have the same strain.) Prior to yielding, all of the stress is carried by the slider, since $\varepsilon = 0$. After yielding, part of the stress is carried by each element. In particular, the stress in the slider is σ_o and that in the spring is $E_2\varepsilon$. Adding gives the total stress as $\sigma = \sigma_o + E_2\varepsilon$, so that

$$\varepsilon = \frac{\sigma - \sigma_o}{E_2} \qquad (\sigma > \sigma_o) \tag{4.33}$$

Adding a spring in series gives the elastic, linear-hardening model of Fig. 4.10(d). The behavior is similar to (c) except that an elastic strain due to spring E_1 is added that depends on the current value of σ. Prior to yielding, $\varepsilon = \sigma/E_1$. After yielding, both springs contribute to the strain. The strain is thus σ/E_1 plus that due to the (E_2, σ_o) parallel combination, which is given by Eq. 4.33 above.

$$\varepsilon = \frac{\sigma}{E_1} + \frac{\sigma - \sigma_o}{E_2} \qquad (\sigma > \sigma_o) \tag{4.34}$$

4.4.3 Unloading and Reloading Response of Models

Consider the situation where the elastic, perfectly plastic model has yielded, and then the stress is returned to zero. This is illustrated in Fig. 4.11(a). The strain in the spring ε_e decreases during unloading, with ε_p remaining unchanged, and with the σ-ε response still being given by Eq. 4.32. Hence, when σ reaches zero, $\varepsilon = \varepsilon_p$. Analogous behavior also occurs for the elastic, linear-hardening model as illustrated in (b). In both cases, the elastic strain in spring E_1 is recovered upon unloading, whereas the strain occurring in the frictional slider is plastic strain that is not recovered.[2] In both cases, the behavior is similar to that of a real material as in (c).

For strains that return to zero after yielding, the behavior of each of the four models is shown in the next-to-last column of Fig. 4.10. Just after the direction of straining is

---

[2]An exception to this is that, for the linear-hardening models, some of the plastic strain will be lost prior to zero stress if the plastic strain is so large that $E_2\varepsilon_p > \sigma_o$. A more complete explanation is given later in Chapter 12.

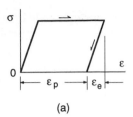 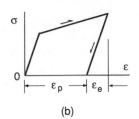 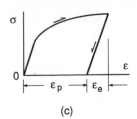

(a)   (b)   (c)

Figure 4.11   Loading and unloading behavior of (a) an elastic, perfectly plastic model, (b) an elastic, linear-hardening model, and (c) a material with nonlinear hardening.

changed, the slope of the stress-strain response is either rigid or elastic, and it is parallel to the initial loading slope. To return the strain to zero, a negative (compressive) stress is required in all cases. The two models with initially rigid response, (a) and (c), must yield again before the strain reaches zero. The two models with initially elastic response, (b) and (d), may or may not be forced to yield prior to reaching zero strain.

Now consider the response of each model to the situation of the last column in Fig. 4.10, where the model is reloaded after elastic unloading to $\sigma = 0$. In all cases, yielding occurs a second time when the strain again reaches the value ε_1 from which unloading occurred. It is obvious that the two perfectly plastic models will again yield at $\sigma = \sigma_o$. But the two linear-hardening models now yield at a value $\sigma = \sigma_1$, which is higher than the initial yield stress. Furthermore, σ_1 is the same value of stress that was present at $\varepsilon = \varepsilon_1$ when the unloading first began. For all four models, the interpretation may be made that the model possesses a *memory* of the point of previous unloading. In particular, yielding again occurs at the same σ-ε point from which unloading occurred, and the subsequent response is the same as if there had never been any unloading. Real materials that deform plastically exhibit a similar memory effect.

The elastic, linear-hardening model is a reasonable one for real materials. It can be extended to model nonlinear hardening by including additional spring-slider parallel combinations as in Fig. 4.12. Note that the memory effect occurs even if there is yielding during unloading, which is true for the simpler models also. Such models provide a useful representation of the behavior of engineering metals where time-dependent effects are small. They will be employed in later chapters in connection with cyclic loading.

Example 4.3

A steel is modeled by the elastic, linear-hardening model of Fig. 4.10 using constants of $E_1 = 200$ GPa, $E_2 = 22.2$ GPa, and $\sigma_o = 400$ MPa. The material is loaded to a strain of 0.007 and then unloaded to a strain of zero. Determine and plot the stress-strain response.

Solution   The response is shown in Fig. E4.3. From O to A, only spring E_1 deforms and the σ-ε response is linear with slope E_1.

$$\sigma = E_1 \varepsilon = 200{,}000\varepsilon \quad (O \text{ to } A)$$

where σ is in units of MPa.

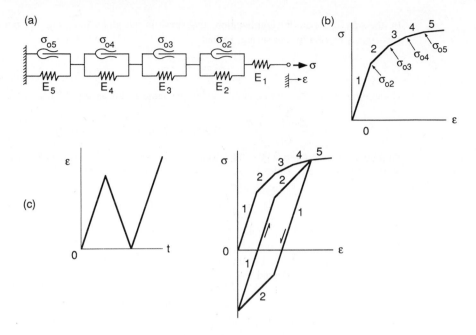

Figure 4.12 Rheological model having an elastic, nonlinear-hardening response. (Adapted from [Dowling 87b]; used with permission; ©Society of Automotive Engineers.)

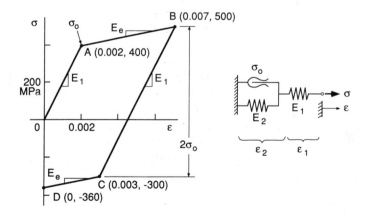

Figure E4.3

Once the stress exceeds $\sigma_o = 400$ MPa, the total strain ε includes a strain ε_2 in the spring and slider parallel combination that adds to the strain $\varepsilon_1 = \sigma/E_1$ in spring E_1.

$$\varepsilon = \varepsilon_1 + \varepsilon_2 = \frac{\sigma}{E_1} + \varepsilon_2$$

In the (E_2, σ_o) parallel combination, the stress in the slider is σ_o and the remainder of $(\sigma - \sigma_o)$ is carried by the spring, so that

$$\varepsilon_2 = \frac{\sigma - \sigma_o}{E_2}$$

Combining the above two equations gives the same result as Eq. 4.34, and solving for σ yields

$$\sigma = \frac{E_1(E_2\varepsilon + \sigma_o)}{E_1 + E_2} = 0.900\,(22{,}200\varepsilon + 400) \quad (A \text{ to } B)$$

At point B, where $\varepsilon_B = 0.007$, the stress is thus $\sigma_B = 500$ MPa. The slope of the resulting straight line on the σ-ε plot is seen to be the same as for the series combination of E_1 and E_2.

$$E_e = \frac{E_1 E_2}{E_1 + E_2} = 20{,}000 \text{ MPa}$$

On unloading, the parallel combination does not deform until point C is reached, where yielding occurs in the opposite direction. Prior to this, only spring E_1 deforms, so that the changes in stress and strain relative to their values at point B, which are $\Delta\sigma = \sigma - \sigma_B$ and $\Delta\varepsilon = \varepsilon - \varepsilon_B$, are related by

$$\Delta\sigma = E_1\Delta\varepsilon, \qquad (\sigma - \sigma_B) = E_1(\varepsilon - \varepsilon_B)$$

so that the stress-strain response follows

$$\sigma = 500 + 200{,}000\,(\varepsilon - 0.007) \quad (B \text{ to } C)$$

Yielding in the opposite direction occurs at C when the stress carried by the slider reaches $-\sigma_o$. At this point, the parallel combination has not deformed since B, so that a stress $(\sigma_B - \sigma_o)$ remains locked in spring E_2. Hence, summing the stresses in spring and slider at the point of reverse yielding gives

$$\sigma_C = (-\sigma_o) + (\sigma_B - \sigma_o) = \sigma_B - 2\sigma_o = -300 \text{ MPa}$$

That is, the stress must change by twice the yield stress for the direction of yielding to reverse. Beyond C, the change in strain due to spring E_1 is

$$\Delta\varepsilon_1 = \frac{\Delta\sigma}{E_1}$$

In the parallel combination, the stress in the slider stays constant at $-\sigma_o$ and all of the change $\Delta\sigma$ is associated with deformation in spring E_2. The change in strain due to the parallel combination is thus

$$\Delta\varepsilon_2 = \frac{\Delta\sigma}{E_2}$$

Adding to obtain the total change in strain gives

$$\Delta\varepsilon = \frac{\Delta\sigma}{E_1} + \frac{\Delta\sigma}{E_2} = \frac{\Delta\sigma}{E_e}, \qquad \text{or} \qquad \varepsilon - \varepsilon_C = \frac{\sigma - \sigma_C}{E_e}$$

Solving for stress gives

$$\sigma = \sigma_C + E_e(\varepsilon - \varepsilon_C) = -300 + 20{,}000\,(\varepsilon - 0.003) \quad (C \text{ to } D)$$

So that at $\varepsilon_D = 0$, we have $\sigma_D = -360$ MPa.

4.5 CREEP DEFORMATION

Significant time-dependent deformations occur in engineering metals and ceramics at elevated temperatures. They also occur at room temperature in low-melting-temperature metals, such as lead, and in many other materials, such as glass, polymers, and concrete. A variety of physical mechanisms are involved, as discussed to an extent in Chapter 2 and considered again in Chapter 15.

4.5.1 Stress-Strain Behavior

Assume that a stress is rapidly applied to a sample of material as in Fig. 4.13. The instantaneous deformation that occurs is a combination of elastic and plastic strain. The plastic portion ε_p could be isolated by immediate unloading, as illustrated by the dashed line. If the stress is instead maintained, creep deformation ε_c may occur that is a combination of transient creep and steady-state creep.

Removal of the stress causes the elastic strain ε_e to be instantly recovered. Some of the creep strain may be recovered after waiting a period of time as indicated by 3-4 in Fig. 4.13. The unrecovered strain is the sum of part of the creep strain and all of the plastic strain. In real materials, the distinctions among the various types of strain blur somewhat. For example, strain in metals due to crystal slip, which is normally regarded as plastic strain, may still be weakly time dependent. In ductile polymers at room temperature, creep is often so prevalent that it is difficult to run a test that does not include it, so that isolated elastic and plastic strains are not observed.

The recovered portion of the creep strain may be quite large in polymers due to chain molecules interfering with one another in such a way that they slowly reestablish their original configuration after removal of the stress, causing the strain to slowly disappear. For example, creep strains as large as 100% in flexible vinyl (plasticized PVC) may be mostly recovered after unloading. In metals and ceramics, large creep strains can occur above about $0.5T_m$ due to motion of lattice vacancies or dislocations, grain

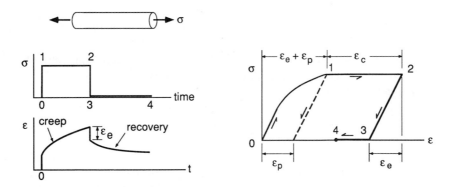

Figure 4.13   Stress-time step applied to a material exhibiting strain response that includes elastic, plastic, and creep components.

boundary sliding, etc. Relatively little recovery occurs for such strains. Viscous flow in glass is similar to the steady-state creep rheological model and is not recovered upon unloading. However, small transient creep strains that are recoverable occur in metals and other materials even at temperatures where there is essentially no steady-state creep. Such strains have the useful effect of aiding in the decay of vibrations and are called *anelastic strains*.

For general three-dimensional states of stress where creep occurs, equations analogous to Hooke's Law, Eq. 4.10, are used except that *rates* of creep strain are obtained.

$$\dot{\varepsilon}_{cx} = \frac{1}{\eta} \left[\sigma_x - 0.5 \left(\sigma_y + \sigma_z \right) \right]$$

$$\dot{\varepsilon}_{cy} = \frac{1}{\eta} \left[\sigma_y - 0.5 \left(\sigma_x + \sigma_z \right) \right] \tag{4.35}$$

$$\dot{\varepsilon}_{cz} = \frac{1}{\eta} \left[\sigma_z - 0.5 \left(\sigma_x + \sigma_y \right) \right]$$

In the simplest application of these equations, the tensile viscosity η is a constant and the creep behavior is of the steady-state type. In more complex applications, η can become a stress-dependent variable. Replacing Poisson's ratio by 0.5 in the above equations specifies that creep strain does not contribute to volume change. Creep strains from the rates of Eq. 4.35 generally need to be added to elastic strains from Eq. 4.10, and plastic strains from Eq. 4.31 may also be present.

4.5.2 Rheological Modeling of Steady-State Creep

Some simple physical models for creep behavior are shown in Fig. 4.14. Note that in (c) and (d) the vertical bar is assumed not to rotate, so that parallel springs and dashpots are subjected to the same strain. These models are expressed in terms of stress and strain, so that springs deform according to $\varepsilon = \sigma/E$ and dashpots according to $\dot{\varepsilon} = \sigma/\eta$. The use of constant viscosities η in these models results in all strain rates and strains being proportional to the applied stress, a situation described by the term *linear viscoelasticity*. Such idealized linear behavior is sometimes a reasonable approximation for real materials, as for polymers, and also for metals and ceramics at high temperature but low-stresses. However, for metals and ceramics at high stresses, strain rates are proportional not to the first power of stress, but to a higher power on the order of five. In such cases, models or equations involving more complex stress-dependence can be used as described later in Chapter 15. However, the simple linear models will suffice for the present to illustrate some of the gross features of creep behavior.

All solid materials exhibit some elastic deformation, which can be included in the steady-state creep model by adding a spring in series with the dashpot as illustrated in Fig. 4.14(b). The response is modified as shown, with the spring simply providing an additional elastic deflection, $\varepsilon_e = \sigma/E_1$. After unloading, this elastic strain is recovered, but the creep strain remains. While the constant stress σ' is applied during 1-2 for model

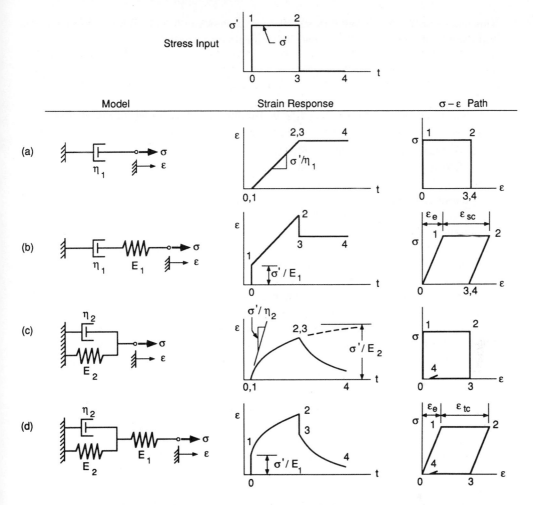

Figure 4.14 Rheological models having time-dependent behavior and their responses to a stress-time step. Both strain-time and stress-strain responses are shown. Model (a) exhibits only steady-state creep, and (b) is similar except that elastic strain is added. Model (c) exhibits only transient creep, and (d) is similar except that elastic strain is added.

(b), the response of the model is given by adding the elastic and creep components of the strain.

$$\varepsilon = \varepsilon_e + \varepsilon_c = \frac{\sigma'}{E_1} + \varepsilon_c \tag{4.36}$$

The rate of creep strain is related to the stress by the dashpot constant.

$$\dot{\varepsilon}_c = \frac{d\varepsilon_c}{dt} = \frac{\sigma'}{\eta_1} \tag{4.37}$$

This represents a very simple differential equation that can be solved for ε_c by integration and combined with Eq. 4.36 to give the strain-time response.

$$\varepsilon = \frac{\sigma'}{E_1} + \frac{\sigma' t}{\eta_1} \tag{4.38}$$

This is the equation of the linear $\varepsilon\text{-}t$ response during 1-2 as shown in Fig. 4.14(b).

4.5.3 Relaxation Behavior of Elastic, Steady-State Creep Model

Now consider imposing a fixed displacement on the elastic, steady-state creep model as illustrated in Fig. 4.15. Initially, all of the strain occurs as elastic strain in the spring. The strain in the dashpot is initially zero since it requires a finite time to respond. As time passes, motion occurs in the dashpot and the strain in the spring decreases, as it must due to the total strain being held constant.

$$\varepsilon' = \varepsilon_e + \varepsilon_c \tag{4.39}$$

where ε' is the constant total strain and ε_c is the creep strain. Hence, elastic strain is being replaced by creep strain.

The stress necessary to maintain the constant strain is related to the elastic strain by

$$\sigma = E_1 \varepsilon_e \tag{4.40}$$

Since ε_e is decreasing, this requires that σ must also decrease. The rate of creep strain is related to the value of σ at any time by

$$\dot{\varepsilon}_c = \frac{d\varepsilon_c}{dt} = \frac{\sigma}{\eta_1} \tag{4.41}$$

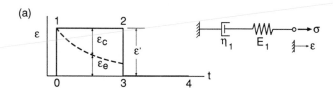

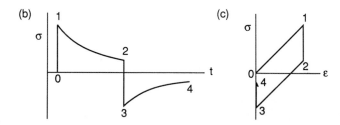

Figure 4.15 Relaxation under constant strain for a model with steady-state creep and elastic behavior. The step in strain (a) causes stress-time behavior as in (b), and stress-strain behavior as in (c).

Combining these equations and solving the resulting differential equation gives the variation of σ as it decreases.

$$\sigma = E_1 \varepsilon' e^{-E_1 t / \eta_1} \tag{4.42}$$

This corresponds to a σ-t response that decays with time as illustrated by curve 1-2 in Fig. 4.15(b). Such behavior is called *relaxation*. Relaxation is the same phenomenon as creep, differing only in that it is observed under constant strain rather than constant stress. Real engineering materials that exhibit creep will also show relaxation behavior in a manner analogous to this model.

Now consider returning the strain of the model to zero after some relaxation has occurred. As indicated by 2-3, a compressive stress is required. If the strain is maintained at zero, the stress will again relax toward zero as shown. The equation for this variation of σ can also be readily obtained.

Relaxation of stress during constant strain should not be confused with time dependent *recovery* at zero stress as in Fig. 4.14(c).

Example 4.4

Derive Eq. 4.42, which describes stress relaxation in the elastic, steady-state creep model as illustrated by 1-2 in Fig. 4.15. Base this on the other equations given just above Eq. 4.42.

Solution   Differentiate both sides of Eqs. 4.39 and 4.40 with respect to time, noting that $d\varepsilon'/dt = 0$ as ε' is held constant.

$$0 = \dot{\varepsilon}_e + \dot{\varepsilon}_c, \qquad \frac{d\sigma}{dt} = \dot{\sigma} = E_1 \dot{\varepsilon}_e$$

Substitute $\dot{\varepsilon}_e$ from the second equation, and also $\dot{\varepsilon}_c$ from Eq. 4.41, into the first equation to obtain

$$\frac{1}{E_1} \frac{d\sigma}{dt} + \frac{\sigma}{\eta_1} = 0$$

Separate the variables σ and t and integrate both sides of the equation.

$$\ln \sigma = -\frac{E_1}{\eta_1} t + C$$

where C is a constant of integration that can be evaluated by noting that the creep strain is initially zero, so that $\sigma = E_1 \varepsilon'$ at $t = 0$. This gives

$$C = \ln E_1 \varepsilon'$$

Substituting for C and solving for σ then gives the desired result.

$$\sigma = E_1 \varepsilon' e^{-E_1 t / \eta_1} \qquad\qquad \textbf{Ans.}$$

4.5.4 Rheological Modeling of Transient Creep

The transient creep model of Fig. 4.14(c) is placed in series with a spring to obtain model (d). The two responses differ only in that an elastic strain $\varepsilon_e = \sigma/E_1$ is added for model (d). This ε_e is labeled, as is ε_{tc}, the transient creep strain. First considering the simple transient creep model (c), this can be analyzed by noting that the stress is the sum of the separate stresses in the spring and the dashpot.

$$\sigma = E_2\varepsilon + \eta_2\dot{\varepsilon} \tag{4.43}$$

This gives

$$\dot{\varepsilon} = \frac{d\varepsilon}{dt} = \frac{\sigma - E_2\varepsilon}{\eta_2} \tag{4.44}$$

Solving this differential equation for the case of a constant stress σ' gives the ε-t response.

$$\varepsilon = \frac{\sigma'}{E_2}\left(1 - e^{-E_2t/\eta_2}\right) \tag{4.45}$$

Study of this equation reveals that ε increases with time as illustrated in Fig. 4.14(c) during time interval 1-2. The time rate of change of ε, that is, $\dot{\varepsilon} = d\varepsilon/dt$, has an initial value of σ'/η_2 and decreases toward zero at infinite time. Also, ε approaches the value σ'/E_2 at infinite time.

The more complex model (d) has a similar ε-t response except that an elastic strain in spring E_1 is added.

$$\varepsilon = \frac{\sigma'}{E_1} + \frac{\sigma'}{E_2}\left(1 - e^{-E_2t/\eta_2}\right) \tag{4.46}$$

Note that the elastic strain appears instantly when the load is applied and disappears instantly when it is removed.

After removal of the load, the strain in the transient creep models varies as shown in Fig. 4.14. In particular, it decreases toward zero at infinite time due to the spring in the parallel arrangement pulling on the dashpot. Equations for this *recovery* response may also be obtained by solving the differential equations involved. In these ideal models, all of the transient creep strain is recovered. If a step strain is applied to the elastic, transient creep model (d), a relaxation behavior occurs that is similar to that of Fig. 4.15. However, the behavior differs in that the stress decays not toward zero, but toward the value

$$\sigma = \frac{\varepsilon'E_1E_2}{E_1 + E_2} \qquad (t = \infty) \tag{4.47}$$

This occurs because the stress in the dashpot decays to zero at infinite time, and the model becomes equivalent to the two springs E_1 and E_2 in series.

4.6 ANISOTROPIC MATERIALS

Real materials are never perfectly isotropic. In some cases, the differences in properties
for different directions are so large that analysis assuming isotropic behavior is no longer
a reasonable approximation. Some examples of significantly anisotropic materials are
shown in Fig. 4.16.

Due to the presence of stiff fibers in particular directions, composite materials can
be highly anisotropic, and engineering design and analysis for these materials requires
the use of a more general version of Hooke's Law than was presented above. In what
follows, we will first discuss Hooke's Law for anisotropic cases in general, and then
we will apply it to in-plane loading of composite materials. Anisotropic plasticity is not
considered in this chapter or even in later chapters. This advanced topic is important in
some cases, but its importance is reduced by the fact that most composite materials fail
prior to the occurrence of large amounts of inelastic strain.

4.6.1 Anisotropic Hooke's Law

In the general three-dimensional case, there are six components of stress, σ_x, σ_y, σ_z, τ_{xy},
τ_{yz}, and τ_{zx}, as illustrated in Fig. 4.4. There are also six corresponding components of
strain ε_x, ε_y, ε_z, γ_{xy}, γ_{yz}, and γ_{zx}. In highly anisotropic materials, any one component of

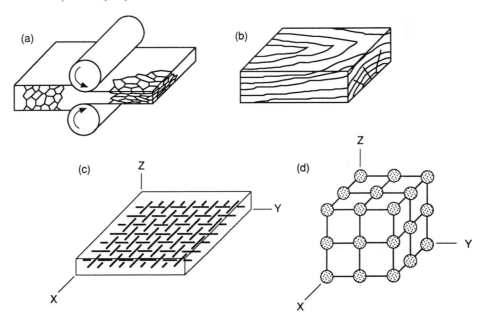

Figure 4.16 Anisotropic materials: (a) rolled metal, (b) wood, (c) glass-fiber
cloth in an epoxy matrix, and (d) a crystal with cubic unit cell. (Adapted from
[Crandall 59] p. 224; used with permission; copyright ©1959 by McGraw-Hill
Pub. Co.)

stress can cause strains in all six components. The general anisotropic form of Hooke's Law is given by the following six equations, here written with the coefficients shown as a matrix:

$$
\begin{Bmatrix} \varepsilon_x \\ \varepsilon_y \\ \varepsilon_z \\ \gamma_{yz} \\ \gamma_{zx} \\ \gamma_{xy} \end{Bmatrix} = \begin{bmatrix} S_{11} & S_{12} & S_{13} & S_{14} & S_{15} & S_{16} \\ S_{12} & S_{22} & S_{23} & S_{24} & S_{25} & S_{26} \\ S_{13} & S_{23} & S_{33} & S_{34} & S_{35} & S_{36} \\ S_{14} & S_{24} & S_{34} & S_{44} & S_{45} & S_{46} \\ S_{15} & S_{25} & S_{35} & S_{45} & S_{55} & S_{56} \\ S_{16} & S_{26} & S_{36} & S_{46} & S_{56} & S_{66} \end{bmatrix} \begin{Bmatrix} \sigma_x \\ \sigma_y \\ \sigma_z \\ \tau_{yz} \\ \tau_{zx} \\ \tau_{xy} \end{Bmatrix}
\tag{4.48}
$$

General anisotropy is considerably more complex than the isotropic case. Not only are there a large number of different materials constants S_{ij}, but their values also change if the orientation of the x-y-z coordinate system is changed.

In the isotropic case, the constants do not depend on the orientation of the coordinate axes, and most of the constants are either zero or have the same values as other ones. For example, $\gamma_{xy} = \tau_{xy}/G$, so that all of the S_{ij} in the γ_{xy} row of the matrix are zero except $S_{66} = 1/G$. This contrasts with the situation for highly anisotropic materials, where γ_{xy} is the sum of contributions due to all six stress components. The matrix of S_{ij} coefficients, specialized to the isotropic case as given by Eq. 4.10, is

$$
[S_{ij}] = \begin{bmatrix} \dfrac{1}{E} & -\dfrac{v}{E} & -\dfrac{v}{E} & 0 & 0 & 0 \\[2ex] -\dfrac{v}{E} & \dfrac{1}{E} & -\dfrac{v}{E} & 0 & 0 & 0 \\[2ex] -\dfrac{v}{E} & -\dfrac{v}{E} & \dfrac{1}{E} & 0 & 0 & 0 \\[2ex] 0 & 0 & 0 & \dfrac{1}{G} & 0 & 0 \\[2ex] 0 & 0 & 0 & 0 & \dfrac{1}{G} & 0 \\[2ex] 0 & 0 & 0 & 0 & 0 & \dfrac{1}{G} \end{bmatrix}
\tag{4.49}
$$

where $G = E/2(1+v)$ from Eq. 4.12, so that there are only two independent constants.

In the most general form of anisotropy, each unique S_{ij} in Eq. 4.48 has a different nonzero value. The matrix is symmetrical about its diagonal in such a way that there are two occurrences of each S_{ij}, where $i \neq j$, so that there are 21 independent constants. An example of a situation with this degree of complexity is the most general form of a single crystal, called a triclinic crystal, where $a \neq b \neq c$ and $\alpha \neq \beta \neq \gamma$, these being the distances and angles defined in Fig. 2.9.

4.6.2 Orthotropic and Cubic Materials

If the material possesses symmetry about three orthogonal planes, that is, about planes oriented 90° to each other, then a special case called an *orthotropic material* exists. In this case, Hooke's Law has a form of intermediate complexity between the isotropic and the general anisotropic cases. To deal with the situation of the S_{ij} values changing with the orientation of the x-y-z coordinate system, it is convenient to define the values for the directions parallel to the planes of symmetry in the material. This special coordinate system will here be identified by capital letters (X, Y, Z) as indicated for two cases in Fig. 4.16. The coefficients for Hooke's Law for an orthotropic material are

$$[S_{ij}] = \begin{bmatrix} \dfrac{1}{E_X} & -\dfrac{v_{YX}}{E_Y} & -\dfrac{v_{ZX}}{E_Z} & 0 & 0 & 0 \\[2mm] -\dfrac{v_{XY}}{E_X} & \dfrac{1}{E_Y} & -\dfrac{v_{ZY}}{E_Z} & 0 & 0 & 0 \\[2mm] -\dfrac{v_{XZ}}{E_X} & -\dfrac{v_{YZ}}{E_Y} & \dfrac{1}{E_Z} & 0 & 0 & 0 \\[2mm] 0 & 0 & 0 & \dfrac{1}{G_{YZ}} & 0 & 0 \\[2mm] 0 & 0 & 0 & 0 & \dfrac{1}{G_{ZX}} & 0 \\[2mm] 0 & 0 & 0 & 0 & 0 & \dfrac{1}{G_{XY}} \end{bmatrix} \tag{4.50}$$

Examples include an orthorhombic single crystal, where $\alpha = \beta = \gamma = 90°$, but $a \neq b \neq c$, and fibrous composite materials with fibers in directions such that there are three orthogonal planes of symmetry.

In Eq. 4.50, there are three moduli E_X, E_Y, and E_Z for the three different directions in the material. These in general have different values. There are also three different shear moduli G_{XY}, G_{YZ}, and G_{ZX} corresponding to three planes. The constants v_{ij} are Poisson's ratio constants.

$$v_{ij} = -\frac{\varepsilon_j}{\varepsilon_i} \tag{4.51}$$

giving the transverse strain in the j-direction due to a stress in the i-direction. Due to the symmetry of S_{ij} values about the matrix diagonal, the following holds:

$$\frac{v_{ij}}{E_i} = \frac{v_{ji}}{E_j} \tag{4.52}$$

where $i \neq j$ and $i, j = X, Y$, or Z. These relationships reduce the number of independent Poisson's ratios to three for a total of nine independent constants. It is important to remember that these constants only apply for the special X-Y-Z coordinate system.

If the material has the same properties in the X, Y, and Z directions, then it is called a *cubic material*. In this case, all three E_i have the same value E_X, all three G_{ij} have the same value G_{XY}, and all six Poisson's ratios have the same value v_{XY}. Thus, there are three independent constants. Examples of such a case include all single crystals with a cubic structure, such as BCC, FCC, and diamond cubic crystals. Note that there is still one more independent constant than for the isotropic case, and the elastic constants still apply only for the special X-Y-Z coordinate system.

For example, for a single crystal of alpha (BCC) iron, $E_X = 129$ GPa. However, consider the direction that is a diagonal of the cubic unit cell, that is, along the direction of one of the arrows in Fig. 2.24(a). The elastic modulus in this direction is $E_{[111]} = 276$ GPa, which is about twice as large as E_X. Considering E values for all possible directions, these two are the largest and smallest that occur. Polycrystalline iron is isotropic, and the value of $E \approx 210$ GPa that applies is the result of an averaging from randomly oriented single crystals. As expected, this E is between the extreme values E_X and $E_{[111]}$ for single crystals.

4.6.3 Fibrous Composites

Many applications of composite materials involve thin sheets or plates that have symmetry corresponding to the orthotropic case, such as simple unidirectional or woven arrangements of fibers as in Fig. 4.16. Also, most laminates (Fig. 3.26) have overall behavior that is orthotropic.

For plates or sheets, the stresses that do not lie in the X-Y plane of the sheet are usually small, so that plane stress with $\sigma_Z = \tau_{YZ} = \tau_{ZX} = 0$ is a reasonable assumption. Although strains ε_Z still occur, these are not of particular interest, so that Hooke's Law can be used in the following reduced form derived from Eq. 4.50:

$$
\left\{ \begin{array}{c} \varepsilon_X \\ \varepsilon_Y \\ \gamma_{XY} \end{array} \right\} = \left[\begin{array}{ccc} \dfrac{1}{E_X} & -\dfrac{\nu_{YX}}{E_Y} & 0 \\ -\dfrac{\nu_{XY}}{E_X} & \dfrac{1}{E_Y} & 0 \\ 0 & 0 & \dfrac{1}{G_{XY}} \end{array} \right] \left\{ \begin{array}{c} \sigma_X \\ \sigma_Y \\ \tau_{XY} \end{array} \right\} \tag{4.53}
$$

where capital letters still indicate that the stresses, strains, and elastic constants are expressed only for directions parallel to the planes of symmetry of the material. (Stresses and strains in other directions can be found by using transformation equations or Mohr's circle as discussed later in Chapter 6 or in textbooks on mechanics of materials.) Equation 4.52 applies, so that

$$
\frac{\nu_{YX}}{E_Y} = \frac{\nu_{XY}}{E_X} \tag{4.54}
$$

with the result that four independent elastic constants are being employed out of the total of nine. Values for some composite materials with unidirectional fibers are given in Table 4.3.

Values of these constants can be obtained from laboratory measurements, but they are also commonly estimated from the separate (and generally known) properties of the reinforcement and matrix materials. The topic of so estimating elastic constants is rather complex, being considered in detail in books on composite materials such as Jones (1975). In the discussion that follows, we will consider only the simple case of unidirectional fibers in a matrix.

TABLE 4.3 ELASTIC CONSTANTS AND DENSITY FOR FIBER-REINFORCED EPOXY WITH 60% UNIDIRECTIONAL FIBERS BY VOLUME

Reinforcement			Composite				
Type	E_r	ν_r	E_X	E_Y	G_{XY}	ν_{XY}	ρ
	GPa (10^3 ksi)		GPa (10^3 ksi)				g/cm³
E-glass	72.3 (10.5)	0.22	45 (6.5)	12 (1.7)	4.4 (0.64)	0.25	1.94
Kevlar 49	124 (18.0)	0.35	76 (11.0)	5.5 (0.8)	2.1 (0.3)	0.34	1.30
Graphite (T-300)	218 (31.6)	0.20	132 (19.2)	10.3 (1.5)	6.5 (0.95)	0.25	1.47
Graphite (GY-70)	531 (77.0)	0.20	320 (46.4)	5.5 (0.8)	4.1 (0.6)	0.25	1.61

Note: For approximate matrix properties, use $E_m = 3.5$ GPa (510 ksi) and $\nu_m = 0.33$.

Sources: Data in [ASM 87] pp. 175-178, and [Kelly 89] p. 262.

Example 4.5

A plate of the epoxy reinforced with unidirectional Kevlar 49 fibers in Table 4.3 is subject to stresses as follows: $\sigma_X = 400$, $\sigma_Y = 12$, and $\tau_{XY} = 15$ MPa, where the coordinate system is that of Fig. 4.17. Determine the in-plane strains ε_X, ε_Y, and γ_{XY}.

Solution   Equation 4.53 applies directly.

$$\varepsilon_X = \frac{\sigma_X}{E_X} - \frac{\nu_{YX}}{E_Y}\sigma_Y, \quad \varepsilon_Y = -\frac{\nu_{XY}}{E_X}\sigma_X + \frac{\sigma_Y}{E_Y}, \quad \gamma_{XY} = \frac{\tau_{XY}}{G_{XY}}$$

Since ν_{YX} is not given in the table, it is convenient to employ Eq. 4.54.

$$\frac{\nu_{YX}}{E_Y} = \frac{\nu_{XY}}{E_X} = \frac{0.34}{76,000} = 4.474 \times 10^{-6} \text{ 1/MPa}$$

Substituting the given stresses, and E_X, E_Y, and G_{XY} converted to MPa from Table 4.3, along with the above quantity that appears in two of the equations, gives the strains.

$$\varepsilon_X = \frac{400}{76,000} - \left(4.474 \times 10^{-6}\right)(12) = 0.00521 \qquad \textbf{Ans.}$$

$$\varepsilon_Y = -\left(4.474 \times 10^{-6}\right)(400) + \frac{12}{5500} = 0.00039 \qquad \textbf{Ans.}$$

$$\gamma_{XY} = \frac{15}{2100} = 0.00714 \qquad \textbf{Ans.}$$

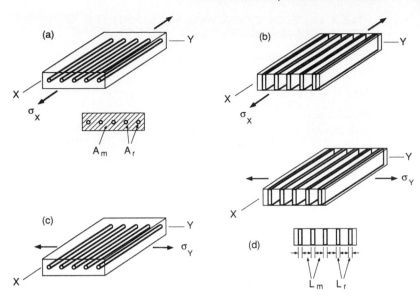

Figure 4.17 Composite materials with various combinations of stress direction and unidirectional reinforcement. In (a) the stress is parallel to fibers, and in (b) to sheets of reinforcement, whereas in (c) and (d) the stresses are normal to similar reinforcement.

4.6.4 Elastic Modulus Parallel to Fibers

Consider a uniaxial stress σ_x parallel to fibers aligned in the X-direction as shown in Fig. 4.17(a). Let the fibers (reinforcement) be an isotropic material with elastic constants E_r, v_r, and G_r, and let the matrix be another isotropic material, E_m, v_m, G_m. Assume that the fibers are perfectly bound to the matrix so that fibers and matrix deform as a unit, resulting in the same strain ε_X in both. Further, let the total cross-sectional area be A, and let the areas occupied by fibers and by matrix be A_r and A_m, respectively.

$$A = A_r + A_m \qquad (4.55)$$

Since the applied force must be the sum of contributions from fibers and matrix, we have

$$\sigma_X A = \sigma_r A_r + \sigma_m A_m \qquad (4.56)$$

where σ_r, σ_m are the differing stresses in fibers and matrix, respectively. The definitions of the various elastic moduli require that

$$\sigma_X = E_X \varepsilon_X, \qquad \sigma_r = E_r \varepsilon_r, \qquad \sigma_m = E_m \varepsilon_m \qquad (4.57)$$

Note that the strain in the composite is the same as that in both fibers and matrix.

$$\varepsilon_X = \varepsilon_r = \varepsilon_m \qquad (4.58)$$

Substitution of Eq. 4.57 into Eq. 4.56 then yields the desired modulus of the composite material.

$$E_X = \frac{E_r A_r + E_m A_m}{A} \tag{4.59}$$

The ratios A_r/A and A_m/A are also the *volume fractions* of fiber and matrix, respectively, denoted V_r and V_m.

$$V_r = \frac{A_r}{A}, \qquad V_m = 1 - V_r = \frac{A_m}{A} \tag{4.60}$$

Thus, Eq. 4.59 can also be written

$$E_X = V_r E_r + V_m E_m \tag{4.61}$$

This result confirms that in this case a simple *rule of mixtures* applies. Note that the same relationship is also valid for a case where the reinforcement is in the form of well-bonded layers as in Fig. 4.17(b).

4.6.5 Elastic Modulus Transverse to Fibers

Now consider uniaxial loading in the other orthogonal in-plane direction, specifically a stress σ_Y as shown in Fig. 4.17(c). An exact analysis of this case is more difficult, but analysis of a transversely loaded layered composite as shown in (d) is a useful approximation. In fact, the E_Y so obtained can be shown by detailed analysis to provide a lower bound on the correct value for case (c). Therefore, let us proceed to analyze case (d).

The stresses in reinforcement and matrix must now be the same and equal to the applied stress.

$$\sigma_Y = \sigma_r = \sigma_m \tag{4.62}$$

As before, we can use the definitions of the various elastic moduli.

$$\sigma_Y = E_Y \varepsilon_Y, \qquad \sigma_r = E_r \varepsilon_r, \qquad \sigma_m = E_m \varepsilon_m \tag{4.63}$$

The total length in the Y direction is the sum of contributions from the layers of reinforcement and the layers of matrix.

$$L = L_r + L_m \tag{4.64}$$

Also, the changes in these lengths give the strains in the overall composite material and in the reinforcement and matrix portions.

$$\varepsilon_Y = \frac{\Delta L}{L}, \qquad \varepsilon_r = \frac{\Delta L_r}{L_r}, \qquad \varepsilon_m = \frac{\Delta L_m}{L_m} \tag{4.65}$$

where

$$\Delta L = \Delta L_r + \Delta L_m \tag{4.66}$$

Substituting for each ΔL in this equation using Eq. 4.65 yields

$$\varepsilon_Y = \frac{\varepsilon_r L_r + \varepsilon_m L_m}{L} \tag{4.67}$$

In this equation, substitute for the strains using Eq. 4.63, and also note that all of the stresses are equal, to obtain

$$\frac{1}{E_Y} = \frac{1}{E_r}\frac{L_r}{L} + \frac{1}{E_m}\frac{L_m}{L} \tag{4.68}$$

The length ratios are equivalent to volume fractions.

$$V_r = \frac{L_r}{L}, \qquad V_m = 1 - V_r = \frac{L_m}{L} \tag{4.69}$$

so that we finally obtain

$$\frac{1}{E_Y} = \frac{V_r}{E_r} + \frac{V_m}{E_m} \tag{4.70}$$

solving for E_Y, this gives

$$E_Y = \frac{E_r E_m}{V_r E_m + V_m E_r} \tag{4.71}$$

4.6.6 Other Elastic Constants and Discussion

Similar logic also leads to an estimate of ν_{XY}, the larger of the two Poisson's ratios, called the *major Poisson's ratio*, and also an estimate of the shear modulus.

$$\nu_{XY} = V_r \nu_r + V_m \nu_m \tag{4.72}$$

$$G_{XY} = \frac{G_r G_m}{V_r G_m + V_m G_r} \tag{4.73}$$

The estimates of composite elastic constants just described are all approximations. Actual values of E_X are usually reasonably close to the estimate, but E_Y may be somewhat higher due to this equation being a lower bound. Books on composite materials contain more accurate, but considerably more complex, derivations and equations. In addition, fibers may occur in two directions, and laminated materials are often employed that consist of several layers of unidirectional or woven composite. Estimates for these more complex cases can also be made.

In a laminate, if equal numbers of fibers occur in several directions, such as the 0°, 90°, +45°, and −45° directions, the elastic constants may be approximately the same for any direction in the X-Y plane, but different in the z-direction. Such a material is said to be *transversely isotropic*. Composite materials made using mats of randomly oriented and intertwined long fibers have similar properties in all in-plane directions and so are also transversely isotropic.

4.7 SUMMARY

Deformations may be classified according to physical mechanisms and analogies with rheological models as to elastic, plastic, or creep deformations. The latter category may be further subdivided into steady-state creep and transient creep. The simplest rheological models for each are shown in Fig. 4.1.

Elastic deformation is associated with stretching the atomic bonds in solids so that the distances between atoms increases. The deformation is not time dependent and is recovered immediately upon unloading. Stress-strain curves for metals especially, but also for many other materials, exhibit a distinct elastic region where the stress-strain behavior is linear.

If a material is both isotropic and homogeneous, the elastic strains for the general three-dimensional case are related to stresses by the generalized Hooke's Law.

$$\varepsilon_i = \frac{1}{E}\left[\sigma_i - \nu\left(\sigma_j + \sigma_k\right)\right], \qquad \gamma_{ij} = \frac{\tau_{ij}}{G}, \qquad \text{etc.} \qquad (4.74)$$

where i is any one of x, y, or z for three orthogonal directions, and j and k are the other two. There are two independent elastic constants: the elastic modulus E, and Poisson's ratio ν. The shear modulus G is related to these by

$$G = \frac{E}{2(1+\nu)} \qquad (4.75)$$

The volumetric strain is the sum of the normal strains.

$$\varepsilon_v = \varepsilon_x + \varepsilon_y + \varepsilon_z \qquad (4.76)$$

For the isotropic, homogeneous case, it is related to the applied stresses by

$$\varepsilon_v = \frac{1-2\nu}{E}\left(\sigma_x + \sigma_y + \sigma_z\right) \qquad (4.77)$$

This equation indicates that the volume change is zero for $\nu = 0.5$. Unless $\nu \geq 0$, lateral expansion for tensile stress is implied, and this is highly unlikely. And unless $\nu \leq 0.5$, a volume decrease occurs for tension stresses, which is also unlikely. Values for virtually all materials lie within these two limits, usually between $\nu = 0.2$ and 0.4.

Plastic deformation is associated with relative movement of planes of atoms or of chain-like molecules that is not strongly time dependent. Volume changes due to such movement are small, so that a constant analogous to Poisson's ratio for plastic strain is close to 0.5. Rheological models containing frictional sliders have behavior analogous to plastic deformation, sharing the following characteristics with plastically deforming materials: (1) Departure from linear behavior occurs that results in permanent deformation if the load is removed. (2) Compressive stressing is required to achieve a return to zero strain after yielding. (3) There is a memory effect on reloading after elastic unloading in that yielding occurs at the same stress and strain from which unloading occurred.

Creep is time-dependent deformation that may or may not be recovered after unloading. The physical mechanisms include vacancy and dislocation motions, grain boundary sliding, and flow as a viscous fluid. Such mechanisms acting in metals, ceramics, and glasses produce creep deformations that are mostly not recovered after unloading. However, considerable recovery of transient (decreasing rate) creep deformation may occur in polymers as a result of interactions among the long carbon-chain molecules. If observed under fixed strain, creep behavior results in a stress decrease toward zero that is termed relaxation.

Rheological models built up of springs and dashpots can be used to study creep behavior. In the simplest form of such models, strain rates are proportional to applied stresses, a situation termed linear viscoelasticity. More complex behavior requiring nonlinear models or equations is often encountered in real materials.

Elastic deformation occurs in all materials at all temperatures. Plastic deformation is important in strengthened metal alloys at room temperature, whereas creep effects are small. Significant creep occurs at room temperature in low-melting-temperature metals, and in many polymers. At sufficiently high temperature, creep becomes an important factor for strengthened metal alloys and even for ceramics.

Some materials, notably fibrous composites, are significantly anisotropic. A particular case of anisotropy that is often encountered is orthotropy, in which the material has symmetry about three orthogonal planes. Such a material has nine independent elastic constants. There are different values of the elastic modulus in three directions, E_X, E_Y, and E_Z, and also three independent values of Poisson's ratio and of shear modulus v_{XY}, G_{XY}, etc., corresponding to the three orthogonal planes. These constants are defined only on the special X-Y-Z coordinate axes that are parallel to the planes of symmetry in the material. If the orientation of the coordinate axes change, the elastic constants change.

For in-plane loading of sheets and plates of composite materials with unidirectional fibers, the elastic constants can be estimated from those of the reinforcement and matrix materials.

$$E_X = V_r E_r + V_m E_m, \qquad E_Y = \frac{E_r E_m}{V_r E_m + V_m E_r} \qquad (4.78)$$

where X is the fiber (reinforcement) direction, Y the transverse direction, and V_r, V_m are the volume fractions of reinforcement and matrix, respectively.

NEW TERMS AND SYMBOLS

(a) Terms

anelastic strain

bulk modulus, B

elastic (Young's) modulus, E

generalized Hooke's Law

homogeneous

hydrostatic stress, σ_h

isotropic

linear elasticity

linear hardening

linear viscoelasticity

orthotropic

perfectly plastic

Poisson's ratio, ν

recovery

relaxation

rheological model

shear modulus, G

steady-state creep

tensile viscosity, η

thermal expansion coefficient, α

transient creep

volumetric strain, ε_v

(b) Nomenclature for Stresses and Strains

x, y, z	Coordinate axes identifying directions for stresses and strains
$\gamma_{xy}, \gamma_{yz}, \gamma_{zx}$	Shear strains
$\varepsilon_x, \varepsilon_y, \varepsilon_z$	Normal strains
$\varepsilon_c; \varepsilon_{cx}$	Creep strain; creep strain in x-direction
$\varepsilon_e; \varepsilon_{ex}$	Elastic strain; elastic strain in x-direction
$\varepsilon_p; \varepsilon_{px}$	Plastic strain; plastic strain in x-direction
$\varepsilon_{sc}, \varepsilon_{tc}$	Steady-state and transient creep strains, respectively
$\sigma_x, \sigma_y, \sigma_z$	Normal stresses
$\tau_{xy}, \tau_{yz}, \tau_{zx}$	Shear stresses

(c) Nomenclature for Orthotropic and Composite Materials

X, Y, Z	The particular x-y-z coordinate axes that are aligned with the planes of material symmetry
E_X, E_Y, E_Z	Elastic moduli in the X-, Y- and Z-directions
G_{XY}, G_{YZ}, G_{ZX}	Shear moduli in X-Y, etc., planes
m, r	Subscripts indicating matrix and reinforcement materials, respectively
V_m, V_r	Volume fractions for matrix and reinforcement materials, respectively
ν_{XY}, etc.	Poisson's ratio giving the transverse strain in the Y-direction due to a stress in the X-direction; others similarly

REFERENCES

FLUGGE, W. 1975 *Viscoelasticity*, 2nd ed., Springer-Verlag, New York.

GOULD, P. L. 1983 *Introduction to Linear Elasticity*, Springer-Verlag, New York.

JONES, R. M. 1975 *Mechanics of Composite Materials*, Hemisphere Pub. Corp., New York.

McCLINTOCK, F. A. and A. S. ARGON, eds. 1966 *Mechanical Behavior of Materials*, Addison-Wesley, Reading, Ma.

NADAI, A. 1950 *Theory of Flow and Fracture of Solids*, McGraw-Hill, New York.

PROBLEMS AND QUESTIONS

Section 4.3

4.1 Simplify Hooke's Law, Eqs. 4.10 and 4.11, for each of the following special cases:
   (a) Plane stress: $\sigma_z = \tau_{yz} = \tau_{zx} = 0$.
   (b) Uniaxial stress: Only σ_x is nonzero.
   (c) Plane strain: $\varepsilon_z = \gamma_{yz} = \gamma_{zx} = 0$.
   For case (c), include the equation that allows the third stress to be calculated if both of the other two are known.

4.2 Consider a flat plate of isotropic material that lies in the x-y plane and which is subject to applied loading in this plane only. Such a plate is under plane stress, so that $\sigma_z = \tau_{yz} = \tau_{zx} = 0$.
   (a) Does the thickness of the plate usually change when the plate is loaded?
   (b) Under what conditions does the thickness not change? That is, when is this state of plane stress also a state of plane strain?

4.3 Consider a plate of isotropic material subject to plane stress where the only nonzero stress component is τ_{xy}. Obtain equations for the three principal normal strains and also for the volumetric strain. Does the thickness of the plate change? Why or why not? Does the volume change? Why or why not?

4.4 A plate of an aluminum alloy is subjected to in-plane x-y loading so that plane stress prevails: $\sigma_z = \tau_{yz} = \tau_{zx} = 0$. Elastic constants for the isotropic material are given in Table 4.2. Strains measured at one point in the plate are $\varepsilon_x = 0.002$, $\varepsilon_y = 0.002$, and $\gamma_{xy} = 0.003$.
   (a) What are the values of all nonzero components of stress at this point?
   (b) What is the strain ε_z normal to the surface of the plate?
   (c) What is the volumetric strain?

4.5 Two strain gages placed 90° apart on the surface of a mild steel part read as follows after loading of the part: $\varepsilon_x = 0.002$ and $\varepsilon_y = 0.003$. The behavior is elastic and the material has elastic constants as given in Table 4.2. What are the normal stresses in the same directions as these strains? If x and y are known to be the directions of the principal normal stresses, do the above strains allow the state of stress to be completely determined?

4.6 Solve Hooke's Law, Eq. 4.10, for σ_x, σ_y, and σ_z, so that these quantities can be calculated directly from known strains. How would you generalize these equations to include the case of Eq. 4.24 where thermal strains are also present?

4.7 A *spherical* pressure vessel contains a pressure p and has average radius r and wall thickness t. It is made of an isotropic, homogeneous material that behaves in a linear-elastic manner. Determine the following as a function of pressure and the geometric dimensions and material constants involved: (a) change in radius, Δr, and (b) change in wall thickness, Δt.

4.8 Consider a pressure vessel that is a thin-walled tube with closed ends and wall thickness t. The volume enclosed by the vessel is determined from the diameter d and length L by

$$V = \frac{\pi d^2 L}{4}$$

The ratio of a small change in the enclosed volume to the original volume can be found by obtaining the differential dV from the above equation, and then dividing by V, which gives

$$\frac{dV}{V} = 2\frac{dd}{d} + \frac{dL}{L}$$

Verify this expression for dV/V. Then derive an equation for dV/V as a function of the pressure p in the vessel, the vessel dimensions, and elastic constants of the isotropic material. Assume that L is large compared to d, so that the details of the behavior of the ends are not important. (Comment: The ratio dV/V in this problem is for the volume enclosed in the pressure vessel; it is not the volumetric strain in the material of the shell.)

4.9 Consider a thin-walled *spherical* pressure vessel of diameter d and wall thickness t. Compute the ratio dV/V, where V is the volume enclosed by the vessel, and dV is the change in V when the vessel is pressurized. (Hint: See the previous problem.)

4.10 A block of material is stressed in the x- and y-directions as shown. The ratio of the magnitudes of the two stresses is a constant, so that $\sigma_y = \lambda\sigma_x$. Determine the following as functions of σ_x, λ, and the elastic constants of the material:
(a) The strain in the z-direction, ε_z.
(b) The stiffness in the x-direction, $E' = \sigma_x/\varepsilon_x$.
(c) Compare this apparent modulus E' with the elastic constant E obtained from a uniaxial test, and comment on the comparison. (Suggestion: Consider λ values of -1, 0, and $+1$.)

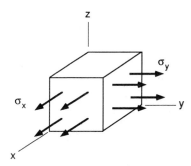

Figure P4.10

4.11 A sample of material subjected to a compressive stress σ_z is confined so that it cannot deform in either the x- or y-directions.
(a) Do stresses occur in the material in the x- and y- directions? If so, how are they related to σ_z?

(b) Determine the stiffness $E' = \sigma_z/\varepsilon_z$ in the direction of the applied stress in terms of the elastic constants E and ν for the material. Is E' equal to the elastic modulus E from uniaxial loading? Why or why not?

(c) What happens if Poisson's ratio for the material approaches 0.5?

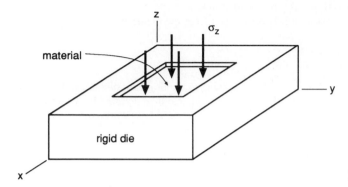

Figure P4.11

4.12 A block of material is stressed in the x- and y-directions as shown, but rigid walls prevent deformation in the z-direction. The ratio of the two applied stresses is a constant, so that $\sigma_y = \lambda\sigma_x$. Answer the following by deriving equations expressed in terms of σ_x, λ, and the elastic constants of the material:

(a) Does a stress develop in the z-direction? If so, how is it related to σ_x and the other constants involved?

(b) Determine the stiffness $E' = \sigma_x/\varepsilon_x$ for the x-direction.

(c) Compare this apparent modulus E' with the elastic modulus E from a uniaxial test. (Suggestion: Consider λ values of -1, 0, and 1 and assume $\nu = 0.3$.)

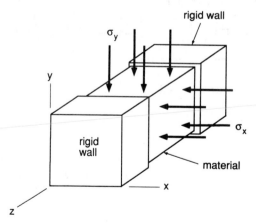

Figure P4.12

Section 4.4

4.13 Consider plastic deformation according to Eq. 4.31 for the situation of Example 4.2, where the material is subjected to a stress σ_z and prevented from deforming in the y-direction. Note that the variable E_p does not change its value if the applied stress is held constant.

Hence, take E_p as constant, and also assume that elastic strains can be neglected as being small compared to plastic strains.

(a) What stress σ_y develops in the y-direction?

(b) What is the ratio $\varepsilon_x/\varepsilon_z$ of the strain in the x-direction to that in the z-direction? Can you rationalize this value?

4.14 Consider the situation of Prob. 4.11, where a stress σ_z is applied to a material that is prevented from deforming in both the x- and y-directions. As in the previous problem, take E_p as constant and neglect elastic strains.

(a) What stresses σ_x and σ_y develop?

(b) What is the strain in the z-direction? Can you rationalize this value?

4.15 A steel is represented by the elastic, perfectly plastic model of Fig. 4.10(b) using constants $E_1 = 200$ GPa and $\sigma_o = 400$ MPa.

(a) Plot the stress-strain response for loading to a strain of $\varepsilon = 0.01$. Of this total strain, how much is elastic, and how much plastic?

(b) Also plot the response following (a) if the strain is now decreased until it reaches zero.

4.16 For the elastic, linear-hardening rheological model of Fig. 4.10(d), determine the slope $d\sigma/d\varepsilon$ after yielding in terms of the constants of the model. What is the meaning of your result? How is the behavior affected by changing E_2 while E_1 remains constant?

4.17 An aluminum alloy is represented by the elastic, linear-hardening model of Fig. 4.10(d) using constants of $E_1 = 70$ GPa, $E_2 = 2$ GPa, and $\sigma_o = 350$ MPa.

(a) Plot the stress-strain response for loading to a strain of $\varepsilon = 0.02$. Of this total strain, how much is elastic, and how much plastic?

(b) Also plot the response following (a) if the strain is decreased until it reaches zero.

4.18 Consider the elastic, linear-hardening model of Fig. 4.10(d), and specifically its response to the second strain input. The model is loaded beyond yielding to point 1, at which let the stress and strain be σ' and ε'. It is then unloaded, proceeding into compression toward zero strain, point 2. Note that during this unloading, there is no deformation in spring E_2 until the portion of the stress carried by the slider reaches a value of $-\sigma_o$, at which point yielding occurs in the opposite direction. Determine the following as functions of the model constants and σ' and ε'.

(a) The stress σ'' where yielding occurs again.

(b) The corresponding strain ε''.

(c) The conditions under which there will be no second yielding event prior to reaching zero strain.

Section 4.5

4.19 A thin-walled *spherical* pressure vessel has diameter 200 mm and wall thickness 5 mm, and it contains a liquid at a pressure of 0.2 MPa above atmospheric pressure. It is made of a borosilicate glass and is used at a temperature of 550°C, where the shear viscosity of this material is $\eta_\tau = 10^{13}$ Pa·s. Note that the tensile viscosity for Eq. 4.35 is $\eta = 3\eta_\tau$.

(a) What is the rate of creep strain in the vessel wall?

(b) How much do the strain in the vessel wall and the vessel diameter increase in one week?

4.20 Show that the shear viscosity, $\eta_\tau = \tau/\dot{\gamma}$, and the tensile viscosity are expected to be related by $\eta = 3\eta_\tau$. (Hint: Follow a procedure parallel to that used to verify Eq. 4.12.)

4.21 At 600°C, a silica glass has an elastic modulus of $E = 60$ GPa and a tensile viscosity of $\eta = 1000$ GPa·s. Assuming that the elastic, steady-state creep model of Fig. 4.14(b) applies, determine the response to a stress of 10 MPa maintained for 1 minute and then removed. Plot both strain versus time and stress versus strain for a total time interval of 2 minutes.

4.22 Derive Eq. 4.45 for the constant stress response of the transient creep model, as in 1-2 in Fig. 4.14(c). Base this on the other equations given just above Eq. 4.45.

4.23 A polymer has constants for the elastic, transient creep model of Fig. 4.14(d) of $E_1 = 6$ GPa, $E_2 = 3$ GPa and $\eta_2 = 10^5$ GPa·s. Determine and plot the strain versus time response to a stress of 15 MPa applied for one day. (PC Problem)

4.24 For the transient creep model of Fig. 4.14(c):
  (a) Derive an equation for the ε-t response after removal of the stress, that is, for the recovery during 3-4. Express the result in terms of the strain ε' reached at the time of stress removal and the time interval Δt since stress removal.
  (b) In Prob. 4.23, assume that the stress is removed after 1 day, and then use your result from (a) of this problem to extend the strain-time plot to a total time of two days. (PC Problem)

4.25 Consider a step strain as in Fig. 4.15, but apply it to the elastic, transient-creep model of Fig. 4.14(d). Derive an equation for the σ-t response during 1-2 while the strain is maintained. From your result, verify that the stress relaxes toward the asymptote given by Eq. 4.47.

4.26 Consider relaxation under constant strain ε' of a model with a spring and dashpot in series as in Fig. 4.15, but let the dashpot behave according to the nonlinear equation

$$\dot{\varepsilon} = B\sigma^m$$

where B and m are material constants, with m being typically in the range 3 to 7. Derive an equation for σ as a function of ε', time t, and the various model constants.

Section 4.6

4.27 Name two materials that fit into each of the following categories: (a) isotropic, (b) transversely isotropic, and (c) orthotropic. Try to think of your own examples rather than using those from the text.

4.28 A composite material is made with an aluminum alloy matrix and 35% by volume of unidirectional SiC fibers. Estimate the elastic constants E_X, E_Y, G_{XY}, ν_{XY}, and ν_{YX}.

4.29 For the epoxy reinforced with E-glass in Table 4.3, use the matrix and reinforcement properties given to estimate the composite properties E_X, E_Y, G_{XY}, ν_{XY}, and ν_{YX}. How well do your estimates compare with the tabulated values? Can you suggest reasons for any discrepancies?

4.30 Proceed as in the previous problem except change the reinforcement material to Kevlar 49.

4.31 A plate of the epoxy reinforced with unidirectional T-300 graphite fibers in Table 4.3 is subject to stresses as follows: $\sigma_X = 300$, $\sigma_Y = -100$, and $\tau_{XY} = 15$ MPa, where the coordinate system is that of Fig. 4.17. Determine the in-plane strains ε_X, ε_Y, and γ_{XY} that result.

4.32 For unidirectional E-glass fibers used to reinforce epoxy, use the matrix and reinforcement properties from Table 4.3 to estimate E_X and E_Y for several volume fractions of reinforcement ranging from zero to 100%. Plot curves of E_X versus V_r and E_Y versus V_r on the same graph and comment on the trends. (PC Problem)

5

Mechanical Testing: Tension Test and Other Basic Tests

5.1 INTRODUCTION

Samples of engineering materials are subjected to a wide variety of mechanical tests to measure their strength or other properties of interest. Such samples, called specimens, are often broken or grossly deformed in testing. Some of the common forms of test specimen and loading situation are shown in Fig. 5.1. The most basic test is simply to break the sample by applying a tensile force as in (a). Compression tests (b) are also common. In engineering, hardness is usually defined in terms of resistance of the material to penetration by a hard ball or point, as in (c). Various forms of bending test are also often used, as is torsion of cylindrical rods or tubes.

The simplest test specimens are smooth (unnotched) ones as illustrated in Fig. 5.2(a). More complex geometries can be used to produce conditions resembling those in actual engineering components. Notches that have a definite radius at the end may be machined into test specimens, as in (b). The term *notch* is used here in a generic manner to indicate any notch, hole, groove, slot, etc., that has the effect of a stress raiser. Sharp notches that behave similar to cracks are also used, as well as actual cracks that are introduced into the specimen prior to testing, as in (c).

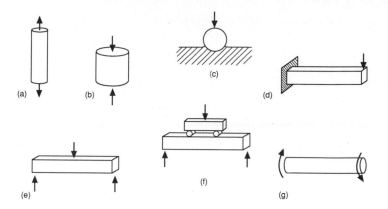

Figure 5.1 Geometry and loading situations commonly employed in mechanical testing of materials: (a) tension, (b) compression, (c) indentation hardness, (d) cantilever bending, (e) three-point bending, (f) four-point bending, and (g) torsion.

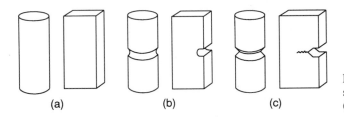

Figure 5.2 Three classes of test specimen: (a) smooth or unnotched, (b) notched, and (c) precracked.

To understand mechanical testing, it is first necessary to briefly consider materials testing equipment and standard test methods. We will then discuss tests involving tension, compression, indentation, notch impact, bending, and torsion. Various more specialized tests are discussed in later chapters in connection with such topics as brittle fracture, fatigue, and creep.

5.1.1 Test Equipment

Equipment of a variety of types is used for applying forces (loads) to test specimens. Test equipment ranges from very simple devices to complex systems that are controlled by digital computer.

Two common configurations of relatively simple devices called *universal testing machines* are shown in Fig. 5.3. These general types of testing machine first became widely used in the period 1900 to 1920, and they are still frequently used today. In the mechanical-screw-driven machine (above), rotation of two large threaded posts (screws) moves a crosshead that applies a load to the specimen. A simple balance system is used to measure the magnitude of the force applied. Loads may also be applied using the pressure of oil pumped into a hydraulic piston (below). In this case, the oil pressure provides a simple means of measuring the force applied. Testing machines of these

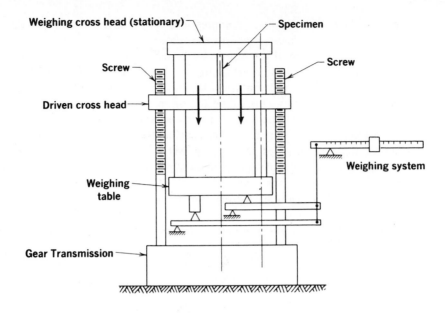

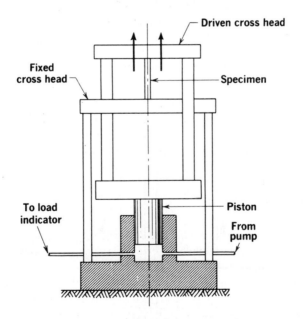

Figure 5.3 Schematics of two relatively simple testing machine designs, called universal testing machines. The mechanical system (above) drives two large screws to apply the load, and the hydraulic system (below) uses the pressure of oil in a piston. (From [Richards 61] p. 114; reprinted by permission of PWS-Kent Publishing Co., Boston.)

types can be used for tension, compression, or bending, and torsion machines based on a similar level of technology are also available.

The introduction of the Instron Corp. testing machine in 1946 represented a major step in that rather sophisticated electronics, based initially on vacuum tube technology, came into use. This is also a screw-driven machine with a moving crosshead, but the electronics, used both in controlling the machine and in measuring loads and displacements, makes the test system much more versatile than its predecessors.

Around 1958, transistor technology and closed-loop automation concepts were used by the forerunner of the present MTS Systems Corp. to develop a high-rate test system using a double-action hydraulic piston as illustrated in Fig. 5.4. The result is called a *closed-loop servohydraulic test system*. Desired variations of load, strain, or testing machine motion (stroke) can be enforced upon a test specimen. Note that the only active motion is that of the actuator rod and piston combination. Hence, the stroke of this actuator replaces the crosshead motion in the older types of testing machines.

The closed-loop servohydraulic concept is the basis of the most advanced test systems in use today. Integrated electronic circuitry has increased the sophistication of

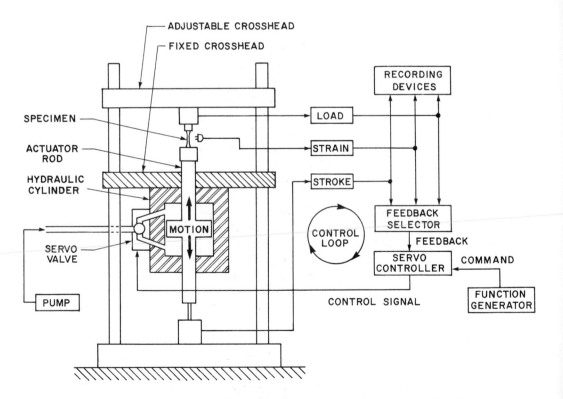

Figure 5.4 Modern closed-loop servohydraulic testing system. (Adapted from [Richards 70]; used with permission.)

these systems. Also, digital computer control and monitoring of such test systems has steadily developed since its introduction around 1965.

Sensors for measuring loads and displacements using electrical signals are important features of testing machines. *Linear variable differential transformers* (LVDTs) were used in this manner relatively early for measuring displacements, which in turn give strains in test specimens. Wire *strain gages* were developed in 1937, and the wire elements were replaced by thin foil elements starting around 1952. Strain gages change their resistance when the material to which they are bonded is deformed, and this change can be used to produce an electrical voltage that is proportional to the strain. They can be used to construct *load cells* for measuring applied load and *extensometers* for measuring displacements on test specimens. The Instron and closed-loop servohydraulic testing machines require electrical signals from such sensors. Strain gages are the type of transducer primarily used at present, but LVDTs are also often used.

Besides the general-purpose test equipment just described, various types of special-purpose test equipment are also employed. Some of these will be discussed in later chapters as appropriate.

5.1.2 Standard Test Methods

The results of materials tests are used for a variety of purposes. One important use is to obtain values of material properties, such as the material's breaking strength in tension, for use in engineering design. Another use is quality control of material that is produced, say plates of steel or batches of concrete, to be sure that they meet established requirements.

Such application of measured values of materials properties requires that everyone who makes these measurements do so in a consistent way. Otherwise, users and producers of materials will not agree as to standards of quality, and much confusion and inefficiency will occur. Perhaps even more important, safety and reliability of engineering design requires that materials properties be well-defined quantities.

Therefore, materials producers and users, and other involved parties, such as practicing engineers, governmental agencies, and research organizations, have worked together to develop *standard test methods.* This activity is often organized by professional societies, with the American Society for Testing and Materials (ASTM) being the most active organization in this area in the United States. Most major industrial nations have similar organizations. Standards activities in Europe have been internationalized to an extent by cooperative activities within the European Economic Community, and the International Organization for Standardization (ISO) coordinates and publishes standards on a worldwide basis.

A wide variety of standard methods have been developed for various materials tests, including all of the basic types of tests discussed in this chapter, and also including other more specialized tests considered in later chapters. The *Annual Book of ASTM Standards* is published yearly and consists of more than sixty volumes, approximately ten of which include a significant number of standards for mechanical tests. The details of the test methods differ depending on the general class of materials involved, such as metals, concrete, plastics, rubber, and glass, and the ASTM Standards are organized according to such classes of materials. The numbers identifying some of the major standards for mechanical testing are given in Table 5.1.

Test standards give the procedures to be followed in detail, but the theoretical basis of the test and background discussion are not generally given. Hence, one purpose of this book is to provide the basic understanding needed to apply materials test standards and to make intelligent use of the results.

TABLE 5.1 SOME OF THE MAJOR ASTM STANDARDS FOR BASIC MECHANICAL TESTS

Class of material (Volume in ASTM Standards)	ASTM Standard Numbers				
	Tension	Compression	Hardness	Impact	Bending
Metals (03.01)	A 370 B 557 E 8 E 602 E 646	E 9 E 209	A 370 E 10 E 18 E 92 E 384 E 448	A 370 E 23 E 208 E 436 E 604	E 290 E 812 E 855
Concrete (04.02)	C 496	C 39 C 469	—	—	C 78 C 293
Stone and rock (04.08)	D 2936 D 3967	C 170 D 2938 D 3148	—	—	C 99 C 120 C 880
Wood and plywood (04.09)	D 143 D 198 D 3500	D 143 D 198 D 3501	D 143	D 143 D 3499	D 143 D 198 D 3043
Plastics (08.01 and .02)	D 638 D 882	D 695 D 1621	D 785 D 2583	D 256 D 746 D 1822 D 3029	D 648 D 747 D 790
Rubber (09.01)	D 412	D 395 D 575 D 1229	D 531 D 1415 D 2240	D 1054 D 2632	D 797
Ceramics and glass (15.02)	—	C 773	C 730 C 849	—	C 158 C 674
Fibers and composites (15.03)	D 3039 D 3379 D 3552 D 4018	C 364 C 365 D 3410	—	—	C 393

5.2 INTRODUCTION TO TENSION TEST

A tension test consists of slowly pulling a sample of material with a tensile load until it breaks. The test specimen used may have either a circular or a rectangular cross section. The ends of tensile specimens are usually enlarged to provide extra area for gripping and to avoid having the sample break where it is being gripped. Specimens both before and after testing are shown for several metals and polymers in Figs. 5.5 and 5.6.

Methods of gripping the ends vary with specimen geometry. A typical arrangement for threaded-end specimens is shown in Fig. 5.7. Note that spherical bearings are used at each end to provide a pure tensile load with no undesirable bending. The usual manner of conducting the test is to deform the specimen at a constant speed. For example, in the universal testing machines of Fig. 5.3, the motion between the fixed and moving crossheads can be controlled at a constant speed. Hence, distance h in Fig. 5.7 is varied so that

$$\frac{dh}{dt} = \dot{h} = \text{constant}$$

The load that must be applied to enforce this displacement rate varies as the test proceeds. This load P may be divided by the cross-sectional area A_i to obtain the stress

Figure 5.5 Tensile specimens of metals (left to right): untested specimen with 9 mm diameter test section, and broken specimens of gray cast iron, aluminum alloy 7075-T651, and hot-rolled AISI 1020 steel. (Photo by R. A. Simonds.)

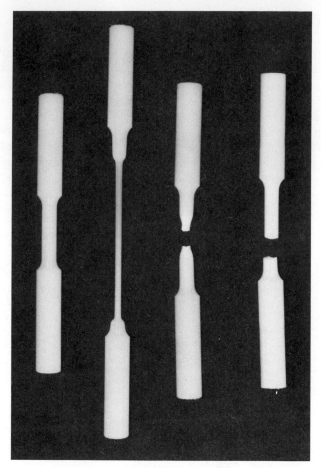

Figure 5.6 Tensile specimens of polymers (left to right): untested specimen with a 7.6 mm diameter test section, a partially tested specimen of high-density polyethylene (HDPE), and broken specimens of nylon 101 and Teflon (PTFE). (Photo by R. A. Simonds.)

in the specimen at any time during the test.

$$\sigma = \frac{P}{A_i} \tag{5.1}$$

Displacements on the specimen are measured within a straight central portion of constant cross section over a *gage length* L_i as indicated in Fig. 5.7. Strain ε may be computed from the change of this length, ΔL.

$$\varepsilon = \frac{\Delta L}{L_i} \tag{5.2}$$

It is sometimes reasonable to assume that all of the grip parts and the specimen ends are nearly rigid. In this case, virtually all of the change in crosshead motion is due to

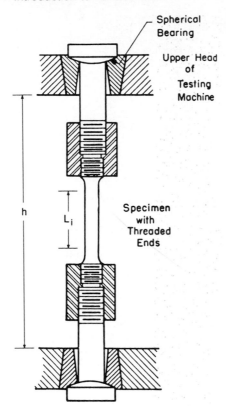

Spherical
Bearing

Upper Head
of
Testing
Machine

Specimen
with
Threaded
Ends

h

L_i

Figure 5.7 Typical grips for a tension test in a universal testing machine. (Adapted from [ASTM 90a] Std. B557; copyright ©ASTM; reprinted with permission.)

deformation within the straight section of the test specimen, so that ΔL is approximately the same as Δh, the change in h. Strain may therefore be estimated from

$$\varepsilon = \frac{\Delta h}{L_i} \qquad (5.3)$$

However, actual measurement of ΔL is preferable where this is feasible. Stress and strain based on the initial (undeformed) dimensions, A_i and L_i, are called *engineering stress and strain*.

The curve giving the relationship between engineering stress and strain during a tension test varies widely for different materials. *Brittle* behavior in a tension test is failure without extensive deformation. Gray cast iron, glass, and some polymers, such as PMMA (acrylic), are examples of materials with such behavior. A stress-strain curve for gray iron is shown in Fig. 5.8. Other materials exhibit *ductile behavior*, failing in tension only after extensive deformation. Stress-strain curves for ductile behavior in engineering metals and some polymers are similar to Figs. 5.9 and 5.10, respectively.

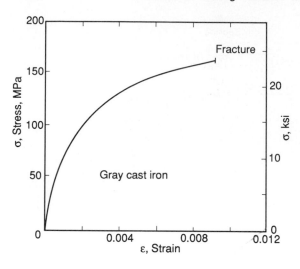

Figure 5.8 Stress-strain curve for gray cast iron in tension showing brittle behavior.

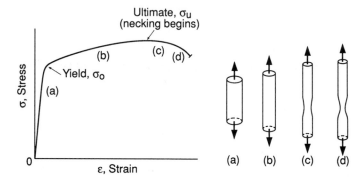

Figure 5.9 Schematic of the engineering stress-strain curve of a typical ductile metal that exhibits necking behavior.

5.3 ENGINEERING STRESS-STRAIN PROPERTIES

Various quantities obtained from the results of tension tests are defined as materials properties. Those based on engineering stress and strain will now be described. In a later portion of this chapter, additional properties based on different definitions of stress and strain, called true stress and strain, will be considered.

5.3.1 Elastic Constants

Initial portions of stress-strain curves from tension tests exhibit a variety of different behaviors for different materials as shown in Fig. 5.11. There may be a well-defined initial straight line, as for many engineering metals, where the deformation is predominantly

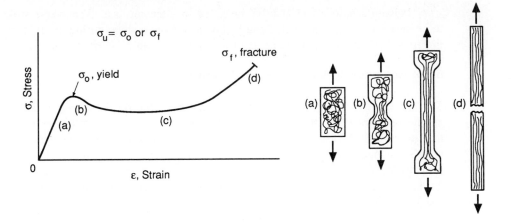

Figure 5.10   Engineering stress-strain curve and geometry of deformation typical of some polymers.

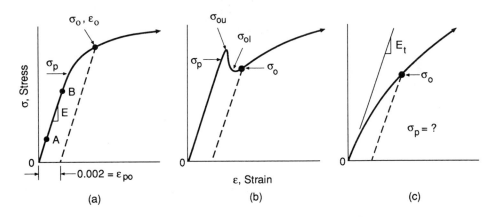

Figure 5.11   Initial portions of stress-strain curves: (a) many metals and alloys, (b) material with yield drop, and (c) material with no linear region.

elastic. The *elastic modulus*, E, also called *Young's modulus*, may then be obtained from the stresses and strains at two points on this line, such as A and B in (a).

$$E = \frac{\sigma_B - \sigma_A}{\varepsilon_B - \varepsilon_A} \tag{5.4}$$

In other cases, there is no well-defined linear region. Here, a *tangent modulus*, E_t, may be employed, which is the slope of a straight line that is tangent to the stress-strain curve at the origin as shown in (c). As a practical matter, obtaining E_t often involves the use of considerable judgment, so that this is not a very well-defined property.

Poisson's ratio ν can also be measured in a tension test. Transverse strains need to be measured so that the value can be obtained from a slope during elastic behavior as in Fig. 4.3. Diameter measurements or a strain gage can be used for this purpose.

5.3.2 Engineering Measures of Strength

The *ultimate tensile strength*, σ_u, also called simply the *tensile strength*, is the highest engineering stress reached prior to fracture. If the behavior is brittle, as for gray cast iron in Fig. 5.8, the ultimate strength occurs at the point of fracture. However, in ductile metals, the load, and hence the engineering stress, reaches a maximum and then decreases prior to fracture, as in Fig. 5.9. In either case, the highest load reached at any point during the test, P_{max}, is used to obtain the ultimate strength by dividing by the original cross-sectional area.

$$\sigma_u = \frac{P_{max}}{A_i} \tag{5.5}$$

In polymers, the ultimate strength may occur prior to a drop in load early in the test, but in other cases it occurs at the point of fracture. See Fig. 5.10.

The *engineering fracture strength*, σ_f, is obtained from the load at fracture, P_f, even if this is not the highest load reached.

$$\sigma_f = \frac{P_f}{A_i} \tag{5.6}$$

Hence, for brittle materials, $\sigma_u = \sigma_f$, whereas for ductile materials, σ_u may exceed σ_f.

The departure from linear-elastic behavior as in Fig. 5.11 is called *yielding* and is of considerable interest. This is simply because stresses that cause yielding result in rapidly increasing deformation due to the contribution of plastic strain. As discussed in the previous chapter, any strain in excess of the elastic strain σ/E is plastic strain and is not recovered on unloading. Hence, plastic strains result in permanent deformation, which manifests itself as changes in size and shape of engineering members, which are generally undesirable. Thus, the first step in engineering design is usually to assure that stresses are sufficiently small that yielding does not occur, except perhaps in very small regions of a component.

The yielding event can be characterized by several methods. The simplest is to identify the stress where the first departure from linearity occurs. This is called the *proportional limit*, σ_p, and is illustrated in Fig. 5.11. Some materials, as in (c), may exhibit a stress-strain curve with a gradually decreasing slope and no proportional limit. Even where there is a definite linear region, it is difficult to precisely locate where this ends. Hence, the value of the proportional limit depends on judgment, so that this is a poorly defined quantity. Another quantity sometimes defined is the *elastic limit*, which is the highest stress that does not cause permanent (i.e. plastic) deformation. Determination of this quantity is difficult, as periodic unloading to check for permanent deformation is necessary.

A third approach is the *offset method*, which is illustrated by dashed lines in Fig. 5.11. A straight line is drawn parallel to the elastic slope, E or E_t, but offset by an arbitrary amount. The intersection of this line with the engineering stress-strain curve is a well-defined point that is not affected by judgment, except in a few cases where E_t is difficult to establish. This is called the *offset yield strength*, σ_o. The most widely used and standardized offset for engineering metals is a strain of 0.002, that is 0.2%, although other values are also used. Note that the offset strain is a plastic strain, such as $\varepsilon_{po} = 0.002$, as unloading from σ_o would follow a dashed line in Fig. 5.11, and this ε_{po} would be the unrecovered strain.

In some engineering metals, notably in low-carbon steels, there is very little non-linearity prior to a dramatic drop in load as illustrated in Fig. 5.11(b). In such cases, one can identify an *upper yield point*, σ_{ou}, and a *lower yield point*, σ_{ol}. The former is the highest stress reached prior to the decrease, and the latter is the lowest stress prior to a subsequent increase. Values of the upper yield point in metals are sensitive to testing rate and to inadvertent small amounts of bending, so that reported values for a given material vary. The lower yield point is generally similar to the 0.2% offset yield strength, with the latter having the advantage of being applicable to other types of stress-strain curve as well.

The offset yield strength is generally the most satisfactory means of defining the yielding event for engineering metals. For polymers, offset yield strengths are also used. However, it is more common for polymers to define a yield point only if there is an early relative maximum or flat region in the curve, in which case σ_o is the stress where $d\sigma/d\varepsilon = 0$ first occurs. In most materials, the proportional limit, elastic limit, and offset yield strength can be considered to be different alternative measures of the beginning of permanent deformation. However, for a nonlinear elastic material such as rubber, the first two of these measure distinctly different events, and the offset yield strength loses its significance. (See Fig. 3.17.)

5.3.3 Engineering Measures of Ductility

Ductility is the ability of a material to accommodate inelastic deformation without breaking. In the case of tension loading, this means the ability to stretch by plastic strain, but with creep strain also sometimes contributing.

The *engineering fracture strain* is one measure of ductility. This is obtained from the length at fracture, L_f, of the gage section that originally had length L_i.

$$\varepsilon_f = \frac{L_f - L_i}{L_i} \tag{5.7}$$

Note that ε_f corresponds to the same point on the stress-strain curve as the engineering fracture strength σ_f. Often, ε_f is expressed as a percentage and is called the *percent elongation*.

$$\% \text{ elongation} = 100\varepsilon_f \tag{5.8}$$

where ε_f is dimensionless according to Eq. 5.7.

The percent elongation in metals is often measured on the sample after it is broken, using marks placed a known distance apart prior to the test. However, it is in general preferable to obtain ε_f from the fracture point on the stress-strain curve, as a value from a broken specimen differs by not including the elastic strain and perhaps some recovered creep strain. The difference can be large in some ductile polymers where a significant amount of recovery of creep strain occurs.

Another measure of ductility is the *percent reduction in area*, called *%RA*, which is obtained by comparing the cross-sectional area after fracture, A_f, with the original area.

$$\%RA = 100\frac{A_i - A_f}{A_i} \qquad \text{(a)}$$

$$\%RA = 100\frac{d_i^2 - d_f^2}{d_i^2} \qquad \text{(b)} \qquad\qquad (5.9)$$

where the second form is derived from the first as a convenience for round cross-sections of initial diameter d_i and final diameter d_f. As for the elongation, a discrepancy may exist between the area after fracture and the area that existed just prior to fracture. This presents little problem for ductile metals, but caution is needed in interpreting area reductions after fracture for polymers.

5.3.4 Discussion of Necking Behavior and Ductility

If the behavior in a tension test is ductile, a phenomenon called *necking* usually occurs as illustrated in Fig. 5.12. The deformation is uniform along the gage length early in the test, but later begins to concentrate in one region, resulting in the diameter there decreasing more than elsewhere. In ductile metals, necking begins at the ultimate strength point, and the decrease in load beyond this is a consequence of the cross-sectional area

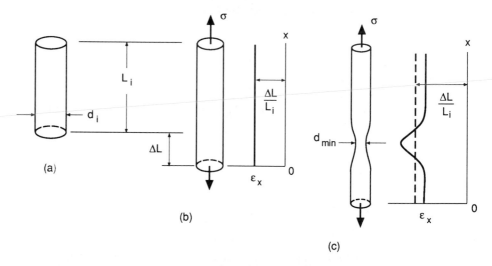

Figure 5.12   Deformation in a tension test of a ductile metal: (a) unstrained, (b) after uniform elongation, and (c) during necking.

rapidly decreasing. Once necking begins, the longitudinal strain becomes nonuniform, as illustrated in Fig. 5.12(c).

Examine the metal samples of Fig. 5.5. Necking occurred in the steel, and to an extent in the aluminum alloy, but not in the brittle cast iron. Enlarged views show the steel and cast iron fractures in more detail in Figs. 5.13 and 5.14.

The percent reduction in area is based on the minimum diameter at fracture and so is a measure of the highest strain along the gage length. In contrast, the percent elongation at fracture is an average over an arbitrarily chosen length. Its value varies

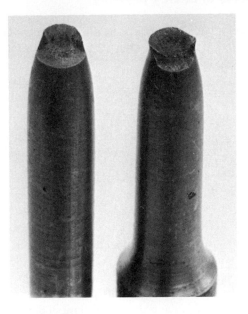

Figure 5.13 Fracture from a tension test on a 9 mm diameter specimen of hot-rolled AISI 1020 steel. (Photo by R. A. Simonds.)

Figure 5.14 Fracture from a tension test on a 9 mm diameter specimen of gray cast iron. (Photo by R. A. Simonds.)

with the ratio of gage length to diameter, L_i/d_i, increasing for smaller values of this ratio. As a consequence, it is necessary to standardize the gage lengths used. For example, $L_i/d_i = 4$ is commonly used in the United States for specimens with round cross sections, but $L_i/d_i = 5$ is specified in most international standards. The reduction in area is not affected by this arbitrariness and is thus a more fundamental measure of ductility than is the elongation.

5.3.5 Engineering Measures of Energy Capacity

In a tension test, let the applied force be P, and let the displacement over gage length L_i be $\Delta L = x$. The amount of work done in deforming the specimen to a value of $x = x'$ is then

$$U = \int_0^{x'} P \, dx \tag{5.10}$$

The volume of material in the gage length is $A_i L_i$. Dividing both sides of the equation by this volume, and using the definitions of engineering stress and strain, Eqs. 5.1 and 5.2, gives

$$u = \frac{U}{A_i L_i} = \int_0^{x'} \frac{P}{A_i} \, d\left(\frac{x}{L_i}\right) = \int_0^{\varepsilon'} \sigma \, d\varepsilon \tag{5.11}$$

Hence, u is the work done per unit volume of material to reach a strain ε', and it is equal to the area under the stress-strain curve up to ε'.

    The work done is equal to the energy absorbed by the material. Within the region of elastic deformation, most of this energy represents potential energy that is released upon unloading. The potential energy at the proportional limit is thus the triangular area shown in Fig. 5.15.

$$u_r = \frac{\sigma_p^2}{2E} \tag{5.12}$$

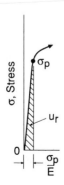
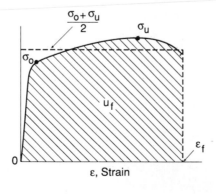

Figure 5.15   Areas under engineering stress-strain curves corresponding to resilience u_r and tensile toughness u_f.

This quantity is called the *resilience*. However, in view of the above-noted difficulty in defining σ_p, it is useful to use this same equation, but with the offset yield strength replacing σ_p.

$$u_r = \frac{\sigma_o^2}{2E} \tag{5.13}$$

In either case, the resilience is a measure of the ability of the material to store elastic energy.

The area under the entire engineering stress-strain curve up to fracture is called the *tensile toughness*, u_f, as also illustrated in Fig. 5.15. This is a measure of the ability of the material to absorb energy without fracture. Where there is considerable plastic strain beyond yielding, as for many engineering metals, some of the energy is stored in the microstructure of the material, but most of it is dissipated as heat.

If the stress-strain curve is relatively flat beyond yielding, it is generally sufficient to approximate the area as that of a rectangle having height equal to the average of the yield and ultimate strengths and width equal to the engineering fracture strain.

$$u_f \approx \varepsilon_f \left(\frac{\sigma_o + \sigma_u}{2} \right) \tag{5.14}$$

This is illustrated by the dashed line in Fig. 5.15. For materials that behave in a brittle manner, the gradually curving stress-strain response may be similar to a parabolic curve with vertex at the origin, in which case

$$u_f \approx \frac{2}{3} \varepsilon_f \sigma_f \tag{5.15}$$

Brittle materials have low tensile toughness, despite perhaps high strength, due to low ductility. In low-strength ductile materials, the converse occurs, and the tensile toughness is also low. To have a high tensile toughness, both the strength and the ductility must be reasonably high, so that a high tensile toughness indicates a "well rounded" material. Tensile toughness as just defined should not be confused with *fracture toughness*, which is the resistance to failure in the presence of a crack.

5.3.6 Strain Hardening

The rise in the stress-strain curve following yielding is described by the term *strain hardening*, as the material is increasing its resistance with increasing strain. A measure of the degree of strain hardening is the ratio of the ultimate strength to the yield strength.

$$\text{Strain hardening ratio} = \frac{\sigma_u}{\sigma_o}$$

Values of this ratio above about 1.4 are considered relatively high, and those below 1.2 relatively low.

Example 5.1

A tension test was conducted on a specimen of AISI 1020 hot-rolled steel having an initial diameter of 9.11 mm. The load at the 0.2% plastic strain offset was 17.21 kN, the highest load reached was 25.75 kN, and the load at fracture was 17.39 kN. After fracture, the broken halves were reassembled and the following measurements were made: (1) Marks on opposite sides of the necked region that were originally 25 mm apart had stretched to 38 mm apart. (2) Similar marks originally 50 mm apart had stretched to 68.5 mm apart. (3) The final diameter was 5.28 mm. Determine the following properties from this test: yield strength, engineering ultimate strength, percent elongation, and percent reduction in area.

Solution   The yield and ultimate strengths are calculated from the yield and maximum loads, respectively, using the original cross-sectional area.

$$\sigma_o = \frac{P_o}{A_i} = \frac{4P_o}{\pi d_i^2} = \frac{4\,(17{,}210\ \text{N})}{\pi\,(9.11\ \text{mm})^2} = 264\ \text{MPa} \qquad \textbf{Ans.}$$

$$\sigma_u = \frac{P_{max}}{A_i} = \frac{4P_{max}}{\pi d_i^2} = \frac{4\,(25{,}750\ \text{N})}{\pi\,(9.11\ \text{mm})^2} = 395\ \text{MPa} \qquad \textbf{Ans.}$$

Two values of percent elongation can be obtained using Eqs. 5.7 and 5.8.

$$\%\ \text{elong.} = 100\frac{L_f - L_i}{L_i} = 100\frac{38 - 25}{25} = 52\%\ \text{(in 25 mm)} \qquad \textbf{Ans.}$$

$$\%\ \text{elong.} = 100\frac{L_f - L_i}{L_i} = 100\frac{68.5 - 50}{50} = 37\%\ \text{(in 50 mm)} \qquad \textbf{Ans.}$$

The percent reduction in area from Eq. 5.9(b) is:

$$\%RA = 100\frac{d_i^2 - d_f^2}{d_i^2} = 100\frac{(9.11)^2 - (5.28)^2}{(9.11)^2} = 66.4\% \qquad \textbf{Ans.}$$

Discussion The elongation over the shorter gage length is higher than the other as this average value is more strongly affected by the concentrated deformation in the neck. Neither gage length corresponds to the international standard value of $L_i/d_i = 5$, but the larger one is reasonably close at $L_i/d_i = 5.5$. Thus, noting that $100\varepsilon_f \approx 37$, the tensile toughness from Eq. 5.14 is

$$u_f \approx \varepsilon_f\left(\frac{\sigma_o + \sigma_u}{2}\right) = 0.37\left(\frac{264 + 395}{2}\right) = 122\ \text{MPa} = 122\frac{\text{MJ}}{\text{m}^3} \qquad \textbf{Ans.}$$

5.4 TRENDS IN TENSILE BEHAVIOR

A wide variety of tensile behaviors occur for different materials. Even for a given chemical composition of a material, the prior processing of the material may have substantial effects on the tensile properties, as may the temperature and strain rate of the test.

5.4.1 Trends for Different Materials

Engineering metals vary widely as to their strength and ductility. This is evident from Table 5.2, where engineering properties from tension tests are given for several metals. Relatively high strength polymers in bulk form are typically only 10% as strong as engineering metals, and their elastic moduli are typically only 3% as large. Their ductilities

TABLE 5.2   TENSILE PROPERTIES FOR SOME ENGINEERING METALS: ENGINEERING PROPERTIES

Material	Elastic Modulus E	0.2% Yield Strength σ_o	Ultimate Strength σ_u	Elongation[1] $100\varepsilon_f$	Reduction in Area $\%RA$
	GPa (10^3 ksi)	MPa (ksi)	MPa (ksi)	%	%
Ductile cast iron A536 (65-45-12)	159 (23)	334 (49)	448 (65)	15	19.8
AISI 1020 steel as rolled	203 (29.4)	260 (37.7)	441 (64)	36	61
ASTM A514, T1 structural steel	208 (30.2)	724 (105)	807 (117)	20	66
AISI 4142 steel as quenched	200 (29)	1619 (235)	2450 (355)	6	6
AISI 4142 steel 205°C temper	207 (30)	1688 (245)	2240 (325)	8	27
AISI 4142 steel 370°C temper	207 (30)	1584 (230)	1757 (255)	11	42
AISI 4142 steel 450°C temper	207 (30)	1378 (200)	1413 (205)	14	48
18 Ni maraging steel (250)	186 (27)	1791 (260)	1860 (270)	8	56
SAE 308 cast aluminum	70 (10.2)	169 (25)	229 (33)	0.9	1.5
2024-T4 aluminum	73.1 (10.6)	303 (44)	476 (69)	20	35
7075-T6 aluminum	71 (10.3)	469 (68)	578 (84)	11	33
AZ91C-T6 cast magnesium	40 (5.87)	113 (16)	137 (20)	0.4	0.4

Note: [1]Typical values from [Boyer 85] are listed in most cases.

Sources: Data in [Conle 84] and [SAE 89].

vary quite widely, some being quite brittle and others quite ductile. Properties of some commercial polymers are given in Table 5.3 to illustrate these trends.

Rubber and rubber-like polymers (elastomers) have very low elastic moduli and relatively low strengths, and they often have extreme ductility. Ceramics and glasses

TABLE 5.3 MECHANICAL PROPERTIES FOR POLYMERS AT ROOM TEMPERATURE[1]

Material	Tensile properties			Rockwell Hardness	Izod Energy[2]	Heat Defl. Temp.
	Modulus E	Ultimate σ_u	Elong. $100\varepsilon_f$			
	GPa (10^3 ksi)	MPa (ksi)	%		J/m (ft·lb/in)	°C
ABS, medium impact	2.4 (0.35)	45 (6.5)	15	R111	240 (4.5)	99
ABS, 30% glass fibers	6.9 (1.0)	90 (13)	1.5	M80	64 (1.2)	102
Acrylic, PMMA	2.8 (0.40)	62 (9)	6	M87	24 (0.45)	87
Epoxy, cast	2.4 (0.35)	59 (8.5)	4.5	M95	32 (0.6)	166
Phenolic, cast	3.8 (0.55)	48 (7)	1.8	M102	17 (0.32)	77
Nylon 6, dry	2.6 (0.38)	81 (11.7)	65	R119	43 (0.8)	77
Nylon 6, 33% glass fibers	9.3 (1.35)	170 (24)	2.9	M95	150 (2.8)	210
Polycarbonate PC	2.4 (0.345)	66 (9.5)	110	M70	850 (16)	132
Polyethylene LDPE	0.23 (0.033)	20 (2.88)	375	R10	No break	42
Polyethylene HDPE	1.1 (0.157)	26.5 (3.85)	600	R65	120 (2.2)	86
Polystyrene PS	2.8 (0.40)	44 (6.35)	1.8	M68	21 (0.4)	86
Polystyrene HIPS	1.9 (0.27)	28 (4.05)	43	R66	210 (4.0)	87
Rigid PVC	3.3 (0.475)	46 (6.7)	60	R115	590 (11)	68

Notes: [1]Properties vary considerably; values are mostly middles of ranges from *Modern Plastics Encyclopedia* [Juran 88].   [2]Energy per unit thickness is tabulated.

Sources: Data in [Juran 88] and [Farag 89] p. 506.

represent the opposite extreme as their behavior is generally so brittle that measures of ductility have little meaning. Strengths in tension are generally lower than for metals but higher than for polymers. The elastic moduli of ceramics are relatively high, often higher than for many metals. Some typical values of ultimate tensile strength and elastic modulus have already been given in Table 3.10.

The tensile behavior of composite materials is of course strongly affected by the details of the reinforcement. For example, hard particles in a ductile matrix increase stiffness and strength but decrease ductility, more so for larger volume percentages of reinforcement. Long fibers have qualitatively similar effects, with the increase in strength and stiffness being especially large for loading directions parallel to large numbers of fibers. Whiskers and short chopped fibers generally produce effects intermediate between those of particles and long fibers. Some of these trends are evident in Table 5.4, where data are given for various SiC reinforcements of an aluminum alloy.

Some stress-strain curves from tension tests of engineering metals are shown in Figs. 5.16 and 5.17. The former gives curves for three steels with contrasting behavior, and the latter gives curves for three aluminum alloys. The curve for gray cast iron in Fig. 5.8 is typical of metals having low ductility, and also of other brittle materials.

TABLE 5.4   TENSILE PROPERTIES FOR VARIOUS SiC REINFORCEMENTS IN A 6061-T6 ALUMINUM MATRIX

Reinforcement[1]	Modulus E GPa (10^3 ksi)	0.2% Yield σ_o MPa (ksi)	Ultimate σ_u MPa (ksi)	Elongation $100\varepsilon_f$ %
None	69 (10)	275 (40)	310 (45)	12
Particles, 20%	103 (15)	414 (60)	496 (72)	5.5
Particles, 40%	145 (21)	448 (65)	586 (85)	2
whiskers, 20%	110 (16)	382 (55)	504 (73)	5
Fibers, 47%, 0°	204 (29.6)	—	1460 (212)	0.9
Fibers, 47%, 90°	118 (17.1)	—	86.2 (12.5)	0.1
Fibers, 47%, 0°/90°	137 (19.8)	—	673 (98)	0.9
Fibers, 47%, 0°/±45°/90°	127 (18.4)	—	572 (83)	1.0

Note: [1]Volume percentage given. For fibers, angles are orientations relative to tensile axis, and there are equal numbers of fibers at each angle given. Fiber properties are $E = 400$ GPa and $\sigma_u = 3950$ MPa.

Source: Data in [ASM 87] pp. 858–901.

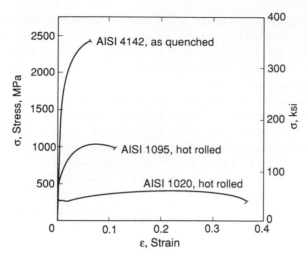

Figure 5.16   Engineering stress-strain curves from tension tests on three steels.

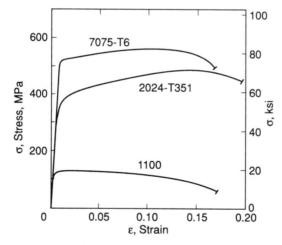

Figure 5.17   Engineering stress-strain curves from tension tests on three aluminum alloys.

Stress-strain curves from tension tests on three ductile polymers are shown in Fig. 5.18. These are in fact the curves for the test specimens shown in Fig. 5.6. An early relative maximum in stress is common for polymers, and this is associated with the distinctive necking behavior evident for HDPE in Fig. 5.6. Necking begins when the stress reaches the early relative maximum, and then it spreads along the specimen length, but with the diameter in the neck remaining approximately constant once the process starts as illustrated in Fig. 5.10. This behavior is due to the chain-like molecules being drawn out of their original amorphous or crystalline structure into an approximately linear and parallel arrangement.

Other polymers, such as Nylon 101, neck in a manner more similar to metals. An additional type of behavior is seen for Teflon (PTFE). This material deformed a

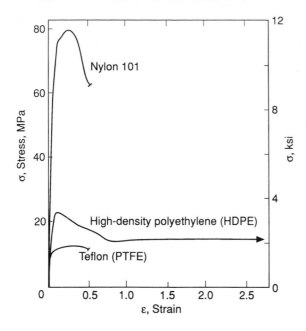

Figure 5.18 Engineering stress-strain curves from tension tests on three polymers.

considerable amount by developing a large number of small tears bridged by filaments of material, a process called *crazing*, followed by failure without necking. Other polymers, such as acrylic (PMMA), behave in a brittle manner and have stress-strain curves that are nearly linear up to the point of fracture.

5.4.2 Effects of Temperature and Strain Rate

If a material is tested in a temperature range where creep occurs, creep strains will contribute to the inelastic deformation in the test. Moreover, the creep strain that occurs is greater if the speed of the test is slower, as a slower test provides more time for the creep strain to accumulate. Under such circumstances, it is important to run the test at a constant value of strain rate, $\dot{\varepsilon} = d\varepsilon/dt$, and to report the value used along with the test results.

For polymers, recall from Chapters 2 and 3 that creep effects are especially large above the particular polymer's glass transition temperature, T_g. As T_g values around and below room temperature are common, large creep effects occur for many polymers. Tension tests on these materials thus require care concerning the effects of strain rate, and it is not uncommon to evaluate the tensile behavior at more than one rate.

For metals and ceramics, creep effects become significant around 0.3 to 0.6 T_m, where T_m is the absolute melting temperature. Thus, creep strains are a factor in room-temperature tension tests for metals with low melting temperatures. Strain rate may also affect the tensile behavior of ceramics at room temperature, but for an entirely different reason unrelated to creep, namely time-dependent cracking due to the detrimental effects of moisture.

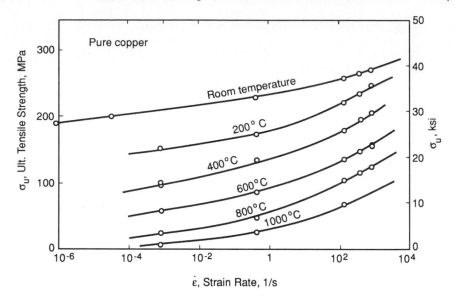

Figure 5.19   Effect of strain rate on the ultimate tensile strength of copper for tests at various temperatures. (Adapted from [Nadai 41]; used with permission of ASME.)

For engineering metals at room temperature, strain-rate effects due to creep exist but are not dramatic. For example, some data for copper are given in Fig. 5.19. In this case, the ultimate tensile strength at room temperature increases about 10% for an increase in strain rate of a factor of 1000. Larger relative effects occur at higher temperatures as creep effects become more important. Also, note that the strength is drastically lowered by increased temperature, especially as $T_m = 1083°C$ is approached.

The following generalizations usually apply to the tensile properties of a given material in a temperature range where creep-related strain-rate effects occur: (1) At a given temperature, increasing the strain rate increases the strength but decreases the ductility. (2) For a given strain rate, decreasing the temperature has the same qualitative effects, specifically of increasing strength and decreasing ductility.

5.5 TRUE STRESS-STRAIN INTERPRETATION OF TENSION TEST

In analyzing the results of tension tests, and in certain other situations, it is useful to work with *true stresses and strains*. Note that engineering stress and strain are most appropriate for small strains where the changes in specimen dimensions are small. True stresses and strains differ in that finite changes in area and length are specifically considered.

5.5.1 Definitions of True Stress and Strain

True stress is simply the load P divided by the current cross-sectional area A, rather than the original area A_i.

$$\tilde{\sigma} = \frac{P}{A} \tag{5.16}$$

Hence, true and engineering stress are related by

$$\tilde{\sigma} = \sigma\left(\frac{A_i}{A}\right) \tag{5.17}$$

For true strain, let the length change be measured in small increments, ΔL_1, ΔL_2, ΔL_3, etc., and let the new gage length, L_1, L_2, L_3, etc., be used to compute the strain for each increment. The total strain is thus

$$\tilde{\varepsilon} = \frac{\Delta L_1}{L_1} + \frac{\Delta L_2}{L_2} + \frac{\Delta L_3}{L_3} + \cdots = \sum \frac{\Delta L_j}{L_j} \tag{5.18}$$

where ΔL is the sum of these ΔL_j. If the ΔL_j are assumed to be infinitesimal, that is, if ΔL is measured in very small steps, the above summation is equivalent to an integral that defines true strain.

$$\tilde{\varepsilon} = \int_{L_i}^{L} \frac{dL}{L} = \ln\frac{L}{L_i} \tag{5.19}$$

where $L = L_i + \Delta L$ is the final length. Noting that $\varepsilon = \Delta L/L_i$ is the engineering strain leads to a relationship between ε and $\tilde{\varepsilon}$.

$$\tilde{\varepsilon} = \ln\frac{L_i + \Delta L}{L_i} = \ln\left(1 + \frac{\Delta L}{L_i}\right) = \ln(1 + \varepsilon) \tag{5.20}$$

5.5.2 Constant Volume Assumption

For materials that behave in a ductile manner, once the strains have increased substantially beyond the yield region, most of the strain that has accumulated is inelastic strain. Recalling from the previous chapter that neither plastic nor creep strains contribute to volume change, the volume change in a tension test is limited to the small change associated with elastic strains. It is therefore reasonable to assume that the volume is constant.

$$A_i L_i = A L \tag{5.21}$$

This gives

$$\frac{A_i}{A} = \frac{L}{L_i} = \frac{L_i + \Delta L}{L_i} = 1 + \varepsilon \tag{5.22}$$

Substitution of the above into Eqs. 5.17 and 5.19 gives two additional equations relating true and engineering stress and strain.

$$\tilde{\sigma} = \sigma \, (1 + \varepsilon) \qquad (5.23)$$

$$\tilde{\varepsilon} = \ln \frac{A_i}{A} \qquad (5.24)$$

For members with round cross sections of original diameter d_i and final diameter d, the last equation may be used in the form

$$\tilde{\varepsilon} = \ln \frac{\dfrac{\pi d_i^2}{4}}{\dfrac{\pi d^2}{4}} = 2 \ln \frac{d_i}{d} \qquad (5.25)$$

It should be remembered that Eqs. 5.22 through 5.25 depend on the constant volume assumption and may be inaccurate unless the inelastic (plastic plus creep) strain is large compared to the elastic strain.

5.5.3 Limitations on True Stress-Strain Equations

Recall that necking begins at the ultimate stress point in a tension test on a ductile metal, or at an early relative maximum stress point for some polymers. Prior to necking, the deformation is reasonably uniform along the gage length, but not thereafter as illustrated in Fig. 5.12. The strain varies along the gage length and is a maximum in the region of necking.

The engineering strain, $\varepsilon = \Delta L / L_i$, is measured between two points spanning the necked region, so that the value obtained is only an average rather than the maximum. Hence, once necking starts, true strain cannot be calculated from engineering strain using Eq. 5.20. Similarly, true stress cannot be obtained from Eq. 5.23 once necking starts, as this also employs the average value ε. Since $\tilde{\sigma}$ and $\tilde{\varepsilon}$ can be evaluated from cross-sectional areas using Eqs. 5.16 and 5.24, points on the true stress-strain curve where necking is occurring can be obtained by measuring cross-sectional areas in the neck. For specimens with a round cross section, this is easily done by making periodic diameter measurements as the test proceeds.

The above-noted limitations of the various true stress-strain equations are summarized in Fig. 5.20. First, note that engineering stress and strain may always be determined from their definitions, Eqs. 5.1 and 5.2. True stress may always be obtained from Eq. 5.16 or 5.17 if areas are directly measured, as from diameters in round cross sections. As just explained, Eqs. 5.20 and 5.23, being based on engineering strain, cannot be used once necking starts.

In addition, Eq. 5.23 is limited by the constant volume assumption. Hence, this conversion to true stress is inaccurate at small strains, such as those below and around the yield stress. An arbitrary lower limit of twice the strain that accompanies the offset yield strength, that is, $2\varepsilon_o$, is suggested in Fig. 5.20. (Note that ε_o is defined in Fig. 5.11.)

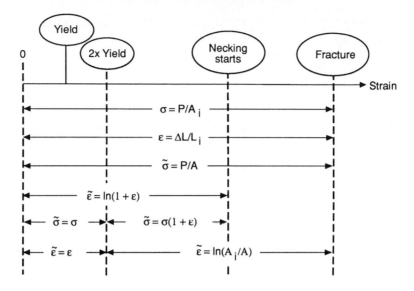

Figure 5.20   Use and limitations of various equations for stresses and strains from a tension test.

Below this limit, the difference between true and engineering stress is generally so small that it can be neglected, so that no conversion is needed. A similar limitation is encountered by Eqs. 5.24 and 5.25, which are otherwise valid at any strain.

5.5.4 Bridgman Correction for Hoop Stress

A complication exists in interpreting tensile results near the end of a test where there is a large amount of necking. As pointed out by P. W. Bridgman in 1944, large amounts of necking result in a tensile hoop stress being generated around the circumference in the necked region. Thus, the state of stress is no longer uniaxial as assumed, and the behavior of the material is affected. In particular, the axial stress is increased above what it would otherwise be. (This arises from Eq. 4.31 and can be understood based on detailed study of plastic deformation later in Chapter 12.)

A rough correction can be made based on the empirical curve developed by Bridgman for steel, which is shown in Fig. 5.21. The curve is entered with the true strain based on area, and it gives a value of the correction factor, B, which is used as follows:

$$\tilde{\sigma}_B = B\tilde{\sigma} \tag{5.26}$$

where $\tilde{\sigma}$ is the true stress simply computed from the area using Eq. 5.16, and $\tilde{\sigma}_B$ is the corrected value of true stress. Rather than using the curve, B may be estimated from the following equation:

$$B = 0.83 - 0.186 \log \tilde{\varepsilon} \qquad (0.15 \leq \tilde{\varepsilon} \leq 3) \tag{5.27}$$

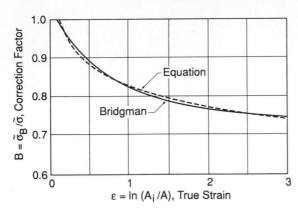

Figure 5.21 Curve giving correction factors on true stress for the effect of hoop stress due to necking. The curve of [Bridgman 44] is shown along with the curve from Eq. 5.27 approximating it.

where the correction is not needed for $\tilde{\varepsilon} < 0.15$, and where *log* is the logarithm to the base 10. This equation closely represents Bridgman's curve, corresponding to the dashed line in Fig. 5.21.

From the curve, note that a 10% correction ($B = 0.9$) corresponds to a true strain of about $\tilde{\varepsilon} = 0.4$. Using Eq. 5.25, this corresponds to a ratio of initial to necked diameter of 1.22. Hence, fairly large strains must occur for the correction to be significant.

Example 5.2

More complete data from the tension test on AISI 1020 steel of Example 5.1 are given in the first three columns of Table 5.5. Length changes over 50 mm have been converted to engineering strains, ε, in the first column. Loads at corresponding times are also given, as are diameters for the large strain portion of the test, with the latter being measured in the neck once this process started. Values of stresses and strains calculated from the data are given in the last four columns of the table. For values of $\varepsilon = 0.01$ and 0.33, verify the values tabulated for engineering stress σ, true strain $\tilde{\varepsilon}$, true stress $\tilde{\sigma}$, and corrected true stress $\tilde{\sigma}_B$.

Solution The engineering stresses are calculated by using the initial diameter with Eq. 5.1.

$$\sigma = \frac{P}{A_i} = \frac{4P}{\pi d_i^2} = \frac{4\,(17{,}210\ \text{N})}{\pi\,(9.11\ \text{mm})^2} = 264\ \text{MPa} \qquad (\varepsilon = 0.01) \qquad \textbf{Ans.}$$

$$\sigma = \frac{4\,(23{,}490)}{\pi\,(9.11)^2} = 360\ \text{MPa} \qquad (\varepsilon = 0.33) \qquad \textbf{Ans.}$$

True strain and stress for $\varepsilon = 0.01$ can be calculated from Eqs. 5.20 and 5.23, respectively.

$$\tilde{\varepsilon} = \ln\,(1 + \varepsilon) = \ln\,(1 + 0.01) = 0.00995 \qquad \textbf{Ans.}$$

$$\tilde{\sigma} = \sigma\,(1 + \varepsilon) = 264\,(1 + 0.01) = 267\ \text{MPa} \qquad \textbf{Ans.}$$

However, $\varepsilon = 0.33$ is beyond the ultimate stress point, where necking begins, so that these equations based on ε from length measurements cannot be used. Fortunately, a diameter

TABLE 5.5 ENGINEERING AND TRUE STRESSES AND STRAINS FOR A TENSION TEST ON AISI 1020 HOT-ROLLED STEEL

Test Data			Calculated Values			
Engr. Strain ε	Load P kN	Diameter d mm	Engineering Stress σ MPa	True Strain[5] $\tilde{\varepsilon}$	Raw True Stress[5] $\tilde{\sigma}$ MPa	Corrected True Stress $\tilde{\sigma}_B$ MPa
0	0	9.11	0	0	0	0
0.0015[1]	19.13	—	293	0.00150	293	—
0.0033[2]	17.21	—	264	0.00329	265	—
0.0050	17.53	—	269	0.00499	270	—
0.0070	17.44	—	268	0.00698	269	—
0.010	17.21	—	264	0.00995	267	—
0.049	20.77	8.89	319	0.0489	335	335
0.218	25.71	8.26	394	0.196	480	461
0.234[3]	25.75	—	395	0.210	488	466
0.306	25.04	7.62	384	0.357	549	501
0.330	23.49	6.99	360	0.530	612	539
0.348	21.35	6.35	328	0.722	674	577
0.360	18.90	5.72	290	0.931	735	615
0.366[4]	17.39	5.28[6]	267	1.091	794	654

Notes: [1]Upper yield. [2]Lower yield and 0.2% offset yield. [3]Ultimate. [4]Fracture. [5]Calculated from $(1 + \varepsilon)$ where d not measured. [6]Measured from the broken specimen.

measurement is available, and so this can be used with Eqs. 5.16 and 5.25.

$$\tilde{\sigma} = \frac{P}{A} = \frac{4P}{\pi d^2} = \frac{4\,(23{,}490\ \text{N})}{\pi\,(6.99\ \text{mm})^2} = 612\ \text{MPa} \qquad \textbf{Ans.}$$

$$\tilde{\varepsilon} = 2\ln\frac{d_i}{d} = 2\ln\frac{9.11}{6.99} = 0.530 \qquad \textbf{Ans.}$$

The true stress does not need to be corrected at $\varepsilon = 0.01$, but the correction is needed at $\varepsilon = 0.33$. Using $\tilde{\sigma}$ and $\tilde{\varepsilon}$ from just above with Eqs. 5.26 and 5.27, we have

$$\tilde{\sigma}_B = \tilde{\sigma}\,(0.83 - 0.186\log\tilde{\varepsilon}) = 539\ \text{MPa} \qquad \textbf{Ans.}$$

Comment   Curves of σ vs. ε and $\tilde{\sigma}$ vs. $\tilde{\varepsilon}$ based on all of the data points in Table 5.5 are plotted in Figure E5.2.

5.5.5 True Stress-Strain Curves

The true stress is always larger than the corresponding engineering stress, and the difference may be a factor of two or more near the end of a tension test on a ductile material. True strain based on a length measurement is somewhat smaller than the corresponding

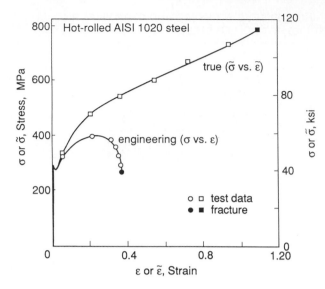

Figure E5.2 Engineering and true stress-strain curves from a tension test on hot-rolled AISI 1020 steel.

engineering strain. However, once necking starts, true strain based on an area measurement becomes larger, dramatically so if a distinct neck forms. These trends are evident in the stress-strain curves for mild steel from Example 5.2.

Consider the true stress-strain curve of a metal in the region well beyond yielding, where most of the strain is plastic strain. Such curves often fit a relationship of the form

$$\tilde{\sigma} = H\tilde{\varepsilon}^n \qquad (5.28)$$

If stress versus strain is plotted on log-log coordinates, this equation gives a straight line. The slope is n, which is called the *strain hardening exponent*. The quantity H, which is called the *strength coefficient*, is the intercept at $\tilde{\varepsilon} = 1$. At large strains during the advanced stages of necking, Eq. 5.28 should be used with true stresses that have been corrected using the Bridgman factor.

$$\tilde{\sigma}_B = H\tilde{\varepsilon}^n \qquad (5.29)$$

5.5.6 True Stress-Strain Properties

Additional materials properties obtained from tension tests may be defined based on true stress and strain. The *true fracture strength*, $\tilde{\sigma}_f$, is obtained simply from the load at fracture and the final area, or from the engineering stress at fracture.

$$\tilde{\sigma}_f = \frac{P_f}{A_f} = \sigma_f \left(\frac{A_i}{A_f}\right) \qquad (5.30)$$

Since the Bridgman correction is generally needed, the value obtained should be converted to $\tilde{\sigma}_{fB}$ using Eq. 5.26.

The *true fracture strain*, $\tilde{\varepsilon}_f$, may be obtained from the final area or from the percent reduction in area.

$$\tilde{\varepsilon}_f = \ln \frac{A_i}{A_f}, \qquad \tilde{\varepsilon}_f = \ln \frac{100}{100 - \%RA} \qquad \text{(a,b)} \qquad (5.31)$$

where the second equation follows readily from the first. Note that $\tilde{\varepsilon}_f$ cannot be computed if only the engineering strain at fracture (percent elongation) is available.

The *true toughness*, $\tilde{u}_f$, is the area under the true stress-strain curve up to fracture. Using Eqs. 5.11 and 5.29, this area is approximately

$$\tilde{u}_f = \int_0^{\tilde{\varepsilon}_f} \tilde{\sigma}_B d\tilde{\varepsilon} = H \int_0^{\tilde{\varepsilon}_f} \tilde{\varepsilon}^n d\tilde{\varepsilon} \qquad (5.32)$$

Evaluating the integral and again using Eq. 5.29 yields

$$\tilde{u}_f = \frac{H\tilde{\varepsilon}_f^{n+1}}{n+1} = \frac{\tilde{\sigma}_{fB}\tilde{\varepsilon}_f}{n+1} \qquad (5.33)$$

The various engineering and true stress-strain properties obtainable from a tension test are summarized by the categorized listing of Table 5.6. Note that the engineering fracture strain ε_f and the percent elongation are only different ways of stating the same quantity. Also, the $\%RA$ and $\tilde{\varepsilon}_f$ can each be calculated from the other using Eq. 5.31(b).

Note that the strength coefficient H determines the magnitude of the true stress in the large strain region of the stress-strain curve, and so it is included as a measure of strength. The strain hardening exponent n is a measure of the rate of strain hardening for

TABLE 5.6   MATERIALS PROPERTIES OBTAINABLE FROM TENSION TESTS

Category	Engineering Property	True Stress-Strain Property
Elastic constants	Elastic modulus, E, E_t Poisson's ratio, ν	—
Strength	Proportional limit, σ_p Yield strength, σ_o Ultimate tensile strength, σ_u Engineering fracture strength, σ_f	True fracture strength, $\tilde{\sigma}_{fB}$ Strength coefficient, H
Ductility	Percent elongation, $100\varepsilon_f$ Reduction in area, $\%RA$	True fracture strain, $\tilde{\varepsilon}_f$
Energy capacity	Resilience, u_r Tensile toughness, u_f	True toughness, $\tilde{u}_f$
Strain hardening	Strain hardening ratio, σ_u/σ_o	Strain hardening exponent, n

the true stress-strain curve. For engineering metals, values above $n = 0.2$ are considered relatively high, and those below 0.1 are considered relatively low.

True stress-strain properties are listed for several engineering metals in Table 5.7. These are the same metals for which engineering properties have already been given in Table 5.2.

TABLE 5.7 TRUE STRESS-STRAIN TENSILE PROPERTIES FOR SOME ENGINEERING METALS, AND ALSO HARDNESS

Material	True Fracture:		Strength Coefficient H	Strain Hardening Exponent n	Brinell Hardness[1] HB
	Strength $\tilde{\sigma}_{fB}$	Strain $\tilde{\varepsilon}_f$			
	MPa (ksi)		MPa (ksi)		
Ductile cast iron A536 (65-45-12)	524 (76)	0.222	456 (66.1)	0.0455	167
AISI 1020 steel as rolled	713 (103)	0.96	737 (107)	0.19	107
ASTM A514, T1 structural steel	1213 (176)	1.08	1103 (160)	0.088	256
AISI 4142 steel as quenched	2580 (375)	0.060	—	0.136	670
AISI 4142 steel 205°C temper	2650 (385)	0.310	—	0.091	560
AISI 4142 steel 370°C temper	1998 (290)	0.540	—	0.043	450
AISI 4142 steel 450°C temper	1826 (265)	0.660	—	0.051	380
18 Ni maraging steel (250)	2136 (310)	0.82	—	0.02	460
SAE 308 cast aluminum	232 (33.6)	0.009	567 (82.2)	0.196	80
2024-T4 aluminum	631 (91.5)	0.43	806 (117)	0.20	120
7075-T6 aluminum	744 (108)	0.41	827 (120)	0.113	150
AZ91C-T6 cast magnesium	137 (20)	0.004	653 (94.7)	0.282	61

Note: [1]Load 3000 kg for irons and steels, 500 kg otherwise; typical values from [Boyer 85] are listed in some cases.

Sources: Data in [Conle 84] and [SAE 89].

Example 5.3

Using the stresses and strains in Table 5.5 for the tension test on AISI 1020 steel, determine the constants H and n for Eq. 5.29, and also the true fracture stress and strain, $\tilde{\sigma}_{fB}$ and $\tilde{\varepsilon}_f$, and the true toughness $\tilde{u}_f$.

Solution   To find H and n, note that Eq. 5.29 can be written

$$\log \tilde{\sigma}_B = n \log \tilde{\varepsilon} + \log H$$

which is a straight line on an x-y plot

$$y = mx + b$$

where

$$y = \log \tilde{\sigma}_B \quad \text{(dependent variable)}$$
$$x = \log \tilde{\varepsilon} \quad \text{(independent variable)}$$
$$m = n$$
$$b = \log H$$

Thus, if $\tilde{\sigma}_B$ is plotted versus $\tilde{\varepsilon}$ on log-log coordinates, a straight line should be formed of slope n and intercept at $\tilde{\varepsilon} = 1$ of $\tilde{\sigma}_B = H$. This is shown in Figure E5.3. A graphical or least squares approach could be used; the latter gives

$$m = n = 0.206 \qquad \qquad \textbf{Ans.}$$

$$b = 2.7967, \qquad H = 10^b = 626 \text{ MPa} \qquad \qquad \textbf{Ans.}$$

where the seven data points at high strains for which diameter measurements are available were used in the fit.

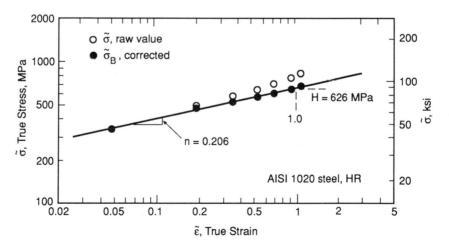

Figure E5.3   Log-log plot of true stress versus true strain from the large strain portion of a tension test on AISI 1020 steel.

Hence, the equation is

$$\tilde{\sigma}_B = 626\,\tilde{\varepsilon}^{0.206} \text{ MPa}$$

The resulting straight line is plotted and is seen to give an excellent fit to the $\tilde{\sigma}_B$ data. The uncorrected $\tilde{\sigma}$ data are also shown and are seen to form a gently upward sweeping curve rather than a straight line.

    The true fracture stress and strain are simply the values from the last line of Table 5.5, as this corresponds to the fracture point.

$$\tilde{\sigma}_{fB} = 654 \text{ MPa}, \qquad \tilde{\varepsilon}_f = 1.091 \qquad\qquad \textbf{Ans.}$$

The true toughness can now be obtained from Eq. 5.33.

$$\tilde{u}_f = \frac{\tilde{\sigma}_{fB}\tilde{\varepsilon}_f}{n+1} = \frac{(654 \text{ MPa})\,(1.091)}{0.206 + 1} = 592 \; \frac{\text{MJ}}{\text{m}^3} \qquad\qquad \textbf{Ans.}$$

5.5.7 True Strain at Necking

An interesting conclusion can be reached concerning the strain that corresponds to the engineering ultimate strength. In particular, the strain hardening exponent is approximately equal to the true strain at this point.

$$n = \tilde{\varepsilon}_u \tag{5.34}$$

To derive this relationship, note that Eq. 5.28 can usually be applied around necking, as the elastic strain there is small compared to the plastic strain.

    To proceed, first write Eq. 5.24 in the form

$$e^{\tilde{\varepsilon}} = \frac{A_i}{A} \tag{5.35}$$

Then substitute this into Eq. 5.17 to obtain an equation for engineering stress.

$$\sigma = \frac{\tilde{\sigma}}{e^{\tilde{\varepsilon}}} \tag{5.36}$$

At the engineering ultimate stress, σ has a maximum value. Thus, differentiate this equation with respect to $\tilde{\varepsilon}$ and equate the result to zero.

$$\frac{d\sigma}{d\tilde{\varepsilon}} = \frac{1}{e^{\tilde{\varepsilon}}}\left(\frac{d\tilde{\sigma}}{d\tilde{\varepsilon}} - \tilde{\sigma}\right) = 0 \tag{5.37}$$

    This equality is satisfied if

$$\frac{d\tilde{\sigma}_u}{d\tilde{\varepsilon}_u} = \tilde{\sigma}_u \tag{5.38}$$

where subscripts are added to indicate the point corresponding to the engineering ultimate stress.

    Next substitute Eq. 5.28 for the right hand side of Eq. 5.38, and also differentiate Eq. 5.28 and substitute the result for the left hand side.

$$H n \tilde{\varepsilon}_u^{n-1} = H \tilde{\varepsilon}_u^n \qquad (5.39)$$

This simplifies to obtain the desired result, $n = \tilde{\varepsilon}_u$.

Using Eq. 5.20, the strain hardening exponent may thus be estimated from the engineering strain that occurs at the engineering ultimate stress.

$$n = \ln(1 + \varepsilon_u) \qquad (5.40)$$

However, it is difficult to precisely identify ε_u, as the engineering stress-strain curve may be quite flat near its maximum. Therefore, where data are available, it is preferable to obtain n from a fit on a log-log plot as in Example 5.3. In that example, note that the fitted value of $n = 0.206$ agrees reasonably, but not precisely, with $\tilde{\varepsilon}_u = 0.210$ from Table 5.5.

5.6 COMPRESSION TEST

Some materials have dramatically different behavior in compression than in tension, and in some cases these materials are used primarily to resist compressive stresses. Examples include concrete and building stone. Data from compression tests are therefore often needed for engineering applications. Compression tests have many similarities to tension tests in the manner of conducting the test and in the analysis and interpretation of the results. Since tension tests have already been considered in detail, the discussion here will focus on areas where these two types of tests differ.

5.6.1 Test Methods for Compression

A typical arrangement for a compression test is shown in Fig. 5.22. Uniform displacement rates in compression are applied in a manner similar to a tension test, except of course for the direction of loading. The specimen is most commonly a simple cylinder having a ratio of length to diameter, L/d, in the range 1 to 3. However, values of L/d up to 10 are sometimes used where the primary objective is to accurately determine the modulus of elasticity in compression. Specimens with square or rectangular cross sections may also be tested.

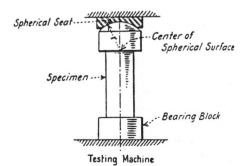

Figure 5.22 Compression test in a universal testing machine using a spherical-seated bearing block. (From [ASTM 90a] Std. E9; copyright ©ASTM; reprinted with permission.)

The choice of a specimen length represents a compromise. Buckling may occur if the L/d ratio is relatively large. If this happens, the test result is meaningless as a measure of the fundamental compressive behavior of the material. Buckling is affected by the unavoidable small imperfections in the geometry of the test specimen and its alignment with respect to the testing machine. For example, the ends of the specimen can be almost parallel, but never perfectly so.

Conversely, if L/d is small, the test result is affected by the details of the conditions at the end. In particular, as the specimen is compressed, the diameter increases due to the Poisson effect, but friction retards this motion at the ends, resulting in deformation into a barrel shape. Although this effect can be minimized by proper lubrication of the ends, it is difficult to avoid entirely. As a result, in materials that are capable of large amounts of deformation in compression, the choice of too small of an L/d ratio may result in a situation where the behavior of the specimen is dominated by the end effects. Again, the test does not measure the fundamental compressive behavior of the material.

Considering both the desirability of small L/d to avoid buckling and large L/d to avoid end effects, a reasonable compromise is $L/d = 3$ for ductile materials. Values of $L/d = 1.5$ or 2 are suitable for brittle materials, where the small amount of deformation that occurs causes less difficulty with end effects.

Some examples of compression specimens of various materials both before and after testing are shown in Figs. 5.23 and 5.24. Mild steel shows typical ductile behavior, specifically large deformation without fracture ever occurring. The gray cast iron and concrete behaved in a brittle manner, and the aluminum alloy deformed considerably but then also fractured. Fracture in compression usually occurs on an inclined plane.

5.6.2 Materials Properties in Compression

The initial portions of compressive stress-strain curves have the same general nature as those in tension. Thus, various materials properties may be defined from the initial portion in the same manner as for tension, such as the elastic modulus E, the proportional limit σ_p, and the yield strength σ_o.

The ultimate strength behavior in compression differs in a qualitative way from that in tension. Note that the decrease in load prior to final fracture in tension is associated

Figure 5.23 Compression specimens of metals (left to right): untested specimen, and tested specimens of gray cast iron, aluminum alloy 7075-T651, and hot-rolled AISI 1020 steel. Diameters before testing were approximately 25 mm, and lengths were 76 mm. (Photo by R. A. Simonds.)

Figure 5.24 Untested and tested 150 mm diameter compression specimens of concrete with Hokie limestone aggregate. (Photo by R. A. Simonds.)

with the phenomenon of necking. This of course does not occur in compression. In fact, an opposite effect occurs, in that the increasing cross-sectional area causes the stress-strain curve to rise rapidly rather than showing a maximum.

As a result, there is no load maximum in compression prior to fracture, and the engineering ultimate strength is the same as the engineering fracture strength. Brittle and moderately ductile materials will fracture in compression. But many ductile metals and polymers simply never fracture. Instead, the specimen deforms into an increasingly larger and thinner pancake shape until the load required for further deformation becomes so large that the test must be suspended.

Ductility may be measured in compression in a manner analogous to that used for tension. Such measures include percentage changes in length and area, and also engineering and true fracture strain. The same measures of energy capacity may also be used, as can constants for true stress-strain curves of the form of Eq. 5.28.

5.6.3 Trends in Compressive Behavior

Ductile engineering metals often have nearly identical initial portions of stress-strain curves in tension and compression, with an example of this being shown in Fig. 5.25. After large amounts of deformation, the curves may still agree if true stresses and strains are plotted.

Many materials that are brittle in tension have this behavior because they contain cracks or voids that grow and combine to cause failures along planes of maximum tension, that is, perpendicular to the specimen axis. Examples are the graphite flakes in gray cast iron, cracks at the aggregate boundaries in concrete, and fissures in natural stone. Such cracks or voids have much less effect in compression, so that materials that behave in a brittle manner in tension usually have considerably higher compressive strengths. For example, compare the strengths in tension and compression given for various ceramics

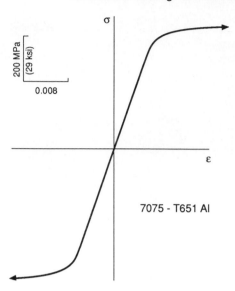

Figure 5.25 Initial portions of stress-strain curves in tension and compression for 7075-T651 aluminum.

in Table 3.10. Quite ductile behavior can even occur for materials that are brittle in tension, as for the polymer in Fig. 5.26.

Where compressive failure does occur, it is generally associated with a shear stress, so that the fracture is inclined about 45° to the specimen axis. This type of fracture is evident for gray cast iron, an aluminum alloy, and concrete in Figs. 5.23 and 5.24. Compare the cast iron fracture plane with that for tension in Fig. 5.14. The tension fracture plane is oriented normal to the applied tension stress, which is typical of brittle behavior in all materials.

5.7 HARDNESS TESTS

In engineering, hardness is most commonly defined as the resistance of a material to *indentation*. Indentation is the pressing of a hard round ball or point against the material sample with a known force, so that a depression is made. The depression, or indentation, results from plastic deformation beneath the indenter as shown in Fig. 5.27. Some specific characteristic of the indentation, such as its size or depth, is then taken as a measure of hardness.

Other principles are also used to measure hardness. For example, the *Scleroscope hardness test* is a rebound test that employs a hammer with a rounded diamond tip. This hammer is dropped from a fixed height onto the surface of the material being tested. The hardness number is proportional to the height of rebound of the hammer, with the scale for metals being set so that fully hardened tool steel has a value of 100. A modified version of this test is also used for polymers.

In mineralogy, the *Mohs hardness scale* is used. Diamond, the hardest known material, is assigned a value of 10. Decreasing values are assigned to other minerals,

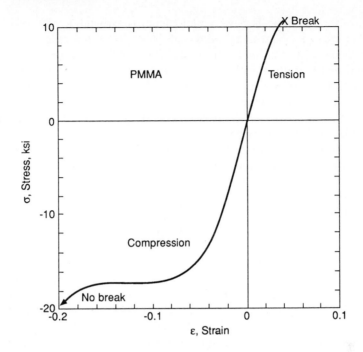

Figure 5.26   Stress-strain curves for plexiglass (acrylic, PMMA) in both tension and compression. (Adapted from [Richards 61] p. 153; reprinted by permission of PWS-Kent Publishing Co., Boston.)

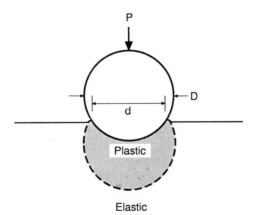

Figure 5.27   Plastic deformation under a Brinell hardness indenter.

down to 1 for the soft mineral talc. Decimal fractions, such as 9.7 for tungsten carbide, are used for materials intermediate between the standard ones. Where a material lies on the Mohs scale is determined by a simple manual scratch test. If two materials are compared, the harder one is capable of scratching the softer one, but not vice versa. This allows materials to be ranked as to hardness, and decimal values between the standard ones are assigned as a matter of judgment.

Very hard steels have a Mohs hardness around 7, and lower strength steels and other relatively hard metal alloys are generally in the range 4 to 5. Soft metals may be below 1, so that their Mohs hardness is difficult to specify. Various materials are compared as to their Mohs hardness in Fig. 5.28. Also shown are values for two of the indentation hardness scales that will be discussed below.

Indentation hardness has an advantage over Mohs hardness in that the values obtained are less a matter of interpretation and judgment. There are a number of different standard indentation hardness tests. They differ from one another as to the geometry of the indenter, the amount of force used, etc. As time-dependent deformations may occur that affect the indentation, loading rates and/or times of application are fixed for each standard test.

Test apparatus for the *Brinell* hardness test is shown in Fig. 5.29. Some of the resulting indentations are shown along with those for *Rockwell* type tests in Fig. 5.30. These two and the *Vickers* test are commonly used for engineering purposes.

5.7.1 Brinell Hardness Test

In this test, a relatively large steel ball, specifically 10 mm in diameter, is used with a relatively high force. The load used is 3000 kg for relatively hard materials, such as steels and cast irons, and 500 kg for softer materials, such as copper and aluminum alloys. For very hard materials, the standard steel ball will deform excessively, and a tungsten carbide ball is used.

The Brinell hardness number, designated *HB*, is obtained by dividing the applied force P, in kilograms, by the curved surface area of the indentation, which is a segment of a sphere. This gives

$$HB = \frac{2P}{\pi D \left[D - \left(D^2 - d^2 \right)^{0.5} \right]} \qquad (5.41)$$

where D is the diameter of the ball and d is the diameter of the indentation, both in millimeters, as illustrated in Fig. 5.27. Brinell hardness numbers are listed for the metals in Table 5.7.

5.7.2 Vickers Hardness Test

The Vickers hardness test is based on the same general principles as the Brinell test. It differs primarily in that the indenter is a diamond point in the shape of a pyramid with a square base. The angle between the faces of the pyramid is $\alpha = 136°$ as shown in Fig. 5.31. This shape results in the depth of penetration, h, being one-seventh of the indentation size, d, measured on the diagonal. The Vickers hardness number HV is obtained by dividing the applied force P by the surface area of the pyramidal depression. This yields

$$HV = \frac{2P}{d^2} \sin \frac{\alpha}{2} \qquad (5.42)$$

where d is in millimeters and P in kilograms.

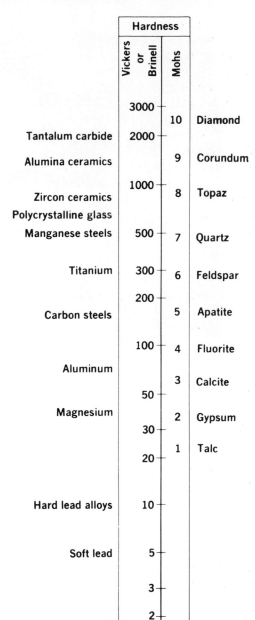

Figure 5.28 Approximate relative hardness of various metals and ceramics. (From [Richards 61] p. 402, as based on data from [Zwikker 54] p. 261; reprinted by permission of PWS-Kent Publishing Co., Boston.)

   Note that the standard pyramidal shape causes the indentation to be geometrically similar regardless of its size. For reasons based on plasticity theory, which are beyond the scope of the present discussion, this geometric similarity is expected to result in the Vickers hardness value being independent of the magnitude of the force used. Hence, a

Figure 5.29 Brinell hardness tester, and indenter being applied to a sample. (Photographs courtesy of Tinius Olsen Testing Machine Co., Inc., Willow Grove, Pa.)

wide range of standard forces usually between 1 and 120 kg are used, so that essentially all solid materials can be included in a single wide-ranging hardness scale.

Approximate Vickers hardness numbers are given for various classes of materials in Fig. 5.28. Also, values for some ceramics are given in Table 5.8. Within the more limited range where the Brinell test can be used, there is approximate agreement with the Vickers scale. This approximate agreement is shown by average curves for steels of various strengths in Fig. 5.32.

Another hardness test that is somewhat similar to the Vickers test is the Knoop test. It differs in that the pyramidal indenter has a diamond-shaped base and in the use of the projected area to calculate hardness.

5.7.3 Rockwell Hardness Test

In the Rockwell test, a diamond point or a steel ball is employed as the indenter. The diamond point, called a Brale indenter, is a cone with an included angle of 120° and a slightly rounded end. Balls of sizes ranging between 1.6 mm and 12.7 mm are also used. Various combinations of indenter and load are applied in the *regular Rockwell test* to accommodate a wide range of materials as listed in Table 5.9. In addition, there is a *superficial Rockwell hardness test* that uses smaller forces and causes smaller indentations.

Figure 5.30   Brinell and Rockwell hardness indentations.   On the left, in hot-rolled AISI 1020 steel, the larger Brinell indentation has a diameter of 5.4 mm, giving $HB = 121$, and the smaller Rockwell B indentation gave $HRB = 72$. On the right, a higher strength steel has indentations corresponding to $HB = 241$ and $HRC = 20$. (Photo by R. A. Simonds.)

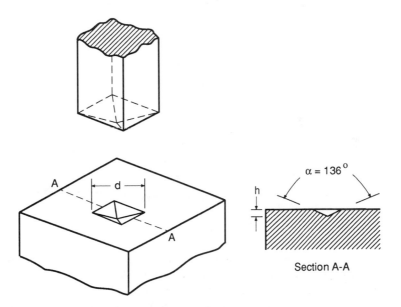

Figure 5.31   Vickers hardness indentation.

TABLE 5.8 VICKERS HARDNESS AND BENDING STRENGTHS FOR SOME CERAMICS AND GLASSES

Material	Hardness HV, 0.1 kg	Bend Strength MPa (ksi)	Elastic Modulus GPa (10^3 ksi)
Soda-lime glass	600	65 (9.4)	74 (10.7)
Fused silica glass	650	70 (10.1)	70 (10.1)
Aluminous porcelain	≈ 800	120 (17.4)	120 (17.4)
Silicon nitride Si_3N_4, hot pressed	1700	600 (87)	400 (58)
Alumina, Al_2O_3 99.5% dense	1750	400 (58)	400 (58)
Silicon carbide SiC, hot pressed	2600	600 (87)	400 (58)
Boron carbide, B_4C hot pressed	3200	400 (58)	475 (69)

Source: Data in [Creyke 82] p. 38.

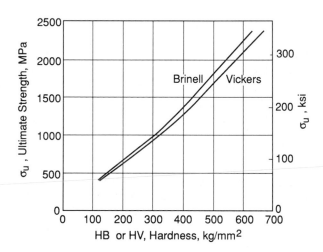

Figure 5.32 Approximate relationship between the ultimate tensile strengths of carbon and alloy steels and Brinell and Vickers hardness. (Data from [Boyer 85] p. 1.61.)

Rockwell tests differ from other hardness tests in that the depth of the indentation is measured, rather than the size. A small initial load called the *minor load* is first applied to establish a reference position for the depth measurement and to penetrate through any surface scale or foreign particles. A minor load of 10 kg is used for the regular test. The *major load* is then applied, and finally the additional penetration due to the major load is measured. This is illustrated by the difference between h_2 and h_1 in Fig. 5.33.

TABLE 5.9   COMMONLY USED ROCKWELL HARDNESS SCALES

Symbol, HRX $X =$	Penetrator Diameter if Ball, mm (in)	Load kg	Typical Application
A	Diamond point	60	Tool materials
D	Diamond point	100	Cast irons, sheet steels
C	Diamond point	150	Steels, hard cast irons, Ti alloys
B	1.588 (0.0625)	100	Soft steels, Cu and Al alloys
E	3.175 (0.125)	100	Al and Mg alloys, other soft metals; reinforced polymers
M	6.35 (0.250)	100	Very soft metals; high modulus polymers
R	12.70 (0.500)	60	Very soft metals; low modulus polymers

Each Rockwell hardness scale has a maximum useful value around 100. An increase of one unit of regular Rockwell hardness represents a decrease in penetration of 0.002 mm. Hence, the hardness number is:

$$HRX = M - \frac{\Delta h}{0.002} \qquad (5.43)$$

where $\Delta h = h_2 - h_1$ is in millimeters and M is the upper limit of the scale. For regular Rockwell hardness, $M = 100$ for all scales using the diamond point (A, C, and D scales), and $M = 130$ for all scales using ball indenters (B, E, M, R, etc., scales). The hardness numbers are designated HRX, where X indicates the scale involved, such as 60 HRC for 60 points on the C scale. Note that a Rockwell hardness number is meaningless unless the scale is specified. In practice, the hardness numbers are read directly from a dial on the hardness tester, rather than being calculated.

5.7.4 Hardness Correlations and Conversions

The deformations caused by a hardness indenter are of similar magnitude to those occurring at the ultimate tensile strength in a tension test. However, an important difference is that the material cannot freely flow outward, so that a complex triaxial state of stress exists under the indenter. Nevertheless, empirical correlations can be established between hardness and tensile properties, primarily the engineering ultimate tensile strength σ_u.

For example, for low- and medium-strength carbon and alloy steels, σ_u can be estimated from Brinell hardness as follows:

$$\sigma_u = 3.45(HB) \text{ MPa}$$
$$\sigma_u = 0.50(HB) \text{ ksi} \qquad (5.44)$$

Note that this equation approximates the curve shown in Fig. 5.32. However, there is considerable scatter in actual data, so that this relationship should be considered to

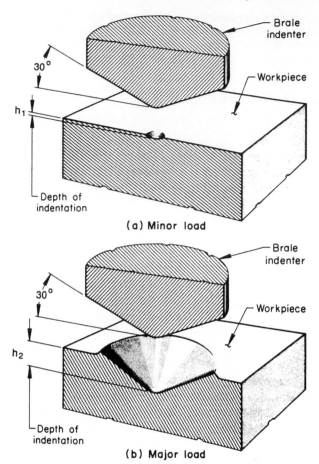

(a) Minor load

(b) Major load

Figure 5.33 Rockwell hardness indentation made by application of (a) the minor load, and (b) the major load, on a diamond Brale indenter. (Adapted from [Boyer 85] p. 34.6; used with permission.)

provide rough estimates only. For other classes of material, the empirical constant will differ, and the relationship may even become nonlinear. Similarly, the relationship will change for a different type of hardness test. Rockwell hardness correlates well with σ_u and with other types of hardness test, but the relationships are usually nonlinear. This situation results from the unique indentation-depth basis of this test. For carbon and alloy steels, a conversion chart for estimating various types of hardness from one another, and also ultimate tensile strength, is given as Table 5.10. More detailed conversion charts for steels and other metals are given in ASTM Standard No. E140, in various handbooks, and in information provided by manufacturers of hardness testing equipment.

5.8 NOTCH-IMPACT TESTS

Notch-impact tests provide information on the resistance of a material to sudden fracture where a sharp stress raiser or flaw is present. In addition to providing information not

TABLE 5.10 APPROXIMATE EQUIVALENT HARDNESS NUMBERS AND ULTIMATE TENSILE STRENGTHS FOR CARBON AND ALLOY STEELS

Brinell HB	Vickers HV	Rockwell HRB	Rockwell HRC	Ultimate, σ_u MPa	Ultimate, σ_u ksi
627	667	—	58.7	2393	347
578	615	—	56.0	2158	313
534	569	—	53.5	1986	288
495	528	—	51.0	1813	263
461	491	—	48.5	1669	242
429	455	—	45.7	1517	220
401	425	—	43.1	1393	202
375	396	—	40.4	1267	184
341	360	—	36.6	1131	164
311	328	—	33.1	1027	149
277	292	—	28.8	924	134
241	253	100	22.8	800	116
217	228	96.4	—	724	105
197	207	92.8	—	655	95
179	188	89.0	—	600	87
159	167	83.9	—	538	78
143	150	78.6	—	490	71
131	137	74.2	—	448	65
116	122	67.6	—	400	58

Note: Load 3000 kg for HB.

Source: Values in [Boyer 85] p. 1.61.

available from any other simple mechanical test, these tests are quick and inexpensive, and so they are frequently employed for engineering purposes.

5.8.1 Types of Test

Various standard impact tests are widely employed in which notched beams are broken by a swinging pendulum or a falling weight. The most common tests of this type are the *Charpy V-notch* and the *Izod tests*. Specimens and loading configurations for these are shown in Fig. 5.34. A swinging pendulum arrangement is used for applying the impact load in both cases, with a device for Charpy tests being shown in Fig. 5.35. The energy required to break the sample is determined from an indicator that measures how high the pendulum swings after breaking the sample. Some broken Charpy specimens are shown in Fig. 5.36. The impact resistance of polymers (plastics) is often evaluated using the Izod test, with some representative data being included in Table 5.3.

Another test that is used fairly often is the *dynamic tear test*. Specimens for this test have a center notch as for the Charpy specimen, and they are impacted in three-point bending, but by a falling weight. These specimens are quite large, being about 180 mm

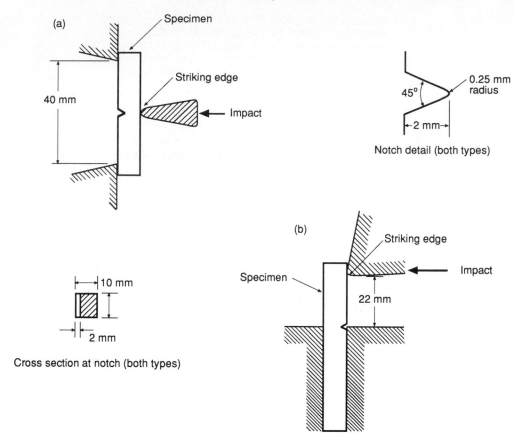

Figure 5.34 Specimens and loading configurations for (a) Charpy V-notch and (b) Izod tests. (Adapted from [ASTM 90a] Std. E23; copyright ©ASTM; reprinted with permission.)

long, 40 mm wide, and 16 mm thick. An even larger size, about 430 mm long, 120 mm wide, and 25 mm thick, is also used.

In notch-impact tests, the energies obtained depend on the details of the specimen size and geometry, including the notch-tip radius. The support and loading configuration used are also important, as is the mass and velocity of the pendulum or weight. Hence, results from one type of test cannot be directly compared with those from another. In addition, all such details of the test must be kept constant as specified in the published standards, notably those of ASTM.

5.8.2 Trends in Impact Behavior and Discussion

Polymers, metals, and other materials with low notch-impact energy are generally prone to brittle behavior and typically have low ductility and low toughness in a tension test.

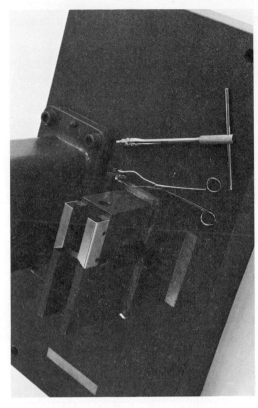

Figure 5.35 Charpy testing machine, and placement of the specimen. (Photographs courtesy of Tinius Olsen Testing Machine Co., Inc., Willow Grove, Pa.)

Figure 5.36 Broken Charpy specimens, left to right, of gray cast iron, AISI 4140 steel tempered to $\sigma_u \approx 1550$ MPa, and the same steel at $\sigma_u \approx 950$ MPa. The specimens are 10 mm in both width and thickness. (Photo by R. A. Simonds.)

However, the correlation with tensile properties is only a general trend, as the results of impact fracture tests are special due to both the high rate of loading and the presence of a notch.

Many materials exhibit marked changes in impact energy with temperature. For example, for plain carbon steels of various carbon contents, Charpy energy is plotted versus temperature in Fig. 5.37. However, even for the same carbon content, and for heat treatment to the same hardness (ultimate strength), there are still differences in the

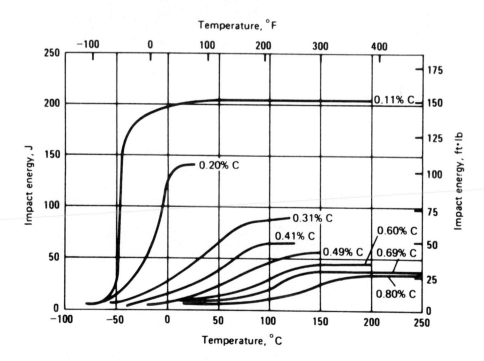

Figure 5.37 Variation in Charpy V-notch impact energy with temperature for normalized plain carbon steels of various carbon contents. (From [Boyer 85] p. 4.85; used with permission.)

impact behavior of steels due to the influence of different percentages of minor alloying elements. This behavior is illustrated by Fig. 5.38.

In Figs. 5.37 and 5.38, there tends to be a region of temperatures over which the impact energy increases rapidly from a lower level that may be relatively constant to an upper level that may also be relatively constant. Such a *temperature-transition* behavior is common in various materials. The fracture surfaces for low-energy (brittle) impact failures are generally relatively smooth, and in metals have a crystalline appearance. But those for high energy (ductile) fractures have regions of shear where the fracture surface is inclined about 45° to the tensile stress, and they have in general a rougher, more highly deformed appearance, called fibrous fracture. These differences can be seen in Fig. 5.36.

The temperature-transition behavior is of some engineering significance as it aids in comparing materials for use at various temperatures. In general, a material should not be severely loaded at temperatures where it has a low impact energy. However, some caution is needed in attaching too much significance to the exact position of the temperature transition. This is because the transition shifts even for different types of impact tests as illustrated in Fig. 5.39. The situation in a particular engineering application represents another set of circumstances and therefore another curve, the position of which cannot be determined with any degree of assurance.

Using fracture mechanics, as described later in Chapter 8, materials containing cracks and sharp notches can be analyzed in a more specific way. In particular, the *fracture toughness* can be quantitatively related to the behavior of an engineering component, and loading-rate effects can be included in the analysis. However, these advantages are achieved at the sacrifice of simplicity and economy. Notch-impact tests have thus re-

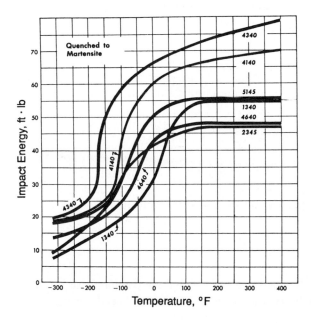

Figure 5.38 Temperature dependence of Charpy *V*-notch impact resistance for different alloy steels of the same carbon content, all quenched and tempered to HRC 34. (Adapted from [French 56]; used with permission.)

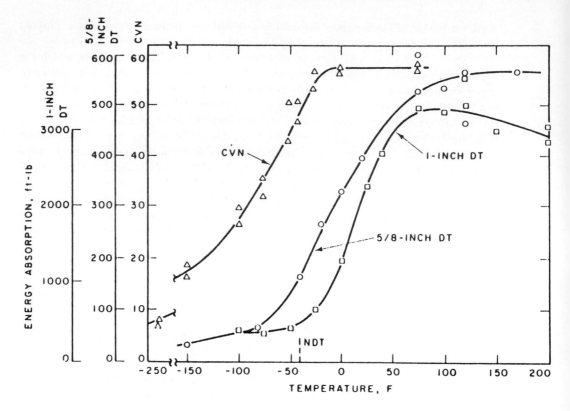

Figure 5.39 Impact test results on A517 steel with $\sigma_o = 690$ MPa for three specimen sizes: Charpy V-notch and two sizes of dynamic tear (DT) specimen having thicknesses of 16 and 25 mm. (From [Barsom 87] p. 111; ©1987 by Prentice Hall, Englewood Cliffs, N.J.; reprinted with permission.)

mained popular despite their shortcomings, as they serve a useful purpose in quickly comparing materials and obtaining general information on their behavior.

5.9 BENDING AND TORSION TESTS

Various bending and torsion tests are widely used for evaluating the elastic modulus, strength, shear modulus, shear strength, and other properties of materials. These differ from tension and compression tests in that the stresses and strains are not uniform in the material being tested.

5.9.1 Bending (Flexure) Tests

Bending tests on smooth (unnotched) bars of material are commonly used, as in various ASTM standard test methods for flat metal spring material, and for concrete, natural building stone, wood, plastics, and glass. Bending tests, also called *flexure tests*, are especially needed to evaluate the tension strength of brittle materials, as such materials are difficult to test in simple uniaxial tension due to cracking in the grips. (Think of trying to grab a flat piece of glass in the jaw-like grips usually used for testing flat pieces of metal.) The specimens usually have rectangular cross sections and may be loaded in either three-point bending or four-point bending as illustrated in Fig. 5.40.

In bending, note that the stress varies through the depth of the beam in such a way that yielding first occurs in a thin surface layer. This results in the load versus deflection curve not being sensitive to the very beginning of yielding. Also, if the stress-strain curve is not linear, as after yielding, the simple elastic bending analysis is not valid. Hence, bending tests are most meaningful for brittle materials that have approximately linear stress-strain behavior up to the point of fracture.

For materials that do have approximately linear behavior, the fracture stress may be estimated from the failure load in the bending test using simple linear elastic beam analysis.

$$\sigma = \frac{Mc}{I} \qquad (5.45)$$

where

M = bending moment
c = half-depth of beam; depth = $2c$
I = moment of inertia of the cross-sectional area about the neutral axis
$I = 2tc^3/3$ for a rectangular cross section of thickness t

Consider three-point bending, due to a force P at midspan, of a beam of length L between supports. In this case, the highest bending moment occurs at midspan and is $M = PL/4$. Equation 5.45 then gives

$$\sigma_f = \frac{3L}{8tc^2} P_f \qquad (5.46)$$

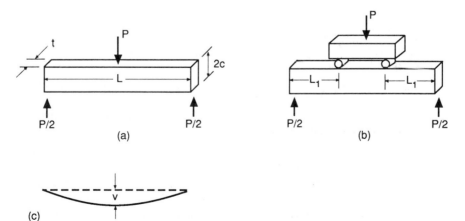

Figure 5.40   Loading configuration for (a) three-point bending and (b) four-point bending. The deflection of the centerline of either beam is similar to (c).

where P_f is the fracture load in the bending test and σ_f is the calculated fracture stress. This is usually identified as the *bend strength* or the *flexural strength*, with the quaint term *modulus of rupture in bending* also being used. Values for some ceramics are given in Table 5.8.

Such values of σ_f should always be identified as being from a bending test. This is because they may not agree precisely with values from tension tests, primarily because of departure of the stress-strain curve from linearity. Note that brittle materials are usually stronger in compression than in tension, so that the maximum tension stress is the cause of the failure in the beam. Corrections for nonlinearity in the stress-strain curve could be made based on methods presented later in Chapter 13, but this is virtually never done.

Yield strengths in bending are also sometimes evaluated. Equation 5.46 is used, but with P_f replaced by a load P_i, corresponding to a strain offset or other means of identifying the beginning of yielding. Such σ_o values are less likely than σ_f to be affected by nonlinear stress-strain behavior, but agreement with values from tension tests is affected by the above-noted insensitivity of the test to the beginning of yielding.

The elastic modulus may also be obtained from a bending test. For example, for three-point bending, and using linear-elastic analysis, the maximum deflection occurs at mid-span.

$$v = \frac{PL^3}{48EI} \tag{5.47}$$

The value of E may then be calculated from the slope dP/dv of the initial linear portion of the load versus deflection curve.

$$E = \frac{L^3}{48I}\left(\frac{dP}{dv}\right) = \frac{L^3}{32tc^3}\left(\frac{dP}{dv}\right) \tag{5.48}$$

Elastic moduli derived from bending are generally reasonably close to those from tension or compression tests of the same material, but several possible causes of discrepancy exist: (1) Local elastic or plastic deformations at the supports and/or points of load application may not be small compared to the beam deflection. (2) In relatively short beams, significant deformations due to shear stress may occur that are not considered by the ideal beam theory used. (3) The material may have differing elastic moduli in tension and compression, so that an intermediate value is obtained from the bending test. Hence, values of E from bending need to be identified as such.

For four-point bending, or for other modes of loading or shapes of cross section, Eqs. 5.46 to 5.48 need to be replaced by the analogous relationships that apply.

5.9.2 Heat-Deflection Test

In this test used for polymers, small beams having rectangular cross sections are loaded in three-point bending using a special apparatus described in ASTM Standard No. D648. Beams $2c = 13$ mm deep, and $t = 3$ to 13 mm thick, are loaded over a span of $L = 100$ mm. A force is applied such that the maximum bending stress, calculated assuming elastic behavior according to Eq. 5.45, is either 0.455 MPa or 1.82 MPa. The temperature is then increased at a rate of 2°C per minute until the deflection of the beam exceeds 0.25 mm, at which point the temperature is noted. This *heat-deflection temperature* is used as an index to compare the resistance of polymers to excessive softening and deformation as a result of heat. It also gives an indication of the temperature range where the material loses its usefulness. Some values are given in Table 5.3.

5.9.3 Torsion Test

Tests of round bars loaded in simple torsion are relatively easy to conduct, and unlike tension tests they are not complicated by the necking phenomenon. The angle of twist θ, which is proportional to shear strain, is generally increased at a constant rate. Torque T, which is of course related to shear stress, is measured as the test proceeds. Within the initial linear-elastic behavior portion of the test, the shear modulus G is proportional to the slope $dT/d\theta$ and so can be evaluated.

Fracture strength values are subject to a similar situation as in bending tests, namely that nonlinear stress-strain behavior may result in stresses calculated based on linear-elastic behavior not being accurate. This limitation can be overcome by testing thin-walled tubes in torsion, for which the error is always small. However, even for solid bars in torsion, the test results may be useful for comparing the strength and ductility of various materials.

In a torsion test, brittle materials fail on planes of maximum tension, which occur at 45° to both the specimen axis and the specimen surface. This produces a helical spiral fracture as shown in Fig. 5.41. In contrast, ductile materials generally fail on planes of

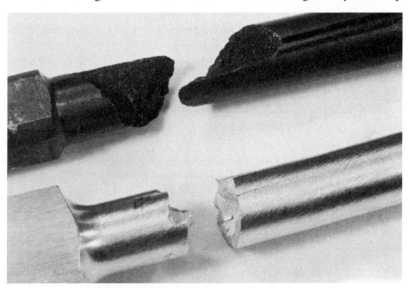

Figure 5.41   Typical torsion failures showing brittle behavior (above) in gray cast iron, and ductile behavior (below) in aluminum alloy 2024-T351. (Photo by R. A. Simonds.)

maximum shear, that is, on planes transverse and longitudinal to the specimen axis, as for the aluminum alloy also shown in Fig. 5.41. (You may wish to refresh your memory on the various stress directions in torsion by looking at Fig. 4.5.)

5.10 SUMMARY

A variety of relatively simple mechanical tests are used to evaluate materials properties. The results are used in engineering design and as a basis for comparing and selecting materials. These include tests involving tension, compression, indentation, impact, bending, and torsion.

Tension tests are frequently used to evaluate stiffness, strength, ductility, and other characteristics of materials as summarized in Table 5.6. One property of special interest is the elastic modulus E, a measure of stiffness and a fundamental elastic constant used in Hooke's Law, Eq. 4.10. Poisson's ratio ν can also be obtained if transverse strains are measured. The yield strength σ_o characterizes resistance to the beginning of plastic deformation, and the ultimate strength σ_u is the highest engineering stress the material can withstand.

Ductility is the ability to resist deformation without fracture. In a tension test, this is characterized by the percent elongation at fracture, $100\varepsilon_f$, and the percent reduction in area. Also, detailed analysis of test results can be done using true stresses and strains, which consider the finite changes in gage length and cross-sectional area that may occur. Additional properties can then be obtained, notably the strain hardening exponent n and the true fracture stress and strain, $\tilde{\sigma}_{fB}$ and $\tilde{\varepsilon}_f$.

If a material is tested under conditions where significant creep strain occurs during the tension test, then the results are sensitive to strain rate. At a given temperature, increasing the strain rate usually increases the strength but decreases the ductility. Qualitatively similar effects occur for a given strain rate if the temperature is decreased. These effects are prevalent for most engineering metals only at elevated temperature, but they occur in many polymers at room temperature.

Compression tests can be used to measure similar properties as tension tests. These tests are especially significant for materials used primarily in compression, such as concrete and building stone, and for other materials that behave in a brittle manner in tension, such as ceramics and glass. In these materials, the strength and ductility are generally greater in compression than in tension, sometimes dramatically so.

Hardness in engineering is usually measured using one of several standard tests that measure resistance to indentation by a ball or sharp point. The Brinell test uses a 10 mm ball, and the Vickers test a pyramidal point. Both evaluate hardness as the average stress in kg/mm^2 on the surface area of the indentation. The values obtained are similar, but the Vickers test is useful for a wider range of materials. Rockwell hardness is based on the depth of indentation by a ball or conical diamond point, with there being several hardness scales to accommodate various materials.

Notch-impact tests evaluate the ability of a material to resist rapid loading where a sharp notch is present. The impact load is applied by a swinging pendulum or a falling weight. Details of specimen size and shape and the manner of loading differ for various standard tests, which include the Charpy, Izod, and dynamic tear tests.

Impact energies often exhibit a temperature transition, below which the behavior is brittle. Thus, impact energy versus temperature curves are useful in comparing the behavior of different materials. However, too much significance should not be attached to the exact position of the temperature transition, as this is sensitive to the details of the test and will not in general correspond to the engineering situation of interest.

Bending tests on unnotched bars are useful for evaluating the elastic modulus and strength of brittle materials. Strengths may differ from those in tension tests, especially if significant nonlinear stress-strain behavior occurs prior to fracture, this arising from the behavior being assumed in analysis to be linear. A special bending test called the heat-deflection test is used to identify the limits of usefulness of polymers with respect to temperature. Torsion tests permit direct evaluation of the shear modulus, G, and also can be used to determine strength and ductility in shear.

NEW TERMS AND SYMBOLS

bending (flexure) test	proportional limit, σ_p
Brinell hardness test	resilience, u_r
Charpy V-notch test	Rockwell hardness test
compression test	strain hardening exponent, n
corrected true stress, $\tilde{\sigma}_B$	strength coefficient, H
elastic limit	tangent modulus, E_t
engineering strain, ε	tensile toughness, u_f
engineering stress, σ	torsion test
heat-deflection temperature	true fracture strain, $\tilde{\varepsilon}_f$
indentation hardness	true fracture strength, $\tilde{\sigma}_f$, $\tilde{\sigma}_{fB}$
Izod test	true strain, $\tilde{\varepsilon}$
necking	true stress, $\tilde{\sigma}$
notch-impact test	true toughness, $\tilde{u}_f$
offset yield strength, σ_o	ultimate tensile strength, σ_u
percent elongation, $100\varepsilon_f$	Vickers hardness test
percent reduction in area, $\%RA$	yielding

REFERENCES

(a) General References

ASTM. 1990 *Annual Book of ASTM Standards*, Am. Soc. for Testing and Materials, Philadelphia, Pa. (Multiple volume set published annually.)

BOYER, H. E. and T. L. GALL, eds. 1985 *Metals Handbook: Desk Edition*, American Society for Metals, Metals Park, Oh.

DIETER, G. E., Jr. 1976 *Mechanical Metallurgy*, 2nd ed., McGraw-Hill, New York.

MCCLINTOCK, F. A. and A. S. ARGON, eds. 1966 *Mechanical Behavior of Materials*, Addison-Wesley, Reading, Ma.

NADAI, A. 1950 *Theory of Flow and Fracture of Solids*, McGraw-Hill, New York.

RICHARDS, C. W. 1961 *Engineering Materials Science*, Wadsworth, Belmont, Ca.

(b) Sources of Materials Properties

ASM. 1987 *Engineered Materials Handbook, Vol. 1: Composites*, ASM International, Metals Park, Oh.

ASM. 1988 *Engineered Materials Handbook, Vol. 2: Engineering Plastics*, ASM International, Metals Park, Oh.

BATTELLE. 1991 *Aerospace Structural Metals Handbook*, 5 vols., Metals and Ceramics Information Center, Battelle Columbus Div., Columbus, Oh.

JURAN, R., et al., eds. 1988 *Modern Plastics Encyclopedia*, Published annually by McGraw-Hill, Inc., New York.

MILHDBK. 1983 *Military Standardization Handbook: Metallic Materials and Elements for Aerospace Vehicle Structures*, MIL-HDBK-5D, 2 Vols., U.S. Dept. of Defense and Federal Aviation Administration, Naval Publications and Forms Ctr., Philadelphia, Pa.

MOREY, G. W. 1954 *The Properties of Glass*, 2nd ed., Reinhold, New York.

PROBLEMS AND QUESTIONS

Section 5.3

5.1 Define the following concepts in your own words: (a) stiffness, (b) strength, (c) ductility, (d) yielding, (e) toughness, and (f) strain hardening.

5.2 Define the following adjectives that might be used to describe the behavior of a material: (a) brittle, (b) ductile, (c) resilient, (d) tough, (e) stiff, and (f) strong.

5.3 The offset yield stress and the proportional limit stress are both used to characterize the beginning of nonlinear behavior in a tension test. Why is the offset method generally preferable? Can you think of any disadvantages of the offset method?

5.4 Using the tensile stress-strain curve for gray cast iron in Fig. 5.8, obtain approximate values for the tangent modulus E_t, the 0.2% offset yield strength σ_o, the ultimate strength σ_u, and the percent elongation, $100\varepsilon_f$. How do these differ from the corresponding values for ductile cast iron in Table 5.2? Based on Section 3.3, explain why these two cast irons have contrasting tensile behavior. (Suggestion: Course instructors may wish to make and distribute enlarged copies of Fig. 5.8.)

5.5 Using the tensile stress-strain curves of Fig. 5.16, obtain approximate values for each steel of the 0.2% offset yield strength σ_o, the ultimate strength σ_u, and the elongation, $100\varepsilon_f$. Do your values agree reasonably well with those in Table 5.2 where data are given for similar materials? (Suggestion: Course instructors may wish to make and distribute enlarged copies of Fig. 5.16.)

5.6 Proceed as in Prob. 5.5 except do so for the three aluminum alloys of Fig. 5.17. Also, add the elastic modulus E to the values to be obtained.

5.7 In a tension test on hot-rolled AISI 1095 steel, the original diameter was 9.07 mm, and the original gage length for longitudinal strain was 50 mm. The highest load observed was 66.7

kN, and at this point the gage length had increased to 54.2 mm. After fracture into two pieces, which occurred at a load of 63.4 kN, the broken halves were fitted back together, and the diameter was measured to be 8.18 mm. Also, the gage length had increased to 55.6 mm. Calculate or estimate the following: (a) engineering ultimate strength, (b) percent elongation at fracture, and (c) percent reduction in area at fracture.

5.8 For the initial portion of a tension test on 1100-O aluminum, a plot is shown below of change in length of the gage section versus load. Before the test, the gage length was 50 mm and the diameter was 9.07 mm. During the large strain portion of the test, additional data as listed in the table that follows were obtained. Diameters were periodically measured, and once necking started, the particular diameter measured was the minimum in the neck.

Use these data to make an accurate plot of engineering stress versus strain. (The result should be similar to the curve in Fig. 5.17.) Also, obtain values for the elastic modulus E and all of the engineering properties that characterize strength and ductility as listed in Table 5.6.

Load P, kN	Length change ΔL, mm	Diameter d, mm	Comment
9.12	1.81	8.89	Highest load
8.93	4.11	8.38	
7.94	5.93	7.11	
6.18	7.36	5.59	
4.00	8.50	4.06	Fracture

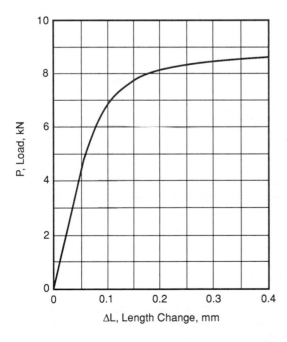

Figure P5.8

Section 5.4

5.9 Based on Tables 3.10, 5.2, and 5.3, write a paragraph that discusses in general terms the differences among engineering metals, polymers, ceramics, and silica glasses as to their tensile strength, ductility, and stiffness.

5.10 Using the data for various steels in Tables 5.2 and 14.1, plot ultimate tensile strength versus percent reduction in area as a data point for each. Use linear coordinates. Also, use two different plotting symbols, one for the steels that are not strengthened by heat treating, which are 1020, 1015, Man-Ten, and 1045 (HR), and one for the remaining which are so strengthened. Then comment on any general trends that are apparent and any exceptions to these trends.

5.11 Based on the data for SiC reinforced aluminum in Table 5.4, write a paragraph discussing the effects of various types and orientations of reinforcement on the stiffness, strength, and ductility. Also, estimate the tensile toughness u_f for each case and include this in your discussion.

Section 5.5

5.12 Explain in your own words, without using equations, the difference between engineering stress and true stress and the difference between engineering strain and true strain.

5.13 Using Figs. 5.16 and 5.17, estimate values of the strain hardening exponent n where this can be done based on the true strain at necking. Where values are given in Table 5.7, compare with your results and comment on the comparison.

5.14 Several values of the strength coefficient H are missing from Table 5.7. Estimate these values.

5.15 A material obeys the relationship of Eq. 5.28 between true stress and true strain in a tension test. Derive an expression for the engineering ultimate strength σ_u in terms of the constants H and n.

5.16 At relatively small strains around the offset yield strength, Eq. 5.28 often applies, but only to the plastic part of the strain. Also, note that true and engineering stresses and strains are numerically similar at such small strains, so that

$$\sigma = H\varepsilon_p^n$$

Based on this, obtain:
(a) An equation for the 0.2% offset yield stress σ_o in terms of H and n.
(b) An equation for the total strain, $\varepsilon = \varepsilon_e + \varepsilon_p$, as a function of σ and material constants.

5.17 For the test described in Prob. 5.7, also determine or estimate the true fracture strain $\tilde{\varepsilon}_f$, the true fracture strength $\tilde{\sigma}_{fB}$, and the constants H and n for Eq. 5.29.

5.18 Using the data given for 1100-O Al in Prob. 5.8, compute true stresses and strains for both the initial and later portions of the test. Then plot true stress versus true strain on linear coordinates. Also, on the same graph, plot the engineering stress-strain curve from your solution to Prob. 5.8 or from Fig. 5.17. Comment on the comparison between the two curves. (PC Problem)

5.19 Using the values of true stresses and strains from your solution to Prob. 5.18, fit a line on log-log coordinates to obtain values of H and n for Eq. 5.29. (Comment: The Bridgman correction factor in the form of Fig. 5.21 was developed for steels and may be inaccurate for this material at the largest strains involved. Hence, do not use the last two data points if omitting them improves the fit.)

5.20 Data are given below from a tension test on Man-Ten steel. Use these to determine the true fracture strain $\tilde{\varepsilon}_f$, true fracture strength $\tilde{\sigma}_{fB}$, strength coefficient H, and strain hardening exponent n. Obtain n both from a straight-line fit on log-log coordinates and approximately from the true strain at the engineering ultimate stress. (PC Problem)

Engineering stress σ, MPa	Engineering strain ε	Diameter d, mm	Comment
0	0	6.32	Before testing
359	0.0017	——	Proportional limit and upper yield point
317	0.0035	— -	0.2% yield
357	0.010	—	
458	0.030	—	
541	0.079	5.99	
576	—	5.72	
558	—	5.33	
531	—	5.08	
476	—	4.45	
379	—	3.50	Fracture

5.21 Based on the data in Tables 5.2 and 5.7 for 2024-T4 aluminum, draw the entire engineering stress-strain curve up to the point of fracture on linear graph paper. Accurately plot the initial elastic slope and the points corresponding to yield, ultimate, and fracture, and approximately sketch the remainder of the curve. How does your result compare to the curve for similar material in Fig. 5.17?

5.22 Proceed as in Prob. 5.21 except draw the true stress-strain curve, and also calculate several additional $(\tilde{\sigma}, \tilde{\varepsilon})$ points along the curve to aid in plotting.

Section 5.6

5.23 Why does gray cast iron fracture normal to the specimen axis in tension and at about 45° to the specimen axis in compression? What type of fractures would you expect in a brittle polymer in tension and compression?

5.24 How would you expect the stress-strain curves for concrete to differ between tension and compression? Give physical reasons for the expected differences.

Section 5.7

5.25 Explain the differences among standard hardness tests based on the following different principles: scratch resistance, rebound, indentation size, and indentation depth. Also, give one example of each.

5.26 Explain why the Brinell and Vickers hardness tests given generally similar results as in Fig. 5.32.

5.27 Using the hardness conversion chart of Table 5.10, plot both Rockwell B and C hardness versus ultimate tensile strength for steel. Comment on the trends observed. Is the relationship approximately linear as for Brinell hardness?

5.28 Using the hardness conversion chart for steels of Table 5.10, plot Vickers hardness versus Brinell hardness. Also plot Rockwell C hardness versus Brinell hardness on the same graph. How do the two relationships differ? Can you think of any reasons why the observed trends occur?

Section 5.8

5.29 Explain in your own words why notch-impact fracture tests are widely used for metals, and also why caution is needed in applying the results to real engineering situations.

5.30 For the polymers listed in Table 5.3, plot data points on linear coordinates of Izod energy versus elongation. Describe and briefly discuss any general trends that are apparent. Also, perform a similar graphical study and discussion of the relationship between Izod energy and ultimate tensile strength.

Section 5.9

5.31 For both three-point bending and four-point bending, as illustrated in Fig. 5.40, draw shear and moment diagrams. Express these in terms of the force P and the various lengths involved. Then use these diagrams to discuss the differences between the two types of test. Can you think of any relative advantages and disadvantages of the two types?

5.32 Equations 5.46 and 5.48 give values of fracture strength and elastic modulus from bending tests, but they apply only to the case of three-point bending. Derive analogous equations for the case of four-point bending as illustrated in Fig. 5.40(b).

5.33 A solid circular shaft of length L and diameter d is to be used in materials testing in torsion.
   (a) Develop equations for calculating the shear fracture stress τ_f and shear modulus G from torque T and angle of twist θ.
   (b) Modify these equations for specimens that are hollow tubes of inner diameter d_1 and outer diameter d_2. Can you think of any advantages or disadvantages of a tubular versus a solid torsion specimen?

Section 5.10

5.34 What characteristics are needed for material for steel ball bearings? What materials tests would be important in judging the suitability of a given steel for this use? Also answer the same questions for the steel race in which the bearings roll.

5.35 You are an engineer designing pressure vessels to hold liquid nitrogen. What general characteristics should the material used have? Of the various types of materials tests described in this chapter, which would you employ to aid in selecting among candidate materials? Explain the reason you need each type of test chosen.

5.36 Answer the questions of Prob. 5.35 where you are instead designing plastic motorcycle helmets.

6

Review of Complex and Principal States of Stress and Strain

6.1 INTRODUCTION

Components of machines, vehicles, and structures are subjected to applied loadings that may include tension, compression, bending, torsion, pressure, or combinations of these. As a result, complex states of normal and shear stress occur that vary in magnitude and direction with location in the component. The designer must assure that the material of the component does not fail as a result of these stresses. To accomplish this, locations where the stresses are the most severe must be identified. Further analysis of the stresses at these locations is then needed.

At any point in a component where the stresses are of interest, it is first necessary to note that the magnitudes of the stresses vary with direction and are highest in certain directions. The highest stresses at a given location are called the *principal stresses*, and the particular directions in which these act are called the *principal axes*. Both the highest normal stresses and the highest shear stresses are of interest. For example, failure in a brittle manner may occur in response to the highest tensile normal stress. However, yielding is caused by shear stress, and so the highest value of this quantity is needed to determine whether or not yielding is likely at a given location in a component. (An alternative approach is to use the value of a special quantity called the *octahedral shear stress*.)

This chapter will review the determination of principal stresses, noting that the topic has already been studied by most readers in standard courses on elementary mechanics of materials. Situations where the stresses acting on one orthogonal plane are zero, called *plane stress*, are considered first. Plane stress is of practical interest as it occurs at any free (unloaded) surface in a component, and surface locations often have the most severe stresses, as in bending of beams and torsion of circular shafts.

Three-dimensional states of stress and strain are also considered, but not in more detail than is needed to prepare the student for later chapters. Fully detailed treatments of three-dimensional cases are given in several books listed at the end of the chapter, and also in numerous other books on advanced mechanics of materials, theory of elasticity, continuum mechanics, and related topics.

The review of principal stresses and directions in this chapter leads into the next chapter, where yielding and fracture of materials due to complex states of stress is considered. The state of stress and strain also affects other modes of material failure, such as brittle fracture from a crack, fatigue due to cyclic loading, and creep under loads maintained for long periods of time. Thus, the later chapters that consider these topics also employ information presented in this chapter.

6.2 PLANE STRESS

Consider any given point in a solid body, such as an engineering component, and assume that an x-y-z coordinate system has been chosen for this point. The material at this point is in general subjected to six components of stress, σ_x, σ_y, σ_z, τ_{xy}, τ_{yz}, and τ_{zx}, as illustrated on a small element of material in Fig. 6.1. If the three components of stress acting on one face of the element are all zero, then a state of plane stress exists. Taking the unstressed plane to be parallel to the x-y plane gives

$$\sigma_z = \tau_{zx} = \tau_{yz} = 0 \tag{6.1}$$

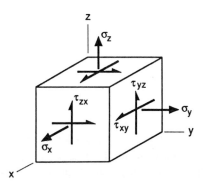

Figure 6.1   The six components needed to completely describe the state of stress at a point.

Equilibrium of forces on the element of Fig. 6.1 requires that the moments must sum to zero about both the x and y axes, requiring in turn that the components of τ_{zx} and τ_{yz} acting on the other two planes must also be zero.

Hence, the components remaining are σ_x, σ_y, and τ_{xy}, as illustrated on a square element of material in Fig. 6.2(a). Note that the square element is simply the cubic element viewed parallel to the z-axis. Positive directions are as shown, the sign convention being as follows: (1) Tensile normal stresses are positive. (2) Shear stresses are positive if the arrows on the positive facing sides of the element are in the directions of the positive x-y coordinate axes.

6.2.1 Rotation of Coordinate Axes

The same state of plane stress may be described on any other coordinate system, such as x'-y' in Fig. 6.2(b). This system is related to the original one by an angle of rotation θ, and the values of the stress components change to σ_x', σ_y', and τ_{xy}' in the new coordinate system. However, it is important to recognize that the new quantities do not represent a new state of stress, but rather an equivalent representation of the original one.

The values of the stress components in the new coordinate system may be obtained by considering the freebody diagram of a portion of the element as indicated by the dashed line in Fig. 6.2(a). The resulting freebody is shown in Fig. 6.3. Equilibrium of forces in both the x- and y-directions provides two equations, which are sufficient to evaluate the unknown normal and shear stress components σ and τ on the inclined plane. The stresses must first be multiplied by the unequal areas of the sides of the triangular element to

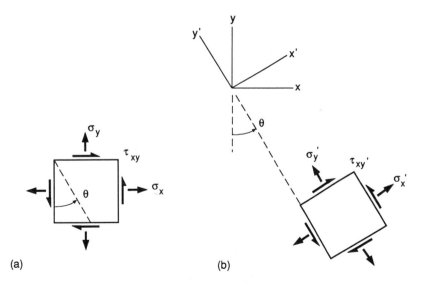

Figure 6.2   The three components needed to describe a state of plane stress (a), and an equivalent representation of the same state of stress for a rotated coordinate system (b).

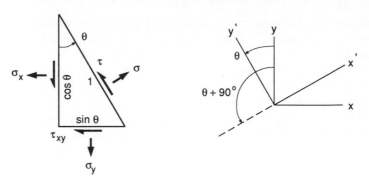

Figure 6.3   Stresses on an oblique plane.

obtain forces. For convenience, the hyopotenuse is taken to be of unit length, as is the thickness of the element normal to the diagram.

Summing forces in the x-direction, and then in the y-direction, gives two equations.

$$\sigma \cos\theta - \tau \sin\theta - \sigma_x \cos\theta - \tau_{xy} \sin\theta = 0 \tag{6.2}$$

$$\sigma \sin\theta + \tau \cos\theta - \sigma_y \sin\theta - \tau_{xy} \cos\theta = 0 \tag{6.3}$$

Solving for the unknowns σ and τ, and also invoking some basic trigonometric indentities, yields

$$\sigma = \frac{\sigma_x + \sigma_y}{2} + \frac{\sigma_x - \sigma_y}{2}\cos 2\theta + \tau_{xy} \sin 2\theta \tag{6.4}$$

$$\tau = -\frac{\sigma_x - \sigma_y}{2}\sin 2\theta + \tau_{xy} \cos 2\theta \tag{6.5}$$

The desired complete state of stress in the new coordinate system may now be obtained. Equations 6.4 and 6.5 give σ_x' and τ_{xy}' directly, and substitution of $\theta + 90°$ gives σ_y'.

The same sign convention for stresses of course applies for the x'-y' system. Also, note that θ is positive in the counterclockwise (CCW) direction, as this was the direction taken as positive in developing these equations. The process of determining the equivalent representation of a state of stress on a new coordinate system is called *transformation of axes*, so that the preceding equations are called the *transformation equations*.

6.2.2 Principal Stresses

The equations just developed give the variation of σ and τ with direction in the material, the direction being specified by the angle θ relative to the originally chosen x-y coordinate system. Maximum and minimum values of σ and τ are of special interest and can be obtained by analyzing the variation with θ.

Taking the derivative $d\sigma/d\theta$ of Eq. 6.4 and equating the result to zero gives the coordinate axes rotations for the maximum and minimum values of σ.

$$\tan 2\theta_n = \frac{2\tau_{xy}}{\sigma_x - \sigma_y} \tag{6.6}$$

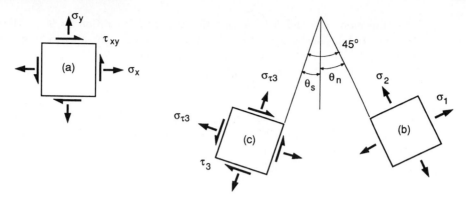

Figure 6.4 A state of plane stress (a), and the special coordinate systems that contain the principal normal stresses (b), and the principal shear stress (c).

Two angles θ_n separated by 90° satisfying this relationship. The corresponding maximum and minimum normal stresses from Eq. 6.4, called the *principal normal stresses*, are

$$\sigma_1, \sigma_2 = \frac{\sigma_x + \sigma_y}{2} \pm \sqrt{\left(\frac{\sigma_x - \sigma_y}{2}\right)^2 + \tau_{xy}^2} \tag{6.7}$$

Also, the shear stress at the θ_n orientation is found to be zero. The resulting equivalent representation of the original state of stress is illustrated by Fig. 6.4(b).

As noted, the shear stress is zero on the planes where the principal normal stresses occur. The converse is also true: If the shear stress is zero, then the normal stresses are the principal normal stresses.

Similarly, Eq. 6.5 and $d\tau/d\theta = 0$ gives the coordinate axes rotation for the maximum shear stress.

$$\tan 2\theta_s = -\frac{\sigma_x - \sigma_y}{2\tau_{xy}} \tag{6.8}$$

The corresponding shear stress from Eq. 6.5 is

$$\tau_3 = \sqrt{\left(\frac{\sigma_x - \sigma_y}{2}\right)^2 + \tau_{xy}^2} \tag{6.9}$$

This is the maximum shear stress in the x-y plane and is called the *principal shear stress*. Also, the two orthogonal planes where this shear stress occurs are found to have the same normal stress of

$$\sigma_{\tau 3} = \frac{\sigma_x + \sigma_y}{2} \tag{6.10}$$

where the special subscript indicates that this is the normal stress that accompanies τ_3. This second equivalent representation of the original state of stress is illustrated by Fig. 6.4(c).

Equations 6.6 and 6.8 indicate that $2\theta_n$ and $2\theta_s$ differ by 90°. Hence, if both $2\theta_n$ and $2\theta_s$ are considered to be limited to the range ±90°, then one of these must be negative, that is, clockwise (CW), and we can write

$$|\theta_n - \theta_s| = 45° \tag{6.11}$$

In addition, τ_3 and the accompanying normal stress can be expressed in terms of the principal normal stresses σ_1 and σ_2 by substituting Eqs. 6.9 and 6.10 into the two relations represented by Eq. 6.7.

$$\tau_3 = \frac{|\sigma_1 - \sigma_2|}{2}, \qquad \sigma_{\tau 3} = \frac{\sigma_1 + \sigma_2}{2} \tag{6.12}$$

The absolute value is necessary for τ_3 due to the two roots of Eq. 6.9

Example 6.1

At a point of interest on the free surface of an engineering component, the stresses with respect to a convenient coordinate system in the plane of the surface are $\sigma_x = 95$, $\sigma_y = 25$, and $\tau_{xy} = 20$ MPa. Determine the principal stresses and the orientations of the principal planes.

Solution   Substitution of the given values into Eq. 6.6 gives the angle to the coordinate axes for the principal normal stresses.

$$\tan 2\theta_n = \frac{2\tau_{xy}}{\sigma_x - \sigma_y} = \frac{4}{7}$$

$$\theta_n = 14.9° \quad \text{(CCW)} \qquad \qquad \textbf{Ans.}$$

Substitution into Eq. 6.7 gives the principal normal stresses.

$$\sigma_1, \sigma_2 = \frac{\sigma_x + \sigma_y}{2} \pm \sqrt{\left(\frac{\sigma_x - \sigma_y}{2}\right)^2 + \tau_{xy}^2}$$

$$\sigma_1, \sigma_2 = 60 \pm 40.3 = 100.3, 19.7 \text{ MPa} \qquad \qquad \textbf{Ans.}$$

The corresponding planes and state of stress are shown in Fig. E6.1. Note that the direction for the larger of the two principal normal stresses is chosen so that it is more nearly aligned with the larger of the original σ_x and σ_y than with the smaller.

Alternatively, a more rigorous procedure is to use $\theta = \theta_n = 14.9°$ in Eq. 6.4, which gives

$$\sigma = \sigma_x' = \sigma_1 = 100.3 \text{ MPa}$$

Use of $\theta = \theta_n + 90° = 104.9°$ in Eq. 6.4 then gives the normal stress in the other orthogonal direction.

$$\sigma = \sigma_y' = \sigma_2 = 19.7 \text{ MPa}$$

The zero value of τ at $\theta = \theta_n$ can also be verified by using Eq. 6.5.

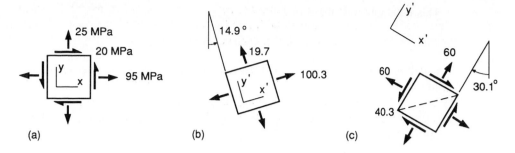

Figure E6.1   Example state of stress (a) and its equivalent representations that contain the principal normal stress (b) and the principal shear stress(c).

For the equivalent representation where the maximum shear stress in the x-y plane occurs, Eq. 6.8 gives

$$\tan 2\theta_s = -\frac{\sigma_x - \sigma_y}{2\tau_{xy}} = -\frac{7}{4}$$

$$\theta_s = -30.1° = 30.1°   \quad \text{(CW)} \qquad \text{\textbf{Ans.}}$$

The stresses for this rotation of the coordinate system may be obtained from σ_1 and σ_2 as calculated above and Eq. 6.12.

$$\tau_3 = \frac{|\sigma_1 - \sigma_2|}{2} = \pm 40.3 \text{ MPa} \qquad \text{\textbf{Ans.}}$$

$$\sigma_{\tau 3} = \frac{\sigma_1 + \sigma_2}{2} = 60 \text{ MPa}$$

This representation of the state of stress is shown as Fig. E6.1(c). The uncertainty as to the sign of the shear stress can be resolved by noting that the positive shear diagonal (dashed line) must be aligned with the larger of σ_1 and σ_2. Alternatively, a more rigorous procedure is to use $\theta = \theta_s = -30.1°$ in Eq. 6.5, which gives

$$\tau_3 = 40.3 \text{ MPa}$$

The positive sign indicates that the shear stress is positive in the new x'-y' coordinate system.

6.2.3 Mohr's Circle

A convenient graphical representation of the transformation equations for plane stress was developed by Otto Mohr in the 1880s. On σ versus τ coordinates, these equations can be shown to represent a circle, called *Mohr's circle*, which is developed as follows:

First, isolate all terms containing θ on one side of Eq. 6.4. Then square both sides of Eqs. 6.4 and 6.5, sum the result, and invoke simple trigonometric identities to eliminate θ, obtaining

$$\left(\sigma - \frac{\sigma_x + \sigma_y}{2}\right)^2 + \tau^2 = \left(\frac{\sigma_x - \sigma_y}{2}\right)^2 + \tau_{xy}^2 \qquad (6.13)$$

This equation is of the form:

$$(\sigma - a)^2 + (\tau - b)^2 = r^2 \tag{6.14}$$

which is the equation of a circle on a plot of σ versus τ with center at coordinates (a, b) and radius r, where

$$a = \frac{\sigma_x + \sigma_y}{2}, \quad b = 0, \quad r = \sqrt{\left(\frac{\sigma_x - \sigma_y}{2}\right)^2 + \tau_{xy}^2} \tag{6.15}$$

Comparison with Eqs. 6.9 and 6.10 reveals that

$$a = \sigma_{\tau 3}, \quad r = \tau_3 \tag{6.16}$$

Mohr's circle is illustrated in Fig. 6.5. It is evident that the radius τ_3 is indeed the maximum shear stress in the x-y plane. Also, the maximum and minimum normal stresses occur along the σ-axis.

$$\sigma_1, \sigma_2 = a \pm r = \frac{\sigma_x + \sigma_y}{2} \pm \tau_3 \tag{6.17}$$

Noting Eq. 6.9, this is seen to be equivalent to Eq. 6.7.

In using Mohr's circle, confusion concerning the signs of shear stresses can be avoided by adopting the convention shown in Fig. 6.6. The complete state of shear stress is considered to be split into two portions as shown. The portion that causes clockwise rotation is considered positive, and the portion that causes counterclockwise rotation is considered negative. For normal stresses, tension is positive, and compression negative.

If this is done, the two ends of a diameter of the circle can be used to represent the stresses on orthogonal planes in the material. This is illustrated by diameter A in Fig. 6.6. Note that the normal and shear stresses that occur together on one orthogonal

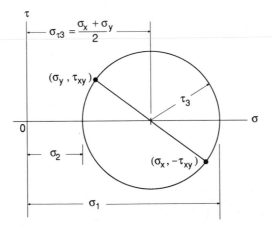

Figure 6.5 Mohr's circle and principal stresses corresponding to a given state of plane stress $(\sigma_x, \sigma_y, \tau_{xy})$.

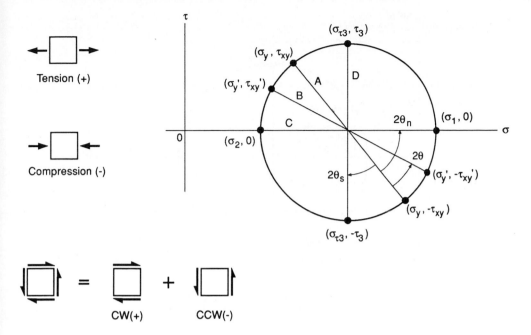

Tension (+)

Compression (-)

CW(+)   CCW(-)

Figure 6.6 Sign convention and diameters of special interest for Mohr's circle.

plane provide the coordinate point for one end of the diameter, and those for the other plane give the opposite end.

A rotation of this diameter by an angle 2θ on the circle gives the state of stress for a coordinate axis rotation of θ in the same direction in the material. This is illustrated by diameter B in Fig. 6.6, which corresponds to the situation of Fig. 6.2(b). If the diameter is rotated by an angle $2\theta_n$, until it becomes horizontal, the principal normal stresses are obtained. Also, if the diameter is rotated by an angle $2\theta_s$, until it becomes vertical, the state of stress obtained contains the principal shear stress. These special choices of coordinate axes are illustrated by diameters C and D in Fig. 6.6, and they correspond to Fig. 6.4(b) and (c), respectively.

Example 6.2

Repeat Example 6.1 using Mohr's circle. Recall that the original state of stress is $\sigma_x = 95$, $\sigma_y = 25$, and $\tau_{xy} = 20$ MPa.

Solution   The circle is obtained by plotting two points that lie at opposite ends of a diameter as shown below (Fig. E6.2).

$$(\sigma, \tau) = \left(\sigma_x, -\tau_{xy}\right) = (95, -20) \text{ MPa}$$

$$(\sigma, \tau) = \left(\sigma_y, \tau_{xy}\right) = (25, 20) \text{ MPa}$$

A negative sign is applied to τ_{xy} for the point associated with σ_x because the shear arrows on the same planes as σ_x tend to cause a counterclockwise rotation. Similarly, a

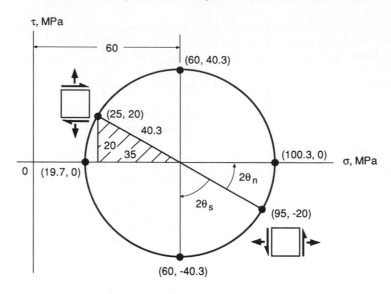

Figure E6.2   Mohr's circle corresponding to the state of stress of Fig. E6.1.

positive sign is used for τ_{xy} when associated with σ_y, due to the clockwise rotation. The center of the circle must lie on the σ axis at a point halfway between σ_x and σ_y, that is, at

$$(\sigma, \tau)_{\text{ctr.}} = \left(\frac{\sigma_x + \sigma_y}{2}, 0 \right) = (60, 0) \ \text{MPa}$$

From the above coordinate points, the shaded right triangle shown has a base of 35 and an altitude of 20 MPa. The hypotenuse is the radius of the circle and is also the principal shear stress.

$$\tau_3 = \sqrt{35^2 + 20^2} = 40.3 \ \text{MPa} \qquad \textbf{Ans.}$$

The angle with the σ-axis is

$$\tan 2\theta_n = \frac{20}{35}$$

$$2\theta_n = 29.74° \quad \text{(CCW)} \qquad \textbf{Ans.}$$

A counterclockwise rotation of the diameter of the circle by this $2\theta_n$ gives the horizontal diameter that corresponds to the principal normal stresses. Their values are obtained from the center location and radius of the circle.

$$\sigma_1 = \frac{\sigma_x + \sigma_y}{2} + \tau_3 = 60 + 40.3 = 100.3 \ \text{MPa} \qquad \textbf{Ans.}$$

$$\sigma_2 = \frac{\sigma_x + \sigma_y}{2} - \tau_3 = 60 - 40.3 = 19.7 \ \text{MPa} \qquad \textbf{Ans.}$$

The resulting state of stress is the same as previously illustrated in Fig. E6.1(b). Note that the counterclockwise rotation $2\theta_n$ on the circle corresponds to a rotation of θ_n in the same direction in the material.

The diameter corresponding to the original state of stress must be rotated clockwise to obtain the equivalent representation that contains the principal shear stress. Since this is 90° from the σ-axis, the angle of rotation is

$$2\theta_s = 90° - 2\theta_n = 60.26°   \text{(CW)} \qquad \textbf{Ans.}$$

So that $\theta_s = 30.1°$ clockwise. The coordinates of the ends of this vertical diameter give the same state of stress as previously shown in Fig. E6.1(c).

6.3 THREE-DIMENSIONAL STATES OF STRESS

In the general three-dimensional case, all six components of stress may be present: σ_x, σ_y, σ_z, τ_{xy}, τ_{yz}, and τ_{zx}. The derivations above may be redone for this more general case to obtain transformation equations that permit any choice of coordinate axes in three dimensions. This is accomplished by considering the freebody of a portion of the unit cube as shown in Fig. 6.7(a).

Assume that the original unit cube is rotated by an angle θ in the x-y plane, followed by rotation out of the x-y plane by an angle ϕ, as shown in (b). These rotations yield a new unit cube whose faces are normal to a new x'-y'-z' coordinate system as illustrated by

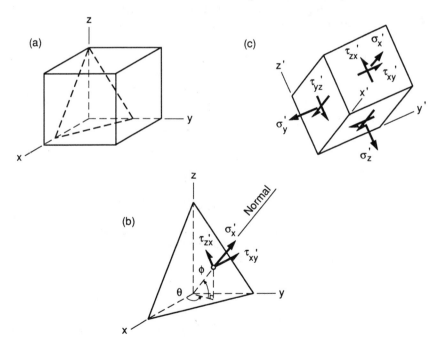

Figure 6.7   An oblique plane in a three-dimensional coordinate system (a), and the stresses acting on this plane (b). The complete state of stress for coordinate axes rotations (θ, ϕ) is shown in (c).

(c). Equilibrium of forces gives the stresses on the oblique plane, specifically the normal stress σ_x' and two shear stress components τ_{xy}' and τ_{zx}'. Similar consideration of the other two faces of the new unit cube gives the remaining stress components, so that the complete state of stress can be obtained. The resulting three-dimensional transformation equations will not be presented here, but can be found in any of the References listed at the end of this chapter.

Certain mathematical functions of the stress components remain constant for a given state of stress for all choices of coordinate system. For example, the sum of the normal stresses is constant.

$$\sigma_x + \sigma_y + \sigma_z = \sigma_x' + \sigma_y' + \sigma_z' = \text{ constant}$$

This is called the first stress invariant. There are two additional invariants, all three being

$$I_1 = \sigma_x + \sigma_y + \sigma_z$$

$$I_2 = \sigma_x \sigma_y + \sigma_y \sigma_z + \sigma_z \sigma_x - \tau_{xy}^2 - \tau_{yz}^2 - \tau_{zx}^2 \qquad (6.18)$$

$$I_3 = \sigma_x \sigma_y \sigma_z + 2\tau_{xy}\tau_{yz}\tau_{zx} - \sigma_x \tau_{yz}^2 - \sigma_y \tau_{zx}^2 - \sigma_z \tau_{xy}^2$$

Since I_1, I_2, and I_3 are constant for a given state of stress, any combined mathematical function of these is also constant, that is, invariant with respect to a change of coordinate system.

$$f(I_1, I_2, I_3) = \text{ constant} \qquad (6.19)$$

6.3.1 Principal Normal Stresses

As for plane stress, a special choice of coordinate system exists where the maximum and minimum normal stresses occur, and also where all shear stresses are absent. These coordinate axes are called the *principal axes*. Of the normal stresses σ_1, σ_2, and σ_3 on these axes, one is the maximum normal stress, another is the minimum normal stress, and the third has an intermediate value. This situation is illustrated in Fig. 6.8(a).

To obtain values for the principal normal stresses, let the state of stress described with respect to any convenient coordinate system be σ_x, σ_y, σ_z, τ_{xy}, τ_{yz}, τ_{zx}. The sign convention employed for these is that tensile normal stresses are positive, and shear stresses are positive if the arrows on the positive-facing side of the cube are in the positive x-y-z-directions as shown in Fig. 6.1. The values of the principal normal stresses (σ_1, σ_2, σ_3) are then given by the solution of a cubic equation.

$$\sigma^3 - \sigma^2 \left(\sigma_x + \sigma_y + \sigma_z\right) + \sigma \left(\sigma_x \sigma_y + \sigma_y \sigma_z + \sigma_z \sigma_x - \tau_{xy}^2 - \tau_{yz}^2 - \tau_{zx}^2\right)$$
$$- \left(\sigma_x \sigma_y \sigma_z + 2\tau_{xy}\tau_{yz}\tau_{zx} - \sigma_x \tau_{yz}^2 - \sigma_y \tau_{zx}^2 - \sigma_z \tau_{xy}^2\right) = 0 \qquad (6.20)$$

The three roots of this equation are the σ_1, σ_2, and σ_3 values, and they are always real numbers. It is often convenient to note that the above equation is the expansion of the following determinant:

$$\begin{vmatrix} (\sigma_x - \sigma) & \tau_{xy} & \tau_{zx} \\ \tau_{xy} & (\sigma_y - \sigma) & \tau_{yz} \\ \tau_{zx} & \tau_{yz} & (\sigma_z - \sigma) \end{vmatrix} = 0 \qquad (6.21)$$

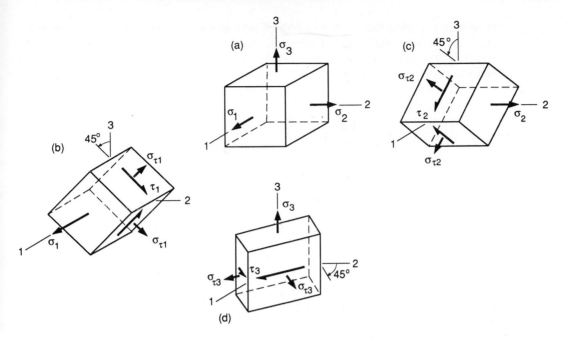

Figure 6.8   Planes of (a) principal normal stresses, and (b, c, d) principal shear stresses. In (b), rotation of the unit cube 45° about the axis of σ_1 gives the planes where τ_1 acts. Similar rotation about σ_2 gives the τ_2 planes (c), and about σ_3 the τ_3 planes (d).

Also, the cubic equation can be expressed in terms of the stress invariants.

$$\sigma^3 - \sigma^2 I_1 + \sigma I_2 - I_3 = 0 \tag{6.22}$$

Since the principal normal stresses are roots of an equation involving the stress invariants as coefficients, their values are also invariant, that is, not dependent on the choice of the original coordinate system.

It is common practice to assign the subscripts 1, 2, and 3 in order to the maximum, intermediate, and minimum values. However, this convention is not a necessity, and it is useful in working numerical problems to relax this requirement and allow the numbers to be assigned as convenient. We will write all equations involving principal stresses in general form, so that it is not necessary to assume that the subscripts are assigned in any particular order.

The orientation of the coordinate system containing the principal normal stresses can also be determined in the general three-dimensional case, but this will not be presented here. (See the References.)

6.3.2 Principal Shear Stresses

Consider the plane containing any two principal normal stresses, such as σ_2 and σ_3. The maximum shear stress in this plane occurs on a unit cube that is rotated 45° about the remaining (σ_1) principal stress axis. These *principal shear planes* are thus oriented 45° away from the σ_2 and σ_3 principal normal stress axes as illustrated by Fig. 6.8(a) and (b). The maximum shear stress in the 2-3 plane is one of three *principal shear stresses* and has the absolute value

$$\tau_1 = \frac{|\sigma_2 - \sigma_3|}{2} \tag{6.23}$$

Similarly, there is a principal shear stress in the plane containing σ_1 and σ_3, and another in the plane containing σ_1 and σ_2 .

$$\tau_2 = \frac{|\sigma_1 - \sigma_3|}{2}, \qquad \tau_3 = \frac{|\sigma_1 - \sigma_2|}{2} \tag{6.24}$$

The planes on which these additional principal shear stresses act are shown in Fig. 6.8(c) and (d). One of τ_1, τ_2, and τ_3 is the maximum shear stress that occurs for all possible choices of coordinate system.

$$\tau_{\max} = \mathrm{MAX}\,(\tau_1, \tau_2, \tau_3) \tag{6.25}$$

Each plane of principal shear stress is also acted upon by a normal stress that is the same in the two orthogonal directions. As for plane stress, these normal stresses are the average of the principal normal stresses in the same plane.

$$\sigma_{\tau 1} = \frac{\sigma_2 + \sigma_3}{2} \qquad (\tau_1 \text{ planes})$$

$$\sigma_{\tau 2} = \frac{\sigma_1 + \sigma_3}{2} \qquad (\tau_2 \text{ planes}) \tag{6.26}$$

$$\sigma_{\tau 3} = \frac{\sigma_1 + \sigma_2}{2} \qquad (\tau_3 \text{ planes})$$

These normal stresses are also shown in Fig. 6.8.

Example 6.3

At a point of interest in an engineering component, the stresses with respect to a convenient coordinate system are:

$$\sigma_x = 100, \qquad \sigma_y = -60, \qquad \sigma_z = 40 \text{ MPa}$$

$$\tau_{xy} = 80, \qquad \tau_{yz} = \tau_{zx} = 0 \text{ MPa}$$

Determine the principal normal and shear stresses.

Solution   Substitution of the stresses into the determinant form of the cubic (Eq. 6.21) gives

$$\begin{vmatrix} (100 - \sigma) & 80 & 0 \\ 80 & (-60 - \sigma) & 0 \\ 0 & 0 & (40 - \sigma) \end{vmatrix} = 0$$

Expanding using the last column yields

$$(40 - \sigma)\,[(100 - \sigma)\,(-60 - \sigma) - 6400] = 0$$

which simplifies partially to

$$(\sigma - 40)\,\left(\sigma^2 - 40\sigma - 12,400\right) = 0$$

or completely to

$$\sigma^3 - 80\sigma^2 - 10,800\sigma + 496,000 = 0$$

which is plotted below.

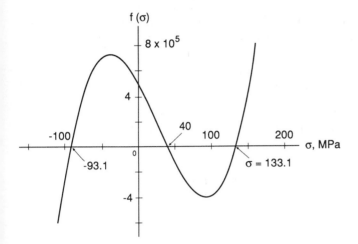

Figure E6.3   Graph of the example cubic equation, showing the three roots that are principal normal stresses.

This particular cubic is easily solved due to the $(\sigma - 40)$ factor in the intermediate form, giving a root that is one of the principal normal stresses.

$$\sigma_3 = 40 \text{ MPa} \qquad\qquad \textbf{Ans.}$$

The quadratic equation that is also a factor now needs to be solved.

$$\sigma^2 - 40\sigma - 12,400 = 0$$

Applying the standard quadratic formula gives the remaining two principal normal stresses.

$$\sigma_1 = 133.1, \qquad \sigma_2 = -93.1 \text{ MPa} \qquad\qquad \textbf{Ans.}$$

The principal shear stresses are then given by Eqs. 6.23 and 6.24.

$$\tau_1 = \frac{|\sigma_2 - \sigma_3|}{2} = \frac{|-93.1 - 40|}{2} = 66.6 \text{ MPa} \qquad\qquad \textbf{Ans.}$$

$$\tau_2 = \frac{|\sigma_1 - \sigma_3|}{2} = \frac{|133.1 - 40|}{2} = 46.6 \text{ MPa} \qquad\qquad \textbf{Ans.}$$

$$\tau_3 = \frac{|\sigma_1 - \sigma_2|}{2} = \frac{|133.1 - (-93.1)|}{2} = 113.1 \text{ MPa} \qquad \textbf{Ans.}$$

Discussion   The maximum and minimum normal stresses for all possible coordinate systems are the largest and smallest principal normal stresses.

$$\sigma_{max} = 133.1, \qquad \sigma_{min} = -93.1 \text{ MPa}$$

Also, the maximum shear stress for all possible coordinate systems is the largest principal shear stress.

$$\tau_{max} = 113.1 \text{ MPa}$$

Comment   More general methods for solving cubic equations are usually needed. For example, a graph similar to Fig. E6.3 can be used to obtain approximate values of the roots. More precise values can then be obtained from these by trial and error or by standard numerical techniques such as Newton's method.

6.3.3 Mohr's Circles for the Principal Planes

Consider a choice of coordinate system such that the description of a state of stress has two components of shear stress equal to zero, such as

$$\tau_{yz} = \tau_{zx} = 0 \qquad (6.27)$$

The normal stress σ_z in the direction perpendicular to the plane of the nonzero component of shear stress (τ_{xy}) is then one of the principal stresses.

$$\sigma_z = \sigma_3 \qquad (6.28)$$

Such a situation is illustrated in a three-dimensional view in Fig. 6.9(a). It can also be illustrated by a diagram in the x-y plane, where the z-direction is normal to the paper, as shown in (b). The freebody of a portion of this unit cube is shown in (c). This freebody is similar to that employed previously for plane stress, specifically Fig. 6.3. The only difference is the presence of the stress σ_z. The equations of equilibrium in the x-y plane are the same as before.

Hence, all of the equations previously developed for the x-y plane apply to this case also. This includes the equations for the principal normal stresses in the x-y plane, σ_1 and σ_2, and also those for the principal shear stress and the accompanying normal stress, τ_3 and $\sigma_{\tau3}$. It is simply necessary to note that $\sigma_z = \sigma_3$ remains unchanged for all rotations of the coordinate axes in the x-y plane. Moreover, since Mohr's circle was also derived from the same equilibrium equations, it can also be employed for the x-y plane.

Now consider a three-dimensional state of stress where the principal normal stresses and their directions are known. For any x-y coordinate system in a plane perpendicular to one of the principal normal stresses, the situation described above exists, that is, the out-of-plane shear components τ_{yz} and τ_{zx} are zero. Mohr's circle can therefore be used for the stresses in a plane perpendicular to any one of the principal normal stresses. For

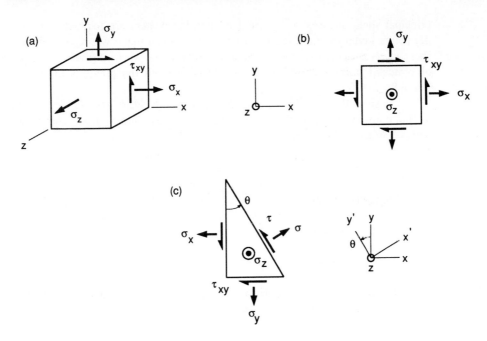

Figure 6.9   State of stress where two components of shear stress are zero.

example, for the plane perpendicular to σ_3, called the 1-2 plane, Mohr's circle is fixed by the (σ, τ) points $(\sigma_1, 0)$ and $(\sigma_2, 0)$. This is illustrated in Fig. 6.10(a).

Mohr's circles can be similarly drawn for the 1-3 and 2-3 planes, with the result that there are three circles, two of which lie inside the largest one. Each circle is also tangent along the σ-axis to each of the other two. The radii of these circles are the

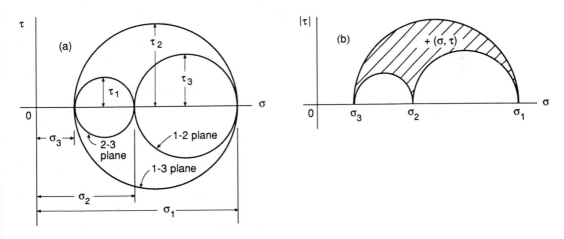

Figure 6.10   Mohr's circles for a three-dimensional state of stress.

principal shear stresses, τ_1, τ_2, and τ_3, and the centers are located along the σ-axis at the points given by the three $\sigma_{\tau i}$ values. Also, each plane where one of these principal shear stresses occurs is seen to be a 45° rotation away from the corresponding planes of principal normal stress, which is consistent with the previous discussion and with Fig. 6.8.

Consider the stresses on the oblique plane of Fig. 6.7(b). Let σ_x' be called simply σ, and combine the two components of shear stress into a single component τ, which still lies in the oblique plane.

$$\sigma = \sigma_x', \quad \tau = \sqrt{\left(\tau_{xy}'\right)^2 + \left(\tau_{zx}'\right)^2} \tag{6.29}$$

For all possible oblique planes, it can be shown that the (σ, τ) combination lies outside the two smaller Mohr's circles, but inside the larger one. Hence, considering only the absolute values of shear stresses, all stress combinations lie within the shaded area of Fig. 6.10(b).

Example 6.4

Repeat Example 6.3 using Mohr's circles. Recall that the original state of stress is

$$\sigma_x = 100, \quad \sigma_y = -60, \quad \sigma_z = 40 \text{ MPa}$$

$$\tau_{xy} = 80, \quad \tau_{yz} = \tau_{zx} = 0 \text{ MPa}$$

Solution   Since there is only one nonzero component of shear stress, the stress normal to the plane of this shear stress is one of the principal normal stresses.

$$\sigma_3 = \sigma_z = 40 \text{ MPa} \qquad\qquad \textbf{Ans.}$$

Mohr's circle may then be employed for the x-y plane just as for a two-dimensional problem. The two ends of a diameter are

$$\left(\sigma_x, -\tau_{xy}\right) = (100, -80) \text{ MPa}$$

$$\left(\sigma_y, \tau_{xy}\right) = (-60, 80) \text{ MPa}$$

The resulting circle is shown below (Fig. E6.4).

Simple geometry as in Example 6.2 is next needed to locate the ends of the horizontal diameter. In particular, the center of the circle is located at a σ value of

$$a = \frac{\sigma_x + \sigma_y}{2} = \frac{100 - 60}{2} = 20 \text{ MPa}$$

From the cross-hatched triangle, the radius of the circle is

$$r = \sqrt{80^2 + 80^2} = 113.1 \text{ MPa}$$

This gives the two remaining principal normal stresses.

$$\sigma_1, \sigma_2 = a \pm r = 133.1, -93.1 \text{ MPa} \qquad\qquad \textbf{Ans.}$$

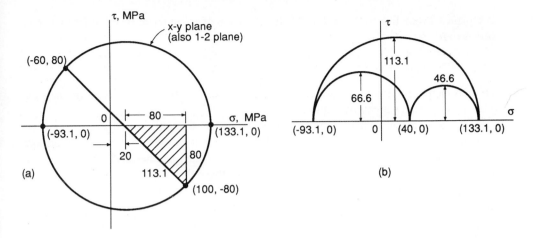

Figure E6.4   Mohr's circles and principal stresses for the three-dimensional state of stress example.

The points $(\sigma_1, 0)$, $(\sigma_2, 0)$, and $(\sigma_3, 0)$ now fix the circles for the three principal planes as shown. The radii of these circles are the principal shear stresses.

$$\tau_1 = \frac{|\sigma_2 - \sigma_3|}{2} = 66.6 \text{ MPa} \qquad \textbf{Ans.}$$

$$\tau_2 = \frac{|\sigma_1 - \sigma_3|}{2} = 46.6 \text{ MPa} \qquad \textbf{Ans.}$$

$$\tau_3 = \frac{|\sigma_1 - \sigma_2|}{2} = 113.1 \text{ MPa} \qquad \textbf{Ans.}$$

with the largest of these being τ_{max}.

Comment   If this example had involved an original state of stress with more than one nonzero component of shear stress, then the use of Mohr's circles would not have been advantageous compared to solving the cubic equation. A method for applying Mohr's circle to any three-dimensional state of stress does exist but is rather complex. It is described in some older advanced textbooks, such as Nadai (1950).

6.4 PLANE STRESS RECONSIDERED AS A THREE-DIMENSIONAL CASE

Plane stress is the situation where the coordinate axes chosen result in the nonzero components of stress all being confined to a plane, such as the x-y plane, so that $\sigma_z = \tau_{yz} = \tau_{zx} = 0$. If the principal normal stresses in the x-y plane are σ_1 and σ_2, then the third principal normal stress σ_3 is zero. Even for this situation, shear stresses are in general present on all of the principal shear planes of Fig. 6.8. This occurs despite the fact that only one of the three sets of principal shear planes involves stresses in the

x-y plane. From Eqs. 6.23 and 6.24, for the case of $\sigma_z = \sigma_3 = 0$, the principal shear stresses are

$$\tau_1 = \frac{|\sigma_2 - \sigma_3|}{2} = \frac{|\sigma_2|}{2}, \qquad \tau_2 = \frac{|\sigma_1 - \sigma_3|}{2} = \frac{|\sigma_1|}{2}, \qquad \tau_3 = \frac{|\sigma_1 - \sigma_2|}{2} \qquad (6.30)$$

These principal shear stresses are accompanied by normal stresses $\sigma_{\tau i}$ acting on the same planes, these being given by Eq. 6.26.

Thus, in a sense, there is no such thing as a state of plane stress, as stresses occur on planes associated with choices of coordinate axes not in the x-y plane. Furthermore, it is hazardous to confine one's attention to the x-y plane, as one of the principal shear stresses τ_1, τ_2, or τ_3 may be larger than the one of these that is the τ_3 of Eq. 6.9 obtained from analysis of the stresses in the x-y plane. From Eq. 6.30, this in fact occurs whenever the two principal normal stresses in the x-y plane are of the same sign.

Mohr's circles further illustrate the situation as shown in Fig. 6.11. The circles are defined by the points σ_1, σ_2, and σ_3 on the σ-axis, where one of these is $\sigma_z = 0$, so that two of the circles must pass through the origin. If the principal normal stresses in the x-y plane are of opposite sign, then the circle for the x-y plane is the largest, and τ_3 for the x-y plane is the maximum shear stress for all possible choices of coordinate axes. This case is illustrated by (a). However, if the principal normal stresses for the x-y plane are of the same sign, as in (b), then one of the other circles is the largest. The radius of the largest circle is the maximum shear stress, and this stress lies not in the x-y plane, but in the plane containing the direction of the largest principal normal stress in the x-y plane and the z-axis.

This situation is illustrated in Fig. 6.12. The principal normal and shear stresses in the x-y plane of the applied stress are shown in (b). If this system of stresses is rotated

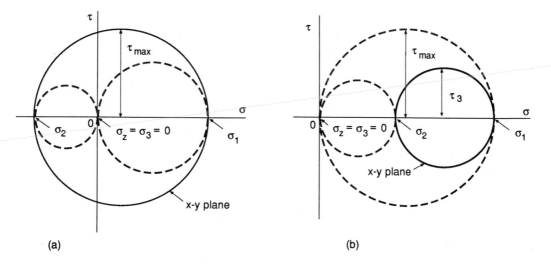

(a)   (b)

Figure 6.11 Plane stress in the x-y plane reconsidered as a three-dimensional state of stress. In case (a), the maximum shear stress lies in the x-y plane, but in case (b) it does not.

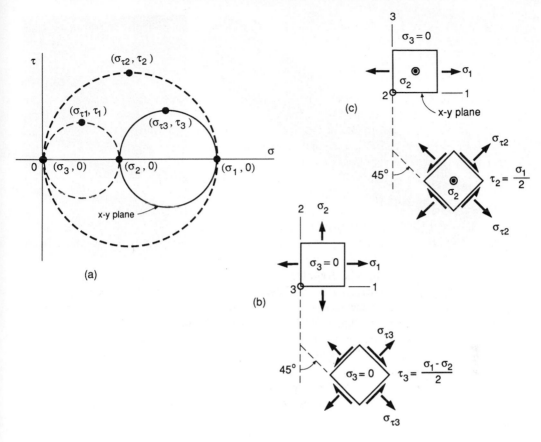

Figure 6.12 Plane stress with Mohr's circle not enclosing the origin, in which case the maximum shear stress is τ_2 oriented at 45° to the x-y plane.

90° about the σ_1 axis so that it is viewed parallel to σ_2, then the stresses in the 1-3 plane appear as shown in (c). The corresponding Mohr's circle has a radius of $\tau_2 = \sigma_1/2$, which is the maximum shear stress in the 1-3 plane. Note that this shear stress is τ_{max} in this case and acts on planes that are inclined 45° to the x-y plane. Failure of a pressurized tube along such an inclined plane of maximum shear stress is shown in Fig. 6.13.

Similarly, 90° rotation about σ_2 and viewing parallel to σ_1 permits the stresses in the 2-3 plane to be analyzed by Mohr's circle. The maximum shear stress in this plane is $\tau_1 = \sigma_2/2$. It also acts on planes that are inclined at 45° to the x-y plane. These are of course not the same set of planes as for τ_2, as study of Fig. 6.8 will indicate.

Example 6.5

What is the maximum shear stress for the situation analyzed in Examples 6.1 and 6.2?

Solution   Recall that the original state of stress is $\sigma_x = 95$, $\sigma_y = 25$, and $\tau_{xy} = 20$ MPa. Also, the principal normal and shear stresses already found by analysis confined to the x-y

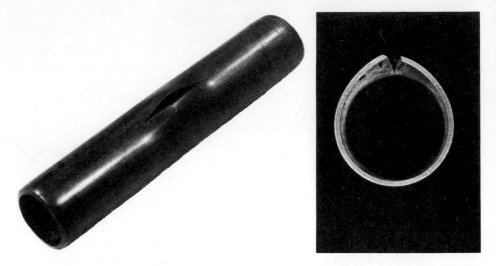

Figure 6.13 Failure of a 15 mm diameter copper water pipe due to excess pressure from freezing. In the cross section on the right, note that failure occurred on a plane inclined 45° to the tube surface, which is the plane of the maximum shear stress. (Photos by R. A. Simonds.)

plane are

$$\sigma_1 = 100.3, \quad \sigma_2 = 19.7 \text{ MPa}$$

$$\tau_3 = 40.3 \text{ MPa}$$

The third principal normal stress is

$$\sigma_3 = \sigma_z = 0$$

Equations 6.23 and 6.24 then give the remaining principal shear stresses.

$$\tau_1 = \frac{|\sigma_2 - \sigma_3|}{2} = \frac{|19.7 - 0|}{2} = 9.8 \text{ MPa}$$

$$\tau_2 = \frac{|\sigma_1 - \sigma_3|}{2} = \frac{|100.3 - 0|}{2} = 50.1 \text{ MPa}$$

The maximum shear stress is thus τ_2, which does not lie in the x-y plane but acts on planes inclined 45° to the x-y plane.

$$\tau_{\text{max}} = 50.1 \text{ MPa} \qquad\qquad \textbf{Ans.}$$

This situation was expected since the principal stresses σ_1 and σ_2 in the x-y plane are of the same sign.

6.5 STRESSES ON THE OCTAHEDRAL PLANES

Consider an oblique plane where the x-y-z axes are chosen to coincide with the principal normal stress axes (1, 2, 3). This is illustrated by Fig. 6.14(a). The normal stress on the oblique plane is σ, and the shear stress is expressed as a single value τ, according to Eq. 6.29. The direction of the normal to the oblique plane is specified by the angles α, β, and γ between this normal and each principal axis.

For the special case where $\alpha = \beta = \gamma$, the oblique plane intersects the principal axes at equal distances from the origin. This special plane is called the *octahedral plane*. Based on equilibrium of forces, the normal stress on this plane can be shown to be the average of the principal normal stresses.

$$\sigma_h = \frac{\sigma_1 + \sigma_2 + \sigma_3}{3} \tag{6.31}$$

The quantity σ_h is called the *octahedral normal stress* or the *hydrostatic stress* and was previously considered in Chapter 4. Equilibrium also permits the shear stress on the same plane, called the *octahedral shear stress*, to be evaluated.

$$\tau_h = \frac{1}{3}\sqrt{(\sigma_1 - \sigma_2)^2 + (\sigma_2 - \sigma_3)^2 + (\sigma_3 - \sigma_1)^2} \tag{6.32}$$

In each octant of the principal axes coordinate system, there is a similar plane where the normal makes equal angles with the axes. The stresses on all eight such planes are the same and are σ_h and τ_h. These planes can be thought of as forming an octahedron as shown in Fig. 6.14. Noting that opposite faces of the octahedron correspond to a single plane, the octahedral stresses act on four planes.

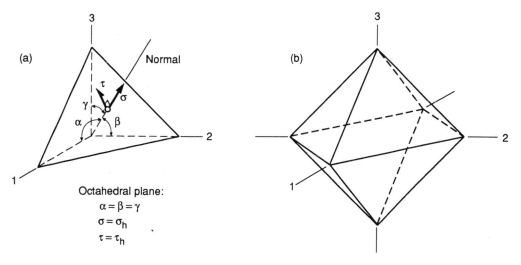

Figure 6.14 Octahedral plane shown relative to the principal normal stress axes (a), and the octahedron formed by the similar such planes in all octants (b).

By evaluating the stress invariants, Eq. 6.18, for the special case of the principal normal stresses, and after some manipulation, σ_h and τ_h can be written in terms of the invariants.

$$\sigma_h = \frac{I_1}{3}, \quad \tau_h = \frac{1}{3}\sqrt{2\left(I_1^2 - 3I_2\right)} \tag{6.33}$$

Substitution of the general form of the invariants and manipulation then gives

$$\sigma_h = \frac{\sigma_x + \sigma_y + \sigma_z}{3} \tag{6.34}$$

$$\tau_h = \frac{1}{3}\sqrt{\left(\sigma_x - \sigma_y\right)^2 + \left(\sigma_y - \sigma_z\right)^2 + \left(\sigma_z - \sigma_x\right)^2 + 6\left(\tau_{xy}^2 + \tau_{yz}^2 + \tau_{zx}^2\right)} \tag{6.35}$$

These more general expressions may be used to compute σ_h and τ_h for stresses described with respect to any coordinate system, so that it is not necessary to first determine the principal stresses. Since the octahedral stresses σ_h and τ_h are functions of the stress invariants, these quantities are themselves invariant. Hence, any equivalent representation of a given state of stress will give the same values of σ_h and τ_h.

The octahedral shear stress is an important quantity as it is used as a basis for predicting yielding and other types of material behavior under complex states of stress. This is considered starting in the next chapter, as is the similar use of the maximum shear stress. Since τ_{max} occurs on only two planes, τ_h occurs twice as frequently as does τ_{max}. (Compare Figs. 6.8 and 6.14.) Also, for all possible states of stress, it can be shown that τ_h is always similar in magnitude to τ_{max}, with the ratio τ_h/τ_{max} being confined to the range 0.866 to unity.

6.6 COMPLEX STATES OF STRAIN

In the discussion above of complex states of stress, equilibrium of forces leads to transformation equations for obtaining an equivalent representation of a given state of stress on a new set of coordinate axes. Of particular interest are two sets of axes, one containing the principal normal stresses and the other the principal shear stresses. The mathematics involved is common to all physical quantities classed as *symmetric second-order tensors*, as distinguished from vectors, which are first-order tensors, or scalars, which are zero-order tensors.

Strain is also a symmetric second-order tensor and so is governed by similar equations. In this case, the basis of the equations is simply the geometry of deformation. Detailed analysis (see the References) gives equations that are identical to those for stress except that shear strains are divided by two. Hence, the various equations developed for stress can be used for strain by changing the variables as follows:

$$\sigma_x, \sigma_y, \sigma_z \rightarrow \varepsilon_x, \varepsilon_y, \varepsilon_z$$

$$\tau_{xy}, \tau_{yz}, \tau_{zx} \rightarrow \frac{\gamma_{xy}}{2}, \frac{\gamma_{yz}}{2}, \frac{\gamma_{zx}}{2} \tag{6.36}$$

These apply in general and also to the special case where the x-y-z axes are axes of principal strain, 1-2-3.

Advanced textbooks on continuum mechanics, theory of elasticity, and similar subjects often redefine shear strain as being half as large as the usual *engineering shear strains* used here, calling these *tensor shear strains*, so that the equations become identical to those for stress. However, we will continue to use engineering shear strains.

6.6.1 Principal Strains

Principal normal strains and *principal shear strains* occur in a similar manner as for stresses. For *plane strain*, where $\varepsilon_z = \gamma_{yz} = \gamma_{zx} = 0$, modifying Eqs. 6.6 and 6.7 according to Eq. 6.36 gives the axis rotations and values for the principal normal strains.

$$\tan 2\theta_n = \frac{\gamma_{xy}}{\varepsilon_x - \varepsilon_y}$$

$$\varepsilon_1, \varepsilon_2 = \frac{\varepsilon_x + \varepsilon_y}{2} \pm \sqrt{\left(\frac{\varepsilon_x - \varepsilon_y}{2}\right)^2 + \left(\frac{\gamma_{xy}}{2}\right)^2} \tag{6.37}$$

Equations 6.8 through 6.10 are similarly modified to obtain the axis rotation and value for the principal shear strain in the x-y plane, and also the accompanying normal strain.

$$\tan 2\theta_s = -\frac{\varepsilon_x - \varepsilon_y}{\gamma_{xy}}$$

$$\gamma_3 = \sqrt{\left(\varepsilon_x - \varepsilon_y\right)^2 + \left(\gamma_{xy}\right)^2} \tag{6.38}$$

$$\varepsilon_{\gamma 3} = \frac{\varepsilon_x + \varepsilon_y}{2}$$

As for the stress equations, θ is positive counterclockwise. Positive normal strains correspond to extension and negative ones to contraction. Positive shear strain causes a distortion corresponding to a positive shear stress in that the long diagonal of the resulting parallelogram has a positive slope. (Look ahead to Fig. E6.6(a) for an example of a positive shear strain.) Direct use of these equations can be replaced by Mohr's circle in a manner similar to its use for stress. In accordance with Eq. 6.36, the σ-axis becomes an ε-axis, and the τ-axis becomes a $\gamma/2$-axis.

For three dimensional states of strain, the principal strains can be obtained by modifying Eqs. 6.21, 6.23, and 6.24 using Eq. 6.36.

$$\begin{vmatrix} (\varepsilon_x - \varepsilon) & \dfrac{\gamma_{xy}}{2} & \dfrac{\gamma_{zx}}{2} \\[3mm] \dfrac{\gamma_{xy}}{2} & (\varepsilon_y - \varepsilon) & \dfrac{\gamma_{yz}}{2} \\[3mm] \dfrac{\gamma_{zx}}{2} & \dfrac{\gamma_{yz}}{2} & (\varepsilon_z - \varepsilon) \end{vmatrix} = 0 \tag{6.39}$$

$$\gamma_1 = |\varepsilon_2 - \varepsilon_3|, \qquad \gamma_2 = |\varepsilon_1 - \varepsilon_3|, \qquad \gamma_3 = |\varepsilon_1 - \varepsilon_2| \tag{6.40}$$

6.6.2 Special Considerations for Plane Stress

For cases of plane stress, $\sigma_z = \tau_{yz} = \tau_{zx} = 0$, the Poisson effect results in normal strains ε_z occurring in the out-of-plane direction, so that the state of strain is three dimensional. If the material is isotropic, or if the material is orthotropic and a material symmetry plane is parallel to the x-y plane, no shear strains γ_{yz} or γ_{zx} occur. This creates a situation analogous to that for stress where σ_z is present but $\tau_{yz} = \tau_{zx} = 0$. Hence, one of the principal normal strains is $\varepsilon_z = \varepsilon_3$, and the other two can be obtained from Eq. 6.37. In addition, Mohr's circle can be used for the x-y plane.

For isotropic materials, ε_z can be obtained from Hooke's Law in the form of Eq. 4.10. Taking $\sigma_z = 0$ and adding Eqs. 4.10(a) and (b) leads to

$$\sigma_x + \sigma_y = \frac{E}{1 - \nu} \left(\varepsilon_x + \varepsilon_y \right) \tag{6.41}$$

Substituting this into Eq. 4.10(c) with $\sigma_z = 0$ gives ε_z in terms of the normal strains in the x-y plane.

$$\varepsilon_z = \frac{-\nu}{1 - \nu} \left(\varepsilon_x + \varepsilon_y \right) \tag{6.42}$$

Since $\tau_{yz} = \tau_{zx} = 0$, Eq. 4.11 gives $\gamma_{yz} = \gamma_{zx} = 0$, and it is confirmed that ε_z is one of the principal normal strains.

Consider an orthotropic material under x-y plane stress where the x-y plane is a plane of symmetry of the material. (This is the situation for most sheets and plates of composite materials.) The strain ε_z is still one of the principal normal strains, as $\gamma_{yz} = \gamma_{zx} = 0$ holds in this case also. Hence, the strains in the x-y plane can still be analyzed using Eqs. 6.37 and 6.38, and Mohr's circle for the x-y plane can still be used. However, ε_z cannot be obtained from Eq. 6.42, the more general form of Hooke's Law for orthotropic materials (Eq. 4.50) being needed.

The principal axes for stress and strain coincide for isotropic materials. Hence, Hooke's Law can be applied to the principal strains, and the resulting stresses are the principal stresses, and vice-versa. However, this is not the case for orthotropic materials unless the principal stresses are parallel to the planes of material symmetry.

Example 6.6

At a point on a free (unloaded) surface of an engineering component made of an aluminum alloy, the following strains exist: $\varepsilon_x = -0.0005$, $\varepsilon_y = 0.0035$, and $\gamma_{xy} = 0.003$. Determine the principal normal and shear strains.

Solution   Since the material is expected to be isotropic, ε_z can be obtained from Eq. 6.42, and this is one of the principal normal strains.

$$\varepsilon_3 = \varepsilon_z = \frac{-0.345}{1 - 0.345}(-0.0005 + 0.0035) = -0.00158 \qquad \textbf{Ans.}$$

where Poisson's ratio from Table 4.2 is used. Substituting the given strains into Eqs. 6.37 and 6.38 gives the axis rotations and values for the other two principal normal strains and for one of the principal shear strains.

$$\tan 2\theta_n = -\frac{3}{4}, \quad \theta_n = -18.4° = 18.4° \quad (\text{CW})$$

$$\varepsilon_1, \varepsilon_2 = 0.004, -0.001 \qquad\qquad\qquad \textbf{Ans.}$$

$$\tan 2\theta_s = \frac{4}{3}, \quad \theta_s = 26.6° \quad (\text{CCW})$$

$$\gamma_3 = 0.005, \quad \varepsilon_{y3} = 0.0015 \qquad\qquad\qquad \textbf{Ans.}$$

The resulting states of strain are shown below as (b) and (c). Signs and directions are determined in a manner similar to that used previously for stresses. In particular, the larger of the two principal normal strains takes a direction such that it is more nearly aligned with the larger of the original ε_x and ε_y than with the smaller. Also, the principal shear strain causes a distortion such that the long diagonal (dashed line) of the resulting parallelogram is aligned with the larger of ε_1 and ε_2.

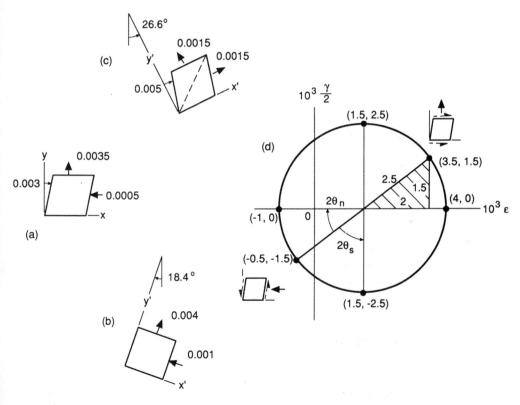

Figure E6.6 A state of strain (a) and the equivalent representations corresponding to principal normal strains (b) and the principal shear strain in the x-y plane (c). Mohr's circle for this case is shown in (d).

The remaining two principal shear strains can be obtained from Eq. 6.40.

$$\gamma_1 = |\varepsilon_2 - \varepsilon_3| = |-0.001 - (-0.00158)| = 0.00058 \qquad \textbf{Ans.}$$

$$\gamma_2 = |\varepsilon_1 - \varepsilon_3| = |0.004 - (-0.00158)| = 0.00558 \qquad \textbf{Ans.}$$

The same result for the in-plane strains can be obtained by using Mohr's circle on a plot of ε versus $\gamma/2$ as shown above. Two ends of a diameter are given by

$$\left(\varepsilon_x, -\frac{\gamma_{xy}}{2}\right) = (-0.0005, -0.0015)$$

$$\left(\varepsilon_y, \frac{\gamma_{xy}}{2}\right) = (0.0035, 0.0015)$$

Analysis on the circle proceeds in a manner similar to that for stress. The special dual sign convention needed for shear strain corresponds to that for the shear stress that would produce the distortion, clockwise rotation being positive as in Fig. 6.6.

6.7 SUMMARY

For a general state of stress, given by components σ_x, σ_y, σ_z, τ_{xy}, τ_{yz}, and τ_{zx}, there is one choice of a new coordinate system where shear stresses are absent and where the maximum and minimum normal stresses occur along with an intermediate normal stress. These special stresses are the principal normal stresses, σ_1, σ_2, and σ_3, and they may be obtained by solving the cubic equation given by a determinant.

$$\begin{vmatrix} (\sigma_x - \sigma) & \tau_{xy} & \tau_{zx} \\ \tau_{xy} & (\sigma_y - \sigma) & \tau_{yz} \\ \tau_{zx} & \tau_{yz} & (\sigma_z - \sigma) \end{vmatrix} = 0 \qquad (6.43)$$

If there is only one nonzero component of shear stress, such as τ_{xy}, the principal normal stresses are

$$\sigma_1, \sigma_2 = \frac{\sigma_x + \sigma_y}{2} \pm \sqrt{\left(\frac{\sigma_x - \sigma_y}{2}\right)^2 + \tau_{xy}^2} \quad \text{(a)}$$
$$\qquad\qquad (6.44)$$

$$\sigma_3 = \sigma_z \qquad\qquad \text{(b)}$$

One method of evaluating Eq. 6.44(a) and obtaining the corresponding axis rotation is to use Mohr's circle.

The principal shear stresses occur on planes inclined 45° with respect to the principal normal stresses. The absolute values of these are given by

$$\tau_1 = \frac{|\sigma_2 - \sigma_3|}{2}, \qquad \tau_2 = \frac{|\sigma_1 - \sigma_3|}{2}, \qquad \tau_3 = \frac{|\sigma_1 - \sigma_2|}{2} \qquad (6.45)$$

One of the values τ_1, τ_2, τ_3 is the maximum shear stress that occurs for all possible choices of coordinate axes. For x-y plane stress, special care is needed that all three of Eq. 6.45

are considered, because the principal shear stress in the x-y plane may not be the largest. It is useful to envision three different Mohr's circles, one for each plane perpendicular to a principal normal stress. The radii of these are the principal shear stresses.

The octahedral normal and shear stresses occur on planes that intercept the principal normal stress axes at equal distances from the origin. Their values are given by

$$\sigma_h = \frac{\sigma_1 + \sigma_2 + \sigma_3}{3} \tag{6.46}$$

$$\tau_h = \frac{1}{3}\sqrt{(\sigma_1 - \sigma_2)^2 + (\sigma_2 - \sigma_3)^2 + (\sigma_3 - \sigma_1)^2} \tag{6.47}$$

where σ_h is also called the hydrostatic stress.

Principal normal strains and principal shear strains occur in a manner analogous to principal stresses. The same equations apply by replacing stresses with strains as follows:

$$\sigma_x, \sigma_y, \sigma_z \rightarrow \varepsilon_x, \varepsilon_y, \varepsilon_z$$

$$\tau_{xy}, \tau_{yz}, \tau_{zx} \rightarrow \frac{\gamma_{xy}}{2}, \frac{\gamma_{yz}}{2}, \frac{\gamma_{zx}}{2} \tag{6.48}$$

Even plane stress causes a three-dimensional state of strain. However, for isotropic materials, and also for orthotropic materials stressed in a plane of material symmetry, the out-of-plane shear strains γ_{yz} and γ_{zx} are zero, permitting two-dimensional analysis to be performed in the x-y plane despite the presence of a nonzero ε_z .

NEW TERMS AND SYMBOLS

axes rotation angles: θ_n and θ_s

Mohr's circle

octahedral normal (hydrostatic) stress, σ_h

octahedral planes

octahedral shear stress, τ_h

plane strain

plane stress

principal axes (1, 2, 3)

principal normal strains: ε_1, ε_2, and ε_3

principal normal stresses: σ_1, σ_2, and σ_3

principal shear planes

principal shear strains: γ_1, γ_2, and γ_3

principal shear stresses: τ_1, τ_2, and τ_3

transformation equations

transformation of axes

REFERENCES

FREDERICK, D. and T. S. CHANG. 1972 *Continuum Mechanics*, Scientific Publishers, Cambridge, Ma.

GOULD, P. L. 1983 *Introduction to Linear Elasticity*, Springer-Verlag, New York.

MENDELSON, A. 1968 *Plasticity: Theory and Applications*, Macmillan, New York. (Reprinted by R. E. Krieger, Malabar, FL, 1983).

NADAI, A. 1950 *Theory of Flow and Fracture of Solids*, McGraw-Hill, New York.

SMITH, J. O. and O. M. SIDEBOTTOM. 1969 *Elementary Mechanics of Deformable Bodies*, Macmillan, New York.

PROBLEMS AND QUESTIONS

Sections 6.2 and 6.4

6.1 A state of stress that occurs at a point on the free surface of a solid body is $\sigma_x = 110$, $\sigma_y = 30$, and $\tau_{xy} = -30$ MPa, where the directions shown in Fig. 6.2(a) are considered positive.
  (a) Evaluate the principal normal stresses and the principal shear stress in the x-y plane, and give the coordinate axes rotations for these.
  (b) Considering shear stresses not in the x-y plane, what is the maximum shear stress?

6.2 Repeat Prob. 6.1 for the following state of stress: $\sigma_x = 28$, $\sigma_y = 0$, $\tau_{xy} = 20.4$ MPa.

6.3 Repeat Prob. 6.1 for the following state of stress: $\sigma_x = 50$, $\sigma_y = 100$, and $\tau_{xy} = -60$ MPa.

6.4 Repeat Prob. 6.1 for the following state of stress: $\sigma_x = 120$, $\sigma_y = 40$, and $\tau_{xy} = -30$ MPa.

6.5 A cylindrical pressure vessel 10 m long has closed ends, a wall thickness of 5 mm, and an average (midthickness) diameter of 3 m. The vessel is filled with a gas to a pressure of 2 MPa. Determine the maximum normal stress and the maximum shear stress, and also describe the orientations of the planes on which these act. Neglect any effects of the discontinuity associated with the end closure.

6.6 A spherical pressure vessel has a wall thickness 2.5 mm and an outside diameter of 150 mm, and it contains a liquid at 0.5 MPa pressure. Determine the maximum normal stress and the maximum shear stress, and also describe the planes on which these act.

6.7 A solid circular shaft with an arm attached to one end is fixed at the other end and is loaded as shown.

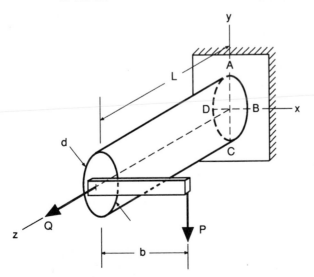

Figure P6.7

   (a) Determine the stresses at points A, B, C, and D in terms of the applied loads P and Q and the geometric dimensions b, d, and L.

   (b) Determine the maximum normal and shear stresses acting in the shaft for the following specific case: $P = 1.8$ kN, $Q = 9$ kN, $d = 100$ mm, $b = 500$ mm, and $L = 750$ mm. Neglect any stress raiser effects where the shaft is attached.

6.8 A weight $W = 45$ kN is hung eccentrically from the end of a cantilevered pipe of length $L = 750$ mm as shown. The outer diameter is $d = 250$ mm, the wall thickness is 6 mm, and a fluid in the pipe has a pressure of 3.5 MPa. Determine the maximum normal and shear stresses at: (a) point A and (b) point B. Neglect any stress raiser effects where the pipe is attached.

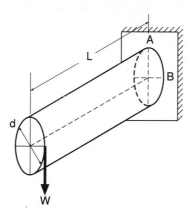

Figure P6.8

Section 6.3

6.9 A three-dimensional state of stress is as follows: $\sigma_x = 110$, $\sigma_y = 30$, $\sigma_z = -40$, $\tau_{xy} = 30$, and $\tau_{yz} = \tau_{zx} = 0$ MPa. Determine the principal normal and shear stresses.

6.10 Repeat Prob. 6.9 for $\sigma_x = 50$, $\sigma_y = 100$, $\sigma_z = 200$, $\tau_{xy} = 60$, and $\tau_{yz} = \tau_{zx} = 0$ MPa.

6.11 Repeat Prob. 6.9 for $\sigma_x = 120$, $\sigma_y = 40$, $\sigma_z = -50$, $\tau_{xy} = 30$, and $\tau_{yz} = \tau_{zx} = 0$ MPa.

6.12 Consider a triaxial state of stress where $\sigma_x = \sigma_y = \sigma_z = S$, and where all shear stresses for the x-y-z coordinate system chosen are zero, so that $\tau_{xy} = \tau_{yz} = \tau_{zx} = 0$. In terms of S, what are the principal normal and shear stresses?

6.13 Consider an axisymmetric state of stress, where $\sigma_x = \sigma_y = T$ and $\sigma_z = S$, and where all shear stresses are zero for the x-y-z coordinate system chosen, so that $\tau_{xy} = \tau_{yz} = \tau_{zx} = 0$. In terms of T and S, what are the principal normal and shear stresses?

6.14 Consider the special case where normal stresses σ_x, σ_y, and σ_z are present, but where the only nonzero shear stress is τ_{xy} for the x-y-z coordinate system chosen, so that $\tau_{yz} = \tau_{zx} = 0$. For determining principal normal stresses, show that the solution of Eq. 6.20 corresponds to the two equations represented by Eq. 6.7 and the third equation $\sigma_z = \sigma_3$.

Section 6.5

6.15 Determine the octahedral normal and shear stresses for the state of stress of Prob. 6.1.

6.16 Determine the octahedral normal and shear stresses for the state of stress of Prob. 6.9.

6.17 Consider a case of plane stress where the only nonzero components for the x-y-z coordinate system chosen are σ_x and τ_{xy}. (For example, this situation occurs at the surface of a shaft under combined bending and torsion.) Develop equations in terms of σ_x and τ_{xy} for the following: maximum normal stress, maximum shear stress, and octahedral shear stress.

6.18 Repeat Prob. 6.17 for the case of plane stress where the only nonzero components for the x-y-z system chosen are σ_x and σ_y. (For example, this occurs in a cylindrical tube with internal pressure and bending and/or axial loads.)

6.19 A thin-walled tube with closed ends has diameter d and wall thickness t, and it is subjected to an internal pressure p. Determine the following in terms of d, t, and p: maximum normal stress, maximum shear stress, and octahedral shear stress.

6.20 Develop an equation for the octahedral shear stress in terms of the principal shear stresses.

6.21 Derive the equations for the octahedral normal and shear stresses, Eqs. 6.31 and 6.32, based on equilibrium of the solid body shown in Fig. 6.14(a). Suggestions: Note that the three faces in the principal planes are acted upon by principal stresses, σ_1, σ_2, and σ_3, and calculate the relative areas of these planes and the octahedral plane. Then sum *forces* normal to the octahedral plane to get σ_h, and parallel to this plane to get τ_h. Also, from the general properties of direction cosines, $\cos^2 \alpha + \cos^2 \beta + \cos^2 \gamma = 1$.

Section 6.6

6.22 For pure planar shear, where only τ_{xy} is nonzero, verify the principal stresses, strains, and planes shown in Fig. 4.5.

6.23 At a point on the free surface of an engineering component made of a low-carbon steel, the following strains exist: $\varepsilon_x = 0.0005$, $\varepsilon_y = 0.00025$, and $\gamma_{xy} = 0.0006$. Determine the principal normal and shear strains.

6.24 By modifying the equations for stress, develop equations for the normal and shear strains on the octahedral plane, ε_h and γ_h. Express these in terms of the components of a general three-dimensional state of strain: ε_x, ε_y, ε_z, γ_{xy}, γ_{yz}, γ_{zx}. What is the significance of ε_h? Under what conditions is the octahedral plane for strain the same as the one for stress?

7

Yielding and Fracture under Combined Stresses

7.1 INTRODUCTION

Engineering components may be subjected to complex loadings in tension, compression, bending, torsion, or pressure, or combinations of these, so that at a given point in the material stresses often occur in more than one direction. If sufficiently severe, such combined stresses can act together to cause the material to yield or fracture. Predicting the safe limits for use of a material under combined stresses requires the application of a *failure criterion*.

A number of different failure criteria are available, some of which predict failure by yielding, and others failure by fracture. The former are specifically called *yield criteria*, and the latter *fracture criteria*. All failure criteria considered in the present chapter will be based on values of stress, so that their application involves calculating a numerical value of stress that characterizes the combined stresses, and then comparing this value with the yield or fracture strength of the material. A given material may fail by either yielding or fracture, depending on its properties and the state of stress, so that in general the possibility of either event occurring first must be considered.

7.1.1 Need for Failure Criteria

The need for careful consideration of failure criteria is illustrated by the examples of Fig. 7.1. For these examples, the material is assumed to be a ductile engineering metal, the behavior of which approximates the ideal elastic, perfectly plastic case. A uniaxial tension test provides the elastic modulus E, and the yield strength σ_o, as shown in (a). Now assume that a transverse compression of equal magnitude to the tensile stress is also applied, as shown in (b). In this case, the tension stress σ_y necessary to cause yielding is experimentally observed to be only about half of the value from the uniaxial test. This result is easily verified by conducting a simple torsion test on a thin-walled tube, where the desired state of stress exists at an orientation of $45°$ to the tube axis. (Recall Fig. 4.5, and note that on the $45°$ axes $\sigma_y = -\sigma_x = |\tau|$.)

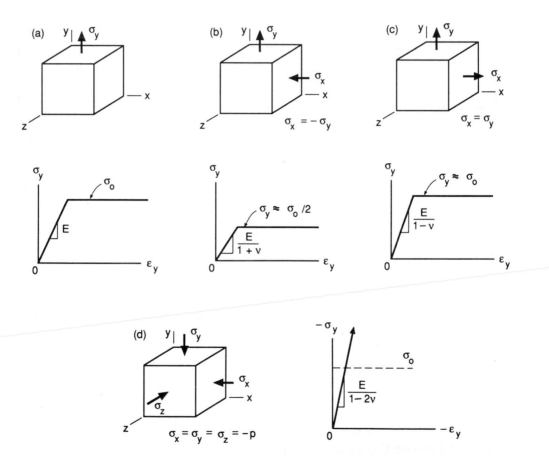

Figure 7.1   Yield strengths for a ductile metal under various states of stress: (a) uniaxial tension, (b) tension with transverse compression, (c) biaxial tension, and (d) hydrostatic compression.

Now consider another example, namely a transverse tension σ_x, of equal magnitude to σ_y, as illustrated in (c). Since transverse compression lowered the yield strength, intuition suggests that transverse tension might increase it. But an experiment shows that the effect of the transverse stress on yielding is small or absent. The experiment could be done by pressurizing a thin-walled spherical vessel until it yielded, or by a combination of pressure and tension on a thin-walled tube. If the material is changed to a brittle one, say gray cast iron, neither tensile nor compressive transverse stresses have much effect on its fracture.

An additional experimental fact of interest is that it is difficult, and perhaps impossible, to yield a ductile material if it is tested under simple hydrostatic stress, where $\sigma_x = \sigma_y = \sigma_z$, in either tension or compression. This is illustrated in Fig. 7.1(d). Hydrostatic tension is difficult to achieve experimentally, but hydrostatic compression consists of simply placing a sample of material in a pressurized chamber.

Hence, failure criteria are needed that are capable of predicting such effects of combined states of stress on yielding and fracture. Although both yield and fracture criteria should in general be employed, materials that typically behave in a ductile manner generally have their usefulness limited by yielding, and those that typically behave in a brittle manner are usually limited by fracture.

7.1.2 Additional Comments

An alternative to failure criteria based on stress is to specifically analyze cracks in the material using the special methods of *fracture mechanics*. Such an approach is not considered in this chapter but is instead the sole topic of the next chapter.

In most of the treatment that follows, materials are assumed to be isotropic and homogeneous. Failure criteria for anisotropic materials is a rather complex topic that is considered only to a limited extent.

Note that the effect of a complex state of stress on deformation prior to yielding has already been discussed in Chapter 4. For example, the initial elastic slopes in Fig. 7.1 are readily obtained from Hooke's Law in the form of Eq. 4.10. The yield criteria considered in this chapter predict the beginning of plastic deformation, beyond which point Hooke's Law ceases to completely describe the stress-strain behavior. Detailed treatment of stress-strain behavior beyond yielding is an advanced topic called *plasticity*, which is considered to an extent later in Chapter 12.

The discussion in this chapter relies rather heavily on the review of complex states of stress in the previous chapter, specifically transformation of axes, Mohr's circle, principal stresses, and octahedral stresses. Readers not familiar with these topics may wish to refer to Chapter 6 as needed.

7.2 GENERAL FORM OF FAILURE CRITERIA

In applying a yield criterion, the resistance of a material is given by its yield strength. Yield strengths are most commonly available as tensile yield strengths σ_o, determined from uniaxial tests using a plastic strain offset as described in Chapter 5. In applying a fracture criterion, the ultimate tensile strength σ_u is usually used. In tension tests on materials that behave in a brittle manner, recall that in most cases yielding is not a well-defined event, and the ultimate strength and fracture events occur at the same point. Hence, using σ_u for brittle materials is the same as using the engineering fracture strength, σ_f.

Failure criteria for isotropic materials can be expressed in the following mathematical form:

$$f(\sigma_1, \sigma_2, \sigma_3) = \sigma_c \quad \text{(at failure)} \tag{7.1}$$

where failure (yielding or fracture) is predicted to occur when a specific mathematical function f of the principal normal stresses is equal to the failure strength of the material, σ_c, from a uniaxial tension test. The failure strength is either the yield strength σ_o, or the ultimate strength σ_u, depending on whether yielding or fracture is of interest.

A requirement for a valid failure criterion is that it must give the same result regardless of the original choice of the coordinate system in a problem. This requirement is met if the criterion can be expressed in terms of the principal stresses. It is also met by any criterion where f is a mathematical function of one or more of the stress invariants given in the previous chapter as Eq. 6.18.

If any particular case of Eq. 7.1 is plotted in *principal normal stress space*, that is, on three-dimensional coordinates of σ_1, σ_2, and σ_3, the function f forms a surface that is called the *failure surface*. A failure surface can be either a *yield surface* or a *fracture surface*. In discussing failure criteria, we will proceed by considering various specific mathematical functions f, hence various types of failure surface.

Consider a point in an engineering component where the applied loads result in particular values of the principal normal stresses, σ_1, σ_2, and σ_3, and where the materials property σ_c is known, and also where a specific function f has been chosen. It is then useful to define an effective stress, $\bar{\sigma}$, which is a single numerical value that characterizes the state of applied stress. In particular,

$$\bar{\sigma} = f(\sigma_1, \sigma_2, \sigma_3) \tag{7.2}$$

where f is the same function as in Eq. 7.1. Thus, Eq. 7.1 states that failure occurs when

$$\bar{\sigma} = \sigma_c \quad \text{(at failure)} \tag{7.3}$$

Failure is not expected if $\bar{\sigma}$ is less than σ_c.

$$\bar{\sigma} < \sigma_c \quad \text{(no failure)} \tag{7.4}$$

Also, the safety factor against failure is

$$X = \frac{\sigma_c}{\bar{\sigma}} \tag{7.5}$$

In other words, the applied stresses can be increased by a factor X before failure occurs. For example, if $X = 2$, the applied stresses can be doubled before failure is expected.[1]

We will now proceed to discuss various specific failure criteria, some of which are appropriate for yielding, and others for fracture. In doing so, unless otherwise noted, the subscripts for σ_1, σ_2, and σ_3 will not be assumed to be assigned in any particular order relative to the magnitudes of these stresses.

7.3 MAXIMUM NORMAL STRESS FRACTURE CRITERION

Perhaps the simplest failure criterion is that failure is expected when the largest principal normal stress reaches the uniaxial strength of the material. Since this approach has its greatest success in predicting fracture of brittle materials, it should be considered to be a fracture criterion, as distinguished from a yield criterion.

The *maximum normal stress fracture criterion* can be specified by a particular function f as follows:

$$\sigma_u = \text{MAX}\,(|\sigma_1|\,,\,|\sigma_2|\,,\,|\sigma_3|) \qquad \text{(at fracture)} \qquad (7.6)$$

where the notation MAX indicates that the largest of the values separated by commas is chosen. Absolute values are used so that compressive principal stresses can be considered, and it is assumed for the present that the ultimate (fracture) strength σ_u of the material is the same in tension and compression.

A particular set of applied stresses can then be characterized by the following effective stress:

$$\bar{\sigma}_N = \text{MAX}\,(|\sigma_1|\,,\,|\sigma_2|\,,\,|\sigma_3|) \qquad (7.7)$$

where the subscript specifies the maximum normal stress criterion. Hence, fracture is expected when $\bar{\sigma}_N$ is equal to σ_u, but not when it is less, and the safety factor against fracture is

$$X = \frac{\sigma_u}{\bar{\sigma}_N} \qquad (7.8)$$

7.3.1 Graphical Representation of the Normal Stress Criterion

For plane stress, such as $\sigma_3 = 0$, this fracture criterion can be graphically represented by a square on a plot of σ_1 versus σ_2 as shown in Fig. 7.2(a). Any combination of σ_1 and σ_2 that plots within the square box is safe, and any on its perimeter corresponds to fracture. Note that the box is the region that satisfies

$$\text{MAX}\,(|\sigma_1|\,,\,|\sigma_2|) \le \sigma_u \qquad (7.9)$$

---

[1] Safety factors are preferably expressed in terms of applied loads according to Eq. 1.1. If loads and stresses are proportional, as is frequently the case, then safety factors on stress are identical to those on load. But caution is needed if such proportionality does not exist, as for problems of buckling and surface contact loading.

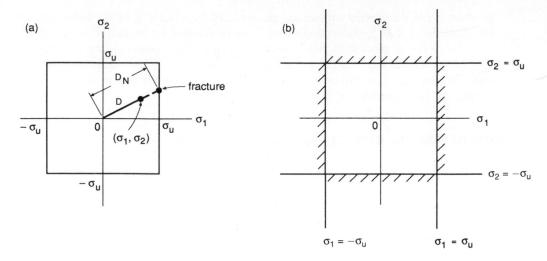

Figure 7.2 Failure locus for the maximum normal stress fracture criterion for plane stress.

Equations for the four straight lines that form the borders of this safe region are obtained as shown in Fig. 7.2(b).

$$\sigma_1 = \sigma_u, \qquad \sigma_1 = -\sigma_u$$

$$\sigma_2 = \sigma_u, \qquad \sigma_2 = -\sigma_u$$

(7.10)

For the general case, where all three principal normal stresses may have nonzero values, Eq. 7.6 indicates that the safe region is bounded by

$$\sigma_1 = \pm\sigma_u, \qquad \sigma_2 = \pm\sigma_u, \qquad \sigma_3 = \pm\sigma_u$$

(7.11)

Each of the above equalities represents a pair of parallel planes normal to one of the principal axes and intersecting each at $+\sigma_u$ and $-\sigma_u$. The failure surface is therefore simply a cube as illustrated in Fig. 7.3. If any one of σ_1, σ_2, or σ_3 is zero, then only the two-dimensional region formed by the intersection of the cube with the plane of the remaining two principal stresses needs to be considered. Such an intersection is shown for the case of $\sigma_3 = 0$, and the result is of course the square of Fig. 7.2.

7.3.2 Graphical Interpretation of the Safety Factor

Consider the situation of a point on the surface of an engineering component, where plane stress prevails, so that $\sigma_3 = 0$. Further assume that increasing the applied load causes σ_1 and σ_2 to both increase with their ratio σ_2/σ_1 remaining constant, a situation called *proportional loading*. For example, for pressure loading of a thin-walled tube with closed ends, the stresses maintain the ratio $\sigma_2/\sigma_1 = 0.5$, where σ_1 is the hoop stress and σ_2 the longitudinal stress.

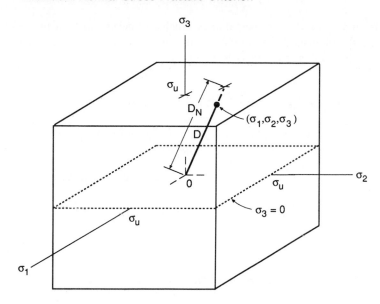

Figure 7.3   Three-dimensional failure surface for the maximum normal stress fracture criterion.

In such a case, a graphical interpretation of the safety factor may be made as illustrated in Fig. 7.2(a). Let D be the straight-line distance from the origin to the (σ_1, σ_2) point corresponding to the applied stress. Then extend this straight line until it strikes the fracture line, and denote the overall distance from the origin as D_N. The safety factor against fracture is the ratio of these lengths.

$$X = \frac{D_N}{D} \tag{7.12}$$

For the particular case of plane stress, and assuming for convenience that σ_1 has the largest absolute value and is positive, Eq. 7.12 can be verified as follows:

$$\bar{\sigma}_N = \text{MAX}\,(|\sigma_1|,\,|\sigma_2|) = \sigma_1 \tag{7.13}$$

By similar triangles on Fig. 7.2(a)

$$\frac{\sigma_u}{\sigma_1} = \frac{D_N}{D} \tag{7.14}$$

Since the safety factor is $X = \sigma_u/\bar{\sigma}_N$, the above two equations combine to give Eq. 7.12. By extending this procedure, Eq. 7.12 is easily seen to apply regardless of the relative magnitude and signs of σ_1 and σ_2.

Such a graphical interpretation of the safety factor, and specifically of Eq. 7.12, also applies in the general three-dimensional case as illustrated in Fig. 7.3. The distances D and D_N are still measured along a straight line, but in this case the line may be inclined relative to all three principal axes. Such an interpretation of the safety factor in terms of lengths of lines for proportional stressing is valid for any physically reasonable failure surface, such as the others to be discussed below.

7.4 MAXIMUM SHEAR STRESS YIELD CRITERION

Yielding of ductile materials is often predicted to occur when the maximum shear stress on any plane reaches a critical value τ_o, which is a material property.

$$\tau_o = \tau_{\max} \quad \text{(at yielding)} \tag{7.15}$$

This is the basis of the *maximum shear stress yield criterion*, also often called the Tresca criterion. For metals, such an approach is logical based on the fact that the mechanism of yielding on a microscopic size scale is the slip of crystal planes, which is a shear deformation that is expected to be controlled by a shear stress. (See Chapter 2.)

7.4.1 Development of the Maximum Shear Stress Criterion

From the previous chapter, recall that the maximum shear stress is the largest of the three principal shear stresses, which act on planes oriented at $45°$ relative to the principal normal stress axes as illustrated in Fig. 6.8. These principal shear stresses may be obtained from the principal normal stresses by Eqs. 6.23 and 6.24, which are repeated here for convenience.

$$\tau_1 = \frac{|\sigma_2 - \sigma_3|}{2}, \quad \tau_2 = \frac{|\sigma_1 - \sigma_3|}{2}, \quad \tau_3 = \frac{|\sigma_1 - \sigma_2|}{2} \tag{7.16}$$

Hence, this yield criterion can be stated as follows:

$$\tau_o = \text{MAX}\left(\frac{|\sigma_1 - \sigma_2|}{2}, \frac{|\sigma_2 - \sigma_3|}{2}, \frac{|\sigma_3 - \sigma_1|}{2}\right) \quad \text{(at yielding)} \tag{7.17}$$

The yield stress in shear, τ_o, for a given material could be obtained directly from a test in simple shear, such as a thin-walled tube in torsion. However, only uniaxial yield strengths σ_o from tension tests are commonly available, so that it is more convenient to calculate τ_o from σ_o. In a uniaxial tension test, at the stress defined as the yield strength, we have

$$\sigma_1 = \sigma_o, \quad \sigma_2 = \sigma_3 = 0 \tag{7.18}$$

Substitution of these values into the yield criterion of Eq. 7.17 gives

$$\tau_o = \frac{\sigma_o}{2} \tag{7.19}$$

In the uniaxial test, note that the maximum shear stress occurs on planes oriented at $45°$ with respect to the applied stress axis. This fact and Eq. 7.19 are easily verified using Mohr's circle as shown in Fig. 7.4.

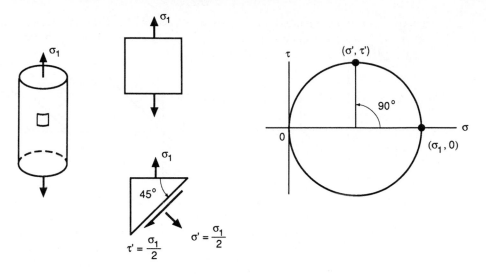

Figure 7.4   The plane of maximum shear in a uniaxial tension test.

Equation 7.17 can thus be written in terms of σ_o.

$$\frac{\sigma_o}{2} = \text{MAX} \left(\frac{|\sigma_1 - \sigma_2|}{2}, \frac{|\sigma_2 - \sigma_3|}{2}, \frac{|\sigma_3 - \sigma_1|}{2} \right) \qquad \text{(at yielding)} \qquad (7.20)$$

or

$$\sigma_o = \text{MAX} \left(|\sigma_1 - \sigma_2|, |\sigma_2 - \sigma_3|, |\sigma_3 - \sigma_1| \right) \qquad \text{(at yielding)} \qquad (7.21)$$

The effective stress is most conveniently defined as in Eq. 7.3, so that it equals the uniaxial strength σ_o at the point of yielding.

$$\bar{\sigma}_S = \text{MAX} \left(|\sigma_1 - \sigma_2|, |\sigma_2 - \sigma_3|, |\sigma_3 - \sigma_1| \right) \qquad (7.22)$$

where the subscript S specifies the maximum shear stress criterion. The safety factor against yielding is then

$$X = \frac{\sigma_o}{\bar{\sigma}_S} \qquad (7.23)$$

7.4.2 Graphical Representation of the Maximum Shear Stress Criterion

For plane stress, such as $\sigma_3 = 0$, the maximum shear stress criterion can be represented on a plot of σ_1 versus σ_2 as shown in Fig. 7.5(a). Points on the distorted hexagon correspond to yielding, and points inside are safe. This failure locus is obtained by substituting $\sigma_3 = 0$ into the yield criterion of Eq. 7.21.

$$\sigma_o = \text{MAX} \left(|\sigma_1 - \sigma_2|, |\sigma_2|, |\sigma_1| \right) \qquad (7.24)$$

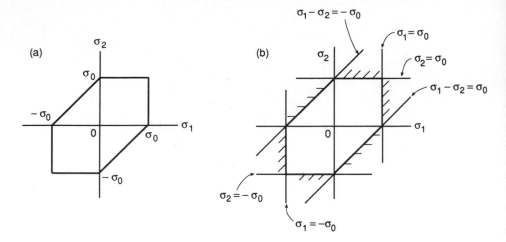

Figure 7.5   Failure locus for the maximum shear stress yield criterion for plane stress.

The region of no yielding, where $\bar{\sigma}_S < \sigma_o$, is thus the region bounded by the lines

$$\sigma_1 - \sigma_2 = \pm\sigma_o, \qquad \sigma_2 = \pm\sigma_o, \qquad \sigma_1 = \pm\sigma_o \tag{7.25}$$

These lines are shown in Fig. 7.5(b). Note that the first equation above gives a pair of parallel lines with a slope of unity, and the other two give pairs of lines parallel to the coordinate axes.

For the general case, where all three principal normal stresses may have nonzero values, the boundaries of the region of no yielding are obtained from Eq. 7.21.

$$\sigma_1 - \sigma_2 = \pm\sigma_o, \qquad \sigma_2 - \sigma_3 = \pm\sigma_o, \qquad \sigma_1 - \sigma_3 = \pm\sigma_o \tag{7.26}$$

Each of the above gives a pair of inclined planes which are parallel to the principal stress direction that does not appear in the equation. For example, the first equation represents a pair of planes parallel to the σ_3 direction.

These three pairs of planes form a tube with a hexagonal cross section as shown in Fig. 7.6. The axis of the tube is the line

$$\sigma_1 = \sigma_2 = \sigma_3 \tag{7.27}$$

This direction corresponds to the normal to the octahedral plane in the octant where the principal normal stresses are all positive, specifically the $\alpha = \beta = \gamma$ line of Fig. 6.14. If the tube is viewed along this line, a regular hexagon is seen as shown in (b).

If any one of σ_1, σ_2, or σ_3 is zero, then the intersection of the tube with the plane of the remaining two stresses give a distorted hexagon failure locus as already shown in Fig. 7.5(a).

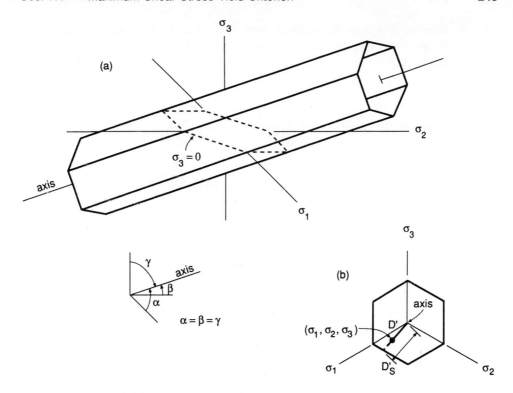

Figure 7.6 Three-dimensional failure surface for the maximum shear stress yield criterion.

7.4.3 Hydrostatic Stresses and the Maximum Shear Stress Criterion

Consider the special case of hydrostatic stress where the principal normal stresses are all equal.

$$\sigma_1 = \sigma_2 = \sigma_3 = S \tag{7.28}$$

For example, the material could be subjected to a simple pressure loading p, so that $S = -p$. This case corresponds to a point along the axis of the hexagonal cylinder of Fig. 7.6. For any such point, the effective stress $\bar{\sigma}_S$ from Eq. 7.22 is always zero, and the safety factor against yielding is thus infinite.

Hence, the maximum shear stress criterion predicts that hydrostatic stress alone does not cause yielding. This seems surprising but is in fact in agreement with experimental results for metals under hydrostatic compression. Testing in hydrostatic tension is essentially impossible, but it is likely that brittle fracture without yielding would occur at a high stress level even in normally ductile materials.

The interpretation of safety factor in terms of lengths of lines from the origin in principal stress space, as discussed earlier, is also valid for the maximum shear stress criterion. For three-dimensional cases, since stresses are expected to affect yielding only to the extent that they deviate from the axis of the hexagonal tube, the projections of lengths normal to this axis can also be used.

$$X = \frac{D_S'}{D'} \tag{7.29}$$

where D_S' is the projected distance corresponding to yielding, and D' to the applied stress, as shown in Fig. 7.6(b).

Example 7.1

A point on the free surface of an engineering component made of 2024-T4 aluminum is subjected to the following state of stress: $\sigma_x = 50$, $\sigma_y = 100$, and $\tau_{xy} = 60$ MPa. What is the safety factor against yielding?

Solution   First, noting that plane stress applies, determine principal normal stresses from either Mohr's circle or Eq. 6.7. Using the latter

$$\sigma_1, \sigma_2 = \frac{\sigma_x + \sigma_y}{2} \pm \sqrt{\left(\frac{\sigma_x - \sigma_y}{2}\right)^2 + \tau_{xy}^2}$$

$$\sigma_1, \sigma_2 = 140, 10 \text{ MPa}$$

The third principal normal stress is

$$\sigma_3 = \sigma_z = 0$$

Equation 7.22 then gives the effective stress for the maximum shear stress yield criterion.

$$\bar{\sigma}_S = \text{MAX}\left(|\sigma_1 - \sigma_2|, |\sigma_2 - \sigma_3|, |\sigma_3 - \sigma_1|\right)$$

$$\bar{\sigma}_S = \text{MAX}\left(|140 - 10|, |10 - 0|, |0 - 140|\right) = 140 \text{ MPa}$$

From Table 5.2, the yield strength of 2024-T4 aluminum is $\sigma_o = 303$ MPa, and the material is quite ductile, having 20% elongation and 35% reduction in area. Therefore, the safety factor against yielding is

$$X = \frac{\sigma_o}{\bar{\sigma}_S} = \frac{303}{140} = 2.16 \qquad \textbf{Ans.}$$

Example 7.2

Consider a thin-walled tube with closed ends and internal pressure p. The wall thickness is t, the radius to mid-thickness is r, and the ductile material has a yield strength σ_o. Derive an equation for the required thickness corresponding to specified values of r and the safety factor X against yielding.

Solution   From elementary mechanics of materials, the stresses in the tube are

$$\sigma_x = \frac{pr}{t} \qquad \text{(hoop direction)}$$

$$\sigma_y = \frac{pr}{2t} \qquad \text{(longitudinal direction)}$$

Since no shear stress is present, these are principal normal stresses.

$$\sigma_1 = \sigma_x = \frac{pr}{t}, \qquad \sigma_2 = \sigma_y = \frac{pr}{2t}$$

Due to the state of plane stress, the third principal stress is

$$\sigma_3 = \sigma_z = 0$$

From Eq. 7.22, the effective stress for the maximum shear stress yield criterion is

$$\bar{\sigma}_S = \text{MAX}\left(|\sigma_1 - \sigma_2|, |\sigma_2 - \sigma_3|, |\sigma_3 - \sigma_1|\right)$$

$$\bar{\sigma}_S = \text{MAX}\left(\left|\frac{pr}{t} - \frac{pr}{2t}\right|, \left|\frac{pr}{2t} - 0\right|, \left|0 - \frac{pr}{t}\right|\right) = \frac{pr}{t}$$

The safety factor against yielding is

$$X = \frac{\sigma_o}{\bar{\sigma}_S} = \frac{\sigma_o t}{pr}$$

which gives the required thickness of

$$t = \frac{Xpr}{\sigma_o} \qquad\qquad\qquad \textbf{Ans.}$$

7.5 OCTAHEDRAL SHEAR STRESS YIELD CRITERION

Another yield criterion often used for ductile metals is that yielding occurs when the shear stress on the octahedral planes reaches a critical value.

$$\tau_h = \tau_{ho} \quad \text{(at yielding)} \tag{7.30}$$

where τ_{ho} is the value of octahedral shear stress τ_h necessary to cause yielding. The resulting *octahedral shear stress yield criterion*, also often called either the von Mises or the distortion energy criterion, represents an alternative to the maximum shear criterion.

To physically justify such an approach, the following argument can be used: Since the hydrostatic stress σ_h is observed not to affect the yielding, it is logical to find the plane where this occurs as the normal stress, and then to use the remaining stress τ_h as the failure criterion. Another justification is to note that although yielding is caused by shear stresses, τ_{max} occurs on only two planes in the material, whereas τ_h is never very much smaller and occurs on four planes. (Compare Figs. 6.8 and 6.14.) Hence, τ_h has a greater chance on a statistical basis of finding crystal planes that are favorably oriented for slip, and this may overcome its disadvantage of being slightly smaller than τ_{max}.

7.5.1 Development of the Octahedral Shear Stress Criterion

From the previous chapter, the shear stress on the octahedral planes is

$$\tau_h = \frac{1}{3}\sqrt{(\sigma_1 - \sigma_2)^2 + (\sigma_2 - \sigma_3)^2 + (\sigma_3 - \sigma_1)^2} \qquad (7.31)$$

so that the failure criterion is

$$\tau_{ho} = \frac{1}{3}\sqrt{(\sigma_1 - \sigma_2)^2 + (\sigma_2 - \sigma_3)^2 + (\sigma_3 - \sigma_1)^2} \quad \text{(at yielding)} \qquad (7.32)$$

As was done for the maximum shear stress criterion, it is useful to express the critical value in terms of the yield strength from a tension test. Substitution of the uniaxial stress state with $\sigma_1 = \sigma_o$ into the octahedral shear criterion gives

$$\tau_{ho} = \frac{\sqrt{2}}{3}\sigma_o \qquad (7.33)$$

From the three-dimensional geometry of the octahedral planes as described in the previous chapter, it can be shown that the plane on which the uniaxial stress acts is related to the octahedral plane by a rotation through the angle α of Fig. 6.14, where

$$\alpha = \cos^{-1}\left(\frac{1}{\sqrt{3}}\right) = 54.7° \qquad (7.34)$$

The same result can also be obtained from Mohr's circle by noting that in uniaxial tension the normal stress on the octahedral plane is

$$\sigma_h = \frac{\sigma_1 + \sigma_2 + \sigma_3}{3} = \frac{\sigma_1}{3} \qquad (7.35)$$

Locating the point that satisfied this on Mohr's circle leads to the above values of α and τ_{ho} as shown in Fig. 7.7.

Combining Eqs. 7.32 and 7.33 gives the yield criterion in the desired form expressed in terms of the uniaxial yield strength.

$$\sigma_o = \frac{1}{\sqrt{2}}\sqrt{(\sigma_1 - \sigma_2)^2 + (\sigma_2 - \sigma_3)^2 + (\sigma_3 - \sigma_1)^2} \quad \text{(at yielding)} \qquad (7.36)$$

As before, the effective stress for this theory is most conveniently defined so that it equals the uniaxial strength σ_o at the point of yielding.

$$\bar{\sigma}_H = \frac{1}{\sqrt{2}}\sqrt{(\sigma_1 - \sigma_2)^2 + (\sigma_2 - \sigma_3)^2 + (\sigma_3 - \sigma_1)^2} \qquad (7.37)$$

where the subscript H specifies that this effective stress is based on the octahedral shear stress criterion. This effective stress may also be determined directly for any state of

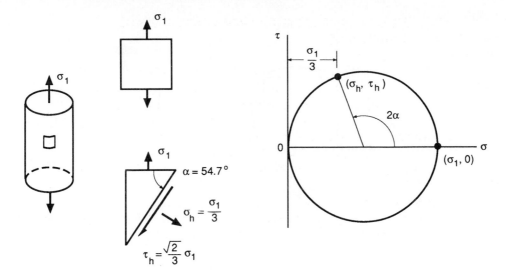

Figure 7.7   The plane of octahedral shear in a uniaxial tension test.

stress, without the necessity of first determining principal stresses, by modifying Eq. 7.37 using Eqs. 6.32 and 6.35.

$$\bar{\sigma}_H = \frac{1}{\sqrt{2}}\sqrt{\left(\sigma_x - \sigma_y\right)^2 + \left(\sigma_y - \sigma_z\right)^2 + \left(\sigma_z - \sigma_x\right)^2 + 6\left(\tau_{xy}^2 + \tau_{yz}^2 + \tau_{zx}^2\right)} \qquad (7.38)$$

7.5.2 Graphical Representation of the Octahedral Shear Stress Criterion

For plane stress, such as $\sigma_3 = 0$, the octahedral shear stress criterion can be represented on a plot of σ_1 versus σ_2 as shown in Fig. 7.8. This elliptical shape can be obtained by substituting $\sigma_3 = 0$ into the failure criterion in the form of Eq. 7.36.

$$\sigma_o = \frac{1}{\sqrt{2}}\sqrt{(\sigma_1 - \sigma_2)^2 + \sigma_2^2 + \sigma_1^2} \qquad (7.39)$$

Manipulation gives

$$\sigma_o^2 = \sigma_1^2 - \sigma_1\sigma_2 + \sigma_2^2 \qquad (7.40)$$

which is the equation of an ellipse with its major axis along the line $\sigma_1 = \sigma_2$ and which crosses the axes at the points $\pm\sigma_o$. Note that the ellipse has the distorted hexagon of the maximum shear criterion inscribed within it as shown.

For the general case, where all three principal normal stresses may have nonzero values, the boundary of the region of no yielding as specified by Eq. 7.36 represents a circular cylindrical surface with its axis along the line $\sigma_1 = \sigma_2 = \sigma_3$. This is illustrated in Fig. 7.9. The view along the cylinder axis, giving simply a circle, is also

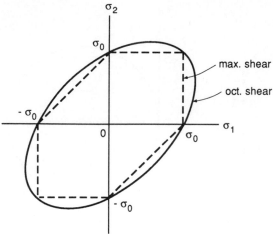

Figure 7.8 Failure locus for the octahedral shear stress yield criterion for plane stress, and comparison with the maximum shear criterion.

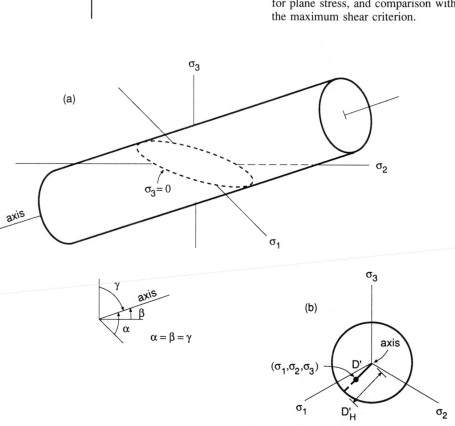

Figure 7.9 Three-dimensional failure surface for the octahedral shear stress yield criterion.

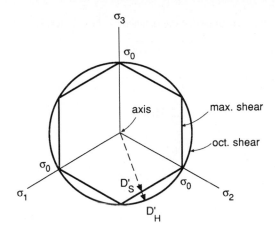

Figure 7.10 Comparison of yield surfaces for the maximum shear and octahedral shear stress criteria.

shown. If any one of σ_1, σ_2, or σ_3 is zero, then the intersection of the cylindrical surface with the plane of the remaining two principal stresses gives an ellipse as in Fig. 7.8.

Thus, we have a situation similar to that for the maximum shear stress criterion where hydrostatic stress is predicted to have no effect on yielding. In particular, substitution of $\sigma_1 = \sigma_2 = \sigma_3 = S$ into Eq. 7.37 gives $\bar{\sigma}_H = 0$, and a safety factor against yielding of infinity. Safety factors against yielding may be similarly interpreted in terms of distances from the cylinder axis as illustrated in Fig. 7.9(b). The hexagonal-tube yield surface of the maximum shear criterion is in fact inscribed within the cylindrical surface of the octahedral shear criterion. A view along the common axis of both gives the comparison of Fig. 7.10.

7.5.3 Energy of Distortion

In applying stresses to an element of material, work must be done, and for an elastic material all of this work is stored as potential energy. This internal strain energy can be partitioned into one portion associated with volume change, and another portion associated with distorting the shape of the element of material. Hydrostatic stress is associated with the energy of volume change, and since hydrostatic stress alone does not cause yielding, the remaining (distortional) portion of the total internal strain energy is a logical candidate for the basis of a failure criterion. When this approach is taken, the resulting failure criterion is found to be the same as the octahedral shear stress criterion. See Nadai (1950) or Seely and Smith (1952) for details.

Example 7.3

Repeat Ex. 7.1 except use the octahedral shear stress yield criterion.

First Solution   The principal normal stresses are as determined in Ex. 7.1.

$$\sigma_1, \sigma_2, \sigma_3 = 140, 10, 0 \text{ MPa}$$

Equation 7.37 gives the effective stress for the octahedral shear criterion.

$$\bar{\sigma}_H = \frac{1}{\sqrt{2}}\sqrt{(\sigma_1 - \sigma_2)^2 + (\sigma_2 - \sigma_3)^2 + (\sigma_3 - \sigma_1)^2}$$

$$\bar{\sigma}_H = \frac{1}{\sqrt{2}}\sqrt{(140 - 10)^2 + (10 - 0)^2 + (0 - 140)^2}$$

$$\bar{\sigma}_H = 135.3 \text{ MPa}$$

The safety factor against yielding is

$$X = \frac{\sigma_o}{\bar{\sigma}_H} = \frac{303}{135.3} = 2.24 \qquad\qquad \textbf{Ans.}$$

Second Solution   Equation 7.38 can be used to solve this problem in a more direct manner as the original state of stress ($\sigma_x = 50$, $\sigma_y = 100$, $\tau_{xy} = 60$ MPa) can be substituted directly, thus avoiding the step of determining principal normal stresses.

$$\bar{\sigma}_H = \frac{1}{\sqrt{2}}\sqrt{\left(\sigma_x - \sigma_y\right)^2 + \left(\sigma_y - \sigma_z\right)^2 + (\sigma_z - \sigma_x)^2 + 6\left(\tau_{xy}^2 + \tau_{yz}^2 + \tau_{zx}^2\right)}$$

$$\bar{\sigma}_H = \frac{1}{\sqrt{2}}\sqrt{(50 - 100)^2 + (100 - 0)^2 + (0 - 50)^2 + 6\left(60^2 + 0 + 0\right)}$$

$$\bar{\sigma}_H = 135.3 \text{ MPa}, \qquad X = \frac{\sigma_o}{\bar{\sigma}_H} = 2.24 \qquad\qquad \textbf{Ans.}$$

which as expected is the same result as above.

Comment   As the octahedral shear criterion is slightly less conservative than the maximum shear criterion, the safety factor obtained is larger than from Ex. 7.1, in this particular case by 4%.

Example 7.4

Repeat Example 7.2 except use the octahedral shear stress criterion.

Solution   The principal normal stresses from Ex. 7.2 can be used.

$$\sigma_1 = \frac{pr}{t}, \qquad \sigma_2 = \frac{pr}{2t}, \qquad \sigma_3 = 0$$

Substitution into Eq. 7.37 gives the effective stress.

$$\bar{\sigma}_H = \frac{1}{\sqrt{2}}\sqrt{\left(\frac{pr}{t} - \frac{pr}{2t}\right)^2 + \left(\frac{pr}{2t} - 0\right)^2 + \left(0 - \frac{pr}{t}\right)^2} = \frac{\sqrt{3}}{2}\frac{pr}{t}$$

so that the factor of safety is

$$X = \frac{\sigma_o}{\bar{\sigma}_H} = \sigma_o\left(\frac{2}{\sqrt{3}}\frac{t}{pr}\right)$$

which gives the required thickness of

$$t = \frac{\sqrt{3}}{2}\frac{Xpr}{\sigma_o} = 0.866\frac{Xpr}{\sigma_o}$$

Comment   The required thickness is about 15% less than from the maximum shear criterion. In this particular case, the difference between the two yield criteria has its maximum value.

7.6 DISCUSSION AND COMPARISON OF THE BASIC FAILURE CRITERIA

The three failure criteria discussed so far, namely the maximum normal stress, maximum shear stress, and octahedral shear stress criteria, may be considered to be the basic ones among a larger number that are available. It is useful at this point to discuss these basic approaches. We will also consider some additional failure criteria that are modifications or combinations of the basic ones.

7.6.1 Comparison of Failure Criteria

Both the maximum shear stress and the octahedral shear stress criteria are widely used to predict yielding of ductile materials, especially metals. Recall that both of these indicate that hydrostatic stress does not affect yielding, and also that the hexagonal-tube yield surface of the maximum shear criterion is inscribed within the circular-cylinder surface of the octahedral shear criterion. Hence, these two criteria never give dramatically different predictions of the yield behavior under combined stress, there being no state of stress where the difference exceeds approximately 15%. This can be seen in Fig. 7.10, where the distance from the cylinder axis to the two yield surfaces differs by a maximum amount at the various points where the circle is farthest from the hexagon. From geometry, the distances at these points have the ratio $2/\sqrt{3} = 1.155$. Hence, safety factors and effective stresses for a given state of stress cannot differ by more than this. For plane stress, $\sigma_3 = 0$, such a maximum deviation occurs for pure shear, where $\sigma_1 = -\sigma_2 = |\tau|$, and also for $\sigma_1 = 2\sigma_2$, as in pressure loading of a thin-walled tube with closed ends.

However, note that in some situations the maximum shear and octahedral shear yield criteria do give dramatically different predictions than a maximum normal stress criterion. Compare the tubular yield surfaces of either to the cube of Fig. 7.3, and consider states of stress near the tube axis ($\sigma_1 = \sigma_2 = \sigma_3$) but well beyond the boundaries of the cube. For plane stress, the three failure criteria compare as shown in Fig. 7.11. Where both principal stresses have the same sign, the maximum shear stress criterion is equivalent to a maximum normal stress criterion. However, if the principal stresses are opposite in sign, the normal stress criterion differs considerably from the other two.

The most convenient method of comparing failure criteria experimentally is to test thin-walled tubes under various combinations of axial, torsion, and pressure loading, thus producing various states of plane stress. Some data obtained in this manner for yielding of ductile metals and fracture of a brittle cast iron are shown in Fig. 7.11. The cast iron data follow the normal stress criterion, whereas the yield data tend to fall between the two yield criteria, perhaps agreeing better in general with the octahedral shear criterion. The maximum shear criterion is more conservative. Based on experimental data for ductile metals similar to that in Fig. 7.10, this criterion seems to represent a lower limit that is infrequently violated.

The maximum difference of 15% between the two yield criteria is relatively small compared to safety factors commonly used and to various uncertainties usually involved

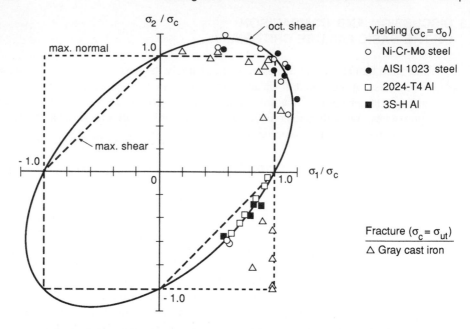

Figure 7.11 Plane stress failure loci for three criteria. These are compared with biaxial yield data for ductile steels and aluminum alloys, and also with biaxial fracture data for gray cast iron. (The steel data are from [Lessells 40] and [Davis 45], the aluminum data from [Naghdi 58] and [Marin 40], and the cast iron data from [Coffin 50] and [Grassi 49].)

in mechanical design, so that a choice between the two is not a matter of major importance. If conservatism is desired, the maximum shear criterion could be chosen.

7.6.2 Yield Criteria for Anisotropic and Uneven Materials

Several empirical modifications have been suggested so that the octahedral shear stress criterion can be used for *anisotropic* or *uneven* materials. Anisotropic materials have different properties in different directions, and uneven materials have different properties in tension versus compression. A material may be either anisotropic or uneven, or both.

Consider anisotropic materials that are orthotropic, that is, possessing symmetry about three planes oriented 90° to each other. For example, such anisotropy can occur in rolled plates of metals where the yield strength may differ somewhat between the rolling, transverse, and thickness directions. The anisotropic yield criterion described in Hill (1983) for this case is:

$$F\left(\sigma_X - \sigma_Y\right)^2 + G\left(\sigma_Y - \sigma_Z\right)^2 + H\left(\sigma_Z - \sigma_X\right)^2 + 2L\tau_{XY}^2 + 2M\tau_{YZ}^2 + 2N\tau_{ZX}^2 = 1 \tag{7.41}$$

where the X-Y-Z axes are aligned with the planes of material symmetry, and F, G, H, L, M, and N are empirical constants for the material. Let σ_{oX}, σ_{oY}, and σ_{oZ} be the uniaxial yield strengths in the three directions, and let τ_{oXY}, τ_{oYZ}, and τ_{oZX} be shear yield strengths on the respective orthogonal planes. The empirical constants can be evaluated from the various yield strengths as follows:

$$F + H = \frac{1}{\sigma_{oX}^2}, \quad F + G = \frac{1}{\sigma_{oY}^2}, \quad G + H = \frac{1}{\sigma_{oZ}^2}$$

$$2L = \frac{1}{\tau_{oXY}^2}, \quad 2M = \frac{1}{\tau_{oYZ}^2}, \quad 2N = \frac{1}{\tau_{oZX}^2}$$

(7.42)

The Hill criterion as just described can also be used with reasonable success as a *fracture* criterion for orthotropic composite materials. The equations are the same except that the various yield strengths are replaced by the corresponding ultimate strengths. However, different values of the constants are generally needed for tension versus compression, and other complexities exist for composite materials that may not be fully predicted by this criterion.

If a material is uneven, having different yield strengths in tension and compression, this suggests that a dependence on hydrostatic stress needs to be added. One proposed yield criterion for this situation is as follows:

$$(\sigma_1 - \sigma_2)^2 + (\sigma_2 - \sigma_3)^2 + (\sigma_3 - \sigma_1)^2 + 2\left(|\sigma_{oc}| - \sigma_{ot}\right)(\sigma_1 + \sigma_2 + \sigma_3) = 2\,|\sigma_{oc}|\,\sigma_{ot}$$

(7.43)

where σ_{ot} and σ_{oc} are the yield strengths in tension and compression, respectively, with the negative sign on σ_{oc} being removed by use of the absolute value.

Polymers often have somewhat higher yield strengths in compression than in tension. This is illustrated by some biaxial test results for several such materials in Fig. 7.12. The ratios $|\sigma_{oc}/\sigma_{ot}|$ for these particular polymers range from 1.20 to 1.33. The behavior expected from Eq. 7.43 using a typical value of $|\sigma_{oc}/\sigma_{ot}| = 1.3$ is plotted. The resulting off-center ellipse is in reasonable agreement with the data.

7.6.3 Fracture in Brittle Materials

The maximum normal stress criterion gives reasonably accurate predictions of fracture in brittle materials as long as the normal stress having the largest absolute value is tensile. However, deviations from this criterion occur if the normal stress having the largest absolute value is compressive. Data illustrating this trend for gray cast iron are shown in Fig. 7.13. The most prominent feature of the deviation is that the ultimate strength in compression is higher than that in tension by more than a factor of three.

Recall from Chapters 2 and 3 that brittle materials, such as ceramics and glasses and some cast metals, commonly contain large numbers of randomly oriented microscopic cracks or other planar interfaces that cannot support significant tensile stress. For example, the numerous fissures in natural stone have this effect, as do the graphite flakes in gray cast iron. Tensile normal stresses are expected to open these flaws and therefore

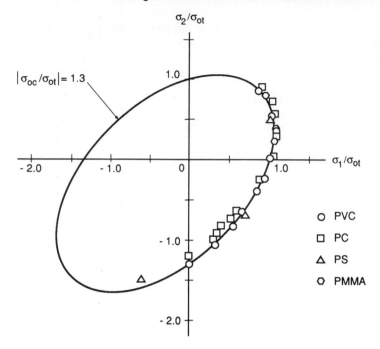

Figure 7.12   Biaxial yield data for various polymers compared to a modified octahedral shear stress theory. (After Raghava, Caddell, and Yeh [Raghava 73]; used with permission.)

to cause them to grow. Thus, failure is expected to occur on the plane where the maximum tensile normal stress occurs and to be controlled by this stress. For example, gray cast iron fails normal to the maximum tensile stress in both tension and torsion, as seen in the photographs of Figs. 5.14 and 5.41.

However, if the dominant stresses are compressive, the planar flaws (cracks, etc.) tend to have their opposite sides pressed together so that they have less effect on the behavior. This explains the higher strengths in compression for brittle materials. Also, failure occurs on planes inclined to the planes of principal normal stress and more nearly aligned with planes of maximum shear. See the compressive fractures of gray iron and concrete in Figs. 5.23 and 5.24.

One possibility for handling the differing behavior of brittle materials in tension and compression is simply to modify the maximum normal stress criterion so that the compressive and tensile ultimate strengths differ. This would give the rectangle shown in Fig. 7.13, which still does not agree with the data. In addition, any successful fracture criterion should predict that even brittle materials do not fail under hydrostatic compression, which is in agreement with both observation and intuition.

Therefore, additional failure criteria need to be considered that are capable of predicting the behavior of brittle materials. A number of such criteria exist, and we will consider two of the simpler ones in the portions of this chapter that follow.

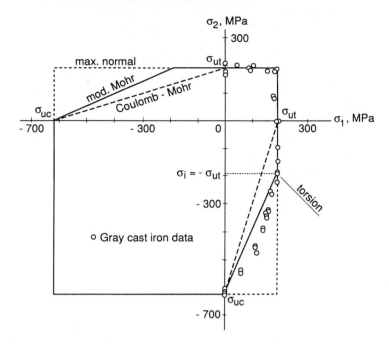

Figure 7.13   Biaxial fracture data of gray cast iron compared to various fracture criteria. (Data from [Grassi 49].)

7.7 COULOMB-MOHR FRACTURE CRITERION

In the Coulomb-Mohr (C-M) criterion, fracture is hypothesized to occur on a given plane in the material when a critical combination of shear and normal stress acts on this plane. In the simplest application of this approach, the mathematical function giving the critical combination of stresses is assumed to be a linear relationship.

$$|\tau| + \mu\sigma = \tau_i \quad \text{(at fracture)} \tag{7.44}$$

where τ and σ are the stresses acting on the fracture plane and μ and τ_i are constants for a given material. This equation forms a line on a plot of σ versus $|\tau|$ as shown in Fig. 7.14. The intercept with the τ axis is τ_i, and the slope is $-\mu$, where both τ_i and μ are defined as positive values.

Now consider a set of applied stresses, which can be specified in terms of the principal stresses, σ_1, σ_2, and σ_3, and plot the Mohr's circles for the principal planes on the same axes as Eq. 7.44. The failure condition is satisfied if the largest of the three circles is tangent to (just touches) the Eq. 7.44 line. If the largest circle does not touch the line, a safety factor greater than unity exists. Intersection of the largest circle and the line is not permissible, as this indicates that failure has already occurred. The line is therefore said to represent a *failure envelope* for Mohr's circle.

The point of tangency of the largest circle to the line occurs at a point (σ', τ') that represents the stresses on the plane of fracture. The orientation of this predicted plane of

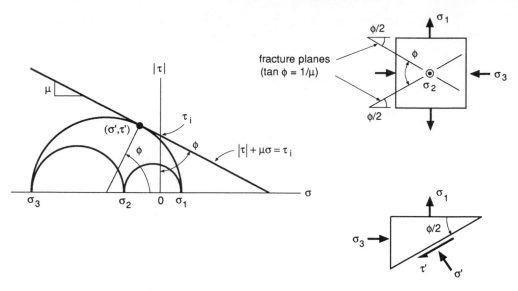

Figure 7.14 Coulomb-Mohr fracture criterion as related to Mohr's circle and predicted fracture planes.

fracture can be determined from the largest circle. In particular, fracture is expected to occur on a plane that is related to the plane where the maximum principal stress acts by a rotation $\phi/2$ in either direction. These planes are illustrated in Fig. 7.14. Also, from the geometry in Fig. 7.14, this angle ϕ can be related to the constant μ.

$$\tan\phi = \frac{1}{\mu} \tag{7.45}$$

The shear stress τ' that causes failure is thus affected by the normal stress σ' acting on the same plane. In particular, τ' increases if σ' is more compressive. Such behavior is logical for materials where a brittle shear fracture is influenced by numerous small and randomly oriented planar flaws. More compressive σ' is expected to cause more friction between the opposite faces of the flaws, thus increasing the τ' necessary to cause fracture.

7.7.1 Development of the Coulomb-Mohr Criterion

The C-M theory can be expressed in terms of the principal normal stresses with the aid of Fig. 7.14.

$$\sigma' = \frac{\sigma_1 + \sigma_3}{2} + \left|\frac{\sigma_1 - \sigma_3}{2}\right|\cos\phi$$

$$|\tau'| = \left|\frac{\sigma_1 - \sigma_3}{2}\right|\sin\phi \tag{7.46}$$

where σ_1 and σ_3 are assumed to be the maximum and minimum principal normal stresses, respectively. Combining Eqs. 7.44 to 7.46 and performing manipulation using simple trigonometric identities leads to

$$|\sigma_1 - \sigma_3| + m\,(\sigma_1 + \sigma_3) = 2\tau_u \tag{7.47}$$

where the new constants m and τ_u are related to the previous ones by

$$m = \frac{\mu}{\sqrt{1+\mu^2}} = \cos\phi \qquad \text{(a)}$$

$$\tau_u = \frac{\tau_i}{\sqrt{1+\mu^2}} = \tau_i \sin\phi \qquad \text{(b)}$$

$$\tag{7.48}$$

Consider a test in pure torsion, where at fracture

$$\sigma_1 = -\sigma_3 = \tau, \qquad \sigma_2 = 0 \tag{7.49}$$

Substitution into Eq. 7.47 yields $\tau = \tau_u$, so that the constant τ_u is the pure shear stress necessary to cause fracture. The corresponding largest Mohr's circle and predicted fracture planes are illustrated in Fig. 7.15.

Similarly applying Eq. 7.47 to uniaxial tension and compression tests gives the following equations for the ultimate strengths in tension and compression, σ_{ut} and σ_{uc}, respectively.

$$\sigma_{ut} = \frac{2\tau_u}{1+m}, \qquad \sigma_{uc} = \frac{-2\tau_u}{1-m} \qquad \text{(a,b)} \tag{7.50}$$

The Mohr's circles and predicted fracture planes for uniaxial tension and compression are shown in Fig. 7.16. Eliminating τ_u between the above two equations gives a predicted relationship between σ_{ut} and σ_{uc}.

$$\sigma_{ut} = -\sigma_{uc}\left(\frac{1-m}{1+m}\right) \tag{7.51}$$

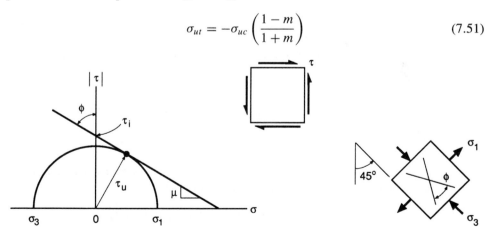

Figure 7.15 Pure torsion and the fracture planes predicted by the Coulomb-Mohr criterion.

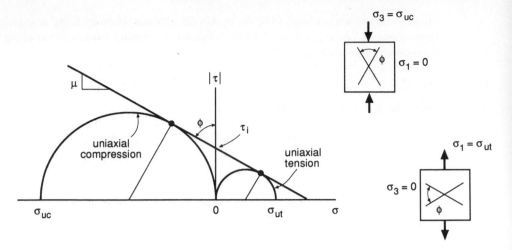

Figure 7.16 Fracture planes predicted by the Coulomb-Mohr criterion for uniaxial tests in tension and compression.

Also, solving for m gives

$$m = \frac{\sigma_{uc} + \sigma_{ut}}{\sigma_{uc} - \sigma_{ut}} \tag{7.52}$$

Thus, the C-M criterion predicts that a single constant m can be used to relate the strengths in tension, compression, and shear. For positive values of m, the strength in tension is predicted to be less than that in compression, which is in agreement with the trend observed for brittle materials.

7.7.2 Graphical Representation of the Coulomb-Mohr Criterion

If the subscripts for the principal stresses are assumed to be arbitrarily assigned, then Eq. 7.47 must be generalized to

$$|\sigma_1 - \sigma_2| + m(\sigma_1 + \sigma_2) = 2\tau_u \quad \text{(a)}$$

$$|\sigma_2 - \sigma_3| + m(\sigma_2 + \sigma_3) = 2\tau_u \quad \text{(b)} \tag{7.53}$$

$$|\sigma_3 - \sigma_1| + m(\sigma_3 + \sigma_1) = 2\tau_u \quad \text{(c)}$$

Note that these actually represent six equations due to the absolute values, fracture being predicted if any one of them is satisfied. For plane stress with $\sigma_3 = 0$, these reduce to

$$|\sigma_1 - \sigma_2| + m(\sigma_1 + \sigma_2) = 2\tau_u$$

$$|\sigma_2| + m\sigma_2 = 2\tau_u$$

$$|\sigma_1| + m\sigma_1 = 2\tau_u \tag{7.54}$$

The six lines represented by the latter equations form the boundaries of a region of no failure as shown in Fig. 7.17. The unequal fracture strengths in tension and compression are related to τ_u by Eq. 7.50.

For the general case of a three-dimensional state of stress, Eq. 7.53 represent six planes that give a failure surface as shown in Fig. 7.18. The surface forms a vertex along the line $\sigma_1 = \sigma_2 = \sigma_3$ at the point

$$\sigma_1 = \sigma_2 = \sigma_3 = \frac{\tau_u}{m} \tag{7.55}$$

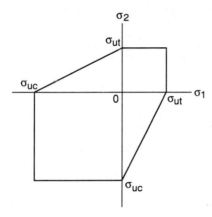

Figure 7.17   Failure locus for the Coulomb-Mohr fracture criterion for plane stress.

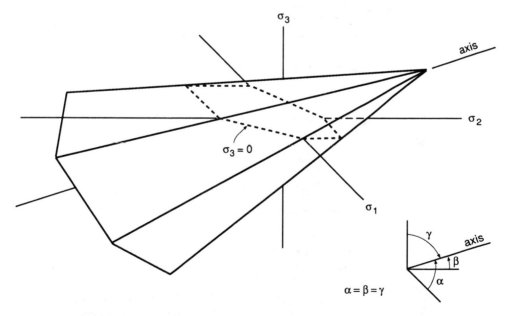

Figure 7.18   Three-dimensional failure surface for the Coulomb-Mohr fracture criterion.

Hence, the value of m, or of the closely related constant μ, determines where the vertex is formed. Higher values of m or μ indicate that the six planes are tilted more abruptly relative to one another and form a vertex closer to the origin. If any one of σ_1, σ_2, or σ_3 is zero, the intersection of this surface with the plane of the remaining two principal stresses forms the shape of Fig. 7.17.

A cross section of the failure surface along a plane normal to the line $\sigma_1 = \sigma_2 = \sigma_3$ forms a six-sided figure. However, due to the tilting of the planes relative to one another, such a cross section is not a regular hexagon, and it changes its size depending on the distance from the origin, that is, depending on the value of the hydrostatic stress σ_h. Several such cross sections are shown in Fig. 7.19 for the relatively large value of $m = 0.5$, which corresponds to $\mu = 0.577$.

From comparison of Eqs. 7.26 and 7.53, it is evident that the C-M criterion with $m = 0$ is equivalent to a maximum shear stress criterion. Figure 7.17 then takes the same more symmetrical shape as Fig. 7.5. The vertex of the failure surface is moved to infinity, and its cross sections become perfect hexagons all of the same size. Thus, the C-M criterion contains the maximum shear criterion as a special case.

7.7.3 Discussion of the Coulomb-Mohr Criterion

The C-M criterion with a positive value of μ is consistent with a number of observations that are typical of brittle materials. First, the fracture strength in compression is greater than that in tension, with the difference increasing with the value of μ. Test data showing different strengths in tension and compression have already been presented in Fig. 7.13 for gray cast iron. Data for the more extreme case of a ceramic material are shown in Fig. 7.20.

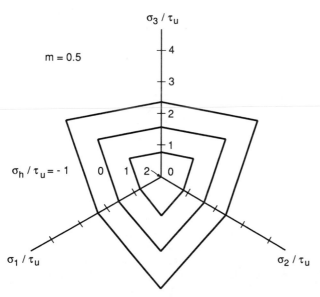

Figure 7.19 Cross sections of the Coulomb-Mohr failure surface normal to its axis for one choice of the constant m.

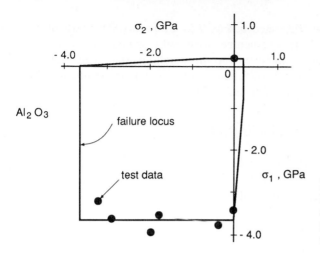

Figure 7.20 Test data and failure locus for biaxial compression of the ceramic alumina, Al_2O_3. Each point plotted is the average from 3 or 4 tests as reported by [Sines 75].

The plane of fracture in compression is often observed to be an acute angle relative to the loading axis on the order of $\phi/2 = 30°$ to $40°$. (See the fractured compression specimens of cast iron and concrete in Figs. 5.23 and 5.24 and compare to Fig. 7.16.) From Eq. 7.45, this corresponds approximately to μ values in the range 0.15 to 0.6.

However, the fracture planes predicted for a tension test are incorrect. Brittle materials generally fail in tension on planes near the plane normal to the maximum tension stress, that is, normal to the specimen axis, not on planes as shown in Fig. 7.16. Failures of brittle materials in torsion generally also occur on planes normal to the maximum tension stress, not on the planes predicted by the C-M theory as in Fig. 7.15. (See the broken tension and torsion specimens of cast iron in Figs. 5.14 and 5.41.) Moreover, the fracture strengths in tension, compression, and shear are not generally related to one another as predicted by a single value of m used with the equations above. This is evident by comparison of the test data with the line for the C-M theory in Fig. 7.13.

The various difficulties with the C-M theory can be avoided by using this theory in combination with the maximum normal stress theory as described next.

7.8 MODIFIED MOHR FRACTURE CRITERION

The difficulties described above for the Coulomb-Mohr fracture criterion can be avoided by using the criterion only where the behavior is dominated by compressive stresses. The maximum normal stress criterion is then used where the behavior is dominated by tension.

Empirical evidence suggests that tension-dominated behavior occurs in torsion tests of brittle materials. In particular, the maximum principal stress at fracture is close to the uniaxial tension strength, as for the data in Figs. 7.11 and 7.13. Also, failure occurs on planes normal to the maximum tensile stress, that is, $45°$ relative to the axis of a round torsion member. In fact, tension-dominated behavior may extend to ratios σ_2/σ_1 below the value of -1 corresponding to pure torsion. The two fracture criteria thus agree at some stress σ_i, as illustrated in Fig. 7.21 by line A. The particular choice of $\sigma_i = -\sigma_{ut}$, line B in Fig. 7.21, is sometimes used and is compared to some test data in Fig. 7.13. The line B choice corresponds to the situation where the limits imposed by the maximum normal stress and the Coulomb-Mohr criteria are simultaneously satisfied for pure torsion, which is illustrated in Fig. 7.21(b).

7.8.1 Details of the Modified Mohr Criterion

The combined criterion is called the *modified Mohr fracture criterion*, and in three dimensions, the resulting failure surface is similar to Fig. 7.22. The three positive faces of the maximum normal stress cube truncate the C-M failure surface. These faces correspond to the three planes

$$\sigma_1 = \sigma_{ut}, \quad \sigma_2 = \sigma_{ut}, \quad \sigma_3 = \sigma_{ut} \tag{7.56}$$

The two failure surfaces intersect, and the two theories agree, along six edges of these faces. (Four of these edges can be seen, and two are hidden, in Fig. 7.22.) For plane stress, a failure locus as in Fig. 7.21(a) is obtained as the intersection of the failure surface with a plane such as $\sigma_3 = 0$.

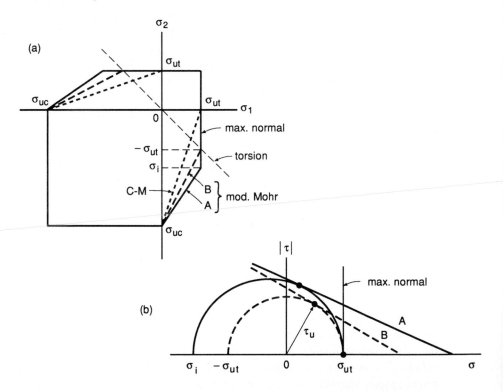

Figure 7.21   Some variations of the modified Mohr fracture criterion.

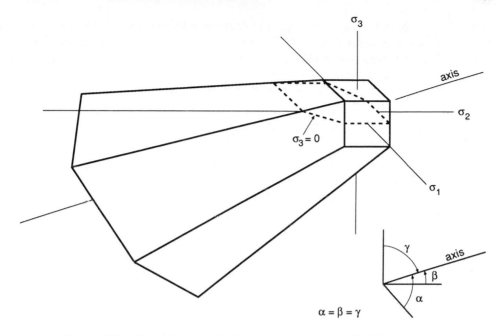

Figure 7.22   Three-dimensional failure surface for the modified Mohr fracture criterion for the particular case of $\sigma_i = -\sigma_{ut}$. The Coulomb-Mohr surface is truncated by three faces of the maximum normal stress cube.

Using Eq. 7.53 as a starting point, the constants for the C-M portion of the modified theory can be developed. Two conditions must be met. First, for uniaxial compression, the failure stress must be σ_{uc}. Appropriate substitution into Eq. 7.53 gives a result analogous to Eq. 7.50(b).

$$2\tau_u' = \sigma_{uc}(m - 1) \tag{7.57}$$

where the prime is introduced for τ_u' as this constant is in general no longer equal to the strength in pure shear. Second, there must be agreement with the maximum normal stress theory at σ_i, so that Eq. 7.53 must be satisfied by the combination

$$\sigma_1 = \sigma_{ut}, \quad \sigma_2 = 0, \quad \sigma_3 = \sigma_i \tag{7.58}$$

Substitution of this into one of the pair of equations represented by Eq. 7.53(c), and also using Eq. 7.57 to evaluate τ_u, permits the constant m to be obtained.

$$m = \frac{\sigma_{uc} + \sigma_{ut} - \sigma_i}{\sigma_{uc} - \sigma_{ut} - \sigma_i} \tag{7.59}$$

for which σ_{uc} and usually σ_i must be used with their appropriate negative signs.

Hence, the constants τ_u' and m for Eq. 7.53, which described the C-M portion of the failure surface, may be determined from three empirical constants, namely σ_{uc}, σ_{ut}, and σ_i. If $\sigma_i = -\sigma_{ut}$ is chosen, one of these becomes known, and Eq. 7.59 reduces to

$$m = 1 + \frac{2\sigma_{ut}}{\sigma_{uc}} \tag{7.60}$$

As before, it is convenient to define an effective stress that can be used to compare applied stresses to those necessary to cause failure. This can be accomplished by defining the following three quantities:

$$C_1 = \frac{\sigma_{ut}}{\sigma_{uc}(m-1)}[|\sigma_1 - \sigma_2| + m(\sigma_1 + \sigma_2)]$$

$$C_2 = \frac{\sigma_{ut}}{\sigma_{uc}(m-1)}[|\sigma_2 - \sigma_3| + m(\sigma_2 + \sigma_3)]$$

$$C_3 = \frac{\sigma_{ut}}{\sigma_{uc}(m-1)}[|\sigma_3 - \sigma_1| + m(\sigma_3 + \sigma_1)] \qquad (7.61)$$

These are obtained from Eq. 7.53 by multiplying by a factor that causes the right hand side to equal σ_{ut} at failure. The largest of C_1, C_2, and C_3 is an effective stress for the C-M portion of the failure surface that can be compared to σ_{ut}. Considering the truncation by the maximum normal stress criterion, the complete definition of the effective stress must be

$$\bar{\sigma}_M = \text{MAX}\,(C_1, C_2, C_3, \sigma_1, \sigma_2, \sigma_3)$$

$$\bar{\sigma}_M = 0 \text{ if } \text{MAX} < 0 \qquad (7.62)$$

The safety factor is obtained by comparing the effective stress to the ultimate strength in tension.

$$X = \frac{\sigma_{ut}}{\bar{\sigma}_M} \qquad (7.63)$$

If the largest of the candidate $\bar{\sigma}_M$ values is negative, this implies that the loading path never intersects the failure surface, explaining why $\bar{\sigma}_M$ must be zero in this situation.

7.8.2 Comments on the Modified Mohr Criterion

If only values of σ_{uc} and σ_{ut} are available, σ_i and hence m cannot in general be determined. If $\sigma_i = \sigma_{ut}$ is thought to be a reasonable choice, then m can be obtained from Eq. 7.60. Hence, such a choice will often be a practical necessity. As σ_i appears generally to be below $-\sigma_{ut}$, this choice (line B in Fig. 7.21) is expected to be conservative compared to the line A choice if at least one principal stress is positive. If all three principal stresses are negative, the $\sigma_i = \sigma_{ut}$ approximation may become nonconservative, and caution is advised in this case.

At least a rough estimate of m may be made from the inclination $\phi/2$ of the fracture plane in a compression test by using Eq. 7.48(a). Such a value could then be used directly for Eq. 7.61 where σ_i is unknown.

In all of the above discussion, we have assumed that m is constant, that is, that the Mohr failure envelope is a straight line. This may not be true over a wide range of stress states in real materials, so that caution is needed in employing this idealization. A more general curved envelope could be used where sufficient test data are available to provide the detailed information needed. Such approaches are described in the literature, for example in the books of Nadai (1950) and Chen and Han (1988).

7.9 ADDITIONAL COMMENTS ON FAILURE CRITERIA

In an attempt to gain a broader perspective on the subject of this chapter, we will engage in some limited additional discussion on three topics. These are brittle versus ductile behavior, stress raiser effects, and time-dependent crack growth.

7.9.1 Brittle Versus Ductile Behavior

Engineering materials that are commonly classed as ductile are those for which the static strength in engineering applications is generally limited by yielding. Many metals and polymers fit into this category. In contrast, the usefulness of materials commonly classed as brittle is generally limited by fracture. In a tension test, brittle materials exhibit no well-defined yielding behavior, and they fail after only a small elongation, on the order of 5% or less. Examples are gray cast iron and certain other cast metals, and also stone, concrete, other ceramics, and glasses.

However, normally brittle materials may exhibit considerable ductility when tested under loading such that the hydrostatic component σ_h of the applied stress is highly compressive. Such an experiment can be conducted by testing the material in a chamber that is already pressurized. The surprising result of large plastic deformations in a normally brittle material is illustrated by some stress-strain curves for limestone in Fig. 7.23.

Also, materials normally considered ductile fail with increased ductility if the hydrostatic stress is compressive, or reduced ductility if it is tensile. For example, although the initial yielding of metals is insensitive to hydrostatic stress, the point of fracture is affected. Data showing this for a steel are given in Fig. 7.24, where the true fracture

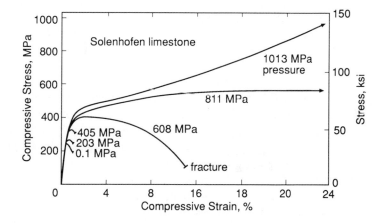

Figure 7.23 Stress-strain data for limestone cylinders tested under axial compression with various hydrostatic pressures ranging from one to 10,000 atmospheres. The applied compressive stress plotted is the stress in the *pressurized laboratory*, that is, the compression in excess of pressure. (Adapted from [Griggs 36]; used with permission; copyright ©The University of Chicago Press.)

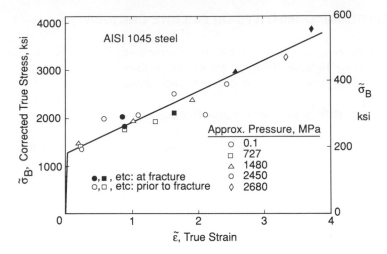

Figure 7.24 Effect of pressures ranging from one to 26,500 atmospheres on the tensile behavior of a steel, specifically AISI 1045 with $HRC = 40$. Stress in the *pressurized laboratory* is plotted. (Data from [Bridgman 52] pp. 47–61.)

stress and strain are seen to increase with pressure, that is, with hydrostatic compression. The fracture event is seen to shift to a later point along a common stress-strain curve.

To explain such behavior, it is useful to adopt the viewpoint that fracture and yielding are separate events and that either one may occur first depending on the combination of material and stress state involved. In three-dimensional principal normal stress space, the limiting surface for yielding (at least for metals) appears to be a cylinder or other prismatic shape symmetrical about the line $\sigma_1 = \sigma_2 = \sigma_3$, such as the surfaces of Figs. 7.6 and 7.9. Limiting surfaces for fracture are generally similar to that for the modified Mohr theory as discussed above and illustrated in Fig. 7.22, although the boundaries may actually be smooth curves.

The situation is illustrated in Fig. 7.25. For certain states of stress, the yield surface is encountered first, whereas for others the fracture surface is encountered first. The relative dimensions of the two surfaces change for different materials. For normally ductile materials, fracture prior to yielding is not expected except for stress states involving a large hydrostatic tension. The stress may be increased by varying amounts beyond yielding before fracture occurs, depending on the amount of hydrostatic compression. However, for normally brittle materials, there is contrasting behavior as fracture occurs prior to yielding except for stress states involving a large hydrostatic compression. Thus, if a wide range of stress states are of interest for any material, it is important to consider the possibility that either yielding or fracture may occur first.

7.9.2 Stress Raiser (Notch) Effects

Geometric discontinuities such as holes, fillets, and grooves, which are collectively known as *stress raisers*, or *notches*, cause the stress to be locally elevated. This is illustrated

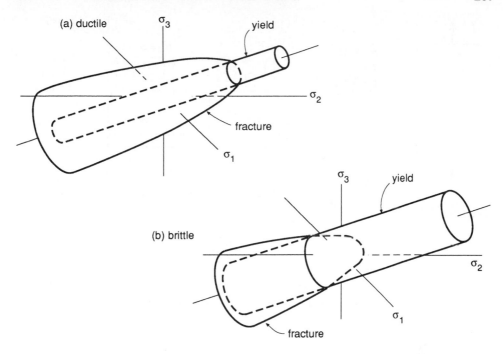

(a) ductile σ_3 yield

σ_2

fracture

σ_1

σ_3 yield

(b) brittle

σ_2

σ_1

fracture

Figure 7.25   Relationships of the limiting surfaces for yielding and fracture for materials that usually behave in a ductile manner, and also for materials that usually behave in a brittle manner.

in Fig. 7.26(a) and (b). A common method of describing the situation is to use stress concentration factors, k_t, as discussed in most books on mechanics of materials or mechanical design. A nominal or average stress S is defined, and the local stress at the notch is considered to be elevated by a factor k_t, so that its value is $\sigma = k_t S$. The handbook of Peterson (1974) is a particularly useful collection of graphs giving k_t values based on linear-elastic stress analysis. (Look ahead to Figs. 10.1 and 10.2 for a sampling of this type of information.)

If a notched member is made of a ductile material, yielding occurs first in a small region near the notch as shown in Fig. 7.26(c). Increased loading causes yielding to spread so that the stress raiser has little effect on the final strength of the member as shown in (d). Thus, for design against failure due to static loads, it is not necessary to consider the effect of a notch if the material is ductile and if the notch effect is localized.

However, if the material does not yield prior to fracture, the locally elevated stress situation prevails up to the point of fracture. The local notch stress $\sigma = k_t S$ should then be compared with the ultimate strength σ_u to determine whether or not fracture is likely. In some cases, the notch may not weaken the material as much as would be expected from k_t, so that use of $\sigma = k_t S$ may be conservative. This occurs where the material itself contains fairly large internal flaws, as in gray cast iron, so that a small geometric notch does not add very much to the stress raisers already present.

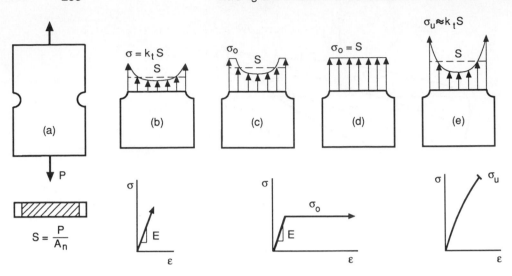

Figure 7.26 Component with a stress raiser (a) and stress distributions for various cases: (b) linear-elastic deformation, (c) local yielding for a ductile material, (d) full yielding for a ductile material, and (e) brittle material at fracture.

Example 7.5

Consider the spherical glass pressure vessel from Probs. 6.6 and 7.3, where the wall thickness is $t = 2.5$ mm, the diameter is $d = 2r = 150$ mm, and the pressure is $p = 0.5$ MPa. The opening in this vessel is made by joining it with a tube as shown in Figure E7.5. The geometric details were carefully chosen so that the stress concentration factor is only $k_t = 1.2$. What is the safety factor against fracture if the effect of this k_t is included?

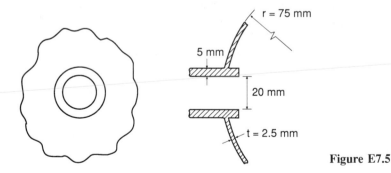

Figure E7.5

Solution   Using a coordinate system attached to the surface of the sphere, with the x-y plane being tangent to the surface, the nominal stresses due to the pressure are

$$S_x = S_y = \frac{pr}{2t} = \frac{(0.5 \text{ MPa}) (75 \text{ mm})}{2 (2.5 \text{ mm})} = 7.5 \text{ MPa}$$

Assuming that the given k_t applies equally to both of these, the local stresses are

$$\sigma_x = \sigma_y = k_t S = 1.2\,(7.5) = 9.0 \text{ MPa}$$

As no shear stress τ_{xy} is present, these are two of the principal normal stresses, and the third is approximately $\sigma_z = \sigma_3 = 0$.

Using the maximum normal stress fracture criterion, the effective stress is thus

$$\bar{\sigma}_N = \text{MAX}\,(|\sigma_1|,\,|\sigma_2|,\,|\sigma_3|) = 9.0 \text{ MPa}$$

and the safety factor is

$$X = \frac{\sigma_{ut}}{\bar{\sigma}_N} = \frac{50 \text{ MPa}}{9 \text{ MPa}} = 5.56 \qquad\qquad \textbf{Ans.}$$

where σ_u in tension is approximated as the value for soda-lime glass from Table 3.10.

7.9.3 Effects of Cracks and Their Time-Dependent Growth

As already noted, normally brittle materials usually contain, or easily develop, small flaws or other geometric features that are equivalent to small cracks. Brittle failure generally occurs as a result of such cracks growing and joining. This process is often time-dependent, principally because it is affected by the presence of moisture (water) or other substances that are present and react chemically with the material. Time-dependent cracking causes the fracture behavior to be dependent on the loading rate. Also, if the stress is held constant, failure can occur after some time has elapsed at a stress that would not cause fracture if maintained for only a short time.

Detailed consideration of the time-dependent growth of cracks would require the use of fracture mechanics, which is the topic of the next chapter. However, that technology is not yet capable of predicting time-dependent behavior for general three-dimensional states of stress. Given this situation, stress-based criteria are still needed, but the fracture strengths should not be considered to be independent of time. Thus, the approaches of this chapter should be used with some caution to assure that the material properties employed are realistic with respect to the time element.

7.10 SUMMARY

For yielding of ductile materials, the safety factor may be obtained by comparing the yield stress with an effective stress.

$$X = \frac{\sigma_o}{\bar{\sigma}} \qquad\qquad (7.64)$$

Two yield criteria that are reasonably accurate are widely used for isotropic materials, namely the maximum shear stress criterion and the octahedral shear stress criterion. The effective stresses for these are respectively as follows:

$$\bar{\sigma}_S = \text{MAX}\,(|\sigma_1 - \sigma_2|,\,|\sigma_2 - \sigma_3|,\,|\sigma_3 - \sigma_1|) \qquad\qquad (7.65)$$

$$\bar{\sigma}_H = \frac{1}{\sqrt{2}} \sqrt{(\sigma_1 - \sigma_2)^2 + (\sigma_2 - \sigma_3)^2 + (\sigma_3 - \sigma_1)^2} \qquad (7.66)$$

Effective stresses, and hence safety factors, from these two criteria never differ by more than 15%. In their basic forms according to these two equations, both predict that hydrostatic stresses have no effect. Modifications of these can be used to predict yielding in anisotropic or uneven materials.

For brittle materials, no single basic failure criterion suffices to describe the fracture behavior. The modified Mohr criterion is a reasonable choice. It is a combination of the maximum normal stress criterion, which is used where the stresses are dominated by tension, and the Coulomb-Mohr criterion. The latter assumes that fracture occurs when the combination of normal and shear stress on any plane in the material reach a critical value given by

$$|\tau| + \mu\sigma = \tau_i \qquad (7.67)$$

where μ and τ_i are material constants. The Coulomb-Mohr criterion can be considered to be a shear stress criterion in which the limiting shear stress increases for greater amounts of hydrostatic compression.

In applying the modified Mohr criterion, values are needed for three material constants. These can be the ultimate strengths in tension and compression, σ_{ut} and σ_{uc}, and one additional constant σ_i related to specifying where the two criteria intersect. The inclination of the fracture plane in compression tests can be used to estimate σ_i, or it can be obtained from biaxial test data, and $\sigma_i = -\sigma_{ut}$ is sometimes used. Expressions for applying the modified Mohr criterion are provided as Eqs. 7.59 to 7.62.

Under high hydrostatic compression, normally brittle materials may behave in a ductile manner, and ductile materials fracture at higher true stresses and strains than otherwise. Such behavior can be explained by considering yielding and fracture to be independent events with different failure surfaces. The possibility of either occurring first should generally be considered.

Where brittle behavior is likely, the stresses used in failure criteria should be elevated by stress concentration factors, k_t, for any localized stress raisers that are present. Also, fracture may be time dependent due to crack growth effects, so that caution is needed in applying failure criteria using materials constants from short-term tests.

NEW TERMS AND SYMBOLS

(a) Terms

anisotropic yield criterion

effective stresses: $\bar{\sigma}_H$, $\bar{\sigma}_M$, $\bar{\sigma}_N$, $\bar{\sigma}_S$

failure criterion (stress based)

failure surface

fracture criteria:
    Coulomb-Mohr
    maximum normal stress
    modified Mohr

principal normal stress space

proportional loading

ultimate strengths:
    compression, σ_{uc}
    shear, τ_u
    tension, σ_{ut}

yield criteria:
    maximum shear stress
    octahedral shear stress

(b) Constants for the Coulomb-Mohr (C-M) and Modified Mohr Criteria

μ, τ_i Slope and intercept, respectively, of the C-M failure envelope line

m, τ_u Constants for the C-M criterion expressed in terms of principal normal stresses

$\phi/2$ Fracture angle; $\phi = \tan^{-1}(1/\mu)$

σ_i For the modified Mohr criterion, stress where the maximum normal and C-M portions of the failure surface agree

REFERENCES

CHEN, W. F. and D. J. HAN. 1988 *Plasticity for Structural Engineers*, Springer-Verlag, New York.

HILL, R. 1983 *The Mathematical Theory of Plasticity*, Oxford University Press, London.

MENDELSON, A. 1968 *Plasticity: Theory and Applications*, Macmillan, New York. (Reprinted by R. E. Krieger, Malabar, Fl., 1983.)

NADAI, A. 1950 *Theory of Flow and Fracture of Solids*, McGraw-Hill, New York.

PAUL, B. 1961 "A Modification of the Coulomb-Mohr Theory of Fracture," *Jnl. of Applied Mechanics, Trans. ASME*, Ser. E., Vol. 28, No. 2, pp. 259-268.

PETERSON, R. E. 1974 *Stress Concentration Factors*, John Wiley, New York.

SEELY, F. B. and J. O. SMITH. 1952 *Advanced Mechanics of Materials*, 2nd ed., John Wiley, New York.

PROBLEMS AND QUESTIONS

Sections 7.3 and 7.4[2]

7.1 What is the safety factor against failure for the stress state of Prob. 6.2 if:
  (a) The material is AISI 1020 steel (as rolled) and yielding is considered failure?
  (b) The material is the ceramic silicon carbide (SiC) and fracture is considered failure?

7.2 For the situation of Prob. 6.5, what is the safety factor against yielding if the material is 18 Ni maraging steel (250 grade)?

7.3 For the situation of Prob. 6.6, what is the safety factor against fracture if the material is borosilicate glass and the pressure is maintained for only a short period of time? The strength of this glass is not expected to be less than that of soda-lime glass.

7.4 For the situation of Prob. 6.1, what yield strength is required for the material if a safety factor of 3.0 against yielding is required?

---

[2]Where specific materials are named, the needed properties can be found in one or more of Tables 3.10, 4.2, 5.2, 5.3, and 5.8.

7.5 A thin-walled circular tube must support a torque of 6 kN·m and a pure bending moment of 4.5 kN·m. If the wall thickness of the tube is 3 mm, what must the diameter be so that the safety factor against yielding is 1.5? The material is ASTM A514 (T1) structural steel.

7.6 A thin-walled cylindrical tube made of 7075-T6 aluminum has a wall thickness of 2.5 mm and a diameter at mid-thickness of 50 mm. An axial load of 60 kN tension is applied along with a torque of 1.0 kN·m. What is the safety factor against yielding?

7.7 A spherical pressure vessel must contain a pressure of 7 MPa and have a diameter of approximately 1.2 m.

  (a) What thickness is required if the material is AISI 1020 steel and if the safety factor against yielding must be 2.0?

  (b) Consider changing the material to ASTM A536 ductile cast iron (grade 65-45-12). Would you decrease the thickness? Would you use the same safety factor?

7.8 Consider a solid circular shaft subjected to bending and torsion so that the state of stress of interest involves only a normal stress σ_x and a shear stress τ_{xy}, with all other stress components being zero. (See Figure P7.8.)

  (a) Determine the safety factor against yielding as a function of the yield strength of the ductile material and the applied stresses.

  (b) Use the results of (a) to develop a design equation for the shaft giving diameter d as a function of yield strength, safety factor, bending moment M, and torque T.

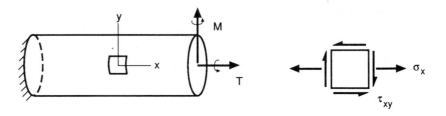

Figure P7.8

7.9 A thin-walled tube with closed ends must withstand both an internal pressure p and a bending moment M. Derive a design formula for the required thickness t if the diameter $d = 2r$, safety factor X, and yield strength σ_o of the ductile material are all considered known. (Suggestion: Note that the polar moment of inertia of the cross sectional area is approximately $J = 2\pi r^3 t$, and that $I = J/2$ for use in the bending stress formula. Using this, and approximating r as the same value in calculating both bending and pressure stresses, will greatly simplify the solution.)

7.10 A piece of a ductile metal is confined on two sides by a rigid die (Figure P7.10). A uniform compressive stress σ_z is applied to the surface of the metal, and it may be assumed that there is no friction between the metal and the die. Determine the value of σ_z necessary to cause yielding in terms of the uniaxial yield strength σ_o and the elastic constants of the material. Is the value of σ_z that causes yielding affected significantly by Poisson's ratio?

7.11 Repeat Prob. 7.10 for the case where the die confines the material on all four sides, that is, in both the x- and y-directions as shown in Figure P7.11.

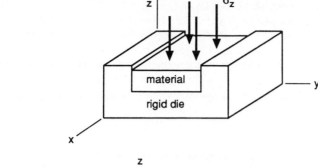

Figure P7.10

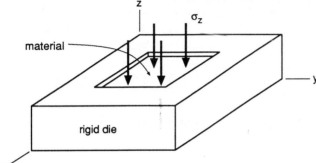

Figure P7.11

Section 7.5[3]

7.12 A solid circular shaft subjected to pure torsion must be designed to avoid yielding with a safety factor X. Find the required diameter as a function of the torque T and the yield strength σ_o using: (a) the maximum shear stress criterion, and (b) the octahedral shear stress criterion. How much do these two sizes differ?

7.13 Work Prob. 7.10 using both the maximum shear and the octahedral shear criteria and compare the results.

Section 7.6

7.14 A block of polycarbonate (PC) is loaded in compression and confined by a rigid die on two sides as in Prob. 7.10. The compressive yield strength is 20% higher than the tensile yield strength and Poisson's ratio is $\nu = 0.42$.

    (a) Estimate the value of σ_z necessary to cause yielding. Note that in a tension test of this material the upper yield strength and ultimate strength points are the same.

    (b) Qualitatively sketch the yield locus for plane stress, in this case $\sigma_x = \sigma_3 = 0$, and show the location of the point corresponding to your answer to (a).

7.15 Specialize the anisotropic yield criterion of Hill, Eq. 7.41, to the case of plane stress. If the yield strengths σ_{oX}, σ_{oY}, and τ_{oXY} are known, can the needed constants be obtained so that the criterion can be used? If not, suggest an additional test on the material and explain how you would use the result to evaluate the needed constants.

---

[3]All of the earlier problems except 7.3 can also be worked by the octahedral shear criterion of Section 7.5. Instructors may allow students to choose the method, or they may specify that one or both be used, as desired.

7.16 Consider the anisotropic yield criterion of Hill, Eq. 7.41.
   (a) Describe the yield locus on a plot of σ_1 vs. σ_2 for plane stress.
   (b) For the general three-dimensional case, what is the shape of the yield surface in principal normal stress space?
   (c) Under what conditions does this criterion reduce to the isotropic octahedral shear stress criterion?

7.17 Answer the questions of Prob. 7.16 for the hydrostatic-stress-sensitive yield criterion of Eq. 7.43.

Section 7.7 and 7.8

7.18 A brittle material has ultimate tensile and compressive strengths of 300 and -1000 MPa, respectively, and it behaves according to the modified Mohr fracture criterion. Biaxial data give a value of the stress $\sigma_i = -500$ MPa, above which the behavior is tension-dominated.
   (a) Draw the biaxial failure locus on σ_1 versus σ_2 coordinates to scale as on graph paper.
   (b) Also draw to scale, on σ versus $|\tau|$ coordinates, both the limiting Coulomb-Mohr and normal stress envelopes.
   (c) Graphically determine the safety factor for biaxial principal stresses as follows:
      (1) $\sigma_1 = 200$, $\sigma_2 = -100$ MPa
      (2) $\sigma_1 = 100$, $\sigma_2 = -600$ MPa
      (3) $\sigma_1 = -300$, $\sigma_2 = -600$ MPa

7.19 Some ultimate strength data from tests on cylinders of two geological materials are given as follows with compression (pressure) considered negative. Construct Coulomb-Mohr envelopes on graphs of σ versus $|\tau|$ for each material by first drawing Mohr's circle for each test. Then evaluate the constants m and τ_u for each material. For the sandstone, comment on the reasonableness of the linear C-M idealization, Eq. 7.44.

Lateral Stress $\sigma_1 = \sigma_2$, MPa	Applied Stress in Excess of Pressure $(\sigma_3 - \sigma_1)$, MPa	Type Test
Limestone		
0	4	Simple tension
0	-83	Simple compression
-100	-419	Compression plus pressure
Siliceous sandstone		
0	≈ 3	Simple tension
0	-100	Simple compression
-100	-600	Compression plus pressure
-200	-1030	Compression plus pressure

Source: Data in [Jaeger 69] pp. 75 and 156.

7.20 Consider the special case of a loose granular material, such as sand, which has zero tensile strength. Sketch the Coulomb-Mohr envelope on coordinates of σ versus $|\tau|$. How can Eq. 7.44 be specialized for this case? If you can, make a physical interpretation of the constant μ.

Section 7.9

7.21 An unusual new material is hypothesized to fail when the absolute value of the hydrostatic stress exceeds a critical value.

$$\left| \frac{\sigma_x + \sigma_y + \sigma_z}{3} \right| = \sigma_c$$

However, there is also a possibility that this material obeys either the maximum normal stress failure criterion or the maximum shear stress failure criterion.

(a) Does the above equation constitute a possible failure theory? Why or why not?

(b) In three-dimensional principal normal stress space, describe the failure surface corresponding to the above equation. Also describe the failure locus for the special case of plane stress.

(c) Describe a critical experiment, consisting of one or a few mechanical tests and a minimum of experimentation, that provides a definitive choice among the three criteria mentioned above. Note that some of the mechanical tests that are feasible are: uniaxial tension and compression, torsion of tubes and rods, internal and external pressure of closed-end tubes, biaxial tension in pressurized diaphragms, and hydrostatic compression.

7.22 Do Fig. 7.25(a) and (b) apply for ductile and brittle polymers, respectively? If not, please sketch what you think the various surface might look like.

8

Fracture of Cracked Members

8.1 INTRODUCTION

The presence of a crack in a component of a machine, vehicle, or structure may weaken it so that it fails by fracturing into two or more pieces. This can occur at stresses below the material's yield strength, where failure would not normally be expected. As an example, photographs from a propane tank truck failure caused in part by pre-existing cracks are shown in Fig. 8.1. Where cracks are difficult to avoid, a special methodology called *fracture mechanics* can be used to aid in selecting materials and designing components to minimize the possibility of fracture.

   In addition to cracks themselves, other types of flaws that are crack-like in form may easily develop into cracks, and these need to be treated as if they were cracks.

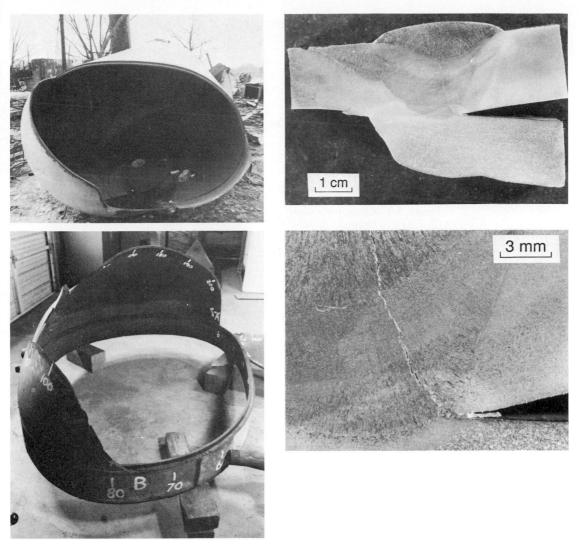

Figure 8.1 Photographs from a propane tank truck that exploded due to fracture from initial environmental cracks in welds. Typical initial cracks are shown from a region that did not participate in the final failure. (Photos courtesy of H. S. Pearson, Pearson Testing Labs, Marietta, Ga: lower left, upper and lower right, published in [Pearson 86]; copyright ©ASTM; reprinted with permission.)

Examples include deep surfaces scratches or gouges, voids in welds, inclusions of foreign substances in cast and forged materials, and delaminations in layered materials. For example, a photograph of a crack starting from a large inclusion in the wall of a forged steel artillery tube is shown in Fig. 8.2.

The study and use of fracture mechanics is of major engineering importance simply because cracks or crack-like flaws occur more frequently than we might at first think. For

2 cm

Figure 8.2 Crack (light area) growing from a large nonmetallic inclusion (dark area within) in an AISI 4335 steel artillery tube. The inclusion was found by inspection, and the tube was not used in service, but rather was tested under cyclic loading to study its behavior. (Photo courtesy of J. H. Underwood, U.S. Army Armament RD&E Center, Watervliet, N.Y.)

example, the periodic inspections of large commercial aircraft frequently reveal cracks, sometimes numerous cracks, that must be repaired. Cracks or crack-like flaws also commonly occur in ship structures, in bridge structures, in pressure vessels and piping, in heavy machinery, and in ground vehicles. They are also a source of concern for various parts of nuclear reactors.

In engineering design, materials and design details can be chosen that are relatively tolerant of the presence of cracks, and design can include *redundancy*, so that fracture of one component does not cause catastrophic failure of the whole machine, vehicle, or structure. Periodic inspections are also sometimes needed, as in aircraft and bridges, so that a crack cannot grow to a dangerous size before it is found and repaired. Methods of inspection for cracks include not only simple visual examination but also sophisticated means such as X-ray photography and ultrasonics. (In the latter method, reflections of high-frequency sound waves are used to reveal the presence of a crack.) Repairs necessitated by cracks may involve replacing a part or modifying it, as by machining away a small crack to leave a smooth surface, or by reinforcing the cracked region in some manner.

In this chapter, we will introduce fracture mechanics and study its application to failure under static loading. Later, in Chapter 11, we will consider growth of cracks due to cyclic loading.

8.2 PRELIMINARY DISCUSSION

Before introducing the details of fracture mechanics, it is useful to make some observations concerning the general nature of cracks and their effects.

8.2.1 Cracks as Stress Raisers

Consider an elliptical hole in a plate of material as illustrated in Fig. 8.3(a). For purposes of discussion, the hole is assumed to be small compared to the width of the plate, and it is aligned with its major axis perpendicular to the direction of a uniform stress, S, applied remotely. The uniform stress field is altered in the neighborhood of the hole as illustrated for one particular case in Fig 8.3(b).

The most notable effect of the hole is its influence on the stress σ_y parallel to S. Far from the hole, this stress is equal to S. If examined along the x-axis of (b), the value of σ_y rises sharply near the hole and has a maximum value at the edge of the hole. This maximum value depends on the proportions of the ellipse and the tip radius ρ of the ellipse as follows:

$$\sigma_y = S\left(1 + 2\frac{c}{d}\right) = S\left(1 + 2\sqrt{\frac{c}{\rho}}\right) \tag{8.1}$$

A stress concentration factor for the ellipse can be defined as the ratio of the maximum stress to the remote stress.

$$k_t = \frac{\sigma_y}{S} = 1 + 2\frac{c}{d} = 1 + 2\sqrt{\frac{c}{\rho}} \tag{8.2}$$

Consider a narrow ellipse where the half-height d approaches zero, so that the tip radius ρ also approaches zero, which corresponds to an ideal slit-like crack. In this case, σ_y from above becomes infinite, as does k_t. Hence, a sharp crack causes a severe concentration of stress and is special in that the stress is theoretically infinite if the crack is ideally sharp.

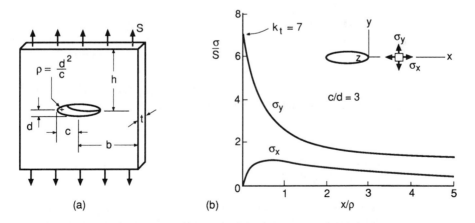

Figure 8.3 Elliptical hole in a wide plate under remote uniform stress, and the stress distribution along the x-axis near the hole for one particular case.

8.2.2 Behavior at Crack Tips in Real Materials

An infinite stress cannot, of course, exist in a real material. If the applied load is not too high, the material can accommodate the presence of an initially sharp crack in such a way that the theoretically infinite stress is reduced to a finite value. This is illustrated in Fig. 8.4. In ductile materials, such as many metals, large plastic deformations occur in the vicinity of the crack tip. The region within which the material yields is called the *plastic zone*. Intense deformation at the crack tip results in the sharp tip being blunted to a small but nevertheless nonzero radius. Hence, the stress is no longer infinite, and the crack is open near its tip by a finite amount, δ, called the *crack-tip opening displacement* (CTOD).

In other types of material, different behaviors occur that have a similar effect of relieving the theoretically infinite stress by modifying the sharp crack tip. In some polymers, a region containing elongated voids develops that contains a fibrous structure bridging the crack faces, which is called a *craze zone*. In brittle materials such as ceramics, a region containing a high density of tiny cracks may develop at the crack tip.

In all three cases, the crack tip experiences intense deformation and develops a finite separation near its tip. The very high stress that would ideally exist near the crack tip is spread over a larger region and is said to be *redistributed*. A finite value of stress that can be resisted by the material thus exists near the crack tip, and the stresses somewhat farther away are higher than they would be for an ideal crack.

8.2.3 Effects of Cracks on Strength

If the load applied to a member containing a crack is too high, the crack may suddenly grow and cause the member to fail by fracturing in a brittle manner, that is, with little

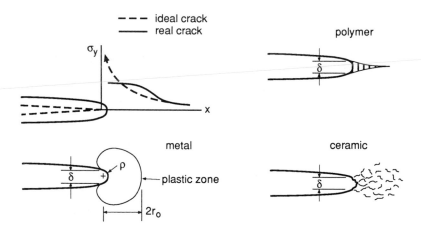

Figure 8.4   Finite stresses and nonzero radii at tips of cracks in real materials. A region of intense deformation forms due to plasticity, crazing, or microcracking.

plastic deformation. From the theory of fracture mechanics, a quantity called the *stress intensity factor*, K, can be defined that characterizes the severity of the crack situation as affected by crack size, stress, and geometry. In defining K, the material is assumed to behave in a linear-elastic manner, according to Hooke's Law, Eq. 4.10, so that the approach being used is called *linear-elastic fracture mechanics* (LEFM).

A given material can resist a crack without brittle fracture occurring as long as this K is below a critical value K_c, which is a property of the material called the *fracture toughness*. Values of K_c vary widely for different materials and are affected by temperature and loading rate, and secondarily by the thickness of the member. Some representative values of K_{Ic}, which is the worst case K_c considering thickness effects, are given in Tables 8.1 and 8.2.

For example, consider a crack in the center of a wide plate of stressed material as illustrated in Fig. 8.5. In this case, K depends on the remotely applied stress S and the crack length a, measured from the centerline as shown.

$$K = S\sqrt{\pi a} \qquad (a \ll b) \tag{8.3}$$

where this equation is accurate only if a is small compared to the half-width b of the member. For a given crack length and a material with fracture toughness K_c, the critical value of remote stress necessary to cause fracture is thus

$$S_c = \frac{K_c}{\sqrt{\pi a}} \tag{8.4}$$

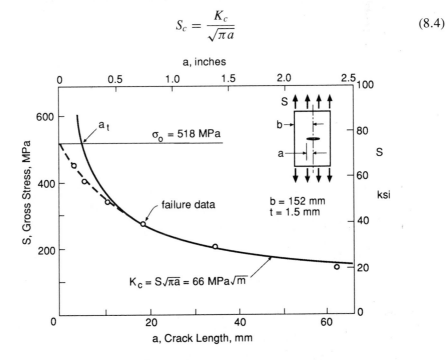

Figure 8.5 Failure data for cracked plates of 2014-T6 Al tested at −195°C. (Data from [Orange 67].)

TABLE 8.1 FRACTURE TOUGHNESS AND CORRESPONDING TENSILE PROPERTIES FOR REPRESENTATIVE STEELS AND ALUMINUM ALLOYS AT ROOM TEMPERATURE

Material	Toughness K_{Ic}	Yield σ_o	Ultimate σ_u	Elong. $100\varepsilon_f$	Red. Area %RA
	MPa$\sqrt{\text{m}}$ (ksi$\sqrt{\text{in}}$)	MPa (ksi)	MPa (ksi)	%	%
(a) Steels					
AISI 1144	66 (60)	540 (78)	840 (122)	5	7
ASTM A470-8 (Cr-Mo-V)	60 (55)	620 (90)	780 (113)	17	45
ASTM A517-F	187 (170)	760 (110)	830 (121)	20	66
AISI 4130	110 (100)	1090 (158)	1150 (167)	14	49
18-Ni maraging air melted	123 (112)	1310 (190)	1350 (196)	12	54
18-Ni maraging vacuum melted	176 (160)	1290 (187)	1345 (195)	15	66
300-M 650°C temper	152 (138)	1070 (156)	1190 (172)	18	56
300-M 300°C temper	65 (59)	1740 (252)	2010 (291)	12	48
(b) Aluminum Alloys					
2014-T651	24 (22)	415 (60)	485 (70)	13	–
2024-T351	34 (31)	325 (47)	470 (68)	20	–
2219-T37	41 (37)	315 (46)	395 (57)	11	–
6061-T651	34 (31)	275 (40)	310 (45)	12	–
7075-T651	29 (26)	505 (73)	570 (83)	11	–

Sources: Data in [Barsom 87] p. 172, [Boyer 85] p. 6.34, [Clark 70], [MILHDBK 83] p. 3.11, and [Ritchie 77].

TABLE 8.2 FRACTURE TOUGHNESS OF SOME POLYMERS AND CERAMICS AT ROOM TEMPERATURE

Material Polymers[1]	K_{Ic} MPa$\sqrt{m}$ (ksi$\sqrt{in}$)		Material Ceramics[2]	K_{Ic} MPa$\sqrt{m}$ (ksi$\sqrt{in}$)	
ABS	3.0	(2.7)	Soda-lime glass	0.76	(0.69)
Acrylic	1.8	(1.6)	Magnesia, MgO	2.9	(2.6)
Epoxy	0.6	(0.55)	Alumina, Al$_2$O$_3$	4.0	(3.6)
PC	2.2	(2.0)	Al$_2$O$_3$, 15% ZrO$_2$	10	(9.1)
PVC	2.4	(2.2)	Silicon carbide	3.7	(3.4)
PVC rubber mod.	3.35	(3.05)	SiC		
			Silicon nitride Si$_3$N$_4$	5.6	(5.1)

Notes: [1,2] See Tables 5.3 and 3.10, respectively, for additional properties of similar materials.

Sources: Data in [Williams 87] p. 243 and [Kelly 86] p. 376.

Hence, longer cracks have a more severe effect on strength than do shorter ones, as might be expected.

Some test data illustrating the effect of different crack lengths on strength are shown in Fig. 8.5. The curve given by Eq. 8.4 is shown, where the particular value of K_c used corresponds to the case at hand, namely 2014-T6 aluminum (1.5 mm thick) at $-195°$C. Fracture strengths from several tests are plotted. Most of these data agree quite nicely with the curve, indicating a degree of success here for LEFM. However, the data fall below the line where the stress approaches the yield strength σ_o. This situation occurs because Eq. 8.3, which assumes linear-elastic behavior, applies only if the plastic zone is small compared to the crack length and other geometric dimensions, and this is not true for the highest stresses corresponding to failure at small crack lengths.

8.2.4 Effects of Cracks on Brittle Versus Ductile Behavior

Consider the crack length where the failure stress predicted by LEFM exceeds the yield strength, identified as a_t in Fig. 8.5. Substituting $S_c = \sigma_o$ into Eq. 8.4 gives its value.

$$a_t = \frac{1}{\pi} \left(\frac{K_c}{\sigma_o} \right)^2 \qquad (8.5)$$

This *transition crack length* can be used as an approximate crack length above which the strength is expected to be limited by brittle fracture. Thus, if cracks of length around or greater than the a_t of a given material are likely to be present, fracture mechanics should be employed in design. Conversely, for crack lengths below a_t, yielding dominated behavior is expected, so that there will be little or no strength reduction due to the crack.

Consider two materials, one with low σ_o and high K_c, and the other with an opposite combination, namely high σ_o and low K_c. These combinations of properties cause a relatively large a_t for the low-strength material, but a small a_t for the high-strength one. Compare Figs. 8.6(a) and (b). Thus, cracks of moderate size may not

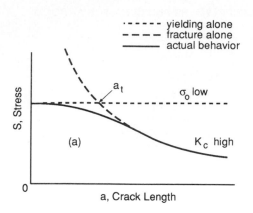

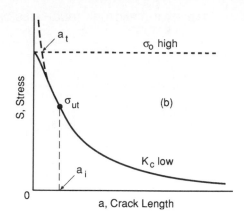

Figure 8.6 Transition crack length a_t for a low-strength, high-toughness material (a), and for high-strength, low-toughness material (b). If (b) contains internal flaws a_i, its strength in tension σ_{ut} is controlled by brittle fracture.

affect the low-strength material, but they may severely limit the usefulness of the high-strength one. Such an inverse trend between yield strength and fracture toughness is fairly common within any given class of materials. Low strength in a tension test is usually accompanied by high ductility and also by high fracture toughness. Conversely, high strength is usually associated with low ductility and low fracture toughness. Trends of this nature for AISI 1045 steel are illustrated in Fig. 8.7.

The relative sensitivity to flaws associated with different a_t values for different materials explains a number of sudden engineering failures that occurred in the 1950s and 1960s. New high-strength materials, such as steels and aluminum alloys developed for the aerospace industry, had sufficiently low fracture toughness that they were sensitive to rather small cracks. One example was the British-made Comet passenger airliner, two of which failed at high altitude in the 1950s, with considerable loss of life in the resulting crashes. Other examples are the late 1950s failures of rocket motor cases for the Polaris missile, and the F-111 aircraft crash in 1969. Such failures accelerated the development of fracture mechanics and led to its adoption by the U.S. Air Force as the basis of their *damage tolerant design* requirements.

Also, some apparently mysterious brittle failures in normally ductile steels occurred in the 1940s and earlier. These were finally understood years later to be due to cracks that were sufficiently large to exceed even the relatively large a_t value of the ductile steel. One example of this is the failure in Boston in 1919 of a large tank, about 90 feet in diameter and 50 feet high, that contained 2 million gallons of molasses. Other examples include welded Liberty Ships and tankers that broke completely in two during and shortly after World War II, and other ship and bridge failures.

8.2.5 Internally Flawed Materials

As already discussed in earlier chapters, many brittle materials naturally contain small cracks or crack-like flaws. This is generally true for glass, natural stones, ceramics, and

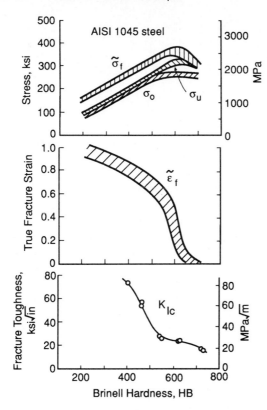

Figure 8.7 Comparison of properties from tension tests and fracture toughness tests for AISI 1045 steel, all plotted as functions of hardness as varied by heat treatment. (Illustration courtesy of R. W. Landgraf, Virginia Tech, Blacksburg, Va.)

some cast metals. The interpretation can be made that these materials have a high yield strength, but this strength can never be reached under tensile loading because of earlier failure due to small flaws and a low fracture toughness. Such a viewpoint is supported by the fact that brittle materials have considerably higher strengths under compression than tension, the flaws simply closing under compression and thus having a much reduced effect.

Denoting the inherent flaw size in such a material as a_i, Eq. 8.4 gives the ultimate strength in tension.

$$\sigma_{ut} = \frac{K_c}{\sqrt{\pi a_i}} \tag{8.6}$$

This situation is illustrated in Fig. 8.6(b). New cracks of size around or below a_i have little effect, as they are no worse than the flaws already present. The material is thus said to be *internally flawed*. Also, since the flaws actually present may vary considerably from sample to sample, there is generally a large statistical scatter in σ_{ut}.

8.3 MATHEMATICAL CONCEPTS

A cracked body can be loaded in any one or a combination of the three displacement modes shown in Fig. 8.8. Mode I is called the *opening mode* and consists of the crack faces simply moving apart. For mode II, the *sliding mode*, the crack faces slide relative

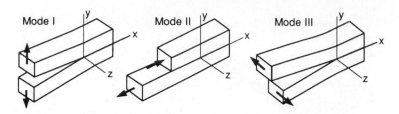

Figure 8.8 The basic modes of crack surface displacement. (Adapted from [Tada 85]; used with permission.)

to one another in a direction normal to the leading edge of the crack. Mode III, the *tearing mode*, also involves relative sliding of the crack faces, but now the direction is parallel to the leading edge. Mode I is caused by tension loading, whereas the other two are caused by shear loading in different directions as shown. Most cracking problems of engineering interest involve primarily Mode I and are due to tension stresses, so we will limit most of our discussion to this case.

Energy methods were employed in the earliest work on fracture mechanics, reported by A. A. Griffith in 1920. This approach is expressed by a concept called the *strain energy release rate*, G. Additional work led to the concept of a stress intensity factor, K, and to the proof that G and K are directly related.

8.3.1 Strain Energy Release Rate, *G*

Consider a cracked member under a Mode I load P, where the crack has length a as shown in Fig. 8.9. Assume that the behavior of the material is linear-elastic, which requires that the load versus displacement behavior also be linear. In a manner similar

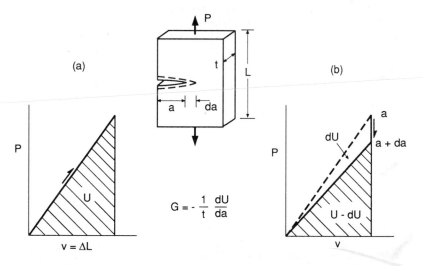

Figure 8.9 Potential energies for two neighboring crack lengths and the energy change dU used to define the strain energy release rate G.

to a linear spring, a potential energy U is stored in the member, as a result of the elastic strains throughout its volume, as shown in Fig. 8.9(a). Note that v is the displacement at the point of loading and $U = Pv/2$ is the triangular area under the P-v curve.

If the crack moves ahead by a small amount da, while the displacement is held constant, the stiffness of the member decreases as shown by (b). This results in the potential energy decreasing by an amount dU, that is, U decreases due to a release of this amount of energy. The rate of change of potential energy with increase in crack area is defined as the strain energy release rate, G.

$$G = -\frac{1}{t}\frac{dU}{da} \tag{8.7}$$

where the change in crack area is $t(da)$, and the negative sign causes G to have a positive value. Thus, G characterizes the energy per unit crack area required to extend the crack, and as such is expected to be the fundamental physical quantity controlling the behavior of the crack.

In the original concept by Griffith, all of the potential energy released was thought to be used in the creation of the new free surface on the crack faces. This is approximately true for materials that crack with essentially no plastic deformation, as for the glass tested by Griffith. However, in more ductile materials, a majority of the energy may be used in deforming the material in the plastic zone at the crack tip. In applying G to metals in the 1950s, G. R. Irwin showed that the concept was applicable even under these circumstances if the plastic zone was small.

As listed in the References, Barsom (1987b) is a collection of papers reporting some of the early work on fracture mechanics by Griffith, Irwin, and others.

8.3.2 Stress Intensity Factor, *K*

The stress intensity factor concept, which has already been introduced, needs to be defined in a rigorous manner. In general terms, K characterizes the magnitude (intensity) of the stresses in the vicinity of an ideally sharp crack tip in a linear-elastic and isotropic material.

A coordinate system for describing the stresses in the vicinity of a crack is shown in Fig. 8.10. The polar coordinates r and θ lie in the x-y plane, which is normal to the

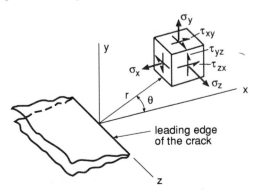

Figure 8.10   Three-dimensional coordinate system for the region of a crack tip. (Adapted from [Tada 85]; used with permission.)

plane of the crack, and the z-direction is parallel to the leading edge of the crack. For any case of Mode I loading, the stresses near the crack tip depend on r and θ as follows:

$$\sigma_x = \frac{K_I}{\sqrt{2\pi r}} \cos\frac{\theta}{2}\left[1 - \sin\frac{\theta}{2}\sin\frac{3\theta}{2}\right] + \cdots \qquad\qquad (a)$$

$$\sigma_y = \frac{K_I}{\sqrt{2\pi r}} \cos\frac{\theta}{2}\left[1 + \sin\frac{\theta}{2}\sin\frac{3\theta}{2}\right] + \cdots \qquad\qquad (b)$$

$$\tau_{xy} = \frac{K_I}{\sqrt{2\pi r}} \cos\frac{\theta}{2}\sin\frac{\theta}{2}\cos\frac{3\theta}{2} + \cdots \qquad\qquad (c) \qquad (8.8)$$

$$\sigma_z = 0 \qquad\qquad\qquad\qquad\qquad\qquad\text{(plane stress)} \qquad (d)$$

$$\sigma_z = \nu\left(\sigma_x + \sigma_y\right) \qquad\qquad\qquad\qquad\text{(plane strain; } \varepsilon_z = 0) \quad (e)$$

$$\tau_{yz} = \tau_{zx} = 0 \qquad\qquad\qquad\qquad\qquad\qquad\qquad (f)$$

These equations are based on the *theory of linear elasticity*, as described in any standard text on that subject, and they are said to describe the *stress field* near the crack tip. Higher order terms are omitted above that are not of significant magnitude near the crack tip. These equations predict that the stresses rapidly increase near the crack tip. Confirmation of this characteristic of the stress field is provided by a photograph of stress contours in a clear plastic specimen in Fig. 8.11.

If the cracked member is relatively thin in the z-direction, plane stress with $\sigma_z = 0$ applies. However, if it is relatively thick, a more reasonable assumption may be plane strain, $\varepsilon_z = 0$, in which case Hooke's Law, specifically Eq. 4.10(c), requires that σ_z depend on the other stresses according to Eq. 8.8(e).

The nonzero stress components in Eq. 8.8 are seen to all approach infinity as r approaches zero, that is, upon approaching the crack tip. Note that this is specifically caused by these stresses being proportional to the inverse of $\sqrt{r}$. Thus, a mathematical singularity is said to exist at the crack tip, and no value of stress at the crack tip can be given. Also, all of the nonzero stresses of Eq. 8.8 are proportional to the quantity K_I, and the remaining factors merely give the variation with r and θ. Hence, the magnitude of the stress field near the crack tip can be characterized by giving the value of the factor K_I. On this basis, K_I is a measure of the severity of the crack. Its definition in a formal mathematical sense is

$$K_I = \lim_{r,\theta\to 0}\left(\sigma_y\sqrt{2\pi r}\right) \qquad\qquad (8.9)$$

It is generally convenient to express K_I as

$$K_I = FS\sqrt{\pi a} \qquad\qquad (8.10)$$

where the factor F is needed to account for different geometries. For example, if a central crack in a plate is relatively long, Eq. 8.3 needs to be modified as the proximity

Figure 8.11   Contours of maximum in-plane shear stress around a crack tip. These were formed by the photoelastic effect in a clear plastic material. The two thin white lines entering from the left are the edges of the crack, and its tip is the point of convergence of the contours. (Photo courtesy of C. W. Smith, Virginia Tech, Blacksburg, Va.)

of the specimen edge causes F to increase above unity. The quantity F is a function of the ratio a/b as shown in Fig. 8.12. Up to $a/b = 0.9$, it is within 2% accuracy for this particular case to use the approximate expression shown there.

8.3.3 Additional Comments on *K* and *G*

For loading in Mode II or III, analogous but distinct stress field equations exist, and stress intensities K_{II} and K_{III} can be defined in a manner analogous to K_I. However,

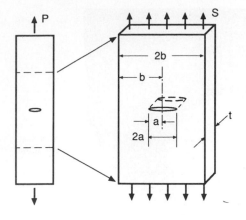

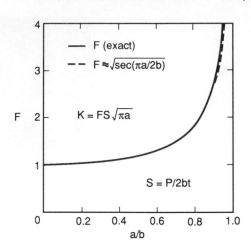

Figure 8.12 Finite width effect on K for a center-cracked plate. The approximate expression for F is within 2% for $a/b \leq 0.9$.

these will not be discussed here due to the emphasis on Mode I. As a convenience, the subscript on K_I will be dropped, and K without such a subscript is understood to denote K_I.

$$K = K_I \tag{8.11}$$

Some exact closed form solutions from theory of elasticity exist for K, one of which is the previously employed Eq. 8.3 for a center crack in a wide plate. However, for many cases of practical interest, it has been necessary to perform numerical analysis, some of the applicable methods being the *finite element*, *boundary integral equation*, and *weight function* methods. See Anderson (1991) and the SAE *Fatigue Design Handbook* (Rice, 1988) for introductions to these methods. Fortunately, many useful solutions in the form of F values for Eq. 8.10 are available from the published literature.

The quantities G and K can be shown to be related as follows:

$$G = \frac{K^2}{E'} \tag{8.12}$$

where

$$E' = E \qquad \text{(plane stress; } \sigma_z = 0\text{)}$$

$$E' = \frac{E}{1 - \nu^2} \qquad \text{(plane strain; } \varepsilon_z = 0\text{)}$$

This relationship and the dependence of G on load versus displacement behavior, Eq. 8.7, can be exploited to evaluate K. Slopes on P-v curves as in Fig. 8.9 are employed in a procedure called the *compliance method*. See any book on fracture mechanics or Tada (1985) for an explanation.

Since G and K are directly related according to Eq. 8.12, only one of these concepts is generally needed. We will primarily employ K, which is consistent with most engineering oriented publications on fracture mechanics.

8.4 APPLICATION OF K TO DESIGN AND ANALYSIS

For fracture mechanics to be put to practical use, values of stress intensity K must be determined for crack geometries that may exist in structural components. Extensive analysis work has been published, and also collected into handbooks, giving equations or plotted curves that enable K values to be calculated for a wide variety of cases. A special section of the References at the end of this chapter lists several such handbooks. What will be done in this portion of the chapter is to give certain fundamental equations for calculating K and also examples of the type of information available from handbooks.

8.4.1 Mathematical Forms Used to Express K

It has already been noted that K can be related to applied stress and crack length by an equation of the form

$$K = F S_g \sqrt{\pi a} \qquad (8.13)$$

The quantity F is a dimensionless function that depends on the geometry and loading configuration, and usually also on the ratio of the crack length to another geometric dimension, such as the member half-width, b, in Fig. 8.12. Hence, for a given type of loading, such as tension or bending

$$F = F(\text{geometry}, a/b) \qquad (8.14)$$

Additional examples are given in Figs. 8.13 and 8.14. In these examples, crack length a is measured from either the surface or the centerline of loading, and the width dimension b is consistently defined as the maximum possible crack length, so that for $a/b = 1$ the member is completely cracked.

Applied loads or bending moments are often characterized by determining a nominal or average stress. In fracture mechanics, it is conventional to use the gross section nominal stress, S_g, calculated assuming that no crack is present. Note that this convention is followed for each case in Figs. 8.13 and 8.14. The subscript g is added merely to avoid any possibility of confusion, as net section stresses, S_n, based on the remaining uncracked area, could be used. Use of S_g rather than S_n is convenient as the effect of crack length is then confined to the F and $\sqrt{a}$ factors. In general, the manner of defining nominal stress S is arbitrary, but consistency with F is necessary. The function F must be redefined and its values changed if the definition of S is changed, and also if the definition of a or b is changed.

It is sometimes convenient to work directly with applied loads (forces), with the following equation being useful for planar geometries:

$$K = F_P \frac{P}{t \sqrt{b}} \qquad (8.15)$$

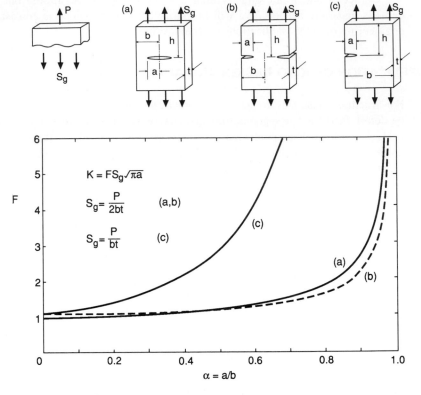

Values for small a/b and limits for 10% accuracy:

(a) $K = S_g \sqrt{\pi a}$
   $(a/b \leq 0.4)$

(b) $K = 1.12 S_g \sqrt{\pi a}$
   $(a/b \leq 0.6)$

(c) $K = 1.12 S_g \sqrt{\pi a}$
   $(a/b \leq 0.13)$

Expressions for any $\alpha = a/b$:

(a) $F = \dfrac{1 - 0.5\alpha + 0.326\alpha^2}{\sqrt{1 - \alpha}}$ $\qquad (h/b \geq 1.5)$

(b) $F = \left(1 + 0.122 \cos^4 \dfrac{\pi\alpha}{2}\right) \sqrt{\dfrac{2}{\pi\alpha} \tan \dfrac{\pi\alpha}{2}}$ $\quad (h/b \geq 2)$

(c) $F = 0.265 (1 - \alpha)^4 + \dfrac{0.857 + 0.265\alpha}{(1 - \alpha)^{3/2}}$ $\qquad (h/b \geq 1)$

Figure 8.13 Stress intensity factors for three cases of cracked plates under tension. An additional expression for (a) is given in Fig. 8.12, and for (c) the load is centered on the uncracked width. (Equations as collected by [Tada 85] pp. 2.2, 2.7, and 2.11.)

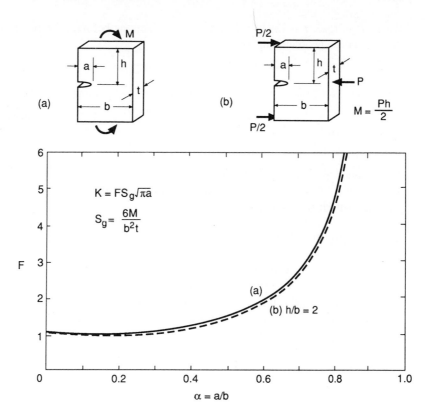

Values for small a/b and limits for 10% accuracy:

$$(a, b) \quad K = 1.12 S_g \sqrt{\pi a} \qquad (a/b \leq 0.4)$$

Expressions for any $\alpha = a/b$:

$$(a) \quad F = \sqrt{\frac{2}{\pi \alpha} \tan \frac{\pi \alpha}{2}} \left[\frac{0.923 + 0.199 \left(1 - \sin \frac{\pi \alpha}{2}\right)^4}{\cos \frac{\pi \alpha}{2}} \right] \qquad \text{(large } h/b)$$

(b) F is within 3% of (a) for $h/b = 4$, and within 6% for $h/b = 2$, at any a/b.

$$F = \frac{1.99 - \alpha \left(1 - \alpha\right) \left(2.15 - 3.93\alpha + 2.7\alpha^2\right)}{\sqrt{\pi} \left(1 + 2\alpha\right) \left(1 - \alpha\right)^{3/2}} \qquad (h/b = 2)$$

Figure 8.14 Stress intensity factors for various cases of bending. Case (b) with $h/b = 2$ is the ASTM standard bend specimen. (Equations from [Tada 85] p. 2.14, and [ASTM 91] Std. E399.)

where P is load, t is thickness, and b is the same as before. The function F_P is a new dimensionless geometry factor.

$$F_P = F_P(\text{geometry}, a/b) \tag{8.16}$$

The examples of Fig. 8.15 are expressed using F_P. Equating K from Eqs. 8.13 and 8.15 allows F_P to be related to the previously defined F.

$$F_P = F \frac{S_g t \sqrt{\pi a b}}{P} \tag{8.17}$$

This relationship can be used to convert any of the expressions for F in Figs. 8.13 and 8.14 to the F_P form. Expressing K in terms of F_P has the advantage that the dependence on crack length is confined to the dimensionless function F_P.

8.4.2 Discussion

Mathematically closed form solutions for K exist primarily for bodies that are of large (ideally infinite) size compared to the crack. These solutions are often reasonably accurate to surprisingly large values of a/b. Corresponding values of F or equations for K are given in Figs. 8.13 through 8.15, along with limits on a/b for 10% accuracy.

    An edge-cracked tension member, Fig. 8.13(c), can be thought of as being similar to a center-cracked plate (a) that has been split in half. Since the crack dimension a is consistently defined for the two cases, the additional relatively modest factor of 1.12 is associated with the effect of the new free surface. In fact, this same factor applies for small a/b to any through-thickness surface crack in a plate under remote stress, thus applying to all of the surface crack cases of Figs. 8.13 and 8.14. A similar difference exists between Figs. 8.15(a) and (b). The free surface factor is in this case 1.30, noting that the load is relatively twice as severe in (b) compared to (a), as bisecting (a) would reduce the load to $P/2$.

    For relatively long cracks where the simple infinite body solutions become inaccurate, polynomial-type equations for calculating F or F_P are given in Figs. 8.13 through 8.15. These are simply convenient expressions fitted to numerical results for a number of different relative crack lengths a/b, including a/b approaching unity. Some of these equations apply ideally to uniform or bending stresses applied an infinite distance h from the crack. However, the errors for finite h are generally negligible at $h/b = 3$ and begin to be significant ($> 5\%$) only at h/b around 1 or 2, which is indicated where the details are known. Where trigonometric functions appear in the expressions given, the arguments for these are in units of radians.

Example 8.1

    A center-cracked plate as in Fig. 8.13(a) has dimensions $b = 50$ mm, $t = 5$ mm, and large h, and a load of $P = 50$ kN is applied.

    (a) What is the stress intensity factor K for a crack length of $a = 10$ mm?

    (b) For $a = 30$ mm?

    (c) What is the critical crack length a_c for fracture if the material is 2014-T651 aluminum?

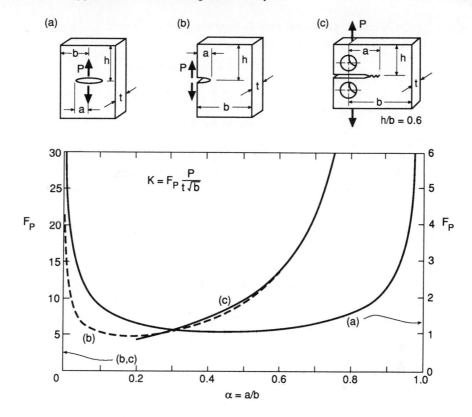

Values for small a/b and limits for 10% accuracy:

$$\text{(a) } K = \frac{P}{t\sqrt{\pi a}} \quad (a/b \le 0.3), \qquad \text{(b) } K = \frac{2.60 P}{t\sqrt{\pi a}} \quad (a/b \le 0.08)$$

Expressions for any $\alpha = a/b$:

$$\text{(a) } F_P = \frac{1.297 - 0.297 \cos \dfrac{\pi \alpha}{2}}{\sqrt{\sin \pi \alpha}} \qquad (h/b \ge 2)$$

$$\text{(b) } F_P = \frac{0.92 + 6.12\alpha + 1.68(1-\alpha)^5 + 1.32\alpha^2(1-\alpha)^2}{\sqrt{\pi\alpha}(1-\alpha)^{3/2}} \qquad (\text{large } h/b)$$

$$\text{(c) } F_P = \frac{(2+\alpha)}{(1-\alpha)^{3/2}}(0.886 + 4.64\alpha - 13.32\alpha^2 + 14.72\alpha^3 - 5.6\alpha^4) \qquad (a/b \ge 0.2)$$

Figure 8.15   Stress intensity factors for three cases of concentrated load. Case (c) is the ASTM standard compact specimen. (Equations from [Tada 85] pp. 2.23 and 2.25, and [Srawley 76].)

Solution   (a) To calculate K using Fig. 8.13(a), we need

$$S_g = \frac{P}{2bt} = \frac{0.050 \text{ MN}}{2(0.050 \text{ m})(0.005 \text{ m})} = 100 \text{ MPa}$$

$$\alpha = \frac{a}{b} = \frac{10 \text{ mm}}{50 \text{ mm}} = 0.2$$

Since $\alpha \leq 0.4$, it is within 10% to use $F = 1$.

$$K = S_g \sqrt{\pi a} = (100 \text{ MPa})\sqrt{\pi(0.010 \text{ m})} = 17.7 \text{ MPa}\sqrt{\text{m}} \qquad \textbf{Ans.}$$

(b) For $a = 30$ mm

$$\alpha = \frac{a}{b} = \frac{30 \text{ mm}}{50 \text{ mm}} = 0.6$$

This does not satisfy $\alpha \leq 0.4$, so that the more general expression for F from Fig. 8.13(a) is needed.

$$F = \frac{1 - 0.5\alpha + 0.326\alpha^2}{\sqrt{1 - \alpha}} = 1.292$$

$$K = F S_g \sqrt{\pi a} = 1.292 (100 \text{ MPa}) \sqrt{\pi(0.030 \text{ m})} = 39.7 \text{ MPa}\sqrt{\text{m}} \qquad \textbf{Ans.}$$

(c) Table 8.1 gives $K_{Ic} = 24$ MPa$\sqrt{\text{m}}$ for 2014-T651 Al. Since a_c is not known, F cannot be determined directly. First, assume that $\alpha \leq 0.4$ is satisfied, in which case $F \approx 1$.

$$K_{Ic} \approx S\sqrt{\pi a_c}$$

Solving for a_c gives

$$a_c \approx \frac{1}{\pi} \left(\frac{K_{Ic}}{S} \right)^2 = \frac{1}{\pi} \left(\frac{24 \text{ MPa}\sqrt{\text{m}}}{100 \text{ MPa}} \right)^2 = 0.0183 \text{ m} = 18.3 \text{ mm} \qquad \textbf{Ans.}$$

This corresponds to

$$\alpha = \frac{a_c}{b} = \frac{18.3}{50} = 0.37$$

which satisfies $\alpha \leq 0.4$, so that the estimated $F \approx 1$ is acceptable and the result obtained is reasonably accurate.

If it is not desired to use the 10% approximation on F, a trial and error or graphical procedure is needed as follows: Use the a_c value from above as a rough estimate and calculate K for several crack lengths in this neighborhood as shown in Table E8.1. Use the same procedure as in (b) above for each crack length. When $K = K_{Ic}$ is obtained, the crack length is the desired value.

Since $K_{Ic} = 24$ MPa$\sqrt{\text{m}}$ was between the calculated values No. 2 and 3, a guess was used to obtain trial a No. 4, which gave K close to K_{Ic}. A linear interpolation then gave trial a No. 5, resulting in $K = K_{Ic}$, so that the final answer is

$$a_c = 16.3 \text{ mm} \qquad \textbf{Ans.}$$

A graphical procedure could be used with fewer calculations to obtain the same result as shown in Figure E8.1. Such a trial and error or graphical solution is not optional, but rather is necessary, if the approximate solution does not satisfy the limits for 10% accuracy of F.

TABLE E8.1

Calc. No.	Trial a mm	$\alpha = a/b$	F	$K = FS_g\sqrt{\pi a}$ MPa$\sqrt{\text{m}}$
1	10	0.20	1.021	18.1
2	15	0.30	1.051	22.8
3	20	0.40	1.100	27.6
4	16.5	0.33	1.063	24.2
5	16.3	0.326	1.0617	24.0

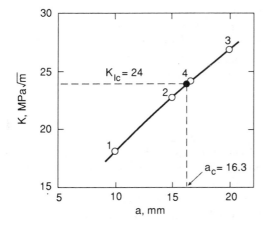

Figure E8.1

8.4.3 Cases of Special Interest for Practical Applications

The handbooks listed in the References contain a wide variety of additional useful cases. These include not only additional situations of cracked plates, but also cracked shafts, tubes, discs, stiffened panels, etc., including three-dimensional cases.

Consider the three-dimensional case of an elliptical crack embedded in an infinite body, where a remote stress is applied normal to the crack plane. This is illustrated in Fig. 8.16. A closed form solution exists and is given in the figure. Note that K now varies with location on the elliptical crack front, having a maximum value where the ellipse is narrowest. The integral $E(k)$ is a standard *elliptic integral of the second kind*, values of which are tabulated in most books of mathematical tables. These values can be approximated quite closely by the additional equation given. For the special case where the ellipse is a circle, $E(k) = \pi/2$ and the solution simplifies to

$$K = \frac{2}{\pi}S\sqrt{\pi a} \tag{8.18}$$

where a is the radius of the circular crack.

A related situation often encountered in practical applications is a surface crack, often of shape near a half circle, as illustrated in Fig. 8.17. For the infinite thickness case, $a/t = 0$, the stress intensity is increased only modestly compared to the embedded

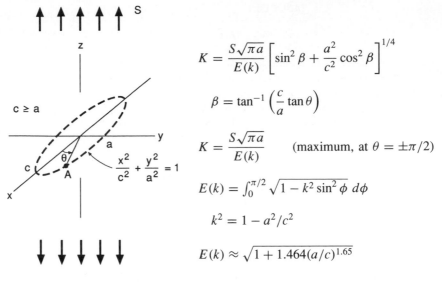

$$K = \frac{S\sqrt{\pi a}}{E(k)} \left[\sin^2 \beta + \frac{a^2}{c^2} \cos^2 \beta \right]^{1/4}$$

$$\beta = \tan^{-1} \left(\frac{c}{a} \tan \theta \right)$$

$$K = \frac{S\sqrt{\pi a}}{E(k)} \qquad (\text{maximum, at } \theta = \pm\pi/2)$$

$$E(k) = \int_0^{\pi/2} \sqrt{1 - k^2 \sin^2 \phi} \; d\phi$$

$$k^2 = 1 - a^2/c^2$$

$$E(k) \approx \sqrt{1 + 1.464(a/c)^{1.65}}$$

Figure 8.16 Exact stress intensity solution for any point A on an elliptical crack embedded in an infinite body under remote stress normal to the crack plane. (Adapted from [Tada 85] p. 26.2; used with permission. The approximation for $E(k)$ is from [Newman 86].)

circular crack case of Eq. 8.18. Where the crack intersects the surface, the effect of bisecting the embedded crack is a maximum and amounts to a factor of about 1.14, giving an overall value of $F = 1.14(2/\pi) = 0.73$. The modification factor decreases smoothly to about 1.04 at the deepest point. An equation and a curve are given that represent this variation, and similar information is also given for finite thickness at $a/t = 0.5$. Various

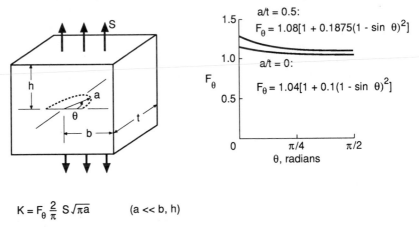

$$K = F_\theta \frac{2}{\pi} S\sqrt{\pi a} \qquad (a \ll b, h)$$

Figure 8.17 Stress intensities for two cases of a half-circular surface crack. (Equations from [Newman 86], which provides more detail.)

other cases of surface and corner cracks are also of practical importance, and these are treated in detail in handbooks and in Newman and Raju as listed in the References.

Another situation that is often of practical interest is a crack growing from a stress raiser, such as a hole, notch, or fillet. The example of a pair of cracks growing from a circular hole in a wide plate is used to illustrate this situation in Fig. 8.18. The solid line shown is from numerical analysis and is closely approximated by the equation given. If the crack is short compared to the hole radius, the solution is the same as for a surface crack in an infinite body, except that the stress is $k_t S$, being amplified by the stress concentration factor, in this case $k_t = 3$ from Eq. 8.2.

$$K_A = 1.12\, k_t S \sqrt{\pi l} \qquad (8.19)$$

where l is the crack length measured from the hole surface. Once the crack has grown far from the hole, the solution is the same as for a single long crack of tip-to-tip length 2a, for which

$$K_B = F S \sqrt{\pi a} \qquad (8.20)$$

where $F = 1$ for this particular case of a wide plate. Hence, for long cracks, the width of the hole acts as part of the crack, and the fact that the material removed to make the hole is missing from the crack faces is of little consequence. The exact K first follows K_A, then falls below it and approaches K_B, agreeing exactly for large l beyond the range of the plot of Fig. 8.18.

Most cases of a crack at an internal or surface notch can be roughly approximated by using K_A for crack lengths up to that where $K_A = K_B$, and then using K_B for all longer crack lengths. Equations 8.19 and 8.20 apply, with the k_t and $F = F(a/b)$ for the

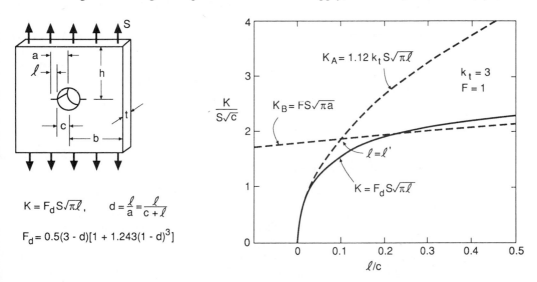

Figure 8.18 Stress intensities for a pair of cracks growing from a circular hole in a remotely loaded wide plate, $a \ll b, h$. (Equation from [Tada 85] p. 19.1.)

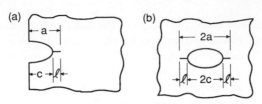

Figure 8.19 Nomenclature for cracks growing from notches.

particular case being used, and with the nomenclature being generalized as in Fig. 8.19. Note that k_t and S in Eq. 8.19 must be consistently defined, and also that F in Eq. 8.20 is usually consistent with S_g based on gross area. Hence, the values k_{tn} used with net section stress S_n that are usually available need to be converted to values k_{tg} that are consistent with S_g, which can be accomplished by using the relationship $k_{tn} S_n = k_{tg} S_g$. The crack length where $K_A = K_B$, labeled as $l = l'$ in Fig. 8.18, can be obtained from the preceding equations.

$$l' = \frac{c}{\left(1.12\dfrac{k_t}{F}\right)^2 - 1} \tag{8.21}$$

The resulting values of l' are typically in the range 0.1ρ to 0.2ρ, where ρ is the notch-tip radius.

More refined estimates of K for cracks at notches may also be made as described in Schijve (1982) and Kujawski (1991) as listed in the References, and solutions for specific cases are given in the various handbooks.

Example 8.2

A pressure vessel made of ASTM A517-F steel operates near room temperature and has a wall thickness of $t = 50$ mm. A crack as shown below was found during an inspection. It has an approximately semi-elliptical shape with surface length $2c = 40$ mm and depth $a = 10$ mm. The stresses in the region of the crack, as calculated without considering the presence of the crack, are approximately uniform through the thickness and are $S_y = 300$ MPa normal to the crack plane and $S_x = 150$ MPa parallel to the crack plane. What is the safety factor against brittle fracture? Would you remove the pressure vessel from service?

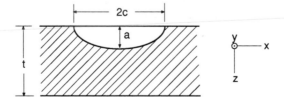

Figure E8.2

Solution From Table 8.1(a), this material has a fracture toughness of $K_{Ic} = 187$ MPa$\sqrt{m}$ and a yield strength of $\sigma_o = 760$ MPa at room temperature. The K for the given stresses and crack can be estimated by applying a correction factor to the value for an embedded elliptical crack from Fig. 8.16. The correction factor is needed due to the free surface created by bisecting the elliptical crack geometry, and due to $a/t = 0.2$ perhaps also having an effect.

$$K = F_s \frac{S\sqrt{\pi a}}{E(k)}$$

where the maximum value at $\theta = \pm\pi/2$ is used and F_s is the added surface factor. The stress needed is $S = S_y = 300$ MPa, as S_x is parallel to the crack plane and therefore does not contribute to K. Noting that $c = 20$ mm, so that $a/c = 0.5$, allows $E(k)$ to be calculated.

$$E(k) \approx \sqrt{1 + 1.464 \left(\frac{a}{c}\right)^{1.65}} = 1.21$$

As an estimate, assume that $a/t = 0.2$ has no effect, which is reasonable considering the curves for the similar half-circular crack in Fig. 8.17. Then take the free surface factor to be the same as for a through-thickness crack, specifically $F_s = 1.12$. Hence, the applied K is

$$K = F_s \frac{S\sqrt{\pi a}}{E(k)} = 1.12 \frac{300 \text{ MPa}\sqrt{\pi (0.010 \text{ m})}}{1.21} = 49.2 \text{ MPa}\sqrt{\text{m}}$$

The stress-based safety factor against brittle fracture can be calculated directly from K values as these are proportional to S for any given crack length.

$$X_S = \frac{K_{Ic}}{K} = \frac{187}{49.2} = 3.80 \qquad\qquad \textbf{Ans.}$$

This is a reasonably high value, so that it would be safe to continue using the pressure vessel until repairs are convenient. However, the crack should be checked frequently to be sure that it is not growing.

Comment   The above estimate of F_s is quite close to the accurate value of $F_s = 1.11$ obtained from Newman and Raju (1986). The procedure used is reasonably accurate for other similar cases of semi-elliptical cracks as long as a/t is fairly small. For large c/a, the semi-elliptical surface crack becomes equivalent to a through-thickness edge crack, Fig. 8.13(c).

8.4.4 Superposition for Combined Loading

Stress intensity solutions for combined loading can be obtained by superposition, that is, by adding the contributions to K from the individual load components. For example, consider an eccentric load applied a distance e from the centerline of a member with a single edge crack as shown in Fig. 8.20. This eccentric load is statically equivalent to the

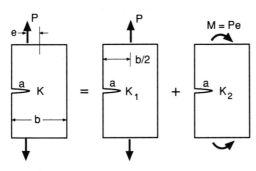

Figure 8.20   Eccentric loading of a plate with an edge crack, and the superposition used to obtain K.

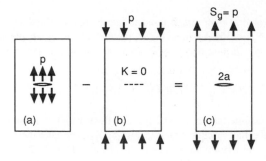

Figure 8.21 Equivalence with respect to K of pressure loading and remote tension for a center-cracked plate.

combination of a centrally applied tension load and a bending moment. The contribution to K from the centrally applied tension may be determined from Fig. 8.13(c).

$$K_1 = F_1 S_1 \sqrt{\pi a}, \qquad S_1 = \frac{P}{bt} \tag{8.22}$$

The contribution from bending may be determined from Fig. 8.14(a).

$$K_2 = F_2 S_2 \sqrt{\pi a}, \qquad S_2 = \frac{6M}{b^2 t} = \frac{6Pe}{b^2 t} \tag{8.23}$$

Hence, the total stress intensity due to the eccentric load is obtained by summing the two solutions and using substitutions from above.

$$K = K_1 + K_2 = \frac{P}{bt} \left(F_1 + \frac{6F_2 e}{b} \right) \sqrt{\pi a} \tag{8.24}$$

where the particular a/b that applies is used to separately determine F_1 and F_2 for tension and bending, respectively.

Ingenious use of superposition sometimes allows handbook solutions to be used for cases not obviously included. For example, consider the case of a central crack in a plate with a uniform pressure, p, prying open the crack as shown in Fig. 8.21(a). This can be shown to have the same solution as the uniform tension case (c), where the stress S_g is equated to the pressure p. The logic is as follows: Consider situation (b) where a uniform compressive stress of the same magnitude p is applied. The crack is present but has no effect as it is forced closed by the uniform compression. Subtracting the stresses on (b) from those on (a), including the compression on the crack faces, which is a pressure p, gives case (c) if $S_g = p$. Since K for case (b) is obviously zero, the K values for (a) and (c) are then equal for any a/b. Similar logic can be used for other cases of pressure-loaded cracks.

8.5 FRACTURE TOUGHNESS VALUES AND TRENDS

In fracture toughness testing, an increasing displacement is applied to an already cracked specimen of the material of interest until it fractures. The arrangement used for a *bend specimen* is shown in Fig. 8.22. Growth of the crack is detected by observing the load versus displacement (*P-v*) behavior as in Fig. 8.23. A deviation from linearity on the *P-v* plot, or a sudden drop in load due to rapid cracking, identifies a point P_Q corresponding to an early stage of cracking. The value of K, denoted K_Q, is then calculated for this point. If there is some tearing of the crack prior to final fracture,

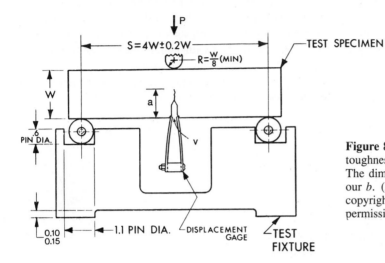

Figure 8.22 Fixtures for a fracture toughness test in a bend specimen. The dimension W corresponds to our b. (Adapted from [ASTM 91]; copyright ©ASTM; reprinted with permission.)

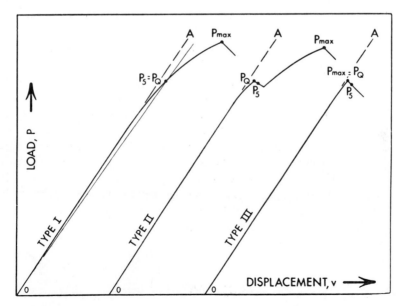

Figure 8.23 Types of load vs. displacement behavior that can occur in a fracture toughness test. (From [ASTM 91]; copyright ©ASTM; reprinted with permission.)

K_Q may be somewhat lower than the value K_c corresponding to the final fracture of the specimen.

In addition to bend specimens, the *compact specimen* geometry of Fig. 8.15(c) is also commonly used, usually with thickness $t = 0.5b$. Figures 8.24 and 8.25 are photographs of untested and tested compact specimens.

Fracture toughness testing of metals based on LEFM principles is governed by several ASTM standards, notably Standard No. E399. Similar tests are also done for other materials, as in Standard No. D5045 on plastics (polymers). A situation addressed in these standards is that K_Q decreases with increasing specimen thickness t, as illustrated by test data in Fig. 8.26. This occurs because the behavior is affected by the plastic zone at the crack tip in a manner that depends on thickness. Once the thickness obeys the following relationship involving the yield strength, no further

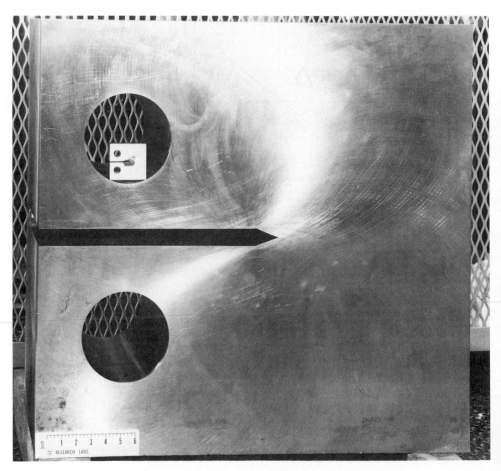

Figure 8.24   Compact specimens of two different sizes, $b = 5.1$ and 61 cm. (Photo courtesy of E. T. Wessel, Haines City, Fla.; used with permission of Westinghouse Electric Corp.)

Figure 8.25   Fracture surfaces from a K_{Ic} test on A533B steel using a compact specimen of dimensions $t = 25$, $b = 51$ cm. (Photo courtesy of E. T. Wessel, Haines City, Fla.; used with permission of Westinghouse Electric Corp.)

decrease is expected:

$$t \geq 2.5 \left(\frac{K_Q}{\sigma_o} \right)^2 \tag{8.25}$$

Values of K_Q meeting this requirement are denoted as K_{Ic} to distinguish them as worst-case values. In engineering design using material of thickness such that K_Q is somewhat greater than K_{Ic}, values of K_{Ic} can be used while recognizing that some extra conservatism is thus included. This is often necessary as only K_{Ic} values are widely available. Note that the values in Table 8.1 and 8.2 are specifically K_{Ic}.

A later section of this chapter will consider plastic zone size effects and other aspects of fracture toughness testing in more detail. We will now proceed to discuss the trends in K_{Ic} with material, temperature, loading rate, and other influences.

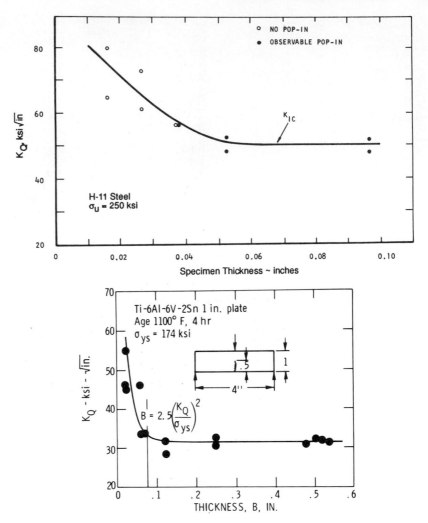

Figure 8.26   Effect of thickness on fracture toughness in two metals. On the bottom σ_{ys} and B correspond to our σ_o and t, respectively. (Top adapted from [Steigerwald 70]; bottom from [Jones 70]; copyright ©ASTM; reprinted with permission.)

8.5.1 Trends in K_{Ic} with Material

Values of K_{Ic} for engineering metals are generally in the range 20 to 200 MPa$\sqrt{m}$. For increasing strength within a given class of engineering metals, it has already been noted that fracture toughness decreases along with tensile ductility. See Figs. 1.6 and 8.7. As a further example, the effect on K_{Ic} of heat treating the alloy steel AISI 4340 to various strength levels is shown in Fig. 8.27.

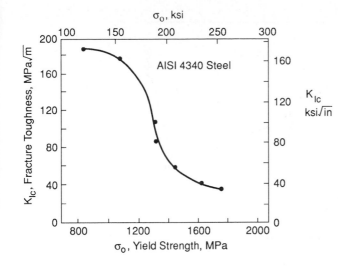

Figure 8.27 Fracture toughness vs. yield strength for AISI 4340 steel quenched and tempered to various strength levels. (Adapted from an illustration courtesy of W. G. Clark, Jr., Westinghouse Science and Technology Ctr., Pittsburgh, Pa.)

Polymers that are useful as engineering materials typically have K_{Ic} values in the range 1 to 5 MPa$\sqrt{m}$. Although these are low values, most polymers are used at low stresses due to their low ultimate strengths, so that under typical usage the likelihood of fracture is roughly similar to that for metals. Modifying a low-ductility polymer with ductile particles such as rubber increases its fracture toughness. The addition of short chopped fibers or other stiff reinforcement may decrease toughness if crack growth paths are available that do not intersect the reinforcement. Conversely, long fibers and especially continuous ones in a polymer matrix composite may obstruct crack growth to the point that the toughness is in the range of that for metals.

Ceramics have low values of fracture toughness, in the range 1 to 5 MPa$\sqrt{m}$, as might be expected from their low ductility. This range of K_{Ic} values is similar to that for polymers, but it is quite low in view of the fact that ceramics are high-strength materials. Indeed, their strengths in tension are usually limited by the inherent flaws in the material as discussed near the beginning of this chapter. Recent efforts in materials development have led to modifications to ceramics that increase toughness somewhat. For example, alumina (Al_2O_3) toughened with a second phase of 15% zirconia (ZrO_2) has $K_{Ic} \approx 10$ MPa$\sqrt{m}$. This occurs because high stress causes a phase transformation (crystal structure change) in the zirconia, which increases its volume by several percent. Thus, when the crack tip encounters a zirconia grain, the increase in volume is sufficient to cause a local compressive stress that retards further extension of the crack.

8.5.2 Effects of Temperature and Loading Rate

Fracture toughness generally increases with temperature, with illustrative test data for a low-alloy steel and a ceramic being shown in Figs. 8.28 and 8.29. An especially abrupt change in toughness over a relatively small temperature range occurs in metals with a BCC crystal structure, notably in steels with ferritic-pearlitic and martensitic structures.

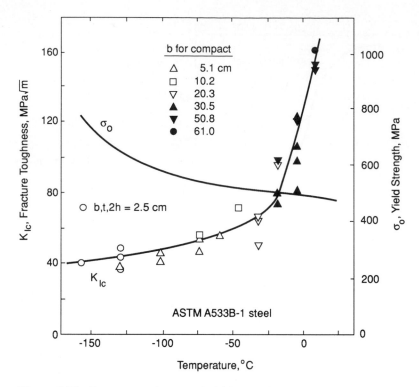

Figure 8.28 Fracture toughness and yield strength vs. temperature for a nuclear pressure vessel steel. Compact specimens and one nonstandard geometry were used in sizes indicated. (Adapted from [Clark 70]; copyright ©ASTM; reprinted with permission.)

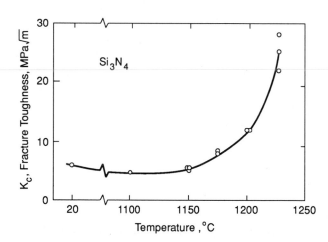

Figure 8.29 Fracture toughness vs. temperature for a silicon nitride ceramic. (Adapted from [Munz 81]; copyright ©ASTM; reprinted with permission.)

The temperature region where the rapid transition occurs varies considerably for different steels as illustrated in Fig. 8.30. There is usually a *lower shelf* of approximately constant K_{Ic} below the transition region, and an *upper shelf* above it, corresponding to a higher approximately constant K_{Ic}. (Only one set of data in Fig. 8.30 covers a sufficient range to exhibit an upper shelf.) Such *temperature-transition* behavior is similar to that observed in Charpy or other notch-impact tests as discussed in Chapter 5. (See Figs. 5.37 and 5.38.)

The distinct temperature-transition behavior in BCC metals is difficult to explain merely on the basis of the increase in ductility associated with the temperature range involved. It is in fact due to a shift in the physical mechanism of fracture. Below the temperature transition, the fracture mechanism is identified as *cleavage*, and above it as *dimpled rupture*. Microphotographs of fracture surfaces exhibiting these mechanisms are shown in Fig. 8.31. Cleavage is fracture with little plastic deformation along specific crystal planes that have low resistance. Dimpled rupture, also called *microvoid coalescence*, involves the plasticity-induced formation, growth, and joining of tiny voids in the material. This process leaves the rough and highly dimpled fracture appearance seen in the photograph.

Data on the K_{Ic} versus temperature behavior of steels and other materials is useful in selecting specific materials for service, as it is generally important to avoid high-stress use of a material at a temperature where its fracture toughness is low. Even the transition region itself should be avoided as there is a large amount of statistical scatter in K_{Ic} there,

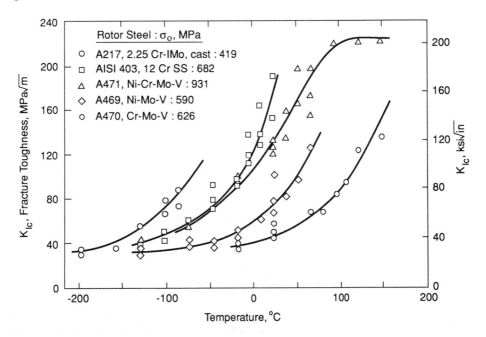

Figure 8.30   Fracture toughness vs. temperature for several steels used for turbine-generator rotors. (Data from [Logsdon 76].)

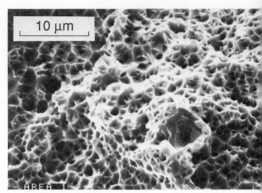

Figure 8.31 Cleavage fracture surface (left) in a 49Co-49Fe-2V alloy, and dimpled rupture (right) in a low-alloy steel. (Photos courtesy of A. Madeyski, Westinghouse Science and Technology Ctr., Pittsburgh, Pa.)

so that a low value may occur. Also, some additional caution is needed as the position of the transition may shift by as much as 50°C for different batches of the same steel.

A higher rate of loading usually lowers the fracture toughness, having an effect similar to decreasing the temperature. The effect can be thought of as causing a temperature shift in the K_{Ic} versus temperature curve as illustrated by test data in Fig. 8.32. Since notch-impact tests involve a high rate of loading, these typically give a higher transition temperature than K_{Ic} tests run at ordinary rates. Some illustrative test data are shown in Fig. 8.33.

8.5.3 Microstructural Influences on K_{Ic}

Seemingly small variations in chemical composition or processing of a given material can significantly affect fracture toughness. For example, sulfide inclusions in steels apparently have effects on a microscopic level that facilitate fracture. The resulting influence of sulfur content on toughness of an alloy steel is shown in Fig. 8.34.

Fracture toughness is generally more sensitive than other mechanical properties to anisotropy and planes of weakness introduced by processing. For example, in forged, rolled, or extruded metal, the crystal grains are elongated and/or flattened in certain directions, and fracture is easier where the crack grows parallel to the planes of the flattened grains. Nonmetallic inclusions and voids may also become elongated and/or flattened so that they also cause the fracture properties to vary with direction. Thus,

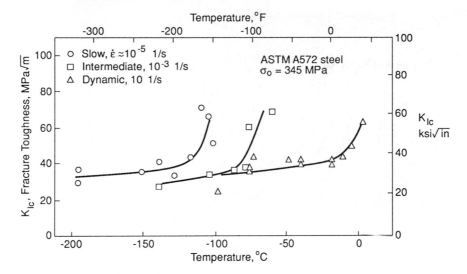

Figure 8.32 Effect of loading rate on the fracture toughness of a structural steel. Approximate strain rates at the edge of the plastic zone are given; the slowest corresponds to an ordinary test. (Adapted from [Barsom 75]; reprinted with permission from *Engineering Fracture Mechanics*, Pergamon Press, Oxford, UK.)

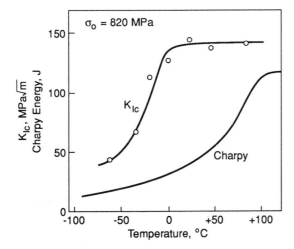

Figure 8.33 Comparisons of temperature-transition behavior for K_{Ic} and Charpy tests on a 2.25Cr-1Mo steel. (Adapted from [Marandet 77]; copyright ©ASTM; reprinted with permission.)

fracture toughness tests are often conducted for various specimen orientations relative to the original piece of material. The six possible combinations of crack plane and direction in a rectangular section of material are shown in Fig. 8.35. Fracture toughness data for three of these possibilities are also given for some typical aluminum alloys.

Neutron radiation affects the pressure vessel steels used in nuclear reactors by introducing large numbers of point defects (vacancies and interstitials) into the crystal

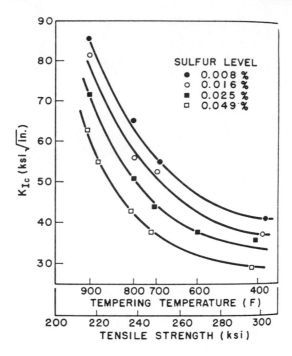

Figure 8.34 Effect of sulfur content on the fracture toughness of AISI 4345 steel. (From [Wei 65]; copyright ©ASTM; reprinted with permission.)

structure of the material. This causes increased yield strength but decreased ductility and decreased fracture toughness. Some test data are shown in Fig. 8.36. Such *radiation embrittlement* is obviously a major concern in nuclear power plants and is an important factor in determining their service life.

8.5.4 Mixed-Mode Fracture

If a crack is not normal to the applied stress, or if a complex state of stress exists, a combination of fracture Modes I, II, and III exists. For example, a situation involving combined Modes I and II is shown in Fig. 8.37. Such a situation is complex because the crack may change direction so that it does not grow in its original plane, and also because the two fracture modes do not act independently, but rather interact. Tests analogous to the K_{Ic} test to determine K_{IIc} or K_{IIIc} are difficult to conduct and are not standardized, so that toughness values for the other modes are generally not known.

A situation exists here that is analogous to the need for a yield criterion for combined stresses. Several combined-mode fracture criteria exist, but there is currently no general agreement on which is best. Any successful theory must predict mixed-mode fracture data of the type shown in Fig. 8.38. These particular data suggest that an elliptical curve could be used as an empirical fit where both K_{Ic} and K_{IIc} are known.

$$\left(\frac{K_I}{K_{Ic}}\right)^2 + \left(\frac{K_{II}}{K_{IIc}}\right)^2 = 1 \tag{8.26}$$

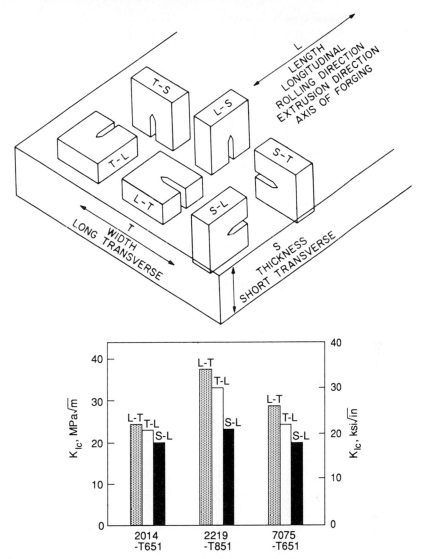

Figure 8.35 Designations for crack plane and growth direction in rectangular sections, and some corresponding effects on fracture toughness in plates of three aluminum alloys. (Top from [ASTM 91]; copyright ©ASTM; reprinted with permission. Bottom from data in [MILHDBK 83] p. 3–11.)

8.6 PLASTIC ZONE SIZE, AND PLASTICITY LIMITATIONS ON LEFM

Near the beginning of this chapter, it was noted that real materials cannot support the theoretically infinite stresses at the tip of a sharp crack, so that upon loading the crack tip becomes blunted and a region of yielding, crazing, or microcracking forms. We

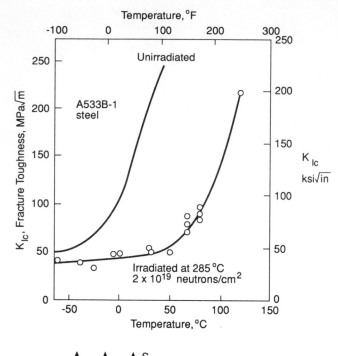

Figure 8.36   Effect of neutron irradiation on fracture toughness of a nuclear pressure vessel steel. (Adapted from [Bush 74]; copyright ©ASTM; reprinted with permission.)

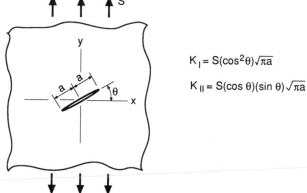

$$K_I = S(\cos^2\theta)\sqrt{\pi a}$$

$$K_{II} = S(\cos\theta)(\sin\theta)\sqrt{\pi a}$$

Figure 8.37   Angled crack in an infinite plate under remote tension, and the resulting stress intensity factors.

will now pursue yielding at crack tips in more detail. It is significant that the region of yielding, called the plastic zone, must not be excessively large if the LEFM theory is to be applied.

8.6.1 Plastic Zone Size for Plane Stress

An equation for estimating plastic zone sizes for plane stress situations can be developed from the elastic stress field equations, Eq. 8.8, with $\sigma_z = 0$. In the plane of the crack,

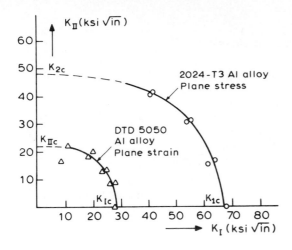

Figure 8.38 Combined mode fracture in two aluminum alloys. (From [Broek 86] p. 378; reprinted by permission of Kluwer Academic Publishers.)

where $\theta = 0$, these simplify to

$$\sigma_x = \sigma_y = \frac{K}{\sqrt{2\pi r}} \qquad (a)$$

$$\sigma_z = \tau_{xy} = \tau_{yz} = \tau_{zx} = 0 \quad (b)$$

$$(8.27)$$

Since all shear stress components along $\theta = 0$ are zero, σ_x, σ_y, and σ_z are principal normal stresses. Both the maximum shear stress and the octahedral shear stress yield criteria of the previous chapter then estimate yielding at $\sigma_x = \sigma_y = \sigma_o$, where σ_o is the yield strength. Substituting this above and solving for r gives

$$r_{o\sigma} = \frac{1}{2\pi} \left(\frac{K}{\sigma_o} \right)^2 \qquad (8.28)$$

This is simply the distance ahead of the crack tip where the elastic stress distribution exceeds the yield criterion for plane stress as illustrated in Fig. 8.39. Note that elastic, perfectly plastic behavior is assumed.

Due to yielding within the plastic zone, the stresses are lower than the values from the elastic stress field equations. The yielded material thus offers less resistance than expected, and large deformations occur, which in turn cause yielding to extend even farther than $r_{o\sigma}$, as also illustrated. The commonly used estimate is that yielding actually extends to about $2r_{o\sigma}$. Hence, the final estimate of plastic zone size for plane stress is

$$2r_{o\sigma} = \frac{1}{\pi} \left(\frac{K}{\sigma_o} \right)^2 \qquad (8.29)$$

As might be expected, the plastic zone size increases if the stress (hence K) is increased, and it is smaller for the same K for materials with higher σ_o.

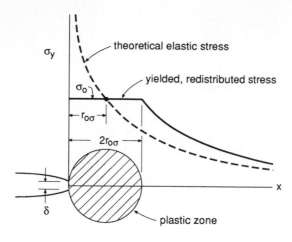

Figure 8.39 Plastic zone size estimate for plane stress, showing the approximate effect of redistribution of stress.

8.6.2 Plastic Zone Size for Plane Strain

For plane strain, the stress σ_z is nonzero, and this elevates the value of $\sigma_x = \sigma_y$ necessary to cause yielding, in turn decreasing the plastic zone size relative to that for plane stress. Irwin's estimate of the resulting size is commonly used.

$$2r_{o\varepsilon} = \frac{1}{3\pi} \left(\frac{K}{\sigma_o} \right)^2 \tag{8.30}$$

This is noted to be one-third as large as the plane-stress value.

Some justification for the smaller value for plane strain is as follows: The elastic stress field equations give the same result as Eq. 8.27 above except that σ_z is not zero.

$$\sigma_z = 2\nu\sigma_y \tag{8.31}$$

The stresses σ_x, σ_y, and σ_z along $\theta = 0$ are still principal normal stresses. Both yield criteria now predict yielding at

$$\sigma_x = \sigma_y = \frac{\sigma_o}{1 - 2\nu} \approx 2.5\sigma_o \tag{8.32}$$

Hence, the triaxial state of stress near the crack tip has the effect of elevating the stress required for yielding by an estimated factor of 2.5, with this particular value being based on a typical $\nu = 0.3$.

In his pioneering work on fracture mechanics and plastic zone effects, Irwin used different logic than that given above, and this led him to select a factor for the elevation of the yield strength due to plane strain that is less than 2.5, in particular about 1.7. Substituting $\sigma_y = 1.7\sigma_o \approx \sqrt{3}\sigma_o$ into Eq. 8.27(a) and solving for r, and also multiplying by two to account for stress redistribution, gives the estimate of plastic zone size for plane strain of Eq. 8.30.

The plastic zone size equations given are based on simple assumptions and do not consider various complexities that exist, so that these should be considered to be rough estimates only.

8.6.3 Plasticity Limitations on LEFM

If the plastic zone is sufficiently small, there will be a region outside of it where the elastic stress field equations (Eq. 8.8) still apply, called the *region of K-dominance*, or the *K-field*. This is illustrated in Fig. 8.40. The existence of such a region is necessary for LEFM theory to be applicable. The K-field surrounds and controls the behavior of the plastic zone and crack tip area, which can be thought of as an incompletely understood "black box." Thus, K continues to characterize the severity of the crack situation, despite the occurrence of some limited plasticity. However, if the plastic zone is so large that it eliminates the K-field, then K no longer applies.

As a practical matter, it is necessary that the plastic zone be small compared to the distance from the crack tip to any boundary of the member, such as distances a, $(b-a)$, and h for a cracked plate as in Fig. 8.41(a). A distance of $8r_o$ is generally considered to

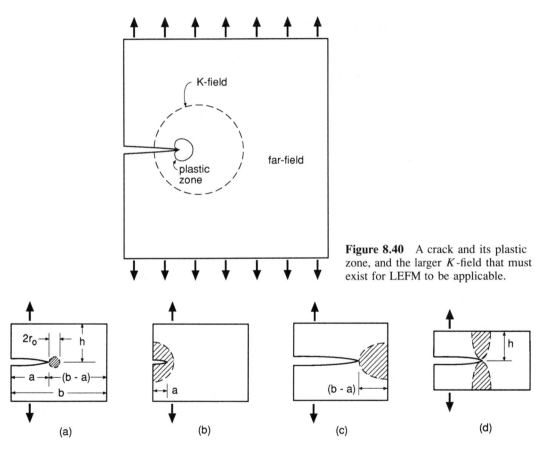

Figure 8.40 A crack and its plastic zone, and the larger K-field that must exist for LEFM to be applicable.

Figure 8.41 Small plastic zone compared to planar dimensions (a), and situations where LEFM is invalid due to the plastic zones being too large compared to (b) crack length, (c) uncracked ligament, and (d) member height.

be sufficient. Note from Eqs. 8.29 and 8.30 that $8r_o$ is four times the plastic zone size, which can be either $2r_{o\sigma}$ or $2r_{o\varepsilon}$, depending on which applies. Since $2r_{o\sigma}$ is larger than $2r_{o\varepsilon}$, an overall limit on the use of LEFM is

$$a, (b - a), h \geq \frac{4}{\pi} \left(\frac{K}{\sigma_o} \right)^2 \quad \text{(LEFM applicable)} \qquad (8.33)$$

This must be satisfied for all three of a, $(b - a)$ and h. Otherwise, the situation too closely approaches gross yielding with a plastic zone extending to one of the boundaries as shown in Fig. 8.41(b), (c), or (d). As discussed below in Section 8.8, a value of K calculated beyond the applicability of LEFM underestimates the severity of the crack.

8.6.4 Plane Stress Versus Plane Strain

If the thickness is not large compared to the plastic zone, Poisson contraction in the thickness direction occurs freely around the crack tip, resulting in the appearance of a dimple as shown in Fig. 8.42. This lack of constraint on deformation in the z-direction results in the stresses in that direction being small, so that x-y plane stress is said to exist. However, if the thickness is large compared to the plastic zone, the material under relatively low stress around the plastic zone tends to prevent the lateral contraction of the material within the plastic zone. This causes ε_z to be small, so that a state of plane strain occurs, and a tensile stress σ_z develops. A situation of *constraint* is said to exist, as the material in the plastic zone is not free to accommodate deformation by Poisson contraction in any direction.

Based on empirical observation of the trends in fracture behavior, especially the thickness effect on toughness, Fig. 8.26, it has become generally accepted that a fully developed situation of plane strain does not occur unless the thickness satisfies the relationship given earlier as Eq. 8.25. In addition, the distances from the crack tip to the in-plane boundaries must be similarly large compared to the plastic zone. Otherwise,

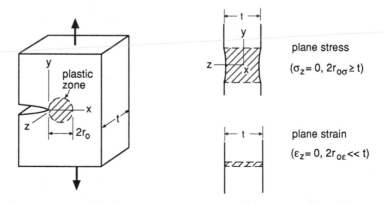

Figure 8.42 Effect of thickness on transverse Poisson contraction in the plastic zone near the crack tip.

deformation in the x- or y-direction can occur that reduces the degree of constraint. Thus, the overall requirement for plane strain is

$$t, a, (b - a), h \geq 2.5 \left(\frac{K}{\sigma_o} \right)^2 \quad \text{(plane strain)} \tag{8.34}$$

Comparison with Eq. 8.30 indicates that this corresponds to the various dimensions all being larger than $47 r_{o\varepsilon}$, or about 24 times the plane strain plastic zone size $2 r_{o\varepsilon}$. Note that the requirements on the in-plane dimensions of Eq. 8.33 are less stringent than Eq. 8.34, so that the limits on the use of LEFM are automatically satisfied if plane strain is satisfied.

Example 8.3

For the situation of Example 8.1:

(a) For $a = 10$ mm, determine whether or not plane strain applies and whether or not LEFM is valid. Also estimate the plastic zone size.

(b) Do the same for the estimated $a_c = 16.3$ mm.

Solution   (a) Plane strain applies if Eq. 8.34 is satisfied. Use $t = 5$ mm, $b = 50$ mm, and K as calculated in Ex. 8.1(a), and also $\sigma_o = 415$ MPa for 2014-T651 Al from Table 8.1.

$$t, a, (b - a), h \geq 2.5 \left(\frac{K}{\sigma_o} \right)^2 = 2.5 \left(\frac{17.7 \text{ MPa}\sqrt{\text{m}}}{415 \text{ MPa}} \right)^2 ?$$

$$5, 10, 40, \text{ large } h \geq 0.0045 \text{ m} = 4.5 \text{ mm}?$$

Yes, the test is successful, so that *plane strain applies and LEFM is applicable.* The plastic zone size is then estimated as the value for plane strain from Eq. 8.30.

$$2 r_{o\varepsilon} = \frac{1}{3\pi} \left(\frac{K}{\sigma_o} \right)^2 = \frac{1}{3\pi} \left(\frac{17.7 \text{ MPa}\sqrt{\text{m}}}{415 \text{ MPa}} \right)^2 = 0.19 \text{ mm} \qquad \textbf{Ans.}$$

(b) For $a_c = 16.3$ mm and $K = K_{Ic} = 24$ MPa$\sqrt{\text{m}}$, the plane strain test is similarly applied.

$$5, 16.3, 33.7, \text{ large } h \geq 2.5 \left(\frac{24 \text{ MPa}\sqrt{\text{m}}}{415 \text{ MPa}} \right)^2 = 8.4 \text{ mm}?$$

No, the test fails and *plane strain does not apply.* LEFM may still be applicable if Eq. 8.33 is satisfied.

$$a, (b - a), h \geq \frac{4}{\pi} \left(\frac{K}{\sigma_o} \right)^2 = \frac{4}{\pi} \left(\frac{24 \text{ MPa}\sqrt{\text{m}}}{415 \text{ MPa}} \right)^2 ?$$

$$16.3, 33.7, \text{ large } h \geq 4.3 \text{ mm}?$$

Yes, the test is successful, and *LEFM is applicable.* The plastic zone size is then estimated as the value for plane stress from Eq. 8.29.

$$2 r_{o\sigma} = \frac{1}{\pi} \left(\frac{K}{\sigma_o} \right)^2 = \frac{1}{\pi} \left(\frac{24 \text{ MPa}\sqrt{\text{m}}}{415 \text{ MPa}} \right)^2 = 1.06 \text{ mm} \qquad \textbf{Ans.}$$

Comment   Due to the state of plane stress in (b), the use of K_{Ic} is conservative. The actual K_c may be somewhat higher than K_{Ic}, and the a_c value therefore larger than estimated.

8.7 DISCUSSION OF FRACTURE TOUGHNESS TESTING

The preceding discussion on plastic zone size and the condition of plane strain permits us to now engage in a more detailed discussion of certain aspects of fracture toughness testing.

8.7.1 Standard Test Methods

Test methods for evaluating fracture toughness based on LEFM, such as ASTM E399 and D5045, require that Eq. 8.34 be satisfied for the toughness to be designated as the plane strain value K_{Ic}. In particular, Eq. 8.34 is explicitly required for t and a. The remaining requirements are then automatically met by specifying a/b near 0.5 and by using specimen geometries where h/a is around unity or greater.

The load versus displacement (P-v) behavior in a fracture toughness test may be similar to Fig. 8.23, Types I, II, or III. Since the plastic zone is required by Eq. 8.34 to be quite small, any nonlinearity in the P-v curve must be due to growth of the crack. A smooth curve as in Type I is caused by a steady tearing type of fracture called *slow-stable crack growth*. In other cases, the crack may suddenly grow a short distance, which is called a *pop-in* (II), or it may suddenly grow to complete failure (III).

The K_{Ic} test standards handle the problem of defining the beginning of cracking by drawing a line with a slope that is 95% as large as the initial elastic slope (O-A) in the test. A load P_Q is identified as the point where this line crosses the P-v curve, or as any larger peak value prior to the crossing point. The stress intensity factor is then evaluated using P_Q and the initial crack length.

$$K_Q = f(a_i, P_Q) \tag{8.35}$$

If this K_Q satisfies Eq. 8.34, it is then considered to be a *valid K_{Ic}* value. However, there is one additional requirement designed to assure that the test involves sudden fracture with little slow-stable crack growth, specifically that the maximum load reached cannot exceed P_Q by 10%.

If the plane strain condition is not satisfied in the test, then it is necessary to use a larger test specimen so that the comparisons of Eq. 8.34 are more favorable. As a result, quite large specimens may be required for materials with relatively low yield strength but high fracture toughness. For example, the large compact specimen of Fig. 8.24 (with $b = 61$ cm) was required to obtain the K_{Ic} value plotted for 10°C for A533B steel in Fig. 8.28.

A successful fracture toughness test requires an initially sharp flaw that is equivalent to a natural crack, otherwise the measured K_{Ic} will be artificially elevated. For metals, this is achieved by using cyclic loading at a low level to start a natural fatigue crack at the end of a machined slot. For plastics (polymers), the usual procedure is to press a razor blade into the material at the end of the machined slot.

8.7.2 Effect of Thickness on Fracture Behavior

Fracture under the highly constrained conditions of plane strain generally occurs rather suddenly with little crack growth prior to final fracture. Also, the fracture surface is quite flat. In contrast, plane stress fractures tend to have sloping or V-shaped surfaces inclined at about $45°$ on planes of maximum shear stress, and the final fracture is usually preceded by considerable slow-stable crack growth. The differences in mechanical behavior and fracture surface are illustrated in Figs. 8.43 and 8.44. These behaviors correlate with the thickness effect on toughness of Fig. 8.26. Flat plane-strain fractures occur where the thickness is sufficient to reach the lower plateau of the curve, where the toughness is the minimum value K_{Ic}. Inclined or V-shaped plane stress fractures occur for relatively thin members, for which the toughness may be well above K_{Ic}. Fracture toughness values K_{Ic} meeting the requirements for plane strain are thus expected to be minimum values that can be safely used in design for any thickness.

8.7.3 Use of Plane Stress Fracture Data

Where a thickness less than that required for plane strain fracture in a given material is used in an engineering application, K_{Ic} may involve an undesirably large degree of conservatism. It may then be useful to use K_Q data for the particular thickness of interest. Also, a toughness K_c can be defined that corresponds to the point of final fracture as illustrated in Fig. 8.43. Since the amount of slow-stable crack growth may be considerable, the crack extension Δa_c from a_i to the final length a_c needs to be measured. The corresponding K can then be calculated from the load P_c at the point of final fracture.

$$K_c = f\,(a_c, P_c) \qquad (8.36)$$

The curve of K versus crack extension Δa up to fracture is called the *resistance curve* or *R-curve*. Failure may occur at different points along the curve if the geometry or loading configuration is changed drastically, and the curve itself may change if the thickness is changed. Hence, K_c is not a material property in a strict sense as it depends on thickness, and to an extent upon other details of the geometry. Some care is therefore needed in using such values. To assure accuracy, it may be necessary to match the laboratory tests fairly closely to the thickness, geometry, and type loading of the engineering application.

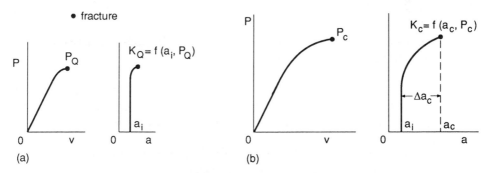

Figure 8.43 Load vs. displacement and load vs. crack extension behavior during fracture toughness tests under plane strain (a), and plane stress (b).

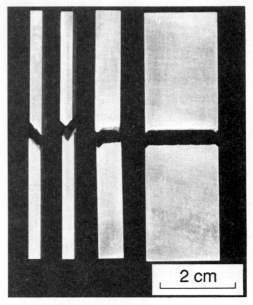

Figure 8.44 Fracture surfaces (left) and cross sections showing profiles of fractures (right) for toughness tests on compact specimens ($b = 51$ mm) of 7075-T651 aluminum. The thinnest specimens shown have typical plane stress fractures on inclined planes; the intermediate thickness has mixed behavior; and the thickest specimens have flat plane strain fractures. (Photos by R. A. Simonds.)

8.8 EXTENSIONS OF FRACTURE MECHANICS BEYOND LINEAR ELASTICITY

If Eq. 8.33 is not satisfied, so that LEFM is not applicable due to excessive yielding, several methods still exist for analyzing cracked members. Excessive yielding causes K to no longer correctly characterize the magnitude of the stress field around the crack tip, and specifically K underestimates the severity of the crack. One possibility is to make an adjustment to K based on the plastic zone size. For large amounts of yielding, the J-integral and CTOD approaches are available.

An introduction to these approaches for extending fracture mechanics is given below starting in Sect. 8.8.2. Before proceeding, it is useful to introduce the concept of a fully plastic yielding load, which is done in Sect. 8.8.1. More detail is available in any recent book on fracture mechanics, and especially in Kumar et al. (1981) and Zahoor (1989) as listed in the References.

8.8.1 Fully Plastic Loads

The plastic zone at a crack tip can extend until the entire uncracked portion of the member has yielded, assuming of course that the combination of material properties, section size, and geometry is such that fracture does not occur first. The situation is the same as that previously illustrated for a notch in Fig. 7.26. A conservative (lower bound) estimate can be made of the load corresponding to full yielding by using an elastic, perfectly plastic stress-strain relationship and by noting that yielding will spread until a uniform stress equal to the yield strength σ_o exists across the entire uncracked portion. The corresponding load is called the *fully plastic load*, P_o, or for bending the *fully plastic moment*, M_o.

For a symmetrically loaded center-cracked plate or double-edge-cracked plate, Fig 8.13(a) or (b), and also for the symmetrical prying load case of Fig. 8.15(a), the average stress on the net (uncracked) portion of the cross-sectional area is

$$S_n = \frac{P}{2t(b-a)} = \frac{S_g}{1 - a/b} \tag{8.37}$$

where the definition $S_g = P/2bt$ is used to obtain the second form. Fully plastic yielding is expected when S_n reaches the yield strength, as illustrated by the freebody diagram of Fig 8.45(a). Hence, substituting $S_n = \sigma_o$ above gives the fully plastic load P_o, or the corresponding gross stress.

$$P_o = 2bt\sigma_o\left(1 - \frac{a}{b}\right), \qquad S_{go} = \sigma_o\left(1 - \frac{a}{b}\right) \tag{8.38}$$

For pure bending, as in Fig. 8.14(a) or (b), the freebody of Fig. 8.45(b) applies. Invoking equilibrium by summing moments leads to the estimate that fully plastic bending occurs at

$$M_o = \frac{b^2 t \sigma_o}{4}\left(1 - \frac{a}{b}\right)^2, \qquad S_{go} = 1.5\sigma_o\left(1 - \frac{a}{b}\right)^2 \tag{8.39}$$

where the definition of S_g for this case is used to obtain the second form. Derivation of the above is similar to the fully plastic bending analysis given in most textbooks on elementary mechanics of materials, and also as presented later in Chapter 13. Equations for fully plastic loads are also given in Fig. 8.45 for two additional situations that involve combined bending and tension, these corresponding to the cases of Figs. 8.15(b) or (c), and Fig. 8.13(c), respectively. They are derived by invoking equilibrium of both forces and moments as done in detail later in Chapter 13.

The P_o and M_o equations given represent lower bound estimates, so that actual values may be somewhat higher, for two reasons. First, strain hardening in actual stress-strain curves will increase the value somewhat. Second, if the member is relatively

(a)
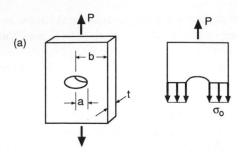

(b)
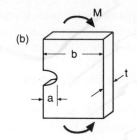
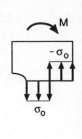

(c)
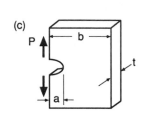
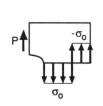

(d)
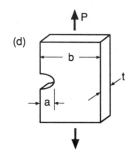
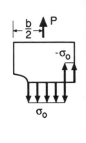

Fully plastic load or moment for given $\alpha = a/b$:

(a) $P_o = 2bt\sigma_o\,(1 - \alpha)$ (b) $M_o = \dfrac{b^2 t\sigma_o}{4}\,(1 - \alpha)^2$

(c) $P_o = bt\sigma_o\left[-\alpha - 1 + \sqrt{2\left(1 + \alpha^2\right)}\,\right]$ (d) $P_o = bt\sigma_o\left[-\alpha + \sqrt{2\alpha^2 - 2\alpha + 1}\,\right]$

Crack length at fully plastic yielding for given load, where for (c) and (d), $P' = P/(bt\sigma_o)$:

(a) $a_o = b\left[1 - \dfrac{P}{2bt\sigma_o}\right]$ (b) $a_o = b\left[1 - \dfrac{2}{b}\sqrt{\dfrac{M}{t\sigma_o}}\right]$

(c) $a_o = b\left[P' + 1 - \sqrt{2P'(P' + 2)}\,\right]$ (d) $a_o = b\left[P' + 1 - \sqrt{2P'(P' + 1)}\,\right]$

Applicability to cases where K is given:

                 (a) Figs. 8.13(a,b), 8.15(a)   (b) Figs. 8.14(a,b)
                 (c) Figs. 8.15(b,c)   (d) Fig. 8.13(c)

Figure 8.45 Freebody diagrams and resulting equations for fully plastic loads or moments for various two-dimensional cases, and also the same equations solved for crack length.

thick, the constrained yielding situation of plane strain can continue into the fully plastic loading region, thus in effect increasing the yield strength in a manner similar to that already discussed for plastic zones. For example, for a center-cracked plate, the second effect can increase P_o by as much as 15%, and for bending the increase in M_o can reach 45%.

8.8.2 Plastic Zone Adjustment

Consider the redistributed stress near the plastic zone as in Fig. 8.39. The stresses outside of the plastic zone are similar to those for the elastic stress field equations for a hypothetical crack of length

$$a_e = a + r_{o\sigma} \tag{8.40}$$

that is, a hypothetical crack with its tip near the center of the plastic zone. This in turn leads to modifying K, specifically increasing it, to account for this yielding by using a_e in place of the actual crack length a.

Where the form $K = FS\sqrt{\pi a}$ is used, the modified value is

$$K_e = FS\sqrt{\pi (a + r_{o\sigma})} \tag{8.41}$$

The F used is the value corresponding to a_e/b, and $r_{o\sigma}$ is calculated using K_e in Eq. 8.28. Making this substitution for $r_{o\sigma}$ and solving for K_e gives

$$K_e = FS\sqrt{\frac{\pi a}{1 - \dfrac{1}{2}\left(\dfrac{FS}{\sigma_o}\right)^2}} \tag{8.42}$$

An iterative calculation is generally involved in using this equation as $F = F(a_e/b)$ cannot be determined in advance, since $r_{o\sigma}$ and hence a_e depend on K_e. If F is not significantly changed for the new crack length a_e, then no iteration is required, and the modified value K_e is related to the unmodified value $K = FS\sqrt{\pi a}$ by

$$K_e = \frac{K}{\sqrt{1 - \dfrac{1}{2}\left(\dfrac{FS}{\sigma_o}\right)^2}} \tag{8.43}$$

In some situations with a high degree of constraint, such as embedded elliptical cracks or half-elliptical surface cracks, it may be appropriate to use the plane strain plastic zone size to make the adjustment. Replacing $r_{o\sigma}$ with $r_{o\varepsilon}$ in the above equations gives

$$K_e = FS\sqrt{\frac{\pi a}{1 - \dfrac{1}{6}\left(\dfrac{FS}{\sigma_o}\right)^2}} \quad \text{(plane strain)} \tag{8.44}$$

An equation for unchanged F analogous to Eq. 8.43 also applies.

Such modified K values allow LEFM to be extended to somewhat higher stress levels than permitted by the limitation of Eq. 8.33. However, large amounts of yielding still cannot be analyzed. The use of even adjusted crack lengths becomes increasingly questionable if the stress approaches a value that would cause yielding fully across the uncracked section of the member. It is suggested that the use of plastic zone adjustments be limited to loads below 80% of the fully plastic value, that is, below $0.8P_o$ or $0.8M_o$.

Example 8.4

Problem 8.21 concerns a test on a double-edge-cracked plate of 7075-T651 Al, for which $a = 5.7$, $b = 15.9$, and $t = 6.35$ mm. A value of $K_Q = 37.3$ MPa$\sqrt{\text{m}}$ is calculated for the load $P_Q = 50.3$ kN, but the test for applicability of LEFM (Eq. 8.33) is not met. Calculate the fully plastic load. If it is reasonable to do so, also apply the plastic zone adjustment to obtain a revised value of K_Q.

Solution   Figure 8.45(a) applies, so that the fully plastic load is

$$P_o = 2bt\sigma_o \left(1 - \frac{a}{b}\right)$$

$$P_o = 2(0.0159 \text{ m})(0.00635 \text{ m})(505 \text{ MPa}) \left(1 - \frac{5.7 \text{ mm}}{15.9 \text{ mm}}\right)$$

$$P_o = 0.0654 \text{ MN} = 65.4 \text{ kN}   \textbf{Ans.}$$

where the yield strength of 7075-T651 Al from Table 8.1 is used. Comparing P_o to P_Q gives

$$\frac{P_Q}{P_o} = \frac{50.3 \text{ kN}}{65.4 \text{ kN}} = 0.77$$

Since this ratio is less than 0.80, it is reasonable to apply the plastic zone adjustment.
    From Fig. 8.13(b), the value $\alpha = a/b = 0.358$ that applies is well within the range $\alpha \leq 0.6$ where $F \approx 1.12$. Hence, F can be taken as unchanged for a_e and Eq. 8.43 applies.

$$K_{Qe} = \frac{K_Q}{\sqrt{1 - \frac{1}{2}\left(\frac{FS}{\sigma_o}\right)^2}} = \frac{37.3 \text{ MPa}\sqrt{\text{m}}}{\sqrt{1 - \frac{1}{2}\left(\frac{1.12 \times 249 \text{ MPa}}{505 \text{ MPa}}\right)^2}} = 40.5 \text{ MPa}\sqrt{\text{m}} \qquad \textbf{Ans.}$$

where $S = S_g = P_Q/2bt$ is used.

Comment   This adjusted K is 40% above the value from Table 8.1 of $K_{Ic} = 29$ MPa$\sqrt{\text{m}}$. The probable explanation for the difference is that K_{Qe} includes an effect of increased toughness for plane stress.

8.8.3 The J-Integral

This advanced approach is capable of handling even large amounts of yielding. The J-integral can be thought of as the generalization of the strain energy release rate, G, to cases of nonlinear-elastic stress-strain curves. This is illustrated in Fig. 8.46. However, for most cases of engineering interest, the nonlinear stress-strain (and hence P-v) behavior is due to elasto-plastic behavior, as in metals. For elasto-plastic materials, J

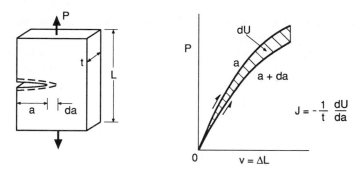

Figure 8.46   Definition of the J-integral in terms of the potential energy difference for cracks of slightly different length.

loses the physical interpretation related to potential energy. But it retains significance as a measure of the intensity of the elasto-plastic stress and strain fields around the crack tip. Values can still be determined experimentally or analytically using P-v curves as in Fig. 8.46. However, the two different P-v curves for crack lengths a and $(a + da)$ need to be determined by independent tests or analysis on two different members, rather than by extending the crack da after a single member is loaded.

The J-integral can be used as the basis of fracture toughness tests according to ASTM Standard E813. Since the plasticity limitations of LEFM can now be exceeded, the need for large test specimens is removed. For example, in A533B steel at room temperature, a fracture toughness J_{Ic} can be obtained from the small specimen in Fig. 8.24, without the need for testing the large one. Thus, J_{Ic} can be used to estimate an equivalent value of K_{Ic} by using Eq. 8.12 with J replacing G.

$$K_{IcJ} = \sqrt{J_{Ic} E'} \qquad (8.45)$$

One complexity encountered in J_{Ic} testing is that nonlinearity in the P-v behavior is now due to a combination of crack growth and plastic deformation. Hence, the beginning of cracking cannot be determined in a straightforward manner from the P-v curve, and special means are needed to more directly measure crack growth.

In engineering applications where crack growth under plastic loading needs to be considered, the J-integral is a candidate for use along with the CTOD approach that is discussed next. One important area of such application is pressure vessels, especially nuclear pressure vessels. Note that an attempt to use K beyond its region of validity will generally cause the results of engineering calculations to be nonconservative, that is, to be in error on the unsafe side. For example, consider a through-thickness center crack in a plate that is small compared to the plate width, and assume that plane stress applies. The J-integral can be used to obtain a modified (equivalent) value, K_J, that includes the plasticity effect. For this case, and also for a similar surface crack

$$\sqrt{JE} = K_J \approx K\sqrt{1 + \frac{\varepsilon_p}{\varepsilon_e \sqrt{n}}} \qquad (8.46)$$

where $K = FS\sqrt{\pi a}$ from LEFM, and ε_e and ε_p are the elastic and plastic strains corresponding to the applied stress. The quantity n is the strain hardening exponent for an elasto-plastic stress-strain curve for the material of the form shown in Fig. 8.47. (See Chapter 12 for detailed discussion of such stress-strain curves.) If the plastic strain ε_p is small, the second term under the radical in Eq. 8.46 disappears, and $K_J = K$. However, beyond yielding, ε_p increases rapidly, and K_J can become much larger than K, as shown in Fig. 8.47 for an n value typical of metals.

8.8.4 Crack-Tip Opening Displacement (CTOD)

The elastic stress field analysis employed for K can also be used to estimate the displacements separating the crack faces. Then pursuing logic that partially parallels that for plastic zone sizes, it is possible to make a rough estimate of the separation of the crack faces near the tip, that is, of the CTOD, which is denoted δ. For ductile materials this estimate is

$$\delta \approx \frac{K^2}{E\sigma_o} \approx \frac{J}{\sigma_o} \tag{8.47}$$

where δ is illustrated in Figs. 8.4 and 8.39. Experimentally determined values of δ are also used as the basis of fracture toughness tests, as in Standard No. BS 5762 of the British Standards Institution and the similar ASTM Standard No. E1290.

For any fracture mechanics approach to be valid, the value of δ must be small compared to the in-plane geometric dimensions. Within the range of LEFM, the ratio of the plane stress plastic zone size to δ can be obtained from the expression above and Eq. 8.29.

$$\frac{2r_{o\sigma}}{\delta} = \frac{E}{\pi\sigma_o} \tag{8.48}$$

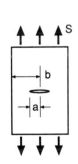
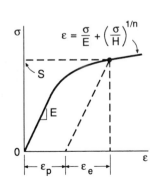
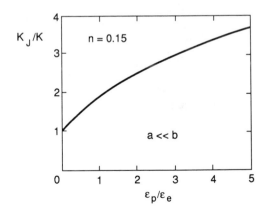

Figure 8.47 Equivalent K from J for a wide center-cracked plate under fully plastic yielding in plane stress, where the material has a nonlinear elasto-plastic stress-strain curve as shown. (See [Dowling 87a] for more detail.)

For engineering materials, the ratio E/σ_o is seldom less than 10, and is more commonly around 100 or more. Therefore, δ is usually very small and essentially never limits the use of fracture mechanics where LEFM is valid. Specimen size and thickness requirements based on δ are included in the J_{Ic} test standard, ASTM E813, but these are likely to require increased specimen sizes only in rather extreme situations of high fracture toughness and low yield strength.

8.9 SUMMARY

Using linear-elastic fracture mechanics (LEFM), the severity of a crack in a component can be characterized by the value of a special variable called the stress intensity factor.

$$K = FS\sqrt{\pi a} \tag{8.49}$$

where S is stress and a is crack length, both consistently defined relative to the dimensionless quantity F. Use of K depends on the behavior being dominated by linear-elastic deformation, so that the zone of yielding (plasticity) at the crack tip must be relatively small. Simple equations and handbooks provide values of F for a wide range of cases of cracked bodies, some examples of which are given in Figs. 8.12 through 8.18. The value of F depends on the crack and member geometry, the loading configuration, such as tension or bending, and on the ratio of the crack length to the width of the member, such as the ratio a/b. Some notable values of F for relatively short cracks under tension stress are as follows:

$$
\begin{aligned}
F &= 1.00 && \text{(center-cracked plate)} \\[4pt]
F &= 1.12 && \text{(through-thickness surface crack)} \\[4pt]
F &= 0.73 && \text{(half-circular surface crack)}
\end{aligned}
\tag{8.50}
$$

It is sometimes convenient to express K in terms of an applied load P using the differently defined dimensionless quantity F_P according to Eq. 8.15.

The value of K where a given material begins to crack significantly is called K_Q, and where it fails K_c. Slow-stable crack growth may follow K_Q until K_c is reached, and both of these may decrease with increased member thickness. If the plastic zone surrounding the crack tip is quite small compared to the thickness and is very well isolated relative to the boundaries of the member, then a state of plane strain is established. Under plane strain, only limited slow-stable crack growth occurs, so that K_Q and K_c have similar values to each other and also to the standard plane strain fracture toughness, K_{Ic}. A value of K_{Ic} thus represents a worst case fracture toughness that can be safely used for any thickness. The flow chart of Fig. 8.48 gives the requirement for plane strain and the plastic zone sizes, and the situation concerning K_{Ic} is also summarized.

Values of K_{Ic} for a given material generally decrease along with ductility if the material is processed to achieve higher strength. For a given material and processing,

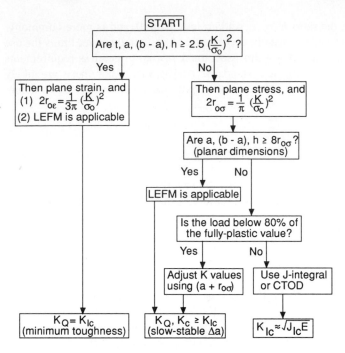

Figure 8.48   Flow chart for distinguishing between plane stress and plane strain, for deciding what fracture mechanics approach is needed, and for identifying what is expected from toughness testing.

K_{Ic} generally increases with temperature, sometimes exhibiting a rather abrupt change over a narrow range of temperatures, and also having relatively constant lower shelf and upper shelf values on opposite sides of the temperature transition. Increased loading rate causes K_{Ic} to decrease, having the effect of shifting the transition to a higher temperature. The microstructure of the material may affect K_{Ic}, as in the detrimental effect of sulfur in some steels, the effect of crystal grain orientation from rolling of aluminum alloys, and radiation embrittlement of pressure vessel steels.

If the plastic zone is too large, LEFM is no longer valid. Modest amounts of yielding can be handled by using adjusted values K_e calculated by adding half the plastic zone size to the crack length. However, above about 80% of the fully plastic load or moment, P_o or M_o, more general methods such as the J-integral or the crack-tip opening displacement (CTOD, δ) are needed. The flow chart of Fig. 8.48 also provides a guide for determining which of the various approaches is required in a given situation.

NEW TERMS AND SYMBOLS

(a) Terms

bend specimen	dimpled rupture
cleavage	fracture modes I, II, and III
compact specimen	fracture toughness: K_c, K_{Ic}
crack-tip opening displacement (CTOD),	fully plastic load, moment
δ	internally flawed material

J-integral, J
K-field
linear-elastic fracture mechanics (LEFM)
mixed-mode fracture
plane strain constraint
plastic zone
pop-in
R-curve

slow-stable crack growth
strain energy release rate, G
stress intensity factor, K
stress redistribution
superposition
temperature transition
transition crack length, a_t

(b) Detailed Nomenclature

a	Crack length, usually measured from the centerline or free surface; minor axis of elliptical crack, or radius of circular crack
a_c	Critical (at fracture) crack length
a_e	Plastic-zone-adjusted crack length
a_i	Inherent flaw size; initial crack length
b	Maximum possible crack length, hence, member width or half-width
c	Major axis of elliptical crack; notch dimension analogous to a
F	Dimensionless function $F(a/b)$ for $K = FS\sqrt{\pi a}$
F_P	Dimensionless function $F_P(a/b)$ for $K = F_P P/(t\sqrt{b})$
h	Distance from crack tip to boundary of member measured normal to the crack plane
k_t	Elastic stress concentration factor for blunt notch
K	Stress intensity factor for a Mode I crack
K_I, K_{II}, K_{III}	Stress intensity factors for the three fracture modes
K_{Ic}	Plane strain fracture toughness
K_c	Fracture toughness
K_e	Plastic-zone-adjusted K
K_J	Plasticity-modified K based on J
K_Q	Provisional fracture toughness
M	Bending moment
M_o	Fully plastic moment
P	Applied load (force)
P_c	Critical (at fracture) load
P_o	Fully plastic load
P_Q	Load for calculating K_Q
r, θ, z	Cylindrical coordinates with axis along leading edge of a crack
$r_{o\varepsilon}$	Half the estimated plastic zone size for plane strain
$r_{o\sigma}$	Half the estimated plastic zone size for plane stress
S	Nominal (average) stress; remotely applied uniform stress
S_g	Gross section S (based on area before cracking)
S_n	Net section S (based on remaining uncracked area)

t	Thickness
U	Potential strain energy
v	Displacement
x, y, z	Rectangular axes aligned parallel and perpendicular to the leading edge and plane of a crack
α	Ratio a/b
σ_x, τ_{xy}, etc.	Stresses at a point

REFERENCES

(a) General References

ANDERSON, T. L. 1991 *Fracture Mechanics: Fundamentals and Applications*, CRC Press, Boca Raton, Fla.

ASTM. 1991 *Annual Book of ASTM Standards*, Am. Soc. for Testing and Materials, Philadelphia, PA. See: No. E399, "Standard Test Method for Plane-Strain Fracture Toughness of Metallic Materials," Vol. 03.01; and Nos. E561, E813, and E1290 in Vol. 03.01; and also No. D5045 in Vol. 08.03.

BARSOM, J. M. and S. T. ROLFE. 1987a *Fracture and Fatigue Control in Structures*, 2nd ed., Prentice-Hall, Englewood Cliffs, N.J.

BARSOM, J. M., ed. 1987b *Fracture Mechanics Retrospective: Early Classic Papers (1913-1965)*, RPS-1, Am. Soc. for Testing and Materials, Philadelphia, Pa.

BROEK, D. 1986 *Elementary Engineering Fracture Mechanics*, 4th ed., Kluwer Academic Pubs., Dordrecht, The Netherlands.

BROEK, D. 1988 *The Practical Use of Fracture Mechanics*, Kluwer Academic Pubs., Dordrecht, The Netherlands.

EWALDS, H. L. and R. J. H. WANHILL. 1984 *Fracture Mechanics*, Edward Arnold Pubs., London.

HERTZBERG, R. W. 1989 *Deformation and Fracture Mechanics of Engineering Materials*, 3rd ed., John Wiley, New York.

RICE, R. C., ed. 1988 *Fatigue Design Handbook*, 2nd ed., SAE Pub. No. AE-10, Society of Automotive Engineers, Warrendale, Pa.

WILLIAMS, J. G. 1984 *Fracture Mechanics of Polymers*, Ellis Horwood Ltd., Chichester, West Sussex, England; dist. by Halsted Press div. of John Wiley, New York.

(b) Handbooks and Other Sources of Stress Intensity Solutions

KUJAWSKI, D., "Estimations of Stress Intensity Factors for Small Cracks at Notches," *Fatigue of Engineering Materials and Structures*, Vol. 14, No. 10, 1991, pp. 953–965.

KUMAR, V., M. D. GERMAN, and C. F. SHIH. 1981 *An Engineering Approach for Elastic-Plastic Fracture Analysis*, EPRI NP-1931, Electric Power Research Institute, Palo Alto, Ca., Jul. 1981.

MURAKAMI, Y., ed. 1987 *Stress Intensity Factors Handbook*, 2 vols., Pergamon Press, Oxford, UK.

NEWMAN, J. C., JR., and I. S. RAJU. 1986 "Stress-Intensity Factor Equations for Cracks in Three-Dimensional Finite Bodies Subjected to Tension and Bending Loads," *Computational Methods in the Mechanics of Fracture*, S. N. Atluri, ed., Elsevier Science Publishers, New York.

RAJU, I. S. and J. C. NEWMAN, JR. 1982 "Stress-Intensity Factors for Internal and External Surface Cracks in Cylindrical Vessels," *Jnl. of Pressure Vessel Technology*, ASME, Vol. 104, Nov. 1982, pp. 293–298.

RAJU, I. S. and J. C. NEWMAN, JR. 1986 "Stress-Intensity Factors for Circumferential Surface Cracks in Pipes and Rods Under Tension and Bending Loads," *Fracture Mechanics, Seventeenth Volume*, J. H. Underwood, et al., eds., ASTM STP 905, American Society for Testing and Materials, Philadelphia, Pa.

ROOKE, D. P. and D. J. CARTWRIGHT. 1976 *Compendium of Stress Intensity Factors*, Her Majesty's Stationery Office, London.

SCHIJVE, J. 1982 "The Stress Intensity Factor of Small Cracks at Notches," *Fatigue of Engineering Materials and Structures*, Vol. 5, No. 1, pp. 77–90.

TADA, H., P. C. PARIS and G. R. IRWIN. 1985 *The Stress Analysis of Cracks Handbook*, 2nd. ed., Paris Productions, Inc., 226 Woodbourne Dr., St. Louis, Mo. (63105).

ZAHOOR, A. 1989 *Ductile Fracture Handbook*, 3 vols., EPRI NP-6301-D, Electric Power Research Institute, Palo Alto, Ca.

(c) Sources for Material Properties

BATTELLE. 1975 *Handbook on Materials for Superconducting Machinery*, Metals and Ceramics Information Center, Battelle Columbus Labs., Columbus, Oh.

BATTELLE. 1991 *Aerospace Structural Metals Handbook*, 5 vols., Metals and Ceramics Information Center, Battelle Columbus Div., Columbus, Oh.

FERGUSON, R. R. and R. C. BERRYMAN. 1976 *Fracture Mechanics Evaluation of B-1 Materials*, AFML-TR-76-137, U. S. Air Force Materials Laboratory, Wright-Patterson AFB, Oh.

GALLAGHER, J.P., ed. 1983 *Damage Tolerant Design Handbook*, 4 vols., Metals and Ceramics Information Ctr., Battelle Columbus Labs., Columbus, Oh.

HUDSON, C. M. and S. K. SEWARD. 1978 "A Compendium of Sources of Fracture Toughness and Fatigue Crack Growth Data for Metallic Alloys," *Int. Jnl. of Fracture*, Vol. 14, No. 4, Aug. 1978, pp. R151–R184.

MARANDET, B. and G. SANZ. 1977 "Evaluation of the Toughness of Thick Medium-Strength Steels ..." *Flaw Growth and Fracture*, ASTM STP 631, Am. Soc. for Testing and Materials, Philadelphia, Pa, pp. 72–95.

MILHDBK. 1983 *Military Standardization Handbook: Metallic Materials and Elements for Aerospace Vehicle Structures*, MIL-HDBK-5D, 2 Vols., U.S. Dept. of Defense and Federal Aviation Administration, Naval Publications and Forms Ctr., Philadelphia, Pa.

PROBLEMS AND QUESTIONS

Section 8.2

8.1 Look ahead to Fig. 8.27, and:

(a) Obtain approximate values of fracture toughness K_{Ic} for AISI 4340 steel heat treated to yield strengths of 800 and 1600 MPa.

(b) For each of these yield strengths, calculate the transition crack length a_t, and comment on the significance of the values obtained.

8.2 Look ahead to Fig. 8.28, and:

   (a) Obtain approximate values of fracture toughness K_{Ic} and yield strength σ_o for A533B steel at temperatures of $-150°C$ and $+10°C$.

   (b) For each temperature, make a plot of stress versus crack length similar to Fig. 8.5, showing the limits imposed by yielding and brittle fracture. (PC Problem)

   (c) Then compare these plots and comment on the engineering use of this steel at these two temperatures.

8.3 For each metal in Table 8.1:

   (a) Calculate the transition crack length a_t .

   (b) Plot these as data points on a logarithmic scale, versus yield strength σ_o on a linear scale, using different symbols for steels versus aluminum alloys.

   (c) Comment on the values obtained and on any trends with yield strength.

8.4 Using Tables 8.1 and 8.2:

   (a) Calculate transition crack lengths a_t for three representative materials from each category of steels, aluminums, polymers, and ceramics. Use ultimate tensile strengths σ_{ut} from Tables 5.3 and 3.10 in place of σ_o for the polymers and ceramics.

   (b) Comment on the values obtained and any trends observed for the different classes of materials. Which particular materials do you think are likely to be internally flawed?

Section 8.4

8.5 A tension member is made of AISI 1045 steel (Fig. 8.7) heat treated to a hardness of $HB = 400$. It has a width of $b = 120$ mm and a thickness of $t = 12$ mm, and there is an edge crack through the full thickness and extending to a length $a = 18$ mm in the width direction. What is the maximum permissible tension load if a safety factor of 3 against brittle fracture is required?

8.6 For the situation of Prob. 8.5, what is the largest permissible crack length if the tension load is 100 kN and if the safety factor of 3 is maintained? (PC Problem)

8.7 A rectangular beam made of ABS plastic is $b = 20$ mm deep and $t = 10$ mm thick. Loads in the plane of the 20 mm depth cause a bending moment of 10 N·m. What is the largest through-thickness edge crack that can be permitted if a safety factor of 2.5 against brittle fracture is required?

8.8 A large part in a turbine-generator unit operates near room temperature and is made of ASTM A470-8 steel. A surface crack has been found that is roughly a semi-ellipse, as in Example 8.2, with $2c = 50$ mm and $a = 15$ mm. The stress normal to the plane of the crack is 250 MPa, and the member width and thickness are large compared to the crack size. What is the safety factor against brittle fracture? Should the power plant continue to operate if failure of this part is likely to cause costly damage to the remainder of the unit?

8.9 A shaft-like component made of 2024-T351 aluminum has a round cross section of diameter 40 mm and is subjected to a bending moment of 600 N·m. What is the largest crack that can be permitted if the crack has a shape that approximates an arc of a circle with center at the shaft surface as shown in Figure P8.9? A safety factor of three against brittle fracture is required. (Suggestion: Approximate K based on a flat surface and a crack under tension equal to the bending stress at the shaft surface. Neither the surface curvature nor the stress gradient due to bending have much effect on K if the ratio a/r is relatively small.)

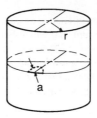

Figure P8.9

8.10 A standard compact fracture specimen, Fig. 8.15(c), has dimensions of $b = 50$ mm and $t = 25$ mm, and it is subjected to an applied load of $P = 22$ kN.
   (a) Calculate several values of K over the interval of crack lengths $a = 15$ to 35 mm. Use these values to plot a smooth curve of K versus a over this interval. (PC Problem)
   (b) If the material is 2219-T37 aluminum, what is the longest crack that would permit the 22 kN load to be applied without brittle fracture occurring?

8.11 For embedded elliptical cracks, Fig. 8.16, it is common to use the form $K = S\sqrt{\pi a/Q}$, where $\sqrt{Q} = E(k)$, with Q being called the *flaw shape parameter*. A plot of a/c versus Q can then be made and used to conveniently evaluate K for any a/c. Make such a plot by first calculating Q values for a number of a/c values between zero and unity. How could you use this plot to also make estimates of K for semi-elliptical surface cracks? (PC Problem)

8.12 Two plates of A533B-1 steel (Fig. 8.28) are butted together and then welded from one side, with the weld only penetrating halfway as shown in Figure P8.12. A uniform tension stress is applied during service at $-75°C$. Estimate the strength of this joint, as limited by brittle fracture from the crack-like flaw, as a value of gross stress S_g, calculated as if the joint were solid. The weld metal has similar properties to the plates.

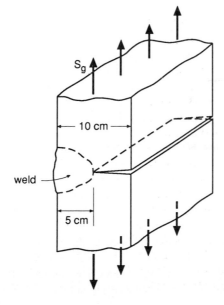

Figure P8.12

8.13 A structural member has dimensions as shown in Figure P8.13 and contains a crack as also shown of length $a = 15$ mm. This member is made of A572 structural steel (Fig. 8.32) and may be subjected in service to dynamic loading at temperatures as low as $-30°$ C.

    (a) What is the maximum permissible bending moment about the x-axis if a safety factor of 1.5 against brittle fracture is required? (Suggestion: Evaluate K approximately by noting that the cracked flange of the beam is essentially an edge-cracked tension member.)

    (b) The standard design code used for such beams permits a moment of 193 kN·m to be applied in service, which is based on a safety factor of 1.5 against yielding. Compare this value with your result to (a) and comment on the difference.

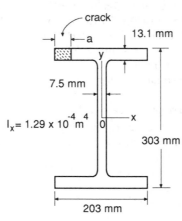

Figure P8.13

Section 8.5

8.14 Assume that each of the rotor steels of Fig. 8.30 except A217 is being considered for use around room temperature (22°C). The design will be such that the highest stress does not exceed half of the respective yield strength in each case. Assuming a flaw geometry that is a half-circular surface crack in an infinite body, Fig. 8.17, determine the largest permissible crack size a for each material if a safety factor of 2.0 against brittle fracture is required. Also comment on how this information might affect the choice among these steels.

8.15 Consider use of the pressure vessel steel of Fig. 8.33 at both +50°C and −50°C, where the stress can be as high as 400 MPa. Assuming a half-circular surface crack in an infinite body, Fig. 8.17, determine the crack size a that can be permitted at each temperature if a safety factor of 3.0 against brittle fracture is required. Comment on the significance of your result relative to use of this material in pressure vessels.

8.16 In Fig. 8.35, the fracture toughness of rolled plates of aluminum alloys is seen to vary with orientation. Explain the physical reasons for this behavior and why the toughness for the L-T orientation is the highest and that for S-L is the lowest. (Suggestion: See Chapter 3.)

8.17 Write a paragraph explaining the significance of the data for unirradiated and irradiated A533B-1 steel of Fig. 8.36.

8.18 Why are the heavy steel turbine-generator rotors in power plants heated as near as possible to their operating temperature before being subjected to the full stresses of service loading?

8.19 Consider the choice of a steel for an oil pipeline in a cold climate, such as Alaska or Siberia. What are the desirable characteristics of a material for this application? What types of test data should be available on candidate materials to serve as a basis for the decision?

Sections 8.6 and 8.7

8.20 For the situation of Example 8.2 under the applied stress given:
(a) Determine whether or not plane strain applies and whether or not LEFM is applicable.
(b) Estimate the plastic zone size, $2r_{o\sigma}$ or $2r_{o\varepsilon}$, whichever applies.

8.21 A double-edge-cracked plate of 7075-T651 aluminum has dimensions, as defined in Fig. 8.13(b), of $b = 15.9$ mm, $t = 6.35$ mm, large h, and sharp precracks with $a = 5.7$ mm. Under tension load, failure by sudden fracture occurred at a load of $P_{max} = 55.6$ kN. Prior to this, there was a small amount of slow-stable crack growth, with the P-v curve being similar to Fig. 8.23, Type I, and crossing the 5% slope deviation at $P_Q = 50.3$ kN.
(a) Calculate K_Q corresponding to P_Q.
(b) At the K_Q point, determine whether or not plane strain applies and whether or not LEFM is applicable.
(c) What is the significance of the K_Q calculated?

8.22 A fracture toughness test was conducted on AISI 4340 steel having a yield strength of 1380 MPa. The standard compact specimen used had dimensions, as defined in Fig. 8.15(c), of $b = 50.8$ mm, $t = 12.95$ mm, and a sharp precrack to $a = 25.4$ mm. Failure occurred suddenly at $P_Q = P_{max} = 15.03$ kN, with the P-v curve resembling Type III of Fig. 8.23.
(a) Calculate K_Q at fracture.
(b) Does this value qualify as a valid (plane strain) K_{Ic} value?
(c) Estimate the plastic zone size at fracture.

8.23 Data are given below for compact specimens of 7075-T651 aluminum in the same sizes as those photographed in Fig. 8.44. All had dimensions, as defined in Fig. 8.15(c), of $b = 50.8$ and $h = 30.5$ mm, and initial sharp precracks and thickness as tabulated below. For each test: (PC Problem)
(a) Calculate K_Q and determine whether or not K_Q qualifies as a valid (plane strain) K_{Ic}.
(b) Estimate the plastic zone size at K_Q, using $2r_{o\sigma}$ or $2r_{o\varepsilon}$ as applicable.
(c) Determine whether analysis by LEFM is applicable.
(d) Plot K_Q versus thickness t and comment on the trend observed and its relationship to the fracture surfaces in Fig. 8.44.

Test No.	a_i mm	t mm	P_Q kN	Basis of P_Q (See Fig. 8.23.)	P_{max} kN
1	24.1	3.18	2.56	P_5 for Type I	3.96
2	24.7	6.86	4.96	Pop-in, Type II	6.16
3	23.3	19.35	12.00	P_{max} for Type III	12.00

8.24 Consider the form $K = F S_g \sqrt{\pi a}$ and the limitation on LEFM of Eq. 8.33.

(a) Develop an equation that gives the largest permissible ratio S_g/σ_o as a function of F.

(b) What is the maximum stress level S_g/σ_o for use of LEFM for $F = 1.00$, $F = 1.12$, and $F = 2/\pi$, corresponding respectively to center, edge, and embedded circular cracks in infinite bodies?

(c) Can you rationalize the trends in Fig. 8.5 on the basis of your results above?

Sections 8.8 and 8.9

8.25 A bending member has dimensions, as defined in Fig. 8.14, of width $b = 50$ mm and thickness $t = 20$ mm. A through-thickness crack in the edge subjected to tension stress may be as long as $a = 10$ mm. What moment is expected to cause failure if the material is AISI 4340 steel (Fig. 8.27) with a yield strength of (a) 800 MPa and (b) 1600 MPa? In each case, consider both brittle fracture and fully plastic yielding as possible failure modes. Then (c) comment on whether or not it is beneficial to use the higher strength steel in this case.

8.26 For the situation of Prob. 8.12, calculate the strength for service temperatures of (a) $-75°C$, and (b) $200°C$. Properties for the former temperature are given in Fig. 8.28, and at the latter temperature the yield strength is $\sigma_o = 400$ MPa and the (upper shelf) fracture toughness is $K_{Ic} = 200$ MPa$\sqrt{m}$. For each temperature, consider both brittle fracture and fully plastic yielding as failure modes. Then (c) comment on the suitability of this steel for use at these two temperatures.

8.27 The combinations of crack length and stress corresponding to failure in Fig. 8.5 are tabulated below. (PC Problem)

(a) Plot these data and the lines for $\sigma_o = 518$ MPa and $K_c = S\sqrt{\pi a} = 66$ MPa$\sqrt{m}$ just as they appear in Fig. 8.5.

(b) Also plot a revised line for $K_c = 66$ MPa$\sqrt{m}$ where the plastic zone adjustment is used.

(c) Comment on the success of curve (b) in predicting the behavior.

Test no.	Crack length a_c, mm	Gross stress at fracture S_g, MPa
1	3.00	453
2	5.35	405
3	10.70	342
4	18.43	276
5	34.60	202
6	62.19	142

Source: Data in [Orange 67].

8.28 A wide plate of 2024-T4 aluminum contains a short center crack ($a \ll b$) and is loaded in tension to a stress S equal to the 0.2% offset yield strength of this material. What is the error in K calculated from LEFM, as determined from the J-integral using K_J? Assume that plane stress applies and use constants from Tables 5.2 and 5.7 for this material, including E, H, and n for the stress-strain curve.

9

Fatigue of Materials: Introduction and Stress-Based Approach

9.1 INTRODUCTION

Components of machines, vehicles, and structures are frequently subjected to repeated loads, also called cyclic loads, and the resulting cyclic stresses can lead to microscopic physical damage to the materials involved. Even at stresses well below a given material's ultimate strength, this damage can accumulate with continued cycling until it develops into a crack or other damage that leads to failure of the component. This process of accumulating damage and finally failure due to cyclic loading is called *fatigue*. Use of this term arose because it appeared to early investigators that cyclic stresses caused a gradual but not readily observable change in the ability of the material to resist stress.

Mechanical failures due to fatigue have been the subject of engineering efforts for more than 150 years. One early study was that of W. A. J. Albert, who tested mine hoist chains under cyclic loading in Germany around 1828. The term *fatigue* was used quite early, as in 1839 in a book on mechanics by J. V. Poncelet of France. Fatigue was further discussed and studied in the mid-1800s by a number of individuals in several countries in response to failures of components such as stagecoach and railway axles, shafts, gears, beams, and bridge girders.

The work in Germany of August Wöhler, starting in the 1850s and motivated by railway axle failures, is especially noteworthy. He began the development of design strategies for avoiding fatigue failure, and he tested irons, steels, and other metals under bending, torsion, and axial loads. Wöhler also demonstrated that fatigue was affected not only by cyclic stresses but also by accompanying steady (mean) stresses. More detailed studies following Wöhler's lead included those of Gerber and Goodman on predicting mean stress effects. The early work on fatigue and subsequent efforts to the 1950s are reviewed in a paper by Mann (1958) that is included in the References.

Fatigue failures continue to be a major concern in engineering design. Recalling from Chapter 1 that the economic costs of fracture and its prevention are quite large, it is noteworthy that an estimated 80% of these costs involve situations where cyclic loading and fatigue are at least a contributing factor. Thus, the annual cost of fatigue of materials to the U.S. economy in 1982 dollars is around $100 billion, corresponding to about 3% of the gross national product (GNP). These costs arise from the occurrence or prevention of fatigue failure for ground vehicles, rail vehicles, aircraft of all types, bridges, cranes, power plant equipment, offshore oilwell structures, and a wide variety of miscellaneous machinery and equipment, including everyday household items, toys, and sports equipment. For example, wind turbines used in power generation, Fig. 9.1, are subjected to cyclic loads due to rotation and wind turbulence, making fatigue a critical aspect of the design of the blade and other moving parts.

At present, there are three major approaches to analyzing and designing against fatigue failures. The traditional approach, which was developed to essentially its present form by 1955, is to base analysis on the nominal (average) stresses in the region of the component being analyzed. The nominal stress that can be resisted under cyclic loading is determined by considering mean stresses and by making adjustments for the effects of stress raisers, such as grooves, holes, fillets, and keyways. We will call this the *stress-based approach*. Another approach is the *strain-based approach*, which involves more detailed analysis of the localized yielding that may occur at stress raisers during cyclic loading. Finally, there is the *fracture mechanics approach*, which specifically treats growing cracks using the methods of fracture mechanics.

The stress-based approach is introduced in this chapter and further considered in Chapter 10, and the fracture mechanics approach is treated in Chapter 11. Discussion of the strain-based approach is postponed until Chapter 14, as it is necessary to first consider plastic deformation in materials and components in Chapters 12 and 13.

9.2 DEFINITIONS AND CONCEPTS

A discussion of the stress-based approach begins with some necessary definitions and basic concepts.

Figure 9.1   Horizontal-axis wind turbine in operation on the Hawaiian Island of Oahu. This is the largest wind turbine in the world, with the blade having a tip-to-tip span of 98 m. (Photo courtesy of the NASA Lewis Research Center, Cleveland, Oh.)

9.2.1 Description of Cyclic Loading

Some practical applications, and also many fatigue tests on materials, involve cycling between maximum and minimum stress levels that are constant. This is called *constant amplitude stressing* and is illustrated in Fig. 9.2.

The *stress range*, $\Delta\sigma$, is the difference between the maximum and the minimum values. Averaging the maximum and minimum values gives the *mean stress*, σ_m. The mean stress may be zero, as in Fig. 9.2(a), but often it is not, as in (b). Half the range is called the *stress amplitude*, σ_a, so that this is the variation about the mean. Mathematical expressions for these basic definitions are

$$\Delta\sigma = \sigma_{max} - \sigma_{min}, \quad \sigma_m = \frac{\sigma_{max} + \sigma_{min}}{2}, \quad \sigma_a = \frac{\Delta\sigma}{2} \tag{9.1}$$

The term *alternating stress* is used by some authors and has the same meaning as stress amplitude. It is also useful to note that

$$\sigma_{max} = \sigma_m + \sigma_a, \quad \sigma_{min} = \sigma_m - \sigma_a \tag{9.2}$$

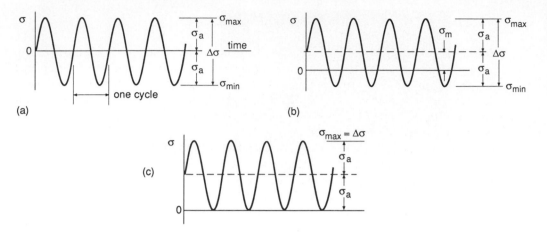

Figure 9.2 Constant amplitude cycling and the associated nomenclature. Case (a) is completely reversed stressing, $\sigma_m = 0$, (b) has a nonzero mean stress σ_m, and (c) is zero-to-tension stressing, $\sigma_{min} = 0$.

The signs of σ_a and $\Delta\sigma$ are always positive, since $\sigma_{max} > \sigma_{min}$, where tension is considered positive. The quantities σ_{max}, σ_{min}, and σ_m can be either positive or negative.

Ratios of certain pairs of the above variables are sometimes used.

$$R = \frac{\sigma_{min}}{\sigma_{max}}, \qquad A = \frac{\sigma_a}{\sigma_m} \tag{9.3}$$

where R is called the *stress ratio* and A the *amplitude ratio*. Some additional relationships derived from the above equations are also useful.

$$\Delta\sigma = 2\sigma_a = \sigma_{max}(1 - R), \qquad \sigma_m = \frac{\sigma_{max}}{2}(1 + R)$$

$$R = \frac{1 - A}{1 + A}, \qquad A = \frac{1 - R}{1 + R} \tag{9.4}$$

Cyclic stressing with zero mean can be specified by giving the amplitude σ_a, or by giving the numerically equal maximum stress, σ_{max}. If the mean stress is not zero, two independent values are needed to specify the loading. Some combinations that may be used are: σ_a and σ_m, σ_{max} and R, $\Delta\sigma$ and R, σ_{max} and σ_{min}, and σ_a and A. The term *completely reversed stressing* is used to describe a situation of $\sigma_m = 0$, hence $R = -1$, as in Fig. 9.2(a). *Zero-to-tension stressing* refers to cases of $\sigma_{min} = 0$, hence $R = 0$, as in Fig. 9.1(c).

The same system of subscripts and the prefix Δ are used in an analogous manner for other variables, such as strain ε, load (force) P, bending moment M, and nominal stress S. For example, P_{max} and P_{min} are maximum and minimum load, ΔP is load range, P_m is mean load, and P_a is load amplitude. If there is any possibility of confusion as to what variable is used with the ratios R or A, a subscript should be used, such as R_ε for strain ratio.

9.2.2 Point Stresses Versus Nominal Stresses

It is important to distinguish between the stress at a point, σ, and the nominal or average stress, S, and for this reason we use two different symbols. *Nominal stress* is calculated from load or moment or their combination as a matter of convenience and is only equal to σ in certain situations. Consider the three cases of Fig. 9.3. For simple axial loading (a), the stress σ is the same everywhere and so is equal to the average value $S = P/A$, where A is the cross-sectional area.

For bending, it is conventional to calculate S from the elastic bending equation, $S = Mc/I$, where c is the distance from neutral axis to edge and I is the area moment of inertia about the bending axis. Hence, $\sigma = S$ at the edge of the bending member, with σ of course being less elsewhere. However, if yielding occurs, the actual stress distribution becomes nonlinear, and σ at the edge of the member is no longer equal to S. This is illustrated in Fig. 9.3(b). In materials testing, and also in engineering applications, it may not be known whether or not yielding occurs, or sometimes values of S are calculated even when yielding is known to occur. It is then essential to distinguish S from the differing edge stress σ. Otherwise, considerable confusion can result. Stresses σ for bending beyond yielding can be calculated by replacing the elastic bending formula with more general analysis as described in Chapter 13.

For notched members, nominal stress S is conventionally calculated from the net area remaining after removal of the notch. (The term *notch* is used in a generic sense to indicate any stress raiser, including holes, grooves, fillets, etc.) If the loading is axial, $S = P/A$ is used, and for bending $S = Mc/I$ is calculated based on bending across the net area. Due to the stress raiser effect, such an S needs to be multiplied by an *elastic stress concentration factor*, k_t, to obtain the peak stress at the notch, $\sigma = k_t S$. However, k_t is based on linear-elastic analysis, and the value does not apply if there is yielding. If yielding occurs even locally at the notch, σ and $k_t S$ are no longer equal as illustrated in Fig. 9.3(c).

Hence, to avoid confusion, we will strictly observe the distinction between the stress σ at a point of interest and nominal stress S. For axial loading of unnotched members, where $\sigma = S$, we will use σ. However, for bending and notched members, we will generally assume that yielding may occur, so that S or $k_t S$ are employed except where it is truly appropriate to use σ. It is also important to be clear as to the area used to define S, as there is sometimes more than one choice, as when S is defined based on the gross area before removal of the notch. If there is any possibility of confusion with gross section nominal stresses, distinguishing subscripts will be applied, specifically the symbols S_n and S_g.

9.2.3 Stress Versus Life (*S-N*) Curves

If a test specimen of a material or an engineering component is subjected to a sufficiently severe cyclic stress, a fatigue crack or other damage will develop, leading to complete failure of the member. If the test is repeated at a higher stress level, the number of

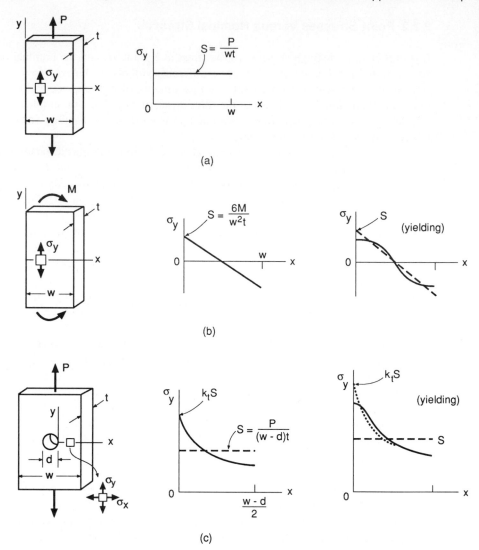

Figure 9.3 Actual and nominal stresses for (a) simple tension, (b) bending, and (c) a notched member. Actual stress distributions σ_y vs. x are shown as solid lines, and hypothetical distributions associated with nominal stresses S as dashed lines. In (c), the stress distribution that would occur if there were no yielding is shown as a dotted line.

cycles to failure will be smaller. The results of such tests from a number of different stress levels may be plotted to obtain a *stress-life curve*, also called an *S-N curve*. The amplitude of stress or nominal stress, σ_a or S_a, is commonly plotted versus the number of cycles to failure N_f, as shown in Figs. 9.4 and 9.5.

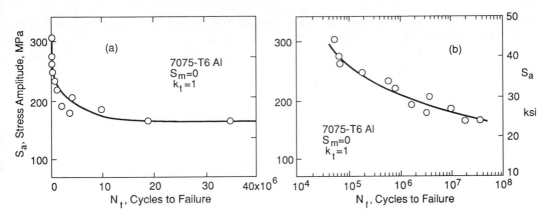

Figure 9.4   Stress versus life (S-N) curves from rotating bending tests of unnotched specimens of an aluminum alloy. Identical linear stress scales are used, but the cycle numbers are plotted on a linear scale in (a), and on a logarithmic one in (b). (Data from [MacGregor 52].)

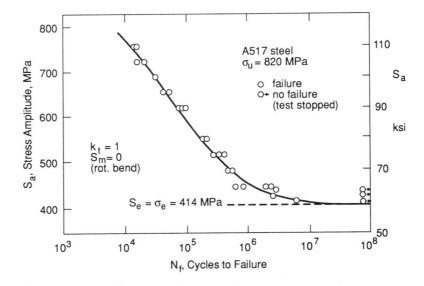

Figure 9.5   Rotating bending S-N curve for unnotched specimens of a steel with a distinct fatigue limit. (Adapted from [Brockenbrough 81]; used with permission.)

A group of such fatigue tests giving an S-N curve may be run all at zero mean stress, or all at some specific nonzero mean stress, σ_m. Also common are S-N curves for a constant value of the stress ratio, R. Although stresses are usually plotted as amplitudes, $\Delta\sigma$ or σ_{max} are also sometimes plotted. Equations 9.2 and 9.4 can be used to convert S-N curves plotted in one form to another.

The number of cycles to failure changes rapidly with stress level and may range over several orders of magnitude. For this reason, the cycle numbers are usually plotted on a logarithmic scale. The difficulty with a linear plot is illustrated in Fig. 9.4, where the same S-N data are plotted on both linear and logarithmic scales of N_f. On the linear plot, the cycle numbers for the shorter lives cannot be read accurately. A logarithmic scale is also often used for the stress axis.

If S-N data are found to approximate a straight line on a log-linear plot, the following equation can be fitted to obtain a mathematical representation of the curve:

$$\sigma_a = C + D \log N_f \tag{9.5}$$

where C and D are fitting constants. For data approximating a straight line on a log-log plot, the corresponding equation is

$$\sigma_a = A N_f^B \tag{9.6}$$

This second equation is often used in a slightly different form.

$$\sigma_a = \sigma_f' \left(2N_f\right)^b \tag{9.7}$$

The fitting constants for the two forms are related by

$$A = 2^b \sigma_f', \quad B = b \tag{9.8}$$

Constants for these equations are given in Table 9.1 for several engineering metals. The constants for Eqs. 9.6 and 9.7 are based on fitting test data for unnotched axial specimens, and those for Eq. 9.5 roughly approximate the same S-N curve in the life range $10^2 < N_f < 10^6$.

At short fatigue lives, the high stresses involved may be accompanied by plastic strains as described in Chapters 12 and 14. Equation 9.7 nevertheless continues to apply for uniaxial test data from unnotched specimens, except that amplitudes of true stress $\tilde{\sigma}_a$ are needed if the strains are quite large. Also, the constant σ_f' is often approximately equal to the true fracture strength $\tilde{\sigma}_f$ from a tension test, which for ductile materials is noted to be a value larger than the engineering ultimate strength σ_u.

In some materials, notably plain-carbon and low-alloy steels, there appears to be a distinct stress level below which fatigue failure does not occur under ordinary conditions. This is illustrated in Fig. 9.5, where the S-N curve appears to become flat and to asymptotically approach the stress amplitude labeled S_e. Such lower limiting stress amplitudes are called *fatigue limits* or *endurance limits*. For test specimens without notches and with a smooth surface finish, these are denoted σ_e and are considered to be material properties. (Since yielding does not generally occur at the low stress levels involved, use of the symbol σ_e is permissible for fatigue limits from bending tests if the member is unnotched. However, σ_e values from bending tend to be 10% to 15% higher than those from axial loading.) For materials where the S-N curve does not appear to approach an

TABLE 9.1   CONSTANTS FOR STRESS-LIFE CURVES FOR VARIOUS DUCTILE ENGINEERING METALS, FROM TESTS AT ZERO MEAN STRESS ON UNNOTCHED AXIAL SPECIMENS

Material	Yield σ_o	Ultimate σ_u	$\sigma_a = \sigma_f'(2N_f)^b = AN_f^B$			$\sigma_a = C + D \log N_f$	
			σ_f'	A	$b = B$	C	D
(a) Steels							
AISI 1015 (normalized)	227 (33)	415 (60.2)	976 (142)	886 (128)	−0.14	545 (79.1)	−69.6 (−10.1)
Man-Ten (hot rolled)	322 (46.7)	557 (80.8)	1089 (158)	1006 (146)	−0.115	703 (102)	−83.0 (−12.0)
RQC-100 (roller Q & T)	683 (99.0)	758 (110)	938 (136)	897 (131)	−0.0648	780 (113)	−68.9 (−10.0)
AISI 4142 (Q & T, 450 HB)	1584 (230)	1757 (255)	1937 (281)	1837 (266)	−0.0762	1529 (222)	−148 (−21.5)
AISI 4340 (aircraft quality)	1103 (160)	1172 (170)	1758 (255)	1643 (238)	−0.0977	1247 (181)	−137 (−19.8)
(b) Other Metals							
2024-T4 Al	303 (44.0)	476 (69.0)	900 (131)	839 (122)	−0.102	624 (90.5)	−69.9 (−10.1)
Ti-6Al-4V (solution treated and aged)	1185 (172)	1233 (179)	2030 (295)	1889 (274)	−0.104	1393 (202)	−157 (−22.8)

Notes: The tabulated values have units of MPa(ksi) except for dimensionless $b = B$. Constants C and D give a rough approximation for $10^2 < N_f < 10^6$ to the σ_f', b curve from test data. See Table 14.1 for sources and additional tensile properties.

asymptote, as for aluminum and copper alloys, a fatigue limit is often arbitrarily defined at a specific long life, say 10^7 or 10^8 cycles.

The term *fatigue strength* is used to specify a stress amplitude value from an *S-N* curve at a particular life of interest. Hence, the *fatigue strength* at 10^5 cycles is simply the stress amplitude corresponding to $N_f = 10^5$. Other terms used with *S-N* curves include *high-cycle fatigue* and *low-cycle fatigue*. The former identifies situations of long fatigue life where the stress is sufficiently low that yielding effects do not dominate the behavior. The life where high-cycle fatigue starts varies with material but is typically in the range 10^2 to 10^4 cycles. In the low-cycle range, the more general strain-based approach of Chapter 14 is particularly useful as this deals specifically with the effects of yielding.

Example 9.1

For an *S-N* curve of the form of Eq. 9.6, two points (N_1, σ_1) and (N_2, σ_2) are known. Develop equations for the constants A and B as a function of these values. Considering a plot of the equation on log-log coordinates, what are the significance of A and B?

Solution   Apply the equation $\sigma_a = AN_f^B$ to the two known points.

$$\sigma_1 = AN_1^B, \qquad \sigma_2 = AN_2^B$$

Then divide the second equation into the first and take logarithms of both sides.

$$\frac{\sigma_1}{\sigma_2} = \left(\frac{N_1}{N_2}\right)^B, \qquad \log\frac{\sigma_1}{\sigma_2} = B\log\frac{N_1}{N_2}$$

Solving for B gives

$$B = \frac{\log\dfrac{\sigma_1}{\sigma_2}}{\log\dfrac{N_1}{N_2}} = \frac{\log\sigma_1 - \log\sigma_2}{\log N_1 - \log N_2} \qquad \textbf{Ans.}$$

Once B is known from this, A can be calculated from either point.

$$A = \frac{\sigma_1}{N_1^B} = \frac{\sigma_2}{N_2^B} \qquad \textbf{Ans.}$$

From the second expression for B, it is evident that this constant is the slope on a plot of $\log N_f$ versus $\log\sigma_a$. However, if the logarithmic coordinates used on the two axes have decades of different lengths, the slope differs from B by the ratio of the lengths of the decades. From the equation $\sigma_a = AN_f^B$, it is seen that A is the value of σ_a corresponding to $N_f = 1$. This situation is illustrated in Figure E9.1 by a plot on log-log coordinates of the line for AISI 1015 steel from Table 9.1.

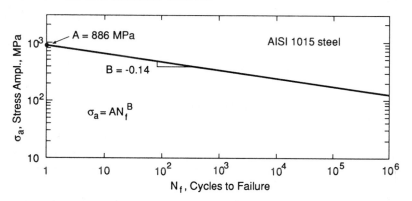

Figure E9.1

9.3 SOURCES OF CYCLIC LOADING

Some practical applications involve cyclic loading at a constant amplitude, but irregular load versus time histories are more commonly encountered. Examples are given in Figs. 9.6 to 9.9. Loads on components of machines, vehicles, and structures can be

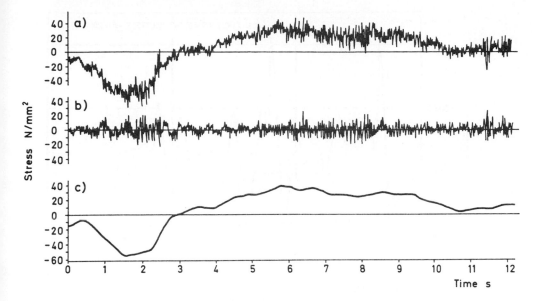

Figure 9.6 Sample record of stresses at the steering knuckle arm of a motor vehicle, including the original stress-time history (a), and the separation of this into vibratory load due to roadway roughness (b), and the working load due to maneuvering the vehicle (c). (From [Buxbaum 73]; used with permission; first published by AGARD/NATO.)

divided into four categories depending on their source. *Static loads* do not vary and are continuously present. *Working loads* change with time and are incurred as a result of the function performed by the component. *Vibratory loads* are relatively high frequency cyclic loads that arise from the environment or as a secondary effect of the function of the component. These are often caused by fluid turbulence or by the roughness of solid surfaces in contact with one another. *Accidental loads* are rare events that do not occur under normal circumstances.

For example, consider highway bridges. Static loads are caused by the always-present weight of the structure and roadway. Cyclic working loads are caused by the weights of vehicles, especially heavy trucks, moving across the bridge. Vibratory loads are added to the working loads and are caused by tires interacting with the roughness of the roadway, including the bouncing of vehicles after hitting potholes. Long-span bridges are also subject to vibratory loading due to wind turbulence. Accidental loading could be caused by a truck hitting an overpass bridge because it was too high for the clearance available, or by an earthquake.

Working loads and vibratory loads, and often their combined effects, are the cyclic loads that can cause fatigue failure. However, the damage due to cyclic loads is greater if the static loads are more severe, so that these also need to be considered. Accidental loads may play an additional role, themselves causing fatigue failure, or damaging a component so that it is more susceptible to fatigue caused by subsequent more ordinary loads.

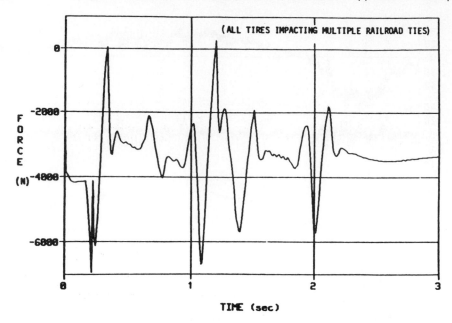

Figure 9.7   Calculated force on the front left lower ball joint in an automobile suspension, for tires impacting railroad ties. (From [Thomas 87]; used with permission; © Society of Automotive Engineers.)

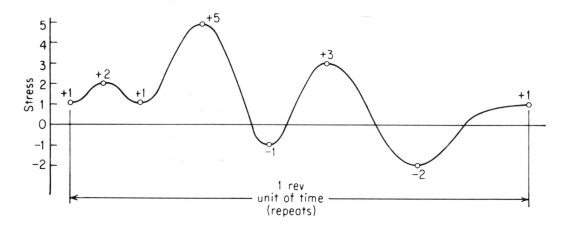

Figure 9.8   Loads during each revolution of a helicopter rotor. Feathering of the blade and interaction with the air cause dynamic loads. (From [Boswell 59]; used with permission.)

S-N curves from constant amplitude testing can be used to estimate fatigue lives for irregular load-time histories. The methodology will be introduced near the end of this chapter.

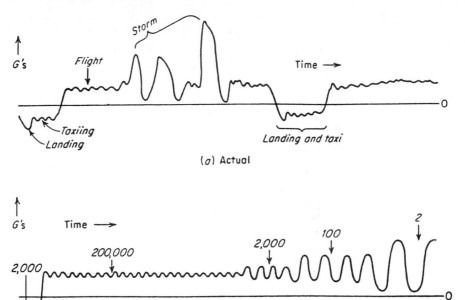

Figure 9.9   Loads for one flight of a fixed-wing aircraft (a), and a simplified version of this loading (b). Working loads occur due to takeoffs, maneuvers, and landings, and there are vibratory loads due to runway roughness and air turbulence, as well as wind gust loads in storms. (From [Waisman 59]; used with permission.)

9.4 FATIGUE TESTING

Materials testing to obtain S-N curves is a widespread practice. Several ASTM Standards address stress-based fatigue testing for metals, especially Standard No. E466, and such tests on plastics (polymers) are covered by Standard No. D671. The resulting data and curves are widely available in the published literature, including various handbooks as listed in a special section of the References. An understanding of the basis of these tests is useful in effectively employing their results for engineering purposes.

9.4.1 Test Apparatus

One of the machines employed by Wöhler tested a pair of rotating test specimens subjected to cantilever bending as shown in Fig. 9.10. Springs supplied a constant force through a bearing, permitting rotation of the specimen, so that the bending moment varied linearly with distance from the spring. In such a *rotating bending test*, any point on the specimen is subjected to a sinusoidally varying stress as it rotates from the tension (top) side of the beam to the compression (bottom) side, completing one cycle each time the specimen rotates 360°.

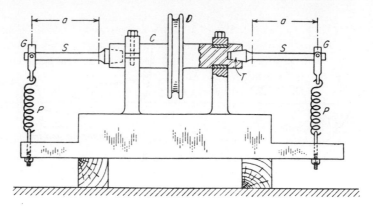

Figure 9.10 Rotating cantilever beam fatigue testing machine used by Wöhler. D, drive pulley; C, arbor; T, tapered specimen butt; S, specimen; a, moment arm; G, loading bearing; P, loading spring. (From [Hartmann 59]; used with permission.)

    Equipment for rotating bending tests operating on similar principles is still in use today. A variation involving four-point bending has probably been more widely used than any other type of fatigue testing machine, this being illustrated in Fig. 9.11. The two bearings near each end of the test specimen permit the load to be applied while the specimen rotates, and two bearings outside of these provide support. A hanging weight usually provides the constant force. Four-point bending has the advantage of providing a constant bending moment and zero shear over the length of the specimen. For all forms of rotating bending test, the applied stress has a mean value of zero. This is because the

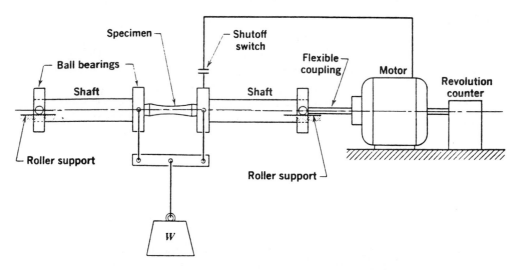

Figure 9.11 The R. R. Moore rotating beam fatigue testing machine. (From [Richards 61] p. 382; reprinted by permission of PWS-Kent Publishing Co., Boston.)

distance from the neutral axis of bending to a given point on the specimen surface varies symmetrically about zero as the circular cross section rotates.

A rotating crank can be used in a *reciprocating bending test* to achieve a nonzero mean stress as shown in Fig. 9.12. Geometric changes in the apparatus that effectively alter the length of the connecting rod from the eccentric drive give different mean deflections, hence different mean stresses. The test specimens are often flat with the width tapered the proper amount to give a constant bending stress despite the linearly varying moment. One of the test specimens shown in Fig. 9.13 is of this type. If yielding occurs in such a test, the stresses cannot be readily determined from the deflections, and the load or specimen strain must be specifically measured. Axial stressing with various mean levels can be achieved by a modification of this device.

A cyclic stress with zero mean level can be achieved by exciting a *resonant vibration* in an clastic system, such as the axial testing machine based on a rotating eccentric mass shown in Fig. 9.14. More complex resonant devices capable of providing mean stresses are also used, and similar principles can be applied to bending or torsion. In addition, the vibration can be induced by other means, such as electromagnetic, piezoelectric, or acoustic effects. Frequencies of cycling up to 100 kHz are possible with some of these special techniques. At such very high frequencies, active cooling of the specimen is required to avoid overheating.

Modifications and elaborations of simple mechanical devices as just described allow fatigue tests to be run in torsion, combined bending and torsion, biaxial bending, etc. Test specimens made of thin-walled tubes may be subjected to cyclic fluid pressure to obtain biaxial stresses. All of the test equipment described so far is best suited to constant amplitude loading at a constant frequency of cycling. However, additional complexity can be added to some of these machines to achieve a slowly changing amplitude or mean level.

Closed loop servo-hydraulic testing machines (Fig. 5.4) are also widely used for fatigue testing. This equipment is expensive and complex, but it has important advantages

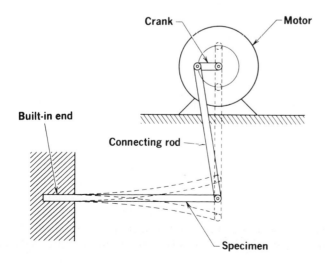

Figure 9.12   A reciprocating cantilever bending fatigue testing machine based on controlled deflections from a rotating eccentric. (From [Richards 61] p. 383; reprinted by permission of PWS-Kent Publishing Co., Boston.)

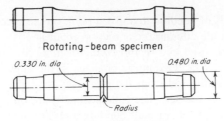

Rotating-beam specimen

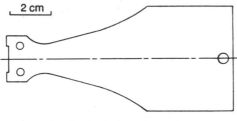

0.330 in. dia   0.480 in. dia

Radius

Notched rotating beam

2 cm

Cantilever specimen from sheet

Axial stress specimen

Figure 9.13   Various fatigue test specimens, all shown to the same scale. (Adapted from [Hartmann 59]; used with permission.)

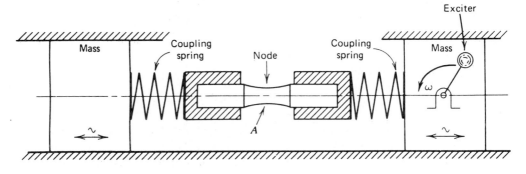

Exciter

Mass   Coupling spring   Node   Coupling spring   Mass

ω

A

Figure 9.14   Axial fatigue testing machine based on a resonant vibration caused by a rotating eccentric mass. (From [Collins 81] p. 179; reprinted by permission of John Wiley & Sons, Inc; copyright ©1981 by John Wiley & Sons, Inc.)

over all other types of fatigue testing equipment. The test specimens can be subjected to constant amplitude cycling with controlled loads, strains, or deflections, and the amplitude, mean, and cyclic frequency of the chosen one of these variables can be set to a desired value using the electronic controls of the machine. Also, any irregular loading history available as an electrical signal can be enforced upon a test specimen. Highly irregular histories similar to Figs. 9.6 to 9.9 can thus be used in testing that closely simulates actual service conditions. Closed-loop machines can be controlled by computers, and the test results monitored by computers.

In most of the test apparatus described, the frequency is fixed by the speed of an electric motor or by the natural frequency of a resonant vibratory device. This fixed frequency is usually in the range 10 to 100 Hz. At the latter value, a test to 10^7 cycles takes 28 hours, one to 10^8 cycles 12 days, and one to 10^9 cycles almost four months. These long test times place a practical limit on the range of lives that can be studied. If very long lives are of interest, one possibility is to use a special high-frequency resonant vibration testing device. However, the frequency may affect the test results, so that it is not clear that an *S-N* curve obtained at say 20 kHz can be applied to service loading at a much lower frequency.

9.4.2 Test Specimens

Specimens for evaluating the fatigue resistance of materials are designed to fit the test apparatus used. Some examples are shown in Fig. 9.13, and two fractures from fatigue tests are shown in Fig. 9.15. The simplest test specimens, called *unnotched* or *smooth specimens*, have no stress raiser in the region where failure occurs. A variety of specimens containing stress raisers, called *notched specimens*, are also used. These are used to evaluate materials using geometries more closely approaching those in an actual component. Notched test specimens are often described by giving the value of the elastic stress concentration factor k_t.

Actual structural components, or portions of components, such as bolted or welded joints, are also often subjected to fatigue testing. Structural assemblies, or even entire structures or vehicles, are also sometimes tested. Examples are tests of aircraft wings or tail sections, or of automobile suspension systems. A test of an entire automobile has already been illustrated by Fig. 1.13.

9.5 THE PHYSICAL NATURE OF FATIGUE DAMAGE

When viewed at a sufficiently small size scale, all materials are anisotropic and inhomogeneous. For example, engineering metals are composed of an aggregate of small crystal grains. Within each grain the behavior is anisotropic due to the crystal planes, and if a grain boundary is crossed, the orientation of these planes changes. Inhomogeneities exist not only due to the grain structure, but also because of the presence of tiny voids or particles of a different chemical composition than the bulk of the material, such as hard silicate or alumina inclusions in steel. Multiple phases, involving grains or other regions of more than one chemical composition, are also common as discussed in Chapter 3. As a result of such nonuniform microstructure, stresses are distributed in a nonuniform manner when viewed at the size scale of this microstructure. Regions where the stresses are severe are usually the points where fatigue damage starts. The details of the behavior at a microstructural level varies widely for different materials due to their different bulk mechanical properties and their different microstructures.

Figure 9.15 Photographs of broken 7075-T6 Al fatigue test specimens: unnotched axial specimen, 7.6 mm diameter (top); and plate 19 mm wide with a round hole (bottom). In the unnotched specimen, the crack started in the flat region with slightly lighter color, and cracks in the notched specimen started on each side of the hole. (Photos by R. A. Simonds.)

    For ductile engineering metals, crystal grains that have an unfavorable orientation relative to the applied stress first develop slip bands. As discussed in Chapter 2, slip bands are regions where there is intense deformation due to shear motion between crystal planes. A sequence of photographs showing this process is presented as Fig. 9.16. Also, the slip band damage previously illustrated in Fig. 2.23 was caused by cyclic loading. Additional slip bands form as more cycles are applied, and their number may become so large that the rate of formation slows, with the number of slip bands approaching a saturation level. Individual slip bands become more severe and some develop into cracks within grains, which then spread into other grains, join with other similar cracks, and produce a large crack that propagates to failure.

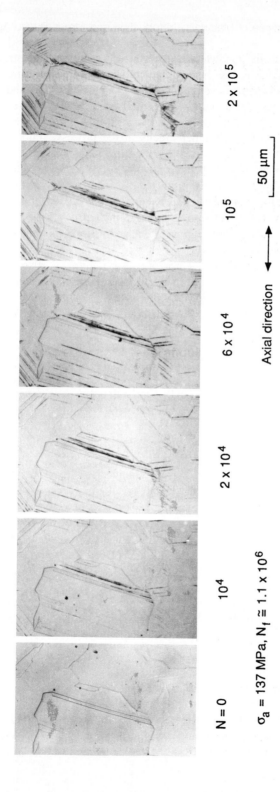

N = 0 10^4 2×10^4 6×10^4 10^5 2×10^5

$\sigma_a = 137$ MPa, $N_f \cong 1.1 \times 10^6$

Axial direction   50 μm

Figure 9.16 The process of slip band damage during cyclic loading developing into a crack in an annealed 70Cu-30Zn brass. (Photos courtesy of Prof. H. Nisitani, Kyushu University, Fukuoka, Japan. Published in [Nisitani 81]; reprinted with permission from *Engineering Fracture Mechanics*, Pergamon Press, Oxford, UK.)

For materials of somewhat limited ductility, such as high-strength metals, the microstructural damage is less widespread, tending to be concentrated at defects in the material. A small crack develops at a void, inclusion, slip band, grain boundary, or scratch, or there may be a sharp flaw initially present that is essentially a crack. This crack then grows in a plane generally normal to the tensile stress until it causes failure, sometimes joining with other cracks in the process. Photographs of progressive damage of this type have already been presented as Fig. 1.8. Thus, the process in limited-ductility materials is characterized by *propagation* of a few defects, in contrast to the more widespread *damage intensification* that occurs in highly ductile materials. In fibrous composite materials, fatigue damage is generally characterized by increasing numbers of fiber breaks and spreading delamination developing over a relatively large area. The final failure involves an irregular geometry of pulled-out fibers and separated layers rather than a distinct crack.

S-N curves can be plotted not only for failure but also for numbers of cycles required to reach various stages of the damage process as illustrated in Fig. 9.17. The curves in one case are for slip-band-dominated damage in an annealed, nearly pure, aluminum alloy. For a precipitation hardened aluminum alloy, *S-N* curves are given for the first detected crack and for failure.

Where failure is dominated by growth of a crack, the resulting fracture, when viewed macroscopically, generally exhibits a relatively smooth area near its origin. This can be seen in Figs. 9.15, 9.18, and 9.19. The portion of the fracture associated with growth of the fatigue crack is usually fairly flat and is oriented normal to the applied tensile stress. Rougher surfaces generally indicate more rapid growth, where the rate of growth usually increases as the crack grows. Curved lines concentric about the crack origin, called *beach marks*, are often seen and mark the progress of the crack at various stages, as seen in Fig. 9.19 and more clearly in Fig. 9.20. Beach marks are seen where the texture of the fracture surface changes as a result of the crack being delayed or accelerated, which may occur due to an altered stress level, or due to an altered temperature or chemical environment. Beach marks may also be caused by discoloration due to greater amounts of corrosion on older portions of the fracture surface.

After the crack has reached a sufficient size, a final failure occurs that may be ductile, involving considerable deformation, or brittle and involving little deformation. The final fracture area is usually rough in texture, and in ductile materials forms a *shear lip*, inclined at approximately 45° to the applied stress. These features can be seen in Figs. 9.15, 9.19, and 9.20. Microscopic examination of fatigue fracture surfaces in ductile materials often reveals the presence of marks left by the progress of the crack on each cycle. These are called *striations* and can be seen in Fig. 9.21.

9.6 TRENDS IN *S-N* CURVES

S-N curves vary widely for different classes of materials, and they are affected by a variety of factors. Any processing that changes the static mechanical properties or microstructure is also likely to affect the *S-N* curve. Additional factors of importance

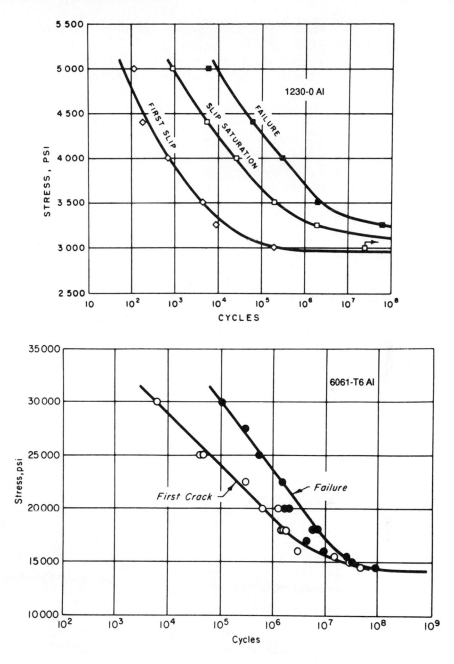

Figure 9.17   Stress-life curves for completely reversed bending of smooth specimens, showing various stages of fatigue damage in an annealed 99% aluminum (1230-0), and in a hardened 6061-T6 aluminum alloy. (Adapted from [Hunter 54] and [Hunter 56]; copyright © ASTM; reprinted with permission.)

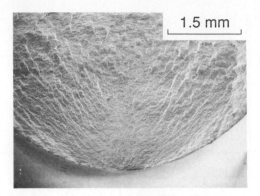

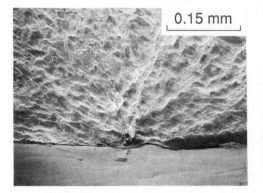

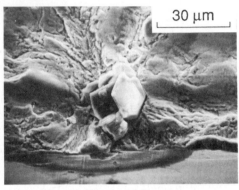

Figure 9.18   Fatigue crack origin in an unnotched axial test specimen of AISI 4340 steel having $\sigma_u = 780$ MPa, tested at $\sigma_a = 440$ MPa with $\sigma_m = 0$. The inclusion that started the crack can be seen at the two higher magnifications. (SEM photos by A. Madeyski, Westinghouse Science and Technology Ctr., Pittsburgh, Pa.; see [Dowling 83] for related data.)

Figure 9.19   Fatigue failure of an aluminum alloy airplane propeller. The failure initiated at a small gouge on the bottom edge approximately 2 cm from the right end of the scale. (Photo by R. A. Simonds; sample loaned for photo by Prof. J. L. Lytton of Virginia Tech., Blacksburg, Va.)

Figure 9.20   Fracture surfaces for fatigue and final brittle fracture in an 18 Mn steel member. (Photo courtesy of A. Madeyski, Westinghouse Science and Technology Ctr., Pittsburgh, Pa.)

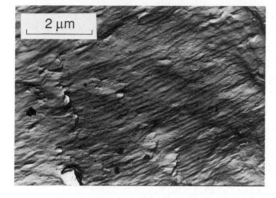

Figure 9.21   Fatigue striations spaced approximately 0.12 μm apart, from a fracture surface of a Ni-Cr-Mo-V steel. (Photo courtesy of A. Madeyski, Westinghouse Science and Technology Ctr., Pittsburgh, Pa. Published in [Madeyski 78]; copyright © ASTM; reprinted with permission.)

include mean stress, member geometry, chemical environment, temperature, cyclic frequency, and residual stress. Some typical *S-N* curves for metals have already been presented, and curves for several polymers and composites are shown in Figs. 9.22 and 9.23.

9.6.1 Trends with Ultimate Strength, Mean Stress, and Geometry

Smooth specimen fatigue limits of steels are often about half of the ultimate tensile strength, which is illustrated in Fig. 9.24. Values drop below $\sigma_e \approx 0.5\sigma_u$ at high-strength levels where most steels have limited ductility. This indicates that a reasonable degree of ductility is helpful in providing resistance to cyclic loading. Lack of appreciation for this fact can lead to fatigue failures in situations where fatigue was not previously a problem, as in substituting high-strength materials to save weight in vehicles. Similar

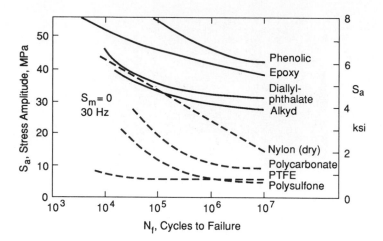

Figure 9.22   Stress-life curves from cantilever bending of mineral and glass-filled thermosets (solid lines) and unfilled thermoplastics (dashed lines). (Adapted from [Riddell 74]; used with permission.)

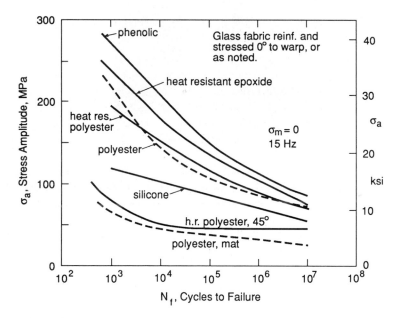

Figure 9.23   Stress-life curves for unnotched axial specimens of various thermosetting plastic laminates, all reinforced with the same (type 181) glass fabric, except for one case of a random fiber mat. The test conditions were 50% relative humidity at 23°C. (Adapted from [Boller 56].)

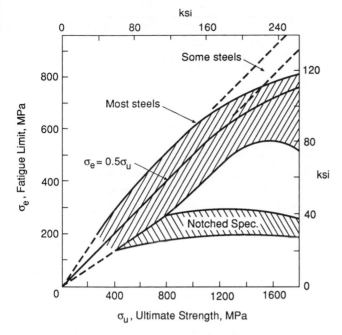

Figure 9.24 Trends in fatigue limits of smoothly polished and notched specimens of steels. (Adapted from [Bullens 48] p. 157; used with permission.)

correlations exist for other metals, but the fatigue limits are generally lower than half the ultimate.

An important influence on *S-N* curves that will be considered in some detail later in this chapter is the effect of mean stress. For a given stress amplitude, tensile mean stresses give shorter fatigue lives than does zero mean stress, and compressive mean stresses give longer lives. Some test data illustrating this are shown in Fig. 9.25. Note that such an effect of mean stress lowers or raises the *S-N* curve, so that for a given life the stress amplitude that can be allowed is lower if the mean stress is tensile, or higher if it is compressive.

Stress raisers (notches) also affect *S-N* curves, the effect being to shorten the life, that is, to lower the *S-N* curve, more so if the elastic stress concentration factor k_t is higher. An example of this effect is shown in Fig. 9.26. Another important trend is that notches have a relatively more severe effect on high-strength, limited-ductility materials, which is apparent in Fig. 9.24. Notch effects are treated in detail in the next chapter, and also later in Chapter 14.

9.6.2 Effects of Environment and Frequency of Cycling

Hostile chemical environments can accelerate the initiation and growth of fatigue cracks. One mechanism is the development of corrosion pits, which then act as stress raisers. Another is a hostile environment causing a crack to grow faster by chemical reactions and dissolution of material at the crack tip. For example, testing in a salt solution similar to seawater lowers the *S-N* curve of one aluminum alloy as shown in Fig. 9.27.

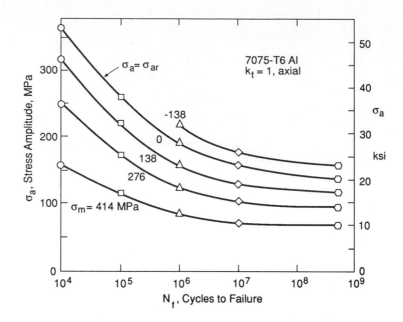

Figure 9.25   Axial loading *S-N* curves at various mean stresses for unnotched specimens of an aluminum alloy. The curves connect average fatigue strengths for a number of lots of material. (Data from [Howell 55].)

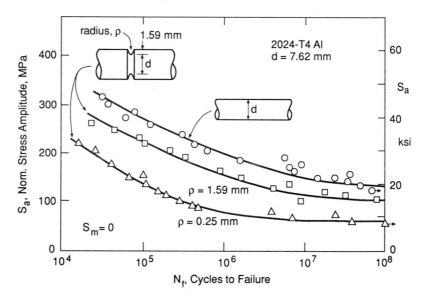

Figure 9.26   Effects of notches having $k_t = 1.6$ and 3.1 on rotating bending *S-N* curves of an aluminum alloy. (Adapted from [MacGregor 52].)

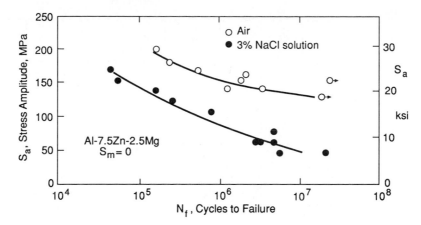

Figure 9.27   Effect of a salt solution similar to seawater on the bending fatigue behavior of an aluminum alloy. (Data from [Stubbington 61].)

Even the moisture and gases in air can act as a hostile environment, especially at high temperature. Time-dependent deformation (creep) is also more likely at high temperature, and when combined with cyclic loading, creep may have a synergistic effect that unexpectedly shortens the life. In general, chemical or thermal effects are greater if more time is available for them to occur. This leads to the fatigue life varying with frequency of cycling in such situations, the life in cycles being shorter for slower frequencies. Such effects are evident in Fig. 9.28.

Polymers may increase in temperature during cyclic loading, as these materials often produce considerable internal energy due to their viscoelastic deformation, that must be dissipated as heat. The effect is compounded because such materials have a poor ability to conduct heat away to their surroundings. A consequence of this is that the *S-N* curve is affected not only by frequency but also by specimen thickness, since thinner test specimens are more efficient at conducting their heat away.

9.6.3 Effects of Microstructure

Any change in the microstructure or surface condition has the potential of altering the *S-N* curve, especially at long fatigue lives. In metals, resistance to fatigue is generally enhanced by reducing the size of inclusions and voids, by small grain size, and by a dense network of dislocations. However, special processing aimed at improvements due to microstructure may not be successful unless it can be accomplished without substantially decreasing the ductility. Some *S-N* curves for brass illustrating effects due to microstructure are shown in Fig. 9.29. In this material, a higher degree of cold work by drawing increases the dislocation density and hence the fatigue strength. Larger grain sizes are obtained by more thorough annealing, thus lowering the fatigue strength. These changes in fatigue strength at long lives are accompanied by generally similar changes in the ultimate tensile strength.

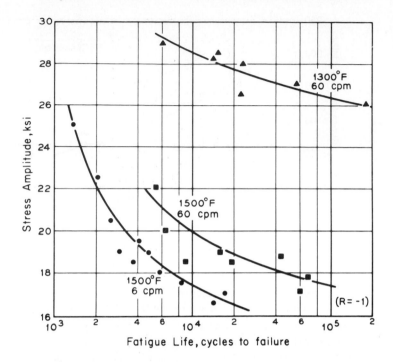

Figure 9.28 Temperature and frequency effects on axial *S-N* curves for the nickel-base alloy Inconel. (Illustration from [Gohn 64] of data in [Carlson 59]; used with permission.)

Microstructures of materials often vary with direction, such as the elongation of grains and inclusions in the rolling direction of metal plates. Fatigue resistance may be lower in directions where the stress is normal to the long direction of such an elongated or layered grain structure. Similar effects are especially pronounced in fibrous composite materials, where the properties and structure are highly dependent on direction. Fatigue resistance is higher where larger numbers of fibers are parallel to the applied stress, and especially low for stresses normal to the plane of a laminated structure.

9.6.4 Residual Stress and Other Surface Effects

Internal stresses in the material, called *residual stresses*, have an effect similar to an applied mean stress. Hence, compressive residual stresses are beneficial. These can be introduced by permanently stretching a thin surface layer by yielding it in tension. The underlying material then attempts to recover its original size by elastic deformation, forcing the surface layer into compression.

One means of doing this is by bombarding the surface with small steel or glass shot, which is called *shot peening*. Another is by sufficient bending to yield a thin surface layer, which is called *presetting*. However, the latter has an opposite (hence harmful) effect on the other side of a bending member, so that the procedure is useful only if

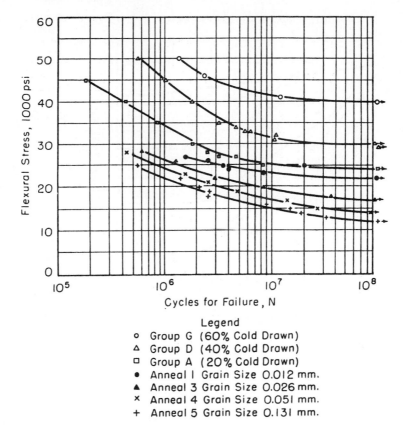

Legend
o   Group G (60% Cold Drawn)
△   Group D (40% Cold Drawn)
□   Group A (20% Cold Drawn)
•   Anneal 1 Grain Size 0.012 mm.
▲   Anneal 3 Grain Size 0.026 mm.
×   Anneal 4 Grain Size 0.051 mm.
+   Anneal 5 Grain Size 0.131 mm.

Figure 9.29   Influence of grain size and cold work on rotating bending *S-N*
curves for 70Cu-30Zn brass. (From [Sinclair 52]; used with permission.)

the bending in service is expected to be primarily in one direction, as for leaf springs
in ground vehicle suspensions. Various combinations of presetting and shot peening
affect the *S-N* curves of steel leaf springs under zero-to-maximum bending as shown in
Fig. 9.30. The effects correlate as expected with measured residual stresses.

Smoother surfaces due to more careful machining generally improve resistance
to fatigue, although some machining procedures are harmful as they introduce tensile
residual stresses. Various surface treatments, such as carburizing or nitriding of steels,
may alter the microstructure, chemical composition, or residual stress of the surface and
therefore affect the fatigue resistance. Plating, such as nickel or chromium plating of
steel, generally introduces tensile residual stresses and is therefore often harmful. Also,
the deposited material itself often has poorer resistance to fatigue than the base material,
so that cracks easily start there and then grow into the base material. Shot peening after
plating can help by changing the residual stress to compression.

Welding results in geometries that involve stress raisers, and residual stresses often
occur as a result of uneven cooling from the molten state. Unusual microstructure may

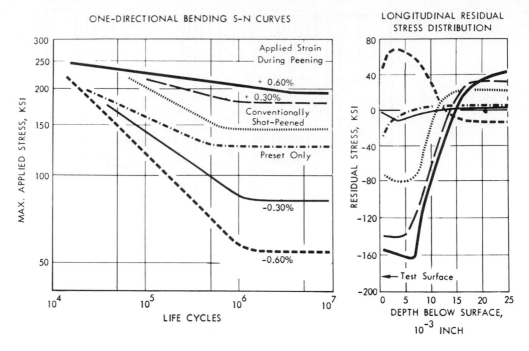

Figure 9.30 *S-N* curves for zero-to-maximum bending, and residual stresses, for variously shot peened steel leaf springs. (From [Mattson 59]; courtesy of General Motors Research Laboratories.)

exist, as well as porosity or other small flaws. Hence, the presence of welds generally reduces fatigue strength and requires special attention.

9.6.5 Statistical Scatter

If multiple fatigue tests are run at one stress level, there is always considerable statistical scatter in the fatigue life. Some *S-N* data illustrating this are shown in Fig. 9.31. If the statistical scatter in cycles to failure N_f is considered, a skewed distribution usually occurs as in Fig. 9.32(left). However, if the logarithm of N_f is treated as the variable, then a reasonably symmetrical distribution is generally obtained as shown on the right. Use of a standard Gaussian (also called normal) statistical distribution of log N_f is then reasonable, which is equivalent to a *lognormal* distribution of N_f. Other statistical models are also used, such as the Weibull distribution. The scatter in log N_f is almost always observed to increase with life, which can be seen in Fig. 9.31.

Statistical analysis of fatigue data permits the average *S-N* curve to be established along with additional *S-N* curves for various probabilities of failure. An example is shown in Fig. 9.33. Such a family of *S-N-P* curves gives detail on the statistical scatter in the life. Since *S-N* curves are affected by a variety of factors, such as surface finish, frequency of cycling, temperature, hostile chemical environments, and residual

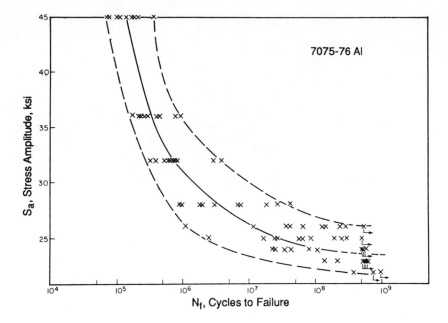

Figure 9.31   Scatter in rotating bending *S-N* data for an unnotched aluminum alloy. (Adapted from [Grover 66] p. 44.)

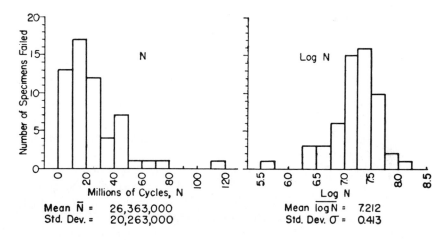

Figure 9.32   Distribution of fatigue lives for 57 small specimens of 7075-T6 Al tested at $S_a = 207$ MPa (30 ksi) in rotating bending. (From [Sinclair 53]; used with permission of ASME.)

stresses, probabilities of failure from *S-N-P* curves based on laboratory data should be considered only as crude estimates. Additional safety margins are usually needed in design to account for complexities and uncertainties that are not included in such data.

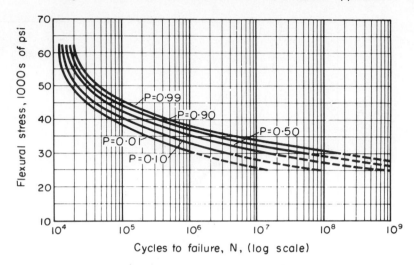

Figure 9.33 Family of rotating bending S-N curves for various probabilities of failure, P, from data for small unnotched specimens of 7075-T6 Al. (From [Sinclair 53]; used with permission of ASME.)

9.7 MEAN STRESSES

S-N curves that include data for various mean stresses are widely available for commonly used engineering metals, and sometimes for other materials. Certain conventions for presenting these data exist, and equations have been developed to estimate the effects of mean stress in situations where specific data are not available.

9.7.1 Presentation of Mean Stress Data

One procedure used for developing data on mean stress effects is to select several values of mean stress and then to run tests at various stress amplitudes for each of these. The results can be plotted as a family of S-N curves, each for a different mean stress, as already illustrated in Fig. 9.25.

An alternate means of presenting the same information is a *constant-life diagram* as shown in Fig. 9.34. This is done by taking points from the S-N curves at various values of life in cycles, and then plotting combinations of stress amplitude and mean stress that produce each of these lives. Interpolation between the lines on either type of plot can be used to obtain fatigue lives for various applied stresses. The constant-life diagram presentation shows clearly that increasing the mean stress in the tensile direction must be accompanied by a decrease in stress amplitude to maintain the same life.

Another procedure often used for developing data on mean stress effect is to choose several values of the stress ratio, $R = \sigma_{min}/\sigma_{max}$, and then to run tests at various stress levels for each of these. A different family of S-N curves is obtained, with each corresponding to a different R value. An example is shown in Fig. 9.35. In this example, the σ_{max} values are plotted, but the curves for each R-ratio can be easily converted to

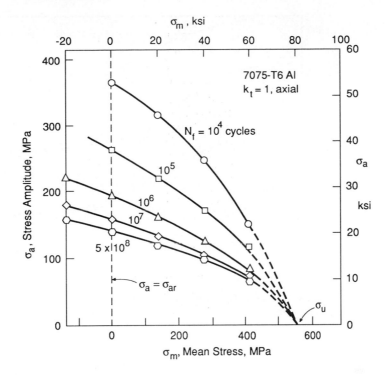

Figure 9.34   Constant-life diagram for 7075-T6 Al taken from the *S-N* curves of Fig. 9.25.

ones plotted in terms of stress amplitude σ_a or range $\Delta\sigma$ by using Eq. 9.4. *S-N* curves for constant values of R provide the same information, but in different form, than *S-N* curves for constant values of mean stress.

   S-N curves for constant R values can be similarly used to construct a constant-life diagram. Since any specific pair of values (σ_m, σ_a) corresponds to a specific pair of values (σ_{min}, σ_{max}) due to Eq. 9.2, the diagram is the same as before. Two sets of coordinate axes 45° apart can be used so that the diagram can be interpreted in either way. An example for a titanium alloy is shown in Fig. 9.36. This particular diagram contains data for both unnotched (smooth) axial test specimens and one case of a notched specimen, as indicated.

   Although all of the examples of mean stress effect presented so far are for engineering metals, families of *S-N* curves and constant-life diagrams can be used for other materials. Generally similar trends in behavior occur, with some data for a composite material being given in Fig. 9.37.

9.7.2 Normalized Amplitude-Mean Diagrams

For an unnotched member, let the stress amplitude for the particular case of zero mean stress be designated σ_{ar}. On a constant-life diagram, σ_{ar} is thus the intercept at $\sigma_m = 0$

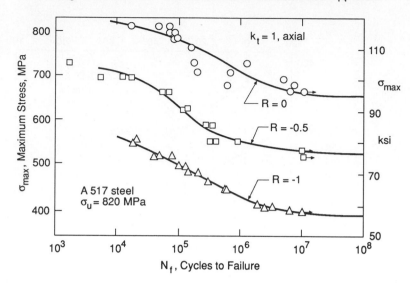

Figure 9.35   Stress-life curves for axial loading of unnotched A517 steel for constant values of the stress ratio R. (Adapted from [Brockenbrough 81]; used with permission.)

of the curve for any particular life. The graph can then be normalized in a useful way by plotting values of the ratio σ_a/σ_{ar} versus the mean stress σ_m. The result of such a normalization for the data of Fig. 9.34 is shown in Fig. 9.38. Such a *normalized amplitude-mean diagram* forces agreement at $\sigma_m = 0$ and tends to consolidate the data at various mean stresses and lives into a single curve. This provides an opportunity to fit a single curve that gives an equation representing the data. For values of stress amplitude approaching zero, the mean stress should approach the ultimate strength of the material. Hence, the amplitude-mean plot must pass through the two points $(\sigma_m, \sigma_a/\sigma_{ar}) = (0, 1)$ and $(\sigma_u, 0)$.

A straight line is often used and is justified by the observation that for tensile mean stresses most data for ductile materials tend to lie near or beyond it, as is the case in Fig. 9.38. Hence, the line is generally conservative, that is, the error is such that it causes extra safety in life estimates. The equation is

$$\frac{\sigma_a}{\sigma_{ar}} + \frac{\sigma_m}{\sigma_u} = 1 \tag{9.9}$$

This equation is used at a particular life by substituting the σ_{ar} value from the completely reversed *S-N* curve. Since fatigue limits are points on *S-N* curves at some specified life, the equation applies to these also.

$$\frac{\sigma_e}{\sigma_{er}} + \frac{\sigma_m}{\sigma_u} = 1 \tag{9.10}$$

where σ_e is the fatigue limit for any mean stress σ_m, and σ_{er} is the fatigue limit for $\sigma_m = 0$. This form of equation and the corresponding straight line on the normalized plot were developed from an early proposal by Goodman, and so they are often called the *modified Goodman equation* and *line*, respectively.

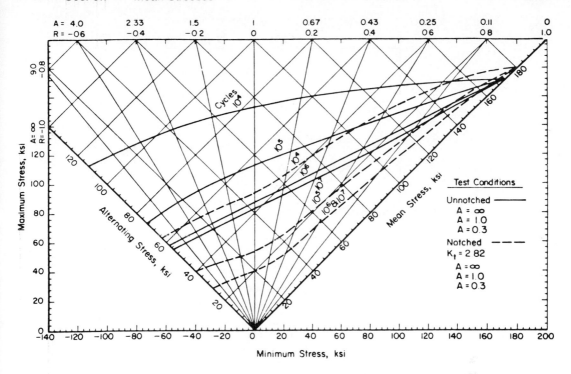

Figure 9.36 Constant-life diagram from axial loading fatigue tests on solution-treated and aged titanium alloy Ti-6Al-4V with $\sigma_u = 1186$ MPa (172 ksi). Net-section nominal stresses are plotted for the notched specimen, which was a plate with a 1.6 mm diameter hole. (From [MILHDBK 83] p. 5-84.)

For compressive mean stresses, Eqs. 9.9 and 9.10 are often somewhat nonconservative, that is, actual data for negative σ_m may be below the line, which is the case in Fig. 9.38. To assure conservatism, it is sometimes assumed that compressive mean stresses provide no benefit. This is equivalent to a horizontal line on the amplitude-mean plot.

$$\frac{\sigma_a}{\sigma_{ar}} = 1, \quad \frac{\sigma_e}{\sigma_{er}} = 1 \quad (\sigma_m \leq 0) \tag{9.11}$$

A variety of more complex equations have been proposed where curved lines are employed to more closely fit the central tendency of data of this type. For example, a relationship called the *Gerber parabola* is sometimes used.

$$\frac{\sigma_a}{\sigma_{ar}} + \left(\frac{\sigma_m}{\sigma_u}\right)^2 = 1 \quad (\sigma_m \geq 0) \tag{9.12}$$

This particular equation is limited to tensile mean stresses as it incorrectly predicts a harmful effect of compressive mean stresses. This is evident from a plot of the above equation, which is also shown on Fig. 9.38.

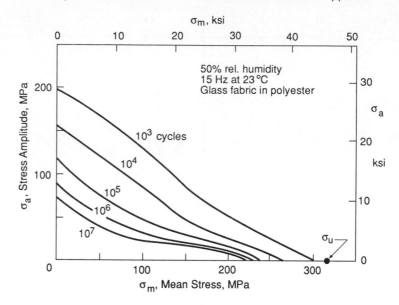

Figure 9.37 Constant-life diagram for unnotched composite material made of 35% heat-resistant polyester resin reinforced with type 181 glass fabric and stressed 0° to the warp. (Adapted from [Boller 56].)

A general trend that occurs on amplitude-mean diagrams is that data for relatively low-ductility metals, such as high-strength steels, tend to lie fairly close to the straight line, Eq. 9.9. In contrast, data for ductile metals may be in better agreement with the Gerber parabola, Eq. 9.12. Improved agreement for ductile metals at both tensile and compressive mean stresses is also often possible by replacing σ_u in Eq. 9.9 with either the true fracture strength $\tilde{\sigma}_f$ from a tension test or the constant σ_f' from the unnotched axial S-N curve in the form of Eq. 9.7. This modification, called the *Morrow parameter*, is used and further discussed in Chapter 14. For brittle metals, such as some cast irons, the data may fall below even the Goodman straight line, so that special equations are often used for these materials. Note that this is also the case for the composite material of Fig. 9.37.

9.7.3 Life Estimates and Families of *S-N* Curves

Let the equation representing the amplitude-mean line or curve, such as Eq. 9.9, be solved for the completely reversed stress σ_{ar}.

$$\sigma_{ar} = \frac{\sigma_a}{1 - \dfrac{\sigma_m}{\sigma_u}} \tag{9.13}$$

Any combination of mean stress σ_m and amplitude σ_a is thus expected to produce the same life as the stress amplitude σ_{ar} applied at zero mean stress. Hence, σ_{ar} may be

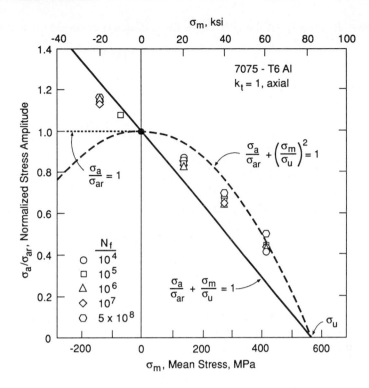

Figure 9.38 Normalized amplitude-mean diagram for 7075-T6 Al based on Fig. 9.34.

thought of as a completely reversed stress that is equivalent, with respect to the life produced, to any (σ_m, σ_a) combination that satisfies the equation. This concept of an *equivalent completely reversed stress* σ_{ar} is useful if the applied stresses involve mean levels that do not correspond to the *S-N* curve that is available. The applied σ_m and σ_a values are used to calculate σ_{ar}, and the life is obtained by entering the *S-N* curve for $\sigma_m = 0$ with this σ_{ar} value.

For example, assume that the *S-N* curve for completely reversed loading is known and has an equation of the form of Eq. 9.6. Since $\sigma_m = 0$, the values correspond to the special case denoted σ_{ar}, so that the equation needs to be rewritten.

$$\sigma_{ar} = A N_f^B \qquad (9.14)$$

Combining this with Eq. 9.13 yields a more general equation that applies for nonzero mean stress

$$\sigma_a = \left(1 - \frac{\sigma_m}{\sigma_u}\right) A N_f^B \qquad (9.15)$$

where this reduces as it should to Eq. 9.6 if $\sigma_m = 0$. On a log-log plot, Eq. 9.15 produces a family of *S-N* curves for different values of mean stress, which are all parallel straight lines. In general, let the *S-N* curve for completely reversed loading be

$$\sigma_{ar} = f\left(N_f\right) \tag{9.16}$$

Combination with Eq. 9.13 then gives

$$\sigma_a = \left(1 - \frac{\sigma_m}{\sigma_u}\right) f\left(N_f\right) \tag{9.17}$$

which is the corresponding family of *S-N* curves.

     If no convenient fitted equation exists for a known *S-N* curve, the equivalent procedure of applying Eq. 9.13 at several different values of N_f can be used to obtain a curve for a given nonzero σ_m. Also, if the *S-N* curve that is available is for a mean stress other than zero, Eq. 9.17 can be used twice, first to obtain the completely reversed curve, and a second time to generalize this curve as needed. Analogous manipulations can be done for *S-N* curves that correspond not to constant values of σ_m but to constant values of *R*.

Example 9.2

     The AISI 4340 steel of Table 9.1 is subjected to cyclic loading with a tensile mean stress of $\sigma_m = 200$ MPa.
     (a) What life is expected if the stress amplitude is $\sigma_a = 450$ MPa?
     (b) Also estimate the σ_a versus N_f curve for this σ_m value.

Solution   **(a)** The *S-N* curve from Table 9.1 for zero mean stress is given by constants $\sigma_f' = 1758$ MPa and $b = -0.0977$ for Eq. 9.7, and the ultimate tensile strength is $\sigma_u = 1172$ MPa. Life estimates may be made for nonzero σ_m by entering Eq. 9.7 with values of equivalent completely reversed stress σ_{ar}.

$$\sigma_{ar} = \sigma_f'\left(2N_f\right)^b = 1758\left(2N_f\right)^{-0.0977} \text{ MPa}$$

Calculating σ_{ar} from Eq. 9.13 gives

$$\sigma_{ar} = \frac{\sigma_a}{1 - \dfrac{\sigma_m}{\sigma_u}} = \frac{450}{1 - \dfrac{200}{1172}} = 543 \text{ MPa}$$

Solving the life equation for N_f and substituting σ_{ar} gives the desired result.

$$N_f = \frac{1}{2}\left(\frac{\sigma_{ar}}{\sigma_f'}\right)^{\frac{1}{b}} = \frac{1}{2}\left(\frac{543}{1758}\right)^{\frac{1}{-0.0977}} = 83{,}400 \text{ cycles} \qquad\qquad \textbf{Ans.}$$

(b) The σ_a vs. N_f curve for $\sigma_m = 200$ MPa is obtained from the same equations with σ_a left as a variable.

$$\sigma_{ar} = \frac{\sigma_a}{1 - \dfrac{\sigma_m}{\sigma_u}} = \sigma_f'\left(2N_f\right)^b, \qquad \sigma_a = \sigma_f'\left(1 - \frac{\sigma_m}{\sigma_u}\right)\left(2N_f\right)^b$$

$$\sigma_a = 1758 \left(1 - \frac{200}{1172}\right)\left(2N_f\right)^b = 1458\left(2N_f\right)^{-0.0977} \text{ MPa} \qquad \textbf{Ans.}$$

The previous calculation of course corresponds to one point that satisfies this equation.

Discussion   This result is probably conservative. An alternate procedure that is likely to be more accurate is to employ σ_f' in place of σ_u in Eq. 9.13. The calculated life is then

$$\sigma_{ar} = \frac{\sigma_a}{1 - \dfrac{\sigma_m}{\sigma_f'}} = \frac{450}{1 - \dfrac{200}{1758}} = 508 \text{ MPa}, \quad N_f = 165{,}000 \text{ cycles} \qquad \textbf{Ans.}$$

The σ_a vs. N_f curve for $\sigma_m = 200$ MPa is now

$$\sigma_a = \sigma_f'\left(1 - \frac{\sigma_m}{\sigma_f'}\right)\left(2N_f\right)^b = \left(\sigma_f' - \sigma_m\right)\left(2N_f\right)^b = 1558\left(2N_f\right)^{-0.0997} \text{ MPa} \qquad \textbf{Ans.}$$

9.8 MULTIAXIAL STRESSES

In engineering components, cyclic loadings that cause complex states of stress are common. Some examples are biaxial stresses due to cyclic pressure in tubes or pipes, combined bending and torsion of shafts, and bending of sheets or plates about more than one axis. Steady applied loads that cause mean stresses may also be combined with such cyclic loads. An additional complexity is that different sources of cyclic loading may differ in phase or frequency or both. For example, if a steady bending stress is applied to a thin-walled tube under cyclic pressure, there are different stress amplitudes and mean stresses in two directions as shown in Fig. 9.39. The axial and hoop directions are the directions of principal stress and remain so as the pressure fluctuates.

If a steady torsion is instead applied, a more complex situation exists as illustrated in Fig. 9.40. At times when the pressure is momentarily at zero, the principal stress directions are controlled by the shear stress and are oriented 45° to the tube axis. However, for nonzero values of pressure, these directions rotate to become more closely aligned with the axial and hoop directions, but never reaching them except for the limiting case where the stresses σ_x and σ_y due to pressure are large compared to the τ_{xy} caused by torsion. Additional complexities could exist. For example, the bending moment in Fig. 9.39 or the torque in Fig. 9.40 could also be cyclic loads, and the frequency of cycling of the bending or torsion could differ from that of the pressure.

9.8.1 One Approach to Multiaxial Fatigue

Consider the simple situation where all cyclic loads are completely reversed and have the same frequency, and further where they are either in-phase or 180° out-of-phase with one another. Also, assume for the present that there are no steady (noncyclic) loads present. For ductile engineering metals, it is reasonable in this case to assume that the fatigue life

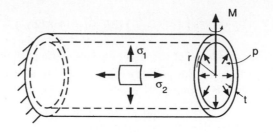

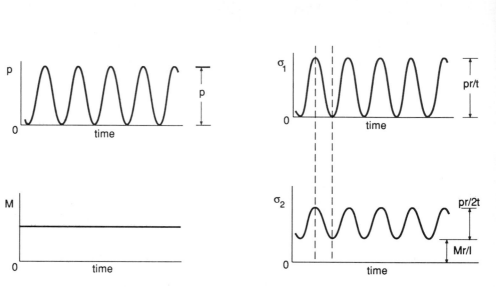

Figure 9.39 Combined cyclic pressure and steady bending of a thin-walled tube with closed ends. The principal directions are constant.

is controlled by the cyclic amplitude of the octahedral shear stress. The amplitudes of the principal stresses, σ_{1a}, σ_{2a}, and σ_{3a} can then be used to compute an *effective stress amplitude* using a relationship similar to that employed for the octahedral shear yield criterion.

$$\bar{\sigma}_a = \frac{1}{\sqrt{2}}\sqrt{(\sigma_{1a} - \sigma_{2a})^2 + (\sigma_{2a} - \sigma_{3a})^2 + (\sigma_{3a} - \sigma_{1a})^2} \tag{9.18}$$

This is identical to Eq. 7.37 except that all stress quantities are amplitudes. In applying this equation, amplitudes considered to be in-phase are positive, and those 180° out-of-phase are negative.

The life may then be estimated by using $\bar{\sigma}_a$ to enter an unnotched *S-N* curve for completely reversed uniaxial stress. Note that the *S-N* curves most commonly available are from bending or axial tests, which do involve a uniaxial state of stress and can therefore be used directly with $\bar{\sigma}_a$ values. (However, difficulties arise for *S-N* curves for bending where yielding occurred in the tests.) For plane stress with $\sigma_3 = 0$, Eq. 9.18

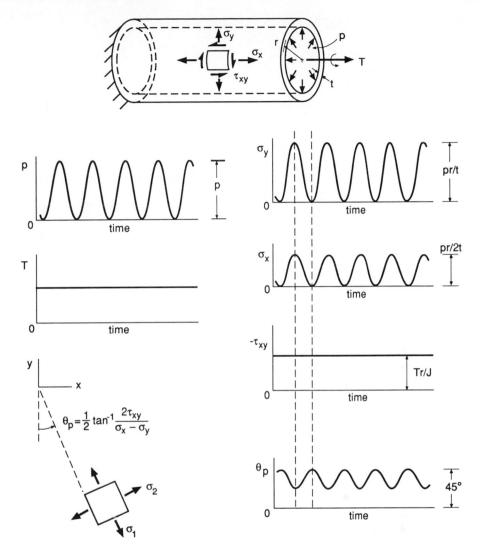

Figure 9.40   Combined cyclic pressure and steady torsion of a thin-walled tube with closed ends. The principal directions oscillate during each cycle.

predicts an elliptical failure locus on a plot of σ_{1a} versus σ_{2a} analogous to the locus for the octahedral shear yield criterion. Some experimental data are compared with such a locus in Fig. 9.41.

If steady (noncyclic) loads are present, these alter the effective stress amplitude $\bar{\sigma}_a$ in a manner analogous to the mean stress effect under uniaxial loading. One approach is to assume that the controlling mean stress variable is the steady value of the hydrostatic stress. Based on this, an *effective mean stress* can be calculated from the mean stresses

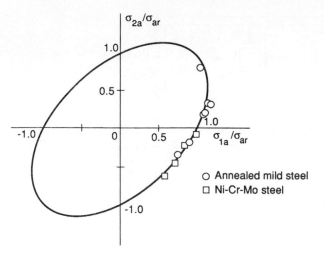

○ Annealed mild steel
□ Ni-Cr-Mo steel

Figure 9.41 Octahedral shear stress criterion compared to 10^7 cycle fatigue strengths for completely reversed biaxial loading. (From the data of Sawert and Gough as compiled by [Sines 59].)

in the three principal directions.

$$\bar{\sigma}_m = \sigma_{1m} + \sigma_{2m} + \sigma_{3m} \tag{9.19}$$

Using stress invariants as discussed in Chapter 6, these effective stress amplitudes and means can also be computed from the amplitudes and means of the stress components for any convenient coordinate axes.

$$\bar{\sigma}_a = \frac{1}{\sqrt{2}} \sqrt{\left(\sigma_{xa} - \sigma_{ya}\right)^2 + \left(\sigma_{ya} - \sigma_{za}\right)^2 + \left(\sigma_{za} - \sigma_{xa}\right)^2 + 6\left(\tau_{xya}^2 + \tau_{yza}^2 + \tau_{zxa}^2\right)}$$

$$\bar{\sigma}_m = \sigma_{xm} + \sigma_{ym} + \sigma_{zm}$$

$$\tag{9.20}$$

where the first equation is analogous to Eq. 7.38, and the various stress components are defined in the conventional manner of Chapter 6 except that additional subscripts a or m indicate amplitudes or means of the particular stress component.

The quantities $\bar{\sigma}_a$ and $\bar{\sigma}_m$ can be combined into an *equivalent completely reversed uniaxial stress* by generalizing Eq. 9.13 or other similar relationship for the amplitude-mean diagram.

$$\sigma_{ar} = \frac{\bar{\sigma}_a}{1 - \dfrac{\bar{\sigma}_m}{\sigma_u}} \tag{9.21}$$

Values of σ_{ar} may be used to enter a completely reversed S-N curve for uniaxial stress, such as one from bending or tension, and the life thus determined.

For uniaxial loading with a mean stress, only σ_{1a} and σ_{1m} are nonzero, and Eq. 9.21 reduces to Eq. 9.13 as it should. For pure shear, as in torsion, only the amplitude and mean of the shear stress, τ_{xya} and τ_{xym}, are nonzero. Hence, Eq. 9.20 gives simply

$$\bar{\sigma}_a = \sqrt{3}\,\tau_{xya}, \quad \bar{\sigma}_m = 0 \tag{9.22}$$

Note that $\bar{\sigma}_m$ is zero even if a mean shear stress is present. This somewhat surprising prediction that mean shear stress has no effect is in fact in agreement with experimental observation. For additional detail on this point and related matters, see Sines and Waisman (1959) in the References, specifically the paper therein by Sines.

9.8.2 Discussion

For situations where the principal axes rotate during cyclic loading, the applicability of the equations just given is questionable. This also applies if cyclic loads occur at more than one frequency, or if there is a difference in phase (other than 180°) between them. A number of other approaches exist. For example, the *critical plane approach* involves finding the maximum shear strain amplitude and the plane on which it acts, and then using the maximum normal stress acting on this plane to obtain a mean stress effect. Some discussion of this approach is given in Chapter 14.

Example 9.3

A tube with closed ends is made of 2024-T4 Al and has a diameter of 120 mm and a wall thickness of 4 mm. It is subjected to a bending moment of 2 kN·m that does not change with time, and also to a pressure of 15 MPa that repeatedly cycles between zero and this value. How many pressure cycles can be applied before fatigue failure is expected?

Solution   This is a particular case of the situation already illustrated in Fig. 9.39. The hoop and longitudinal directions in the tube are the (1-2) principal directions, as no shear stresses occur on the planes normal to these. In the hoop (1) direction, both the amplitude and the mean stress are half the zero-to-tension stress that occurs.

$$\sigma_{1a} = \sigma_{1m} = \frac{1}{2}\left(\frac{pr}{t}\right) = \frac{(15 \text{ MPa})(60 \text{ mm})}{2(4 \text{ mm})} = 112.5 \text{ MPa}$$

For the longitudinal (2) direction, the zero-to-tension pressure stress is half as large, and the mean stress must be increased by the contribution of the bending moment.

$$\sigma_{2a} = \frac{1}{2}\left(\frac{pr}{2t}\right) = \frac{(15 \text{ MPa})(60 \text{ mm})}{4(4 \text{ mm})} = 56.3 \text{ MPa}$$

$$\sigma_{2m} = \frac{1}{2}\left(\frac{pr}{2t}\right) + \frac{Mr}{I} = \frac{1}{2}\left(\frac{pr}{2t}\right) + \frac{M}{\pi r^2 t}$$

$$\sigma_{2m} = 56.3 \text{ MPa} + \frac{0.002 \text{ MN·m}}{\pi (0.06 \text{ m})^2 (0.004 \text{ m})} = 100.5 \text{ MPa}$$

where the area moment of inertia of the cross section of the thin-walled tube is approximated as $I = \pi r^3 t$. The effective stress amplitude and mean can now be calculated from Eqs. 9.18 and 9.19.

$$\bar{\sigma}_a = \frac{1}{\sqrt{2}}\sqrt{(\sigma_{1a} - \sigma_{2a})^2 + (\sigma_{2a} - \sigma_{3a})^2 + (\sigma_{3a} - \sigma_{1a})^2}$$

$$\bar{\sigma}_a = \frac{1}{\sqrt{2}}\sqrt{(112.5 - 56.3)^2 + (56.3 - 0)^2 + (0 - 112.5)^2} = 97.4 \text{ MPa}$$

$$\bar{\sigma}_m = \sigma_{1m} + \sigma_{2m} + \sigma_{3m} = 112.5 + 100.5 + 0 = 213 \text{ MPa}$$

These combine to give an equivalent completely reversed uniaxial stress amplitude of

$$\sigma_{ar} = \frac{\bar{\sigma}_a}{1 - \dfrac{\bar{\sigma}_m}{\sigma_u}} = \frac{97.4}{1 - \dfrac{213}{476}} = 176.3 \text{ MPa}$$

The value $\sigma_u = 476$ MPa is obtained for 2024-T4 Al from Table 9.1, where constants $A = 839$ MPa and $B = -0.102$ are also given for a stress-life curve of the form of Eq. 9.6. Noting that these constants give the completely reversed stress-life curve, it is appropriate to use them with σ_{ar} from above. Hence, the estimated life is

$$N_f = \left(\frac{\sigma_{ar}}{A}\right)^{\frac{1}{B}} = \left(\frac{176.3 \text{ MPa}}{839 \text{ MPa}}\right)^{\frac{1}{-0.102}} = 4.39 \times 10^6 \text{ cycles} \qquad \textbf{Ans.}$$

Discussion   Due to the high mean stresses involved, yielding at peak pressure is a possibility. This would be considered failure at $N_f = 1$ in many engineering situations, and so we will estimate the safety factor against yielding. The principal stresses at peak pressure are

$$\sigma_{1\,\text{max}} = \sigma_{1a} + \sigma_{1m} = 112.5 + 112.5 = 225 \text{ MPa}$$

$$\sigma_{2\,\text{max}} = \sigma_{2a} + \sigma_{2m} = 56.3 + 100.5 = 156.8 \text{ MPa}$$

$$\sigma_{3\,\text{max}} \approx 0$$

The effective stress for the octahedral shear yield criterion is obtained by substituting these values into Eq. 7.37.

$$\bar{\sigma}_H = \frac{1}{\sqrt{2}} \sqrt{(\sigma_1 - \sigma_2)^2 + (\sigma_2 - \sigma_3)^2 + (\sigma_3 - \sigma_1)^2}$$

$$\bar{\sigma}_H = \frac{1}{\sqrt{2}} \sqrt{(225 - 156.8)^2 + (156.8 - 0)^2 + (0 - 225)^2} = 199.8 \text{ MPa}$$

Comparing this to the yield strength of $\sigma_o = 303$ MPa from Table 9.1 gives the safety factor.

$$X = \frac{\sigma_o}{\bar{\sigma}_H} = \frac{303}{199.8} = 1.52 \qquad \textbf{Ans.}$$

Therefore, no yielding failure occurs.

9.9 VARIABLE AMPLITUDE LOADING

As discussed earlier in this chapter, fatigue loadings in practical applications usually involve stress amplitudes that change in an irregular manner. We will now consider methods of making life estimates for such loadings.

9.9.1 The Palmgren-Miner Rule

Consider a situation of variable amplitude loading as illustrated in Fig. 9.42. A certain stress amplitude σ_{a1} is applied for a number of cycles N_1, where the number of cycles to

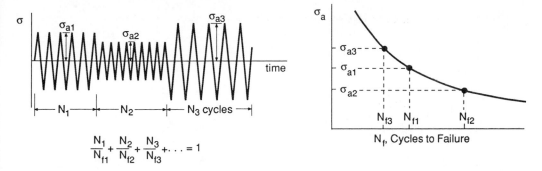

Figure 9.42   Use of the Palmgren-Miner rule for life prediction for variable amplitude loading which is completely reversed.

failure from the S-N curve for σ_{a1} is N_{f1}. The fraction of the life used is then N_1/N_{f1}. Now let another stress amplitude σ_{a2}, corresponding to N_{f2} on the S-N curve, be applied for N_2 cycles. An additional fraction of the life N_2/N_{f2} is then used. The *Palmgren-Miner rule* simply states that fatigue failure is expected when such life fractions sum to unity, that is, when 100% of the life is exhausted.

$$\frac{N_1}{N_{f1}} + \frac{N_2}{N_{f2}} + \frac{N_3}{N_{f3}} + \ldots = \sum \frac{N_j}{N_{fj}} = 1 \qquad (9.23)$$

This simple rule was first used by A. Palmgren in Sweden in the 1920s for predicting the life of ball bearings. However, it was not widely known or used until it was reinvented in 1945 by M. A. Miner.

Often, a sequence of variable amplitude loading is repeated a number of times. Under these circumstances, it is convenient to sum cycle ratios over one repetition of the history, and then to multiply this by the number of repetitions required for the summation to reach unity.

$$B_f \left[\sum \frac{N_j}{N_{fj}} \right]_{\text{one rep.}} = 1 \qquad (9.24)$$

where B_f is the number of repetitions to failure. This is illustrated in Fig. 9.43.

Some cycles of the variable amplitude loading may involve mean stresses. Equivalent completely reversed stresses then need to be calculated, using an amplitude-mean relationship similar to Eq. 9.13, before N_f values are obtained from an S-N curve for completely reversed loading. In addition, the stress ranges caused by changing the mean level also need to be considered in summing cycle ratios. For example, in Fig. 9.43 the cycles at amplitude-mean combinations $(\sigma_{a1}, \sigma_{m1})$ and $(\sigma_{a2}, \sigma_{m2})$ are obvious. However, one additional cycle $(\sigma_{a3}, \sigma_{m3})$ also needs to be considered. In fact, this cycle will cause most of the fatigue damage if σ_{a1} and σ_{a2} are small, so that omitting it could result in a seriously nonconservative life estimate. The task of identifying cycles can be handled in a comprehensive manner by *cycle counting* as described in the next subsection.

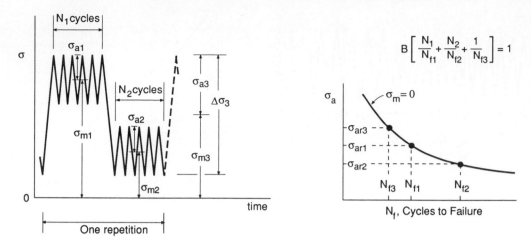

Figure 9.43   Life prediction for a repeating stress history with mean level shifts.

Example 9.4

The stress history shown in Figure E9.4 is repeatedly applied as a uniaxial stress to an unnotched member made of the AISI 4340 steel of Table 9.1. Estimate the number of repetitions required to cause fatigue failure.

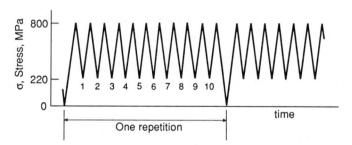

Figure E9.4

Solution   In one repetition of the history, the stress rises from zero to 800 MPa and later returns to zero, forming one cycle with $\sigma_{min} = 0$ and $\sigma_{max} = 800$ MPa. There are also 10 cycles with $\sigma_{min} = 220$ and $\sigma_{max} = 800$ MPa. A table as given below is useful, with all stress values being in MPa.

j	N_j	σ_{min}	σ_{max}	σ_a	σ_m	σ_{ar}	N_{fj}
1	1	0	800	400	400	607	26,600
2	10	220	800	290	510	513	148,000

Each of the two levels of cycling forms a line in the table. The values of stress amplitude σ_a, mean stress σ_m, equivalent completely reversed stress amplitude σ_{ar}, and the corresponding life N_f, are calculated from Eqs. 9.1, 9.6, and 9.13.

$$\sigma_a = \frac{\sigma_{max} - \sigma_{min}}{2}, \qquad \sigma_m = \frac{\sigma_{max} + \sigma_{min}}{2}$$

$$\sigma_{ar} = \frac{\sigma_a}{1 - \dfrac{\sigma_m}{\sigma_u}}, \qquad N_f = \left(\frac{\sigma_{ar}}{A}\right)^{\frac{1}{B}}$$

Constants $\sigma_u = 1172$ MPa, $A = 1643$ MPa, and $B = -0.0977$ from Table 9.1 are used. The life N_f corresponding to each $(\sigma_{min}, \sigma_{max})$ combination is calculated from σ_{ar} using the constants A and B as these correspond to test data with zero mean stress.

The life in repetitions to failure B_f may be estimated from the N and N_f values in the table using the Palmgren-Miner rule in the form of Eq. 9.24.

$$B_f = \frac{1}{\left[\sum \dfrac{N_j}{N_{fj}}\right]_{one\ rep.}} = \frac{1}{\dfrac{1}{26,600} + \dfrac{10}{148,000}} = 9510 \text{ repetitions} \qquad \textbf{Ans.}$$

9.9.2 Cycle Counting for Irregular Histories

For highly irregular variations of load with time, such as those in Figs. 9.6 to 9.9, it is not obvious how individual events should be isolated and defined as cycles so that the Palmgren-Miner rule can be employed. In past years, there has been considerable uncertainty and debate concerning the proper procedure, and a number of different methods have been proposed and used. However, a consensus has recently emerged that the best approach is to use a procedure called *rainflow cycle counting*, developed by Prof. T. Endo and his colleagues in Japan around 1968, or certain other procedures that are essentially equivalent.

An irregular stress history consists of a series of *peaks* and *valleys*, which are points where the direction of loading changes as illustrated in Fig. 9.44. Also of interest are *ranges*, that is, stress differences, measured between peaks and valleys, or between valleys and peaks. A *simple range* is measured between a peak and the next valley, or between a valley and the next peak. An *overall range* is measured between a peak and a valley that is not the next one, but is one that occurs later, or between a valley and a later peak. In Fig. 9.44, $\Delta\sigma_{AB}$ and $\Delta\sigma_{BC}$ are simple ranges, and $\Delta\sigma_{AD}$ and $\Delta\sigma_{DG}$ are overall ranges.

In performing rainflow cycle counting, a cycle is identified or *counted* if it meets the criterion illustrated in Fig. 9.45. A peak-valley-peak or valley-peak-valley combination X-Y-Z in the loading history is considered to contain a cycle if the second range, $\Delta\sigma_{YZ}$, is greater than or equal to the first range, $\Delta\sigma_{XY}$. If the second range is indeed larger or equal, then a cycle equal to the first range ($\Delta\sigma_{XY}$) is counted. The mean value for this cycle, specifically the average of σ_X and σ_Y, is also of interest.

The complete procedure is described as follows using the example of Fig. 9.46: Assume that the stress history given is to be repeatedly applied, so that it can be taken

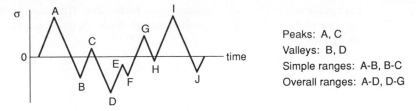

Peaks: A, C
Valleys: B, D
Simple ranges: A-B, B-C
Overall ranges: A-D, D-G

Figure 9.44   Definitions for irregular loading.

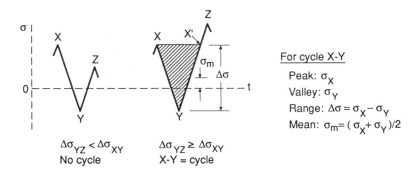

For cycle X-Y
 Peak: σ_X
 Valley: σ_Y
 Range: $\Delta\sigma = \sigma_X - \sigma_Y$
 Mean: $\sigma_m = (\sigma_X + \sigma_Y)/2$

$\Delta\sigma_{YZ} < \Delta\sigma_{XY}$
No cycle

$\Delta\sigma_{YZ} \geq \Delta\sigma_{XY}$
X-Y = cycle

Figure 9.45   Condition for counting a cycle using the rainflow method.

to begin at any any peak or valley. On this basis, it is convenient to move a portion of the history to the end, so that a sequence is obtained that begins and ends with the peak or valley having the highest absolute value of stress. Figures 9.46(a) and (b) illustrate this. Cycle counting then proceeds, starting at the beginning of the rearranged history and using the criterion of Fig. 9.45. If a cycle is counted, this information is recorded, and its peak and valley are assumed not to exist for purposes of further cycle counting, as illustrated for cycle *E-F* in Fig. 9.46(c). If no cycle can be counted at the current location, one then moves ahead until a count can be made. For example, in (d), after *E-F* is counted, the counting condition is next met for *A-B*.

Counting is complete when all of the history is exhausted. For this example, the cycles counted are *E-F*, *A-B*, *H-C*, and *D-G*, and they have ranges and means as tabulated in Fig. 9.46. Note that some of the cycles counted correspond to simple ranges in the original history, specifically *E-F* and *A-B*, and others to overall ranges, specifically *H-C* and *D-G*. The largest range counted is always the one between the highest peak and the lowest valley, *D-G* for this example.

For lengthy histories, it is convenient to present the results of rainflow cycle counting as a matrix giving the numbers of cycles occurring at various combinations of range and mean. An example of this is shown in Fig. 9.47. Note that values of range and mean are rounded off to discrete values to give a matrix of manageable size. Cycle counting is often applied directly to load histories, with this being the case for Fig. 9.47.

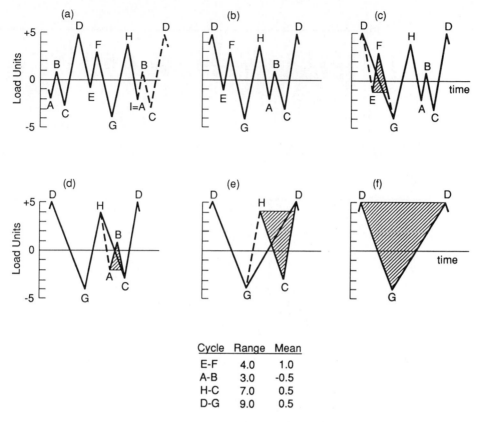

Cycle	Range	Mean
E-F	4.0	1.0
A-B	3.0	-0.5
H-C	7.0	0.5
D-G	9.0	0.5

Figure 9.46   Example of rainflow cycle counting. (Adapted from [ASTM 90b] Std. E1049; copyright © ASTM; reprinted with permission.)

Additional information on cycle counting can be found in ASTM Standard No. E1049, in Bannantine et al. (1990), and also in the second edition of the *SAE Fatigue Design Handbook* (Rice, 1988). The latter reference also contains relevant computer programs.

Example 9.5

The load history of Fig. 9.46(a) is repeatedly applied as a uniaxial stress to an unnotched member of 2024-T4 Al, with stress values σ being given by 1 unit = 60 MPa. Estimate the number of repetitions required to cause fatigue failure.

Solution   It is convenient to mark each cycle on a plot of the history as it is counted, subsequently ignoring any cycles already marked as if they were removed. The process is shown in Figure E9.5.

This history is first reordered to start with either the highest peak or lowest valley as shown in Fig. 9.46(a) and (b). Starting at the beginning of the reordered history, the cycle counting condition of Fig. 9.45 is not met by event *D-E-F*. Moving ahead one point to *G* and considering *E-F-G*, a cycle *E-F* is counted and marked as shown in (a). Then moving

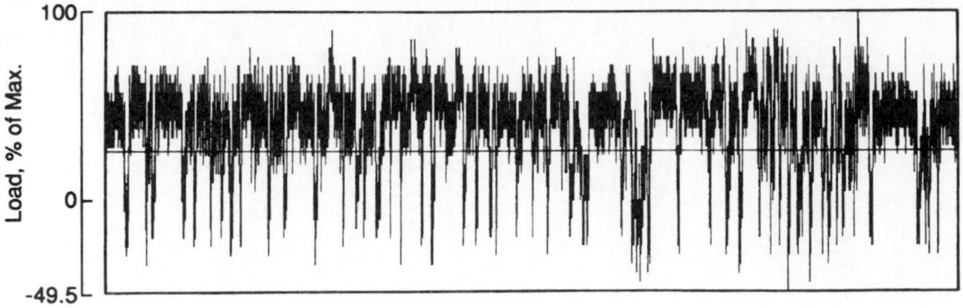

	Mean																			
Range	−15	−10	−5	0	5	10	15	20	25	30	35	40	45	50	55	60	65	70	75	All
20	4	1	5	2	2	5	—	—	3	6	15	27	29	32	22	12	6	2	—	173
25	2	4	3	9	8	10	4	6	2	7	17	37	36	43	33	13	7	1	2	244
30	1	1	5	3	1	1	4	3	—	4	13	20	20	23	20	8	6	1	—	134
35	1	1	4	2	3	2	—	1	3	2	8	17	16	11	11	7	2	—	—	91
40	—	1	1	1	2	1	1	—	—	4	7	15	16	9	8	2	—	—	—	68
45	—	1	—	4	3	—	—	—	—	2	1	9	7	2	3	1	—	—	—	33
50	—	—	2	2	2	1	—	—	—	2	2	3	3	1	1	1	1	—	—	21
55	—	—	1	1	—	—	—	—	—	2	2	4	4	2	—	1	—	1	—	18
60	—	1	1	—	—	—	—	—	—	1	1	3	2	1	—	—	—	—	—	10
65	—	—	—	—	—	—	—	—	—	—	2	1	—	—	—	—	—	—	—	3
70	—	—	—	—	—	—	—	—	—	—	2	—	1	—	—	—	—	—	—	3
75	—	—	—	—	—	—	1	—	—	—	1	2	—	—	—	—	—	—	—	4
80	—	—	—	—	—	—	—	—	—	—	—	—	—	—	—	—	—	—	—	—
85	—	—	—	—	1	—	1	3	3	—	—	—	—	—	—	—	—	—	—	8
90	—	—	—	—	—	—	—	—	4	—	—	—	—	—	—	—	—	—	—	4
95	—	—	—	—	—	1	—	1	4	1	—	—	—	—	—	—	—	—	—	7
100	—	—	—	—	—	—	—	5	3	1	—	—	—	—	—	—	—	—	—	9
105	—	—	—	—	—	—	—	3	3	3	—	—	—	—	—	—	—	—	—	9
110	—	—	—	—	—	—	—	—	2	3	—	—	—	—	—	—	—	—	—	5
115	—	—	—	—	—	—	—	—	3	—	—	—	—	—	—	—	—	—	—	3
120	—	—	—	—	—	—	1	—	1	1	—	—	—	—	—	—	—	—	—	3
125	—	—	—	—	—	—	—	—	2	—	—	—	—	—	—	—	—	—	—	2
130	—	—	—	—	—	—	—	—	—	—	—	—	—	—	—	—	—	—	—	—
135	—	—	—	—	—	—	—	1	—	—	—	—	—	—	—	—	—	—	—	1
140	—	—	—	—	—	—	—	—	—	—	—	—	—	—	—	—	—	—	—	—
145	—	—	—	—	—	—	—	—	—	—	—	—	—	—	—	—	—	—	—	—
150	—	—	—	—	—	—	—	—	1	—	—	—	—	—	—	—	—	—	—	1

Figure 9.47 An irregular load vs. time history from a ground vehicle transmission, and a matrix giving numbers of rainflow cycles at various combinations of range and mean. The range and mean values are percentages of the peak load, and these were rounded to the discrete values shown in constructing the matrix. (Load history from [Wetzel 77] pp. 15-18.)

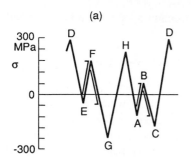 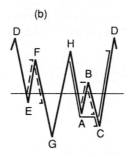 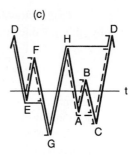

Figure E9.5

ahead one point at a time and checking for cycles, none are found for D-G-H, for G-H-A, or for H-A-B. Cycle A-B is finally found for A-B-C and is marked as also shown in (a). Moving ahead to D and considering H-C-D, cycle H-C is found and marked as shown in (b). Only event D-G-D remains, and this constitutes a cycle D-G as marked in (c).

The σ_{min} and σ_{max} values for each cycle are entered into a table as they are counted, each occurring only once in this case.

Cycle	N_j	σ_{min}	σ_{max}	σ_a	σ_m	σ_{ar}	N_{fj}
E-F	1	-60	180	120	60	137.3	5.09×10^7
A-B	1	-120	60	90	-30	84.7	5.83×10^9
H-C	1	-180	240	210	30	224.1	4.17×10^5
D-G	1	-240	300	270	30	288.2	3.55×10^4

Units of MPa are used for all stresses above, and the values of σ_a, σ_m, σ_{ar}, and N_f are calculated as for Example 9.4 using constants σ_u, A, and B for 2024-T4 Al from Table 9.1. The number of repetitions to failure can now be estimated using Eq. 9.24.

$$B_f = \frac{1}{\left[\sum \dfrac{N_j}{N_{fj}}\right]_{\text{one rep.}}} = \frac{1}{\dfrac{1}{5.09 \times 10^7} + \dfrac{1}{5.83 \times 10^9} + \dfrac{1}{4.17 \times 10^5} + \dfrac{1}{3.55 \times 10^4}}$$

$$B_f = 32{,}700 \text{ repetitions} \qquad\qquad \textbf{Ans.}$$

9.10 SUMMARY

Fatigue of materials is the process of accumulating damage and failure due to cyclic loading. Engineering efforts over more than 150 years, aimed at preventing fatigue failure, led first to the development of a stress-based approach. This approach emphasizes stress versus life curves and nominal (average) stresses. More sophisticated approaches, namely the strain-based approach and the fracture mechanics approach, have arisen in recent years.

Cyclic stressing between constant maximum and minimum values can be described by giving values of the stress amplitude and mean.

$$\sigma_a = \frac{\Delta\sigma}{2} = \frac{\sigma_{\max} - \sigma_{\min}}{2}, \quad \sigma_m = \frac{\sigma_{\max} + \sigma_{\min}}{2} \tag{9.25}$$

Alternatively, one can specify $\sigma_{\max}$ and the ratio R, or $\Delta\sigma$ and R, where $R = \sigma_{\min}/\sigma_{\max}$. Additional useful relationships among these quantities are given by Eqs. 9.1 to 9.4.

It is important to distinguish between the actual stress σ at a point and any nominal (average) stress S that might be calculated. For simple axial loading of unnotched members, these two quantities are the same. If bending occurs, σ at the edge of the member is equal to S only if there is no yielding. Notches (stress raisers) are commonly characterized by an elastic stress concentration factor k_t, so that the peak stress at the notch is $\sigma = k_t S$, except that the equality fails if yielding occurs even locally at the notch.

Stress versus life (S-N) curves are commonly plotted in terms of stress amplitude versus cycles to failure. Such curves can be obtained from a variety of test apparatus ranging from relatively simple rotating bending machines to sophisticated closed loop servo-hydraulic equipment. The most basic S-N curve is considered to be the one for zero mean stress, which is the only case that can be tested using rotating bending.

S-N curves vary with the material and its prior processing. They are also affected by mean stress and geometry, especially the presence of notches, and also by surface finish, chemical and thermal environment, frequency of cycling, and residual stress. Some equations for representing S-N curves are

$$\sigma_a = A N_f^B, \quad \sigma_a = C + D \log N_f \tag{9.26}$$

The first of these gives a straight line on log-log coordinates, and the second a straight line on log-linear coordinates.

Mean stress effects can be represented by plotting stress amplitude versus mean stress for various specific lives on a *constant-life diagram*. Estimates of mean stress effects for unnotched members may be made by using equations similar to the one for the modified Goodman line.

$$\sigma_{ar} = \frac{\sigma_a}{1 - \dfrac{\sigma_m}{\sigma_u}} \tag{9.27}$$

In such an equation, the applied combination of stress amplitude σ_a and mean stress σ_m is expected to result in the same life as the stress amplitude σ_{ar} applied at zero mean stress. Values of σ_{ar}, called the *equivalent completely reversed stress amplitude*, may be used to enter a curve of σ_a versus N_f from completely reversed loading. The above equation may also be employed in making life estimates for simple cases of multiaxial loading. The quantities σ_a and σ_m are replaced by an effective stress amplitude $\bar{\sigma}_a$, which is proportional to the amplitude of the octahedral shear stress, and an effective mean stress $\bar{\sigma}_m$, which is proportional to the hydrostatic stress due to mean stresses in three directions.

If more than one amplitude or mean level occurs, life estimates may be made by summing cycle ratios, called the Palmgren-Miner rule.

$$\sum \frac{N_j}{N_{fj}} = 1 \qquad (9.28)$$

The alternate from given by Eq. 9.24 is useful for sequences of loading that occur repeatedly. Cycles introduced by changing mean levels need to be considered in making such life estimates, and rainflow cycle counting is needed for highly irregular load versus time histories.

NEW TERMS AND SYMBOLS

beach marks
completely reversed
constant amplitude
constant-life diagram
elastic stress concentration factor, k_t
equivalent completely reversed stress
   amplitude, σ_{ar}
fatigue (of materials)
fatigue limit, σ_e
fatigue strength
fracture mechanics approach
Gerber parabola
high-cycle fatigue
low-cycle fatigue
maximum stress, σ_{max}
mean stress, σ_m
minimum stress, σ_{min}
modified Goodman line
nominal stress, S
normalized amplitude-mean diagram

notch (stress raiser)
Palmgren-Miner rule
peak, valley
point stress, σ
rainflow cycle counting
reciprocating bending test
residual stress
rotating bending test
shear lip
S-N curve
static load
strain-based approach
stress amplitude, σ_a
stress-based approach
stress range, $\Delta\sigma$
stress ratio, R
striations
vibratory load
working load
zero-to-tension

REFERENCES

(a) General References

ASTM. 1990 *Annual Book of ASTM Standards*, Am. Soc. for Testing and Materials, Philadelphia, Pa. See: No. E466, "Conducting Constant Amplitude Axial Fatigue Tests of Metallic Materials," Vol. 03.01; and No. E1049, "Cycle Counting in Fatigue Analysis," Vol. 03.01; and also No. D671,"Flexural Fatigue of Plastics by Constant Amplitude of Force," Vol. 08.01.

BANNANTINE, J. A., J. J. COMER, and J. L. HANDROCK. 1990 *Fundamentals of Metal Fatigue Analysis*, Prentice Hall, Englewood Cliffs, N.J.

FORREST, P. G. 1962 *Fatigue of Metals*, Pergamon Press, Oxford, UK, and Addison-Wesley, Reading, Ma.

FROST, N. E., K. J. MARSH, and L. P. POOK. 1974 *Metal Fatigue*, Clarendon Press, Oxford, UK.

GRAHAM, J. A., J. F. MILLAN, and APPL F. J., eds. 1968 *Fatigue Design Handbook*, SAE Pub. No. AE-4, Society of Automotive Engineers, Warrendale, Pa.

HEYWOOD, R. B. 1962 *Designing Against Fatigue of Metals*, Reinhold, New York.

MANN, J. Y. 1958 "The Historical Development of Research on the Fatigue of Materials and Structures," *The Jnl. of the Australian Inst. of Metals*, Nov. 1958, pp. 222–241.

RICE R. C., ed. 1988 *Fatigue Design Handbook*, 2nd ed., SAE Pub. No. AE-10, Society of Automotive Engineers, Warrendale, Pa.

SINES, G. and J. L. WAISMAN. 1959 *Metal Fatigue*, McGraw-Hill, New York.

SURESH, S. 1991 *Fatigue of Materials*, Cambridge University Press, Cambridge, UK.

(b) Sources of Material Properties

BATTELLE. 1975 *Handbook on Materials for Superconducting Machinery*, Metals and Ceramics Information Center, Battelle Columbus Labs, Columbus, Oh.

BATTELLE. 1991 *Aerospace Structural Metals Handbook*, 5 vols., Metals and Ceramics Information Center, Battelle Columbus Div., Columbus, Oh.

BOYER, J. E. 1986 *Atlas of Fatigue Curves*, American Society for Metals, Metals Park, Oh.

HALLOWELL, J. B., ed. 1989 *Structural Alloys Handbook*, 3 vols., Metals and Ceramics Information Center, Battelle Columbus Labs, Columbus, Oh.

HOWELL, F. M. and J. L. MILLER. 1955 "Axial Stress Fatigue Strengths of Several Structural Aluminum Alloys," *Proc. of the Am. Soc. for the Testing and Materials*, Vol. 55, pp. 955–968.

JSMS. 1982 *Data Book on Fatigue Strength of Metallic Materials*, 3 vols., The Society of Materials Science (JSMS), Kyoto, Japan, 1982.

NRIM. 1978 *NRIM Fatigue Data Sheets*, National Research Institute for Metals, Tokoyo, Japan, starting in 1978 and continuing.

SAE. 1989 "Technical Report on Low Cycle Fatigue Properties of Wrought Metals," SAE J1099, Information Report, Society of Automotive Engineers, Warrendale, Pa. See also L. E. Tucker, R. W. Landgraf, and W. R. Brose, "Proposed Technical Report on Fatigue Properties for the SAE Handbook," SAE Paper No. 740279, Automotive Engineering Congress, Detroit, Mich. 1974.

PROBLEMS AND QUESTIONS

Section 9.2

9.1 Which of the following are true for completely reversed cycling? Which ones are true for zero-to-tension cycling?

    (a) $R = 0$
    (b) $R = -1$
    (c) $\sigma_a = \sigma_{max}$
    (d) $\Delta\sigma = \sigma_{max}$
    (e) $\sigma_m = \sigma_a$
    (f) $\sigma_m = 0$

9.2 Constant amplitude cycling occurs at a stress amplitude of 20 MPa about a mean stress level of 60 MPa. Identify which of the following correctly describe this situation, and make changes where needed to provide a correct description:
    (a) $\sigma_a = 20$, $\sigma_m = 60$ MPa
    (b) $\sigma_{max} = 80$ MPa, $R = 0.33$
    (c) $\Delta\sigma = 40$ MPa, $R = 0.50$
    (d) $\sigma_{max} = 80$, $\sigma_{min} = 40$ MPa
    (e) $\sigma_a = 20$ MPa, $A = 0.5$

9.3 Verify each item of Eq. 9.4 starting from any of Eqs. 9.1 to 9.3 or preceding items of Eq. 9.4.

9.4 Write definitions in your own words for point stress σ, nominal stress S, and the product $k_t S$. Give an example of a situation where $\sigma = S$, and also give an example where $\sigma \neq S$. Similarly, give an example for $\sigma = k_t S$ and $\sigma \neq k_t S$.

9.5 For an S-N curve of the form of Eq. 9.5, two points (N_1, σ_1) and (N_2, σ_2) are known.
    (a) Develop equations for the constants C and D as a function of these values.
    (b) Using log-linear coordinates, plot the Eq. 9.5 line for AISI 1015 steel using constants from Table 9.1. The plot should cover $N_f = 1$ to 10^6, but use a dotted line for $N_f < 10^2$ where the equation is not accurate.
    (c) How does the line plotted in (b) relate to the constants C and D?

9.6 For the Ti-6Al-4V alloy of Table 9.1, plot the S-N curves for both the Eq. 9.5 and 9.6 forms. Use either log-linear or log-log coordinates covering $N_f = 1$ to 10^6. Then comment on the types of lines obtained and on their agreement.

9.7 Hot-rolled and normalized AISI 1045 steel can be assumed to have an S-N curve of the form of Eq. 9.6. Some test data for unnotched specimens under axial stress with zero mean are given below.
    (a) Plot these data on log-log coordinates and use the graph to obtain approximate values for the constants A and B.
    (b) Obtain refined values for A and B using a linear least-squares fit to $\log N_f$ versus $\log \sigma_a$.

σ_a, MPa	N_f, cycles
524	257
459	1 494
410	6 749
352	19 090
315	36 930
270	321 500
241	2 451 000

Source: Data in [Leese 85].

9.8 Proceed as in Prob. 9.7, but use the following data for RQC-100 steel.

σ_a, MPa	N_f, cycles
631	227
574	1 030
505	6 450
472	22 250
455	110 000

Section 9.3

9.9 Describe the likely sources of cyclic loading for a sailboat rudder. Consider static loads, working loads, vibratory loads, and accidental loads.

9.10 For a bicycle front fork, answer the same question as in Prob. 9.9.

9.11 For the plow on a small garden tractor, answer the same question as in Prob. 9.9.

9.12 For the *fork* of a fork-lift truck, which is placed under boxes to lift them, answer the same question as in Prob. 9.9.

Section 9.7

9.13 The steel AISI 4142, heat treated to a hardness of 450 HB, is subjected to cyclic loading at a stress amplitude of $\sigma_a = 800$ MPa. Estimate the life N_f for mean stresses σ_m of (a) zero, (b) 200 MPa tension, and (c) 200 MPa compression.

9.14 For RQC-100 steel:
   (a) Obtain an estimated equation relating stress amplitude σ_a and life N_f for a mean stress of $\sigma_m = 100$ MPa.
   (b) Plot this σ_a vs. N_f equation along with the one for $\sigma_m = 0$ and comment on the difference.
   (c) Also estimate the σ_a vs. N_f equation for $\sigma_m = -100$ MPa (compression), add this to the plot from (b), and further comment.

9.15 Proceed as in Prob. 9.14 except change the material to 2024-T4 aluminum.

9.16 A material has a zero mean stress S-N curve of the form of Eq. 9.5. Write an equation for estimating life N_f as a function of stress amplitude σ_a, mean stress σ_m, and constants C and D obtained from tests at $\sigma_m = 0$.

9.17 One of the various mean stress equations sometimes used in place of those of Goodman or Gerber is Eq. 9.9 with σ_u replaced by the true fracture strength, $\tilde{\sigma}_f$.
   (a) Apply this equation to the data of Fig. 9.38, obtaining $\tilde{\sigma}_f$ for the alloy involved from Table 5.7.
   (b) Is a reasonable fit to the data obtained? Can you think of any advantages or disadvantages of this equation compared to other two?

9.18 In a paper by E. K. Walker, it is proposed that mean stress effects can be estimated by $\sigma^* = \sigma_{max}(1 - R)^\gamma$, where γ is a constant for the material and σ^* is an equivalent zero-to-maximum ($R = 0$) stress range that causes the same fatigue life as the actual σ_{max}, R combination.

(a) Modify the above equation so that it is expressed in terms of the equivalent completely reversed stress amplitude. In particular, develop an equation for σ_{ar} as a function of the stress amplitude and mean, σ_a and σ_m, and γ.

(b) Based on (a), write the equation for σ_{ar} for the special case where $\gamma = 0.5$.

Section 9.8

9.19 The alloy Ti-6Al-4V of Table 9.1 is used to make a cylindrical pressure vessel having closed ends with a diameter 250 mm and wall thickness of 2.5 mm.

(a) What repeatedly applied pressure will cause fatigue failure in 10^5 cycles? (Neglect the stress raiser effect of the end closure or other geometric discontinuities.)

(b) For the pressure from (a), what is the safety factor against yielding?

9.20 A unnotched solid circular shaft is made of AISI 4142 steel heat treated to a hardness of 450 HB. It has a diameter of 50 mm and is loaded in cyclic torsion between zero and a torque of 20 kN·m.

(a) How many torsion cycles are expected to result in fatigue failure?

(b) What is the safety factor against yielding?

(c) What shaft diameter is required if 100,000 cycles are expected in service and the safety factor on life must be 20?

(d) What shaft diameter is required if the safety factor against yielding must be 1.5?

(e) What diameter is required if requirements (c) and (d) must both be met?

Section 9.9

9.21 An unnotched member of the AISI 4340 steel of Table 9.1 is subjected to uniaxial cyclic stressing at zero mean stress. The amplitude is at first $\sigma_a = 650$ MPa for 2000 cycles, followed by $\sigma_a = 575$ MPa for 10,000 cycles. If the stress is then changed to $\sigma_a = 700$ MPa, how many cycles can be applied at this third level before fatigue failure is expected?

9.22 A uniaxial stress history with a changing mean level similar to Fig. 9.43 is repeatedly applied to an unnotched member of the Ti-6Al-4V alloy of Table 9.1. Amplitudes, means, and cycle numbers as follows apply: $\sigma_{a1} = \sigma_{a2} = 210$ MPa, $\sigma_{m1} = 730$ MPa, $\sigma_{m2} = 210$ MPa, and $N_1 = N_2 = 100$ cycles. Estimate the number of repetitions necessary to cause fatigue failure.

9.23 An unnotched member of the AISI 4142 (450 HB) steel of Table 9.1 is repeatedly subjected to the uniaxial stress history shown in Figure P9.23. Estimate the number of repetitions necessary to cause fatigue failure.

9.24 Assume that the history of Fig. 9.8 is a bending stress on the main rotor shaft of the helicopter. The shaft is made of the alloy Ti-6Al-4V of Table 9.1, and no yielding occurs even locally at a notch where fatigue cracking is a possibility. The stress in the notch is approximately uniaxial, and with the stress concentration factor of the notch included, one stress unit in Fig. 9.8 is equivalent to 150 MPa.

(a) How many rotor revolutions can be applied before fatigue failure is expected? What is the estimated life in hours of flight if each rotor revolution requires 0.3 seconds?

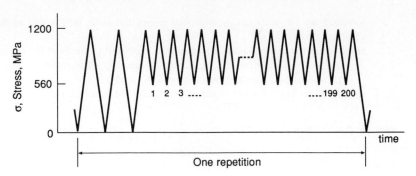

Figure P9.23

(b) A life of 1000 flight hours and a safety factor of 1.5 on stress are required. Assuming that the relative magnitudes of the peaks and valleys are unchanged, what is the highest permissible peak stress? (PC Problem).

10

Stress-Based Approach to Fatigue: Notched Members

10.1 INTRODUCTION

Geometric discontinuities that are unavoidable in design, such as holes, fillets, grooves, and keyways, cause the stress to be locally elevated and so are called *stress raisers*. Stress raisers, here generically termed *notches* for brevity, require special attention as their presence reduces the resistance of a component to fatigue failure. This is simply a consequence of the locally higher stresses causing fatigue cracks to start at such locations.

Values of the elastic stress concentration factor k_t are used to characterize notches, where $k_t = \sigma/S$ is the ratio of the local notch (point) stress σ to the nominal (average) stress S. Values of k_t are widely available in various textbooks and handbooks. A good collection can be found in Peterson (1974), and examples are given in Figs. 10.1 and 10.2.

Any particular definition of S is arbitrary, as in basing S on net area rather than gross area, and the exact choice made affects the value of k_t. Hence, it is important to be consistent with the definition of S being used when k_t values are given. In the *S-N* approach, it is conventional to define S in terms of the net area, that is, the area after

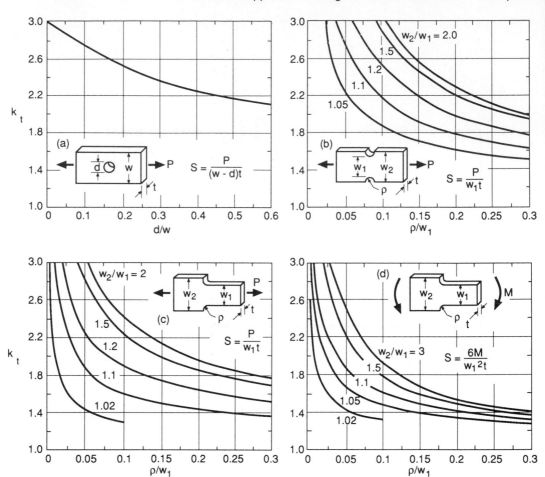

Figure 10.1 Elastic stress concentration factors for various cases of notched plates. (Values from [Peterson 74] pp. 35, 89, 98, and 150.)

the notch has been removed. Note that this convention is observed for each example in Figs. 10.1 and 10.2. Also, the equality $\sigma = k_t S$ holds at the notch only if there is no yielding. (The quantities k_t, S, and σ were defined near the beginning of the previous chapter and are illustrated in Fig. 9.3. The reader may wish to review that discussion.)

If a simplistic view is taken, we would expect unnotched (smooth) and notched members to have the same fatigue life if the stress $\sigma = S$ in the smooth member is the same as the stress $\sigma = k_t S$ at the notch in the notched member. Hence, on a plot of S versus life N_f, the effect of a notch should be to reduce the stress amplitude corresponding to any given life by the factor k_t. An example of such an estimate is the lower line in Fig. 10.3. However, it is seen that the actual test data lie above this estimate, so that the notch has less effect than expected based on k_t. The actual reduction

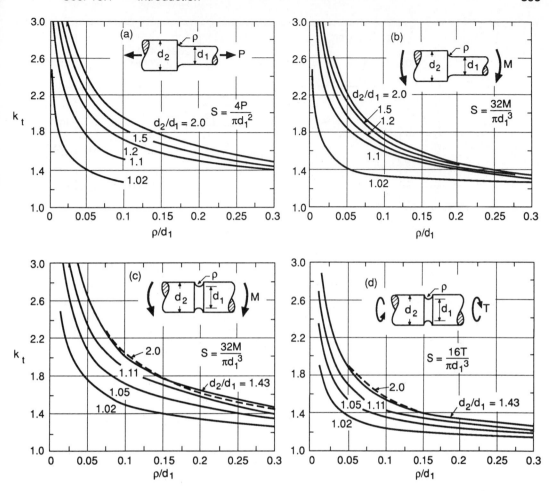

Figure 10.2 Elastic stress concentration factors for various cases of notched circular shafts. (Values from [Peterson 74] pp. 96 and 103 for (a) and (b), [Nisitani 84] for (c) and (d).)

factor at long fatigue lives, specifically at $N_f = 10^6$ to 10^7 cycles or greater, is called the *fatigue notch factor* and is denoted k_f.

$$k_f = \frac{\sigma_{ar}}{S_{ar}} \tag{10.1}$$

where k_f is formally defined only for completely reversed stresses, σ_{ar} for the smooth member and S_{ar} for the notched member.

  If the notch has a large radius ρ at its tip, k_f may be essentially equal to k_t. However, for small ρ, the above noted discrepancy may be quite large, so that k_f is considerably smaller than k_t. Some values of k_f illustrating this variation with ρ for notched bars of mild steel are shown in Fig. 10.4. As will be discussed in the following

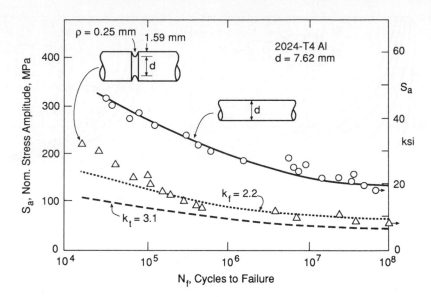

Figure 10.3 Effect of a notch on the rotating bending *S-N* behavior of an aluminum alloy, and comparisons with strength reductions using k_t and k_f. (Data from [MacGregor 52].)

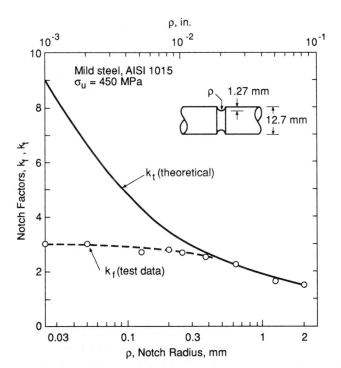

Figure 10.4 Fatigue notch factors for various notch radii based on fatigue limits from rotating bending of mild steel. (Data from [Frost 59].)

section, more than one physical cause of this behavior may exist. Due to the complexity involved, empirical estimates are generally used to obtain k_f values for use in design.

10.2 CAUSES OF THE $k_f < k_t$ EFFECT

The stress in a notched member decreases rapidly with increasing distance from the notch as illustrated in Fig. 10.5. The slope $d\sigma/dx$ of the stress distribution is called the *stress gradient*, and the magnitude of this quantity is especially large near sharp notches. It is generally agreed that the $k_f < k_t$ effect is associated with this stress gradient, and several detailed explanations have been suggested on this basis.

One argument based on stress gradients is that the material is not sensitive to the peak stress, but rather to the average stress that acts over a region of small but finite size. In other words, some finite volume of material must be involved for the fatigue damage process to proceed. The size of the active region can be characterized by a dimension δ, called the *process zone size*, as illustrated in Fig. 10.5. Thus, the stress that controls the initiation of fatigue damage is not the highest stress at $x = 0$, but rather the somewhat lower value that is the average out to a distance $x = \delta$. This average stress is then expected to be the same as the smooth specimen fatigue limit σ_e, so that k_f is estimated by

$$k_f = \frac{\left(\text{average } \sigma_y \text{ out to } x = \delta\right)}{S_a} = \frac{\sigma_e}{S_a} \tag{10.2}$$

which is less than k_t. The ratio k_f/k_t falls further below unity, that is, the discrepancy increases, if the notch radius ρ is smaller. This is because the drop in stress with increasing distance x away from the notch is more abrupt if ρ is smaller. Such a trend is consistent with observations, as in Fig. 10.4.

An effect of this type might be expected due to discrete microstructure, such as crystal grains, having the effect of equalizing the stress over a small dimension, so that the peak stress is actually lowered. However, no generally applicable correlation of trends in k_f with microstructural features has been established.

Another possible stress gradient effect involves a *weakest-link* argument based on statistics. Recall that the fatigue damage process may initiate in a crystal grain that has

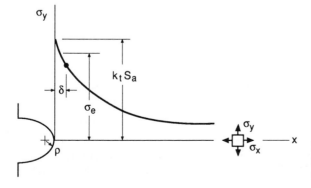

Figure 10.5 Interpretation of the fatigue limit as the average stress over a finite distance δ ahead of the notch.

an unfavorable orientation of its slip planes relative to the planes of applied shear stress, or in other cases at an inclusion, void, or other microscopic stress raiser. Many potential damage initiation sites occur within the volume of a smooth specimen. However, at a sharp notch, there is a possibility that no such damage initiation site occurs in the small region where the stress is near its peak value. Hence, on the average, the notched member will be more resistant to fatigue if the comparison is made on the basis of the local notch stress, $\sigma_a = k_t S_a$.

A third possible effect is related to the fact that a crack may start at a sharp notch but then be unable to grow to failure. Consider cyclic loading of a smooth member under a stress σ_s, and also of a notched member under a nominal stress S, such that the stress at the notch has the same value as the smooth specimen stress, $k_t S = \sigma_s$. Therefore, based on k_t, the same fatigue life is expected in smooth and notched members. Now assume that a through-thickness crack of the same length l is present in both cases as illustrated in Fig. 10.6. In each case, consider the stresses at the crack-tip location prior to the introduction of the crack. This stress is still σ_s in the smooth member, but it is lower in the notched member. The stress decrease away from the notch thus has the effect of lowering k_t. The crack in the notched member is less likely to grow, and if it does grow it will do so more slowly, than the one in the smooth member.

This argument can be further rationalized by noting that for equal smooth and notched specimen stresses, the stress intensity K of fracture mechanics for a finite crack

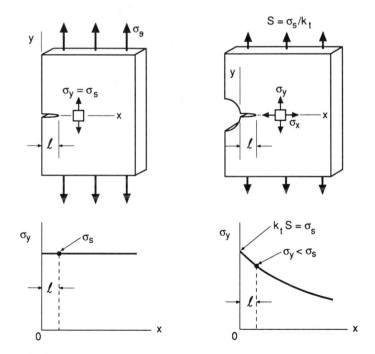

Figure 10.6 Smooth and notched members with the same local stress prior to the introduction of a crack.

length l is lower in the notched member. The following logic applies: In the smooth member, if l is relatively short

$$K_s = 1.12\ \sigma_s\sqrt{\pi l} = 1.12\ k_t S\sqrt{\pi l} \tag{10.3}$$

For the notched member, K varies qualitatively as discussed previously in connection with Fig. 8.18.

$$K = F_d S\sqrt{\pi l} \tag{10.4}$$

where F_d is defined as in the example of Fig. 8.18. For small l, we have $F_d = 1.12\ k_t$, but otherwise F_d is less than this value. Thus, equal stress intensities occur in smooth and notched members for small l, but otherwise K is lower in the notched member. The difference becomes substantial beyond the crack length l' of Eq. 8.21, which is typically only 0.1ρ to 0.2ρ, where ρ is the notch-tip radius. Since K controls the behavior of the crack, growth is less likely (or slower) in the notched member.

Finally, if there is *reversed yielding* at the notch during each loading cycle, plastic strains occur and the actual stress amplitude σ_a is less than $k_t S_a$ as illustrated in Fig. 10.7. This is simply a consequence of the fact that the calculation based on assuming elastic behavior, $\sigma_a = k_t S_a$, is not valid if yielding occurs. Such an effect occurs at high stress levels corresponding to short fatigue lives in most engineering metals, and it occurs even at long lives in a few very ductile metals. However, there is little or no yielding at long lives, say around 10^6 or 10^7 cycles, for most engineering metals, so that this explanation alone is insufficient. (Yielding effects at short lives are considered in detail later in Chapter 14 using the strain-based approach.)

Cracks that have failed to grow in long life fatigue tests, called *nonpropagating cracks*, are commonly found at sharp notches. Analysis of these indicates that the presence of cracks is probably a major cause of the $k_f < k_t$ effect. Reversed yielding

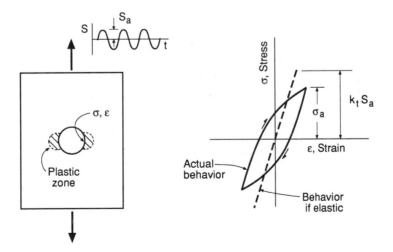

Figure 10.7   Effect of reversed yielding in a small region near the notch on the stress amplitude.

is clearly also a factor at short lives, and the process zone and weakest link effects may play a further role. In general, however, the situation is quite complex and is not completely understood. This circumstance has resulted in several empirical approaches being developed.

10.3 NOTCH SENSITIVITY AND EMPIRICAL ESTIMATES OF k_f

A useful concept in dealing with notch effects is the *notch sensitivity*, q.

$$q = \frac{k_f - 1}{k_t - 1} \tag{10.5}$$

If the notch has its maximum possible effect, so that $k_f = k_t$, then $q = 1$. The value of q decreases from unity if $k_f < k_t$, having a minimum value of $q = 0$ where $k_f = 1$, that is, where the notch has no effect. The value of q between 0 and 1 is therefore a convenient measure of how severely a given member is affected by a notch. Examples of the variation of q with material and notch radius are shown in the upper parts of Figs. 10.8 and 10.9. For a given material, q increases with notch radius, and within a given class of materials, q increases with ultimate tensile strength. Hence, the discrepancy between k_f and k_t is greatest for highly ductile materials and for sharp notches, and least for low-ductility materials and blunt notches.

Values of q and hence k_f may be estimated from empirical material constants that are independent of notch radius. Peterson (1974) employs

$$q = \frac{1}{1 + \dfrac{\alpha}{\rho}} \tag{10.6}$$

where α is a material constant having dimensions of length, with some typical values being as follows:

$\alpha = 0.51$ mm (0.02 in.)   (aluminum alloys)
$\alpha = 0.25$ mm (0.01 in.)   (annealed or normalized low-carbon steels)   (10.7)
$\alpha = 0.064$ mm (0.0025 in.)   (quenched and tempered steels)

For relatively high-strength steels, a more specific estimate than the above single value is

$$\alpha = 0.025 \left(\frac{2070 \text{ MPa}}{\sigma_u} \right)^{1.8} \text{ mm} \qquad (\sigma_u \geq 550 \text{ MPa})$$

$$\tag{10.8}$$

$$\alpha = 0.001 \left(\frac{300 \text{ ksi}}{\sigma_u} \right)^{1.8} \text{ in.} \qquad (\sigma_u \geq 80 \text{ ksi})$$

The graph in the lower part of Fig. 10.8 may also be used. The above values are for axial or bending loading, and approximate values of α for torsion are obtained by multiplying those from above by 0.6.

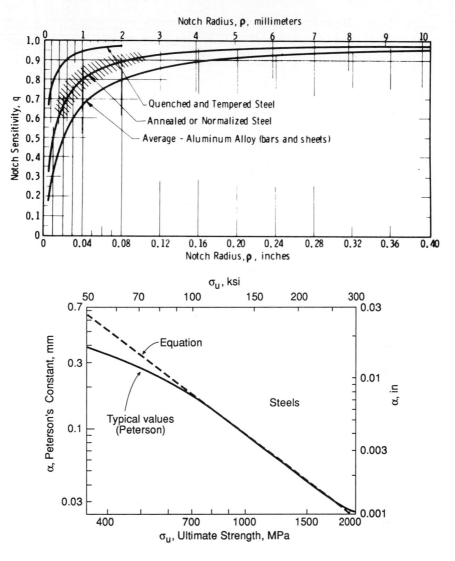

Figure 10.8 Empirical constants for k_f as employed by Peterson for axial or bending loading. Notch sensitivities q are shown above, and the length constant α for steels below, where typical values are compared with the approximation of Eq. 10.8. (Above adapted from [Peterson 74] p. 10; reprinted by permission of John Wiley & Sons, Inc.; copyright ©1974 by John Wiley & Sons, Inc. Values below from [Peterson 59a].)

Empirical curves giving either q or α can therefore be used to obtain k_f. It is convenient in making such calculations to solve Eq. 10.5 for k_f.

$$k_f = 1 + q(k_t - 1) \qquad (10.9)$$

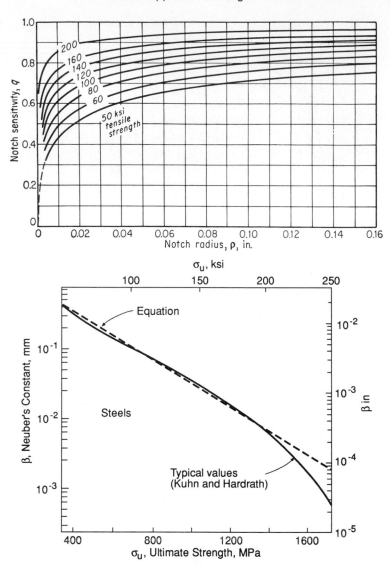

Figure 10.9 Empirical constants for k_f for steels based on the Neuber equation. Notch sensitivities q are shown above, and the length constant β for axial or bending loading below, where typical values are compared with the approximation of Eq. 10.12. (Above adapted from [Peterson 59b]; used with permission. Values below from [Kuhn 52].)

Combining this with Eq. 10.6 gives k_f directly from α.

$$k_f = 1 + \frac{k_t - 1}{1 + \dfrac{\alpha}{\rho}} \qquad (10.10)$$

Another frequently used empirical formulation for q and the resulting equation for k_f are

$$q = \frac{1}{1 + \sqrt{\dfrac{\beta}{\rho}}}, \qquad k_f = 1 + \frac{k_t - 1}{1 + \sqrt{\dfrac{\beta}{\rho}}} \tag{10.11}$$

where β is a different material constant. These particular expressions represent a simplification of an equation developed by H. Neuber. Typical values of β for steels are shown in the lower part of Fig. 10.9. The following empirical equation may also be used for steels over the range indicated:

$$\log \beta = -\frac{\sigma_u - 134 \text{ MPa}}{586} \text{ (mm)} \qquad (\sigma_u \leq 1520 \text{ MPa})$$

$$\log \beta = -\frac{\sigma_u + 100 \text{ ksi}}{85} \text{ (in.)} \qquad (\sigma_u \leq 220 \text{ ksi})$$

$$\tag{10.12}$$

Equations 10.10 and 10.11, and a number of other analogous equations in the literature, are based on the process zone, weakest link, and similar hypothesis as reviewed by Peterson (1959). However, no single physical model or combination of models fully satisfies all of the complexities involved, so that the equations for estimating k_f should be regarded only as estimates that are useful for engineering purposes. All k_f equations are empirical in nature in that they involve a material constant that is adjusted for agreement with test data. Further, the equations for estimating k_f are intended for relatively mild notches such as those intentionally used in mechanical design. If the notch is relatively deep and sharp, so that it is crack-like in form, it will generally be more accurate to simply assume that the notch is already a crack. Its behavior can then be predicted using fracture mechanics.

Example 10.1

For the rotating bending situation of Fig. 10.4, estimate k_t and then k_f for a notch radius of $\rho = 0.4$ mm.

Solution   The geometry and loading correspond to Fig. 10.2(c). Values of the ratios d_2/d_1 and ρ/d_1 are determined and used to obtain k_t from this graph.

$$d_2 = 12.7 \text{ mm}, \qquad \rho = 0.4 \text{ mm}, \qquad d_1 = 12.7 - 2(1.27) = 10.16 \text{ mm}$$

$$\frac{d_2}{d_1} = \frac{12.7}{10.16} = 1.25, \qquad \frac{\rho}{d_1} = \frac{0.4}{10.16} = 0.0394$$

$$k_t = 2.7 \qquad\qquad\qquad \textbf{Ans.}$$

where a judgmental interpolation between the curves for $d_2/d_1 = 1.11$ and 1.43 is needed.

A value of k_f could be obtained from the Peterson constant α, the Neuber constant β, or from the notch sensitivity q. The value of $\sigma_u = 450$ MPa given in Fig. 10.4 for this material allows $\alpha \approx 0.3$ mm to be obtained from the lower part of Fig. 10.8, which with Eq. 10.10 gives k_f.

$$k_f = 1 + \frac{k_t - 1}{1 + \dfrac{\alpha}{\rho}} = 1 + \frac{2.7 - 1}{1 + \dfrac{0.3}{0.4}} = 1.97 \qquad\qquad \textbf{Ans.}$$

To illustrate the use of q, the top part of Fig. 10.9 is entered with $\rho = 0.4$ mm (0.0157 in.) and $\sigma_u = 450$ MPa (65 ksi). Interpolating between the curves for $\sigma_u = 60$ and 80 ksi gives $q \approx 0.56$, and applying Eq. 10.9 then gives k_f.

$$k_f = 1 + q(k_t - 1) = 1 + 0.56(2.7 - 1) = 1.95 \qquad \textbf{Ans.}$$

Comment   Such close agreement between the two differently calculated k_f values is ideally expected, but often does not occur. Note that these values agree only roughly with the experimental value of $k_f \approx 2.5$ from Fig. 10.4.

10.4 NOTCH EFFECTS AT INTERMEDIATE AND SHORT LIVES

At intermediate and short fatigue lives in ductile materials, the reversed yielding effect of Fig. 10.7 becomes increasingly important as higher stresses, and therefore shorter lives, are considered. One consequence of this behavior is that the ratio of the smooth specimen to notched specimen fatigue strengths becomes even less than k_f, so that it is useful to define a fatigue notch factor k'_f that varies with life.

$$k'_f = f(N_f) = \frac{\sigma_{ar}}{S_{ar}} \tag{10.13}$$

Data illustrating this effect are shown in Fig. 10.10. As is typical for ductile metals, k'_f decreases from k_f at long lives to a value near unity at short lives.

Considering completely reversed loading, these trends can be rationalized by idealizing the behavior of the material as elastic, perfectly plastic with yield strength σ_o as illustrated in Fig. 10.11. There are three possible situations: (a) no yielding, (b) local yielding, and (c) full yielding. If the stress at the notch never exceeds the yield strength, which is satisfied if $k_t S_a \le \sigma_o$, there is *no yielding*. In this case, ignoring effects other than yielding for the present, k'_f is expected to be equal to k_t.

$$k'_f = k_t \qquad \text{(no yielding; } k_t S_a \le \sigma_o) \tag{10.14}$$

Assume that the loading is sufficiently severe to cause yielding at least locally at the notch, that is, assume that $k_t S_a > \sigma_o$. Completely reversed loading will then cause yielding to occur in both tension and compression on each loading cycle as illustrated in Fig. 10.11(b). If this reversed yielding does not spread over the entire cross section, the situation is described as *local yielding*. Since the stress amplitude at the notch is equal to σ_o, the value of k'_f is expected to be

$$k'_f = \frac{\sigma_o}{S_a} \qquad \text{(local yielding; } k_t S_a > \sigma_o) \tag{10.15}$$

In the extreme case of *full yielding*, the reversed yielding spreads across the entire cross section as in Fig. 10.11(c). This causes the stress amplitude over the net section to be uniform and equal to σ_o, so that the nominal stress amplitude is also equal to σ_o. Hence, the notch has little effect, and

$$k'_f \approx 1 \qquad \text{(full yielding; } S_a \approx \sigma_o) \tag{10.16}$$

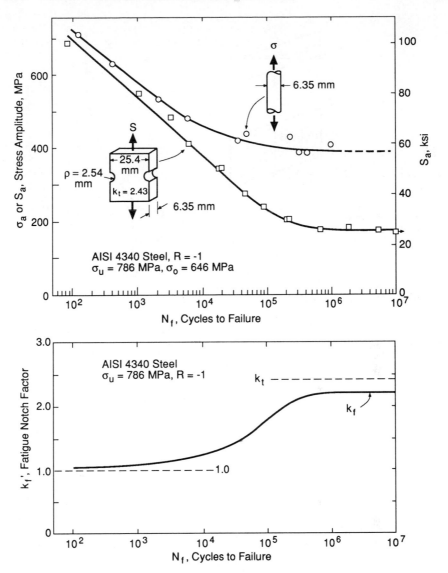

Figure 10.10   Test data for a ductile metal illustrating variation of the fatigue notch factor with life. The *S-N* data above are used to obtain $k'_f = \sigma_a / S_a$ below. The notches are half-circular cutouts.

Such behavior is similar to the spreading of yielding across the cross section in static loading as in Fig. 7.26, except that now reversed yielding in both tension and compression on each cycle of loading is involved.

The expected trends in k'_f for the three situations are summarized by the preceding equations as shown in Fig. 10.11(d). If a smooth specimen *S-N* curve as

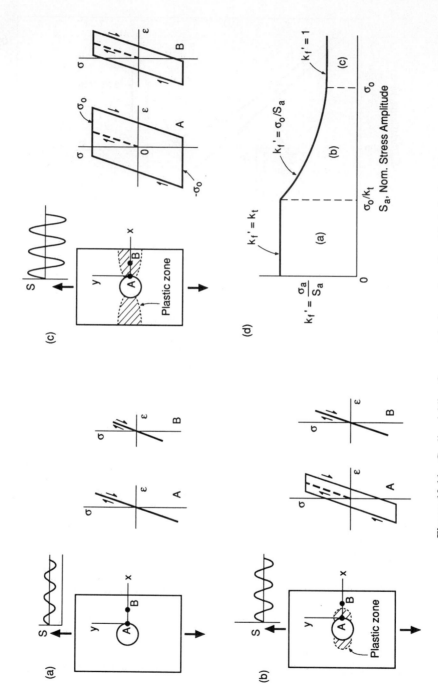

Figure 10.11 Cyclic yielding for a notched member of an ideal elastic, perfectly plastic material. There are three possibilities: (a) no yielding, (b) local yielding, and (c) full yielding. The fatigue notch factor is thus expected to vary with the stress level as in (d).

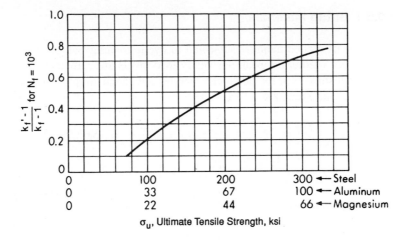

Figure 10.12   Curve based on empirical data for estimating the fatigue notch factor k'_f at $N_f = 1000$ cycles. (Adapted from [Juvinall 67] p. 260; used with permission; copyright ©1967 by McGraw-Hill Publishing Co.)

in Fig. 10.10 is adjusted using k'_f values from Eqs. 10.14 to 10.16, the result will be qualitatively similar to the notched member curve. However, the difference between k_f and k_t at long lives indicates that at least one effect in addition to yielding is acting, and some additional adjustment would be needed to obtain close agreement.

In applying the stress-based approach, *S-N* curves for notched members may need to be estimated. An empirical approach rather than analysis as just described is generally used to make adjustments for yielding effects at short and intermediate lives. For example, once k_f for long life has been estimated, k'_f at $N_f = 10^3$ cycles can be estimated using curves based on test data as in Fig. 10.12. A graphical interpolation is then used to obtain k'_f for lives between $N_f = 10^3$ and 10^6. Study of the particular curves of Fig. 10.12 indicates that k'_f at $N_f = 10^3$ approaches k_f for high-strength (usually brittle, low ductility) metals, and k'_f approaches unity for low-strength (usually ductile) metals. This is precisely what is expected based on a reversed yielding effect as just discussed. Empirical estimates of entire *S-N* curves for notched members are discussed in some detail in a later section of this chapter.

10.5 COMBINED EFFECTS OF NOTCHES AND MEAN STRESS

The empirical expressions and curves for k_f and k'_f are based on trends and data observed under completely reversed loading. Hence, these values cannot be applied directly if mean stresses are present. We will now consider the additional procedures necessary to handle notches under mean stress.

10.5.1 Brittle Materials

Consider a notched member subjected to cyclic loading with a nonzero mean level, and assume that neither the peak nor valley load levels are sufficient to cause yielding at the notch. All stresses at the notch can then be calculated simply from $k_t S$, so that the stress amplitude and mean are

$$\sigma_a = k_t S_a, \quad \sigma_m = k_t S_m \quad (k_t |S|_{\max} < \sigma_o) \tag{10.17}$$

where the notation $|S|_{\max}$ denotes the maximum absolute value of S, so that the limitation $k_t |S|_{\max} < \sigma_o$ requires both $k_t S_{\max} < \sigma_o$ and $k_t S_{\min} > \sigma_o$, forbidding yielding in either tension or compression.

The effect of this mean stress on life can be estimated by applying the modified Goodman expression, Eq. 9.13, or other analogous equation, to the stresses σ at the notch.

$$\sigma_{ar} = \frac{\sigma_a}{1 - \dfrac{\sigma_m}{\sigma_u}} \tag{10.18}$$

where σ_{ar} is the equivalent completely reversed stress amplitude for smooth specimens of the same material. Combining the above equations gives

$$\sigma_{ar} = \frac{k_t S_a}{1 - \dfrac{k_t S_m}{\sigma_u}} \quad (k_t |S|_{\max} < \sigma_o) \tag{10.19}$$

However, in view of the notch sensitivity effect discussed above, this equation is usually applied in modified form with k_t replaced by k_f.

General application of Eq. 10.19, or a similar form using k_f, is a reasonable approach for notched members made of brittle (low-ductility) materials, for which the yielding limitation is not easily violated. Also, for finite lives in brittle materials, it was noted previously that $k_f' \approx k_f$, making it reasonable to apply such an equation to brittle materials even at finite lives. Equation 10.19 used with k_f is plotted in Fig. 10.13 as the line marked *brittle*. Compared to the modified Goodman line for smooth members, it intercepts both axes at stresses that are smaller due to the values being divided by k_f. The intercept with the stress amplitude axis occurs at σ_{ar}/k_f, which agrees with the definition of k_f in Eq. 10.1. The intercept with the mean stress axis corresponds to a static load and is σ_u/k_f. Noting that for brittle materials k_f and k_t are similar values, this is consistent with the fact that static strengths of notched members of such materials are reduced by the stress concentration factor. (See Section 7.9.2 and Fig. 7.26, and note the exception mentioned for small notches in severely internally flawed materials.)

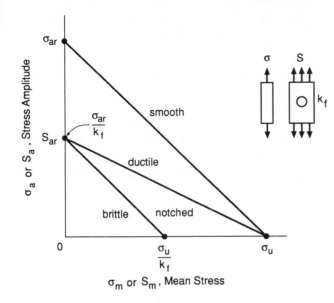

Figure 10.13 Approximate amplitude-mean plots for smooth and notched members of brittle and ductile materials.

10.5.2 Ductile Materials

A quite different situation occurs for ductile materials that are capable of accommodating considerable yielding. In this case, the ultimate strength of the notched member is not severely affected by the notch. This is because yielding can spread into the material and eventually across the entire cross section as illustrated in Fig. 7.26. For simple axial loading, the ultimate strength of the notched member is therefore roughly the same as for a smooth member, $S_u = \sigma_u$. This in turn implies that the intercept of the amplitude-mean line with the mean stress axis should not be reduced by k_f as was done for brittle materials, but rather should be the same as for an unnotched member. This gives the line marked *ductile* in Fig. 10.13. The equation is

$$\sigma_{ar} = \frac{k'_f S_a}{1 - \dfrac{S_m}{\sigma_u}} \tag{10.20}$$

where the more general variable k'_f of Eq. 10.13 is substituted to permit application at finite life. Aside from this change, the equation differs from Eq. 10.19 only in that S_m is not multiplied by a notch factor. Hence, Eq. 10.20 can be interpreted as stating that for ductile materials there is no stress concentration effect on the mean stress.

Equation 10.19, used with k_f, and Eq. 10.20 can be written as one by defining a fatigue notch factor for the mean stress, k_{fm}.

$$\sigma_{ar} = \frac{k'_f S_a}{1 - \dfrac{k_{fm} S_m}{\sigma_u}}, \qquad S_{ar} = \frac{S_a}{1 - \dfrac{k_{fm} S_m}{\sigma_u}} \qquad \text{(a,b)} \tag{10.21}$$

where

$$k_{fm} = k'_f = k_f \quad \text{(brittle materials)}$$
$$k_{fm} = 1 \quad \text{(ductile materials)}$$

The first form of the equation may be used to compute σ_{ar}, the equivalent completely reversed stress amplitude for use with a smooth member S-N curve for which $\sigma_m = 0$. Alternatively, invoking $S_{ar} = \sigma_{ar}/k'_f$ gives the second form, where S_{ar} is the equivalent completely reversed stress amplitude for the notched member, appropriate for use with an S-N curve for the notched member for which $S_m = 0$. The latter use of the equation for ductile materials with $k_{fm} = 1$ corresponds simply to applying the modified Goodman equation (Eq. 9.13) to the nominal stress. Brittle materials can be considered to be those with less than 5% elongation in a tension test. However, this definition is obviously arbitrary, and borderline cases do exist where classification is difficult.

10.5.3 Discussion

If there is no yielding at a notch, the local mean stress can be estimated directly from k_t. However, in a ductile material, yielding may occur at high mean stresses even at long lives, where there is little or no yielding for completely reversed loading. This causes the mean stress at the notch to be less than would be estimated from k_t as illustrated in Fig. 10.14. The concentration factor for the mean stress can be considered to be a variable.

$$k_{fm} = \frac{\sigma_m}{S_m} \tag{10.22}$$

where S_m is the mean level for the nominal stress and σ_m is the mean level for the local stress at the notch.

Let the behavior of the material be approximated as being elastic, perfectly plastic, and assume that a mean stress is present. There are three possible situations as illustrated in Fig. 10.14: (a) no yielding, (b) initial yielding, and (c) reversed yielding. There is *no yielding* if neither the peak nor the valley of the load causes the stress $k_t S$ at the notch to exceed the yield strength. In this case, ignoring effects other than yielding for the present, we have the same situation as for a brittle material, specifically Eq. 10.19, so that $k_{fm} = k_t$ is expected.

Assume that the loading is such that either the peak or valley load causes yielding, that is, $k_t|S|_{\max} > \sigma_o$, but also that it is not sufficiently severe for reversed yielding to occur. We then have the situation shown in Fig. 10.14(b) where there is only *initial yielding*. Since the cyclic stressing is elastic, the notch stress amplitude may be calculated from $\sigma_a = k_t S_a$. On each stress cycle, the maximum stress will return to the same yield strength value σ_o that it had at the first load peak. Hence, for initial yielding in tension, the mean stress is

$$\sigma_m = \sigma_{\max} - \sigma_a = \sigma_o - k_t S_a \tag{10.23}$$

Finally, if the loading is sufficiently severe that $k_t \, \Delta S > 2\sigma_o$, *reversed yielding* occurs as shown in Fig. 10.14(c). The stresses at the notch are

$$\sigma_{\max} = \sigma_o, \quad \sigma_{\min} = -\sigma_o \tag{10.24}$$

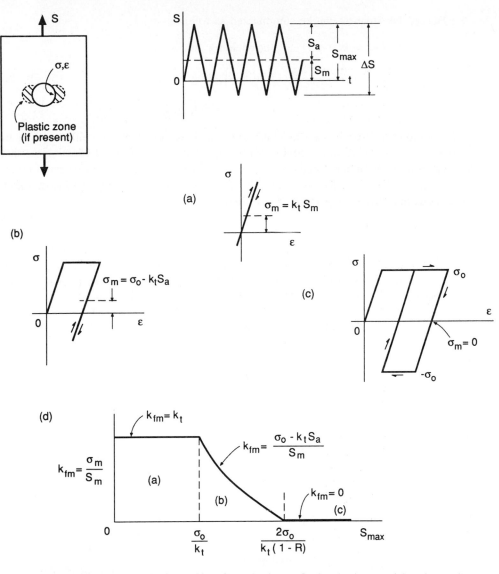

Figure 10.14   A notched member of an elastic, perfectly plastic material under cyclic loading with nonzero mean level. There are three possible stress-strain behaviors at the notch: (a) no yielding, (b) initial yielding but elastic cycling, and (c) reversed yielding. The concentration factor k_{fm} for mean stress is thus expected to vary with S_{max} as shown in (d).

so that

$$\sigma_m = \frac{\sigma_{max} + \sigma_{min}}{2} = 0 \tag{10.25}$$

which gives $k_{fm} = 0$.

Hence, the situation can be summarized as follows:

$$k_{fm} = k_t \qquad \text{(no yielding; } k_t |S|_{\max} < \sigma_o\text{)}$$

$$k_{fm} = \frac{\sigma_o - k_t S_a}{|S_m|} \qquad \text{(initial yielding; } k_t |S|_{\max} > \sigma_o\text{)} \qquad (10.26)$$

$$k_{fm} = 0 \qquad \text{(reversed yielding; } k_t \Delta S > 2\sigma_o\text{)}$$

where absolute values are used to make the equations applicable for either tensile or compressive S_m. The variation of k_{fm} with $S_{\max}$ is similar to Fig. 10.14(d). Note that local yielding at the notch causes k_{fm} to be less than k_t, even zero in extreme cases. The $k_{fm} = 0$ behavior occurs beyond the $S_{\max}$ value indicated in Fig. 10.14(d), which corresponds to $k_t \Delta S > 2\sigma_o$, as can be verified from Eq. 9.4.

In view of this analysis, Eq. 10.21 could perhaps be improved for ductile materials by making k_{fm} a continuous variable according to Eq. 10.26. The value $k_{fm} = 1$ is indeed within the range of zero to k_t given by these equations, but this or any other single value is seen to represent only a crude approximation. This question is discussed in Juvinall (1967) and Juvinall and Marshek (1991), and one of the alternatives suggested there is similar to the use of Eq. 10.26, except that k_t is replaced with k_f.

$$k_{fm} = k_f \qquad \text{(no yielding; } k_f |S|_{\max} < \sigma_o\text{)}$$

$$k_{fm} = \frac{\sigma_o - k_f S_a}{|S_m|} \qquad \text{(initial yielding; } k_f |S|_{\max} > \sigma_o\text{)} \qquad (10.27)$$

$$k_{fm} = 0 \qquad \text{(reversed yielding; } k_f \Delta S > 2\sigma_o\text{)}$$

The simpler approach of using $k_{fm} = 1$ for ductile materials is nevertheless widely employed.

Example 10.2

The RQC-100 steel of Table 9.1 is to be used in the form of a plate with a width change under bending as in Fig. 10.1(d). The dimensions are $w_2 = 88$, $w_1 = 80$, $\rho = 4$, and $t = 10$ mm. What amplitude of bending moment M_a will result in a life of 10^6 cycles if cycling is applied at mean moment of $M_m = 4$ kN·m?

First Solution   One approach is to use Eq. 10.21 with $k_{fm} = 1$ for this ductile material. Since 10^6 cycles is a long life, we have $k'_f = k_f$, so that

$$\sigma_{ar} = \frac{k_f S_a}{1 - \dfrac{S_m}{\sigma_u}}$$

The quantities k_f, S_m, and σ_{ar} need to be evaluated, and then we can solve for S_a, which gives M_a.

To estimate k_f, first determine k_t from Fig. 10.1(d).

$$\frac{w_2}{w_1} = \frac{88 \text{ mm}}{80 \text{ mm}} = 1.1, \qquad \frac{\rho}{w_1} = \frac{4 \text{ mm}}{80 \text{ mm}} = 0.05, \qquad k_t = 1.85$$

The Peterson constant α is given by Eq. 10.8 used with $\sigma_u = 758$ MPa from Table 9.1.

$$\alpha = 0.025 \left(\frac{2070 \text{ MPa}}{\sigma_u} \right)^{1.8} = 0.025 \left(\frac{2070}{758} \right)^{1.8} = 0.153 \text{ mm}$$

Equation 10.10 then gives k_f .

$$k_f = 1 + \frac{k_t - 1}{1 + \dfrac{\alpha}{\rho}} = 1 + \frac{1.85 - 1}{1 + \dfrac{0.153 \text{ mm}}{4 \text{ mm}}} = 1.82$$

We next calculate S_m from the definition of S in Fig. 10.1(d).

$$S_m = \frac{6M_m}{w_1^2 t} = \frac{6 (0.004 \text{ MN·m})}{(0.08 \text{ m})^2 (0.01 \text{ m})} = 375 \text{ MPa}$$

Constants A and B from Table 9.1 give the completely reversed stress amplitude σ_{ar} at $N_f = 10^6$ for smooth specimens of this material.

$$\sigma_{ar} = AN_f^B = 897(10^6)^{-0.0648} = 366 \text{ MPa}$$

We can now solve the first equation above for S_a and substitute the values determined, and finally use the definition of S to get M_a.

$$S_a = \frac{\sigma_{ar}}{k_f} \left(1 - \frac{S_m}{\sigma_u} \right) = \frac{366 \text{ MPa}}{1.82} \left(1 - \frac{375 \text{ MPa}}{758 \text{ MPa}} \right) = 102 \text{ MPa}$$

$$M_a = \frac{w_1^2 t S_a}{6} = \frac{(0.08 \text{ m})^2 (0.01 \text{ m}) (102 \text{ MPa})}{6} = 0.00109 \text{ MN·m}$$

$$M_a = 1.09 \text{ kN·m} \qquad \qquad \textbf{Ans.}$$

Second Solution   Another alternative is to more specifically account for possible local yielding using k_{fm} from Eq. 10.27. First, assume that there is no yielding, hence $k_{fm} = k_f$, so that Eq. 10.21 becomes

$$\sigma_{ar} = \frac{k_f S_a}{1 - \dfrac{k_f S_m}{\sigma_u}}$$

Solving for S_a and substituting quantities obtained above gives

$$S_a = \frac{\sigma_{ar}}{k_f} \left(1 - \frac{k_f S_m}{\sigma_u} \right) = 20.0 \text{ MPa}$$

However, this produces a local notch stress (based on k_f) above the yield strength of $\sigma_o = 683$ MPa.

$$k_f S_{\max} = k_f (S_a + S_m) = 1.82 (20.0 + 375) = 719 \text{ MPa}$$

$$k_f S_{\max} > \sigma_o \quad \text{(yielding occurs)}$$

Hence, the calculation is invalid, and Eq. 10.27 is tried assuming initial yielding.

$$k_{fm} = \frac{\sigma_o - k_f S_a}{S_m}$$

which is substituted into Eq. 10.21 to obtain

$$\sigma_{ar} = \frac{k_f S_a}{1 - \dfrac{\sigma_o - k_f S_a}{\sigma_u}}$$

Solving for S_a and substituting the known values gives

$$S_a = \frac{\sigma_{ar}(\sigma_u - \sigma_o)}{k_f(\sigma_u - \sigma_{ar})} = 38.5 \text{ MPa}$$

This calculation is valid since no reversed yielding occurs.

$$k_f \Delta S = 2k_f S_a = 2(1.82)(38.5 \text{ MPa}) = 140 \text{ MPa}$$

$$k_f \Delta S < 2\sigma_o = 2(683 \text{ MPa}) \qquad \text{(no reversed yielding)}$$

Hence the result is

$$M_a = \frac{w_1^2 t S_a}{6} = \frac{(0.08 \text{ m})^2(0.01 \text{ m})(38.5 \text{ MPa})}{6} = 0.41 \text{ kN·m} \qquad \textbf{Ans.}$$

Comment   The two solutions give answers for M_a that differ considerably due to the different handling of mean stress in this problem where the mean stress level is relatively high.

10.6 ESTIMATING LONG LIFE FATIGUE STRENGTHS (FATIGUE LIMITS)

In applications involving relatively low stresses applied for large numbers of cycles, design against fatigue may require only that the fatigue strength at long lives on the order of 10^6 to 10^8 cycles be known. Such values are often available in the literature from rotating bending fatigue tests, and they may be adjusted for nonzero mean stress by the methods already described. For situations that differ from standard test conditions as to type loading, size, surface finish, etc., values may often be estimated based on trends observed in existing data.

10.6.1 Estimation of Smooth Specimen Fatigue Limits

Distinct fatigue limits, where the S-N curve appears to become horizontal at long lives, are observed for many low-strength carbon and alloy steels, for some stainless steels, irons, molybdenum alloys, titanium alloys, and polymers, but not for many other materials. For aluminum, magnesium, copper, and nickel alloys, and for some stainless steels and high-strength carbon and alloy steels, *S-N* curves generally continue to decrease slowly at the longest lives that have been studied. Fatigue strengths at long lives from rotating bending tests on smoothly polished specimens are commonly tabulated as materials properties. These fatigue strengths are often loosely termed *fatigue limits*, even if there is no distinct horizontal region on the *S-N* curve.

It is convenient to consider the ratio of the fatigue limit to the ultimate tensile strength.

$$m_e = \frac{\sigma_{erb}}{\sigma_u} \tag{10.28}$$

where σ_{erb} is the polished specimen fatigue limit for completely reversed loading in bending, often rotating bending. A value around $m_e = 0.5$ is common for low and intermediate strength steels as shown in Fig. 10.15. A similar plot for fatigue strengths of wrought aluminum alloys at $N_f = 5 \times 10^8$ cycles is given in Fig. 10.16. A value of $m_e = 0.4$ applies for the lower strength levels. For both steels and aluminums, m_e decreases beyond a certain ultimate tensile strength level, that is, the fatigue limit fails to keep up with the increased static strength. The fatigue limit appears to level off around 700 MPa for many steels and around 130 MPa for wrought aluminum alloys. This trend is associated with the fact that a degree of ductility is helpful in providing fatigue resistance, and high-strength alloys generally have limited ductility.

Some other typical m_e values are 0.4 for cast irons at $N_f = 10^7$ cycles, 0.35 for wrought magnesium alloys at $N_f = 10^8$ cycles, and 0.5 for titanium alloys at $N_f = 10^7$

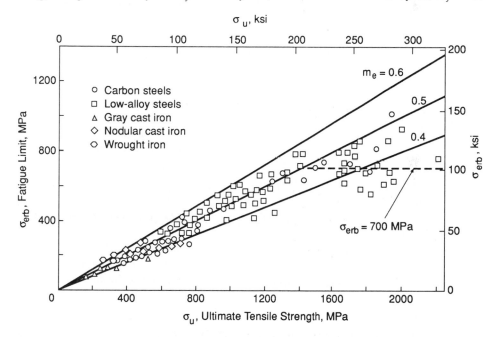

Figure 10.15 Rotating bending fatigue limits from polished specimens of various ferrous metals. (From data compiled by [Forrest 62].)

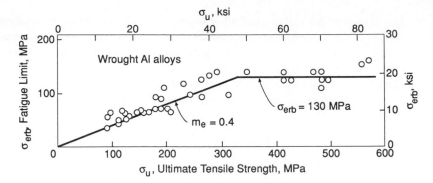

Figure 10.16 Fatigue strengths in rotating bending at 5×10^8 cycles for various tempers of common wrought aluminum alloys, including 1100, 2014, 2024, 3003, 5052, 6061, 6063, and 7075 alloys. (Adapted from [Juvinall 67] p. 215; used with permission; copyright ©1967 by McGraw-Hill Publishing Co.)

cycles. Forrest (1962) and Heywood (1962) in the Chapter 9 References contain data compilations for the metals in common use at that time, and additional data are given in the materials property sources also listed in the Chapter 9 References. Systematic collections of long-life fatigue strength data are not generally available for the less commonly used metals or for polymers and composites, but data on particular materials of these types can sometimes be found in the literature.

10.6.2 Factors Affecting Long Life Fatigue Strength

If a notch is present, the fatigue strength at long lives is reduced by the factor k_f as discussed in detail earlier in this chapter. A variety of additional factors may also affect the long life fatigue strength, often reducing it. For example, axial loading produces a lower fatigue strength than bending, typically by 10% or more. This is thought to be because the process zone, weakest link, or related effects due to a stress gradient act to a limited extent in bending as described earlier for notches. Such effects are beneficial and can occur in bending (or torsion) due to the stress variation with depth in the material, but not for axial loading, where the stress is uniform. Another factor is that a slight but unknown eccentricity of the axial load may cause some bending that is not included in the stress calculation, $S = P/A$.

The state of stress also has an effect on fatigue strength as described in Chapter 9, where an octahedral shear stress criterion is suggested for ductile materials. For example, for pure torsion, such an approach gives an estimate of the fatigue limit in shear from Eq. 9.22.

$$\tau_{er} = \frac{\sigma_{erb}}{\sqrt{3}} = 0.577\sigma_{erb} \tag{10.29}$$

For large size members in bending or torsion, stress gradient effects would be expected to cause fatigue strengths to decrease with member size, and such a *size effect* is in fact observed. A size effect is expected specifically because the decrease of stress with depth is less abrupt in larger cross sections, so that a larger volume of material is subjected to relatively high stress. Some test data for steel shafts are shown in Fig. 10.17. As a result of this effect, fatigue limits from small (typically 8 mm diameter) rotating bending test specimens need to be decreased for application to larger sizes.

If the *surface finish* is rougher than the polished surface of a typical smooth test specimen, the long life fatigue strength is likely to be reduced. Careful grinding reduces the fatigue limit around 10%, and more ordinary machining by 20% or more. Relatively rough surfaces that are unmodified after forging or casting may cause the fatigue limit to be less than half of the smooth specimen value. Some typical reduction factors for various surface conditions for steel are given in Fig. 10.18. Note that the reductions are greater for increased ultimate tensile strength. This occurs because surface roughness acts as a stress raiser (notch), and as previously discussed, higher strength materials are relatively more sensitive to notches. Surface finish effects are complicated by other factors that may accompany them, such as residual stresses from machining or heat treating, and also by surface compositional or microstructural changes that may occur during some types of processing, such as hot-rolling or forging.

10.6.3 Reduction Factors for the Fatigue Limit

Combinations of effects such as those just described are common in engineering situations. What is usually done is to multiply reduction factors for the various effects to

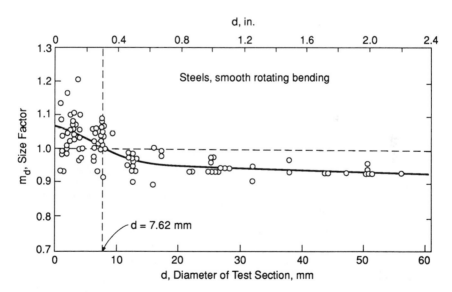

Figure 10.17   Effect of size on the fatigue limit of smoothly polished specimens of steels tested in rotating bending. Values are plotted of m_d, the ratio of the fatigue limit to that for the frequently used 7.62 mm (0.3 in.) specimen diameter. (Data from [Heywood 62] p. 23.)

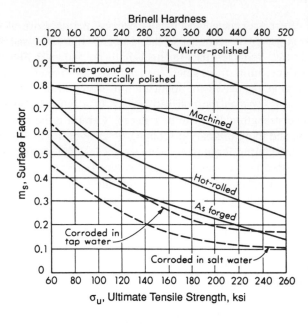

Figure 10.18 Effect of various surface finishes on the fatigue limit of steel. Values are plotted of m_s, the ratio of the fatigue limit to that for polished specimens. (Adapted from [Juvinall 67] p. 234; used with permission; copyright ©1967 by McGraw-Hill Publishing Co.)

obtain an adjusted fatigue limit σ_{er}, which is lower than σ_{erb}.

$$\sigma_{er} = m_t m_d m_s m_o \sigma_{erb} = m\sigma_u \tag{10.30}$$

where m is a combined reduction factor that includes m_e from Eq. 10.28.

$$m = m_e m_t m_d m_s m_o \tag{10.31}$$

The various factors account for the effects of type of loading (m_t), size (m_d), surface finish (m_s), and any other effects (m_o) that may be involved, such as elevated temperature, corrosion, etc. Any one of these factors obviously has no effect if the value is unity, and a value of 0.9 corresponds to a 10% reduction, etc. Some examples are $m_t = 0.58$ for torsion from Eq. 10.29, $m_d = 0.95$ for diameters around 25 mm from Fig. 10.17, and $m_s = 0.8$ for a machined surface in low-strength steel from Fig. 10.18.

If the starting point for the estimate is the ultimate tensile strength, then the fatigue limit for a notched member, S_{er}, is obtained by applying Eq. 10.30 along with k_f.

$$S_{er} = \frac{\sigma_{er}}{k_f} = \frac{m\sigma_u}{k_f} \tag{10.32}$$

10.7 ESTIMATING *S-N* CURVES

Estimates of fatigue limits can be used as part of a procedure for estimating entire *S-N* curves, with most mechanical engineering design books including such a procedure. Three typical and widely used methods are those described in Collins (1981), Juvinall and Marshek (1991), and Shigley and Mischke (1989). These methods involve using an estimated fatigue limit and one or two additional points at shorter lives as illustrated in Fig. 10.19. Straight line segments are then used on either log-linear or log-log coordinates to make estimates for other lives.

Actual data approximating the situation of interest are always preferable to an estimate. Thus, such data should be used where possible to aid in estimating the *S-N* curve, or even to replace the estimate entirely. Also, where data are not found in the literature, it will sometimes be appropriate to expend the time and funds necessary to obtain them. Estimated *S-N* curves are nevertheless useful for preliminary calculations, and they may be all that is needed if a rough evaluation of the fatigue resistance is

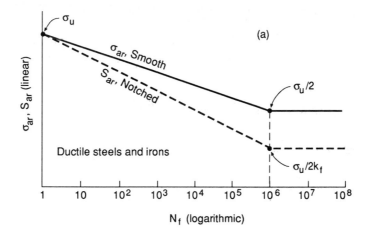

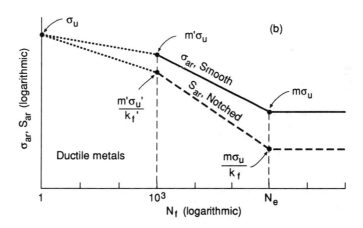

Figure 10.19 Estimating completely reversed *S-N* curves for smooth and notched members according to procedures suggested by (a) Collins and (b) Juvinall or Shigley.

sufficient. Generic S-N curves for particular classes of materials are also often available and may be useful. An example is given as Fig. 10.20.

10.7.1 Estimates by the Method of Collins

Collins (1981) suggests the following procedure for estimating the completely reversed S-N curve for ductile steels and irons with σ_u below about 1200 MPa: Draw a straight line on log-linear coordinates between the ultimate tensile strength at 1 cycle and half of this value at one million cycles. If a notch is present, use the same procedure except divide the stress value at $N_f = 10^6$ cycles by k_f. Thus, Eq. 10.32 with $m = 0.5$ is applied at $N_f = 10^6$ cycles as illustrated in Fig. 10.19(a). Since no change is made at $N_f = 1$, the value $k_f' = 1$ is in effect being used there, which is consistent with the effect of reversed yielding as discussed above. This procedure can be viewed as a means of estimating values of k_f' that vary between 1 and k_f for any desired life between 1 and 10^6 cycles.

The straight S-N line on log-linear coordinates has the form of Eq. 9.5.

$$S_{ar} = C + D \log N_f \tag{10.33}$$

The constants C and D can be determined from the two known points, which leads to

$$S_{ar} = \sigma_u \left(1 - \frac{2k_f - 1}{12k_f} \log N_f \right) \quad \left(N_f \le 10^6 \right)$$
$$S_{er} = \frac{\sigma_u}{2k_f} \qquad\qquad\qquad \left(N_f > 10^6 \right) \tag{10.34}$$

where a constant fatigue strength (fatigue limit) is used beyond 10^6 cycles. This equation includes unnotched members as a special case by substitution of $k_f = 1$. Nonzero mean stresses are handled by simply applying Eq. 10.21 with $k_{fm} = 1$ for ductile materials.

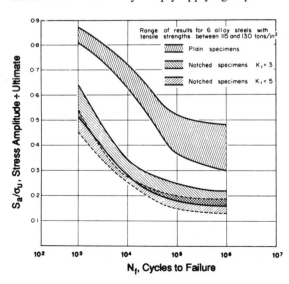

Figure 10.20 Generic S-N curves from completely reversed axial loading of various low-alloy quenched and tempered steels with σ_u in the range 1700 to 2100 MPa. The notch radii were $\rho = 0.23$ mm for $k_t = 3$, and 0.08 mm for $k_t = 5$. (Adapted from [Forrest 62] p. 160, which summarizes data in [Muvdi 57].)

The approach of Collins can be easily extended to other materials, and additional modification factors can be included. In particular, a value of m other than 0.5 could be used, and this could be applied at a long-life point $N_f = N_e$ other than 10^6 cycles as appropriate to the material. For example, noting Fig. 10.16, $m_e = 0.4$ at $N_e = 5 \times 10^8$ cycles would estimate an *S-N* curve for aluminum alloys having $\sigma_u \le 325$ MPa. Also, a logical extension for brittle materials would be to use σ_u / k_f instead of σ_u for the point at $N_f = 1$ cycle.

10.7.2 Estimates by the Methods of Juvinall or Shigley

In Juvinall and Marshek (1991) a procedure is suggested that can be applied to a variety of engineering metals, and a similar approach is used by Shigley and Mischke (1989) for steels. These approaches are illustrated by Fig. 10.19(b). Both use log-log coordinates, establish a point at 1000 cycles, and then draw a straight line to a fatigue limit at a specified number of cycles N_e. The value of N_e varies with material in Juvinall's method, and $N_e = 10^6$ is used for the Shigley method.

The point at 1000 cycles is estimated by taking advantage of the fact that the completely reversed fatigue strength at this life for unnotched members, called σ'_{ar}, is only a little below the ultimate tensile strength, generally ranging from $0.75\sigma_u$ to $0.9\sigma_u$ for various situations. Also, many of the factors that strongly affect the fatigue limit, such as size and surface finish, have little or no effect at this life. It is therefore useful to use a new factor m' that is generally in the range 0.75 to 0.9.

$$m' = \frac{\sigma'_{ar}}{\sigma_u} \quad \left(N_f = 1000\right) \tag{10.35}$$

For a notched member, the estimated stress amplitude at 1000 cycles is

$$S'_{ar} = \frac{m'\sigma'_u}{k'_f} \quad \left(N_f = 1000\right) \tag{10.36}$$

For bending or tension, σ'_u is the ultimate tensile strength σ_u, but for torsion it is the ultimate strength in shear, τ_u. Juvinall uses the estimates $\tau_u = 0.8\sigma_u$ for steels and $\tau_u = 0.7\sigma_u$ for other ductile materials.

The quantity k'_f is a life-dependent fatigue notch factor as discussed previously. Recall that values between k_f and unity are expected. Juvinall nevertheless uses $k'_f = k_f$, noting that this is sometimes overly conservative, but justifying the choice on the basis that it simplifies calculations for multiaxial loading. In Juvinall (1967), specific empirical estimates of k'_f were used based on Fig. 10.12. In contrast, Shigley uses $k'_f = 1$.

Beyond N_e, the estimated *S-N* curve is assumed to be horizontal. Neither book makes a specific estimate for lives less than 1000 cycles. For ductile materials, it would be consistent with the spirit of either to use a straight line to connect the point at 1000 cycles with a point obtained by plotting the ultimate strength σ_u at $N_f = 1$. The two methods differ as to the detailed estimates of the various reduction factors, such as m_e, m_t, m_d, and m_s. Some of these details are given in Table 10.1. (Consult the books

TABLE 10.1   SOME PARAMETERS USED BY TWO AUTHORS IN ESTIMATING S-N CURVES

Parameter	Applicability	Juvinall (1991)	Shigley (1989)
Bending fatigue limit factor: m_e	Steels, $\sigma_u \leq 1400$ MPa	0.5	0.5
	Higher strength steels	≤ 0.5	$\sigma_{erb} = 700$ MPa
	Cast irons; Al alloys if $\sigma_u \leq 325$ MPa	0.4	—
	Higher strength Al	$\sigma_{erb} = 130$ MPa	—
	Magnesium alloys	0.35	—
Load type factor: m_t	Bending	1.0	1.0
	Axial	1.0	0.92[1]
	Torsion	0.58	0.58
Size (stress gradient) factor: m_d	Bending or torsion[2]		
	$d < 10$ mm	1.0	$(d/7.62 \text{ mm})^{-0.1133}$
	$d = 10$ to 50 mm	0.9	$(d/7.62 \text{ mm})^{-0.1133}$
	$d = 50$ to 100 mm	0.8	0.6 to 0.75
	$d = 100$ to 150 mm	0.7	0.6 to 0.75
	Axial	0.7 to 0.9[3]	1.0
Surface finish factor: m_s	Polished	1.0	1.0
	Ground[4]	See Fig. 10.18	$1.58 \, \sigma_u^{-0.085}$
	Machined[4]	See Fig. 10.18	$4.51 \, \sigma_u^{-0.265}$
Life for fatigue limit point: N_e, cycles	Steels, cast irons	10^6	10^6
	Aluminum alloys	5×10^8	—
	Magnesium alloys	10^8	—
Constants for point at $N_f = 10^3$: m', k_f'	Bending or torsion	$m' = 0.9$ $k_f' = k_f$	$m' = 0.9$ $k_f' = 1$
	Axial, small eccentricity	$m' = 0.75$ $k_f' = k_f$	$m' = 0.9$ $k_f' = 1$

Notes: [1]If $\sigma_u > 1520$ MPa, use $m_t = 1.0$. [2]For Shigley, for reversed (nonrotating) bending, replace d with $d_e = 0.37d$ for round sections, and with $0.81\sqrt{wt}$ for rectangular sections, where w is beam depth and t is thickness. [3] Use for $d < 50$ mm only, and use the higher value if the eccentricity is small. [4] For Shigley, enter equations with σ_u in MPa.

involved for additional information.) Note that the factor $m_t = 0.58$ is used for torsion in both cases, so that σ_u at the long life point does not need to be adjusted as is done at $N_f = 10^3$ by using σ_u'.

The two authors use factors m_d for size effect that vary with diameter d as indicated in Table 10.1. For bending of cross-sectional shapes that are not round, the dimension d is interpreted by Juvinall as the depth of the bending member, so that stress gradients are matched. Shigley calculates the volume subjected to stresses within 95% of the maximum value, and on this basis determines the size factor using an equivalent diameter corresponding to a circular rotating bending member with the same volume of highly stressed material. In general, caution is needed for materials other than steels, as most of the values of the various factors are based on data for steels.

Once all of the numerical values needed are obtained, the *S-N* curve between $N_f = 10^3$ and N_e is fixed by Eq. 10.36 and by Eq. 10.32 applied at $N_f = N_e$. Hence, it must connect the two points

$$\left(S'_{ar}, N_f\right) = \left(\frac{m' \sigma'_u}{k'_f}, 10^3\right)$$

$$\left(S_{er}, N_f\right) = \left(\frac{m \sigma_u}{k_f}, N_e\right) \tag{10.37}$$

Since log-log coordinates are used, the equation has the form of Eq. 9.6.

$$S_{ar} = A N_f^B \tag{10.38}$$

Evaluating the constants as in Example 9.1 to require the line to pass through the two points gives

$$B = -\frac{\log \dfrac{m' \sigma'_u k_f}{m \sigma_u k'_f}}{\log N_e - 3}$$

$$A = \frac{m \sigma_u}{k_f N_e^B} = \left[\frac{\left(\dfrac{m' \sigma'_u}{k'_f}\right)^{\frac{\log N_e}{3}}}{\dfrac{m \sigma_u}{k_f}}\right]^{\frac{3}{\log N_e - 3}} \tag{10.39}$$

The second expression for A is generally the most convenient to use, being obtained by substituting for B in the first expression and performing several steps of manipulation.

These equations for A and B can be simplified considerably once N_e is chosen. For example, consider the Juvinall method where $k'_f = k_f$ is used, and apply it to steels under axial or bending load, so that $\sigma'_u = \sigma_u$ and $N_e = 10^6$.

$$B = -\frac{1}{3} \log \frac{m'}{m}, \qquad A = \frac{(m')^2 \sigma_u}{m k_f} \tag{10.40}$$

Similarly, applying the Shigley method to steels under axial or bending load, we substitute $k'_f = 1$, $m' = 0.9$, $\sigma'_u = \sigma_u$, and $N_e = 10^6$.

$$B = -\frac{1}{3} \log \frac{0.9 k_f}{m}, \qquad A = \frac{0.81 \sigma_u k_f}{m} \tag{10.41}$$

It should not be forgotten that constants A and B from Eqs. 10.39 to 10.41 are intended for use only in the interval $10^3 \leq N_f \leq 10^6$.

For handling nonzero mean stresses, Juvinall uses $k_{fm} = k_f$ and accounts for yielding effects to calculate a local mean stress σ_m, then entering the unnotched member S-N curve with the resulting σ_{ar} value from Eq. 10.18. In terms of nominal stress, this is equivalent to using Eq. 10.27. In contrast, Shigley uses $k_{fm} = 1$ for ductile materials and $k_{fm} = k_f$ for brittle ones in Eq. 10.21.

10.7.3 Discussion

The three methods discussed for estimating S-N curves, and other similar ones, should not be regarded as providing anything more than very rough curves for use in design. For any particular case, there may be large differences among the estimates, and close agreement with test data should not be expected. Practical use of the Collins procedure will generally require that the fatigue limit be reduced below the $m = 0.5$ level by using additional reduction factors for size, surface finish, etc., as in the other procedures. The Juvinall method is likely to be excessively conservative compared to real test data as a result of using $k'_f = k_f$ even at short lives. This can be avoided if desired by making a more detailed estimate of k'_f at $N_f = 10^3$ as from Fig. 10.12.

Of the three procedures discussed, Juvinall's is the most complete. It incorporates nonferrous metals and uses a lower m' factor at $N_f = 10^3$ for tension than for bending and torsion. These different m' values reflect trends that do occur in S-N data for ductile materials, being due to the behavior at short lives approaching fully plastic yielding. In particular, a lower S-N curve occurs where the net section nominal stress S_n at fully plastic yielding is lower. For example, elastic, perfectly plastic behavior occurs at $S_{no} = \sigma_o$ for axial loading, but at $S_{no} = 1.5\sigma_o$ for bending of a rectangular cross section. As a consequence, S-N curves at short lives for axial loading are lower than those for bending.

Concerning the handling of mean stresses, the use of $k_{fm} = 1$ for ductile metals, as by Collins and Shigley, is a crude approximation to the variation of k_{fm} between zero and k_t that is expected due to yielding. (Recall Fig. 10.14.) The choice of $k_{fm} = 1$ tends to be nonconservative for tensile mean stresses at long lives, especially for materials of moderate ductility that might not be classed as brittle. Juvinall's approach of considering yielding effects on mean stress in more detail, as in Eq. 10.27, is advantageous as this in effect provides a k_{fm} value that varies in a rational manner with stress level and with the yield strength of the material.

Where a choice among the three procedures needs to be made for design purposes, Juvinall's is generally the most satisfactory. It is likely to be conservative, but perhaps excessively so at relatively short lives. In any stress-based approach, the crude assumptions necessarily made regarding k'_f and k_{fm} indicate that such approaches lack the generality needed for detailed handling of yielding effects. The strain-based approach of Chapter 14 does consider yielding effects in a more complete and rational manner and should be considered for use where a degree of refinement is needed. For example, the rough elastic, perfectly plastic idealization of the stress-strain curve is replaced by a nonlinear hardening curve, and the different S-N curves for various geometry and loading configurations can all be estimated from a single strain versus life curve.

Example 10.3

A round bar of the aircraft quality AISI 4340 steel of Table 9.1 is subject to bending and contains a circumferential groove with a ground surface. The dimensions, as defined in Fig. 10.2(c), are $d_1 = 32$, $d_2 = 35$, and $\rho = 1.5$ mm. Estimate the completely reversed *S-N* curve, and also estimate the life for cyclic loading at a nominal stress amplitude of $S_a = 240$ MPa with mean of $S_m = 300$ MPa.

First Solution   One approach is to use the estimate of Collins, which is applicable to a steel of this strength ($\sigma_u = 1172$ MPa) in the form of Eq. 10.34.

$$S_{ar} = \sigma_u \left(1 - \frac{2k_f - 1}{12k_f} \log N_f \right) \qquad \left(N_f \leq 10^6 \right)$$

$$S_{er} = \frac{\sigma_u}{2k_f} \qquad\qquad\qquad \left(N_f > 10^6 \right)$$

The notch factor k_f is estimated from k_t by using α, β, or q as in previous examples. Figure 10.2(c) provides k_t

$$\frac{d_2}{d_1} = 1.094, \quad \frac{\rho}{d_1} = 0.047, \quad k_t = 2.35$$

If k_f is obtained from Eqs. 10.8 and 10.10, it is

$$\alpha = 0.070 \text{ mm}, \quad k_f = 2.29$$

The *S-N* curve from Eq. 10.34 is then

$$S_{ar} = (1172 \text{ MPa}) \left[1 - \frac{2(2.29) - 1}{12(2.29)} \log N_f \right]$$

$$S_{ar} = 1172 \left(1 - 0.130 \log N_f \right) \text{ MPa} \qquad \left(N_f \leq 10^6 \right) \qquad\qquad \textbf{Ans.}$$

$$S_{er} = \frac{1172 \text{ MPa}}{2(2.29)} = 256 \text{ MPa} \qquad \left(N_f > 10^6 \right) \qquad\qquad \textbf{Ans.}$$

This relationship is plotted in Figure E10.3a, forming a straight line on log-linear coordinates.

For the given stresses, the life can be estimated by entering the *S-N* curve with a value of equivalent completely reversed stress from Eq. 10.21, for which we use $k_{fm} = 1$ for this ductile material.

$$S_{ar} = \frac{S_a}{1 - \dfrac{S_m}{\sigma_u}} = \frac{240 \text{ MPa}}{1 - \dfrac{300 \text{ MPa}}{1172 \text{ MPa}}} = 323 \text{ MPa}$$

This value is above the estimated fatigue limit of $S_{er} = 256$ MPa and so corresponds to a finite life. Hence, it can be used with the *S-N* equation above to obtain N_f.

$$\log N_f = \frac{1}{0.130} \left(1 - \frac{S_{ar}}{1172} \right) = \frac{1}{0.130} \left(1 - \frac{323 \text{ MPa}}{1172 \text{ MPa}} \right)$$

$$N_f = 3.74 \times 10^5 \text{ cycles} \qquad\qquad \textbf{Ans.}$$

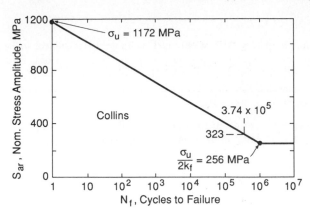

$\sigma_u = 1172$ MPa

Collins

3.74×10^5

323

$\dfrac{\sigma_u}{2k_f} = 256$ MPa

S_{ar}, Nom. Stress Amplitude, MPa

N_f, Cycles to Failure

Figure E10.3a

Second Solution   The estimates of either Juvinall or Shigley could also be used; let us pursue the former. The various m_i factors are evaluated by following the Juvinall column of Table 10.1.

$$m_e = 0.5, \quad m_t = 1.0, \quad m_d = 0.9, \quad m_s = 0.88$$

where m_s is obtained by entering Fig. 10.18 with $\sigma_u = 1172$ MPa (170 ksi.) Hence, the combined reduction factor for the fatigue limit is

$$m = m_e m_t m_d m_s = 0.396$$

so that the estimated fatigue limit is

$$S_{er} = \frac{m\sigma_u}{k_f} = \frac{0.396\,(1172\ \text{MPa})}{2.29} = 203\ \text{MPa}$$

where k_f from the previous solution is used.

Further following Table 10.1 gives the additional values needed.

$$N_e = 10^6, \quad m' = 0.9, \quad k'_f = k_f$$

The point on the estimated S-N curve at $N_f = 10^3$ cycles is thus

$$S'_{ar} = \frac{m'\sigma'_u}{k'_f} = \frac{0.9(1172\ \text{MPa})}{2.29} = 461\ \text{MPa}$$

Thus the estimated S-N curve on log-log coordinates connects the two points:

$$(S_{ar}, N_f) = (461, 10^3) \text{ and } (203, 10^6)$$

The equation for this line can be obtained from Eq. 10.39, or in this particular case the simplified form of Eq. 10.40 applies. The latter gives

$$B = -\frac{1}{3} \log \frac{m'}{m} = -\frac{1}{3} \log \frac{0.9}{0.396} = -0.119$$

$$A = \frac{(m')^2 \sigma_u}{m k_f} = \frac{(0.9)^2 (1172\ \text{MPa})}{0.396(2.29)} = 1047\ \text{MPa}$$

so that the equation is

$$S_{ar} = AN_f^B = 1047\ N_f^{-0.119}\ \text{MPa} \quad (10^3 \le N_f \le 10^6) \qquad \textbf{Ans.}$$

This forms a straight line on log-log coordinates and is plotted in Figure E10.3b.

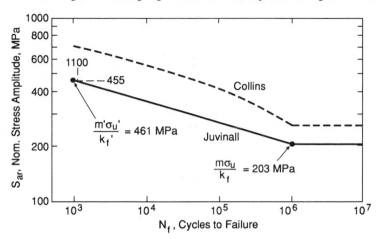

Figure E10.3b

To continue following the Juvinall procedure, Eq. 10.21 is used with k_{fm} from Eq. 10.27 to evaluate S_{ar} for the given stresses. First, noting the three cases for Eq. 10.27, use $\sigma_o = 1103$ MPa from Table 9.1 to determine whether or not initial yielding or reversed yielding occur.

$$k_f S_{max} = k_f(S_a + S_m) = 2.29(240 + 300) = 1237\ \text{MPa}$$

$$1237 > \sigma_o = 1103\ \text{MPa} \quad \text{(initial yielding)}$$

$$k_f\,\Delta S = 2k_f S_a = 2(2.29)(240) = 1099\ \text{MPa}$$

$$1099 < 2\sigma_o = 2206\ \text{MPa} \quad \text{(no reversed yielding)}$$

Hence, the initial yielding case of Eq. 10.27 is used in Eq. 10.21, and S_{ar} is obtained as follows:

$$S_{ar} = \frac{S_a}{1 - \dfrac{\sigma_o - k_f S_a}{\sigma_u}} = \frac{240}{1 - \dfrac{1103 - 2.29(240)}{1172}} = 455\ \text{MPa}$$

Substituting into the estimated *S-N* equation from above and solving for N_f finally gives the life.

$$N_f = \left(\frac{S_{ar}}{A}\right)^{\frac{1}{B}} = \left(\frac{455\ \text{MPa}}{1047\ \text{MPa}}\right)^{\frac{1}{-0.119}} = 1100\ \text{cycles} \qquad \textbf{Ans.}$$

Discussion The drastic difference between the two estimated lives results from the m_i factors in the Juvinall estimate giving a lower fatigue limit, and also from the different handling of k_f' and k_{fm}, all of which combine to push the Juvinall estimate toward shorter lives. Note that the two *S-N* curves differ considerably where they are compared in the second graph. (The Collins estimate is no longer a straight line due to the log-log scales.)

10.8 DESIGNING TO AVOID FATIGUE FAILURE

In analyzing and designing against fatigue failure, there are two general classes of problems. First, for situations involving low stresses applied for large numbers of cycles, the expected stresses in service must be below the fatigue limit. Second, relatively high stresses may occur at least occasionally, creating a situation where fatigue failure is possible in a finite number of cycles. An appropriate S-N curve from an estimate or from test data is then needed, with the effects of any notches being included in the curve, as by use of k_f, or by use of test data obtained directly on notched specimens.

Design details determine k_f and thus affect the fatigue limit and the S-N curve. Safety factors in stress and/or life are of course needed. These may be assigned based on engineering judgment, or they may be specified by design codes. Safety factors may be fixed values, or they may be based on statistics, perhaps with a particular probability of failure in mind. (Recall Figs. 9.31 to 9.33.)

10.8.1 Design Details

In design of engineering components, resistance to fatigue failure can be improved by careful attention to detail. For notches such as grooves, fillets, and noncircular holes, stress concentration factors are decreased if the radius of the notch is increased, as study of Figs. 10.1 and 10.2 will confirm. Other aspects of the geometry also have an effect, such as the relative width of a notched plate, or the ratio of the two diameters in a stepped shaft. Hence, within the constraints imposed by functional requirements, geometries can be adjusted to minimize the elastic stress concentration factor k_t. For a given material, k_f will then also generally be minimized. If more than one material is being considered, the different notch sensitivities q of these also need to be examined.

Consider the example of Fig. 10.21. A relatively small radius occurs in (a) at the diameter step in a shaft. The stress concentration factor k_t can be decreased by increasing the fillet radius, ρ from Fig. 10.2, while keeping the other dimensions the same. Even better, the radius can be essentially eliminated by using a taper as in (b). If functional requirements preclude a simple taper, the geometry of (c) has a lower k_t than (a) and could perhaps be used.

Similar principles apply to other design details, such as keyways, as illustrated in Fig. 10.22. As a further example, the common *fir tree* design for connecting the roots of turbine blades is shown in Fig. 10.23. Smooth curves with radii as large as permitted by the tight spaces involved are used, and the overall taper tends to distribute the load fairly evenly among the projections (lugs) on the blade root.

Where small motions occur between tightly fitting metal parts, a problem called *fretting* may occur. A metal oxide in the form of a powder is usually present, and surface damage occurs that can cause cracks to start and grow. Hence, considerable care is needed in designing certain mechanical connections, such as the press-fitted shaft of Fig. 10.24. Altered geometry to reduce the stresses and to make a more gradual

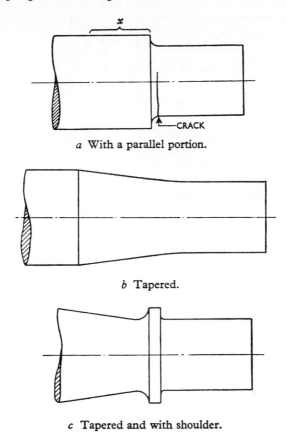

a With a parallel portion.

b Tapered.

c Tapered and with shoulder.

Figure 10.21   The common location (a) of fatigue cracks in a stepped shaft, and (b) reducing the stress raiser effect by using a taper, or (c) a taper with a shoulder. (From [Cottell 56]; reprinted by permission of the Council of the Institution of Mechanical Engineers, London, UK.)

transition into the press fit is helpful, with some possibilities being shown. Since certain combinations of materials are particularly susceptible to this problem, it is often helpful to change one of the materials, or to use a bushing or a surface coating in the joint. Intentional introduction of beneficial compressive residual stresses, as by shot peening, is also often used.

Bolts involve severe stress raisers in the threads and elsewhere, and the tightly fitting surfaces in a bolted connection may also be subject to fretting. Locations where fatigue cracks are likely to start on a bolt, and some changes that can be used to improve the fatigue resistance, are shown in Fig. 10.25. Highly specialized bolt and rivet designs are sometimes used in critical applications. Some bolted connections that might be found in metal aircraft structure are shown in Fig. 10.26. Improved resistance to fatigue is provided by symmetrical geometries that minimize bending stresses in the members

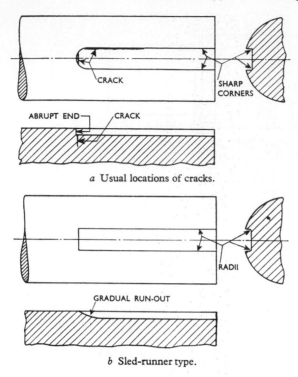

a Usual locations of cracks.

Figure 10.22   Usual location (a) of fatigue cracks in keyways, and an improved design (b) called a *sled-runner* keyway. (From [Cottell 56]; reprinted by permission of the Council of the Institution of Mechanical Engineers, London, UK.)

b Sled-runner type.

being connected, and thus also in the bolts. Tapered or *scarf* joints cause the loads to be more evenly distributed among the bolts, and so this feature is also usually beneficial. Special care is required in designing not only bolted joints, but also joints that are riveted, welded, bonded, etc.

Additional information on design details for various mechanical elements, such as joints, springs, gears, bearings, shafts, etc., may be found in textbooks on mechanical design, such as those listed at the end of this chapter.

10.8.2 Surface Residual Stresses

Consider a notched member as in Fig. 10.27 that is subjected to a tensile overload sufficient to cause local yielding. Upon removal of the load, the unyielded material around the plastic zone attempts to recover its original shape, and in so doing forces the yielded material into compression. Other regions away from the notch are in tension, so that there is a distribution of stress that sums to zero as required by equilibrium and the now zero applied load. These locked-in stresses are called *residual stresses*. If compressive at the notch, they retard fatigue cracking by biasing the mean stress in the compressive direction during subsequent cyclic loading.

Assuming that the material is an elastic, perfectly plastic one, logic similar to that used previously for mean stresses applies. The residual stress remaining after removal

Figure 10.23   Steam turbine rotor with blades attached and the *fir tree* type of connection at the blade root. On the lower right, a fatigue crack can be seen running across the blade root just above the base of the blade. (Photos courtesy of Neville F. Rieger, Stress Technology, Inc., Rochester, N.Y. Reprinted with permission of the Electric Power Research Institute, EPRI, from *Failure Analysis of Fossil Low Pressure Turbine Blade Group.*)

of a nominal stress S' is thus

$$
\begin{aligned}
\sigma_r &= \sigma_o - k_t S' \quad \left(k_t S' \leq 2\sigma_o \right) \\
\sigma_r &= -\sigma_o \qquad\quad \left(k_t S' > 2\sigma_o \right)
\end{aligned}
\tag{10.42}
$$

In the first case, corresponding to Fig. 10.27(a), no compressive yielding occurs during unloading, but in the second case (b) it does, resulting in a residual stress equal to the yield strength in compression. If the overload is compressive, an analogous but opposite

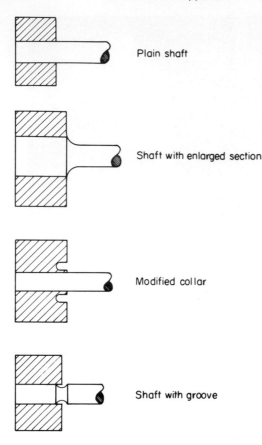

Plain shaft

Shaft with enlarged section

Modified collar

Shaft with groove

Figure 10.24 Some designs to alleviate stress concentration at a press-fitted shaft. The plain shaft involves a severe stress raiser and is susceptible to fretting. Some possible improvements are enlarging the shaft end, modifying the collar, or grooving the shaft. (Adapted from [Grover 66] p. 211.)

effect occurs, giving a tensile residual stress. The above equations can still be used if σ_o is replaced by $-\sigma_o$ and S' is used with its negative value.

Compressive surface residual stresses are often intentionally introduced into mechanical components to improve the fatigue strength at long lives. Any method of yielding the surface in tension will result in a compressive residual stress in a manner similar to the notch case above. Methods used in addition to tensile overloading of notches include *shot peening*, *cold rolling* of the surface, and overloading in bending, called *presetting*. (See Fig. 9.30.) Shot peening is the most commonly used method and involves bombarding the surface with small, hard, often steel balls. These cause biaxial yielding in tension under each point of impact, hence a biaxial compressive residual stress occurs due to the elastic recovery of the unyielded material beneath. Components commonly shot peened include leaf springs, gears, crankshafts, and turbine blades.

Compressive surface residual stresses can also be produced by surface hardening treatments, as in carburizing or nitriding of steels, or by thermal treatments, notably rapid quenching of steel shafts. Other surface treatments, such as abusive grinding and chrome plating, need to be used with care as they may introduce harmful tensile residual

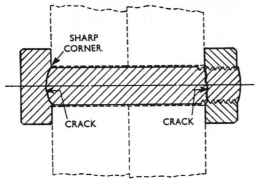

a Usual locations of cracks.

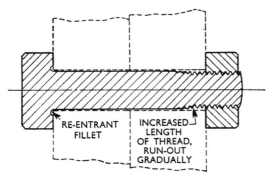

b Improved type of bolt.

Figure 10.25   Usual location (a) of fatigue cracks in a bolt, and (b) some measures to improve fatigue resistance. (From [Cottell 56]; reprinted by permission of the Council of the Institution of Mechanical Engineers, London, UK.)

stresses. Tensile residual stresses may also remain after welding, leading to the common practice of a subsequent *stress relief* heat treatment to remove them.

Surface residual stresses are beneficial only where subsequent yielding does not occur due to loads that occur in service, as this may remove the compressive residual stress or even change it to a harmful tensile one. Additional discussion of residual stress effects on fatigue can be found in Fuchs and Stephens (1980) and in the SAE *Fatigue Design Handbook* (Rice, 1988).

10.8.3 Variable Amplitude Loading

Since variable amplitude loading is common in practical applications, the Palmgren-Miner (P-M) rule as discussed in Chapter 9 is often needed to estimate fatigue lives. The *S-N* curve used with the P-M rule should of course include all relevant effects, such as those of notches and surface finish. Where some stress levels occur that are above the fatigue limit, those below would be assigned zero fatigue damage if the fatigue limits (horizontal lines) beyond N_e in estimated *S-N* curves are taken seriously. However, test results indicate that the presence of even a few relatively high stresses in a load history causes those below the fatigue limit to contribute to the fatigue damage. This occurs

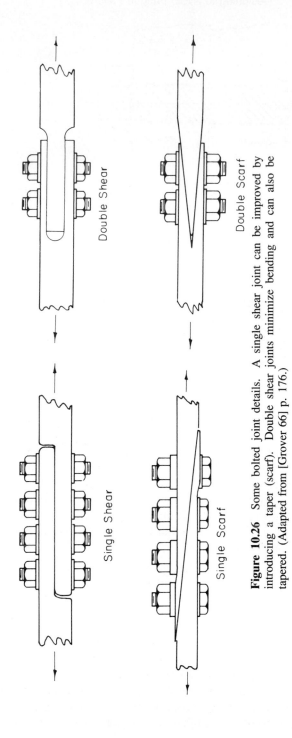

Figure 10.26 Some bolted joint details. A single shear joint can be improved by introducing a taper (scarf). Double shear joints minimize bending and can also be tapered. (Adapted from [Grover 66] p. 176.)

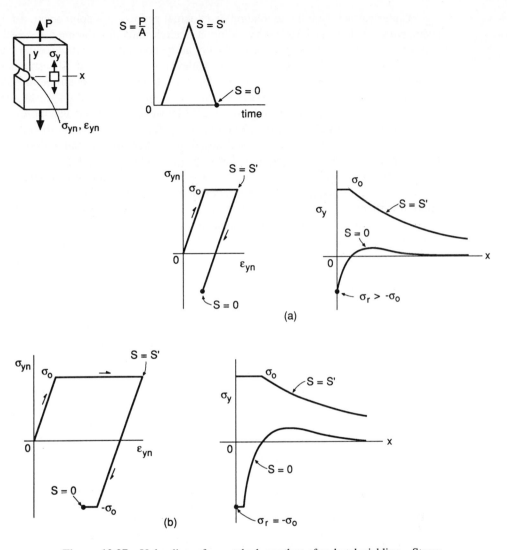

Figure 10.27 Unloading of a notched member after local yielding. Stress-strain behavior at the notch and residual stress distributions are shown for (a) elastic behavior during unloading, and (b) compressive yielding during unloading.

even for materials such as many steels that exhibit a distinct fatigue limit in the *S-N* curve for constant amplitude loading. Thus, the concept of a fatigue limit is valid only if all stresses in the load history are below this level. An extrapolation of the *S-N* curve below the fatigue limit on a log-log plot is a reasonable and usually conservative method of handling this situation. See Example 10.5 and the additional discussion of this matter given later in Section 14.6.

Under variable amplitude loading as in actual service of a component, a relatively severe stress cycle can cause local yielding at a notch, thus introducing a residual stress. This affects the mean stress that occurs locally at the notch during subsequent lower level cycles, hence also affecting the fatigue life. Unless this is considered, life estimates based on the P-M rule can be highly inaccurate. Such a *sequence effect* could be included in a rough way in a stress-based approach by extending the logic based on elastic, perfectly plastic behavior of Fig. 10.14 and Eq. 10.27. However, a related but more complete methodology has already been developed as a key feature of the strain-based approach of Chapter 14. Use of the strain-based approach is therefore recommended where the degree of refinement needed requires that sequence effects of this type be analyzed.

Further, the specific logic of Fig. 10.14 is seen to apply only to constant amplitude loading. Thus, use of Eq. 10.27 in a direct manner for variable amplitude loading is not valid. It is therefore suggested that the following simpler alternative be used: If there is little or no yielding, use $k_{fm} = k_f$, otherwise use $k_{fm} = 1$. For relatively high strength (but still ductile) metals at intermediate and long lives, this will result in the choice of $k_{fm} = k_f$ as for a brittle material.

10.8.4 Safety Factors

In designing components that must resist fatigue loading, various uncertainties exist. These include the actual values of the service loads, the amount of statistical variation in the fatigue strength of the component, and the magnitudes of other effects such as those of surface finish and chemical and thermal environment. Since these items are generally difficult to quantify, reasonable estimates must be made, and then a safety factor applied to take care of inaccurate estimates or unforeseen circumstances. Larger safety factors are needed where the uncertainties are greater or where the consequences of failures are severe, as when loss of life could occur. One approach is to lower the *S-N* curve in the stress direction by a factor X_S, and another is to shift it by a factor of X_N in the life direction, as illustrated in Fig. 10.28. In either case, a new *design S-N curve* is obtained that provides a margin of safety relative to the curve corresponding to failure.

It is useful to designate the stress and number of cycles actually expected in service as $\hat{S}_a$ and $\hat{N}$, respectively. Also, let the *S-N* curve for failure be

$$S_a = f(N_f) \tag{10.43}$$

The safety factor in stress is then

$$X_S = \frac{S_a}{\hat{S}_a} \tag{10.44}$$

where $S_a = f(\hat{N})$ is the stress corresponding to failure at the end of the service life $\hat{N}$, as at point A in Fig. 10.28. Thus, the design *S-N* curve is

$$\hat{S}_a = \frac{f(\hat{N})}{X_S} \tag{10.45}$$

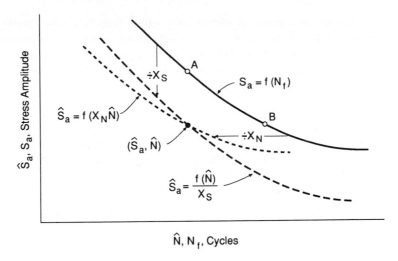

Figure 10.28   Safety factors in stress X_S, and in life X_N, applied to obtain S-N curves for design.

Alternatively, a safety factor in life can be employed.

$$X_N = \frac{N_f}{\hat{N}}$$

(10.46)

where N_f is the life to failure corresponding to the service stress, $\hat{S}_a = f(N_f)$, as at point B in Fig. 10.28. The design S-N curve is in this case

$$\hat{S}_a = f(X_N \hat{N})$$

(10.47)

Safety factors on stress are typically 1.5 to 2.0 or more. Safety factors on life must be larger, typically 10 to 20 or more. This is because a relatively small change in stress corresponds to a large change in life on most S-N curves. A combination of both is sometimes employed, where the more conservative result from either a specified X_S or a specified X_N is used. In other words, the X_N curve as in Fig. 10.28 is used where it is lower than the X_S curve, and the X_S curve is used elsewhere.

Another approach is to use a design curve that corresponds to a specific small probability of failure, such as one curve from the family in Fig. 9.33. It is sometimes observed and often assumed that such curves for given probabilities of failure correspond approximately to a constant percent reduction in stress at all values of life. In this case, a given probability of failure corresponds to a specific safety factor in stress, X_S. Such a probabilistic approach to choosing safety factors is in general preferable as it provides a rational basis for the decision. However, sufficient information is seldom available to calculate all of the statistical uncertainties involved, such as those for the applied loads that might actually occur, so that the input information to the statistical analysis may represent only rough estimates.

The above comments address only safety factors against fatigue. Safety factors against yielding are also required and apply to the maximum and minimum stresses that occur during cyclic loading. For a ductile material, recalling Section 7.9.2 and Fig. 7.26, yielding can spread over the entire net area before failure occurs, so that the limiting conditions are approximately

$$S_{max} = \sigma_o, \quad S_{min} = -\sigma_o \tag{10.48}$$

where σ_o is the yield strength and S is defined by using the net (notch removed) area. If X_o is the safety factor against yielding, the permissible values of the maximum and minimum nominal stresses are

$$\hat{S}_{max} = \frac{\sigma_o}{X_o}, \quad \hat{S}_{min} = -\frac{\sigma_o}{X_o} \tag{10.49}$$

where X_o may have a different numerical value than the safety factor on stress for fatigue, X_S. The combined limitations for yielding and fatigue can be illustrated by adding lines from the equations above to a constant-life diagram as shown in Fig. 10.29. For bending, such an approach is noted to be conservative, as fully plastic yielding occurs beyond $S = \sigma_o$, as at $S_{no} = 1.5\sigma_o$ for a rectangular section.

Example 10.4

The circumferentially grooved bar of Ex. 10.3, made of the AISI 4340 steel of Table 9.1, is expected to be subjected in service to 5000 cycles at a nominal bending stress amplitude of $S_a = 200$ MPa, with this applied at a mean level of $S_m = 150$ MPa. What are the safety factors on both stress and life?

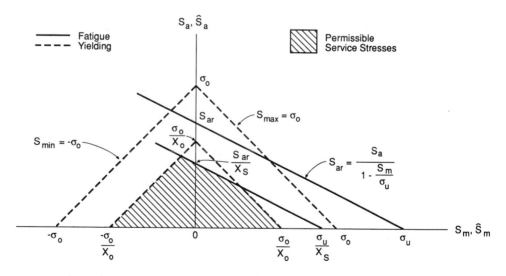

Figure 10.29 Diagram for design at a given life N_f corresponding to a given stress level S_{ar}. The combined limitations of failure due to fatigue and fully plastic yielding are shown, as are the permissible service stresses as determined by safety factors.

Solution   The S-N curve estimated in Ex. 10.3 using the Juvinall procedure can be used. Recall that this curve is written in terms of the equivalent completely reversed nominal stress S_{ar}, with the notch effect already included by the use of $k_f = 2.29$ in developing the curve.

$$S_{ar} = AN_f^B = 1047N_f^{-0.119} \text{ MPa} \qquad \left(10^3 \leq N_f \leq 10^6\right)$$

Further following Juvinall, the S_{ar} corresponding to the given stresses is obtained using Eq. 10.27 with Eq. 10.21. Noting that $\sigma_o = 1103$ MPa from Table 9.1, there is no yielding since

$$k_f S_{max} = k_f (S_a + S_m) = 2.29(200 + 150) = 802 \text{ MPa} < 1103 \text{ MPa}$$

Hence, $k_{fm} = k_f$ is used in Eq. 10.21.

$$S_{ar} = \frac{S_a}{1 - \dfrac{k_f S_m}{\sigma_u}} = \frac{200 \text{ MPa}}{1 - \dfrac{2.29(150 \text{ MPa})}{1172 \text{ MPa}}} = 283 \text{ MPa}$$

Entering the S-N curve with this value gives the estimated life to failure.

$$N_f = \left(\frac{S_{ar}}{A}\right)^{\frac{1}{B}} = \left(\frac{283 \text{ MPa}}{1047 \text{ MPa}}\right)^{\frac{1}{-0.119}} = 59{,}500 \text{ cycles}$$

The safety factor on life for $\hat{N} = 5000$ cycles in service is thus

$$X_N = \frac{N_f}{\hat{N}} = \frac{59{,}500}{5000} = 11.9 \qquad\qquad \textbf{Ans.}$$

This safety factor is shown on the S-N curve in Figure E10.4.

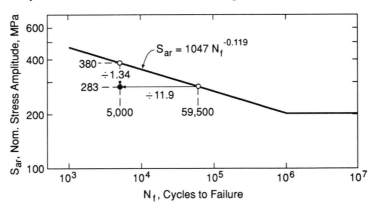

Figure E10.4

To determine the safety factor on stress, the service stress level of $\hat{S}_{ar} = 283$ MPa is compared to the stress S_{ar} corresponding to $\hat{N} = 5000$ cycles.

$$S_{ar} = A\hat{N}^B = 1047(5000)^{-0.119} = 380 \text{ MPa}.$$

Ans.

$$X_S = \frac{S_{ar}}{\hat{S}_{ar}} = \frac{380 \text{ MPa}}{283 \text{ MPa}} = 1.34$$

This factor is also illustrated on the S-N curve above. However, yielding failure is also a possibility, so that Eq. 10.49 needs to be checked.

$$X_o = \frac{\sigma_o}{\hat{S}_{max}} = \frac{\sigma_o}{\hat{S}_a + \hat{S}_m} = \frac{1103 \text{ MPa}}{(200 + 150) \text{ MPa}} = 3.15$$

Since the smaller of the two values is the controlling one, the safety factor in stress is confirmed to be $X_S = 1.34$ from the fatigue calculation.

Example 10.5

The nominal bending stress history shown in Figure E10.5a is applied to the circumferentially grooved bar of AISI 4340 steel from Ex. 10.3.

(a) Estimate the number of repetitions of this history necessary to cause fatigue failure.

(b) In the service usage of this component, 800 repetitions of this history are expected, and safety factors of 10 on life and 1.5 on stress are required. Are these factors satisfied? If not, by what ratio must the stresses be scaled down by redesign or by modified usage of the component?

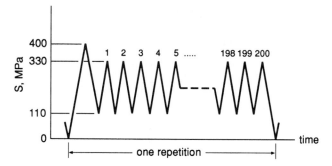

Figure E10.5a

Solution (a) The P-M rule can be used with the S-N curve estimated in Ex. 10.3 using the Juvinall procedure. Recall that this curve is written in terms of the equivalent completely reversed nominal stress S_{ar}, with the notch effect already being included by the previous use of $k_f = 2.29$ in developing the curve.

$$S_{ar} = AN_f^B = 1047N_f^{-0.119} \text{ MPa} \qquad (10^3 \le N_f \le 10^6)$$

Each repetition of the load history contains one cycle with $S_{min} = 0$ and $S_{max} = 400$ MPa, and 200 cycles with $S_{min} = 110$ and $S_{max} = 330$ MPa. (This is apparent by inspection and can be verified by rainflow cycle counting.) There is no yielding at the notch even for the most severe stress peak.

$$k_f S_{max} = 2.29 \,(400 \text{ MPa}) = 916 \text{ MPa}$$

which is well below $\sigma_o = 1103$ MPa.

The following calculations are needed for each stress level:

$$S_a = \frac{S_{max} - S_{min}}{2}, \quad S_m = \frac{S_{max} + S_{min}}{2}$$

$$S_{ar} = \frac{S_a}{1 - \dfrac{k_f S_m}{\sigma_u}}, \quad N_f = \left(\frac{S_{ar}}{A}\right)^{\frac{1}{B}}$$

Since there is no yielding, $k_{fm} = k_f$ from Eq. 10.27 is used in Eq. 10.21. The results, with all stresses in MPa, are

j	N_j	S_{min}	S_{max}	S_a	S_m	S_{ar}	N_{fj}
1	1	0	400	200	200	328	1.709×10^4
2	200	110	330	110	220	193	1.488×10^6

The P-M rule in the form of Eq. 9.24 is then used to estimate the number of repetitions to failure.

$$B_f = \frac{1}{\left[\displaystyle\sum \frac{N_j}{N_{fj}}\right]_{\text{one rep.}}} = \frac{1}{\dfrac{1}{1.709 \times 10^4} + \dfrac{200}{1.488 \times 10^6}} = 5180 \text{ repetitions} \qquad \textbf{Ans.}$$

Comment   The second stress level is below the fatigue limit of $S_{er} = 203$ MPa from Ex. 10.3 (second solution) but is nevertheless expected to contribute to the fatigue damage as discussed in the text. Calculation of the life from the S-N equation used is equivalent to the dotted-line extrapolation (A) shown in Figure E10.5b.

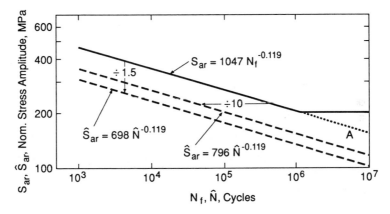

Figure E10.5b

(b) The safety factor in life for $\hat{B} = 800$ repetitions in service and the given stresses is

$$X_N = \frac{N_f}{\hat{N}} = \frac{B_f}{\hat{B}} = \frac{5180}{800} = 6.5$$

which is less than the required value of $X_N = 10$. The existing safety factor on stress could also be calculated by comparing the current stresses with those that would cause failure at a life of $\hat{B} = 800$ repetitions. However, the calculation is unnecessary as the inadequate X_N is alone sufficient to require lowering the stresses.

Using the S-N curve from above as $S_{ar} = f(N_f)$ with Eqs. 10.43 to 10.47, the design S-N curve as required by $X_S = 1.5$ is

$$\hat{S}_{ar} = \frac{f\left(\hat{N}\right)}{X_S} = \frac{1047\hat{N}^{-0.119}}{1.5} = 698\hat{N}^{-0.119} \text{ MPa}$$

and that required for $X_N = 10$ is

$$\hat{S}_{ar} = f\left(X_N\hat{N}\right) = 1047\left(10\hat{N}\right)^{-0.119} = 796\hat{N}^{-0.119} \text{ MPa}$$

These are both plotted on the graph above. Due to the resulting parallel straight lines on this log-log plot, the first equation always gives the lowest S-N curve, so that X_S always controls in this problem, X_N automatically being exceeded if X_S is satisfied. Hence, the stresses needed are those that give a life of $\hat{B} = 800$ repetitions using the first equation.

The needed stresses can be obtained by programming the life calculation on a digital computer and then varying a reduction factor on stress until the life $\hat{B} = 800$ is found by trial and error. The procedure is the same as before except for use of the design S-N curve. The iterations and their results are as follows, with all stresses being in MPa:

Ratio	$\hat{S}_{\min 1}$	$\hat{S}_{\max 1}$	$\hat{S}_{\min 2}$	$\hat{S}_{\max 2}$	$\hat{B}$
1.00	0	400	110	330	172
0.90	0	360	99	297	746
0.89	0	356	98	294	866
0.895	0	358	98	295	804

Ans.

$\hat{B} = 804$ repetitions is judged to be sufficiently close, so the stresses on the last line are the required lower values for the peaks and valleys in the load history.

10.8.5 Load Line Approach to Safety Factors

Consider a cyclic loading situation where a safety factor in stress X_S has been selected and where mean stresses are present. The constant-life diagram corresponding to the desired service life is shown in Fig. 10.30(a). If the stress amplitude and mean stress are proportional, then the stresses applied in service are related to the constant-life diagram by the line labeled *load line*. All possible magnitudes of loading form a straight line through the origin that has a known slope, and both the service and failure stresses, ($\hat{S}_m$, $\hat{S}_a$) and (S_m, S_a), respectively, must lie on this line. The same safety factor applies to both the stress amplitude and mean.

$$X_S = \frac{S_a}{\hat{S}_a} = \frac{S_m}{\hat{S}_m} \tag{10.50}$$

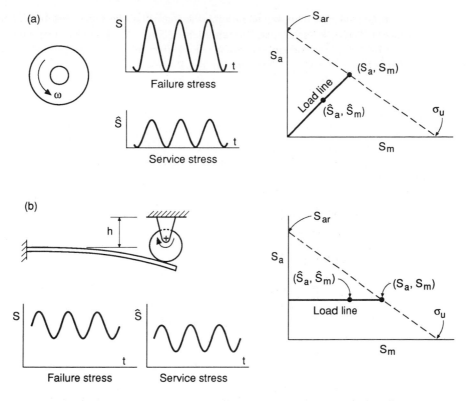

Figure 10.30 Load lines for two design situations. For the rotating disc (a), the same safety factor should be applied to both the stress amplitude and mean, but for the cam follower spring (b), the safety factor should be applied only to the mean stress.

For example, this situation exists in fatigue due to the centrifugal stresses caused by spinning a disc. One stress cycle occurs each time the rotation is stopped and started, with the magnitude of the stress for a disc of given mass and dimensions depending on the square of the rotation speed. The stress returns to zero whenever the rotation stops, so that $S_a = S_m = S_{max}/2$. Hence, in this case, the load line in Fig. 10.30(a) has a slope of unity and passes through the origin.

But this simple situation often does not apply. For example, consider a leaf spring that acts as a cantilever beam and is in contact with a rotating eccentric roller (called a cam) as shown in Fig. 10.30(b). The amplitude of the cyclic load is fixed by the dimensions of the cam and therefore cannot vary. However, assume that the dimension h between the cam and the spring's unstressed position may change during operation of this device, thus changing the mean stress. The load line describing the applied stresses that are possible is in this case horizontal due to the amplitude being fixed. A safety factor is necessary only for the mean stress, a safety factor of unity being adequate for the stress amplitude if it is truly impossible for it to vary.

In general, for complex situations, the load line can have various slopes and may not pass through the origin, and it can even be curved. The subject of safety factors in cases with complex load lines becomes somewhat involved. Additional detail is found in Mitchell and Vaughn (1975) listed in the References.

10.9 SUMMARY

Stress raisers (notches) reduce fatigue strength and require careful attention to detail in design. The strength reduction is often not as great as would be expected from the elastic stress concentration factor k_t, so that special fatigue notch factors are used. These are calculated from k_t and empirical curves giving the notch sensitivity value q for the material and notch radius of interest.

$$k_f = 1 + q\,(k_t - 1) \tag{10.51}$$

Alternatively, the same empirical information may be expressed in terms of a material constant. For example, Peterson employs a material constant α and the notch radius ρ to estimate k_f.

$$k_f = 1 + \frac{k_t - 1}{1 + \dfrac{\alpha}{\rho}} \tag{10.52}$$

In ductile materials, yielding causes even further reductions in k_f at short lives. A more general variable k_f', that varies with life, is then needed. Limiting values of k_f' are k_f and unity, at long and short lives, respectively.

Where mean stresses are applied to notched members, combinations of stress amplitude and mean stress can be used to calculate an equivalent completely reversed stress.

$$S_{ar} = \frac{S_a}{1 - \dfrac{k_{fm} S_m}{\sigma_u}} \tag{10.53}$$

where

$$k_{fm} = 1 \qquad \text{(ductile materials)}$$

$$k_{fm} = k_f' = k_f \qquad \text{(brittle materials)}$$

The fatigue notch factor k_{fm} for the mean stress is actually expected to approach zero at short life due to yielding effects, and at long life it may approach k_f even for reasonably ductile materials. Thus, use of the single value of unity as above for ductile materials is a common but crude approximation. A more sophisticated approach is to treat k_{fm} as a variable, thus including the effects of yielding using Eq. 10.27.

For engineering situations involving low stresses applied large numbers of times, design may be based on a long-life fatigue strength, often called the fatigue limit, σ_{er}. Where specific data are not available, σ_{er} for commonly used metals may be estimated from correlations with the ultimate tensile strength σ_u. In applying σ_{er} values to engineering components, the fatigue notch factor k_f is employed along with additional modifying factors, such as those for type of load, size, and surface finish.

Where relatively high stresses occur at least occasionally, fatigue failure in a finite number of cycles is a possibility, and an entire S-N curve is needed. A curve based on data approximating the situation of interest is preferable, but estimates may also be made. The typical estimation methods of Collins, Juvinall, and Shigley are described in the text and illustrated by Fig. 10.19, and some details for the latter two are also given in Table 10.1. Juvinall's method is the most complete.

Efficient design against fatigue requires keeping stress concentration factors as low as possible, while also avoiding such problems as fretting and tensile residual stresses from processing. Beneficial compressive residual stresses are often introduced intentionally. Safety factors may be applied either to the stress or to the life, or to both. Safety factors on life need to be on the order of a factor of ten or larger, which is a consequence of the life being sensitive to small changes in the applied stress or in the S-N curve.

NEW TERMS AND SYMBOLS

fatigue limit:
    polished bend, σ_{erb}
    adjusted, σ_{er}
    notch, S_{er}
fatigue limit life, N_e
fatigue limit ratio, $m_e = \sigma_{erb}/\sigma_u$
fatigue limit reduction factors:
    load type, m_t
    size, m_d
    surface, m_s
fatigue notch factor:
    long life, k_f
    not long life, k_f'
    mean stress, k_{fm}
fretting
initial yielding
load line
local yielding

Neuber constant, β
nonpropagating crack
notch sensitivity, q
Peterson constant, α
process zone effect
residual stress
reversed yielding
safety factor in life, X_N
safety factor in stress, X_S or X_o
sequence effect
service life, $\hat{N}$
service stresses, $\hat{S}_a$, $\hat{S}_m$
shot peening
size effect
stress gradient, $d\sigma/dx$
stress raiser
surface finish
weakest-link effect

REFERENCES

(a) General References

COLLINS, J. A. 1981 *Failure of Materials in Mechanical Design*, John Wiley, New York.

FUCHS, H. O. and R. I. STEPHENS. 1980 *Metal Fatigue in Engineering*, John Wiley, New York.

JUVINALL, R. C. 1967 *Stress, Strain and Strength*, McGraw-Hill, New York.

JUVINALL, R. C. and K. M. MARSHEK. 1991 *Fundamentals of Machine Component Design*, 2nd ed., John Wiley, New York.

MITCHELL, L. D. and D. T. VAUGHAN. 1975 "A General Method for the Fatigue-Resistant Design of Mechanical Components, Part 1, Graphical, and Part 2, Analytical," *Journal of Engineering for Industry*, ASME, Aug. 1975, pp. 965–975.

PETERSON, R. E. 1959 "Notch-Sensitivity," *Metal Fatigue*, G. Sines and J. L. Waisman, eds., McGraw-Hill, New York, pp. 293-306.

RICE, R. C., ed. 1988 *Fatigue Design Handbook*, 2nd ed., SAE Pub. No. AE-10, Society of Automotive Engineers, Warrendale, Pa.

SHIGLEY, J. E. and C. R. MISCHKE. 1989 *Mechanical Engineering Design*, 5th ed., McGraw-Hill, New York.

(b) Sources for Stress Concentration Factors

NISITANI, H. and N. NODA. 1984 "Stress Concentration of a Cylindrical Bar with a V-Shaped Circumferential Groove Under Torsion, Tension, or Bending," *Engineering Fracture Mechanics*, Vol. 20, No. 5/6, pp. 743-766.

PETERSON, R. E. 1974 *Stress Concentration Factors*, John Wiley, New York.

SORS, L. 1971 *Fatigue Design of Machine Components*, Pergamon Press, Oxford, UK.

PROBLEMS AND QUESTIONS

Section 10.3

10.1 Define the following terms in your own words: (a) elastic stress concentration factor, k_t; (b) stress intensity factor, K; (c) fatigue notch factor, k_f; and (d) notch sensitivity, q. You may use equations to supplement, but not to replace, your word definitions.

10.2 Determine k_t and then estimate k_f for the notched member of Fig. 10.3. Do you agree with the values given? Why might there be a discrepancy in k_f?

10.3 For the notched member of Fig. 10.10:
   (a) Determine your own value of k_t and then estimate k_f. (Note that the half-circular notch shape gives $w_1 = w_2 - 2\rho$.)
   (b) For 10^6 cycles, use your k_f value with the smooth specimen fatigue strength from the S-N curve to estimate the notched specimen strength. How good is the agreement with the test data?

10.4 A shaft with a step-down in diameter has dimensions, as defined in Fig. 10.2(b), of $d_1 = 50$, $d_2 = 55$, and $\rho = 1.3$ mm. The shaft is subjected to bending and is made of a quenched and tempered low-alloy steel having $\sigma_u = 1100$ MPa.
   (a) Determine k_t and then estimate k_f.
   (b) The 10^6 cycle fatigue strength in bending of this steel in unnotched form is $\sigma_{ar} = 500$ MPa. What completely reversed bending moment amplitude M_a can be applied to the notched shaft for 10^6 cycles before fatigue failure is expected?

Sections 10.4 and 10.5

10.5 For notched members of ductile materials, explain in your own words why k'_f tends to approach unity at short fatigue lives, and also why k_{fm} tends to approach zero at short fatigue lives.

10.6 Verify that the initial yielding case of Eq. 10.26 is correct for loading with a compressive value of S_m. (Suggestion: sketch the local stress-strain response at the notch.)

10.7 For the notched member of Fig. 10.10, estimate the repeatedly applied zero-to-maximum nominal stress S_{max} that corresponds to a life of $N_f = 10^5$ cycles: (a) using Eq. 10.21 and (b) using Eq. 10.27. (Suggestion: Use one or more values read from the graphs in Fig. 10.10.)

10.8 For the same situation as in Example 10.2, determine the mean value M_m of bending moment that will result in a life of 10^6 cycles if cycling is applied at a moment amplitude of $M_a = 1.5$ kN·m: (a) using Eq. 10.21 and (b) using Eq. 10.27.

10.9 Aluminum alloy 2024-T4 as in Table 9.1 is to be used in the form of a plate with a central hole under axial loading as in Fig. 10.1(a). The dimensions are $w = 50$, $d = 10$, and $t = 20$ mm. For a life of $N_f = 10^7$ cycles, estimate the load amplitude P_a if the mean load is $P_m = 80$ kN: (a) using Eq. 10.21 and (b) using Eq. 10.27.

Sections 10.6 and 10.7

10.10 According to Fig. 10.15, most steels with ultimate tensile strengths beyond $\sigma_u = 1400$ MPa have fatigue limits below half the ultimate. However, a few high-strength steels ($\sigma_u = 1700$ to 2000 MPa) have higher fatigue limits than is typical, with the values still being near the $\sigma_{erb} = 0.5\sigma_u$ line. Speculate on how the tensile properties other than σ_u for such steels might compare to those for more ordinary high-strength steels.

10.11 Generalize the S-N curve estimate of Collins, Eq. 10.34, to depend on m and N_e, so that it can be used for cases other than steels where $m = 0.5$ and $N_e = 10^6$.

10.12 The notched plate in bending of Ex. 10.2, which is made of RQC-100 steel with $\sigma_u = 758$ MPa, is noted to have a value of $k_f = 1.82$.
(a) Estimate the S-N curve according to Collins.
(b) What life is expected for a moment amplitude of $M_a = 2.0$ kN·m applied at a mean level of $M_m = 2.5$ kN·m?

10.13 Repeat Prob. 10.12 but use the S-N curve estimate of Shigley, and assume that the notch has a machined surface.

10.14 The axially loaded plate with a hole of Prob. 10.9, made of 2024-T4 aluminum as in Table 9.1, has a value of $k_f = 2.36$, and the hole is smoothly polished.
(a) Estimate the S-N curve according to the procedure of Juvinall.
(b) What life is expected if a nominal stress amplitude of $S_a = 60$ MPa is applied at a mean level of $S_m = 30$ MPa?

10.15 An S-N curve is given for rotating bending of unnotched, smoothly polished samples of A517 steel in Fig. 9.5. Additional curves are given for polished unnotched samples of the same material, but under axial load, including a curve for completely reversed loading, in Fig. 9.35.
(a) Apply the estimate of Collins to both sets of completely reversed data and comment on its success. (Suggestion: Obtain approximate stress values from the curves at $N_f = 10^4$,

10^5, 10^6, and 10^7, and plot them on log-linear scales using graph paper or computer graphics.)

(b) Proceed as in (a), except use the Juvinall procedure, which requires log-log scales.

(c) Proceed as in (a), except use the Shigley procedure, which requires log-log scales.

(d) Based on (a), (b), and (c), comment on the relative success for these data of the three estimates.

10.16 Man-Ten steel (Table 9.1) is made into axially loaded notched test specimens in the form of plates with central holes. The dimensions, as defined in Fig. 10.1(a), are $w = 19.05$, $d = 5.08$, and $t = 2.54$ mm, and the inside of the hole is smoothly polished.

(a) Estimate the S-N curve for completely reversed ($R = -1$) loading using the procedure of Collins. Compare your estimate to the actual test data listed below by plotting them together on the same graph, and also comment on the success of the estimate.

(b) Repeat (a), but use the procedures of both Juvinall and Shigley.

S_a, MPa	N_f, cycles
517	12
421	108
421	330
338	683
286	5.73×10^3
245	2.52×10^4
217	4.79×10^4
207	1.50×10^5
172	3.66×10^5
162	1.68×10^6
138	1.00×10^7 (no failure)

Source: Data in [Rosenberger 68].

10.17 For the AISI 4340 steel notched member of Fig. 10.10, some points on the S-N curve are given below. Proceed for this data as in Prob. 10.16(a) and (b).

S_a, MPa	N_f, cycles
696	10^2
617	$10^{2.5}$
538	10^3
459	$10^{3.5}$
379	10^4
300	$10^{4.5}$
231	10^5
193	$10^{5.5}$
176	10^6 to 10^7

10.18 Aluminum alloy 2024-T351 is made into axially loaded notched test specimens which are plates with central holes. The dimensions, as defined in Fig. 10.1(a), are $w = 50.8$, $d = 12.7$, and $t = 2.22$ mm, and the inside of the hole is smoothly polished. This material has properties from a tension test of $\sigma_u = 469$ MPa, $\sigma_o = 379$ MPa, and reduction in area 25%. Estimate the S-N curve for zero-to-maximum $(R = 0)$ loading using the procedure of Juvinall. Compare your estimate to the actual test data below by plotting them together on the same graph, and also comment on the success of the estimate. (Suggestion: Estimate the S-N curve for $R = -1$, and then use S_{ar} values at several lives on this curve to estimate corresponding S_{max} values applied at $R = 0$.)

S_{max}, MPa	N_f, cycles
365	2.4×10^3
313	4.5×10^3
261	1.5×10^4
209	3.2×10^4
209	3.6×10^4
157	1.3×10^5
115	4.0×10^5
105	7.2×10^5
101	7.0×10^6 (no failure)

Source: Data in [Wetzel 68].

Section 10.8

10.19 Examine a bicycle front fork.
  (a) Draw a sketch of it and point out any design features that are beneficial in avoiding fatigue failure.
  (b) Suggest any design changes that might improve fatigue resistance.
  (c) Of what material does this part appear to be made? Try to explain the choice of material from the viewpoints of function, cost, and resistance to fatigue, wear, and other possible failure causes.

10.20 Examine the leaf or coil springs in the axle suspension system of a small boat trailer, and then answer (a), (b), and (c) as in Prob. 10.19.

10.21 Examine an airplane propeller, and then answer (a), (b), and (c) as in Prob. 10.19.

10.22 Assume that the notched member of Fig. 10.10 is a component that is required to resist 2000 cycles of a nominal stress amplitude of $S_a = 220$ MPa, applied at a mean level of $S_m = 140$ MPa. Obtain approximate values for the safety factors in both stress and life.

10.23 A circular rod made of the AISI 4142 (450 HB) steel of Table 9.1 is loaded axially and has a step change in diameter. The dimensions, as defined in Fig. 10.2(a), are $d_1 = 15$, $d_2 = 18$, and $\rho = 1$ mm, and the fillet radius is ground.
  (a) Using the S-N curve estimate of Juvinall, evaluate the safety factors in both stress and life if the expected service loading is 30,000 cycles at a zero-to-tension load of $P_{max} = 70$ kN.
  (b) A safety factor in stress of $X_S = 1.7$ is required. Is the design adequate? If not, try to redesign the shaft by increasing the fillet radius.

10.24 The notched rod of Prob. 10.23 with $\rho = 1$ mm is subjected to the load history shown in Figure P10.24. Assume that the S-N curve with the notch effect of $k_f = 1.92$ included is

$$S_{ar} = 2250 \, N_f^{-0.172} \text{ MPa} \quad \left(10^3 \leq N_f \leq 10^6\right)$$

In the expected service of this component, 500 repetitions of the load history are anticipated, and a safety factor of 10 on life is required.

(a) Estimate the number of repetitions of this load history that can be applied before fatigue failure occurs.

(b) Is the design adequate? If not, estimate the ratio by which the stresses should be lowered by redesign or less severe usage.

(c) What is the safety factor in stress for the original design?

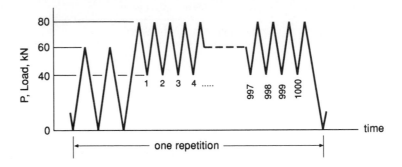

Figure P10.24

10.25 In an engineering failure that actually occurred, the shaft shown in Figure P10.25 was supported by bearings at its ends and was used in a hoist for lifting ore out of a mine. It was loaded in rotating bending by the loads shown and failed after 15 years of service and 2×10^7 rotations. An *impact factor* of 1.5 has already been included in the loads given to roughly account for loads exceeding the dead weight of the ore, cable, and bucket. (An impact factor is needed due to such effects as bouncing during loading at the bottom of the mine and acceleration to start moving the ore upward.) The shaft was made of AISI 1040 steel with tensile properties of $\sigma_u = 620$ MPa, $\sigma_o = 310$ MPa, and reduction in area 25%. The design drawings showed a fillet radius of 6.35 mm at the failure location, but measurements on the shaft after failure indicated that it had been machined at only 2 mm. Estimate the safety factor in stress for both the intended and actual fillet radii. Does this error in manufacture explain the failure?

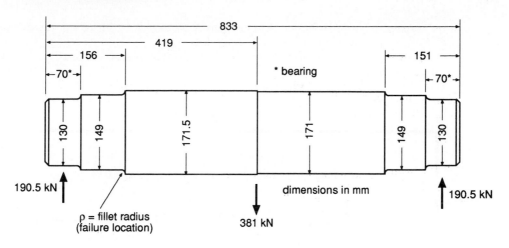

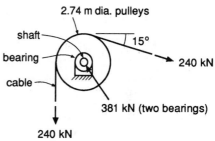

Figure P10.25

11

Fatigue Crack Growth

11.1 INTRODUCTION

The presence of a crack can significantly reduce the strength of an engineering component due to brittle fracture as already discussed in Chapter 8. However, it is unusual for a crack of dangerous size to exist initially, although this can occur, as when there is a large defect in the material used to make a component. A more common situation is that a small flaw that was initially present develops into a crack and then grows until it reaches the critical size for brittle fracture.

Crack growth can be caused by cyclic loading, a behavior called *fatigue crack growth*. However, if a hostile chemical environment is present, even a steady load can cause *environmental crack growth*. Both types of crack growth can occur if cyclic loads are applied in the presence of a hostile environment, especially if the cycling is slow or if there are periods of steady load interrupting the cycling. This chapter primarily considers fatigue crack growth, but limited discussion of environmental crack growth is included near the end in Section 11.10.

Engineering analysis of crack growth is often required and can be done using the stress intensity concept, K, of fracture mechanics. Recall from Chapter 8 that K

quantifies the severity of the combination of crack length, loading, and geometry that exists, being expressed by

$$K = FS\sqrt{\pi a} \tag{11.1}$$

where a is crack length, and S is nominal stress, usually defined based on the gross area of the uncracked member, and F is a dimensionless function of geometry. The value of F is affected by the relative crack length, $\alpha = a/b$, where b is a width dimension of the member such that $\alpha = 1$ for complete cracking.

The rate of fatigue crack growth is controlled by K. Hence, under constant amplitude cyclic loading, the dependence of K on a and F causes cracks to accelerate as they grow. The variation of crack length with cycles is thus similar to Fig. 11.1(a).

11.2 PRELIMINARY DISCUSSION

Before proceeding in detail, it is useful to describe the general nature of crack growth analysis and the need for it, and further to present some definitions.

11.2.1 Need for Crack Growth Analysis

It has been found from experience that careful inspection of certain types of hardware often reveals the presence of cracks. For example, this is the case for large welded components, such as pressure vessels and bridge and ship structure, for metal structure in large aircraft, and for large forgings, as in the rotors of turbines and generators in power plants. Cracks are especially likely to be found in such hardware after some actual service usage has occurred. The likely presence of cracks strongly suggests that specific analysis based on fracture mechanics is appropriate.

Let us assume that a certain structural component may contain cracks, but none are larger than a known *minimum detectable length* a_d. This situation could be the result of an inspection that is capable of finding all cracks larger than a_d, so that all such cracks have been repaired, or the parts scrapped. (Inspections for cracks are done by a variety of means, including visual examination, x-ray photography, reflection of ultrasonic waves, and application of electric currents, where in the latter case a crack causes a detectable disturbance in the resulting voltage field.) This worst-case crack of initial length a_d then grows until it reaches a critical length a_c, where brittle fracture occurs after N_{if} cycles of loading. If the number of cycles expected in actual service is $\hat{N}$, then the safety factor on life is

$$X_N = \frac{N_{if}}{\hat{N}} \tag{11.2}$$

This situation is illustrated in Fig. 11.1(a). Such a safety factor is needed because uncertainties exist as to the actual stress that will occur in service, the exact a_d that can be reliably found, and the cracks growth rates in the material.

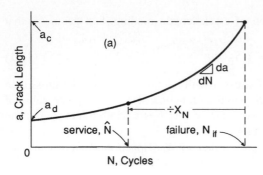
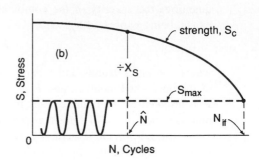

Figure 11.1 Growth of a worst-case crack from the minimum detectable length a_d to failure (a), and the resulting variation in worst-case strength (b).

The critical strength for brittle fracture of the member is determined by the current crack length and the fracture toughness K_c for the material and thickness involved.

$$S_c = \frac{K_c}{F\sqrt{\pi a}} \qquad (11.3)$$

As the worst-case crack grows, its length increases, causing the worst-case strength S_c to decrease, failure occurring when S_c reaches S_{max}, the maximum value for the cyclic loading applied in actual service. This is illustrated in Fig. 11.1(b). The safety factor on stress against sudden brittle fracture due to the applied cyclic load is

$$X_S = \frac{S_c}{S_{max}} \qquad (11.4)$$

Such a safety factor is generally needed in addition to X_N because of the possibility of an unexpected high load that exceeds the normal cyclic load. Within the expected actual service life, X_S decreases and has its minimum value at the end of this service life.

It sometimes occurs that the combination of minimum detectable crack length a_d and cyclic stress is such that the safety margin, as expressed by X_N and X_S, is insufficient. Predicted failure prior to reaching the actual service life, $X_N < 1$, may even be the case. *Periodic inspections* for cracks are then necessary, following which any cracks exceeding a_d are repaired, or the part replaced. This assures that after each inspection no cracks larger than a_d exist. Assuming that inspections are done at intervals of N_p cycles, the length of the worst-case crack increases due to growth between inspections, varying as shown in Fig. 11.2. The safety factor on life is then determined by the inspection period.

$$X_N = \frac{N_{if}}{N_p} \qquad (11.5)$$

After each inspection, the worst-case strength of the member increases as shown in Fig. 11.2(b). The safety factor on stress is lowest just prior to each inspection.

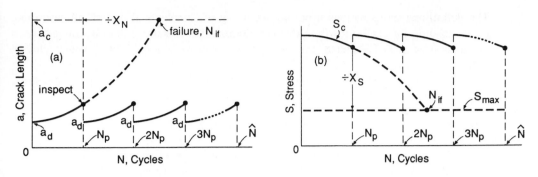

Figure 11.2   Variation of worst-case crack length (a) and strength (b) where periodic inspections are required.

Analysis based on fracture mechanics allows the variations in worst-case crack length and strength to be estimated so that safety factors can be evaluated. Where periodic inspections are necessary, fracture mechanics analysis thus permits a safe inspection interval to be set. For example, for large military and civilian aircraft, cracks are so commonly found during periodic inspections that safe operation and economic maintenance are both critically dependent on fracture mechanics analysis. The term *damage tolerant design* is used to identify this approach of requiring that structures be able to survive even in the presence of growing cracks.

In addition to design applications, analysis of crack growth life is also useful in situations where an unexpected crack has been found in a component of a machine, vehicle, or structure. The remaining life can be calculated to determine whether the crack may be ignored, or whether repair or replacement is needed immediately, or whether this can be postponed until a more convenient time. Situations of this sort have arisen in steel-mill machinery, where an immediate shutdown would disrupt operations and perhaps cause a large employee layoff. Similar situations have also occurred in turbine-generator units in major electrical power plants, where fracture of a large steel component could cause a power outage and expenditures of millions of dollars.

11.2.2 Definitions for Fatigue Crack Growth

Consider a growing crack that increases its length by an amount Δa due to the application of a number of cycles ΔN. The rate of growth with cycles can be characterized by the ratio $\Delta a/\Delta N$, or for small intervals, by the derivative da/dN. The *fatigue crack growth rate, da/dN*, is of course the slope at a point on an a versus N curve as in Fig. 11.1(a).

Assume that the applied loading is cyclic with constant values of the loads P_{max} and P_{min}, hence also with constant values of the nominal stresses S_{max} and S_{min}. For fatigue crack growth, it is conventional to use the range and stress ratio.

$$\Delta S = S_{max} - S_{min}, \qquad R = \frac{S_{min}}{S_{max}} \tag{11.6}$$

The definitions and equations from the beginning of Chapter 9 also apply in describing cyclic loading for crack growth. However, nominal stresses are generally defined based on gross area as in Chapter 8, this being convenient to avoid having S values change with crack length.

The primary variable affecting the growth rate of a crack is the range of the stress intensity factor. This is calculated from the stress range ΔS.

$$\Delta K = F \ \Delta S \sqrt{\pi a} \tag{11.7}$$

The value of F depends only on the geometry and the relative crack length, $\alpha = a/b$, just as if the loading were not cyclic. Since K and S are proportional for a given crack length according to Eq. 11.1, the maximum, minimum, range, and R-ratio for K during a loading cycle are given by

$$K_{\max} = F S_{\max} \sqrt{\pi a}, \qquad K_{\min} = F S_{\min} \sqrt{\pi a}$$

$$\Delta K = K_{\max} - K_{\min}, \qquad R = \frac{K_{\min}}{K_{\max}} \tag{11.8}$$

Also, it may be convenient, especially for laboratory test specimens, to use the alternate expression of K in terms of applied load P as discussed in Chapter 8 relative to Eq. 8.15.

$$\Delta K = F_P \frac{\Delta P}{t \sqrt{b}}, \qquad R = \frac{P_{\min}}{P_{\max}} \tag{11.9}$$

Recall that F_P is also a function of a/b, with some F_P values being given in Fig. 8.15, and others being obtainable from F using Eq. 8.17.

11.2.3 Describing Fatigue Crack Growth Behavior of Materials

For a given material and set of test conditions, the crack growth behavior can be described by the relationship between cyclic crack growth rate da/dN and stress intensity range ΔK. Test data and a fitted curve for one material are shown on a log-log plot in Fig. 11.3. At intermediate values of ΔK, there is often a straight line on the log-log plot as in this case. A relationship representing this line is

$$\frac{da}{dN} = C(\Delta K)^m \tag{11.10}$$

where C is a constant and m is the slope on the log-log plot, assuming of course that the decades on both log scales are the same length. This equation is identified with P.C. Paris, who first used it and who was influential in the first application of fracture mechanics to fatigue in the early 1960s.

At low growth rates, the curve generally becomes steep and appears to approach a vertical asymptote denoted ΔK_{th}, which is called the *fatigue crack growth threshold*.

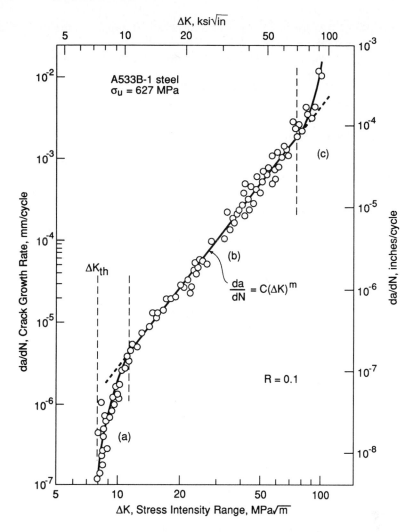

Figure 11.3   Fatigue crack growth rates over a wide range of stress intensities for a ductile pressure vessel steel. Three regions of behavior are indicated: (a) slow growth near the threshold ΔK_{th}, (b) intermediate region following a power equation, and (c) unstable rapid growth. (Plotted from the original data for the study of [Paris 72].)

This quantity is interpreted as a lower limiting value of ΔK below which crack growth does not ordinarily occur. At high growth rates, the curve may again become steep. This is due to rapid unstable crack growth just prior to final failure of the test specimen. Such behavior can occur where the plastic zone is small, in which case the curve asymptotically approaches the fracture toughness K_c for the material and thickness of interest. Rapid unstable growth at high ΔK sometimes involves fully plastic yielding. In such cases, the use of ΔK for this portion of the curve is improper as the theoretical limitations of the K concept are exceeded.

The value of the stress ratio R affects the growth rate in a manner analogous to the effects observed in S-N curves for different values of R or mean stress. For a given ΔK, increasing R increases the growth rate, and vice-versa. Some data illustrating this effect for a steel are shown in Fig. 11.4.

11.2.4 Discussion

The logical path involved in evaluating the crack growth behavior of a material and using the information is summarized in Fig. 11.5. First, a convenient test specimen geometry is

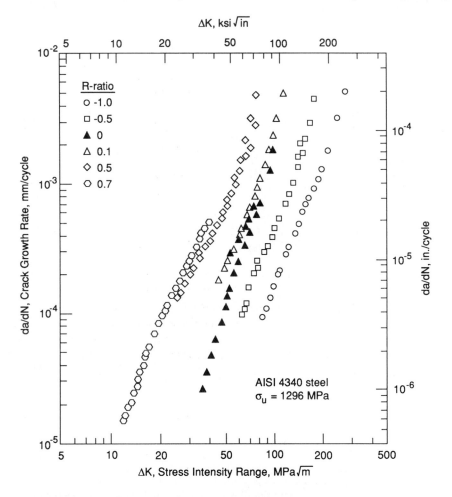

Figure 11.4 Effect of R-ratio on crack growth rates for an alloy steel. For $R < 0$, the compressive portion of the load cycle is here included in calculating ΔK. (Data from [Dennis 86].)

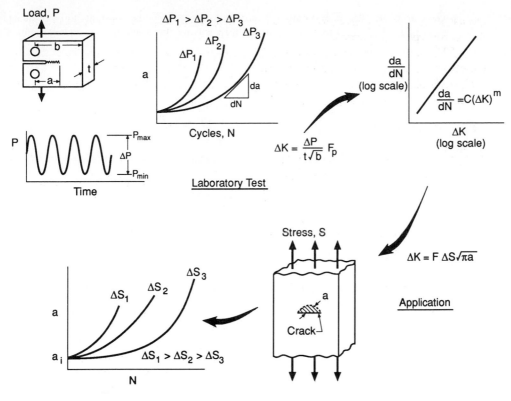

Figure 11.5 Steps in obtaining da/dN vs. ΔK data and using it for an engineering application. (Adapted from [Clark 71]; used with permission.)

employed in tests at each of several different load levels, so that a wide range of fatigue crack growth rates are obtained. Growth rates are then evaluated and plotted versus ΔK to obtain the da/dN versus ΔK curve. This curve can later be used in an engineering application, with ΔK values being calculated as appropriate for the particular component geometry of interest. Crack length versus cycles curves for a specific initial crack length can be predicted for the component, leading to life estimates and the determination of safety factors and inspection intervals as discussed earlier.

Example 11.1

Obtain approximate values of constants C and m of Eq. 11.10 for the data at $R = 0.7$ in Fig. 11.4.

Solution These data appear to fall along a straight line on this log-log plot, so that it is reasonable to apply Eq. 11.10. Aligning a straight edge with the data gives a line that passes near two points as follows:

$$\left(\Delta K, \frac{da}{dN}\right) = \left(10, 10^{-5}\right) \quad \text{and} \quad \left(110, 10^{-2}\right)$$

where units of MPa$\sqrt{m}$ and mm/cycle are used. These allow the constants to be evaluated by applying Eq. 11.10 to each point and then solving the resulting two equations for C and m.

$$10^{-5} = C \ 10^m, \quad 10^{-2} = C \ 110^m$$

Dividing the second equation by the first allows C to be eliminated.

$$\frac{10^{-2}}{10^{-5}} = \left(\frac{110}{10}\right)^m, \quad 10^3 = 11^m$$

Taking logarithms of both sides of the second equation just above and solving for m gives

$$m = \frac{\log \ 10^3}{\log \ 11} = 2.88 \qquad \textbf{Ans.}$$

Finally, we obtain C by substituting this m for either point above.

$$10^{-5} = C \ 10^{2.88}, \quad C = 1.32 \times 10^{-8} \qquad \textbf{Ans.}$$

Discussion If a more accurate fit is desired, the original source of the data should be consulted for numerical values of the data points, and a log-log least squares line obtained of the form

$$y = mx + b$$

where

$$y = \log \frac{da}{dN}, \quad x = \log (\Delta K), \quad m = m, \quad b = \log \ C$$

The procedure has already been illustrated in Example 5.3 for an equation of the same mathematical form.

11.3 FATIGUE CRACK GROWTH RATE TESTING

Standard methods for conducting fatigue crack growth tests have been developed, notably ASTM Standard No. E647. The most commonly used test specimen geometries are the standard compact specimen, Fig 8.15(c), and center cracked plates, Fig. 8.13(a), but others may be used.

11.3.1 Test Methods and Data Analysis

In a typical test, constant amplitude cyclic loading is applied to a specimen of a size such that its width dimension b (as defined in Chapter 8) is perhaps 50 mm. Before starting the test, a precrack is necessary. This is accomplished by first machining a sharp notch into the specimen and then starting a crack by cyclic loading at a low level. Cyclic loading is then applied at the higher level to be used for the remainder of the test. The progress of the crack is recorded in terms of the numbers of cycles required for its length to reach each of ten to twenty or more different values, with these being on the order of

1 mm apart for a specimen of size $b \approx 50$ mm. The resulting crack length data may then be plotted as discrete points versus the corresponding cycle numbers as in Fig. 11.6.

To measure these crack lengths, one approach is simply to note by visual observation, through a low-power (20 to 50X) microscope, when the crack reaches various lengths that have been previously marked on the specimen. An arrangement for such a test is shown in Fig. 11.7. More sophisticated means may be used to measure crack lengths. For example, as the crack grows, the deflection of the specimen increases,

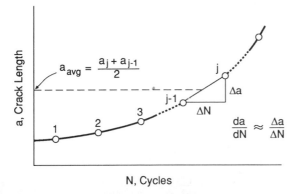

Figure 11.6 Crack growth rates obtained from adjacent pairs of a vs. N data points.

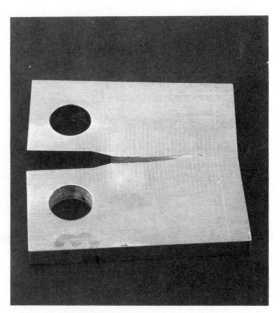

Figure 11.7 Crack growth rate test underway (left) on a compact specimen ($b = 51$ mm), with a microscope and a strobe light being used to visually monitor crack growth. Cycle numbers are recorded when the crack reaches each of a number of scribe lines (right.) (Photos by R. A. Simonds.)

resulting in a decreased stiffness. This stiffness change may be measured and used to calculate the crack length. Another approach is to pass an electrical current through the specimen, measuring changes in the voltage field due to growth of the crack, from which one can obtain its length. Ultrasonic waves can also be reflected from the crack and used to measure its progress.

To obtain growth rates from crack length versus cycles data, a simple and generally suitable approach is to calculate straight-line slopes between the data points as shown in Fig. 11.6. If the data points are numbered, $1, 2, 3 \ldots j$, then the growth rate for the segment ending at point number j is

$$\left(\frac{da}{dN}\right)_j \approx \left(\frac{\Delta a}{\Delta N}\right)_j = \frac{a_j - a_{j-1}}{N_j - N_{j-1}} \qquad (11.11)$$

The corresponding ΔK is calculated from the average crack length during the interval.

$$\Delta K_j = F \, \Delta S \sqrt{\pi a_{\text{avg}}}, \qquad \Delta K_j = F_P \frac{\Delta P}{t\sqrt{b}} \qquad (11.12)$$

whichever is more convenient, where

$$a_{\text{avg}} = \frac{a_j + a_{j-1}}{2} \qquad (11.13)$$

The geometry factor, $F = F(\alpha)$, or $F_P = F_P(\alpha)$, where $\alpha = a/b$, is evaluated at the same crack length using

$$\alpha_{\text{avg}} = \frac{a_{\text{avg}}}{b} = \frac{a_j + a_{j-1}}{2b} \qquad (11.14)$$

The above procedure is valid only if the crack length is measured at fairly short intervals. Otherwise, the growth rate and K may differ so much between adjacent observations that the averaging involved causes difficulties. Detailed requirements are given in the ASTM Standard. Also, curve-fitting methods of evaluating da/dN, which are more sophisticated than simple point-to-point slopes, are sometimes used to smooth the scatter in the a versus N data. Fitting a polynomial over all of the data from a test usually does not work very well, but such a fit applied in an incremental manner to portions of the data works well as described in the ASTM Standard.

Example 11.2

Crack length versus cycles data are given in Table E11.2(a) from a test on a center-cracked plate of 7075-T6 Al. The specimen had dimensions, as defined in Fig. 8.13(a), of $h = 445$, $b = 152.4$, and $t = 2.29$ mm. The load was cycled between zero and a maximum value of $P_{\text{max}} = 48.1$ kN. Obtain da/dN and ΔK values corresponding to the first five data points.

Solution The average growth rate between points No. 1 and 2 is obtained by applying Eq. 11.11 with $j = 2$.

$$\left(\frac{da}{dN}\right)_2 = \frac{a_2 - a_1}{N_2 - N_1} = \frac{7.62 - 5.08}{18,300 - 0} = 1.388 \times 10^{-4} \text{ mm/cycle} \qquad \textbf{Ans.}$$

TABLE E11.2(a)

No.	a, mm	N, kilocycles
1	5.08	0
2	7.62	18.3
3	10.16	28.3
4	12.70	35
5	15.24	40
6	17.78	43
7	20.32	47
8	22.86	50
9	25.40	52
10	30.48	57
11	35.56	59
12	40.64	61
13	45.72	62

Source: Data in [Hudson 69].

The corresponding ΔK is evaluated using the average crack length, hence using Eq. 11.13 with $j = 2$.

$$a_{\text{avg}} = \frac{a_2 + a_1}{2} = 6.35 \text{ mm}$$

To evaluate F, the expression of Fig. 8.12 is convenient.

$$F = \sqrt{\sec \frac{\pi a_{\text{avg}}}{2b}} = \sqrt{\sec \frac{\pi (6.35 \text{ mm})}{2(152.4 \text{ mm})}} = 1.001$$

Hence, using $\Delta S = \Delta P / 2bt$ in Eq. 11.12, we have

$$(\Delta K)_2 = F \ \Delta S \sqrt{\pi a_{\text{avg}}}$$

$$(\Delta K)_2 = 1.001 \frac{48,100 \text{ N}}{2(152.4 \text{ mm})(2.29 \text{ mm})} \sqrt{\pi (0.00635 \text{ m})} = 9.74 \text{ MPa}\sqrt{\text{m}} \qquad \textbf{Ans.}$$

Similarly applying Eqs. 11.11 to 11.13 with $j = 3$, and then with $j = 4$, etc., gives the additional values seen in Table E11.2(b).

TABLE E11.2(b)

j	a, mm	N, kilocycles	$\dfrac{da}{dN}, \dfrac{\text{mm}}{\text{cycle}}$	ΔK, MPa$\sqrt{\text{m}}$
1	5.08	0	—	—
2	7.62	18.3	1.388×10^{-4}	9.74
3	10.16	28.3	$2.54 \ \times 10^{-4}$	11.54
4	12.70	35	$3.79 \ \times 10^{-4}$	13.10
5	15.24	40	$5.08 \ \times 10^{-4}$	14.51

Ans.

11.3.2 Test Variables

Crack growth tests are most commonly conducted using zero-to-tension loading, $R = 0$, or tension-to-tension loading with a small R value, such as $R = 0.1$. Variations in R in the range 0 to 0.2 have little effect on most materials, and tests in this range are accepted by convention as the standard basis for comparing the effects of various materials, environments, etc. It is usually necessary to test several specimens at different load levels to obtain data over a wide range of growth rates. Such results for a steel are shown in Figs. 11.8 and 11.9. For more complete data, groups of several tests at each of several R values can be conducted to obtain data similar to Fig. 11.4. Also, if data are desired in the ΔK_{th} region, a special decreasing load test is needed as described in ASTM Standard E647.

A wide range of variables may affect fatigue crack growth rates in a given material, so that test conditions may be selected to include situations that resemble the anticipated service use of the material. Some of these variables are temperature, frequency of the cyclic load, and hostile chemical environments. Minor variations in the processing or composition of materials may affect fatigue crack growth rates due to the different microstructures that result. Hence, tests on different variations of a material may be conducted to aid in developing materials that can best resist fatigue crack growth.

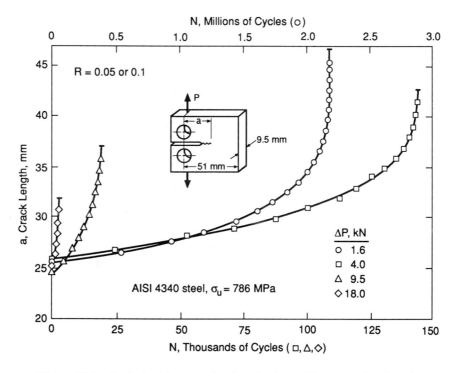

Figure 11.8 Crack length vs. cycles data for four different levels of cyclic load applied to compact specimens of an alloy steel.

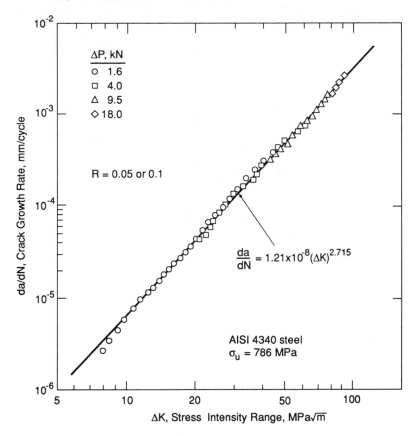

Figure 11.9 Data and least squares fitted line for da/dN vs. ΔK from the a vs. N data of Fig. 11.8.

11.3.3 Geometry Independence of *da/dN* vs. *ΔK* Curves

For a given material and set of test conditions, such as a particular R value, test frequency, and environment, the growth rates should depend only on ΔK. This arises simply from the fact that K characterizes the severity of a combination of loading, geometry, and crack length, and ΔK serves the same function for cyclic loading. Hence, regardless of the load level, crack length, and specimen geometry, all da/dN vs. ΔK data for a given set of test conditions should fall together along a single curve, except that some statistical scatter is of course expected. This occurs for the different load levels and crack lengths involved in Figs. 11.8 and 11.9. A single curve should occur even if more than one specimen geometry is included in the tests. Some data demonstrating geometry independence are shown in Fig. 11.10.

Such uniqueness of the da/dN vs. ΔK curve for different geometries is a crucial test of the applicability of the K concept to both materials testing and engineering

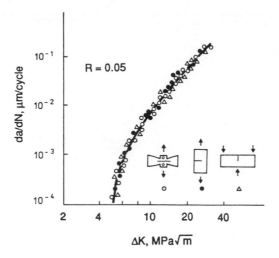

Figure 11.10 Fatigue crack growth rate data for a 0.65% carbon steel, demonstrating geometry independence. (Adapted from [Klesnil 80] p. 111; used with permission.)

applications. (Recall Fig. 11.5.) This uniqueness has been sufficiently verified, so that it is not generally necessary to include more than one test specimen geometry in obtaining materials data. However, difficulty with the applicability of ΔK can occur if there is excessive yielding, or for very small cracks, as discussed in Section 11.9 near the end of this chapter.

11.4 EFFECTS OF $R = S_{min}/S_{max}$ ON FATIGUE CRACK GROWTH

An increase in the R-ratio of the cyclic loading causes growth rates for a given ΔK to be larger, which has already been illustrated by Fig. 11.4. The effect is generally more pronounced for more brittle materials. For example, the granite rock of Fig. 11.11 shows an extreme effect, being sensitive to increasing R from 0.1 to only 0.2. In contrast, mild steel and other relatively low-strength, highly ductile, structural metals exhibit only a weak R effect in the intermediate growth rate region of the da/dN vs. ΔK curve.

11.4.1 Effects on ΔK_{th}

The R-ratio generally has a strong effect on the behavior at low growth rates, hence also on the threshold value ΔK_{th}. This occurs even for low-strength metals where there is little effect at intermediate growth rates. Some values of ΔK_{th} for various steels over a range of R-ratios are shown in Fig. 11.12. The lower limit of the scatter shown corresponds to ΔK_{th} as follows:

$$\Delta K_{th} = 6.0 \text{ MPa}\sqrt{\text{m}}, \qquad \Delta K_{th} = 5.5 \text{ ksi}\sqrt{\text{in}} \qquad (R \leq 0.17)$$

$$\Delta K_{th} = 7.0(1 - 0.85R) \text{ MPa}\sqrt{\text{m}} \qquad\qquad (R \geq 0.17) \qquad (11.15)$$

$$\Delta K_{th} = 6.4(1 - 0.85R) \text{ ksi}\sqrt{\text{in}} \qquad\qquad (R \geq 0.17)$$

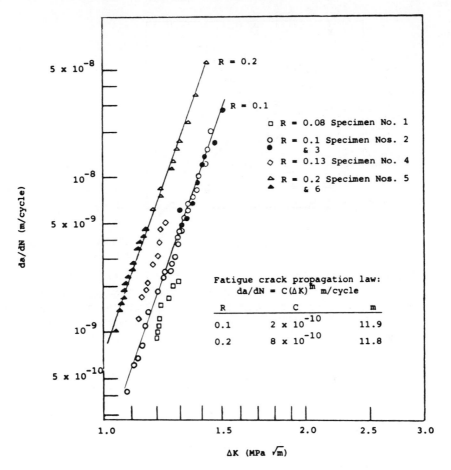

Figure 11.11   Effect of R-ratio on fatigue crack growth rates for Westerly granite, tested in the form of three-point bend specimens. (From [Kim 81]; copyright ©ASTM; reprinted with permission.)

Based on the discussion in Barsom and Rolfe (1987), these equations appear to represent a reasonable worst-case estimate for a wide range of steels. However, lower values of ΔK_{th} may apply for highly strengthened steels, which will be illustrated later. Similar trends occur for other classes of metals, but the numerical constants differ.

11.4.2 The Walker Equation

Various empirical equations are used for estimating the effect of R on da/dN vs. ΔK curves. One of the most widely used equations is based on a proposal by E. K. Walker that the effect can be estimated by

$$\overline{\Delta K} = K_{max}(1 - R)^{\gamma} \tag{11.16}$$

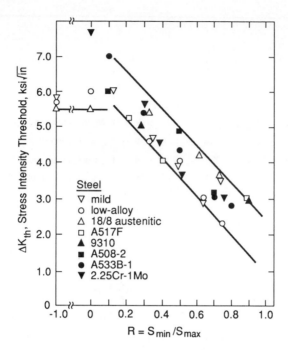

Figure 11.12 Effect of R-ratio on the threshold ΔK_{th} for various steels. For $R = -1$, the compressive portion of the loading cycle is here excluded in calculating ΔK_{th}. (Adapted from [Barsom 87] p. 285; ©1987 by Prentice Hall, Englewood Cliffs, N.J.; reprinted with permission.)

where γ is a constant for the material and $\overline{\Delta K}$ is an equivalent zero-to-maximum ($R = 0$) stress intensity that causes the same growth rate as the actual $K_{\max}$, R combination. By applying Eq. 9.4 to K, which gives $\Delta K = K_{\max}(1 - R)$, the above expression is seen to be equivalent to

$$\overline{\Delta K} = \frac{\Delta K}{(1 - R)^{1-\gamma}} \tag{11.17}$$

Let the constants C and m in Eq. 11.10 be denoted C_1 and m_1 for the special case of $R = 0$.

$$\frac{da}{dN} = C_1 (\Delta K)^{m_1} \qquad (R = 0) \tag{11.18}$$

Since $\overline{\Delta K} = \Delta K$ for $R = 0$, this substitution can be made above, giving

$$\frac{da}{dN} = C_1 \left[\frac{\Delta K}{(1 - R)^{1-\gamma}} \right]^{m_1} \tag{11.19}$$

This equation represents a family of da/dN vs. ΔK curves, which on a log-log plot are all parallel straight lines of slope m_1. Some manipulation gives

$$\frac{da}{dN} = \frac{C_1}{(1 - R)^{m_1(1-\gamma)}} (\Delta K)^{m_1} \tag{11.20}$$

Constants C and m for Eq. 11.10 are thus

$$C = \frac{C_1}{(1 - R)^{m_1(1-\gamma)}}, \quad m = m_1 \tag{11.21}$$

Hence, the intercept constant C is predicted to be a function of R, but the slope m is unaffected by R.

A useful interpretation arising from Eq. 11.19 is that $\overline{\Delta K}$, the equivalent zero-to-maximum $(R = 0)$ stress intensity, can be plotted versus da/dN, and a single straight line should result. The data of Fig. 11.4 are plotted in this manner in Fig. 11.13 using constants as listed in Table 11.1. Since all of the data lie quite close to the single line, the equation is reasonably successful. However, it was necessary to handle loadings

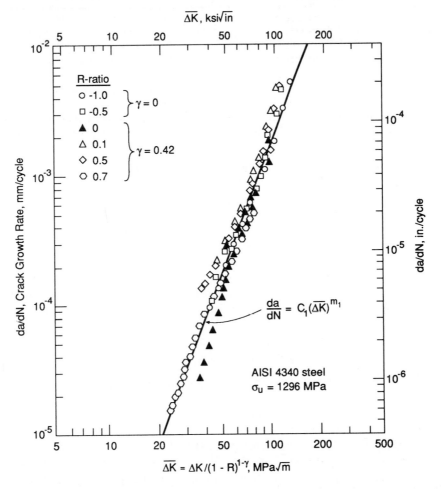

Figure 11.13   Representation of the data of Fig. 11.4 by a single relationship based on the Walker equation. (Data from [Dennis 86].)

TABLE 11.1   FATIGUE CRACK GROWTH CONSTANTS FOR THREE METALS

Material	Walker Equation[1]				Forman Equation[1]		
	C_1	m_1	γ $(R \geq 0)$	γ $(R < 0)$	C_2	m_2	K_c
AISI 4340 steel[2] ($\sigma_u = 1296$ MPa)	5.11×10^{-10} (2.73×10^{-11})	3.24	0.42	0	—	—	—
7075-T6 Al	2.71×10^{-8} (1.51×10^{-9})	3.70	0.64	0	5.29×10^{-6} (2.56×10^{-7})	3.21	78.7^3 (71.6)
2024-T3 Al	1.42×10^{-8} (7.85×10^{-10})	3.59	0.68	—	2.31×10^{-6} (1.14×10^{-7})	3.38	110^3 (100)

Notes: [1]Values of C_1, C_2, and K_c not in parentheses should be used with units of MPa$\sqrt{\text{m}}$ for K and mm/cycle for da/dN; those in parentheses are used with units of ksi$\sqrt{\text{in}}$ and in/cycle.   [2]Additional properties for this steel are yield strength $\sigma_o = 1255$ MPa and fracture toughness $K_{Ic} = 130$ MPa$\sqrt{\text{m}}$.   [3]For 2.3 mm thick sheet material.

Sources: Values for Al alloys adapted from [Hudson 69], including modification for use there of k, where $K = k\sqrt{\pi}$. Values for steel from [Dennis 86].

involving compression, $R < 0$, by assuming that the compressive portion of the cycle had no effect, which is accomplished by using $\gamma = 0$ where $R < 0$, so that $\overline{\Delta K} = K_{\max}$. This is reasonable based on the logic that the crack closes at zero load and no longer acts as a crack below this. In more ductile metals, the compressive portion of the loading may contribute to the growth, so that this approach is not universally applicable.

Values of the constant γ for various metals are typically around 0.5 but vary from around 0.3 to nearly 1.0. A value of $\gamma = 1$ gives simply $\overline{\Delta K} = \Delta K$, that is, no effect of R. Decreasing values of γ imply a stronger effect of R. Constants for the Walker equation are also given for two aluminum alloys in Table 11.1.

A value of γ can be obtained from data at various R values by a trial and error procedure. The desired γ is the one that best consolidates the data along a single straight line on a plot of da/dN vs. $\overline{\Delta K}$. Where a straight line on a log-log plot is expected, a good initial estimate of γ can be obtained by using the data for two different and contrasting R values as illustrated below in Example 11.3.

The Walker equation can be used for da/dN vs. ΔK equations of forms other than Eq. 11.10. Let the equation for $R = 0$ be $da/dN = f(\Delta K)$. The generalized equation which includes other R values is then

$$\frac{da}{dN} = f\left(\overline{\Delta K}\right) = f\left[\frac{\Delta K}{(1 - R)^{1-\gamma}}\right] \tag{11.22}$$

where as before γ is the value that consolidates the data for various R all onto a single line on the da/dN vs. $\overline{\Delta K}$ plot. However, the Walker equation is primarily used for intermediate growth rates for which Eq. 11.10 does generally apply. Attempts to use it over a wider range, as into the low growth rate region, may not be successful because there is generally an increased sensitivity to R at low growth rates. In other words, for a given material, there is often no single constant value of γ that works at both intermediate and low growth rates.

Example 11.3

Parallel straight lines on a log-log plot of da/dN vs. ΔK provide a reasonable fit to data for 2024-T3 Al at two different R-ratios as follows:

$$\frac{da}{dN} = 1.60 \times 10^{-8} (\Delta K)^{3.59} \quad (R = 0.1)$$

$$\frac{da}{dN} = 3.15 \times 10^{-8} (\Delta K)^{3.59} \quad (R = 0.5)$$

where units of MPa$\sqrt{m}$ and mm/cycle are used. Determine values of the constants C_1 and γ that permit this behavior to be described by the Walker equation.

Solution   Since the two exponents $m = m_1$ are the same, we can proceed. The coefficients C for the equations given represent two particular cases of C in Eq. 11.21.

$$C = \frac{C_1}{(1 - R)^{m_1(1-\gamma)}}$$

Substituting the given C values provides two equations.

$$1.60 \times 10^{-8} = \frac{C_1}{(1 - 0.1)^{3.59(1-\gamma)}}, \quad 3.15 \times 10^{-8} = \frac{C_1}{(1 - 0.5)^{3.59(1-\gamma)}}$$

Dividing the second equation by the first gives an expression involving only γ.

$$\frac{3.15 \times 10^{-8}}{1.60 \times 10^{-8}} = \left(\frac{0.9}{0.5}\right)^{3.59(1-\gamma)}, \quad 1.97 = 1.8^{3.59(1-\gamma)}$$

Taking logarithms of both sides leads to γ.

$$3.59(1 - \gamma) = \frac{\log 1.97}{\log 1.8}, \quad (1 - \gamma) = 0.321, \quad \gamma = 0.679 \qquad \textbf{Ans.}$$

Substituting this γ into Eq. 11.21 gives C_1.

$$C_1 = 1.60 \times 10^{-8}(1 - 0.1)^{3.59(0.321)} = 1.42 \times 10^{-8} \qquad \textbf{Ans.}$$

Comment   These constants agree precisely with those in Table 11.1, due to the two da/dN vs. ΔK equations that were given being members of the family of curves described by those constants.

11.4.3 The Forman Equation

Another proposed generalization to include R effects is that of Forman.

$$\frac{da}{dN} = \frac{C_2 (\Delta K)^{m_2}}{(1 - R) K_c - \Delta K} = \frac{C_2 (\Delta K)^{m_2}}{(1 - R)(K_c - K_{max})} \tag{11.23}$$

where K_c is the fracture toughness for the material and thickness of interest. The second form arises from the first simply by applying Eq. 9.4 to ΔK in the denominator. As K_{max} approaches K_c, the denominator approaches zero, and da/dN becomes large. In particular, there is an asymptote at $\Delta K/(1 - R) = K_{max} = K_c$. The equation thus has the attractive feature of predicting accelerated growth as the final toughness failure is approached, while conforming to Eq. 11.10 at low ΔK. Hence, it can be used to fit data that cover both the intermediate and high growth rate regions.

Assuming that crack growth data are available for various R values, these can be fitted to Eq. 11.23 by computing the following quantity for each data point.

$$Q = \frac{da}{dN}[(1 - R)K_c - \Delta K] \qquad (11.24)$$

If these Q values are plotted versus the corresponding ΔK values on a log-log plot, a straight line is expected. This is illustrated for 7075-T6 Al in Fig. 11.14. The slope of the Q vs. ΔK line on the log-log plot is m_2, assuming of course that the decades on both log scales are the same length. Also, the value of Q at $\Delta K = 1$ gives C_2.

For a given material, the success of the Forman equation can be judged by the extent to which data for various ΔK, R combinations all fall together on a straight line on a log-log plot of Q versus ΔK. For the data of Fig. 11.14(a), the consolidation

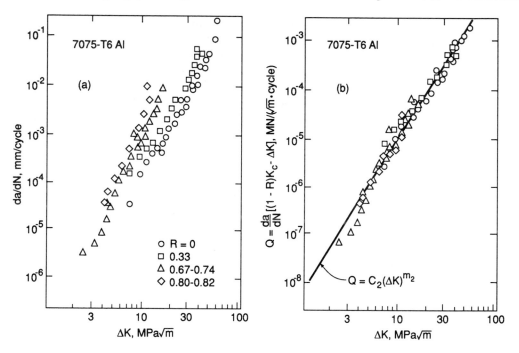

Figure 11.14 Effect of R-ratio on growth rates in 7075-T6 aluminum (a), and correlation of these data (b) based on the Forman equation, using constants as listed in Table 11.1. (Data from [Hudson 69].)

onto a straight line in (b) is reasonably successful. Constants for the Forman equation corresponding to these data, and also constants for one additional aluminum alloy, are given in Table 11.1.

11.4.4 Other Equations

A variety of other mathematical expressions, some of them quite complex, have been used to represent da/dN vs. ΔK curves. Some of these are not merely empirical, but are based on attempts to include modeling of the closing of the crack, and other physical phenomena that affect crack growth. Many give a curve shape similar to Fig. 11.3, where the curve steepens at both low and high growth rates. An example of this is the following modification of the Forman equation:

$$\frac{da}{dN} = \frac{C_3 (\Delta K)^{m_3} (\Delta K - \Delta K_{th})^{0.5}}{(1 - R) K_c - \Delta K} \tag{11.25}$$

This equation has an asymptote not only at high growth rates as $K_{\max}$ approaches K_c, but also one at low growth rates as ΔK approaches ΔK_{th}. As ΔK_{th} is sensitive to R, a specific value of this parameter in the equation is generally needed for any given R value.

The use of K_c as a constant in a crack growth equation is necessary for accurate representation of behavior at high growth rates. However, some care is needed. First, K_c varies with thickness unless the behavior is plane strain, where K_{Ic} applies. In addition, the severe cyclic loading that occurs just prior to brittle fracture at the end of a crack growth test may alter K_c, increasing it for certain materials and decreasing it for others. Further, for ductile, high-toughness materials such as mild steel, fatigue crack growth tests may be terminated by gross yielding instead of brittle fracture. It is then not appropriate to obtain a K_c value from such data, as the theoretical limitations of K are exceeded near the end of the test. The apparent K_c obtained will be lower than one from a larger size test specimen, where higher K values can be reached prior to yielding failure.

11.5 TRENDS IN FATIGUE CRACK GROWTH BEHAVIOR

Fatigue crack growth behavior differs considerably for different classes of material. It is also affected, sometimes to a large extent, by changes in the environment, such as temperature or hostile chemicals.

11.5.1 Trends with Material

The crack growth behavior in air at room temperature may vary only modestly within a narrowly defined class of materials. For example, data for $R \approx 0$ for several ferritic-pearlitic steels are shown in Fig. 11.15. An equation of the form of Eq. 11.10 for the intermediate growth rate region is shown that represents the worst case of the several

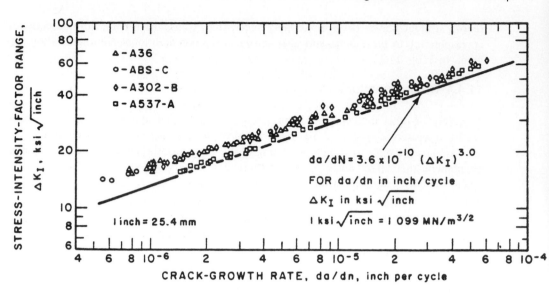

Figure 11.15 Fatigue crack growth rate data at $R \approx 0$ for four ferritic-pearlitic steels, and a line giving worst-case growth rates. Note that the axes are reversed compared to the other da/dN vs. ΔK plots given. (From [Barsom 71]; used with premission of ASME.)

steels tested. Recall from Chapter 3 that ferritic-pearlitic steels have low carbon contents and are relatively low-strength steels used for structural members, pressure vessels, and similar applications.

Additional worst-case da/dN vs. ΔK equations are given in Barsom and Rolfe (1987) for two additional classes of steel, namely martensitic steels and austenitic stainless

TABLE 11.2 CONSTANTS FOR WORST-CASE da/dN VS. ΔK CURVES FOR VARIOUS CLASSES OF STEELS FOR $R \approx 0$

	Constants[1] for $da/dN = C(\Delta K)^m$	
Class of Steel	C	m
Ferritic-pearlitic	6.89×10^{-9} (3.6×10^{-10})	3.0
Martensitic	1.36×10^{-7} (6.6×10^{-9})	2.25
Austenitic	5.61×10^{-9} (3.0×10^{-10})	3.25

Note: [1]Values of C not in parentheses should be used with units of MPa$\sqrt{m}$ for K and mm/cycle for da/dN; those in parentheses are used with units of ksi$\sqrt{in}$ and in/cycle.

Source: Values in [Barsom 87] pp. 288-291.

steels. The constants are given in Table 11.2. Martensitic steels are distinguished as being steels that are heat treated using quenching and tempering, so that this group includes many low-alloy steels, and also those 400-series stainless steels with less than 15% Cr. Austenitic steels are primarily the 300-series stainless steels, which are used where corrosion resistance is critical. These equations apply for R values near zero, say up to $R = 0.2$. However, for ductile steels that are only weakly affected by R at intermediate growth rates, they may apply even at higher values of R.

These general-purpose equations need to be used with some care, as exceptions do exist where they are not very accurate. For example, if the widely used martensitic steel AISI 4340 is heat treated to various strength levels, including very high strength levels, the crack growth rates may exceed the suggested worst case trend as illustrated in Fig. 11.16. In addition, the ΔK_{th} value may be considerably below the typical behavior, test data showing the trend of ΔK_{th} with strength level in this material being given in Fig. 11.17. The decrease in ΔK_{th} with strength parallels a similar trend in fracture toughness for this material. (See Fig. 8.27.)

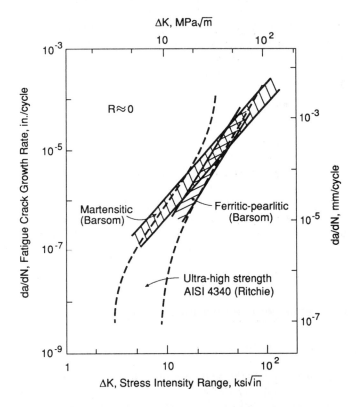

Figure 11.16 Ranges of fatigue crack growth behavior for various high-strength steels. (Illustration courtesy of R. W. Landgraf, Virginia Tech, Blacksburg, Va. See [Barsom 87] pp. 287 and 290, and [Ritchie 77], for more detail.)

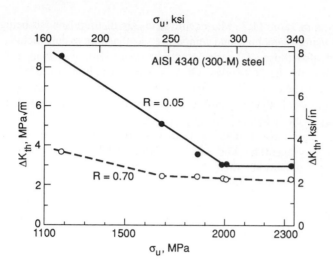

Figure 11.17 Effect of strength level of an alloy steel on ΔK_{th} at two R-ratios. (Adapted from [Ritchie 77]; used with permission of ASME.)

If different classes of metals are considered, crack growth rates differ considerably when compared on a da/dN vs. ΔK plot as shown in Fig. 11.18. However, the ΔK values corresponding to a given growth rate scale roughly with the elastic modulus E. Hence, a plot of da/dN vs. $\Delta K/E$ removes much of the difference as also shown. Polymers exhibit a wide range of growth rates when compared on the basis of ΔK as shown in Fig. 11.19. For any given ΔK level, the growth rates are considerably higher than for most metals.

One generalization that may be made is that the crack growth rate exponent m is higher for lower ductility (more brittle) materials. For ductile metals, m is typically in the range 2 to 4 and is often around 3. Higher exponents occur for more brittle cast metals, for short-fiber reinforced composites, and for ceramics. For example, m is near 12 for the granite rock of Fig. 11.11.

Despite the generalizations that may be made as to similar behavior within classes of materials, surprisingly small differences can sometimes have a significant effect. For example, decreasing the grain size in steels has the detrimental effect of lowering ΔK_{th}, while the behavior outside of the low growth rate region is relatively unaffected. Also, as might be expected, variations in reinforcement often have a significant effect on crack growth in composites materials, an example of which is given in Fig. 11.20.

11.5.2 Trends with Temperature and Environment

Changing the temperature usually affects the fatigue crack growth rate, with higher temperature often causing faster growth. Data illustrating such behavior for the austenitic (FCC) stainless steel AISI 304 are shown in Fig. 11.21. However, an opposite trend can occur in BCC metals due to the cleavage mechanism contributing to fatigue crack growth at low temperature. (See Section 8.5.) Such a trend for an Fe-21Cr-6Ni-9Mn alloy is also illustrated in Fig. 11.21. This alloy is austenitic at room temperature, but at low temperatures it is martensitic (BCC), and hence subject to cleavage, so that the more

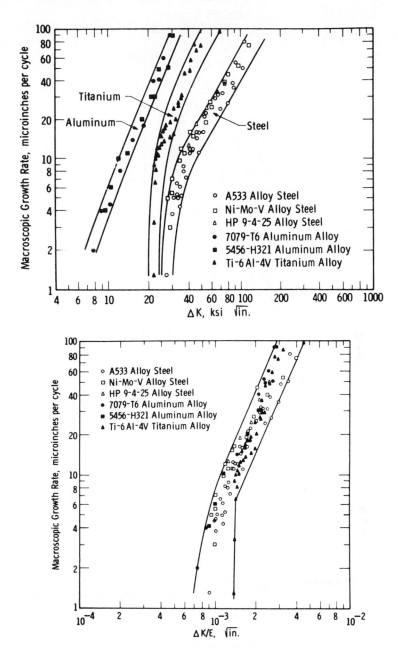

Figure 11.18 Fatigue crack growth trends for various metals, top, and correlation of the behavior, bottom, by plotting $\Delta K/E$. (From [Bates 69]; used with permission.)

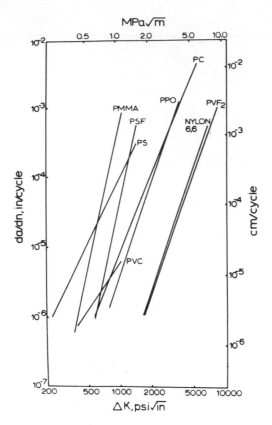

Figure 11.19 Fatigue crack growth trends for various crystalline and amorphous polymers. (From [Hertzberg 75]; used with permission.)

usual temperature effect is reversed. The effect of this cleavage contribution in BCC irons and steels can have a large effect on the fatigue crack growth exponent m for Eq. 11.10 as shown in Fig. 11.22. Suppressing this effect by adding sufficient nickel avoids high growth rates at low temperature.

Hostile chemical environments often increase fatigue crack growth rates, with certain combinations of material and environment causing especially large effects. The term *corrosion fatigue* is often used when the environment involved is a corrosive medium, such as seawater. Such behavior is illustrated by the effect of a saltwater solution similar to seawater on two strength levels of AISI 4340 steel in Fig. 11.23. The effect is considerably greater for the higher strength level of this steel. The effect on growth rate per cycle, da/dN, of a given hostile environment is usually greater at slower frequencies, where the environment has more time to act. This trend is apparent in the data of Fig. 11.24.

Even the gases and moisture in air can act as a hostile environment, which can be demonstrated by comparing test data in vacuum or an inert gas with data in air. Such comparisons for a metal and a ceramic are shown in Fig. 11.25. This circumstance results in frequency effects occurring in ambient air for some materials. Since chemical activity increases with temperature, the general trend of increasing growth rate with temperature is explained at least in part by the ambient air having a hostile effect.

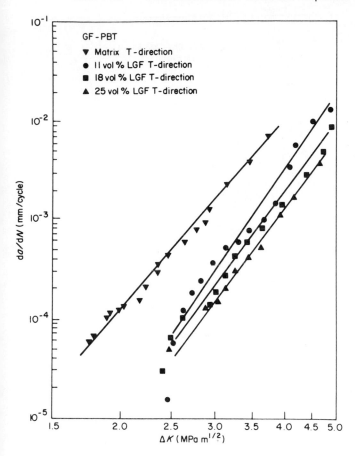

Figure 11.20 Effect on crack growth rates at $R = 0.2$ of various amounts of short glass fibers, in a matrix of the thermoplastic polymer PBT, with crack propagation perpendicular to the mold-fill direction. (From [Voss 88]; used with permission.)

11.6 LIFE ESTIMATES FOR CONSTANT AMPLITUDE LOADING

Since ΔK increases with crack length during constant amplitude stressing ΔS, and since the crack growth rate da/dN depends on ΔK, the growth rate is not constant, but increases with crack length. In other words, the crack accelerates as it grows, as for the data of Fig. 11.8. This situation of changing da/dN necessitates the use of an integration procedure to obtain the life required for crack growth.

Crack growth rates da/dN for a given combination of material and R-ratio are given as a function of ΔK by one of Eqs. 11.10, 11.19, and 11.23, or other similar equation, which may be represented in general by

$$\frac{da}{dN} = f(\Delta K, R) \tag{11.26}$$

where any effects of environment, frequency, etc., are assumed to be included in the material constants involved. The life in cycles required for crack growth may be calculated

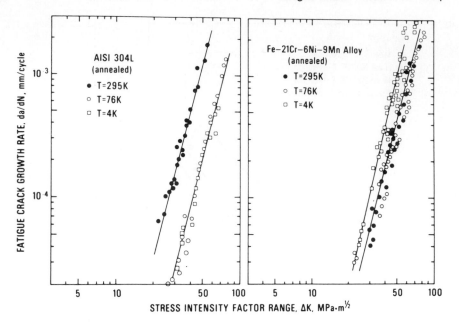

Figure 11.21   Effect of temperature on fatigue crack growth rates in two metals. (From [Tobler 78]; used with permission.)

by solving this equation for dN and integrating both sides.

$$\int_{N_i}^{N_f} dN = N_f - N_i = N_{if} = \int_{a_i}^{a_f} \frac{da}{f(\Delta K, R)} \qquad (11.27)$$

This integral gives the number of cycles required for the crack to grow from an initial size a_i at cycle number N_i, to a final size a_f at cycle number N_f. It is convenient to use the symbol N_{if} to represent the number of elapsed cycles, $N_f - N_i$.

To perform the integration for a particular case, it is necessary to substitute the specific da/dN equation for the material and R of interest, and also the specific equation for ΔK for the geometry of interest. Some useful closed form solutions exist, but numerical integration is necessary in many cases.

11.6.1 Closed Form Solutions

Consider a situation where growth rates are given by Eq. 11.10 and where $F = F(a/b)$ in Eq. 11.7 is constant, or can be approximated as constant, over the range of crack lengths a_i to a_f.

$$\frac{da}{dN} = f(\Delta K, R) = C(\Delta K)^m, \qquad \Delta K = F \, \Delta S \sqrt{\pi a} \qquad (11.28)$$

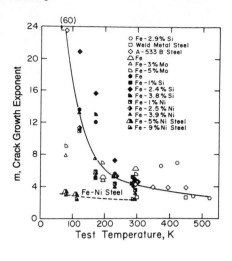

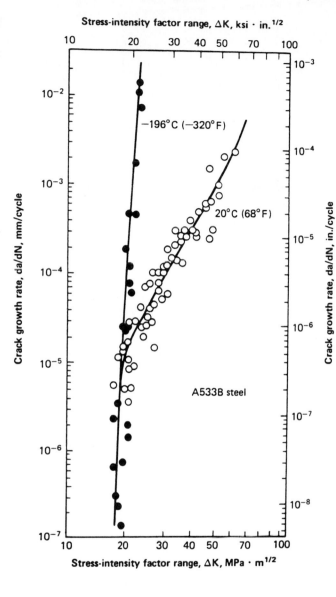

Figure 11.22 Variation (above) of the exponent m for the Paris equation with test temperature, for iron and various steels at R-ratios near zero, and (left) the associated drastic increase in growth rates at low temperature for A533B steel tested at $R = 0.1$. (Above from [Gerberich 79]; copyright ©ASTM; reprinted with permission. Left from [Campbell 82] p. 83, as based on data from [Stonesifer 76]; used with permission.)

The value of C used can include the effect of the ratio $R = S_{min}/S_{max}$, as from the Walker approach using Eq. 11.21. Assume that S_{max} and S_{min} are constant, so that ΔS and R are also both constant.

Substituting this particular $f(\Delta K, R)$ into Eq. 11.27, and then substituting for ΔK, gives

$$N_{if} = \int_{a_i}^{a_f} \frac{da}{C(\Delta K)^m} = \int_{a_i}^{a_f} \frac{da}{C(F\,\Delta S\sqrt{\pi a})^m} = \int_{a_i}^{a_f} \frac{1}{C(F\,\Delta S\sqrt{\pi})^m}\frac{da}{a^{m/2}} \quad (11.29)$$

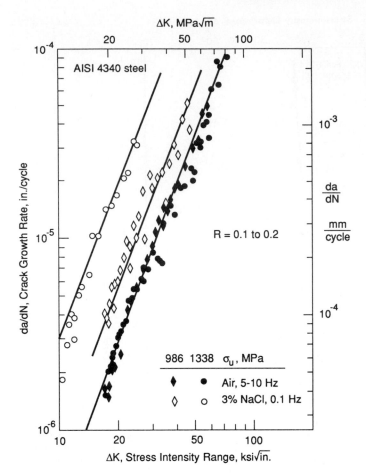

Figure 11.23 Contrasting sensitivity to corrosion fatigue crack growth of two strength levels of an alloy steel. (Adapted from [Imhof 73]; copyright ©ASTM; reprinted with permission.)

Since C, F, ΔS, and m are all constant, the only variable is a, and integration is straightforward, giving

$$N_{if} = \frac{a_f^{1-m/2} - a_i^{1-m/2}}{C\left(F\,\Delta S\sqrt{\pi}\right)^m (1 - m/2)} \quad (m \neq 2) \tag{11.30}$$

If $m = 2$, this equation is mathematically indeterminate. Manipulation yields a convenient equivalent form.

$$N_{if} = \left(\frac{1 - \left(a_i/a_f\right)^{m/2-1}}{C\left(F\,\Delta S\sqrt{\pi}\right)^m (m/2 - 1)}\right)\left(\frac{1}{a_i^{m/2-1}}\right) \quad (m \neq 2) \tag{11.31}$$

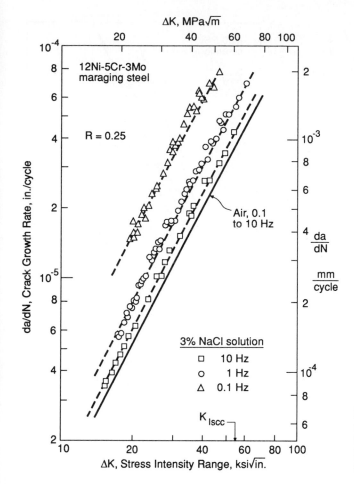

Figure 11.24 Frequency effects on corrosion fatigue crack growth rates in a maraging steel. (Adapted from [Imhof 73]; copyright ©ASTM; reprinted with permission.)

It is apparent from this second form that the life is highly sensitive to the value of a_i and less sensitive to a_f. In particular, for typical m values of say 3 or greater, and if the ratio a_i/a_f is small, the numerator approaches unity and the life depends mainly on a_i.

Additional closed form solutions exist that may be useful, such as one for the case of $m = 2$, with derivations of some of these being included as Problems at the end of this chapter. However, where $F = F(a/b)$ varies, the variety of these is severely limited due to the appearance of m as an exponent on F in the denominator of Eq. 11.29.

The preceding equations assume constant amplitude loading, that is, constant P_{max} and P_{min}, hence constant gross section stresses S_{max} and S_{min}. If these change, the integral of Eq. 11.27, and any equations obtained from it, can be used for separate calculations for periods of crack growth during which the load levels are constant. The cycle numbers for each of these periods can then be summed to get the total life. However, see the additional discussion of variable amplitude loading given later in Section 11.7.

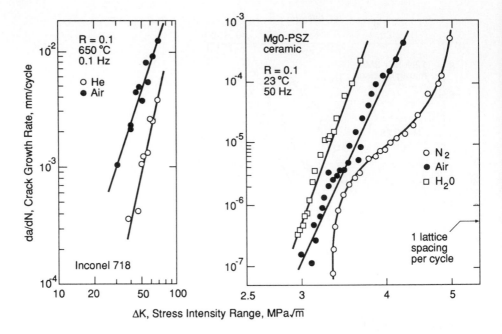

Figure 11.25 Faster fatigue crack growth in air than in inert gas for (left) the Ni-base alloy Inconel 718 at elevated temperature, and (right) for a magnesia, partially stabilized zirconia ceramic. (Left adapted from [Floreen 79]; used with permission. Right adapted from [Dauskardt 90]; reprinted by permission of the American Ceramic Society.)

11.6.2 Crack Length at Failure

In employing Eq. 11.27 to estimate crack growth life, the final crack length a_f is often unknown and must be determined before the equation can be applied. In addition, if F is taken as constant as in Eq. 11.30, it is also necessary to determine $F_f = F(a_f/b)$, so that it can be confirmed that this value does not differ excessively from $F_i = F(a_i/b)$. If F_f and F_i differ by more than about 10 to 15%, the resulting error in N_{if} due to using a constant value will generally be unacceptably large. Numerical integration as described below is then usually needed.

Under constant amplitude cyclic loading, the value K_{max} corresponding to S_{max} increases as crack growth proceeds. When K_{max} reaches the fracture toughness K_c for the material and thickness of interest, failure is expected at the length a_c that is critical for brittle fracture.

$$a_c = \frac{1}{\pi} \left(\frac{K_c}{F S_{max}} \right)^2 \tag{11.32}$$

Since F varies, a graphical or iterative solution as already illustrated by Example 8.1(c) is generally needed to obtain a_c.

In addition, crack growth causes a loss of cross-sectional area, and thus an increase in the stress on the remaining uncracked (net) area. Depending on the material and the member geometry and size, fully plastic yielding may be reached prior to $K_{max} = K_c$. This is most likely for ductile materials with low strength and high fracture toughness. Hence, a_f is in general the smaller of two possibilities, a_c and a_o, where the latter is the crack length corresponding to fully plastic yielding. Values of a_o may be estimated based on fully plastic behavior as discussed previously in Section 8.8 and later in more detail in Section 13.5. For simple two-dimensional situations, some useful equations for a_o obtained in this manner are given in Fig 8.45.

Use of linear-elastic fracture mechanics up to the crack length a_o corresponding to fully plastic yielding violates the plastic zone size limitations of LEFM as discussed in Chapter 8. The effect of yielding just prior to reaching a_o will generally be to increase growth rates to higher values than those calculated, so that the actual life is shorter than calculated. However, as cracks accelerate during their growth, most cycles are exhausted while the crack is short, and few are spent while the crack is near its final length. The error in life from this source is thus generally small, so that the procedure suggested of choosing the smaller of a_c and a_o is useful and appropriate as an approximation for engineering purposes. A more detailed approach would be much more complex, and for most situations of practical interest would not give life estimates that differ greatly.

Another source of possible error in life estimates is that the fracture toughness K_c at the end of cyclic loading may differ from standard values obtained in static tests. Clark (1981) discusses this issue. However, recalling Eq. 11.31, if the crack length increases prior to failure sufficiently for the ratio a_i/a_f to be small, then the effect of changing K_c on the estimated life will generally be small.

Example 11.4

A center-cracked plate of the AISI 4340 steel ($\sigma_u = 1296$ MPa) of Table 11.1 has dimensions, as defined in Fig. 8.13(a), of $b = 38$ and $t = 6$ mm, and it contains an initial crack of length $a_i = 1$ mm. It is subjected to tension-to-tension cyclic loading between constant values of minimum and maximum load, $P_{min} = 80$ and $P_{max} = 240$ kN.

  (a) At what crack length a_f is failure expected? Is the cause of failure yielding or brittle fracture?

  (b) How many cycles can be applied before failure occurs?

Solution   (a) The crack length at fully plastic yielding can be estimated from Fig. 8.45(a).

$$a_o = b \left(1 - \frac{P_{max}}{2bt\sigma_o} \right)$$

$$a_o = (38 \text{ mm}) \left(1 - \frac{240,000 \text{ N}}{2(38 \text{ mm})(6 \text{ mm})(1255 \text{ MPa})} \right) = 22.1 \text{ mm}$$

where the yield strength (and also K_{Ic}) is obtained from the footnote in Table 11.1.

The crack length a_c at brittle fracture is given by Eq. 11.32

$$a_c = \frac{1}{\pi} \left(\frac{K_{Ic}}{F S_{max}} \right)^2$$

Since F may vary, a trial and error solution as illustrated in Example 8.1(c) may be needed. With reference to Fig. 8.13(a), an initial estimate of a_c may be made by assuming that $a_c/b \le 0.4$, so that $F \approx 1$.

$$S_{max} = \frac{P_{max}}{2bt} = \frac{240,000 \text{ N}}{2(38 \text{ mm})(6 \text{ mm})} = 526 \text{ MPa}$$

$$a_c \approx \frac{1}{\pi}\left(\frac{K_{Ic}}{S_{max}}\right)^2 = \frac{1}{\pi}\left(\frac{130 \text{ MPa}\sqrt{\text{m}}}{526 \text{ MPa}}\right)^2 = 0.0194 \text{ m} = 19.4 \text{ mm}$$

This corresponds to $a_c/b = 0.51$, which is beyond the region of 10% accuracy for $F \approx 1$. A trial and error solution is thus needed using F from Fig. 8.13(a).

Calc. No.	Trial a mm	$\alpha = a/b$	F	$K_{max} = F S_{max}\sqrt{\pi a}$ MPa$\sqrt{\text{m}}$
1	15	0.395	1.097	125.3
2	16	0.421	1.114	131.3
3	15.8	0.416	1.110	130.1

The final K value is close enough to $K_{Ic} = 130$ MPa$\sqrt{\text{m}}$, so that $a_c = 15.8$ mm. Since this is smaller than a_o, it is the controlling value a_f.

$$a_f = 15.8 \text{ mm} \qquad\qquad \textbf{Ans.}$$

(b) The value of F increases from from $F \approx 1$ for a_i to $F = 1.11$ at a_f, and is thus approximately constant. Equation 11.30 can be used to estimate N_{if} by using an average value of $F \approx 1.05$.

$$N_{if} = \frac{a_f^{1-m/2} - a_i^{1-m/2}}{C\left(F \, \Delta S\sqrt{\pi}\right)^m (1 - m/2)}$$

Noting that Table 11.1 gives constants for the Walker equation, the nonzero R-ratio for the applied load can be handled by calculating a C value from Eq. 11.21.

$$R = \frac{S_{min}}{S_{max}} = \frac{P_{min}}{P_{max}} = \frac{80}{240} = 0.333$$

$$C = \frac{C_1}{(1-R)^{m_1(1-\gamma)}} = \frac{5.11 \times 10^{-10} \text{ mm/cycle}}{(0.667)^{3.24(1-0.42)}} = 1.095 \times 10^{-9} \text{ mm/cycle}$$

Two additional calculations are useful before computing N_{if}.

$$\Delta S = S_{max}(1 - R) = 526(0.667) = 351 \text{ MPa}$$

$$\left(1 - \frac{m}{2}\right) = \left(1 - \frac{3.24}{2}\right) = -0.62$$

where $m = m_1 = 3.24$ from Table 11.1 is used. Substituting the various numerical values finally gives N_{if}.

$$N_{if} = \frac{(0.0158 \text{ m})^{-0.62} - (0.001 \text{ m})^{-0.62}}{(1.095 \times 10^{-12} \text{ m/cycle}) \left(1.05 \times 351 \text{ MPa} \times \sqrt{\pi}\right)^{3.24} (-0.62)}$$

$$N_{if} = 66{,}200 \text{ cycles} \qquad \textbf{Ans.}$$

Comment   To avoid difficulty with units, it was necessary to use units for all quantities that correspond to the units of MPa$\sqrt{\text{m}}$ used for ΔK in defining C_1. Hence MPa and *meters* were used throughout, which necessitated converting C to units of m/cycle. Changing the units of K is not recommended as this would necessitate also changing the value of C to maintain consistency.

11.6.3 Solutions by Numerical Integration

As already discussed, Eq. 11.30 and related equations that might be derived for calculating crack growth life assume that F is constant, so that these cannot be used if F changes excessively between the initial and final crack lengths, a_i and a_f. Since closed-form integration of Eq. 11.27 is seldom possible if F is treated as a variable, numerical integration becomes necessary. Also, some elaborate mathematical forms used to fit da/dN vs. ΔK curves lead to equations that cannot be integrated in closed form even for constant F, again necessitating numerical integration.

To estimate crack growth life using numerical integration, note that equations 11.26 and 11.27 are together equivalent to

$$N_{if} = \int_{a_i}^{a_f} \left(\frac{dN}{da}\right) da \qquad (11.33)$$

where

$$\frac{dN}{da} = \frac{1}{da/dN} = \frac{1}{f(\Delta K, R)}$$

The quantity dN/da is the rate of accumulation of cycles, N, per unit of increase in crack length, a. Equation 11.33 suggests the use of a graphical solution on a plot of dN/da versus a, which is illustrated by Fig. 11.26. The life N_{if} is simply the area under the curve between a_i and a_f. Curves of dN/da vs. a generally decrease as in Fig. 11.26 due to ΔK, and hence da/dN, increasing with crack length. A large portion of the area under the curve is associated with the early growth just beyond a_i, and a smaller portion with the later stages of growth near a_f.

The most practical way to proceed is to pick a number of crack lengths between a_i and a_f.

$$a_i, a_1, a_2, a_3, \ldots a_f$$

For each of these, and for the material, geometry, and loading of interest, we calculate ΔK, and then da/dN, inverting the latter to get dN/da. We then plot the resulting dN/da values versus the corresponding a values, and find the area under the curve

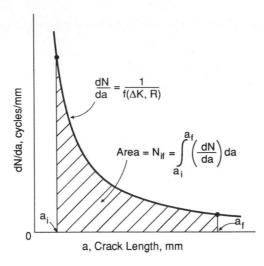

Figure 11.26 Area under the dN/da vs. a curve used to estimate the number of cycles to grow a crack from initial size a_i to final size a_f.

between a_i and a_f. This can be done for any mathematical form of the ΔK and da/dN equations. For example, for the forms of Eq. 11.28 with F allowed to vary, the dN/da for any given crack length a_j is

$$\left(\frac{dN}{da}\right)_j = \frac{1}{C\left(\Delta K_j\right)^m} = \frac{1}{C\left(F_j\ \Delta S\sqrt{\pi a_j}\right)^m} \qquad (11.34)$$

where F_j needs to be specifically calculated for each a_j. Also, R-ratio effects can be included in C, as from the Walker approach.

The intervals Δa between these can be made equal, but this is not necessary. It is important that Δa be sufficiently small for accurate representation of the dN/da curve. This is most likely to be a problem for the shorter crack lengths where the curve is generally steepest. One alternative that gives small Δa only where needed is to increase a by a fixed percentage for each interval. Ten percent, that is, a factor of 1.1, increase in a for each interval is sufficiently small for typical values of m.

$$a_{j+1} = 1.1a_j \qquad (11.35)$$

A manual solution for N_{if} may be done using graph paper. It is also straightforward to program an approximate area calculation on a digital computer. Standard methods and computer programs for numerical integration also apply.

A relatively simple method of numerical integration usually described in books on numerical analysis is Simpson's rule. To use this, consider three neighboring crack lengths a_j, a_{j+1}, and a_{j+2}, as shown in Fig. 11.27. Between a_j and a_{j+2}, the area under the curve $y = dN/da$ is estimated by assuming that a parabola passes through the three points (a_j, y_j), (a_{j+1}, y_{j+1}), and (a_{j+2}, y_{j+2}). If the points are equally spaced Δa apart, the area estimate is

$$\int_{a_j}^{a_{j+2}} y\ da = \frac{\Delta a}{3}\left(y_j + 4y_{j+1} + y_{j+2}\right) \qquad (11.36)$$

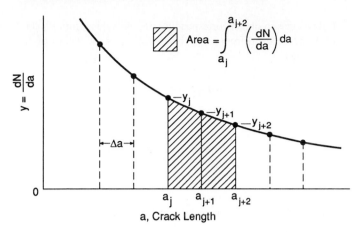

Figure 11.27   Area under the dN/da vs. a curve over two intervals Δa as estimated by Simpson's rule.

This equation is applied for each of $j = 0, 2, 4, 6 \ldots n$, where n is even. Adding the contributions to the area from each calculation gives an approximate value of the total area under the curve between a_i and a_f, where $a_f = a_n$.

However, for our present purpose, it is better if the a values are not evenly spaced, but instead differ by a constant factor r, such as $r = 1.1$ in Eq. 11.35.

$$a_i, \qquad a_1 = r a_i, \qquad a_2 = r^2 a_i, \qquad \ldots a_n = r^n a_i \qquad (11.37)$$

The area for a parabola through three such points is given by

$$\int_{a_j}^{a_{j+2}} y \, da = \frac{a_j \left(r^2 - 1 \right)}{6r} \left[y_j r (2 - r) + y_{j+1} (r + 1)^2 + y_{j+2} (2r - 1) \right] \qquad (11.38)$$

The integration up to a_n points can be performed in a manner analogous to a Simpson's rule calculation, except for the use of the new area formula.

Example 11.5

Refine the life estimate of Example 11.4 by using numerical integration.

Solution   The modified Simpson's rule of Eq. 11.38 can be used. A factor for incrementing a is first chosen to be near $r = 1.1$, such that the integration will end at $a_f = 15.8$ mm. A preliminary calculation indicates that approximately $n = 30$ increments are required, so that Eq. 11.37 is used to obtain

$$r^{30} = \frac{a_{30}}{a_i} = \frac{15.8 \text{ mm}}{1 \text{ mm}}, \qquad r = 1.09637$$

For the first integration step, the crack lengths from Eq. 11.37 are

$$a_i = 1 \times 10^{-3}, \qquad a_1 = 1.096 \times 10^{-3}, \qquad a_2 = 1.202 \times 10^{-3} \text{ m}$$

Calculations to obtain corresponding dN/da values from Eq. 11.34 are given below. Values of F are obtained from Fig. 8.13(a), and the same values of ΔS, C, and m are used as in Example 11.4.

No.	a, m	$F = F\left(\dfrac{a}{b}\right)$	$y = \dfrac{dN}{da}$, $\dfrac{\text{cycles}}{\text{m}}$
0	1.000×10^{-3}	1.00	5.87×10^7
1	1.096×10^{-3}	1.00	5.06×10^7
2	1.202×10^{-3}	1.00	4.35×10^7

The first increment of area (cycles) from substitution of the resulting $y = dN/da$ values into Eq. 11.38 is

$$\int_{a_i}^{a_2} y \, da = 10{,}201 \text{ cycles}$$

Similar computations for additional integration steps up to a_f can be programmed on a personal computer. The values for the end of each step are as follows:

j	a, m	$y = \dfrac{dN}{da}$, $\dfrac{\text{cycles}}{\text{m}}$	$N_{ij} = \int_{a_i}^{a_j} y \, da$, cycles
0	1.000×10^{-3}	5.87×10^7	0
2	1.202	4.35	10 201
4	1.445	3.23	19 297
6	1.737	2.39	27 406
8	2.088	1.774	34 632
10	2.509	1.314	41 068
12	3.016	9.73×10^6	46 795
14	3.626	7.19	51 886
16	4.358	5.30	56 404
18	5.239	3.90	60 403
20	6.297	2.85	63 928
22	7.569	2.07	67 019
24	9.099	1.488	69 705
26	1.094×10^{-2}	1.052	72 008
28	1.315	7.25×10^5	73 942
30	1.580	4.79	75 514

The intermediate (odd numbered) a and y values are not shown, and each new increment in area (cycles) is added to the previous total to obtain the cumulative values shown. The final N_{ij} value is thus the calculated life.

$$N_{if} = 75{,}500 \text{ cycles} \qquad\qquad \textbf{Ans.}$$

11.7 LIFE ESTIMATES FOR VARIABLE AMPLITUDE LOADING

If the stress levels vary during crack growth, life estimates may still be made. One simple approach is to assume that growth for a given cycle is not affected by the prior history, that is, *sequence effects* are absent. Large sequence effects do occur in special situations, but it is often useful and sufficiently accurate to neglect these.

11.7.1 Summation of Crack Increments

The crack growth Δa in each individual cycle of variable amplitude loading can be estimated from the da/dN vs. ΔK curve of the material. Summing these Δa, while keeping track of the number of cycles applied, leads to a life estimate. Such a procedure is equivalent to a numerical integration where a is the dependent variable, rather than N as above.

Hence, if the current crack length is a_j and the increment is Δa_j, the new value of crack length a_{j+1} for the next cycle is

$$a_{j+1} = a_j + \Delta a_j = a_j + \left(\frac{da}{dN} \right)_j \tag{11.39}$$

where the Δa are numerically equal to da/dN, since $\Delta N = 1$ for one cycle. Denoting the initial crack length as a_i, the crack length after N cycles is

$$a_N = a_i + \sum_{j=1}^{N} \left(\frac{da}{dN} \right)_j \tag{11.40}$$

Each da/dN is calculated from the ΔK and R for that particular cycle, where ΔK is obtained from the current crack length a_j and the ΔS for the particular cycle. Any form of expression for varying $F = F(a/b)$, and any form of a da/dN vs. ΔK relationship, can be readily used with this procedure. For highly irregular loading, rainflow cycle counting as described in Chapter 9 can be used to identify the cycles.

The summation is continued until a load peak is encountered that is sufficiently severe to cause either fully plastic yielding or brittle fracture. At this point, the calculation is terminated, and the number of cycles accumulated is the estimated crack growth life.

11.7.2 Special Method for Repeating or Stationary Histories

In some cases, it may be reasonable to approximate the actual service load history by assuming that it is equivalent to repeated applications of a loading sequence of finite length. This can be useful where some repeated operation occurs, such as lift cycles for a crane, or flights of an aircraft, and also for random loading with characteristics that are constant with time, called *stationary* loading. The crack growth life can then be estimated by an alternate procedure that is equivalent to summing crack increments. The necessary mathematical derivation follows.

First, assume that the da/dN vs. ΔK behavior obeys a power relationship of the form of Eq. 11.10. The increment in crack length for any cycle ($\Delta N = 1$) is then

$$\Delta a_j = C\left(\overline{\Delta K_j}\right)^m \tag{11.41}$$

where different R-ratios are handled by calculating an equivalent zero-to-maximum ($R = 0$) value $\overline{\Delta K}$, as from the Walker approach using Eq. 11.16. If the repeating load history contains N_B cycles, the increase in crack length during one repetition is obtained by summing.

$$\Delta a_B = \sum_{j=1}^{N_B} \Delta a_j = \sum_{j=1}^{N_B} C\left(\overline{\Delta K_j}\right)^m \tag{11.42}$$

The average growth rate per cycle during one repetition of the history is thus

$$\left(\frac{da}{dN}\right)_{\text{avg.}} = \frac{\Delta a_B}{N_B} = \frac{C \sum_{j=1}^{N_B} \left(\overline{\Delta K_j}\right)^m}{N_B} \tag{11.43}$$

Note that C is constant due to the use of $\overline{\Delta K}$ and so can be factored from the summation. Manipulation gives

$$\left(\frac{da}{dN}\right)_{\text{avg.}} = C\left(\left[\frac{\sum_{j=1}^{N_B} \left(\overline{\Delta K_j}\right)^m}{N_B}\right]^{1/m}\right)^m = C\left(\Delta K_e\right)^m \tag{11.44}$$

where

$$\Delta K_e = \left[\frac{\sum_{j=1}^{N_B} \left(\overline{\Delta K_j}\right)^m}{N_B}\right]^{1/m} \tag{11.45}$$

The quantity ΔK_e can be interpreted as an equivalent zero-to-maximum stress intensity range that is expected to cause the same crack growth as the variable amplitude history when applied for the same number of cycles N_B.

Since K and nominal stress S are proportional for any given crack length, an equivalent zero-to-maximum stress level can also be defined.

$$\Delta S_e = \frac{\Delta K_e}{F\sqrt{\pi a}} = \left[\frac{\sum_{j=1}^{N_B} \left(\overline{\Delta S_j}\right)^m}{N_B}\right]^{1/m} \tag{11.46}$$

where the $\overline{\Delta S}$ for each cycle in the history are equivalent zero-to-maximum values corrected for R effect. If this is done based on the Walker approach using Eq. 11.16, these values are obtained from

$$\overline{\Delta S} = S_{\text{max}}(1 - R)^\gamma \tag{11.47}$$

A life estimate can be made using ΔS_e just as if it were a constant amplitude loading at $R = 0$, for example, by using Eq. 11.30. However, to determine the final crack length

a_f as caused by either fully plastic yielding or brittle fracture, the actual peak stress S_{max} in one repetition of the history should be employed.

The above procedure is numerically equivalent to summing crack increments if the crack length a does not change by a large amount during one repetition of the sample load history that is used to represent the service loading. If the repeating load history is extremely long, the calculation becomes approximate where the severity of the loading varies markedly during different portions of the repeating history.

Some authors recommend a similar approach, called the *root-mean-square approach*, where ΔK_e is calculated by an equation that can be obtained from Eq. 11.45 by substituting $m = 2$. Such a simplification seems incorrect in view of the above derivation and is likely to lead to error if the actual value differs greatly from $m = 2$.

11.7.3 Sequence Effects

In all of the treatment so far of variable amplitude loading, it has been assumed that the crack growth in a given cycle is unaffected by prior events in the load history. However, this assumption may sometimes lead to significant error. Consider the situation of Fig. 11.28. After a high tensile overload is applied as in case C, the growth rate during the lower level cycles is decreased. Slower than normal growth continues for a large number of cycles until the crack grows beyond the region affected by the overload, where the size of the affected region is related to the size of the crack-tip plastic zone caused by the overload. For the case illustrated, the overall effect of only three overloads was to increase the life by about a factor of 10. This beneficial effect of tensile overloads is called *crack growth retardation*.

A tensile overload introduces a compressive residual stress around the crack tip in a manner similar to the notched member of Fig. 10.27. This compression tends to keep the

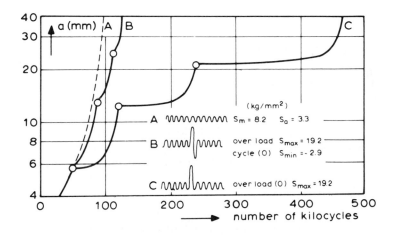

Figure 11.28 Effect of overloads on crack growth in center-cracked plates ($b = 80$, $t = 2$ mm) of 2024-T3 Al. (From [Broek 86] p. 273, as based on data in [Schijve 62]; reprinted by permission of Kluwer Academic Publishers.)

crack tip closed during the subsequent lower level cycles, retarding crack growth. The magnitude of the effect is related to the ratio S_{max2}/S_{max1} where S_{max2} is the overload stress and S_{max1} is the peak value of the lower level. For ratios greater than about 2.0, crack growth may be *arrested*, that is, stopped entirely. Conversely, if the ratio is less than about 1.4, the effect is small. Compressive overloads have an opposite but lesser effect. The effect is not as great because the crack tends to close during the overload, so that the faces of the crack support much of the compressive load and shield the crack tip from its effect. Also, the effect of a tensile overload is much reduced if it is followed by a compressive one, as in case *B* of Fig. 11.28.

Several methods have been developed to incorporate sequence effects due to overloads into life calculations for crack growth. The general approach used is to base the life estimate on calculating crack growth increments for each cycle as described above in connection with Eqs. 11.39 and 11.40. However, the da/dN values used are each modified in a manner that is determined by the prior history of overloads. This is generally done by determining da/dN from an effective ΔK that is modified based on logic related to residual stress fields or crack closure levels. More detailed explanation can be found in Chang and Rudd (1984), Broek (1986) and (1988), and Suresh (1991), which are listed as References.

Overload sequence effects are likely to be important where high overloads occur predominantly in one direction. This occurs in the service of some aircraft, where occasional severe wind gust loadings or maneuver loadings may introduce sequence effects. However, less effect is expected if overloads occur in both directions, if the history is highly irregular, or if the overloads are relatively mild. Noting that the effect is mainly to retard crack growth, neglecting sequence effects generally provides conservative estimates of crack growth life that will be sufficient for engineering purposes in many cases. Load histories that include severe compressive overloads then need to be handled with caution, due to the possibility of these causing faster crack growth than predicted.

11.8 DESIGN CONSIDERATIONS

It is becoming increasingly common to assure adequate service life for components of machines, vehicles, and structures based on crack growth calculations as described above. This is appropriate for large structures subjected to cyclic loading, especially where personal safety or high costs are factors, and especially if cracks are commonly found in the type of hardware involved. Examples include bridge structure, large aircraft, the space shuttle, and nuclear and other pressure vessels. Such a *damage tolerant* approach is critically dependent on initial and sometimes periodic inspections for cracks as described near the beginning of this chapter. (See Figs. 11.1 and 11.2.)

Inspection for cracks, especially small ones, is an expensive process and is not generally feasible for inexpensive components that are made in large numbers. If the service stresses are relatively high, the cracks that would need to be found to use a damage tolerant approach can be so small that the inspection would greatly increase the cost of the item. Periodic inspections would allow a larger crack to be tolerated initially, but the component may not be available for periodic inspection. Examples of parts that fall in this category are automobile engine, steering, and suspension parts, bicycle front forks and pedal cranks, and parts for home appliances. Here, fatigue life estimates are based on an *S-N* approach, or on the related strain-based approach, neither of which specifically consider cracks. Where personal safety is involved, safety factors reflect this fact and are generally larger than if a damage tolerant approach could be used. Failures are minimized by careful attention to design detail and to manufacturing quality control, including inspection to eliminate any obviously flawed parts.

Regardless of the approach used, a finite probability of failure always exists. For the damage tolerant approach, this arises because the minimum detectable crack length a_d is difficult to establish and is never precisely known. The a_d value used is often treated as a statistical variable. For example, for critical parts of the space shuttle, there is 95% confidence that the size a_d for that particular part can be found 90% of the time. For the *S-N* and related approaches, a finite probability of failure arises due to the possibility that a part passing inspection still contains a flaw which, though small, nevertheless leads to early failure. Also, all approaches to assuring adequate life are subject to additional uncertainties, such as: (1) estimates of the service loading being too low, (2) accidental substitution during manufacturing of the wrong material, (3) undetected manufacturing quality control problems, and (4) hostile environmental effects that are more severe than forecast, with the latter including both ordinary corrosion and environmental crack growth.

Where a damage tolerant approach is used, critical components must be designed so that they are accessible for inspection. For example, cracks at fastener (rivet or bolt) holes are of concern in aircraft structure, and access to the interior of the skin of the fuselage or wing structure may be needed for situations such as that in Fig. 11.29. If periodic inspections are required, then the design must accommodate disassembly if this is necessary for inspection. For example, in large aircraft, passenger seats, interior panels, and even paint are removed, and some structural parts are disassembled, for costly but necessary periodic inspections.

Specific measures can also be taken by the designer to allow structures to function without sudden failure even if a large crack does develop. Some examples for aircraft structure are illustrated in Figs. 11.30 and 11.31. Stiffeners retard crack growth, and

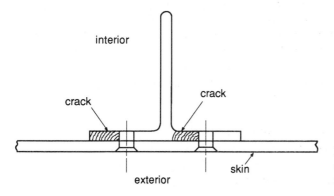

Figure 11.29   Cracks in the interior of an aircraft skin structure. (Adapted from [Chang 78].)

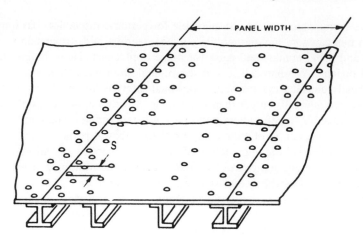

Figure 11.30 Stiffened panel in aircraft structure with a crack delayed before growing into adjacent panels. The rivet spacing dimensioned is 38 mm. (From the paper by J. P. Butler in [Wood 70] p. 41.)

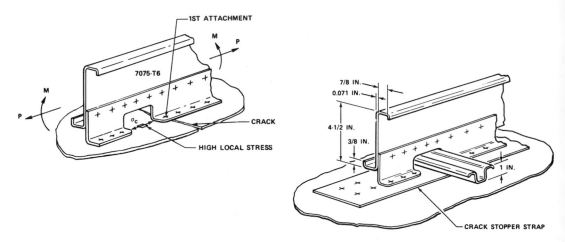

Figure 11.31 Crack (left) in a DC-10 fuselage in the longitudinal direction, due to cabin pressure loading, and (right) a crack stopper strap. Rivet locations are indicated by (+), and the longeron member with a hat-shaped cross section is omitted on the left for clarity. (From [Swift 71]; copyright ©ASTM; reprinted with permission.)

joints in skin panels may be intentionally introduced so that a crack in one panel has difficulty growing into the next. Similarly, a crack stopper strap may lower stresses in a critical area and provide some strength even if a crack does start.

Recall from the early part of this chapter and Eq. 11.2 that the safety factor on life X_N is the ratio of the failure life for crack growth N_{if} to the expected service life $\hat{N}$. The value of N_{if} depends not only on the detectable crack length a_d, but also on the stress level and the material. If the safety factor is insufficient, perhaps even less than unity, several different options exist to resolve the situation. Obviously, the design could be changed to lower the stress, thus increasing the calculated life N_{if} and with it X_N. Another possibility is to make a more careful initial inspection for cracks, decreasing a_d, and thus increasing the worst-case failure life N_{if}. Alternatively, the material could be changed to one with slower fatigue crack growth rates, as judged by comparing da/dN vs. ΔK curves. Depending on whether failure occurs by brittle fracture or by yielding, increasing either the fracture toughness or the yield strength of the material also increases the life by increasing the final crack length a_f, but the effect is usually small as the life is generally insensitive to the value of a_f.

If design changes or improved initial inspection do not suffice, it may be necessary to perform periodic inspections, making it permissible to calculate the safety factor from the inspection period N_p using Eq. 11.5.

11.9 PLASTICITY ASPECTS AND LIMITATIONS OF LEFM FOR FATIGUE CRACK GROWTH

During cyclic loading, a region of reversed yielding exists at the crack tip, and the size of this region can be estimated using a procedure similar to that applied to static loading in Section 8.6. Based on this, plasticity limitations on LEFM for fatigue crack growth can be explored. Limitations are also needed if the crack is so small that its size is comparable to that of the microstructural features of the material.

11.9.1 Plasticity at Crack Tips

In the immediate vicinity of the crack tip, there is a finite separation δ between the crack faces as discussed in Chapter 8. Behavior on the size scale of δ determines how the crack advances through the material during cyclic loading. Details are not fully understood and vary with material, and they even vary with the K level for a given material. In ductile metals, the process of crack advance during a cycle is thought to be similar to Fig. 11.32. Localized deformation by slip of crystal planes occurs and is most intense in bands above and below the crack plane. The crack tip moves ahead and becomes blunt as the maximum load is reached, and it is resharpened during decreasing load. This process results in striations on the fracture surface as previously illustrated by Fig. 9.21.

Another mechanism is crack growth by small increments of brittle cleavage during each cycle. It is not uncommon in metals for the fracture surface to have regions of striation growth mixed with regions of cleavage, especially at high growth rates where K_{max} approaches K_c. In other cases, the boundaries between grains are the weakest regions in the material, so that the crack grows along grain boundaries. This is called *intergranular fracture*, to distinguish it from the more usual *transgranular fracture* by striation formation or cleavage. For example, intergranular fatigue cracking occurred for the granite rock of Fig. 11.11. In metals, intergranular cracking is likely to occur if there is a hostile environmental influence.

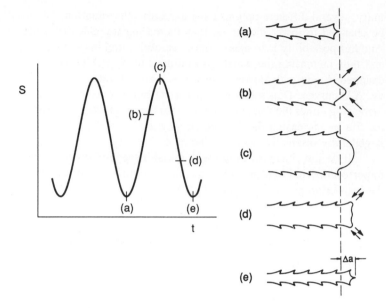

Figure 11.32 Hypothesized plastic deformation behavior at the tip of a growing fatigue crack during a loading cycle. Slip of crystal planes along directions of maximum shear occurs as indicated by arrows, and this plastic blunting process results in one striation (Δa) being formed for each cycle. (Adapted from the paper by J. C. Grosskreutz in [Wood 70] p. 55.)

If the material is relatively ductile, a crack-tip plastic zone will exist that is considerably larger than δ. The peak stress in the cyclic loading determines K_{max}, which can be substituted into Eq. 8.29 or 8.30 to estimate the extent of yielding ahead of the crack. For example, for plane stress

$$2r_{o\sigma} = \frac{1}{\pi} \left(\frac{K_{max}}{\sigma_o} \right)^2 \tag{11.48}$$

This is called the *monotonic plastic zone*. As the minimum load in a cycle is approached, yielding in compression occurs in a region of smaller size called the *cyclic plastic zone*, as illustrated in Fig. 11.33.

Consider an ideal elastic, perfectly plastic material and the behavior during unloading following $K = K_{max}$. For compressive yielding to occur as K changes by an amount ΔK, the stress of σ_o near the crack tip must change to $-\sigma_o$, which is a change of $2\sigma_o$, or twice the yield strength. In effect, for changes relative to K_{max}, the yield strength is doubled. The size of the cyclic plastic zone where yielding occurs not only in tension but also in compression can therefore be approximated by using ΔK for K, and $2\sigma_o$ for σ_o, in the monotonic plastic zone estimate.

$$2r'_{o\sigma} = \frac{1}{\pi} \left(\frac{\Delta K}{2\sigma_o} \right)^2 \tag{11.49}$$

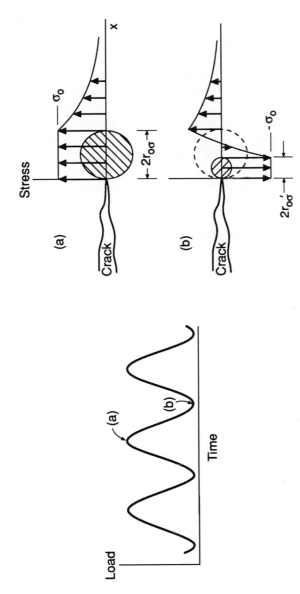

Figure 11.33 Monotonic (a) and cyclic (b) plastic zones. (Adapted from [Paris 64]; used with permission.)

For plane strain, monotonic and cyclic plastic zone sizes $r_{o\sigma}$ and $r'_{o\sigma}$ can be similarly estimated. Using the logic from Chapter 8, these are one-third as large as the corresponding plane stress zones. For zero-to-tension loading, where $\Delta K = K_{max}$ and $R = 0$, the cyclic plastic zone is thus estimated as above to be one-fourth as large as the monotonic one. More generally, since $\Delta K = K_{max}(1 - R)$, the estimated sizes compare as follows:

$$\frac{2r'_o}{2r_o} = \frac{(1 - R)^2}{4} \tag{11.50}$$

where subscript σ is dropped to indicate that this equation applies for either plane stress or plane strain.

These two plastic zone types can be further understood by considering the stress-strain history at a point in the material as the crack approaches as illustrated in Fig. 11.34. When the point being observed is still outside the monotonic plastic zone, no yielding occurs. Yielding begins, but only in the tensile direction, when the monotonic plastic zone boundary passes the point. Once the cyclic plastic zone boundary passes, yielding in both compression and tension occurs during each loading cycle.

11.9.2 Thickness Effects and Plasticity Limitations

If the monotonic plastic zone is not small compared to the thickness, then plane stress exists, and fatigue cracks may grow in a shear mode, with the fracture inclined about

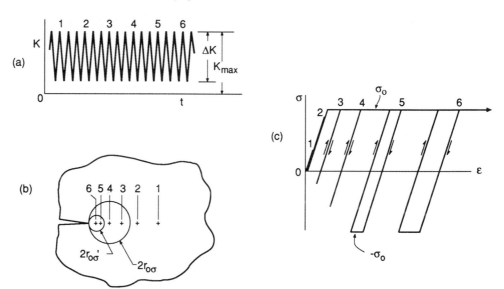

Figure 11.34 Stress-strain behavior at a point as the tip of a growing fatigue crack approaches. For selected cycles (a), relative positions of the point and the crack tip are shown in (b), and the stress-strain responses in (c). (Adapted from [Dowling 77]; copyright ©ASTM; reprinted with permission.)

45° to the surface. Since K and hence the plastic zone size increase with crack length, a transition to this behavior can occur during the growth of a crack as illustrated in Fig. 11.35. Crack growth rates can be affected somewhat by member thickness as a result of different behavior in plane stress and plane strain. However, the effect is sufficiently small that it can generally be ignored, so that crack growth data for one thickness can be used for any other thickness.

If large amounts of plasticity occur during cyclic loading, crack growth rates rapidly increase and exceed what would be expected from the da/dN vs. ΔK curve. This circumstance arises from the fact that the theory supporting the use of K requires that plasticity occur only in a region that is small compared to the planar dimensions of the member, as discussed previously in Sections 8.6 and 8.8. Large effects occur only where the maximum load exceeds about 80% of fully plastic yielding, so that this level represents a sufficient plasticity limitation in most cases. Modest effects may occur at somewhat lower levels. If a fairly strict limitation is desired, the limitation of Eq. 8.33 on the in-plane dimensions, as previously employed for static loading, can be applied to the peak stress.

$$a, (b-a), h \geq 8r_{o\sigma} = \frac{4}{\pi}\left(\frac{K_{max}}{\sigma_o}\right)^2 \qquad (11.51)$$

For fatigue crack growth, thickness effects and plasticity limitations are not generally issues of major importance as they are for fracture toughness applications. This is because nominal stresses around or exceeding yielding are rare in engineering situations except near the very end of the life, when the fatigue crack growth phase is essentially complete. However, local yielding at stress raisers is fairly common, so that difficulties may be encountered if it is necessary to use fracture mechanics for cracks growing from

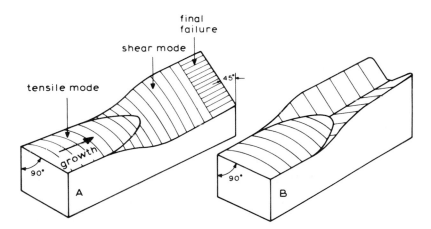

Figure 11.35 Schematic of surfaces of fatigue cracks showing transition from a flat tensile mode to an angular shear mode. The shear growth can (A) occur on a single sloping surface, or (B) form a V-shape. (From [Broek 86] p. 269; reprinted by permission of Kluwer Academic Publishers.)

notches while they are still small, and thus may be affected by the local plasticity there. Fortunately, a crack is under the influence of the local stress field of a notch only if its length is quite small, specifically less than about 10 to 20% of the notch radius. See Eq. 8.21 and Fig. 8.18.

11.9.3 Limitations for Small Cracks

Fracture mechanics in the form considered so far is based on stress analysis in an isotropic and homogeneous solid. The microstructural features of the material are in effect assumed to occur on such a small scale that only the average behavior needs to be considered. However, if a crack is sufficiently small, it can interact with the microstructure in ways that cause the behavior to differ from what would otherwise be expected. In engineering metals, small cracks tend to grow faster than estimated from the usual da/dN vs. ΔK curves from test specimens with long cracks, as illustrated in Fig. 11.36.

It is useful to distinguish between *small cracks* and *short cracks*. For a small crack, all of its dimensions are similar to or smaller than the dimension of greatest microstructural significance, such as the average crystal grain size or the average reinforcement particle spacing. However, a short crack has one dimension that is large compared to the microstructure. The behavior of a small crack can be profoundly affected by the microstructure. For example, while the crack is within a single crystal grain in a metal, the growth rate is much higher than expected from the usual da/dN vs. ΔK curve as illustrated by Fig. 11.36(a). Upon encountering a grain boundary, the growth is temporarily retarded. Until the crack becomes several times larger than the grain size, the average growth rate as affected by lattice planes within grains and grain boundaries is considerably above the usual da/dN vs. ΔK curve.

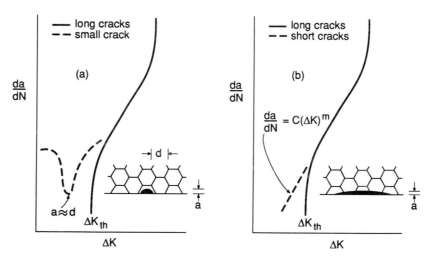

Figure 11.36   Behavior for a crack that is small in all dimensions (left), vs. that for a crack with one dimension which is large compared to the microstructure (right).

A less drastic effect occurs if the crack is merely short in one dimension, being large compared to the microstructure in the other dimension as illustrated by Fig. 11.36(b). Growth rates for such cracks in metals are similar to the da/dN vs. ΔK curve except at low ΔK, where a reasonable estimate of the behavior can be obtained by extrapolating Eq. 11.10 from the intermediate region of the curve. The cause of the special behavior in this case appears to be associated with the fact that the faces of a crack normally interfere behind the tip during part of the stress cycle. In particular, the crack opens and closes, and the portion of the stress cycle that occurs while the crack is closed does not contribute to its growth. Note that this cannot occur if there is not sufficient length behind the tip for the interference to occur. Thus, for low ΔK where crack closure effects are especially important, short cracks grow faster than expected.

An approximate method for identifying crack sizes below which the usual da/dN vs. ΔK curve may not apply is illustrated in Fig 11.37. Note that the unnotched-specimen fatigue limit is the stress level below which the small naturally occurring flaws in the material will not grow, even with their growth being enhanced by the small crack effect as just discussed. Considering members containing cracks of various sizes, the fatigue limit decreases with crack length, and for relatively long cracks follows the behavior expected from LEFM and the threshold ΔK_{th} from the long crack da/dN vs. ΔK curve.

The crack length a_s where the ΔK_{th} prediction exceeds the unnotched-specimen fatigue limit is the intersection of the lines for the two equations

$$\Delta S = \Delta\sigma_e, \qquad \Delta K_{th} = \Delta S \sqrt{\pi a} \qquad (11.52)$$

where the completely reversed $(R = -1)$ fatigue limit is given as a stress range, $\Delta\sigma_e = 2\sigma_{er}$, where ΔK_{th} is the value for $R = -1$, and where the geometry factor is approximated as $F = 1$. Combining these and solving for a gives

$$a_s = \frac{1}{\pi}\left(\frac{\Delta K_{th}}{\Delta\sigma_e}\right)^2 \qquad (11.53)$$

For cracks larger than a_s in all dimensions, fracture mechanics based on long crack data is expected to be reasonably accurate. For example, approximate values of $\Delta\sigma_e$ and ΔK_{th}

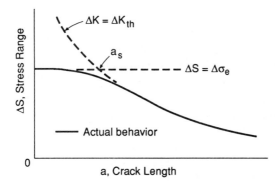

Figure 11.37 Fatigue limit stress as a function of crack length, and the transition length a_s, below which special small crack effects are expected.

for two steels with contrasting ultimate tensile strengths σ_u give a_s values as follows:

σ_u, MPa	$\Delta\sigma_e = 2\sigma_{er}$, MPa	ΔK_{th}, MPa$\sqrt{\text{m}}$	a_s, mm
500	500	12	0.18
1500	1400	6	0.006

For the lower strength steel, a_s is relatively large and could be within a range of crack sizes that is of engineering interest. The opposite is true for the higher strength steel, where a_s is so small that unusual short crack behavior would probably never affect the use of fracture mechanics for engineering applications.

A discussion of small crack effects, with references to additional literature, is given in Suresh (1991).

11.10 ENVIRONMENTAL CRACK GROWTH

Similar considerations of inspection for cracks, and a similar need for life estimates, exist where crack growth is caused by a hostile chemical environment. There are several physical mechanisms that occur. One of these is *stress corrosion cracking*, where material removal by corrosion in water, salt water, or other liquid assists in growing the crack. In other cases, no corrosion is involved, as in cracking of steels due to *hydrogen embrittlement*, or cracking of aluminum alloys due to *liquid metal embrittlement* caused by mercury. In these cases, the embrittling substance appears to enhance the breaking of chemical bonds in the highly stressed region of the crack tip. Embrittlement and hence crack growth can occur even where the harmful substance is not present as an external environment, but is instead in solid solution in the material, which is sometimes the case for hydrogen cracking of metals. Also, even the moisture and gases in air can cause environmental crack growth in some materials, for example, in silica glass.

11.10.1 Life Estimates for Static Loading

In situations of environmental crack growth during an unchanging static load, the crack growth life can be estimated based on fracture mechanics, in a manner analogous to the procedures described above for fatigue crack growth under constant amplitude loading. The parameter controlling crack growth is simply the static value K of the stress intensity factor, as determined from the applied static stress and the current crack length. Growth rates for the material are characterized by the use of a da/dt versus K curve, where da/dt is the time-based growth rate, or *crack velocity*, also denoted $\dot{a}$. For example, the $\dot{a}$ vs. K relationship sometimes fits a straight line on a log-log plot, so that it has the form

$$\dot{a} = \frac{da}{dt} = AK^n \tag{11.54}$$

where A and n are material constants that depend on the particular environment and are affected by temperature. Data for two glasses that obey such a relationship are shown in Fig. 11.38.

Once the $\dot{a}$ vs. K relationship is known, life estimates can proceed as for fatigue crack growth using either closed-form expressions or numerical integration. For example, if $F = F(a/b)$ does not change substantially during crack growth, a relationship similar to Eq. 11.30 is obtained due to the mathematical forms of Eqs. 11.10 and 11.54 being the same.

$$t_{if} = \frac{a_f^{1-n/2} - a_i^{1-n/2}}{A\left(FS\sqrt{\pi}\right)^n (1 - n/2)} \qquad (n \neq 2) \qquad (11.55)$$

where t_{if} is the time required for a crack to grow from an initial size a_i to a final size a_f. As before, a_f can be estimated as the smaller of a_o due to fully plastic yielding or a_c due to brittle fracture.

Where the behavior follows Eq. 11.54, the exponent n may be quite high. For example, for silica glasses in various environments, it is usually at least 10 and may be considerably higher. A high value of n indicates that cracks accelerate rapidly, and also that growth rates da/dt are highly sensitive to the value of K, so that modest increases in stress can have a large effect.

A different behavior is sometimes observed where the growth rate is constant over a range of K values as in Fig. 11.39. At low K, the growth rate may drop abruptly, so that the curve approaches an asymptote at the value K_{Iscc}, called the *stress*

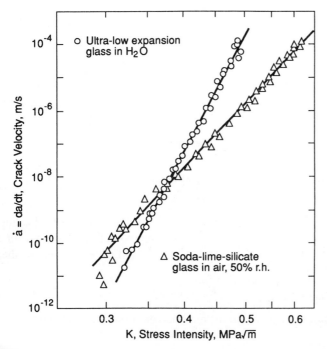

Figure 11.38 Crack velocity data for two silica glasses in room temperature environments as indicated. (Data from [Wiederhorn 77].)

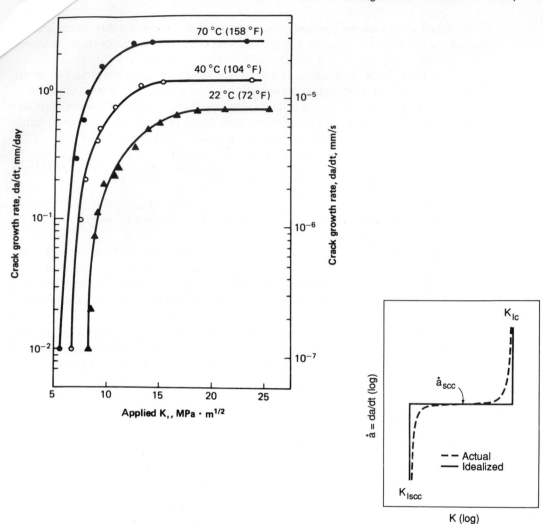

Figure 11.39 Crack velocity data (left) for 7075-T6 Al in a 3.5% NaCl solution similar to seawater, and approximation of such behavior (right) using a constant $\dot{a}$ between K_{Iscc} and K_{Ic}. (Left from [Campbell 82] p. 20; used with permission.)

corrosion cracking threshold, below which no crack growth occurs under static loading. A reasonable engineering approach in such cases is to approximate the curve with a constant rate $\dot{a}_{scc}$, except that no growth occurs below K_{Iscc}. Life estimates are then given simply by

$$t_{if} = \frac{a_f - a_i}{\dot{a}_{scc}} \quad (K > K_{Iscc}) \tag{11.56}$$

The value of $\dot{a}_{scc}$ must of course be specific to the material, environment, and temperature of interest.

Values of K_{Iscc} are generally determined from long-term static loading tests. One approach is to hang weights on previously cracked cantilever beams as shown in Fig. 11.40. A number of different initial values of K are obtained by using various weights. As also shown, the value of K below which no failure occurs after a long period of time is then identified as K_{Iscc}.

More complex forms of $\dot{a}$ vs. K relationship that do not fit either Eq. 11.54 or constant $\dot{a}$ are often encountered. For example, the lower curve of Fig. 11.39 does not fit the former and is not represented very well by constant $\dot{a}$ for the lower half of the range of K involved. If necessary in such cases, numerical integration can be employed, with any appropriate $\dot{a}$ vs. K relationship, in a manner parallel to the procedure given above for fatigue crack growth.

11.10.2 Additional Comments

Environmental cracking problems occur for certain particular combinations of material and environment. Small changes in the processing or composition of a material, hence in its microstructure, may eliminate or introduce the problem. For example, AISI 4340 steel is susceptible to environmental cracking in H_2S gas as illustrated by K_{Iscc} data in Fig. 11.41. The K_{Iscc} value is sensitive to the gas pressure and especially to the strength level (heat treatment) of the steel. Similar trends occur in this steel for other environments, such as seawater, and also in other alloy steels. In such cases, a modest decrease in strength may solve a cracking problem

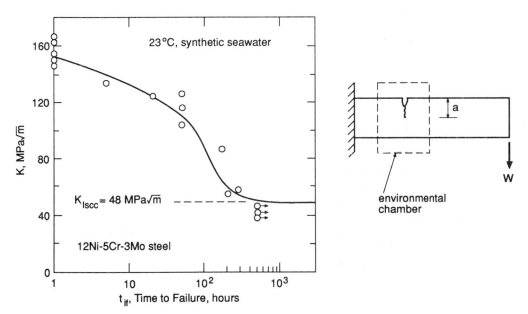

Figure 11.40   Determination of K_{Iscc} from cantilever beams loaded with dead weights. (Data from [Novak 69].)

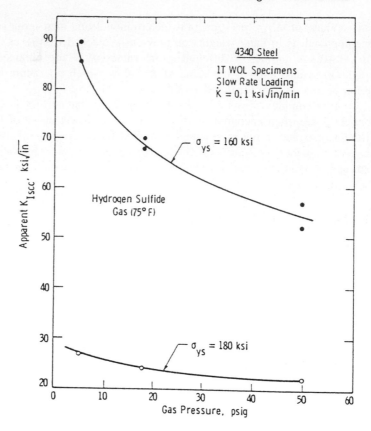

Figure 11.41   Effect of H$_2$S gas pressure on K_{Iscc} for two yield strength levels (σ_{ys}) of AISI 4340 steel. (From [Clark 76]; copyright ©ASTM; reprinted with permission.)

by increasing K_{Iscc}, despite the safety factor against yielding decreasing somewhat. Also, a seemingly small change in the environment can have a large effect. For example, alloy steels similar to AISI 4340 also crack in pure hydrogen gas, but the effect is considerably decreased if a small amount of oxygen is added to the hydrogen.

As an additional example, grain boundary cracking and the resulting intergranular fracture surface from an actual pressure vessel cracking problem are shown in Fig. 11.42. The environment in the 2.25Cr-1Mo steel vessel contained hydrogen gas at a partial pressure of 10 MPa, and the temperature was 420°C. However, some of the weld rods used in fabricating the vessel were of the wrong type, resulting in some welds having much less Cr and Mo content than specified for this steel, in turn causing a loss of resistance to environmental cracking in the affected welds. Grain boundary cracking as shown then led to large cracks, necessitating a costly program of inspection and repair. Intergranular cracking is caused by the segregation of impurities at

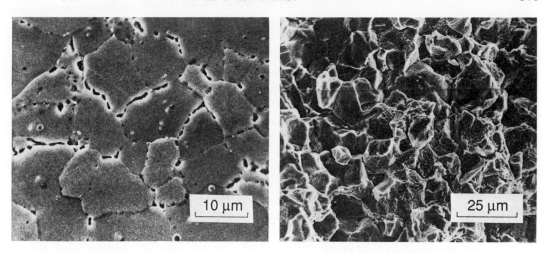

Figure 11.42   Grain boundary damage (left) and resulting intergranular fracture surface (right) in steel used in a pressure vessel containing H_2 gas at elevated temperature. (Photos courtesy of K. Rahka, Technical Research Center of Finland, Espoo, Finland. Published in [Rahka 86]; copyright ©ASTM; reprinted with permission.)

grain boundaries, or other metallurgical differences there, that make these boundaries susceptible to environmental attack. In this particular case, the Cr and Mo are needed to form carbides and thus limit the amount of iron carbide formed, and also to stabilize the iron carbides (such as Fe_3C) that do form. Note that iron carbides are the source of difficulty at the grain boundaries, probably by reacting with the hydrogen to form methane gas.

Opportunities thus exist for eliminating environmental cracking problems, making life estimates as described above unnecessary. This is, of course, the preferred solution where it is feasible. As suggested by the above examples, a detailed knowledge of the material and environment combination involved is required to aid in choosing the correct course of action, so that relevant literature or expert advice is generally needed. Some information along these lines can be found in the more detailed but still introductory treatments in Campbell et al. (1982), Courtney (1990), and Hertzberg (1989), and some recent research and review papers are given in the workshop proceedings of Thompson et al. (1988). Also, relevant materials data for $\dot{a}$ vs. K and K_{Iscc} are sometimes available for commonly encountered environments, such as water and saltwater, as in Batelle (1991) and Gallagher (1983).

Environmental and fatigue crack growth may occur in combination when cyclic loading is applied in a hostile environment, as in corrosion fatigue. A simple approach to making life estimates in such cases is to add the two contributions to crack growth by using both da/dN vs. ΔK and da/dt vs. K curves. However, this is not always sufficiently accurate, and a number of complexities exist that are difficult to incorporate into life estimates. Some discussion is given in Suresh (1991), and recent research is reviewed in Wei and Gangloff (1989).

11.11 SUMMARY

The resistance of a material to fatigue crack growth under a given set of conditions can be characterized by a da/dN versus ΔK curve. At intermediate growth rates, the behavior can often be represented by the Paris equation.

$$\frac{da}{dN} = C (\Delta K)^m \tag{11.57}$$

Values of the exponent m are typically in the range 2 to 4 for ductile materials, but are higher for brittle materials, with values above 10 not being unusual.

Growth rates are affected by the value of the stress ratio $R = S_{min}/S_{max}$. For a given ΔK, increasing R increases da/dN in a manner analogous to the effect of mean stress on S-N curves. More general equations for da/dN have thus been developed that include this effect. For example, if the Walker equation is used, C in the above equation depends on R as in Eq. 11.21. Hostile chemical environments may increase da/dN, especially at slow frequencies of cyclic loading. At low growth rates, da/dN vs. ΔK curves generally exhibit a lower limiting or threshold value, ΔK_{th}, below which crack growth does not usually occur. The low growth rate region of the curve is especially sensitive to the effects of R-ratio and material variables, such as grain size and heat treatment in metals.

For a given applied stress, material, and component geometry, the crack growth life N_{if} depends on both the initial crack size a_i and the final crack size a_f. The life N_{if} is generally quite sensitive to the value of a_i, and considerably less sensitive to a_f. To make a calculation of N_{if}, it is necessary to have an equation for the da/dN vs. ΔK curve for the material, and a mathematical expression for the stress intensity for the geometry and loading case of interest, such as $K = FS\sqrt{\pi a}$. For example, where F is constant or approximately so, and for behavior according to Eq. 11.57, the life is

$$N_{if} = \frac{a_f^{1-m/2} - a_i^{1-m/2}}{C \left(F \, \Delta S \sqrt{\pi} \right)^m (1 - m/2)} \quad (m \neq 2) \tag{11.58}$$

In design applications, the initial crack size a_i is often the minimum size a_d that can be reliably detected by inspection. The final crack size a_f is either a_c or a_o, whichever is smaller, as either brittle fracture or fully plastic yielding may occur first.

As F often varies, and as mathematical complexities may occur for certain forms of the da/dN vs. ΔK equation, a closed form expression for N_{if} may not be obtainable. It is then necessary to perform numerical integration by first evaluating ΔK and then da/dN for a number of different crack lengths. The crack growth life can be interpreted as the area under the dN/da versus a plot between a_i and a_f as illustrated in Fig. 11.26.

$$N_{if} = \int_{a_i}^{a_f} \left(\frac{dN}{da} \right) da \tag{11.59}$$

For variable amplitude loading, the da/dN vs. ΔK curve can be used to estimate increments in crack length Δa for each cycle. The end of the crack growth life occurs when the crack length increases such that a stress peak is expected to cause either brittle fracture or fully plastic yielding. An alternate procedure is to identify a representative sample of the load history, applying Eq. 11.46 to obtain an equivalent zero-to-maximum stress level ΔS_e, and then using ΔS_e to make a life estimate as for constant amplitude loading. If isolated severe overloads occur, these may cause sequence effects that need to be included in life estimates.

The estimated crack growth life from the minimum detectable crack length a_d must be longer than the expected actual service life by a sufficient safety factor X_N. If X_N is inadequate, it may be possible to resolve the situation by redesign that lowers stresses, by improving the initial inspection to decrease a_d, by changing materials, or by resorting to periodic inspections. Special crack-stopping design features as used in aircraft structure also contribute to safety.

Limitations on the use of LEFM due to excessive plasticity can be set based on plastic zone sizes according to Eq. 11.51. However, the looser restriction of 80% of fully plastic yielding is generally sufficient. If a crack is so small that all of its dimensions are similar to or smaller than the microstructural features of the material, then its growth is likely to be significantly faster than expected from the usual da/dN vs. ΔK curve. Equation 11.53 can be employed to estimate a crack length above which such behavior is not expected.

For static loading in a hostile chemical environment, time-dependent crack growth may occur. Life estimates may be made by employing a da/dt vs. K curve for the particular combination of material and environment. Since environmental cracking problems are sensitive to the exact combination of material and environment, it may be possible to make a modest change in the material or the environment that eliminates the problem.

NEW TERMS AND SYMBOLS

crack growth life, N_{if}

crack growth retardation

crack velocity, $\dot{a} = da/dt$

cyclic plastic zone size, $2r'_o$

damage tolerant design

environmental crack growth

fatigue crack growth rate, da/dN

fatigue crack growth threshold, ΔK_{th}

Forman equation constants: C_2, m_2, K_c

initial and final crack lengths: a_i, a_f

inspection period, N_p

intergranular fracture

minimum detectable crack length, a_d

monotonic plastic zone size, $2r_o$

Paris equation constants: C, m

sequence effects

small crack; short crack

small crack transition length, a_s

stationary loading

stress corrosion cracking threshold, K_{Iscc}

stress intensity range, ΔK

transgranular fracture

Walker equation constants: C_1, m_1, γ

REFERENCES

(a) General References

ASTM. 1990 *Annual Book of ASTM Standards*, Vol. 03.01, Am. Soc. for Testing and Materials, Philadelphia, Pa. See No. E647, "Measurement of Fatigue Crack Growth Rates."

BARSOM, J. M. and S. T. ROLFE. 1987 *Fracture and Fatigue Control in Structures*, 2nd ed., Prentice-Hall, Englewood Cliffs, N.J.

BROEK, D. 1986 *Elementary Engineering Fracture Mechanics*, 4th ed., Kluwer Academic Pubs., Dordrecht, The Netherlands.

BROEK, D. 1988 *The Practical Use of Fracture Mechanics*, Kluwer Academic Pubs., Dordrecht, The Netherlands.

CAMPBELL, J. E., W. W. GERBERICH, and J. H. UNDERWOOD, eds. 1982 *Application of Fracture Mechanics for Selection of Metallic Structural Materials*, Am. Soc. for Metals, Metals Park, Oh.

CHANG, J. B. and J. L. RUDD, eds. 1984 *Damage Tolerance of Metallic Structures*: *Analysis Methods and Applications*, ASTM STP 842, Am. Soc. for Testing and Materials, Philadelphia, Pa.

CLARK, W. G., Jr. 1981 "Some Problems in the Application of Fracture Mechanics," *Fracture Mechanics: Thirteenth Conference*, ASTM STP 743, R. Roberts, ed., Am. Soc. for Testing and Materials, Philadelphia, Pa., pp. 269–287.

COURTNEY, T. H. 1990 *Mechanical Behavior of Materials*, McGraw-Hill, New York.

FISHER, J. W. 1984 *Fatigue and Fracture in Steel Bridges: Case Studies*, John Wiley, New York.

HERTZBERG, R. W. 1989 *Deformation and Fracture Mechanics of Engineering Materials*, 3rd ed., John Wiley, New York.

HUDSON, C. M. and T. P. RICH, eds. 1986 *Case Histories Involving Fatigue and Fracture Mechanics*, ASTM STP 918, Am. Soc. for Testing and Materials, Philadelphia, Pa.

RICE, R. C., ed. 1988 *Fatigue Design Handbook*, 2nd ed., SAE Pub. No. AE-10, Soc. of Automotive Engineers, Warrendale, Pa.

SURESH, S. 1991 *Fatigue of Materials*, Cambridge University Press, Cambridge, UK.

THOMPSON, R. B., et al., eds. 1988 *Mechanics and Physics of Crack Growth: Application to Life Prediction*, Elsevier Applied Science Pubs., London. (See also *Materials Science and Engineering*, Vol. A103, No. 1, 1988.)

WEI, R. P. and R. P. GANGLOFF. 1989 "Environmentally Assisted Crack Growth in Structural Alloys: Perspectives and New Directions," *Fracture Mechanics: Perspectives and Directions (Twentieth Symposium)*, R. P. Wei and R. P. Gangloff, eds., ASTM STP 1020, Am. Soc. for Testing and Materials, Philadelphia, Pa., pp. 233–264.

(b) Sources of Material Properties

BATTELLE. 1991 *Aerospace Structural Metals Handbook*, 5 vols., Metals and Ceramics Information Center, Battelle Columbus Div., Columbus, Oh.

BOYER, H. E. 1986 *Atlas of Fatigue Curves*, Am. Soc. for Metals, Metals Park, Oh.

FREIMAN, S. W. 1980 "Fracture Mechanics of Glass," *Glass: Science and Technology; Vol. 5: Elasticity and Strength in Glasses*, D. R. Uhlmann and N. J. Kreidl, eds., Academic Press, New York, pp. 21–78.

GALLAGHER, J. P., ed. 1983 *Damage Tolerant Design Handbook*, 4 vols., Metals and Ceramics Information Ctr., Battelle Columbus Labs., Columbus, Oh.

JSMS. 1983 *Data Book on Fatigue Crack Growth Rates of Metallic Materials*, 2 vols., The Society of Materials Science (JSMS), Kyoto, Japan.

MILHDBK. 1983 *Military Standardization Handbook: Metallic Materials and Elements for Aerospace Vehicle Structures*, MIL-HDBK-5D, 2 Vols., U.S. Dept. of Defense and Federal Aviation Administration, Naval Publications and Forms Ctr., Philadelphia, Pa.

SPEIDEL, M. O. 1975 "Stress Corrosion Cracking of Aluminum Alloys," *Metallurgical Transactions A*, Vol. 6A, No. 4, Apr. 1975, pp. 631–651.

TAYLOR, D. 1985 *A Compendium of Fatigue Thresholds and Crack Growth Rates*, Engineering Materials Advisory Services Ltd., Cradley Heath, Warley, West Midlands, UK. (See also *Computer Database on Fatigue Thresholds and Crack Growth Rates*, same publisher.)

PROBLEMS AND QUESTIONS

Section 11.2

11.1 For martensitic steels at $R \approx 0$, Eq. 11.10 is sometimes used with the constants

$$C = 1.36 \times 10^{-7}, \quad m = 2.25$$

where units of MPa$\sqrt{\text{m}}$ and mm/cycle are used.

(a) Plot the resulting equation on log-log coordinates for ΔK between 1 and 100 MPa$\sqrt{\text{m}}$. What type of curve is obtained? (Suggestions: Two log cycles on the horizontal scale, and five on the vertical scale, are required. If one of the scales on your graph paper runs the wrong way, turn the paper over, look through to see the grid lines, and plot on the back.)

(b) What is the significance of the value of da/dN at $\Delta K = 1$ MPa$\sqrt{\text{m}}$? What is the significance of the slope of the line?

(c) Also plot this equation on linear-linear coordinates for ΔK from zero to 100 MPa$\sqrt{\text{m}}$. How does the curve differ from (a)?

11.2 Estimate the constants C and m for the straight line portion of the data of Fig. 11.3.

11.3 A fatigue crack growth test was done at $R = 0.032$ on a compact specimen of a hard ($HRC = 60$) tool steel having tensile properties of $\sigma_u = 2200$ MPa and elongation 1.7%. Some of the da/dN vs. ΔK data points are listed in Table P11.3.

(a) Plot these points on log-log coordinates and obtain approximate values of constants C and m for Eq. 11.10.

(b) Use a log-log least squares fit to obtain refined values of C and m.

11.4 A center-cracked plate made of 7075-T651 aluminum alloy has dimensions, as defined in Fig. 8.13(a), of $b = 19.05$, $t = 6.60$ mm, and large h. A zero-to-maximum ($R = 0$) cyclic load of $P = 18.8$ kN is applied, causing the crack to grow. Crack growth rates measured at various crack lengths are given in Table P11.4.

TABLE P11.3

da/dN, mm/cycle	ΔK, MPa$\sqrt{\text{m}}$
4.24×10^{-6}	6.84
9.12×10^{-6}	8.76
1.75×10^{-5}	10.35
3.51×10^{-5}	13.30

Source: Data in [Luken 87].

(a) Calculate ΔK for each crack length, and then make a da/dN vs. ΔK plot for these data.
(b) Obtain approximate values for constants C and m for Eq. 11.10.
(c) Use a log-log least squares fit to obtain refined values of C and m.

TABLE P11.4

a, mm	da/dN, mm/cycle
7.32	1.76×10^{-4}
9.53	5.08×10^{-4}
12.07	1.27×10^{-3}
14.94	3.18×10^{-3}

11.5 A center-cracked plate is made of a high-strength martensitic steel with C and m as in Prob. 11.1 and a fracture toughness of $K_{Ic} = 80$ MPa$\sqrt{\text{m}}$. The plate dimensions, as defined in Fig. 8.13(a), are $b = 50$, $t = 10$ mm, and large h. A fatigue crack of length a grows from a small initial size to failure during zero-to-maximum ($R = 0$) cyclic loading at $P_{max} = 140$ kN. (PC Problem)
   (a) For several crack lengths between $a = 2$ and 45 mm, calculate ΔK and plot the values versus a. What is the nature of the variation of K with a?
   (b) If K_{Ic} is exceeded prior to $a = 45$ mm, determine the approximate crack length $a = a_c$ where failure occurs.
   (c) Estimate crack growth rates da/dN for several ΔK values from (a) that do not exceed a_c, and plot these on a log scale versus a on a linear scale. What is the nature of the variation of da/dN with a?

Section 11.3

11.6 Based on the data given in Example 11.2, determine the remaining da/dN and ΔK values and make a plot of all the results. (PC Problem)

11.7 A center-cracked plate of 7075-T6 Al was tested as in Example 11.2. All details were the same except that the load was cycled between $P_{min} = 48.1$ and $P_{max} = 96.2$ kN. The data obtained are listed in Table P11.7. Calculate da/dN and ΔK for these data and make a da/dN vs. ΔK plot of the results. (PC Problem)

11.8 Complete crack length versus cycles data are given in Table P11.8 from the test on a hard tool steel of Prob. 11.3. A standard compact specimen was used with dimensions, as defined

TABLE P11.7

a, mm	N, kilocycles
5.08	0
7.62	9.5
10.16	14.3
12.70	17.1
15.24	19.1
17.78	20.5
20.32	21.5
22.86	22.3
25.40	22.9
30.48	23.5
35.56	24.0

Source: Data in [Hudson 69].

in Fig. 8.15(c) of $b = 50.8$ and $t = 6.35$ mm. The load was cycled at a frequency of 30 Hz between values of $P_{min} = 44.5$ and $P_{max} = 1379$ N. Crack lengths a, measured from the centerline of the pin holes, are given in Table P11.8 with the corresponding cycle numbers. (PC Problem)

(a) Determine the da/dN and ΔK values from these data and prepare a da/dN vs. ΔK plot of the results.

(b) Obtain approximate values of C and m for Eq.11.10.

(c) Refine the C and m values from (b) by using a log-log least squares fit.

TABLE P11.8

a, mm	N, kilocycles
19.86	0
21.13	300
22.15	508
22.94	658
23.80	792
24.61	892
25.37	976
26.54	1070
27.43	1122
27.89	1148
28.32	1167
29.08	1191
29.85	1221
30.71	1241
31.01	1251
31.29	1259

Source: Data in [Luken 87].

11.9 Crack length versus cycles data are given in Table P11.9 from a test on 7075-T651 Al. A standard compact specimen was used with dimensions, as defined in Fig. 8.15(c), of $b = 50.8$ and $t = 6.63$ mm. The load was cycled between $P_{min} = 267$ and $P_{max} = 2669$ N, and crack lengths a are measured from the centerline of the pin holes. (PC Problem)

    (a) Determine the da/dN and ΔK values from these data and prepare a plot of the results.

    (b) Similarly reduce the remaining data of Example 11.2 and add the data from that test to your da/dN vs. ΔK plot from (a). Is the behavior independent of test specimen geometry? (Comment: The -T6 and -T651 alloys are nearly identical, and the difference in behavior between $R = 0$ and 0.1 is expected to be small.)

TABLE P11.9

a, mm	N, cycles
13.97	480
15.24	10 460
16.51	21 170
17.78	29 940
19.05	36 910
20.32	43 150
21.59	48 490
22.86	52 940
24.13	56 350
25.40	59 370
26.67	61 410
27.94	63 120
29.21	64 460
30.48	65 470
31.75	65 900

Section 11.4

11.10 A martensitic steel has constants C and m for Eq. 11.10 as given in Prob. 11.1, which values are noted to apply for $R = 0$.

    (a) Assuming that the constant for the Walker equation is $\gamma = 0.4$, estimate da/dN vs. ΔK equations of the form of Eq. 11.10 for both $R = 0.5$ and $R = 0.75$.

    (b) Plot the equations from (a) along with the $R = 0$ equation on log-log coordinates.

    (c) By about what factor does da/dN increase for a given ΔK if R is increased from 0 to 0.5? From 0 to 0.75? Are these factors constant for different ΔK values?

11.11 Use the data of Fig. 11.4 to obtain your own approximate value of the Walker constant γ for this AISI 4340 steel ($\sigma_u = 1296$ MPa). You may take the C_1 and m_1 values in Table 11.1 as given. (Suggestion: Start by plotting a line of slope corresponding to m_1 that passes through the $R = 0.7$ data.)

11.12 Apply the Walker equation to the data for granite rock in Fig. 11.11, evaluating C_1, m_1, and γ. Suggestion: First get two parallel lines by slightly adjusting the $R = 0.2$ line to have the same slope as the $R = 0.1$ line, which gives

$$\frac{da}{dN} = 7.8 \times 10^{-10} \, (\Delta K)^{11.9} \quad (R = 0.2)$$

Also, compare the γ value obtained with those for the metals in Table 11.1 and comment on the significance of its magnitude.

11.13 Using the Forman equation and the constants in Table 11.1 for 2024-T3 Al:

(a) Obtain the particular da/dN vs. ΔK equations that apply for both $R = 0$ and $R = 0.5$.

(b) Then plot these on log-log coordinates for growth rates between 10^{-6} and 10^{-2} mm/cycle. Are the lines straight? (PC Problem)

(c) By about what factor is da/dN for a given ΔK typically increased by changing R from 0 to 0.5? Its the factor constant for different ΔK values?

11.14 Using the da/dN vs. ΔK data for 7075-T6 Al at $R = 0.5$ obtained from Prob. 11.7: (PC Problem)

(a) Calculate Q from Eq. 11.24 for each data point, and using K_c from Table 11.1, plot Q vs. ΔK on log-log coordinates.

(b) On the plot from (a), add the straight line from the Forman equation using constants from Table 11.1. Also add Q vs. ΔK data for $R = 0$ based on the data of Example 11.2. Do the data for both R values fall along the line?

11.15 The SAE *Fatigue Design Handbook* gives the following equation as being useful in fitting da/dN vs. ΔK curves:

$$\frac{da}{dN} = \frac{C_4 (\Delta K - \Delta K_{th})^{m_4}}{(1 - R)K_c - \Delta K}$$

The quantities C_4 and m_4 are new material constants, and K_c and ΔK_{th} are material constants as previously defined, with ΔK_{th} being in general a function of R.

(a) Describe the shape of the resulting curve on a log-log plot of da/dN vs. ΔK. Qualitatively, how is it affected by changing R? By changing ΔK_{th}? By changing K_c?

(b) Assume that K_c is known, and also that ΔK_{th} is known for various R. How could values for C_4 and m_4 then be obtained by using a log-log plot of data from crack growth tests at several different R-ratios?

Section 11.6

11.16 Derive the equation for crack growth life N_{if} that is analogous to Eq. 11.30, but which is applicable to the special case of $m = 2$.

11.17 Consider da/dN vs. ΔK behavior according to Eq. 11.10 with ΔK given by

$$K = \frac{F'P}{t\sqrt{\pi a}}$$

where F' is approximately constant as for small a/b for two cases of prying loads in Fig. 8.15. Derive an equation for crack growth life N_{if} as a function of material constants C and m, load range ΔP, the constant F', geometric dimensions, and initial and final crack sizes, a_i and a_f.

11.18 Consider closed-form equations for crack growth life N_{if} that are analogous to Eq. 11.30, but for variable F according to

$$F = \sqrt{\frac{2b}{\pi a} \tan \frac{\pi a}{2b}}$$

(a) Derive such an equation for the particular case of $m = 2$.

(b) Are closed form solutions possible for other values of m? If so, which ones? (Suggestion: Consult a table of integrals.)

(This expression for F is within 10% out to a $a/b = 0.7$ for a center-cracked plate under uniform stress, and it is the exact solution for an array of colinear cracks in an infinite plate under remote uniform stress normal to the crack plane.)

11.19 Derive an equation for calculating crack growth life N_{if} that is analogous to Eq. 11.30, where F is constant, but where da/dN is given by the Forman equation, Eq. 11.23.

11.20 A center-cracked plate made of 2024-T3 Al has dimensions, as defined in Fig. 8.13(a), of $b = 50$, $t = 4$ mm, and large h, and an initial crack length of $a_i = 2$ mm.

(a) How many cycles of loading in tension between $P_{min} = 6$ and $P_{max} = 24$ kN are required to grow the crack to a length of $a_f = 12$ mm?

(b) Does either fully plastic yielding or brittle fracture occur prior to reaching this a_f? (Suggestion: Use properties for 2024-T351 Al in Table 8.1).

11.21 A bending member has a rectangular cross section of dimensions, as defined in Fig. 8.14, of depth $b = 60$ and thickness $t = 12$ mm. It is made of the AISI 4340 steel of Table 11.1 and is subjected to a cyclic moment between $M_{min} = 0.8$ and $M_{max} = 4.0$ kN·m. Failure occurs after 70,000 cycles of this loading by brittle fracture from a through-thickness edge crack extending 15 mm in the depth direction. Estimate the initial crack length present at the beginning of the cyclic loading.

11.22 A center-cracked plate loaded in tension has dimensions, as defined in Fig. 8.13(a), of $b = 50$, $t = 5$ mm, and large h, and an initial crack length of $a_i = 5$ mm. The material is the ferritic-pearlitic structural steel ASTM A572 with yield strength $\sigma_o = 430$ MPa and a plane stress fracture toughness in this thickness of at least $K_c = 200$ MPa$\sqrt{m}$. What zero-to-maximum cyclic load will cause the crack to grow to $a_f = 25$ mm in 200,000 cycles?

11.23 A single-edge-cracked plate is loaded in tension and has dimensions, as defined in Fig. 8.13(c), of $b = 75$, $t = 5$ mm, and large h, and it contains an initial crack of length $a_i = 4$ mm. A zero-to-maximum cyclic load of 20 kN is applied, and the material is the same as in Prob. 11.22. How many cycles can be applied before failure is expected? (PC Problem: Numerical integration is needed.)

11.24 For the situation of Prob. 11.20, estimate the number of cycles to failure using numerical integration. (PC Problem)

Section 11.7

11.25 For the material and member of Prob. 11.20, replace the loading with repeated applications of the load history shown in Figure P11.25.

(a) How many repetitions are required to grow the crack from $a_i = 2$ mm to $a_f = 12$ mm? (PC Problem)

(b) Does your life estimate need to be revised due to either fully plastic yielding or brittle fracture occurring prior to reaching this a_f?

11.26 For the same material and member as in Prob. 11.21, the initial crack length is $a_i = 0.5$ mm, and the load history is replaced by repeated applications of the sequence shown in Figure P11.26. Estimate the number of repetitions to failure.

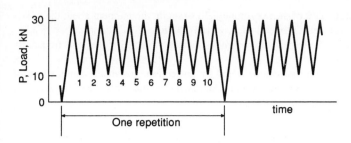

Figure P11.25

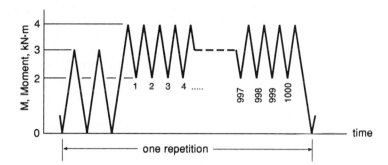

Figure P11.26

Section 11.8

11.27 A bending member with a rectangular cross section is made of 7075-T6 Al and has dimensions, as defined in Fig. 8.14, of $b = 50$ and $t = 12$ mm. Any through-thickness edge crack larger than $a = 0.25$ mm can be found by inspection.

    (a) What is the safety factor on life if the expected actual service life is 80,000 cycles, and the applied loading is a zero-to-maximum cyclic bending moment of 500 N·m.

    (b) What inspection period is necessary to achieve a safety factor of five on life?

11.28 A plate of the AISI 4340 steel of Table 11.1 is loaded in tension and may contain an edge crack. It has dimensions, as defined in Fig. 8.13(c), of width $b = 250$ and thickness $t = 25$ mm. Cyclic loading occurs between loads of $P_{min} = 1.7$ and $P_{max} = 3.4$ MN.

    (a) Estimate the number of cycles required to grow a crack to failure starting from a minimum detectable size of $a_d = 1.3$ mm.

    (b) The desired actual service life is 60,000 cycles, and a safety factor of 3.0 on life is required. Is the design adequate?

    (c) If not, what a_d would have to be found by improved inspection?

    (d) If $a_d = 1.3$ mm cannot be improved, what interval of periodic inspections is required?

    (e) If neither improved inspection nor periodic inspection is possible, how much must the stresses be lowered while maintaining $R = 0.5$?

11.29 The structural member of Prob. 8.13 is made of the ferritic-pearlitic steel ASTM A572. The specified minimum yield strength is $\sigma_o = 345$ MPa, and for the service conditions of interest, the fracture toughness from Fig. 8.32 may be as low as $K_{Ic} = 55$ MPa$\sqrt{m}$. Based on the stress not exceeding 55% of the minimum yield strength, the moment about the x-axis in actual service is not allowed to exceed $M_{max} = 161$ kN·m.

(a) At what crack length a of Fig. P8.13 is brittle fracture expected if the load frequently comes close to M_{max}?

(b) If cracks of length $a = 2.5$ mm can be found by inspection, how many applications of loading from zero to M_{max} can be permitted before the member must be inspected? A safety factor of five on life is required.

Section 11.10

11.30 Obtain approximate values of the constants A and n of Eq. 11.54 for both sets of data in Fig. 11.38. Comment on the values of n obtained. In each case, by what factor is $\dot{a}$ increased if K is increased by 25%?

11.31 The constants A and n of Eq. 11.54 for soda-lime-silicate glass are approximately $A = 1.67$ and $n = 20.3$, where these values apply for K and $\dot{a}$ in units of MPa$\sqrt{m}$ and m/s, respectively. Assume that this glass contains initial cracks of length $a = 10$ μm, and that these are half-circular surface cracks as in Fig. 8.17, so that $F \approx 0.73$ applies for $K = FS\sqrt{\pi a}$.

(a) Use Eq. 11.55 to derive a relationship between stress and time to failure for this situation. (Suggestion: Manipulate Eq. 11.55 to be in the same form as Eq. 11.31. Is the life affected significantly by a_f?)

(b) Develop similar equations for both $a_i = 5\mu$m and $a_i = 20$ μm. Then plot all three S vs. *time* equations on log-log coordinates and briefly discuss the dependence of life on stress level and initial crack length.

12

Plastic Deformation Behavior and Models for Materials

12.1 INTRODUCTION

Deformation beyond the point of yielding that is not strongly time dependent, called *plastic deformation*, frequently occurs in engineering components and may need to be analyzed in design, or to determine the cause of a failure. During plastic deformation, stresses and stains are no longer proportional, so that relationships more general than Hooke's Law (Eq. 4.10) are needed to provide an adequate description of the stress-strain behavior.

12.1.1 Significance of Plastic Deformation

Plastic deformation can impair the usefulness of an engineering component by causing large permanent deflections. Also, as already noted in Chapter 10, plastic deformation commonly causes residual stresses to remain after unloading. (See Fig. 10.27.) Residual stresses can either decrease or increase the subsequent resistance of a component to fatigue or environmental cracking, depending on whether the residual stress is tensile or compressive, respectively. Furthermore, the stress-based approach to fatigue as in Chapter 10 is based primarily on elastic analysis. As a result of this limitation, rough empirical adjustments are needed to account for the occurrence of plastic deformation at short lives and at high mean stresses.

Improved understanding and analysis of permanent deflections, of residual stresses, and of yielding during cyclic loading, is made possible through a study of plastic deformation. Before analysis of components is considered, it is first necessary to characterize the plastic deformation behavior of materials in more detail than is provided by the brief introduction in Chapter 4. Such a treatment is undertaken in this chapter, and the topic is then extended in Chapter 13 to stress-strain analysis of beams, shafts, and notched members. Information from both this chapter and Chapter 13 is then used in Chapter 14 in presenting the strain-based approach to fatigue, which considers plasticity effects on fatigue in a fairly complete and rigorous manner.

Plastic deformation is generally considered to be time independent. This simplifying assumption is often a reasonable one for engineering purposes, as long as the always present time-dependent (creep) deformations are relatively small. Further consideration of time-dependent behavior is postponed until Chapter 15.

12.1.2 Preview of Chapter

In characterizing the plastic deformation behavior of materials, the obvious starting point is to consider stress-strain curves for *monotonic* loading, that is, for loading that proceeds in only one direction as in Fig. 12.1(a). Mathematical representation $\varepsilon = f(\sigma)$ of such curves is needed, as for later use in Chapter 13 for component analysis. Recalling from Chapter 7 that yielding is affected by the state of stress, it might be expected that the

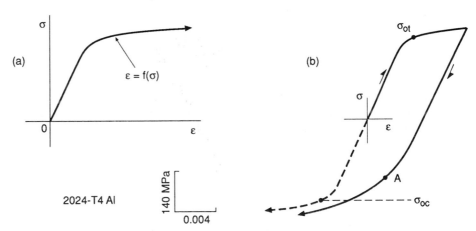

Figure 12.1 Monotonic stress-strain curve (a), and unloading stress-strain curve (b), where the Bauschinger effect causes yielding at A prior to the yield strength σ_{oc} from monotonic compression.

stress-strain curve beyond yielding is also affected. This is indeed the case, and we will consider state-of-stress effects on stress-strain curves in some detail.

If the direction of straining is reversed after yielding has occurred, the stress-strain path that is followed differs from the initial monotonic one as illustrated in Fig. 12.1(b). Yielding on unloading generally occurs prior to the stress reaching the yield strength σ_{oc} for monotonic compression, as at point A. This early yielding behavior is called the *Bauschinger effect*, after the German engineer who first studied it in the 1880s. Unloading stress-strain paths need to be described mathematically for use in predicting the behavior of components after unloading from a severe load, as in estimating residual stresses.

Stress-strain behavior during cyclic loading exhibits a number of complexities. Some example test data are shown in Fig. 12.2. In this case, an aluminum alloy specimen under axial loading was subjected to cyclic straining between the strains levels $\varepsilon_{\max} = 0.01$ and $\varepsilon_{\min} = -0.01$. Yielding occurs on each half-cycle of loading, and the behavior is observed to gradually change with the number of applied cycles. At least approximate modeling of such behavior is needed in implementing the strain-based approach to fatigue. Rheological models composed of linear springs and frictional sliders, as introduced in Chapter 4, are found to be particularly useful for this purpose.

12.1.3 Additional Comments

Before proceeding in more detail, it will be noted that the present chapter and the next are intended only as a brief introduction to the subject of *plasticity*. Furthermore, the emphasis on rheological modeling and cyclic loading is somewhat unconventional compared to the treatments found in standard (graduate level) textbooks on the subject, such as those of Chen and Han (1988), Hill (1983), and Mendelson (1968) in the References for this chapter. Here, we will consider primarily the total strain theory of plasticity,

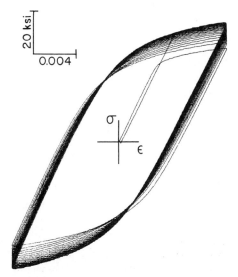

Figure 12.2 Stress-strain response in 2024-T4 aluminum for twenty cycles of completely reversed strain at $\varepsilon_a = 0.01$. (From [Dowling 72]; copyright ©ASTM; reprinted with permission.)

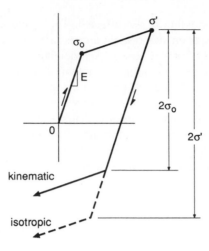

Figure 12.3 Differing unloading behavior for kinematic and isotropic hardening.

also called *deformation theory*, rather than the more advanced *incremental theory*. Also, the rheological models used are consistent with the behavior called *kinematic hardening*, the alternative choice of *isotropic hardening* not being employed as it is a poor model for real materials.

These two hardening rules are illustrated for unloading behavior in Fig. 12.3. Kinematic hardening predicts that yielding in the reverse direction occurs when the stress change from the unloading point is twice the monotonic yield strength, $\Delta\sigma = 2\sigma_o$. In contrast, isotropic hardening predicts yielding later at $\Delta\sigma = 2\sigma'$, where σ' is the highest stress reached prior to unloading. Thus, kinematic hardening predicts a Bauschinger effect as observed in real materials, but isotropic hardening predicts the opposite.

12.2 STRESS-STRAIN CURVES

Two simple elasto-plastic stress-strain curves and the corresponding rheological models are shown in Fig. 12.4. Additional curves involving nonlinear hardening are shown is Fig. 12.5. Either of these nonlinear curves may be represented using a multistage model as in Fig. 12.6. Note that these rheological models follow the conventions of Chapter 4 and contain only linear springs and frictional sliders, there being no time-dependent dashpot elements. Forces on such models are proportional to stress in the material being modeled, and displacements are proportional to strains.

12.2.1 Elastic, Perfectly Plastic Relationship

An elastic, perfectly plastic stress-strain relationship is flat beyond yielding as illustrated by Fig. 12.4(a).

$$\sigma = E\varepsilon \quad (\sigma \leq \sigma_o)$$

$$\sigma = \sigma_o \quad \left(\varepsilon \geq \frac{\sigma_o}{E}\right)$$

(12.1)

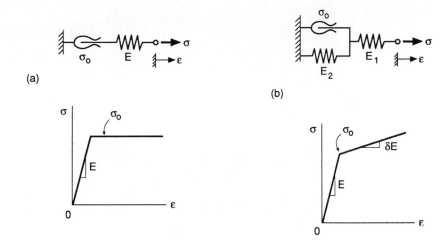

Figure 12.4   Stress-strain curves and rheological models for (a) elastic, perfectly plastic behavior and (b) elastic, linear-hardening behavior.

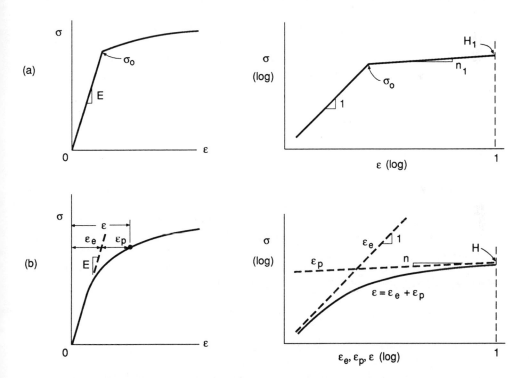

Figure 12.5   Stress-strain curves on linear and logarithmic coordinates for (a) an elastic, power-hardening relationship and (b) the Ramberg-Osgood relationship.

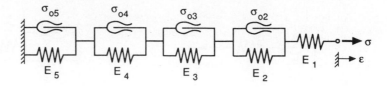

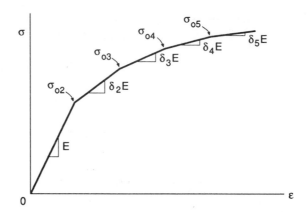

Figure 12.6 Multistage spring and slider model for nonlinear-hardening stress-strain curves.

where σ_o is the yield strength. This form is a reasonable approximation for the initial yielding behavior of certain metals and other materials. Also, it is often used as a simple idealization to make rough estimates even where the stress-strain curve has a more complex shape.

Beyond yielding, the strain is the sum of elastic and plastic parts.

$$\varepsilon = \varepsilon_e + \varepsilon_p = \frac{\sigma}{E} + \varepsilon_p \qquad \left(\varepsilon > \frac{\sigma_o}{E}\right) \tag{12.2}$$

In the rheological model, the elastic strain ε_e is analogous to the deflection of the linear spring of stiffness E, and the plastic strain ε_p is analogous to the movement of the frictional slider.

12.2.2 Elastic, Linear-Hardening Relationship

Elastic, linear-hardening behavior, Fig. 12.4(b), is useful as a rough approximation for stress-strain curves that rise appreciably following yielding. Such a relationship requires an additional constant, δ. This is the *reduction factor* for the slope following yielding, the slope before yielding being the elastic modulus E, and that after yielding being δE. The value of δ can vary from zero to unity, with smaller values giving flatter post-yielding behavior. Also, $\delta = 0$ gives a special case that corresponds to the elastic, perfectly plastic relationship.

An equation for the post-yielding portion can be obtained by taking the slope between any point on this part of the curve and the yield point.

$$\delta E = \frac{\sigma - \sigma_o}{\varepsilon - \varepsilon_o} \tag{12.3}$$

Noting that the yield strain is given by $\varepsilon_o = \sigma_o/E$, and solving for stress, allows the entire relationship to be specified.

$$\begin{aligned} \sigma &= E\varepsilon & (\sigma \leq \sigma_o) \\ \sigma &= (1 - \delta)\,\sigma_o + \delta E\varepsilon & (\sigma \geq \sigma_o) \end{aligned} \tag{12.4}$$

It is sometimes convenient to solve the second equation for strain.

$$\varepsilon = \frac{\sigma_o}{E} + \frac{(\sigma - \sigma_o)}{\delta E} \quad (\sigma \geq \sigma_o) \tag{12.5}$$

The response of the rheological model is the sum of the elastic strain in spring E_1 and any plastic strain in the spring-slider (E_2, σ_o) parallel combination. No plastic strain occurs until the stress exceeds the slider yield strength σ_o, and beyond this point the deflection of spring E_2 is also equal to the plastic strain. Noting that beyond yielding spring E_2 is subjected to a stress of $(\sigma - \sigma_o)$, the deflections in the two springs can be added to obtain the total strain.

$$\varepsilon = \frac{\sigma}{E_1} + \frac{(\sigma - \sigma_o)}{E_2} \quad (\sigma \geq \sigma_o) \tag{12.6}$$

which was previously developed as Eq. 4.34. Equations 12.5 and 12.6 are equivalent if the constants are related by

$$E = E_1, \quad \delta E = \frac{E_1 E_2}{E_1 + E_2} \tag{12.7}$$

Where it is noted that the slope δE corresponds to the stiffness of the two springs E_1 and E_2 in series.

12.2.3 Elastic, Power-Hardening Relationship

Reasonably accurate representation of the stress-strain curves of real materials generally requires a more complex mathematical relationship than those described so far. One form that is sometimes used assumes that stress is proportional to strain raised to a power, with this being applied only beyond a yield strength σ_o.

$$\begin{aligned} \sigma &= E\varepsilon & (\sigma \leq \sigma_o) & \quad \text{(a)} \\ \sigma &= H_1 \varepsilon^{n_1} & (\sigma \geq \sigma_o) & \quad \text{(b)} \end{aligned} \tag{12.8}$$

The term *strain hardening exponent* is used for n_1, and H_1 is an additional constant that is needed.

The most convenient means of fitting this relationship to a particular set of stress-strain data is to make a log-log plot of stress versus strain, where for the post-yield portion a straight line is expected. This is illustrated on the right in Fig. 12.5(a). The value of σ at $\varepsilon = 1$ is H_1. Assuming that the logarithmic decades are the same length in both directions, the slope of the line is n_1. On the same graph, the elastic region equation, $\sigma = E\varepsilon$, also forms a straight line, but with a slope of unity, and the two lines intersect at $\sigma = \sigma_o$. Values of the exponent n_1 are typically in the range 0.05 to 0.4 for metals where this equation fits well.

Equation 12.8(b) can be easily expressed in terms of strain.

$$\varepsilon = \left(\frac{\sigma}{H_1}\right)^{\frac{1}{n_1}} \quad (\sigma \geq \sigma_o) \tag{12.9}$$

Also, the yield strength is not an independent constant, as any two of σ_o, H_1, and n_1 may be used to calculate the remaining one. An equation relating these can be obtained by applying both Eqs. 12.8(a) and (b) at the point $(\varepsilon_o, \sigma_o)$ and combining the results.

$$\sigma_o = E\left(\frac{H_1}{E}\right)^{\frac{1}{1-n_1}} \tag{12.10}$$

12.2.4 Ramberg-Osgood Relationship

A relationship similar to that proposed in a report by Ramberg and Osgood in 1943 is frequently used. Here, elastic and plastic strains, ε_e and ε_p, are considered separately and summed. An exponential relationship is used, but it is applied to the plastic strain, rather that to the total strain as above.

$$\sigma = H\varepsilon_p^n \tag{12.11}$$

where this n is also called a *strain hardening exponent*, despite the fact that it is defined differently than n_1 above. The total strain is the sum of the elastic and plastic strains, where the elastic strain is related to stress by E and the plastic strain is the deviation from the elastic slope as shown in Fig. 12.5(b).

$$\varepsilon = \varepsilon_e + \varepsilon_p = \frac{\sigma}{E} + \varepsilon_p \tag{12.12}$$

Combining the above equations gives a relationship between stress and total strain.

$$\varepsilon = \frac{\sigma}{E} + \left(\frac{\sigma}{H}\right)^{\frac{1}{n}} \tag{12.13}$$

This equation cannot be solved explicitly for stress. It provides a single smooth curve for all values of σ and does not exhibit a distinct yield point. Thus, it contrasts with the previously described elastic, power-hardening form, which is discontinuous at a distinct yield point σ_o.

Constants for Eq. 12.13 for a particular set of stress-strain data are obtained by making a log-log plot of stress versus *plastic* strain, σ vs. ε_p, as illustrated on the right in Fig. 12.5(b). The constant H is the value of σ at $\varepsilon_p = 1$, and n is the slope on the log-log plot if the logarithmic decades in the two directions are of equal length. A plot of σ versus *total* strain ε is a curve on the log-log plot. At small strains, this curve approaches the line of unity slope corresponding to elastic strains, and at large strains it approaches the plastic strain line of slope n.

The Ramberg-Osgood equation and the power-hardening relationship are essentially equivalent if the strains are sufficiently large that the plastic portion dominates, so that the elastic portion can be considered to be negligible. The first term of Eq. 12.13 is then negligible, and values of H and n fitted to data at large strains for ductile materials will be similar to values of H_1 and n_1 fitted to the same data. Recall that a power-hardening relationship was used in this manner in Chapter 5 for true stresses and strains in tension tests. See Eq. 5.28.

Example 12.1

Some test data points on the monotonic stress-strain curve of 7075-T651 Al for uniaxial stress are given below. Obtain values of the constants for a stress-strain curve of the Ramberg-Osgood form, Eq. 12.13, that fits these data. Use $E = 71$ GPa.

Test Data		Calculation
σ, MPa	ε	$\varepsilon_p = \varepsilon - \dfrac{\sigma}{E}$
433	7.40×10^{-3}	1.301×10^{-3}
451	8.95×10^{-3}	2.598×10^{-3}
469	1.28×10^{-2}	6.194×10^{-3}
487	2.29×10^{-2}	1.604×10^{-2}
505	4.57×10^{-2}	3.859×10^{-2}

Solution   In addition to E as given, constants H and n for Eq 12.13 are needed. These can be found by fitting the data to Eq. 12.11.

$$\sigma = H\varepsilon_p^n$$

Plastic strains are required to proceed, values of which are calculated for each data point as indicated above. Taking logarithms of both sides of Eq. 12.11 gives

$$\log \sigma = n \log \varepsilon_p + \log H$$

This is a straight line on a log-log plot

$$y = mx + b$$

where

$$y = \log \sigma \quad \text{(dependent variable)}$$
$$x = \log \varepsilon_p \quad \text{(independent variable)}$$
$$m = n, \qquad b = \log H$$

Performing a linear least-squares fit on this basis gives

$$m = n = 0.04453$$ **Ans.**

$$b = 2.7675, \quad H = 10^b = 585.5 \text{ MPa}$$ **Ans.**

Discussion   Equation 12.11 with the constants evaluated is thus

$$\sigma = 585.5\varepsilon_p^{0.04453} \text{ MPa}$$

The resulting straight line on a log-log plot of ε_p versus σ is shown in Figure E12.1(a) along with the test data.

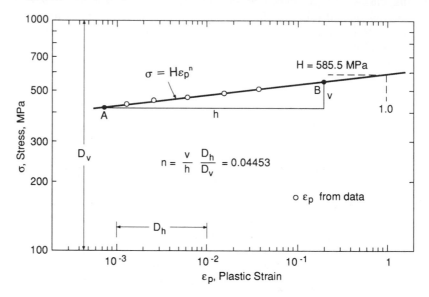

Figure E12.1(a)

An approximate graphical evaluation of the constants could be done by drawing a straight line through the ε_p versus σ data on this plot. As shown, the value of σ at $\varepsilon_p = 1$ is H. The slope of the line determines n, being equal to n if the decades of both logarithmic scales have the same length. To graphically determine n, distances h, v, D_h, and D_v need to be measured as with a ruler, where h and v correspond to any two convenient points A and B on the line. The raw slope v/h is not equal to n, as the decade lengths D_h and D_v are not equal, but n can be obtained by multiplying v/h by the ratio of the decade lengths as shown.

Substituting the constants obtained into Eq 12.13 gives an equation for the total strain.

$$\varepsilon = \frac{\sigma}{71{,}000} + \left(\frac{\sigma}{585.5}\right)^{\frac{1}{0.04453}}$$

where σ is assumed to be in units of MPa. Entering a number of values of σ into this equation and calculating the corresponding total strains ε gives the curve plotted in Figure E12.1(b). The original data are also plotted, and there is good agreement with the fitted curve.

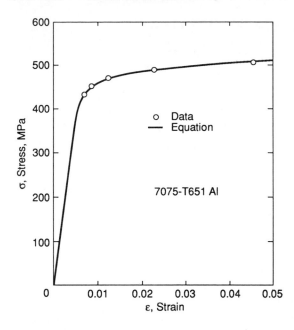

12.2.5 Rheological Modeling of Nonlinear Hardening

A stress-strain curve of either the elastic, power-hardening or Ramberg-Osgood types can be modeled by approximating it as a series of straight line segments as illustrated in Fig. 12.6. The first segment ends at the yield strength for the elastic, power-hardening case, and at a low stress where the plastic strain is small for the Ramberg-Osgood case. The corresponding rheological model has a linear spring that gives an initial elastic slope, and then a series of spring and slider parallel combinations that cause nonlinear behavior. The yield stresses for the various sliders have increasingly higher values and correspond to the stresses at the ends of the straight line segments on the stress-strain curve. The slope of any segment corresponds to the stiffness of all the springs in series for which the associated sliders have yielded.

$$\delta_j E = \frac{1}{\dfrac{1}{E_1} + \dfrac{1}{E_2} + \dfrac{1}{E_3} + \cdots + \dfrac{1}{E_j}} \tag{12.14}$$

where δ_j is the slope reduction factor for the segment that starts at σ_{oj}.

The strain in the model is the sum of the strains in each stage.

$$\varepsilon = \varepsilon_1 + \varepsilon_2 + \varepsilon_3 + \cdots + \varepsilon_j \tag{12.15}$$

Until a slider yields, it absorbs all of the stress and the associated spring none, as the strain in that stage is zero. Once a slider has moved, say the ith one, any stress in excess of that slider's yield stress must be resisted by the associated spring. Hence, the total stress is

$$\sigma = \sigma_{oi} + E_i \varepsilon_i \quad (\sigma > \sigma_{oi}) \qquad (12.16)$$

Solving for strain gives

$$\varepsilon_i = \frac{\sigma - \sigma_{oi}}{E_i} \quad (\sigma > \sigma_{oi}) \qquad (12.17)$$

This equation applies to each stage that has yielded, so that if all sliders up to the jth one have yielded, the strain is

$$\varepsilon = \frac{\sigma}{E_1} + \frac{\sigma - \sigma_{o2}}{E_2} + \frac{\sigma - \sigma_{o3}}{E_3} + \cdots + \frac{\sigma - \sigma_{oj}}{E_j} \qquad (12.18)$$

Evaluating $d\sigma/d\varepsilon$ to get the slope verifies Eq. 12.14.

12.3 THREE-DIMENSIONAL STRESS-STRAIN RELATIONSHIPS

From Chapters 4 and 7, the presence of stress components in more than one direction affects both a material's elastic stiffness and its yield strength. During plastic deformation, the state of stress continues to affect the behavior. Relationships between stress and strain are therefore needed for plastic deformation for the general three-dimensional case.

The generalized Hooke's Law for elastic strains was previously developed in Chapter 4 as Eqs. 4.10 and 4.11. These relationships apply not only prior to yielding but also after yielding, except that in the latter case they give only the elastic portions of the strains.

$$\varepsilon_{ex} = \frac{1}{E} \left[\sigma_x - \nu \left(\sigma_y + \sigma_z \right) \right] \qquad \text{(a)}$$

$$\varepsilon_{ey} = \frac{1}{E} \left[\sigma_y - \nu \left(\sigma_x + \sigma_z \right) \right] \qquad \text{(b)}$$

$$\qquad (12.19)$$

$$\varepsilon_{ez} = \frac{1}{E} \left[\sigma_z - \nu \left(\sigma_x + \sigma_y \right) \right] \qquad \text{(c)}$$

$$\gamma_{exy} = \frac{\tau_{xy}}{G}, \quad \gamma_{eyz} = \frac{\tau_{yz}}{G}, \quad \gamma_{ezx} = \frac{\tau_{zx}}{G} \qquad \text{(d)}$$

Subscripts e have been added to indicate that these are elastic strains, and the elastic constants E, ν, and G are defined in the usual manner.

The various strain components also include plastic portions that must be added to the elastic portions to obtain total strains.

$$\varepsilon_x = \varepsilon_{ex} + \varepsilon_{px}, \qquad \varepsilon_y = \varepsilon_{ey} + \varepsilon_{py}, \qquad \varepsilon_z = \varepsilon_{ez} + \varepsilon_{pz}$$
$$\gamma_{xy} = \gamma_{exy} + \gamma_{pxy}, \qquad \gamma_{yz} = \gamma_{eyz} + \gamma_{pyz}, \qquad \gamma_{zx} = \gamma_{ezx} + \gamma_{pzx} \qquad (12.20)$$

Shear strains, γ_{xy}, etc., are *engineering shear strains* as employed in Chapters 4 and 6 and in most elementary mechanics of materials textbooks. These values are twice the *tensor shear strains* often used in advanced textbooks.

12.3.1 Deformation Plasticity Theory

In Chapter 7, it was noted that yielding for a given material occurs at a value of *effective stress* $\bar{\sigma}$ that is approximately the same for all states of stress. Equation 7.37 gives $\bar{\sigma}$ as a function of the principal normal stresses.

$$\bar{\sigma} = \frac{1}{\sqrt{2}}\sqrt{(\sigma_1 - \sigma_2)^2 + (\sigma_2 - \sigma_3)^2 + (\sigma_3 - \sigma_1)^2} \qquad (12.21)$$

Recall that $\bar{\sigma}$ is proportional to the octahedral shear stress and reduces to $\bar{\sigma} = \sigma_1$ for the uniaxial case.

Corresponding strain quantities are needed. The *effective plastic strain* is a function of the plastic strains in the principal directions.

$$\bar{\varepsilon}_p = \frac{\sqrt{2}}{3}\sqrt{\left(\varepsilon_{p1} - \varepsilon_{p2}\right)^2 + \left(\varepsilon_{p2} - \varepsilon_{p3}\right)^2 + \left(\varepsilon_{p3} - \varepsilon_{p1}\right)^2} \qquad (12.22)$$

where subscripts (1, 2, 3) indicate the (x, y, z) axes that are the principal stress directions. This $\bar{\varepsilon}_p$ is proportional to the plastic shear strain on the octahedral planes and reduces to $\bar{\varepsilon}_p = \varepsilon_{p1}$ for the uniaxial case. Equation 12.22 may also be written in an equivalent but somewhat simpler form.

$$\bar{\varepsilon}_p = \sqrt{\frac{2}{3}\left(\varepsilon_{p1}^2 + \varepsilon_{p2}^2 + \varepsilon_{p3}^2\right)} \qquad (12.23)$$

The *effective total strain* is the sum of elastic and plastic parts.

$$\bar{\varepsilon} = \frac{\bar{\sigma}}{E} + \bar{\varepsilon}_p \qquad (12.24)$$

When defined in this manner, $\bar{\varepsilon}$ reduces to ε_1 for the uniaxial case. A key feature of deformation theory is its prediction that a single curve relates $\bar{\sigma}$ and $\bar{\varepsilon}$ for all states of stress. Some test data on thin-walled copper tubes that approximately verify this are shown in Fig. 12.7(a). (Note that the octahedral shear stress τ_h is proportional to $\bar{\sigma}$, and the plastic octahedral shear strain γ_{ph} is proportional to $\bar{\varepsilon}_p$.)

Equations analogous to Hooke's Law are used to relate stresses and plastic strains.

$$\varepsilon_{px} = \frac{1}{E_p}\left[\sigma_x - 0.5\left(\sigma_y + \sigma_z\right)\right] \qquad (a)$$

$$\varepsilon_{py} = \frac{1}{E_p}\left[\sigma_y - 0.5\left(\sigma_x + \sigma_z\right)\right] \qquad (b)$$

$$(12.25)$$

$$\varepsilon_{pz} = \frac{1}{E_p}\left[\sigma_z - 0.5\left(\sigma_x + \sigma_y\right)\right] \qquad (c)$$

$$\gamma_{pxy} = \frac{3}{E_p}\tau_{xy}, \qquad \gamma_{pyz} = \frac{3}{E_p}\tau_{yz}, \qquad \gamma_{pzx} = \frac{3}{E_p}\tau_{zx} \qquad (d)$$

Comparing these to Hooke's Law, Eq. 12.19, the elastic modulus E is replaced by a *plastic modulus* E_p.

$$E_p = \frac{\bar{\sigma}}{\bar{\varepsilon}_p} \qquad (12.26)$$

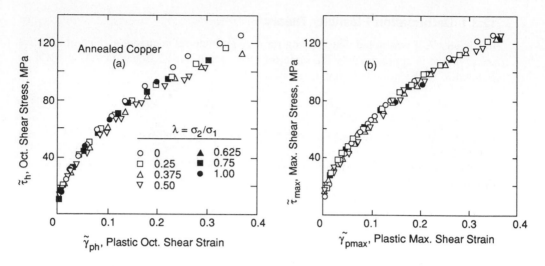

Figure 12.7 Correlation of true stresses and true plastic strains from combined axial and pressure loading of thin-walled copper tubes in terms of (a) octahedral shear stress and strain, and (b) maximum shear stress and strain. (Adapted from [Davis 43]; used with permission of ASME.)

Graphically, E_p corresponds to a secant modulus drawn to a point on the $\bar{\sigma}$ vs. $\bar{\varepsilon}_p$ curve as shown in Fig. 12.8. Hence, E_p is a variable that decreases as plastic deformation progresses along the $\bar{\sigma}$ vs. $\bar{\varepsilon}_p$ curve for the material.

12.3.2 Discussion

Poisson's ratio ν in Hooke's Law is replaced by 0.5 in Eq. 12.25, which is equivalent to the assumption that plastic strains do not contribute to volume change. This is supported by experimental evidence in metals, and is consistent with the physical mechanism of plastic strain being slip of crystal planes as discussed in Chapter 2. Hence, the volumetric strain ε_v is given by Eq. 4.19 even in the presence of plastic strains. This can be verified by obtaining the sum $\varepsilon_v = \varepsilon_x + \varepsilon_y + \varepsilon_z$ from Eq. 12.20, with the expres-

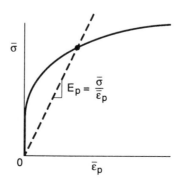

Figure 12.8 Definition of the plastic modulus as the secant modulus to a point on the effective stress versus effective plastic strain curve.

sions from Eqs. 12.19 and 12.25 substituted for the various elastic and plastic strain components.

Further comparing Eqs. 12.19 and 12.25, the elastic shear modulus G is replaced by $E_p/3$. This can be verified by using Eq. 12.25 to derive a plastic shear modulus G_p in a manner parallel to the derivation leading to $G = E/2(1 + v)$ as in Eq. 4.12. The presence of 0.5 in place of v in the equation is seen to give $G_p = E_p/3$. The equations given above constitute stress-strain relationships that can be used beyond the point of yielding. Although elastic and plastic strains are treated separately, Eqs. 12.19 and 12.25 can be substituted into Eq. 12.20 to write relationships between stress and total strain. An important use of the above equations is in predicting the effects of state of stress on stress-strain curves. To do so for a particular material, it is necessary to have the stress-strain curve for one state of stress, such as the uniaxial one. The procedure needed will be described shortly.

A similar plasticity theory may be formulated based on the maximum shear stress and plastic shear strain. The particular test results of Fig. 12.7 correlate even better on this basis as shown in (b). However, the use of $\bar{\sigma}$ and $\bar{\varepsilon}_p$ as given above is more conventional and will be the only approach pursued here.

12.3.3 The Effective Stress-Strain Curve

Assume that the uniaxial stress-strain curve for a given material is known.

$$\varepsilon_1 = f(\sigma_1) \quad (\sigma_2 = \sigma_3 = 0) \tag{12.27}$$

The equations of deformation plasticity theory given above can be used with any particular uniaxial σ-ε curve to estimate stress-strain curves for other states of stress that might be of interest. Such estimates are based on the concept of an effective stress-strain curve.

To proceed, first consider the uniaxial case in detail. Equation 12.21 gives

$$\bar{\sigma} = \sigma_1 \quad (\sigma_2 = \sigma_3 = 0) \tag{12.28}$$

Since the (x, y, z) axes are in this case the principal stress $(1, 2, 3)$ axes, Eq. 12.25 gives the plastic strains.

$$\varepsilon_{p1} = \frac{\sigma_1}{E_p}, \quad \varepsilon_{p2} = \varepsilon_{p3} = -\frac{\sigma_1}{2E_p} \tag{12.29}$$

Since $E_p = \bar{\sigma}/\bar{\varepsilon}_p$, Eqs. 12.28 and 12.29 give

$$\bar{\varepsilon}_p = \varepsilon_{p1} \quad (\sigma_2 = \sigma_3 = 0) \tag{12.30}$$

Alternatively, substitution of $\varepsilon_{p2} = \varepsilon_{p3} = -\varepsilon_{p1}/2$ from Eq. 12.29 into either Eq. 12.22 or 12.23 gives the same result.

Summing the elastic and plastic parts of the effective strain (Eq. 12.24) gives the total as

$$\bar{\varepsilon} = \frac{\sigma_1}{E} + \varepsilon_{p1} = \varepsilon_1 \quad (\sigma_2 = \sigma_3 = 0) \tag{12.31}$$

Hence, for the uniaxial case $\bar{\sigma} = \sigma_1$ and $\bar{\varepsilon} = \varepsilon_1$, so that a curve relating effective stress and strain (as defined above) is the same as the uniaxial stress-strain curve.

$$\bar{\varepsilon} = f(\bar{\sigma}) \tag{12.32}$$

Since the effective stress-strain curve is hypothesized to be independent of state of stress, it can thus be used to estimate stress-strain curves for other states of stress.

12.3.4 Application to Plane Stress

Consider the case of plane stress where the coordinate axes chosen are the principal axes and thus result in zero shear components.

$$\sigma_x = \sigma_1, \qquad \sigma_y = \sigma_2 = \lambda\sigma_1, \qquad \sigma_z = \sigma_3 = 0 \tag{12.33}$$

where to ratio $\lambda = \sigma_2/\sigma_1$ is used as a convenience. Equation 12.21 gives the effective stress for this situation to be

$$\bar{\sigma} = \sigma_1\sqrt{1 - \lambda + \lambda^2} \tag{12.34}$$

The total strain in the direction of σ_1 is composed of elastic and plastic parts.

$$\varepsilon_1 = \varepsilon_{e1} + \varepsilon_{p1} \tag{12.35}$$

The elastic part may be evaluated from Eq. 12.19(a), and the plastic part is most conveniently obtained from Eq. 12.25(a) with $E_p = \bar{\sigma}/\bar{\varepsilon}_p$ substituted.

$$\varepsilon_{e1} = \frac{\sigma_1}{E}(1 - \nu\lambda), \qquad \varepsilon_{p1} = \frac{\sigma_1\bar{\varepsilon}_p}{\bar{\sigma}}(1 - 0.5\lambda) \tag{12.36}$$

Taking $\bar{\varepsilon} = f(\bar{\sigma})$ as any particular uniaxial curve, Eq. 12.24 gives

$$\bar{\varepsilon}_p = \bar{\varepsilon} - \frac{\bar{\sigma}}{E} = f(\bar{\sigma}) - \frac{\bar{\sigma}}{E} \tag{12.37}$$

Substituting this for $\bar{\varepsilon}_p$ and then combining Eqs. 12.35 and 12.36 gives

$$\varepsilon_1 = \frac{1 - \nu\lambda}{E}\sigma_1 + \frac{(1 - 0.5\lambda)\sigma_1}{\bar{\sigma}}\left[f(\bar{\sigma}) - \frac{\bar{\sigma}}{E}\right] \tag{12.38}$$

Finally, substituting $\bar{\sigma}$ from Eq. 12.34 gives the desired result.

$$\varepsilon_1 = \frac{(0.5 - \nu)\lambda}{E}\sigma_1 + \frac{1 - 0.5\lambda}{\sqrt{1 - \lambda + \lambda^2}}f\left(\sigma_1\sqrt{1 - \lambda + \lambda^2}\right) \tag{12.39}$$

Hence, for any particular uniaxial curve, $\bar{\varepsilon} = f(\bar{\sigma})$, and for any particular biaxial state of stress as given by λ, this equation provides a relationship between σ_1 and the strain ε_1 in the same direction. Equations can be similarly developed for the other strain components, ε_2 and ε_3, as functions of σ_1 and λ. More general equations for the case of nonzero σ_3 can also be developed by a similar procedure.

Equation 12.39 does not apply for an elastic, perfectly plastic material beyond yielding, as no $\bar{\varepsilon} = f(\bar{\sigma})$ can be defined. In this case, Eq. 12.19 applies up to yielding, and the new yield stress can be estimated by substituting $\bar{\sigma} = \sigma_o$ into Eq. 12.34, where σ_o is the uniaxial yield strength. This gives

$$\sigma_1 = \frac{E}{1 - \nu\lambda}\varepsilon_1 \qquad (\bar{\sigma} \le \sigma_o) \qquad \text{(a)}$$

$$\sigma_1 = \frac{\sigma_o}{\sqrt{1 - \lambda + \lambda^2}} \qquad \left(\bar{\varepsilon} \ge \frac{\sigma_o}{E}\right) \qquad \text{(b)}$$

$$(12.40)$$

For other discontinuous stress-strain curves with a distinct yield point, apply Eq. 12.40(a) for the elastic portion, and beyond this employ Eq. 12.39 with the particular post-yield $\bar{\varepsilon} = f(\bar{\sigma})$ that is of interest.

As an example of the use of Eq. 12.39, consider a uniaxial curve that fits the Ramberg-Osgood form, Eq. 12.13, so that

$$\bar{\varepsilon} = \frac{\bar{\sigma}}{E} + \left(\frac{\bar{\sigma}}{H}\right)^{\frac{1}{n}} \tag{12.41}$$

Substitution of this particular $\bar{\varepsilon} = f(\bar{\sigma})$ into Eq. 12.39, followed by manipulation, gives

$$\varepsilon_1 = (1 - \nu\lambda)\frac{\sigma_1}{E} + (1 - 0.5\lambda)\left(1 - \lambda + \lambda^2\right)^{\frac{1-n}{2n}}\left(\frac{\sigma_1}{H}\right)^{\frac{1}{n}} \tag{12.42}$$

An example of the use of Eq. 12.42 to estimate the effects of various transverse stresses on the σ_1 vs. ε_1 curve is given in Fig. 12.9. Note that transverse tension, $\lambda > 0$, raises the stress-strain curve, whereas transverse compression, $\lambda < 0$, lowers it.

Example 12.2

A material has the Ramberg-Osgood type uniaxial stress-strain curve given by the constants of Fig. 12.9. Write the equation relating the principal stress and strain σ_1 and ε_1 for the case of plane stress, $\sigma_3 = 0$, with $\lambda = \sigma_2/\sigma_1 = 1.0$.

Solution   Equation 12.39 in the particular form of Eq. 12.42 applies. The following constants for the uniaxial (same as effective) stress-strain curve are given in Fig. 12.9:

$$E = 69{,}000 \text{ MPa}, \qquad H = 690 \text{ MPa}, \qquad n = 0.15, \qquad \nu = 0.3$$

Substituting these values and $\lambda = 1.0$ into Eq. 12.42 gives numerical factors preceding each term.

$$\varepsilon_1 = 0.700\frac{\sigma_1}{69{,}000} + 0.500\left(\frac{\sigma_1}{690}\right)^{\frac{1}{0.15}}$$

Consolidating the numerical constants then gives

$$\varepsilon_1 = \frac{\sigma_1}{98{,}570} + \left(\frac{\sigma_1}{765.6}\right)^{\frac{1}{0.15}} \qquad\qquad \textbf{Ans.}$$

A plot of this equation corresponds to the $\lambda = 1.0$ line in Fig. 12.9. Note that this equation has the same form as the original uniaxial one, Eq. 12.13, except that new values appear in place of E and H.

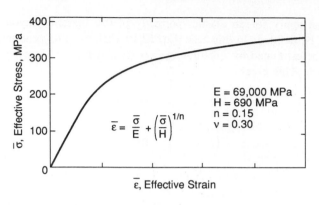

$$\bar{\varepsilon} = \frac{\bar{\sigma}}{E} + \left(\frac{\bar{\sigma}}{H}\right)^{1/n}$$

E = 69,000 MPa
H = 690 MPa
n = 0.15
ν = 0.30

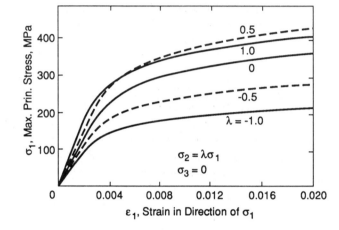

$\sigma_2 = \lambda\sigma_1$
$\sigma_3 = 0$

Figure 12.9 Estimated effect of biaxial stress on a Ramberg-Osgood stress-strain curve. (The constants correspond to a fictitious aluminum alloy.)

Example 12.3

A thin-walled tubular pressure vessel of radius r and wall thickness t has closed ends and is made of a material having a uniaxial stress-strain curve of the Ramberg-Osgood form, Eq. 12.13. If the internal pressure p is increased monotonically, derive an equation for the relative change in radius, $\Delta r/r$, as a function of p and the various constants involved.

Solution   The principal stresses are

$$\sigma_1 = \frac{pr}{t}, \qquad \sigma_2 = \frac{pr}{2t}, \qquad \sigma_3 \approx 0$$

where σ_1 is in the hoop direction and σ_2 is in the longitudinal direction. The strain ε_1 in the hoop direction is the ratio of the change in circumference to the original circumference, so that

$$\varepsilon_1 = \frac{\Delta(2\pi r)}{2\pi r} = \frac{\Delta r}{r}$$

Since we have plane stress, $\lambda = \sigma_2/\sigma_1 = 0.5$ can be substituted into Eq. 12.42, which gives

$$\frac{\Delta r}{r} = \varepsilon_1 = (1 - 0.5\nu)\frac{\sigma_1}{E} + (0.75)(0.75)^{\frac{1-n}{2n}}\left(\frac{\sigma_1}{H}\right)^{\frac{1}{n}}$$

Substitution of σ_1 from above and manipulation gives the desired result.

$$\frac{\Delta r}{r} = \left(1 - \frac{\nu}{2}\right)\frac{pr}{tE} + \frac{\sqrt{3}}{2}\left(\frac{\sqrt{3}\,pr}{2tH}\right)^{\frac{1}{n}}$$ **Ans.**

12.3.5 Deformation Versus Incremental Plasticity Theories

Experimental results indicate that plastic strains depend not only on the values of the stresses reached but also on the history of stressing. For example, consider a thin-walled tube loaded to particular values of axial load $P = P'$ and torque $T = T'$, either of which is sufficient to cause yielding by itself. If the axial loading beyond yielding is applied first and then the torsion, the plastic strains that result differ from those that occur if the torsion is instead applied first. Also, a third result is obtained if the tension and torsion are increased proportionally, so that the ratio P/T remains constant until P' and T' are simultaneously reached.

These sequences of stressing correspond to variations in principal stresses as shown in Fig. 12.10. The lines shown are called *loading paths*, and any loading path that is a straight line through the origin is termed *proportional loading*. The situation of the plastic strains differing despite the final stresses being the same is said to represent *loading path dependence*. To analyze such path-dependent behavior, an *incremental plasticity theory* is needed, which is applied by following the loading path in small steps. The equations of incremental plasticity theory are similar to those given above for deformation theory, Eq. 12.21 to 12.26, except that all plastic strains, $\bar{\varepsilon}_p$, ε_{px}, etc., are replaced by the corresponding differential quantities, $d\bar{\varepsilon}_p$, $d\varepsilon_{px}$, etc.

However, if all loads are applied so that their magnitudes are proportional, and if no unloading occurs, then incremental plasticity theory gives the same result as deformation theory. The discussion above is thus restricted to cases of *monotonic proportional*

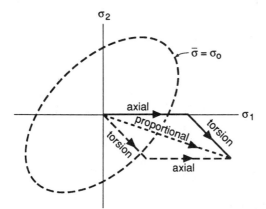

Figure 12.10 Three possible paths for combined axial and torsional loading of a thin-walled tube.

loading. Proportional loading is defined mathematically at a given point in a deforming solid if the principal stresses maintain constant ratios as their values increase.

$$\frac{\sigma_2}{\sigma_1} = \lambda_2, \qquad \frac{\sigma_3}{\sigma_1} = \lambda_3 \tag{12.43}$$

where λ_2 and λ_3 may vary from point to point in the deforming solid, but for any given point must not change.

As a practical matter, modest variations in λ_2 and λ_3 can occur without causing significant problems with deformation theory. Also, if proportionality is preserved but unloading does occur, as in cyclic loading, deformation theory can still be used as discussed later. However, some practical problems involve obviously nonproportional loading and thus require the use of incremental theory. Details on incremental theory and its application are given in textbooks on plasticity, such as those listed in the References.

12.4 UNLOADING AND CYCLIC LOADING BEHAVIOR FROM RHEOLOGICAL MODELS

Spring and slider rheological models do not include time-dependent effects, nor do they exhibit certain other complexities that are observed for real materials. They nevertheless provide a useful idealization, even for cyclic loading, as their behavior is basically similar to that of engineering metals and some other materials.

Some features of the behavior of these models are illustrated in Fig. 12.11. First consider the simple elastic, perfectly plastic model (a). If the direction of loading is reversed, the behavior may be elastic until yielding again occurs on reloading. However, if the strain excursion for the unloading–reloading event is sufficiently large, reversed yielding occurs. More specifically, reversed yielding occurs when the stress change since unloading reaches

$$\Delta\sigma = 2\sigma_o \tag{12.44}$$

For completely reversed loading, the response for a $\Delta\varepsilon$ sufficient to cause reversed yielding is also shown. The behavior is symmetrical about the origin and repeats itself for each cycle of loading.

If the model has one or more stages of parallel spring-slider combinations, then the behavior is analogous but more complex as shown in Fig. 12.11 by model (b). An unloading-reloading event similarly causes reversed yielding when $\Delta\sigma = 2\sigma_o$, which is similar to kinematic hardening as in Fig. 12.3. For the second strain history illustrated, reversed yielding occurs, causing a small loop to be formed, following which the stress-strain path rejoins the original path, then proceeding just as if the small loop had never occurred. This special behavior is called the *memory effect* and is similar to the behavior of real materials. For completely reversed straining, a loop that is symmetrical about the origin is formed that resembles the material behavior of Fig. 12.2.

We will now consider the behavior of multistage spring-slider models in more detail, starting with unloading behavior and then proceeding to cyclic loading behavior.

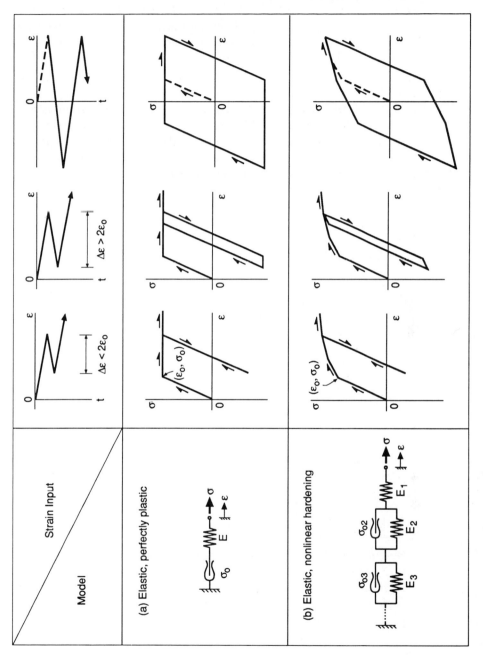

Figure 12.11 Unloading and reloading behavior for two rheological models. The first strain history causes only elastic deformation during unloading, but the second one is sufficiently large to cause compressive yielding. The third history is completely reversed and causes a hysteresis loop that is symmetrical about the origin.

12.4.1 Unloading Behavior

Consider any one stage (no. i) of a multistage parallel spring-slider model as shown in Fig. 12.12. After this stage has yielded on monotonic loading, its stress and strain are related by Eq. 12.16. At a particular later stress-strain point (ε', σ'), we thus have

$$\sigma' = \sigma_{oi} + E_i \varepsilon_i' \tag{12.45}$$

where ε_i' is the strain in the ith stage only. If the direction of loading is reversed at the point (ε', σ'), no change in ε_i' occurs until the stress on the slider reaches $-\sigma_{oi}$. Hence, ε_i' at first remains unchanged, and thus so does the stress $E_i \varepsilon_i'$ in the spring, so that this stage can be said to be *locked* until the resistance of the slider is overcome. At the point of reversed yielding of this ith stage, the stress is thus

$$\sigma'' = -\sigma_{oi} + E_i \varepsilon_i' \tag{12.46}$$

The change in stress necessary to cause reversed yielding in the ith stage is then

$$\Delta \sigma'' = \sigma' - \sigma'' = 2\sigma_{oi} \tag{12.47}$$

Beyond the point of reversed yielding of the ith stage, the stress on the slider remains at $-\sigma_{oi}$, and the stress change beyond σ'', which is the difference $\Delta\sigma - 2\sigma_{oi}$, is

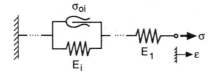

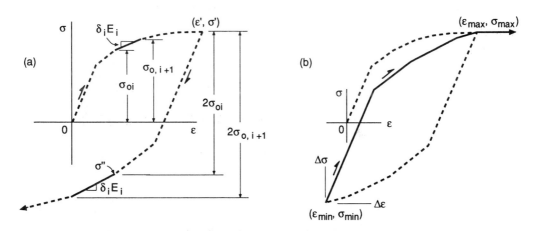

Figure 12.12 Unloading behavior of spring and slider rheological models: (a) showing doubling of segment lengths with the slope unchanged and (b) similar behavior on reloading.

applied to the spring and causes its elastic deflection. The contribution of the ith stage to the strain change is thus

$$\Delta\varepsilon_i = \frac{\Delta\sigma - 2\sigma_{oi}}{E_i} \qquad (12.48)$$

The total strain change is the sum for all yielded stages.

$$\Delta\varepsilon = \frac{\Delta\sigma}{E_1} + \frac{\Delta\sigma - 2\sigma_{o2}}{E_2} + \frac{\Delta\sigma - 2\sigma_{o3}}{E_3} + \cdots + \frac{\Delta\sigma - 2\sigma_{oj}}{E_j} \qquad (12.49)$$

where all sliders through the jth one have reverse yielded. The values of stress and strain relative to the original (σ, ε) axes are

$$\sigma = \sigma' - \Delta\sigma, \qquad \varepsilon = \varepsilon' - \Delta\varepsilon \qquad (12.50)$$

where (ε', σ') is the point of load reversal. Combining the above two equations and evaluating the derivative $d\sigma/d\varepsilon$ gives the slope of the stress-strain response.

$$\frac{d\sigma}{d\varepsilon} = \frac{1}{\dfrac{1}{E_1} + \dfrac{1}{E_2} + \dfrac{1}{E_3} + \cdots + \dfrac{1}{E_j}} \qquad (12.51)$$

Thus, noting Eq. 12.14, the slope is the same as for the interval in the monotonic response where the same number of sliders have yielded.

Furthermore, since Eq. 12.47 states that the $\Delta\sigma$ at each slider's point of reversed yielding is twice its monotonic yield stress, the intervals between reversed yielding events are twice as large as the intervals between the corresponding monotonic yieldings. In other words, the stress-strain path relative to a shifted origin at the point of unloading (ε', σ') has the same shape as the monotonic curve, differing in that it is expanded by a scale factor of two. Each straight-line segment of the unloading path has the same slope as the corresponding one in the monotonic response, but its length is doubled.

If the monotonic response of Eq. 12.18 is represented as

$$\varepsilon = f(\sigma) \qquad (12.52)$$

then the unloading response of Eq. 12.49 is

$$\frac{\Delta\varepsilon}{2} = f\left(\frac{\Delta\sigma}{2}\right) \qquad (12.53)$$

which can be verified from the two equations. The factor of two in the denominator is the mathematical expression of the factor-of-two expansion noted above.[1] The unloading stress-strain path relative to the original coordinate axes is obtained by combining Eqs. 12.50 and 12.53.

$$\varepsilon = f\left(\sigma'\right) - 2f\left(\frac{\sigma' - \sigma}{2}\right) \qquad (12.54)$$

---

[1]The appearance of the two in the denominator may seem incorrect, but this does represent a twofold expansion of the curve, as half the values $\Delta\sigma$ and $\Delta\varepsilon$ now fall on the original curve of σ vs. ε. If this still seems confusing, try the following example: Plot y versus x for the first half-cycle of both $y = \sin x$ and $y/2 = \sin(x/2)$, and compare the two curves.

For initial loading in compression to a point (ε', σ'), followed by loading in the tensile direction, analogous behavior occurs. Equation 12.53 still applies, but sign changes are needed in Eq. 12.50, and $\varepsilon' = -f(-\sigma')$, so that Eq. 12.54 must be replaced by

$$\varepsilon = -f\left(-\sigma'\right) + 2f\left(\frac{\sigma - \sigma'}{2}\right) \tag{12.55}$$

12.4.2 Discussion of Unloading

Since the point of reversed yielding of the first slider stage must conform to Eq. 12.47, the first reversed yielding occurs at $\Delta\sigma = 2\sigma_o$, where $\sigma_o = \sigma_{o2}$ is the monotonic yield strength of the model. Hence, the first reversed yielding obeys kinematic hardening as in Fig. 12.3. If unloading proceeds to such a degree that more sliders yield during unloading than previously yielded on loading, then the equations given above no longer apply. What happens is that Eq. 12.53 is followed until the stress-strain path joins the one that would be traced during monotonic compression. Beyond this point, the monotonic compression path is followed.

If the number of stages in the model is made relatively large, then its response approximates a smooth stress-strain curve. Equations 12.52 to 12.55 can thus be thought of as representing smooth curves.

Stress-strain paths for unloading are shown for laboratory tests of two engineering metals in Fig. 12.13. Also shown are estimated paths for unloading based on Eq. 12.53. Fair agreement is obtained for the aluminum alloy, but not for the steel. Materials that yield abruptly on monotonic loading seldom have similar behavior or unloading. A more gentle curve occurs instead, as for the steel in this case. For both experimental curves, note that yielding in compression begins at a lower absolute value of stress than the initial tensile yield strength, that is, a Bauschinger effect is observed. For the aluminum alloy, this is estimated reasonably well by the rheological model.

12.4.3 Cyclic Loading Behavior

Let unloading be terminated at a point $(\varepsilon_{min}, \sigma_{min})$, and then let reloading proceed in the tensile direction as illustrated in Fig. 12.12(b). The rheological model follows the same factor-of-two expanded curve as during unloading, with the origin now being at $(\varepsilon_{min}, \sigma_{min})$. This behavior persists until the original point of unloading (ε', σ') is reached, where a closed stress-strain loop is formed. At this point, the previously mentioned memory effect acts, and the stress-strain path beyond follows the monotonic one, namely Eq. 12.18, that is, $\varepsilon = f(\sigma)$.

Now assume that reloading is stopped at a point $(\varepsilon_{max}, \sigma_{max})$ that is the same as (ε', σ'), and that the strain is cycled between the two values ε_{max} and ε_{min}. During the resulting constant amplitude cycling, the stress-strain loop between ε_{max} and ε_{min}, called a *hysteresis loop*, is retraced for each cycle. This is further illustrated by Fig 12.14. In this case, it is assumed that there are a large number of models stages, so that smooth curves occur.

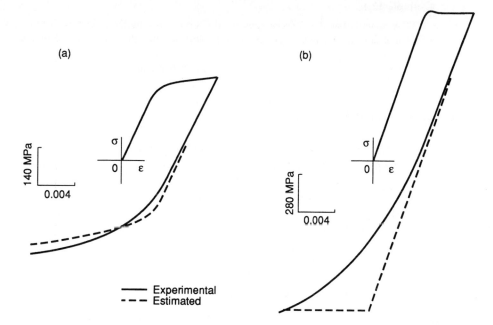

Figure 12.13   Monotonic tension followed by loading into compression for (a) aluminum alloy 2024-T4 and (b) quenched and tempered AISI 4340 steel. Unloading curves estimated from a factor-of-two expansion of the monotonic curve are also shown.

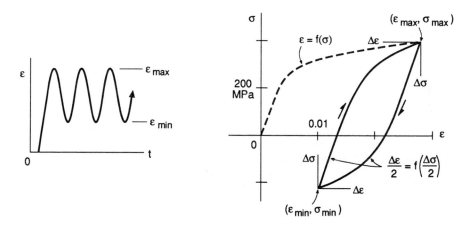

Figure 12.14   Stress-strain unloading and reloading behavior consistent with a spring and slider rheological model. The example curves plotted correspond to a Ramberg-Osgood stress-strain curve with constants as in Fig. 12.9.

Example 12.4

A material has the uniaxial stress-strain curve given by the constants in Fig. 12.9. Assume that this curve also applies for cyclic loading and that the stress-strain behavior is similar to that of a multistage spring-slider model. Then estimate the stress-strain response for starting from zero and then cycling between $\varepsilon_{max} = 0.028$ and $\varepsilon_{min} = 0.01$.

Solution   The stress-strain response for this example is the one plotted in Fig. 12.14. For the initial monotonic loading from zero to ε_{max}, we use Eq. 12.13 with E, H, and n values as in Fig. 12.9.

$$\varepsilon = f(\sigma) = \frac{\sigma}{69,000} + \left(\frac{\sigma}{690}\right)^{\frac{1}{0.15}}$$

We first need to solve this equation for σ_{max} given $\varepsilon_{max} = 0.028$, which gives

$$\sigma_{max} = 390.1 \text{ MPa} \qquad \qquad \textbf{Ans.}$$

An approximation to this value could be obtained simply by plotting $\varepsilon = f(\sigma)$ and reading σ_{max} from the graph. For a more accurate result, a trial and error solution could be done, as could iteration using Newton's method, as described in books on numerical methods. A good initial guess for Newton's method can be obtained by noting that the first term is dominant at low strains and the second at high strains. (See Fig. 12.5(b).)

For unloading from the point $(\varepsilon_{max}, \sigma_{max})$, Eq. 12.53 applies.

$$\Delta\varepsilon = 2f\left(\frac{\Delta\sigma}{2}\right) = \frac{\Delta\sigma}{69,000} + 2\left(\frac{\Delta\sigma}{1380}\right)^{\frac{1}{0.15}}$$

This equation can be used to plot the unloading response relative to an origin at $(\varepsilon_{max}, \sigma_{max})$ as shown in Fig. 12.14. At $(\varepsilon_{min}, \sigma_{min})$ the direction of straining reverses, where

$$\Delta\varepsilon = \varepsilon_{max} - \varepsilon_{min} = 0.018$$
$$\Delta\sigma = \sigma_{max} - \sigma_{min}$$

Entering the equation above with $\Delta\varepsilon$ and performing a second graphical or iterative solution gives $\Delta\sigma$.

$$\Delta\sigma = 614.5 \text{ MPa}$$

so that

$$\sigma_{min} = \sigma_{max} - \Delta\sigma = 390.1 - 614.5 = -224.4 \text{ MPa} \qquad \textbf{Ans.}$$

For reloading back to ε_{max}, the same Eq. 12.53 path is followed for the $\Delta\sigma$ versus $\Delta\varepsilon$ response relative to an origin at $(\varepsilon_{min}, \sigma_{min})$. Hence, the estimated curve reaches the same point $(\sigma_{max}, \varepsilon_{max})$ as before, and subsequent cycles are estimated to retrace the hysteresis loop thus formed.

12.4.4 Application to Irregular Strain Versus Time Histories

The behavior of a multistage spring and slider rheological model (as in Fig. 12.6) can be summarized by a set of rules that apply to any irregular variation of strain with time. These rules are stated as follows in terms of the behavior of the straight-line segments that approximate the monotonic loading curve, these being numbered starting from the origin as shown in Fig. 12.15(a):

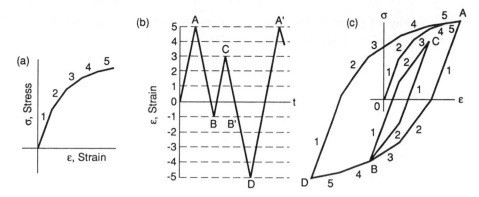

Figure 12.15   Behavior of a multistage spring-slider rheological model for an irregular strain history. A model having the monotonic stress-strain curve (a) is subjected to strain history (b), resulting in stress-strain response (c). (Adapted from [Dowling 79b]; used with permission of Elsevier Science Publishers.)

1. Initially, and after each reversal of strain direction, segments are used in order, starting with the first.
2. Each segment may be used once in either direction with its original length. Thereafter, the length is twice the monotonic value and the segment retains the same slope.
3. An exception to rule (1) is that a segment (or portion thereof) must be skipped if its most recent use was not in the opposite direction of its impending use.

Figure 12.15 illustrates the application of these rules, the strain history of (b) resulting in the stress-strain response of (c). Segments 1 through 5 are each used once in reaching point A, and no additional segments are required by the example strain history, so that only double-length segments are used during the remainder of the history. At point B' in the strain history, segment 3 must be skipped as its most recent use was not in the opposite direction of its impending use, so that the sequence of segments between C and D is 1-2-4-5.

Only double-length segments are used after the strain first reaches its largest absolute value. Thus, beyond this point, the model behaves such that all stress-strain curves follow a single shape, which is the monotonic curve expanded with a scale factor of two according to Eq. 12.53. Rule (3) above causes that memory effect to act, the skipping of segments (or portions thereof) resulting in a return to the stress-strain path previously established. The origins for the various $\Delta\sigma$ vs. $\Delta\varepsilon$ curves are of course located at points where the direction of straining changes, and the particular origin that applies for a given section of curve is determined by the memory effect. The skipping of a portion of a segment does not occur in Fig. 12.15, but this can be seen for the second case of Fig. 12.11(b).

If the strain history subsequently returns to its largest absolute value, the stress-strain paths traced form a set of closed stress-strain hysteresis loops. This is the case in Fig. 12.15, where the loops correspond to events B-C-B' and A-D-A'. If the same strain history is applied again, the same loops are retraced.

12.5 CYCLIC LOADING BEHAVIOR OF REAL MATERIALS

A special laboratory test methodology has been developed for characterizing cyclic stress-strain behavior, which is done during low-cycle fatigue testing. This is described in ASTM Standard No. E606, and also in Landgraf et al. (1969). As a result, considerable data are available for engineering metals, and limited data from other materials.

12.5.1 Cyclic Stress-Strain Tests and Behavior

The most common test involves completely reversed ($R = -1$) cycling between constant strain limits as illustrated in Fig. 12.16. A strain amplitude, $\varepsilon_a = \Delta\varepsilon/2$, is selected, and an axial test specimen is loaded until the tensile strain reaches a value of $\varepsilon_{\max} = +\varepsilon_a$. Then the direction of loading is reversed until the strain reaches $\varepsilon_{\min} = -\varepsilon_a$, and the test is continued, with the direction of loading being reversed each time the strain reaches $+\varepsilon_a$ or $-\varepsilon_a$. The rate of straining between these limits may be held constant, or sometimes a fixed-frequency sinusoidal variation of strain with time is used. The rate or frequency of the test affects the behavior of materials that exhibit significant creep at the test

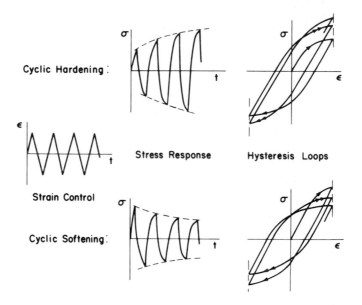

Figure 12.16 Completely reversed controlled strain test and two possible stress responses, namely cycle-dependent hardening and softening. (From [Landgraf 70]; copyright ©ASTM; reprinted with permission.)

temperature used. For engineering metals tested at room temperature, rate effects are generally small.

Such cyclic strain tests are continued until fatigue failure occurs. The stresses that are needed to enforce the strain limits generally change as the test progresses. Some materials exhibit *cycle-dependent hardening*, which means that the stresses increase as shown in Fig. 12.16. Others exhibit *cycle-dependent softening*, or a decrease in stress with increasing numbers of cycles, as also illustrated. In engineering metals, the cyclic hardening or softening is usually rapid at first, but the change from one cycle to the next decreases with increasing numbers of cycles. Often, the behavior becomes approximately stable in that further changes are small. These trends can be seen in Fig. 12.2, which shows hardening in an aluminum alloy.

If the stress-strain variation during stable behavior for one cycle is plotted, a closed hysteresis loop is formed on each cycle as shown in Fig. 12.17. After the direction of loading changes at either the positive or the negative strain limit, the slope of the stress-strain path is at first constant and close to the elastic modulus, E, as from a tension test. Then the path gradually deviates from linearity as plastic strain occurs. One can think of each branch of the hysteresis loop as being a separate stress-strain curve that begins at an origin that is shifted to one of the loop tips, with the axes being inverted for the lower branch. Note that this hysteresis looping behavior is similar to that of the rheological models discussed earlier. Compare the completely reversed case in Fig. 12.11(b) with Fig. 12.17.

The maximum deviation from linearity reached during a cycle is the plastic strain range, labeled $\Delta\varepsilon_p$ in Fig. 12.17. The stress range is $\Delta\sigma$, and the elastic portion of the strain range is related to $\Delta\sigma$ by the elastic modulus E. Summing the elastic and plastic portions gives the total strain range, $\Delta\varepsilon$.

$$\Delta\varepsilon = \frac{\Delta\sigma}{E} + \Delta\varepsilon_p \qquad (12.56)$$

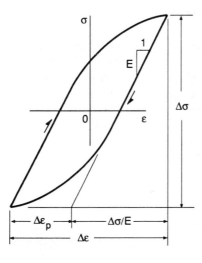

Figure 12.17 Stable stress-strain hysteresis loop.

It is often useful to work with the amplitudes, that is, half-ranges, of these quantities, $\varepsilon_a = \Delta\varepsilon/2$, $\sigma_a = \Delta\sigma/2$, and $\varepsilon_{pa} = \Delta\varepsilon_p/2$, so that

$$\varepsilon_a = \frac{\sigma_a}{E} + \varepsilon_{pa} \tag{12.57}$$

In most engineering metals, the stable hysteresis loops are nearly symmetrical with respect to tension and compression. One exception is gray cast iron, where the different behavior of graphite flakes in tension and compression causes asymmetric behavior. Ductile polymers and their composites also often have asymmetric hysteresis loops, with an example shown in Fig. 12.18.

12.5.2 Cyclic Stress-Strain Curves and Trends

Hysteresis loops from near half the fatigue life are conventionally used to represent the approximately stable behavior. Such loops from tests at several different strain amplitudes can be plotted on one set of axes as shown in Fig. 12.19. A line from the origin that passes through the tips of the loops, such as *O-A-B-C,* is called *cyclic stress-strain curve.* Where the branches in tension and compression do not differ greatly, which is often the case, their average is used. The cyclic stress-strain curve is thus the relationship between stress amplitude and strain amplitude for cyclic loading.

Cyclic stress-strain curves for several engineering metals are compared with monotonic tension curves in Fig. 12.20. Where the cyclic curve is above the monotonic one, the material is one that cyclically hardens, and vice versa. A mixed behavior may also occur, with crossing of the curves indicating softening at some strain levels and hardening at others. The cyclic curves virtually always deviate smoothly from linearity, even for materials where the monotonic curve has a distinct yield point or even a yield drop. A *cyclic yield strength* σ_o' may be defined that corresponds to a 0.2% plastic strain offset on the cyclic curve.

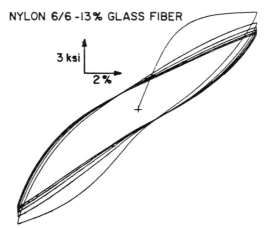

NYLON 6/6 -13% GLASS FIBER

3 ksi

2%

Figure 12.18   Cycle-dependent softening and asymmetric hysteresis loops in Nylon 6-6 reinforced with 13% glass fiber, which was cycled at a constant strain rate of $\dot{\varepsilon} = 0.01$ 1/s. (From [Beardmore 75]; used with permission.)

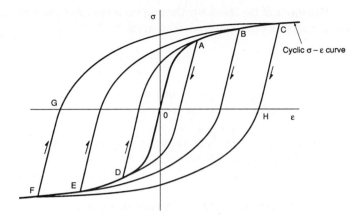

Figure 12.19 Cyclic stress-strain curve defined as the locus of tips of hysteresis loops. Three loops are shown, *A-D*, *B-E*, and *C-F*. The tensile branch of the cyclic stress-strain curve is *O-A-B-C*, and the compressive branch is *O-D-E-F*.

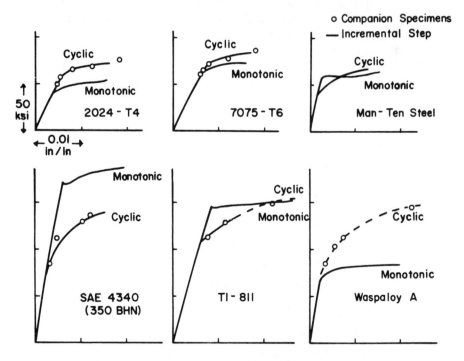

Figure 12.20 Cyclic and monotonic stress-strain curves for several engineering metals. (From [Landgraf 69]; copyright ©ASTM; reprinted with permission.)

Equations of the Ramberg-Osgood form have this character and are thus commonly used to represent cyclic stress-strain curves.

$$\varepsilon_a = \frac{\sigma_a}{E} + \left(\frac{\sigma_a}{H'}\right)^{\frac{1}{n'}} \tag{12.58}$$

where primes are used to specify that the constants for the plastic term are from fitting cyclic rather than monotonic stress-strain data. For engineering metals, n' is often in the range of 0.1 to 0.2, so that 0.15, or about 1/7, is a typical value. A low value of n from monotonic tension, say 0.05, corresponds to a fairly flat stress-strain curve. A metal with low n is likely to cyclically soften to a lower but steeper curve, with the resulting n' generally being around 0.1 to 0.2. Conversely, a metal with high n, that is, a steeply rising monotonic stress-strain curve, is likely to harden to a higher but flatter curve, with n' again being around 0.1 to 0.2. Values of H' and n' for four engineering metals are given in Table 12.1, and more are given later in Chapter 14, specifically Table 14.1.

The alloy composition and processing of engineering metals affects the cyclic stress-strain behavior, sometimes differently than it affects the monotonic tension properties. For example, strength achieved by cold working is often substantially reduced by cycle-dependent softening. Conversely, metals softened by annealing generally harden considerably. Hardening due to a fine precipitate, as in many aluminum alloys, is usually preserved and often increases under cyclic loading. This is the case for the two aluminum alloys included in Fig. 12.20. In medium carbon steels that are hardened by heat treatment using quenching and tempering, some of the effect is generally lost if cyclic loading occurs. An example is provided by the curve for SAE 4340 steel in Fig. 12.20. For such steels, the average variation with hardness of the monotonic and cyclic yield strengths, σ_o and σ_o', is shown in Fig. 12.21. Cyclic softening, indicated by σ_o' being lower the σ_o, occurs except at very high hardness.

Ductile polymers usually soften under cyclic loading. Recall from Section 7.6.2 that polymers have monotonic yield strengths that are typically 20% to 30% higher in

TABLE 12.1 CONSTANTS FOR CYCLIC STRESS-STRAIN CURVES FOR FOUR ENGINEERING METALS

Material	Yield	Ultimate	Cyclic σ-ε curve		
	σ_o	σ_u	E	H'	n'
RQC-100 steel	683	758	200,000	903	0.0905
	(99)	(110)	(29,000)	(131)	
AISI 4340 steel	1103	1172	207,000	1655	0.131
	(160)	(170)	(30,000)	(240)	
2024-T351 Al	379	455	73,100	662	0.070
	(55)	(66)	(10,600)	(96)	
7075-T6 Al	469	578	71,000	977	0.106
	(68)	(84)	(10,300)	(142)	

Notes: Units are MPa(ksi) except for dimensionless n'. See Table 14.1 for additional properties and sources.

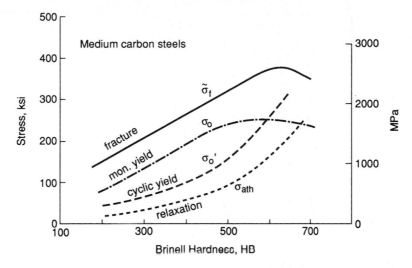

Figure 12.21 Average property trends for SAE 1045 and other medium-carbon steels as a function of hardness, including the true fracture strength, the monotonic and cyclic yield strengths, and the threshold stress amplitude for relaxation of mean stress. (Adapted from [Landgraf 88]; copyright ©ASTM; reprinted with permission.)

compression than in tension, indicating a sensitivity to hydrostatic stress. A similar ratio of compressive to tensile yield strengths is maintained in the cyclic stress-strain curves, such behavior being evident for polycarbonate in Fig. 12.22. The cyclic stress-strain curves for ductile polymers are affected by the strain rate even at room temperature.

12.5.3 Hysteresis Loop Curve Shapes

One item of interest is the shape of hysteresis loop curves, such as C-H-F and F-G-C in Fig. 12.19. If the stable behavior obeys a multistage spring and slider rheological model of the type discussed earlier, the stress-strain path for the hysteresis loops should have the same shape as the cyclic stress-strain curve except for expansion by a scale factor of two according to Eq. 12.53. Thus, for a cyclic stress-strain curve of the Ramberg-Osgood form, Eq. 12.58, the equation of the loop curves should be

$$\Delta\varepsilon = \frac{\Delta\sigma}{E} + 2\left(\frac{\Delta\sigma}{2H'}\right)^{\frac{1}{n'}} \qquad (12.59)$$

Here, the variables $\Delta\sigma$ and $\Delta\varepsilon$ represent changes relative to coordinates axes at either loop tip.

This expected behavior can be compared to actual behavior by expanding the cyclic stress-strain curve with a scale factor of two and comparing the result to actual loop curves. Such a comparison for a steel is shown in Fig. 12.23. To enable this comparison, three actual hysteresis loops from experimental data are plotted with shifted

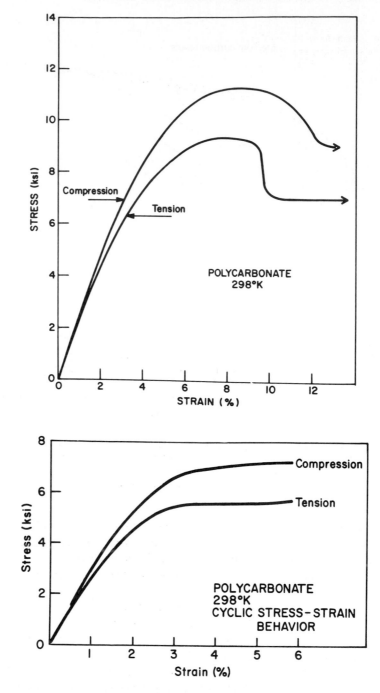

Figure 12.22 Monotonic (top) and cyclic (bottom) stress-strain curves for polycarbonate, the latter being for $\dot{\varepsilon} = 0.01$ 1/s. (From [Beardmore 75]; used with permission.)

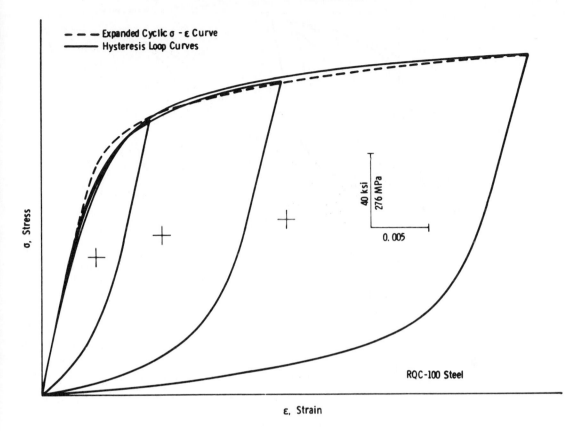

Figure 12.23 Stable hysteresis loops for a steel plotted with shifted axes so that their compressive tips coincide. The loop curves fall near the dashed line, which is obtained by expanding the cyclic stress-strain curve with a scale factor of two, $\Delta\varepsilon/2 = f(\Delta\sigma/2)$. (From [Dowling 78]; used with permission of ASME.)

origins so that their tips fall at the origin used to plot Eq. 12.59. Reasonably good agreement of the actual and estimated loop curves exists in this case. Similar behavior generally occurs for other engineering metals, but in some cases the agreement is not as good. Nevertheless, Eq. 12.59 provides a reasonable estimate for engineering purposes of loop shape in metals.

For irregular variations of strain with time, the hysteresis loop curves for stable behavior still generally follow shapes close to the cyclic stress-strain curve expanded with a scale factor of two, Eq. 12.59. Hence, the behavior is similar to that predicted by the spring-slider rheological model as discussed above and illustrated in Fig. 12.15. Some supporting test data for an alloy steel are provided by Fig. 12.24. The stress-strain response for a short irregular history is shown in (a) as predicted by the rheological model. The actual measured response for stable behavior shown in (b) is almost identical.

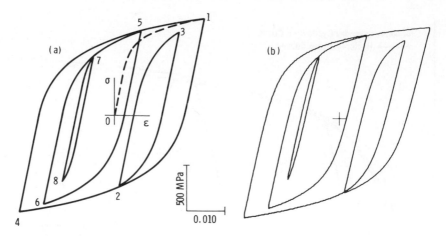

Figure 12.24   Stable stress-strain response of AISI 4340 steel ($\sigma_u = 1158$ MPa) subjected to a repeatedly applied irregular strain history. The predicted response is shown in (a) and actual test data in (b). (From [Dowling 79b]; used with permission of Elsevier Science Publishers.)

For nonuniaxial states of stress, Eq. 12.42 and related more general equations that are not limited to plane stress can be used to estimate cyclic stress-strain curves. The constants E, H', and n' for the uniaxial cyclic stress-strain curve of the material of interest are of course employed. It is then reasonable to apply the same assumption of hysteresis loops having shapes according to $\Delta \varepsilon / 2 = f(\Delta \sigma / 2)$, where $\varepsilon_a = f(\sigma_a)$ is now the cyclic stress-strain curve for a particular direction of the nonuniaxial state of stress. Such a procedure thus allows deformation plasticity theory to be used for cyclic loading, but the limitation of at least approximately proportional loading is still needed.

12.5.4 Transient Behavior; Mean Stress Relaxation

The spring and slider rheological models exhibit only stable behavior for cyclic loading, that is, continuous repetition of an identical hysteresis loop for each cycle. Cycle-dependent hardening or softening thus represents a class of transient behavior that is not predicted by such models. Also, these models still exhibit stable behavior even if the mean stress is not zero, that is, if the loop is biased in either the tensile or compressive directions.

However, if a real material is subjected to loading with a nonzero mean stress, an additional class of transient behavior may be observed. Examples of such behavior for a steel are shown in Fig. 12.25. If biased strain limits are imposed, and if the strain range is sufficiently large to cause some cyclic plasticity, as on the left, the resulting mean stress will gradually shift toward zero as increasing numbers of cycles are applied. Sometimes a stable nonzero value is reached, or if the degree of plasticity is large, the mean stress may shift essentially to zero. This behavior is called *cycle-dependent relaxation*. Behavior of this type is further illustrated by Fig. 12.26.

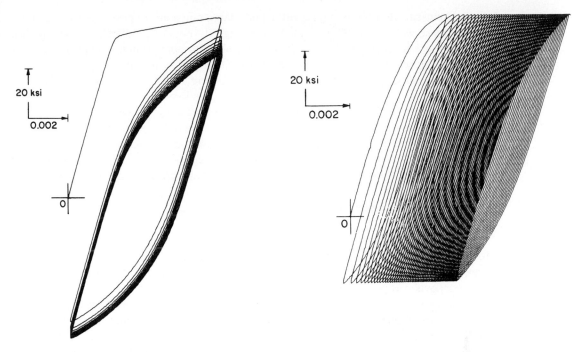

Figure 12.25 Cycle-dependent relaxation of mean stress (left) and cycle-dependent creep (right), both for an AISI 1045 steel. On the right, the specimen was previously yielded, so that the monotonic curve does not appear. (From [Landgraf 70]; copyright ©ASTM; reprinted with permission.)

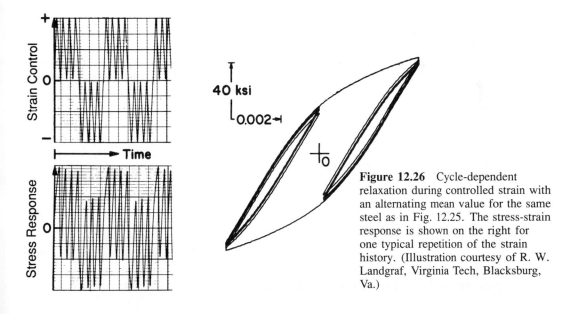

Figure 12.26 Cycle-dependent relaxation during controlled strain with an alternating mean value for the same steel as in Fig. 12.25. The stress-strain response is shown on the right for one typical repetition of the strain history. (Illustration courtesy of R. W. Landgraf, Virginia Tech, Blacksburg, Va.)

If biased stress limits with a sufficiently large range are imposed, the mean strain will increase with cycles as illustrated on the right in Fig. 12.25. The mean strain shift may decrease its rate and stop, or it may establish an approximately constant rate, or it may accelerate and lead to a failure somewhat similar to that in a tension test. This behavior is called *cycle-dependent creep* or *ratchetting*. Cyclic creep and relaxation are a single phenomenon viewed under two different situations.

Cyclic creep-relaxation may occur at the same time as cyclic hardening or softening, and the two phenomena may interact. Relaxation effects are especially obvious when they are enhanced by cyclic softening occurring at the same time, which is the case for Fig. 12.25(left). Cyclic creep-relaxation behavior occurs for combinations of material and temperature where time-dependent effects are usually considered small, as well as in situations where there is an obvious combination of cycle-dependent and time-dependent effects. However, subtle time dependency may play a role even where the behavior is thought to be mainly cycle dependent.

A paper by Martin et al. (1971) listed in the References describes a fairly straightforward approach of modifying the spring and slider model to handle cycle-dependent transient behavior. More general three-dimensional incremental plasticity models can also be devised that exhibit cycle-dependent transient behavior, such as the model of Dafalias and Popov (1976). Other models can handle time-dependent transients, such as the model of Krempl and Yao (1987). Predicting all effects simultaneously is difficult given the current state of knowledge.

A detailed study of cycle-dependent relaxation in SAE 1045 steel is described by Landgraf and Chernenkoff (1988). The following equation is used to determine the mean stress σ_{mN} after N cycles of relaxation:

$$\sigma_{mN} = \sigma_{mi} N^r \tag{12.60}$$

where σ_{mi} is the initial ($N = 1$) value of mean stress. The exponent r has values typically in the range 0 to -0.2 and varies with the level of cyclic strain and the material. For SAE 1045 steel heat treated to various strength levels, r was found to vary as follows:

$$r = 0.085 \left(1 - \frac{\varepsilon_a}{\varepsilon_{ath}} \right) \qquad (\varepsilon_a > \varepsilon_{ath})$$

$$r = 0 \qquad (\varepsilon_a \leq \varepsilon_{ath}) \tag{12.61}$$

$$\varepsilon_{ath} = e^{-8.41 + 0.00536(HB)}$$

where ε_a is the applied strain amplitude, ε_{ath} is a threshold value below which no relaxation occurs, and HB is the Brinell hardness. The stress value σ_{ath} corresponding to ε_{ath} on the cyclic stress-strain curve is somewhat lower than the cyclic yield strength. This trend is included in Fig. 12.21.

Equations 12.60 and 12.61 also provide reasonable estimates for mean stress relaxation in other heat-treated steels. For other engineering metals, similar equations may be useful if employed with numerical constants based on test data for the particular material.

12.6 SUMMARY

Some relationships commonly used to fit stress-strain curves involve linear-elastic behavior up to a distinct yield point σ_o. If the curve is flat beyond σ_o, the relationship is said to be elastic, perfectly plastic, and Eq. 12.1 applies. A rising linear behavior beyond σ_o is called an elastic, linear-hardening relationship, and this is given by Eq. 12.4. A power relationship beyond σ_o may also be used and is given by Eq. 12.8. The Ramberg-Osgood relationship differs from the others described in that there is no distinct yield point.

$$\varepsilon = \frac{\sigma}{E} + \left(\frac{\sigma}{H}\right)^{\frac{1}{n}} \tag{12.62}$$

At all values of stress, elastic and plastic strain are summed to obtain total strain, resulting in a smooth continuous curve. The plastic (not total) strain and stress have a power relationship. Rheological models consisting of linear springs and frictional sliders can be used to model any of these stress-strain curves.

Considering three-dimensional states of stress, Hooke's Law still applies beyond yielding, but gives only the elastic portion of the strain. The plastic portion can be related to the stress by deformation plasticity theory, which employs equations analogous to Hooke's Law.

$$\varepsilon_{px} = \frac{1}{E_p} \left[\sigma_x - 0.5\left(\sigma_y + \sigma_z\right)\right], \qquad \varepsilon_{py}, \ \varepsilon_{pz}, \ \text{similarly}$$

$$\gamma_{pxy} = \frac{3}{E_p}\tau_{xy}, \qquad\qquad\qquad\qquad \gamma_{pyz}, \ \gamma_{pzx}, \ \text{similarly} \tag{12.63}$$

where Poisson's ratio is replaced by 0.5, so that plastic strains do not contribute to volume change. The elastic modulus is replaced by the variable $E_p = \bar{\sigma}/\bar{\varepsilon}_p$, which is the secant modulus to a point on the effective stress versus plastic strain curve, $\bar{\sigma}$ vs. $\bar{\varepsilon}_p$. These effective quantities are proportional to the corresponding octahedral shear stresses and strains and are related to the principal stresses and plastic strains by Eqs. 12.21 to 12.23. Total strains are obtained by adding elastic and plastic portions.

A key feature of deformation plasticity theory is that the $\bar{\sigma}$ vs. $\bar{\varepsilon}_p$ curve is assumed to be independent of state of stress, thus permitting a stress-strain curve from one state of stress to be used to estimate those for other states of stress. As employed here, the effective stress-strain curve, $\bar{\sigma} = f(\bar{\varepsilon})$, is identical to the uniaxial curve, $\sigma = f(\varepsilon)$. For plane stress, estimates of stress-strain curves as a function of $\lambda = \sigma_2/\sigma_1$ are given in general by Eq. 12.39, and more specific relationships for elastic, perfectly plastic and Ramberg-Osgood materials are given by Eqs. 12.40 and 12.42, respectively.

For unloading following yielding, and for cyclic loading, spring and frictional slider rheological models suggest that yielding should occur when the stress range reaches twice the yield stress from the monotonic stress-strain curve, $\Delta\sigma = 2\sigma_o$. Furthermore, stress-strain paths for these situations are predicted to follow a path that is given by a factor-of-two expansion of the monotonic stress-strain curve.

$$\frac{\Delta\varepsilon}{2} = f\left(\frac{\Delta\sigma}{2}\right) \tag{12.64}$$

where $\varepsilon = f(\sigma)$ is the monotonic curve. For the above equation, $\Delta\sigma$ and $\Delta\varepsilon$ are measured from origins at points where the loading direction changes, resulting in symmetrical stress-strain hysteresis loops being formed. If the strain varies in an irregular manner with time, the stress-strain paths still form such hysteresis loops obeying Eq. 12.64. After completion of a loop, the stress-strain behavior exhibits a memory effect in that it returns to the path previously established.

For stable cyclic loading following completion of most cycle-dependent hardening or softening, the behavior predicted by a multistage spring and slider rheological model is reasonably accurate for many engineering metals. It is necessary to replace the monotonic stress-strain curve with a special cyclic stress-strain curve, with the Ramberg-Osgood form often being used.

$$\varepsilon_a = \frac{\sigma_a}{E} + \left(\frac{\sigma_a}{H'}\right)^{\frac{1}{n'}} \tag{12.65}$$

where stress and strain amplitudes appear and the constants H' and n' differ from those for the monotonic curve. In addition to hardening-softening, a second type of transient, cycle-dependent behavior occurs in engineering metals, namely cyclic creep-relaxation.

NEW TERMS AND SYMBOLS

(a) Terms

Bauschinger effect

cycle-dependent creep (ratchetting)

cycle-dependent hardening

cycle-dependent relaxation

cycle-dependent softening

cyclic stress-strain curve

cyclic yield strength, σ_o'

deformation plasticity theory

hysteresis loop

incremental plasticity theory

isotropic hardening

kinematic hardening

memory effect

monotonic loading

power hardening σ-ε curve: σ_o, H_1, n_1

proportional loading

Ramberg-Osgood σ-ε curve

    monotonic: E, H, n

    cyclic: E, H', n'

slope reduction factor, δ

strain hardening exponent: n_1, n, n'

(b) Nomenclature for Three-Dimensional Stresses and Strains

E_p	Plastic modulus
$\gamma_{xy}, \gamma_{yz}, \gamma_{zx}$	Total shear strains on orthogonal planes
$\gamma_{exy}, \gamma_{eyz}, \gamma_{ezx}$	Elastic shear strains
$\gamma_{pxy}, \gamma_{pyz}, \gamma_{pzx}$	Plastic shear strains
$\varepsilon_x, \varepsilon_y, \varepsilon_z$	Total normal strains in orthogonal directions
$\varepsilon_{ex}, \varepsilon_{ey}, \varepsilon_{ez}$	Elastic normal strains
$\varepsilon_{px}, \varepsilon_{py}, \varepsilon_{pz}$	Plastic normal strains
$\varepsilon_1, \varepsilon_2, \varepsilon_3$	Normal strains in the principal directions
$\varepsilon_{e1}, \varepsilon_{e2}, \varepsilon_{e3}$	Elastic strains in the principal directions
$\varepsilon_{p1}, \varepsilon_{p2}, \varepsilon_{p3}$	Plastic strains in the principal directions
$\bar{\varepsilon}$	Effective total strain
$\bar{\varepsilon}_p$	Effective plastic strain
λ	Ratio σ_2/σ_1 for plane stress ($\sigma_3 = 0$)
$\sigma_x, \sigma_y, \sigma_z$	Normal stresses in orthogonal directions
$\sigma_1, \sigma_2, \sigma_3$	Principal normal stresses
$\bar{\sigma}$	Effective stress
$\tau_{xy}, \tau_{yz}, \tau_{zx}$	Shear stresses on orthogonal planes

REFERENCES

ASTM. 1990 *Annual Book of ASTM Standards*, Vol. 03.01, Am. Soc. for Testing and Materials, Philadelphia, PA. See No. E606, "Constant Amplitude Low-Cycle Fatigue Testing."

CHEN, W. F. and D. J. HAN. 1988 *Plasticity for Structural Engineers*, Springer-Verlag, New York.

DAFALIAS, Y. F. and E. P. POPOV. 1976 "Plastic Internal Variables Formalism of Cyclic Plasticity," Paper No. 76-WA/APM-21, *Journal of Applied Mechanics*, ASME, Vol. 43, Dec. 1976, pp. 645–651.

HILL, R. 1983 *The Mathematical Theory of Plasticity*, Oxford University Press, London.

KREMPL, E. and D. YAO. 1987 "The Viscoplasticity Theory Based on Overstress Applied to Ratchetting and Cyclic Hardening," *Low Cycle Fatigue and Elasto-Plastic Behavior of Materials*, K. T. Rie, ed., Elsevier Applied Science Pubs., London.

LANDGRAF, R. W., J. MORROW and T. ENDO. 1969 "Determination of the Cyclic Stress-Strain Curve," *Journal of Materials*, ASTM, Vol. 4, No. 1, Mar. 1969, pp. 176–188.

LANDGRAF, R. W. and R. A. CHERNENKOFF. 1988 "Residual Stress Effects on Fatigue of Surface Processed Steels," *Analytical and Experimental Methods for Residual Stress Effects in Fatigue*, ASTM STP 1004, R. L. Champous et al., eds., Am. Soc. for Testing and Materials, Philadelphia, Pa., pp. 1–12.

MARTIN, J. F., T. H. TOPPER and G. M. SINCLAIR. 1971 "Computer Based Simulation of Cyclic Stress-Strain Behavior with Applications to Fatigue," *Materials Research and Standards*, ASTM, Vol. 11, No. 2, Feb. 1971, pp. 23–29.

MENDELSON, A. 1968 *Plasticity: Theory and Applications*, Macmillian, New York. (Also reprinted by R. E. Krieger. Malabar, Fl., 1983.)

NADAI, A. 1950 *Theory of Flow and Fracture of Solids*, McGraw-Hill, New York.

PROBLEMS AND QUESTIONS

Section 12.2

12.1 The monotonic stress-strain curve of RQC-100 steel under uniaxial stress can be approximated by an elastic, linear-hardening relationship. Two points on this curve are as follows, with the first of these being the point of yielding:

σ, MPa	ε
703	3.52×10^{-3}
738	2.50×10^{-2}

Plot the curve and write its equation in the form of Eq. 12.4, with numerical values substituted for all constants.

12.2 Proceed as in Prob. 12.1 except use the following two points for AISI 4340 steel:

σ, MPa	ε
1103	5.33×10^{-3}
1214	4.00×10^{-2}

12.3 Some test data points on the monotonic stress-strain curve of 2024-T351 Al for uniaxial stress are given below.
   (a) Obtain values of the constants H and n for a stress-strain curve of the Ramberg-Osgood form, Eq. 12.13, that fits these data. Use $E = 73.1$ GPa, and evaluate the constants approximately using an intercept and slope on a log-log plot.
   (b) Plot the resulting equation $\varepsilon = f(\sigma)$ along with the test data. How well does your equation fit the data? What is the 0.2% offset yield strength implied by your equation?
   (c) Obtain refined values of H and n from a least-squares fit of $\log \varepsilon_p$ versus $\log \sigma$. (PC Problem)

σ, MPa	ε
317	4.74×10^{-3}
341	6.07×10^{-3}
366	9.50×10^{-3}
390	1.91×10^{-2}
414	3.29×10^{-2}
439	5.23×10^{-2}

12.4 Based on the data of Prob. 12.3, obtain constants H_1, n_1, and σ_o for an elastic, power-hardening relationship, Eq. 12.8, that provides a reasonable fit.

Section 12.3

12.5 Given constants E, H, and n for a uniaxial stress-strain curve of the Ramberg-Osgood form, Eq. 12.13, estimate the stress-strain curve $\gamma = f_\tau(\tau)$ for a state of pure planar shear stress. Note that the principal stresses and strains for this state of stress are:

$$\sigma_1 = -\sigma_2 = \tau, \quad \varepsilon_1 = -\varepsilon_2 = \gamma/2, \quad \sigma_3 = \varepsilon_3 = 0$$

(a) Express $\gamma = f_\tau(\tau)$ using the constants E, H, n from the uniaxial curve.
(b) Consider the form

$$\gamma = \frac{\tau}{G} + \left(\frac{\tau}{H_\tau}\right)^{\frac{1}{n_\tau}}$$

Based on (a), give expressions for the new constants G, H_τ and n_τ in terms of the ones from the uniaxial curve.

12.6 For a uniaxial stress-strain curve of linear-hardening form, Eq. 12.5:
(a) Proceed as in Prob. 12.5(a) to find $\gamma = f_\tau(\tau)$ for a state of pure planar shear stress. (Suggestion: Use Eq. 12.39.)
(b) Does the new equation still exhibit linear hardening? What is its slope $d\tau/d\gamma$? If the new slope is denoted $\delta_\tau G$, where G is the shear modulus, is δ_τ the same as δ for the uniaxial curve?

12.7 Consider a material with a uniaxial stress-strain curve of the Ramberg-Osgood form, Eq. 12.13, and the case of axisymmetric loading, where

$$\sigma_2 = \sigma_3 = \lambda' \sigma_1$$

(a) Derive an equation analogous to Eq. 12.42 for ε_1 as a function of σ_1 and the constants from the uniaxial curve.
(b) Write the particular equation that applies for $\lambda' = 0.5$ for the material of Fig. 12.9. Plot this equation along with the uniaxial one and comment on the comparison. (PC Problem)

12.8 Proceed as in Example 12.3, except change the pressure vessel to a thin-walled spherical one of radius r and wall thickness t.

12.9 A thin-wall tubular pressure vessel of radius r, wall thickness t, and length L has closed ends and is made of a material having a uniaxial stress-strain curve of the Ramberg-Osgood form, Eq. 12.13. If the internal pressure p is increased monotonically, derive an equation for the relative change in the enclosed volume, $\Delta V/V$, as a function of the pressure p and the various constants involved. (Suggestion: See Prob. 4.8.)

12.10 Consider a state of plane stress with $\sigma_2/\sigma_1 = \lambda$ and $\sigma_3 = 0$.
(a) Derive equations analogous to Eq. 12.39 for the other two principal strains, ε_2 and ε_3.
(b) Specialize your result (a) to the Ramberg-Osgood form, obtaining equations for ε_2 and ε_3 analogous to Eq. 12.42.

Section 12.4

12.11 An elastic, linear-hardening material has elastic modulus $E = 200$ GPa, yield strength $\sigma_o = 500$ MPa, and a value of $\delta = 0.1$ for Eq. 12.4. Assuming behavior according to the rheological model of Fig. 12.4(b), estimate and plot the stress versus strain response for:
  (a) Completely reversed cyclic straining at $\varepsilon_a = 0.006$.
  (b) Straining from zero to $\varepsilon = 0.012$, followed by decreasing strain to $\varepsilon = 0.005$, and then increasing strain to $\varepsilon = 0.015$.

12.12 A multistage spring-slider rheological model as in Fig. 12.6 has a spring in series with three parallel spring-slider combinations. For model constants as given below, determine and plot the stress-strain response as the strain increases from zero to $\varepsilon = 0.016$ and then returns to zero.

Stage No.	E_i, GPa	σ_{oi}, MPa
1	200	0
2	120	800
3	150	1100
4	50	1300

12.13 For the rheological model of Prob. 12.12, determine and plot the stress-strain response for starting at zero and then following the sequence given below.

Peak or Valley	A	B	C	D	E	F	G
σ, Stress, MPa	1400	−800	800	−1400	1200	−1000	1400

12.14 Aluminum alloy 7075-T6 has constants for a stress-strain curve of the Ramberg-Osgood form, Eq. 12.13, of $E = 71,000$ MPa, $H = 977$ MPa, and $n = 0.106$. This particular curve applies for cyclic loading and the behavior can be assumed to be similar to that of a multistage spring-slider model. Estimate and plot the stress-strain response for starting from zero and then cycling between $\varepsilon_{max} = 0.025$ and $\varepsilon_{min} = 0.005$. (PC Problem)

Section 12.5

12.15 For a cyclic stress-strain curve of the Ramberg-Osgood form, Eq. 12.58, consider a cyclic yield strength σ_o' that corresponds to a plastic strain offset of 0.2%.
  (a) Derive an equation for σ_o' that gives its value in terms of other material constants.
  (b) What is the significance of σ_o'? How does it differ from the monotonic yield strength σ_o?

12.16 For aluminum alloy 2024-T351, the monotonic and cyclic stress-strain curves both fit Ramberg-Osgood forms, Eqs. 12.13 and 12.58, respectively. Constants for the monotonic curve are $E = 73.1$ GPa, $H = 527$ MPa, and $n = 0.0663$, and those for the cyclic curve are given in Table 12.1.
  (a) Plot both of these curves on the same linear-linear axes out to a strain of $\varepsilon = 0.02$. (PC Problem)
  (b) Does this material harden or soften cyclically? How do the yield strengths from the two curves compare?

12.17 Various properties are given in Chapter 14, specifically Table 14.1, for four strength levels of SAE 4142 steel.

(a) Use the constants given to plot all four cyclic stress-strain curves on the same graph. (PC Problem)

(b) Then comment on the trends in these curves and how they correlate with the strength and ductility from tension tests. Include in your comparison the monotonic yield strength versus the 0.2% offset yield strength from the cyclic curves.

12.18 Four strain amplitudes, and the corresponding cyclically stable stress amplitudes from test data, are given below for RQC-100 steel under uniaxial stress.

(a) Evaluate constants H' and n' for a cyclic stress-strain curve of the Ramberg-Osgood form, Eq. 12.58. The elastic modulus is $E = 200$ GPa.

(b) Plot both the fitted curve and the data on the same graph and compare them. (PC Problem)

(c) The initial portion of the monotonic stress-strain curve for this material is given in Prob. 12.1. Add this curve to the graph from (b). Does this steel cyclically harden or soften? How much does cyclic loading change the yield strength?

ε_a	σ_a, MPa
3.00×10^{-3}	472
4.50×10^{-3}	505
1.00×10^{-2}	574
2.02×10^{-2}	631

Source: The author's data on the ASTM Committee E9 material.

12.19 Proceed as in Prob. 12.18, except use the data below for AISI 4340 steel ($\sigma_u = 1158$ MPa). The monotonic stress-strain curve is in this case given in Prob. 12.2, and $E = 207$ GPa applies.

ε_a	σ_a, MPa
6.43×10^{-3}	827
9.80×10^{-3}	886
1.97×10^{-2}	1000
3.93×10^{-2}	1119

Source: Data from [Dowling 79b].

12.20 For each test data point in Prob. 12.19, estimate the hysteresis loop curve shapes, and plot the four resulting loops all together on one set of axes. (PC Problem)

13

Stress-Strain Analysis of Plastically Deforming Members

13.1 INTRODUCTION

It is often useful for engineering purposes to analyze plastic deformation in components of machines, vehicles, or structures. This occurs in two types of situations. First, it may be desirable to know the load necessary to cause gross plastic deformation, sometimes called *plastic collapse*. The safety factor for failure due to accidental overload can then be calculated by comparing the failure load to the loads expected during service use of the component. Second, the stresses and strains that accompany plastic deformation in localized areas, such as at the edge of a beam or at a stress raiser, may be of interest. Such deformations introduce residual stresses, perhaps intentional ones, that can be evaluated by analyzing the plastic deformation. Localized plastic deformation caused by cyclic loading is of considerable importance as this is often associated with fatigue cracking.

Local plasticity at stress raisers (notches) may be analyzed by applying an approximate method such as *Neuber's rule*. In some cases that are frequently of practical interest, such as bending of beams and torsion of circular shafts, plastic deformation can be analyzed in a fairly straightforward manner by a mechanics of materials type of approach. Three steps are needed: (1) Assume a strain distribution. (2) Apply equilibrium

570

of forces. (3) Choose a particular stress-strain relationship, and use this to complete the analysis.

We will apply such approaches not only to static loading, but also to unloading and to cyclic loading. The application to cyclic loading will be employed later in the next chapter on the strain-based approach to fatigue. Detailed analysis of many complex situations of loading and geometry is not possible using the relatively simple methods of this chapter. Analysis by *finite elements* or another numerical method may then be needed. (Chapter-length introductions to numerical stress-strain analysis are provided by Dhalla and Gallagher (1982) and in the SAE *Fatigue Design Handbook* (Rice, 1988), which are listed in the References.) The treatment given nevertheless provides some useful engineering tools, and it can also serve as an introduction for those who will later pursue more advanced study of plasticity.

In what follows, it is assumed that the reader is familiar with Section 12.2 on stress-strain curves. Also, for the discussions on analysis of residual stresses and on cyclic loading, familiarity is needed with additional relevant portions of Chapter 12.

13.2 PLASTICITY IN BENDING

Bending of beams where the material deforms in a linear-elastic manner is quite thoroughly covered in elementary and advanced texts on mechanics of materials. Plastic bending is also often introduced, with this generally being limited to elastic, perfectly plastic behavior. In this portion of the chapter, bending involving plastic deformation will be considered for various stress-strain curves. Procedures for developing closed form solutions are described, and some completed solutions are given, for cases where the applied loads lie in a plane of symmetry of the cross section.

13.2.1 Review of Elastic Bending

As a brief review of linear-elastic bending, consider the simple case of three-point bending of a beam having a rectangular cross section as illustrated in Fig. 13.1. The applied loads lie in the x-y plane, which is a plane of symmetry of the beam. Elastic stress distributions in the beam involve a zero stress or *neutral axis* N-N that coincides with a centroidal axis of the cross-sectional area.

Normal stresses for linear-elastic bending are given by

$$\sigma = \frac{My}{I_z} \tag{13.1}$$

where M is the bending moment at the cross section of interest, such as $M = Px/2$ for the cross section illustrated in Fig. 13.1. The quantity I_z is the moment of inertia of the cross-sectional area about the neutral axis, and the distance y is measured from the neutral axis. The positive direction for y is chosen as the direction that gives positive σ on the side of the neutral axis where tensile stresses occur. The equation gives a linear distribution of the normal stress, $\sigma = \sigma_x$, with distance y from the neutral axis, which

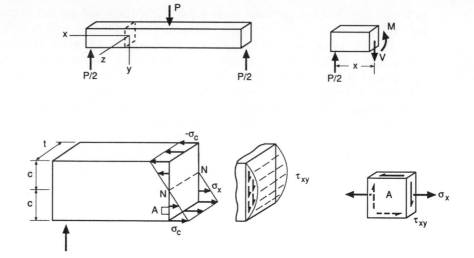

Figure 13.1 Elastic bending and shear in a rectangular beam. For a cross section as indicated above, the stresses are distributed as shown.

arises from two key assumptions: (1) linear-elastic stress-strain behavior, and (2) plane sections remaining plane after deformation, which requires a linear strain distribution. For rectangular cross sections of thickness t and depth $2c$, Eq. 13.1 gives the maximum tensile stress to be

$$\sigma_c = \frac{3M}{2tc^2} \tag{13.2}$$

where the subscript indicates that this stress occurs at $y = c$. Since in this case the cross section is symmetrical above and below the neutral axis, an equal compressive stress $-\sigma_c$ occurs at $y = -c$.

Except for cases of pure bending, these normal stresses are accompanied by shear stresses that can be computed from an additional equation found in any text on elementary mechanics of materials. For a rectangular cross section, the shear stress $\tau = \tau_{xy}$ for elastic deformation varies as a parabola with a maximum value at the neutral axis.

$$\tau = \frac{3V\left(c^2 - y^2\right)}{4tc^3} \tag{13.3}$$

where V is the shear force, such as $V = P/2$ for the cross section illustrated in Fig. 13.1.

Equation 13.1 applies even if the cross-sectional area is not symmetrical about the neutral axis, provided that the requirement of symmetry about the plane of loading is still met. Such a case is illustrated in Fig. 13.2(a). However, if symmetry about the plane of loading does not exist, then the stresses can still be determined by considering bending about two axes, as for Fig. 13.2(b). In addition, if the plane of unsymmetrical bending does not pass through the centroid of the cross-sectional area, torsional stresses must be considered.

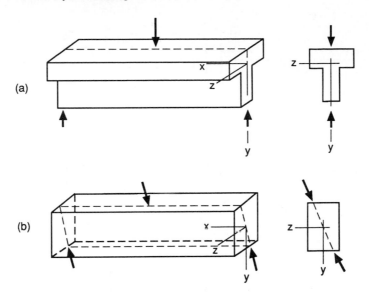

Figure 13.2   Bending due to loads in a plane of symmetry of the cross section (a), and bending due to loads not in a plane of symmetry (b).

13.2.2 Plastic Bending Analysis by Integration

Let us restrict our attention to cases of bending in a plane of symmetry of the beam and generalize the problem to permit plastic deformation. Also, assume that shear stresses are absent, or if present, that their effects are small. Under these circumstances, a reasonably accurate physical assumption even for plastic deformation is that originally plane cross sections remain plane. This results in a linear variation of strain with distance from the neutral axis.

$$\frac{\varepsilon}{y} = \frac{\varepsilon_c}{c} \tag{13.4}$$

where ε is strain in the longitudinal (x) direction, and ε_c is its value at $y = c$, the edge of the beam. Hence, if yielding occurs, the nonlinear stress-strain curve causes the stress distribution to become nonlinear as illustrated in Fig. 13.3. Note that, due to the linear strain distribution, the stress distribution has the same shape as the portion of the stress-strain curve up to $\varepsilon = \varepsilon_c$.

The bending moment can be computed by considering the contribution of a differential element of area as shown in Fig. 13.4.

$$dM = (\text{stress})(\text{area})(\text{distance}) = (\sigma)(t\, dy)(y) \tag{13.5}$$

where the thickness t can in general vary with y. Integrating to obtain the moment gives:

$$M = \int_{-c_2}^{c_1} \sigma t y \, dy \tag{13.6}$$

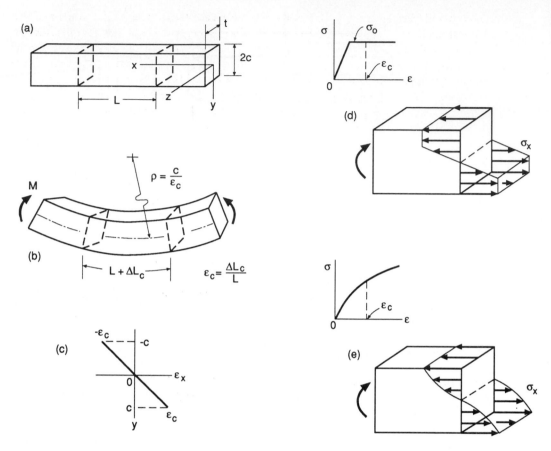

Figure 13.3 A rectangular beam (a) subjected to pure bending (b) which causes yielding. Plane sections remaining plane results in a linear strain distribution (c), but the stress distribution is nonlinear as in (d) or (e), having the same shape as that portion of the stress-strain curve up to ε_c.

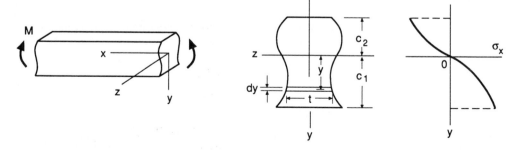

Figure 13.4 Area element ($t\,dy$) and stress distribution needed for integration to relate bending moment M to stresses and strains.

The cross-sectional area may not be symmetrical above and below the x-axis, and the stress-strain curve may not be symmetrical with respect to tension and compression. If either of these asymmetries exists, then the neutral axis shifts somewhat from the centroid of the cross-sectional area as plastic deformation progresses, so that it may need to be located before the integration can proceed. The principal to be followed in doing so is that the volumes under the tensile and compressive portions of the stress distribution must give equal and opposite forces, that is, a sum of zero, corresponding to an axial force P of zero.

$$P = \int_{-c_2}^{c_1} \sigma t \, dy = 0 \tag{13.7}$$

Equations 13.4, 13.6, and 13.7, and a stress-strain curve $\varepsilon = f(\sigma)$, are needed to solve any specific problem.

13.2.3 Rectangular Cross Sections

Consider the simple case of a rectangular cross section and a material with a symmetrical stress-strain curve. Hence, t is constant, $c_1 = c_2 = c$, and the neutral axis remains at the centroid. The symmetry that exists is such that the integral can be computed for one side of the neutral axis and then doubled.

$$M = 2t \int_0^c \sigma y \, dy \tag{13.8}$$

This equation is easily integrated for various forms of stress-strain curve. Stress σ first needs to be written as a function of y for integration to proceed.

For example, let

$$\varepsilon = f(\sigma) = \left(\frac{\sigma}{H_2}\right)^{\frac{1}{n_2}} \tag{13.9}$$

which is a simple power-hardening stress-strain curve with no linear-elastic region. Eliminating ε between Eqs. 13.4 and 13.9, and solving for σ as a function of y and constants c and ε_c, gives

$$\sigma = H_2 \left(\frac{y \varepsilon_c}{c}\right)^{n_2} \tag{13.10}$$

This is the equation of the positive part of a nonlinear stress distribution as illustrated in Fig. 13.3(e). We then substitute this σ into Eq. 13.8 and perform the integration, obtaining

$$M = \frac{2tc^2 H_2 \varepsilon_c^{n_2}}{n_2 + 2} = \frac{2tc^2 \sigma_c}{n_2 + 2} \tag{13.11}$$

Since the stress-strain curve holds for the edge of the beam where $\sigma = \sigma_c$ and $\varepsilon = \varepsilon_c$, Eq. 13.9 permits the result to be expressed in terms of stress, giving the second form. The special case of $n_2 = 1$ corresponds to the linear-elastic case with $H_2 = E$, so that Eq. 13.11 then gives the same result as Eq. 13.2.

13.2.4 Discontinuous Stress-Strain Curves

Consider a stress-strain curve that is discontinuous in that the mathematical relationship changes at the end of a distinct linear-elastic region. Three such relationships are described in Chapter 12, specifically behavior that is perfectly plastic, linear-hardening, or power-hardening. A beam made of such a material has a distinct elastic-plastic boundary and a region on each side of the neutral axis where only elastic deformation occurs. For example, for perfectly plastic behavior beyond yielding, the stress distribution is similar to Fig. 13.3(d).

As a result of the discontinuity, integration must be performed in two steps. For a symmetrical stress-strain curve and a rectangular cross section, Eq. 13.8 applies. The integration step occurs at y_b, the distance from the neutral axis to the point where yielding begins.

$$M = 2t \left[\int_0^{y_b} \sigma y \, dy + \int_{y_b}^c \sigma y \, dy \right] \tag{13.12}$$

To evaluate the integral, y_b must first be found by applying Eq. 13.4 at $y = y_b$.

$$\frac{\varepsilon_o}{y_b} = \frac{\varepsilon_c}{c} \tag{13.13}$$

Noting that the yield stress and strain are related by $\sigma_o = E\varepsilon_o$ gives

$$y_b = \frac{\sigma_o c}{E\varepsilon_c} \tag{13.14}$$

Between $y = 0$ and $y = y_b$, the stress distribution is linear as a consequence of the linear-elastic stress-strain relationship, $\varepsilon = \sigma/E$. To obtain σ as a function of y in this interval, apply Eq. 13.4 at any $y < y_b$.

$$\frac{\sigma/E}{y} = \frac{\varepsilon_c}{c} \qquad (0 \le y \le y_b) \tag{13.15}$$

Then combine this with Eq. 13.13 and solve for σ.

$$\sigma = \frac{E\varepsilon_o y}{y_b} \qquad (0 \le y \le y_b) \tag{13.16}$$

The second step of the integration is affected by the type of hardening beyond yielding. As an example, consider an elastic, perfectly plastic stress-strain curve. In this case, the stress beyond $y = y_b$ is simply equal to the yield strength.

$$\sigma = \sigma_o \qquad (y_b \le y \le c) \tag{13.17}$$

To obtain a solution, first substitute Eqs. 13.16 and 13.17 into the first and second terms, respectively, of Eq. 13.12, perform the integration, and then use Eq. 13.14 to eliminate

y_b from the equation. After some manipulation, the result is

$$M = tc^2\sigma_o \left[1 - \frac{1}{3}\left(\frac{\sigma_o}{E\varepsilon_c}\right)^2 \right] \quad (\varepsilon_c \geq \sigma_o/E) \tag{13.18}$$

The case where yielding is just beginning at the edge of the beam corresponds to $\varepsilon_c = \sigma_o/E$. Equation 13.18 then gives the same result as the elastic solution, Eq. 13.2.

$$M_i = \frac{2tc^2\sigma_o}{3} \quad (\varepsilon_c = \sigma_o/E) \tag{13.19}$$

which is called the *initial yielding moment*. For smaller values of M, the elastic solution applies in the form of Eq. 13.2 with $\sigma_c = E\varepsilon_c$. For large values of the maximum strain, Eq. 13.18 approaches a limiting value called the *fully plastic moment*.

$$M_o = tc^2\sigma_o \quad (\varepsilon_c \gg \sigma_o/E) \tag{13.20}$$

Note that $M_o/M_i = 1.5$, which ratio changes if the cross-sectional shape is other than a rectangle. The variation of moment with strain according to Eq. 13.18, and also the accompanying changes in the stress distribution, are shown in Fig. 13.5. As the fully plastic moment is approached, large deformation occurs and a *plastic hinge* is said to develop. This corresponds to the elastic region of the beam, $y \leq y_b$, shrinking and approaching zero. The development of increased plastic deformation and finally a plastic hinge in three-point bending is illustrated in Fig. 13.6.

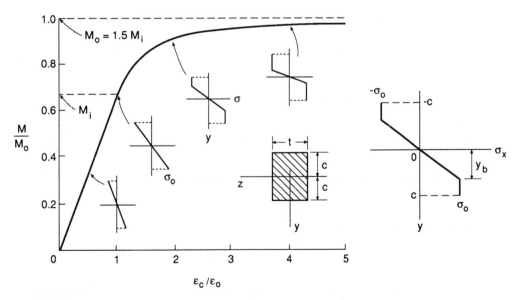

Figure 13.5 Moment vs. strain behavior for a rectangular beam of an elastic, perfectly plastic material. As loading progresses, the stress distribution changes as shown.

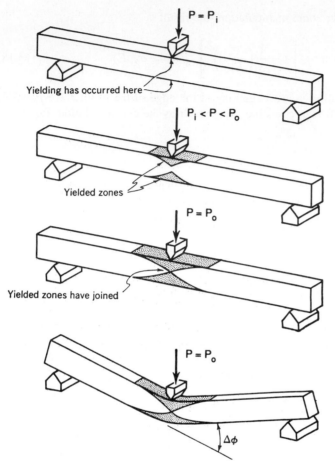

Figure 13.6 Development of a plastic hinge in three-point bending. The initial yield load P_i and the fully plastic load P_o correspond to moments M_i and M_o, respectively. Large deflections occur when P_o is reached. (Adapted from [Crandall 59] p. 332; used with permission; copyright 1959 by McGraw-Hill Publishing Co.)

Additional analysis similar to that given above can be done for various other combinations of stress-strain curve and cross-sectional shape, some of which are given as exercises at the end of this chapter. It is not always possible to perform the integration analytically, in which case numerical integration is needed.

13.2.5 Ramberg-Osgood Stress-Strain Curve

The Ramberg-Osgood stress-strain curve, Eq. 12.13, has the advantage that it can be used to accurately represent the stress-strain curves of many materials.

$$\varepsilon = \frac{\sigma}{E} + \left(\frac{\sigma}{H}\right)^{\frac{1}{n}} \tag{13.21}$$

Although this relationship is not explicitly solvable for stress, closed-form integration can still be performed in certain cases by changing the variable of integration from y to σ. This is demonstrated just below for a rectangular cross section.

To begin, use the linear strain distribution, Eq. 13.4, to obtain y in terms of strain, and differentiate to obtain dy.

$$y = \frac{c}{\varepsilon_c}\varepsilon, \quad dy = \frac{c}{\varepsilon_c}\, d\varepsilon \tag{13.22}$$

Substitute these into Eq. 13.8 to express M in terms of an integral with both stress and strain as variables.

$$M = 2t \left(\frac{c}{\varepsilon_c}\right)^2 \int_0^{\varepsilon_c} \sigma\varepsilon\, d\varepsilon \tag{13.23}$$

Strain ε is given as a function of stress by Eq. 13.21, and $d\varepsilon$ can be obtained from this by differentiation and manipulation to be

$$d\varepsilon = \left[\frac{1}{E} + \frac{1}{n\sigma}\left(\frac{\sigma}{H}\right)^{\frac{1}{n}}\right] d\sigma \tag{13.24}$$

Substituting both Eqs. 13.21 and 13.24 into the integral of Eq. 13.23 gives

$$M = 2t \left(\frac{c}{\varepsilon_c}\right)^2 \int_0^{\sigma_c} \sigma \left[\frac{\sigma}{E} + \left(\frac{\sigma}{H}\right)^{\frac{1}{n}}\right]\left[\frac{1}{E} + \frac{1}{n\sigma}\left(\frac{\sigma}{H}\right)^{\frac{1}{n}}\right] d\sigma \tag{13.25}$$

This integral can be evaluated in a straightforward manner by first obtaining the product of the two quantities in brackets. Doing so and performing some manipulation gives

$$M = 2t\sigma_c \left(\frac{c}{\varepsilon_c}\right)^2 \left[\frac{1}{3}\left(\frac{\sigma_c}{E}\right)^2 + \frac{n+1}{2n+1}\left(\frac{\sigma_c}{E}\right)\left(\frac{\sigma_c}{H}\right)^{\frac{1}{n}} + \frac{1}{n+2}\left(\frac{\sigma_c}{H}\right)^{\frac{2}{n}}\right] \tag{13.26}$$

This result may be written with the beam-edge stress σ_c as the only variable by substituting $\varepsilon_c = f(\sigma_c)$ using Eq. 13.21. A useful form obtained after some manipulation is

$$M = \frac{2tc^2\sigma_c}{3}\left[\frac{1 + \dfrac{3n+3}{2n+1}\beta + \dfrac{3}{n+2}\beta^2}{(1+\beta)^2}\right] \tag{13.27}$$

where

$$\beta = \frac{\varepsilon_{pc}}{\varepsilon_{ec}}, \quad \varepsilon_{pc} = \left(\frac{\sigma_c}{H}\right)^{\frac{1}{n}}, \quad \varepsilon_{ec} = \frac{\sigma_c}{E}, \quad \varepsilon_c = \varepsilon_{ec} + \varepsilon_{pc}$$

The quantities ε_{pc} and ε_{ec} are the edge-of-beam values of plastic and elastic strain, respectively, and β is their ratio. Using the above quantities, the moment can be related to either stress or strain, with the relationship with strain being implicit.

Equation 13.27 gives a smooth variation of moment with strain as shown in Fig. 13.7. Also shown for comparison are the trends for elastic analysis, Eq. 13.2, and also for Eq. 13.11, which is the result derived using a simple power-hardening stress-strain curve with $H_2 = H$ and $n_2 = n$. If the ratio of plastic to elastic strain β is small, Eq. 13.27 reduces to the elastic solution. Conversely, if this ratio is large,

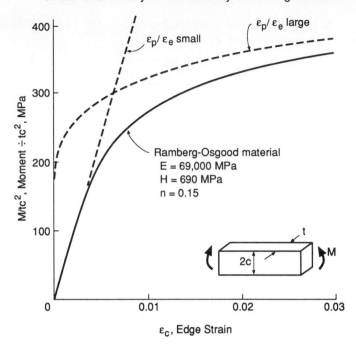

Figure 13.7 Moment vs. strain relationship for a material having a particular Ramberg-Osgood stress-strain curve. The curve approaches the limiting case of the elastic solution for small strains, and another limiting case corresponding to simple power hardening for large strains.

Eq. 13.27 gives the same result as the simple power-hardening case, corresponding to the elastic term in the Ramberg-Osgood stress-strain equation being negligible compared to the plastic term.

13.2.6 Combined Axial and Bending Loads

The analysis methods just applied to pure bending are readily extended to combined bending and axial loads. If the loads all lie in a plane of symmetry of the member, Eq. 13.6 applies for calculating the bending moment. This particular equation is needed, rather than Eq. 13.8, due to the asymmetry in the stress distribution arising from the axial load. The integral sum of the stresses in the axial direction is no longer zero but must be equal to the axial load P.

$$P = \int_{-c_2}^{c_1} \sigma t \, dy \tag{13.28}$$

As an example, consider the simple case of an eccentric load applied to a member with a rectangular cross section as shown in Fig. 13.8. The eccentricity relative to the centroid of the cross-sectional area is y_e, so that the moment applied is $M = P y_e$. Let the

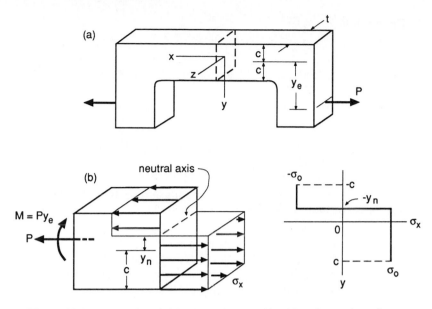

Figure 13.8 Eccentric axial load causing combined bending and tension on a rectangular cross section (a). The stress distribution for fully plastic yielding is shown in (b).

material be an elastic, perfectly plastic one, and consider only the extreme limiting case of fully plastic loading. The stress distribution must then be as shown in Fig. 13.8(b). The full cross-sectional area is subject to stresses equal to the yield strength in either tension or compression. For a tensile load as shown, the neutral axis is shifted so that the area under tension is larger than that under compression. This problem can be solved by first using Eq. 13.28 to locate the neutral axis, and then integrating to evaluate the bending moment.

Letting the distance from the centroid to the neutral axis be y_n, Eq. 13.28 gives

$$P = t \int_{-c}^{c} \sigma \, dy = t \int_{-c}^{-y_n} (-\sigma_o) \, dy \; + \; t \int_{-y_n}^{c} \sigma_o \, dy \qquad (13.29)$$

where a two step integration broken at $y = -y_n$ is necessary due to the discontinuous stress distribution. Evaluating this integral and solving the result for y_n gives

$$y_n = \frac{P}{2t\sigma_o} \qquad (13.30)$$

We then apply the integral of Eq. 13.6 to evaluate the bending moment, proceeding in a similar manner as before.

$$M = P y_e = t \int_{-c}^{c} \sigma y \, dy$$

$$M = t \int_{-c}^{-y_n} (-\sigma_o) y \, dy \; + \; t \int_{-y_n}^{c} \sigma_o y \, dy \qquad (13.31)$$

The result is

$$P = \frac{\sigma_o t}{y_e} \left(c^2 - y_n^2 \right) \tag{13.32}$$

Substituting y_n from Eq. 13.30 and solving the resulting quadratic equation gives

$$M_o = P_o y_e = 2t\sigma_o y_e \left[-y_e + \sqrt{y_e^2 + c^2} \right] \quad (P > 0) \tag{13.33}$$

$$M_o = P_o y_e = -2t\sigma_o y_e \left[y_e + \sqrt{y_e^2 + c^2} \right] \quad (P < 0) \tag{13.34}$$

where positive P corresponds to tension. Subscripts have been added to M and P to make it clear the values obtained are limiting values for the fully plastic situation. Note that these equations do not apply for $y_n = 0$, that is, simple bending, which solution has already been given.

If the combined axial and bending loads are given in terms of a combination of a *centric* load P and a bending moment M, the above equations can still be applied. This is done by first calculating $y_e = M/P$, where y_e is the location of P for the statically equivalent system of forces consisting of P alone.

Example 13.1

A simply supported beam with a rectangular cross section is subjected to a load P at mid-span. The material has a simple power-hardening stress-strain curve, Eq. 13.9. Derive an equation for the deflection at mid-span in terms of P, beam dimensions, and material constants. (See Figure E13.1.)

Solution   Consider a short length m of the deflected beam, over which the radius of curvature ρ is approximately constant. The angle θ measured at the center of curvature as shown is related to m and ρ by

$$\theta = \frac{m}{\rho} = \frac{m + \Delta m_c}{\rho + c}$$

where Δm_c is the change in m at the tensile edge of the beam, which is located at a distance $\rho + c$ from the center of curvature. Substituting $\varepsilon_c = \Delta m_c/m$ into this relationship gives

$$\frac{1}{\rho} = \frac{\varepsilon_c}{c}$$

For small curvature, where $y = f(x)$ is the beam deflection, we have

$$\frac{1}{\rho} = \frac{d^2 y}{dx^2}$$

which is the same assumption as used for elastic bending in any elementary mechanics of materials book.

Hence, for our nonlinear material, Eq. 13.9, we have

$$\frac{d^2 y}{dx^2} = \frac{\varepsilon_c}{c} = \frac{1}{c} \left(\frac{\sigma_c}{H_2} \right)^{\frac{1}{n_2}}$$

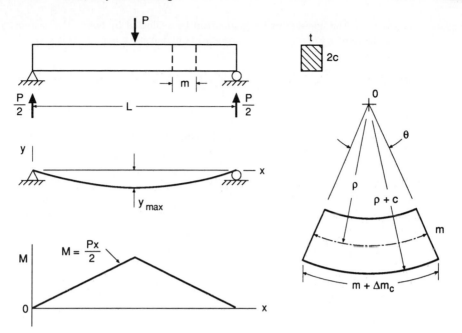

Figure E13.1

Equation 13.11 gives the relationship between bending moment M and either ε_c or σ_c for a rectangular beam of such a material, allowing the equation above to be expressed in terms of M.

$$\frac{d^2 y}{dx^2} = \frac{1}{c}\left[\frac{M(n_2 + 2)}{2tc^2 H_2}\right]^{\frac{1}{n_2}} = \left(\frac{M}{K H_2}\right)^{\frac{1}{n_2}}$$

The constant K is employed for convenience where

$$K = \frac{2tc^{n_2+2}}{n_2 + 2}$$

The symmetry involved permits consideration of only one-half of the beam, say the left half, for which $M = Px/2$ can be substituted.

$$\frac{d^2 y}{dx^2} = \left(\frac{Px}{2K H_2}\right)^{\frac{1}{n_2}}$$

Since x is now the only variable, we can integrate, obtaining

$$\frac{dy}{dx} = \left(\frac{P}{2K H_2}\right)^{\frac{1}{n_2}}\frac{n_2 x^{\frac{n_2+1}{n_2}}}{n_2 + 1} + C_1$$

The constant of integration can be evaluated by noting that symmetry requires that the slope dy/dx of the beam be zero at mid-span, $x = L/2$. This gives

$$\frac{dy}{dx} = \frac{n_2}{n_2 + 1} \left(\frac{P}{2KH_2} \right)^{\frac{1}{n_2}} \left[x^{\frac{n_2+1}{n_2}} - \left(\frac{L}{2} \right)^{\frac{n_2+1}{n_2}} \right]$$

Integrating again and evaluating the constant of integration by noting that $y = 0$ at $x = 0$ gives

$$y = \frac{n_2}{n_2 + 1} \left(\frac{P}{2KH_2} \right)^{\frac{1}{n_2}} \left[\frac{n_2 x^{\frac{2n_2+1}{n_2}}}{2n_2 + 1} - x \left(\frac{L}{2} \right)^{\frac{n_2+1}{n_2}} \right]$$

This equation provides the beam deflection y at any x on the left half of the beam. For the mid-span deflection, substituting $x = L/2$ and K from above gives the desired result.

$$y_{max} = -\frac{n_2}{c(2n_2 + 1)} \left[\frac{P(n_2 + 2)}{4H_2tc^2} \right]^{\frac{1}{n_2}} \left(\frac{L}{2} \right)^{\frac{2n_2+1}{n_2}} \qquad \textbf{Ans.}$$

Comment   For the elastic case, which corresponds to $n_2 = 1$ and $H_2 = E$, the above equation reduces as expected to

$$y_{max} = -\frac{PL^3}{32Etc^3} = -\frac{PL^3}{48EI_z}$$

13.3 RESIDUAL STRESSES AND STRAINS FOR BENDING

Consider a beam that is plastically deformed, and then the bending moment removed, as illustrated in Fig. 13.9. Let the material exhibit not only plastic deformation but also elastic deformation that is recovered during unloading. Upon unloading of the beam, the strains will decrease, but not to zero, so that a permanent deflection remains in the beam. Locked-in or *residual* stresses will also exist in the beam.

Since plane sections are still expected to remain plane during unloading, the greatest changes in strain are at the edge of the beam. Here, the residual stress is opposite in sign to the stress previously present at the maximum load. What happens is that the material at the beam-edge that was permanently stretched in tension is forced into compression as the beam springs back upon removal of the load. Conversely, the material at the beam-edge subjected to compressive yielding is forced into tension upon unloading. Closer to the neutral axis, there are regions where the residual stresses have the same sign as the previously applied stress.

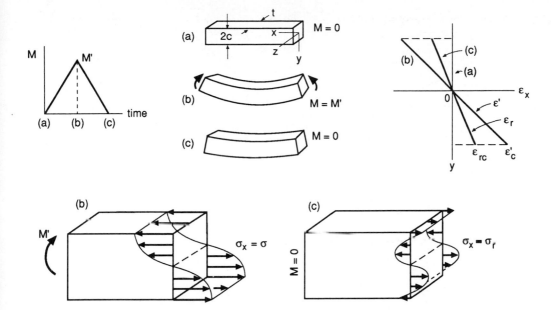

Figure 13.9 Loading of a rectangular beam beyond the point of yielding followed by unloading. Loading starts from zero moment at time (a) and proceeds to the maximum moment M' at time (b). When unloading is complete at time (c), residual strains ε_r having a linear distribution remain, and residual stresses σ_r are distributed as shown.

13.3.1 Rectangular Beam of Elastic, Perfectly Plastic Material

Consider an elastic, perfectly plastic material and the case of a rectangular cross section that has been loaded beyond yielding. This situation is illustrated in Fig. 13.10. At the highest moment reached, M', the stress distribution is similar to Fig. 13.10(a). The moment M' is related to the strain at the edge of the beam, ε_c', by Eq. 13.18.

$$M' = tc^2\sigma_o \left[1 - \frac{1}{3} \left(\frac{\sigma_o}{E\varepsilon_c'} \right)^2 \right] \tag{13.35}$$

The material deforms elastically upon unloading unless the compressive yield strength is exceeded. Analysis is simplified if there is no compressive yielding in the beam after the load is removed. Let us make this assumption and proceed, later checking the result to be sure that the assumption is confirmed. Since plane sections are expected to remain plane during unloading, the change in strain $\Delta\varepsilon$ at any y during unloading is proportional to that at $y = c$, denoted $\Delta\varepsilon_c$.

$$\Delta\varepsilon = \frac{y}{c} \Delta\varepsilon_c \tag{13.36}$$

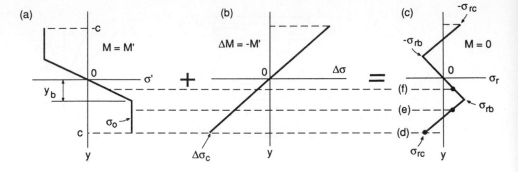

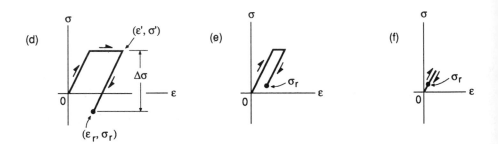

Figure 13.10 For a rectangular beam of an elastic, perfectly plastic material, stresses at the maximum moment are shown in (a), stress changes during unloading in (b), and residual stresses in (c). Depending on the location, residual stresses may be opposite in sign to the maximum stress (d) or of the same sign (e, f). The particular case illustrated corresponds to $M' = 0.95M_o$.

Also, if only elastic unloading occurs, then the changes in stress and strain are related by the elastic modulus.

$$\Delta\sigma = E\,\Delta\varepsilon, \qquad \Delta\sigma_c = E\,\Delta\varepsilon_c \qquad (13.37)$$

Equation 13.36 thus requires

$$\Delta\sigma = \frac{y}{c}\Delta\sigma_c \qquad (13.38)$$

The change in stress is negative for the side of the neutral axis where positive σ' previously occurred, the distribution of $\Delta\sigma$ being as shown in Fig. 13.10(b). For various locations in the beam, the changes in stress and strain during unloading occur on stress-strain paths as shown by (d), (e), and (f).

Since only elastic changes are assumed to occur, $\Delta\sigma$ can be related to the change in moment ΔM by the elastic bending formula.

$$\Delta M = \frac{I_z\Delta\sigma_c}{c} = \frac{2tc^2\Delta\sigma_c}{3} \qquad (13.39)$$

Since the beam is unloaded to zero moment, we have $\Delta M = -M'$, and Eqs. 13.35 and 13.39 can be combined. Doing so and solving for $\Delta\sigma_c$ gives

$$\Delta\sigma_c = -\frac{\sigma_o}{2}\left[3 - \left(\frac{\sigma_o}{E\varepsilon_c'}\right)^2\right] \tag{13.40}$$

The residual stress at any value of y in the beam, denoted σ_r, is obtained by adding the change $\Delta\sigma$ during unloading to the value σ' that existed at $M = M'$, and similarly for residual strains.

$$\sigma_r = \sigma' + \Delta\sigma, \qquad \varepsilon_r = \varepsilon' + \Delta\varepsilon = \varepsilon' + \frac{\Delta\sigma}{E} \qquad \text{(a,b)} \tag{13.41}$$

The residual stress at the edge of the beam, σ_{rc}, can thus be obtained by substituting Eq. 13.40 into Eq. 13.41(a), noting that at $y = c$ the stress before unloading was $\sigma' = \sigma_o$.

$$\sigma_{rc} = -\frac{\sigma_o}{2}\left[1 - \left(\frac{\sigma_o}{E\varepsilon_c'}\right)^2\right] \tag{13.42}$$

The residual strain at the edge of the beam can be similarly obtained by substituting $\Delta\sigma_c$ from Eq. 13.40 into Eq. 13.41(b).

$$\varepsilon_{rc} = \frac{\sigma_o}{2E}\left[-3 + 2\left(\frac{E\varepsilon_c'}{\sigma_o}\right) + \left(\frac{\sigma_o}{E\varepsilon_c'}\right)^2\right] \tag{13.43}$$

The quantity $E\varepsilon_c'/\sigma_o = \varepsilon_c'/\varepsilon_o$ is the ratio of the highest strain reached to the yield strain. This quantity must be greater than unity for an initially yielded beam. Noting that this ratio and its inverse appears in both Eqs. 13.42 and 13.43, it is readily concluded that σ_{rc} is negative and ε_{rc} is positive at $y = +c$, that is, for the edge of the beam that was yielded in tension. Appropriate sign changes give values equal in magnitude but opposite in sign for $y = -c$.

For the extreme case where the highest strain reached is large compared to the yield strain, Eq. 13.42 gives a residual stress of magnitude half the yield strength, and Eq. 13.43 indicates that essentially none of the strain is lost upon unloading.

$$\sigma_{rc} = -\frac{\sigma_o}{2}, \qquad \varepsilon_{rc} = \varepsilon_c' \qquad \left(\varepsilon_c'/\varepsilon_o \gg 1\right) \tag{13.44}$$

The corresponding stress distributions are shown in Fig. 13.11. Conversely, if the edge of the beam just reaches the yield stress, $\varepsilon_c'/\varepsilon_o = 1$, these equations give $\sigma_{rc} = \varepsilon_{rc} = 0$ as expected. Values of σ_{rc} and ε_{rc} vary smoothly between the above extremes, with the trend for σ_{rc} being shown as the solid line in Fig.13.12. Note that the initial assumption of no yielding in compression is confirmed for the most extreme case by Eq. 13.44, so that the analysis is valid.

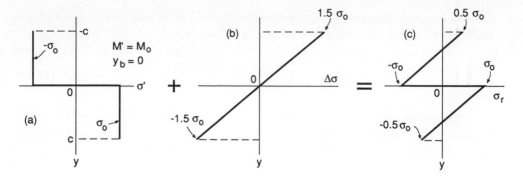

Figure 13.11   For the special case of fully plastic yielding of a rectangular beam, maximum stresses prior to unloading are shown in (a), stress changes during unloading in (b), and residual stresses in (c).

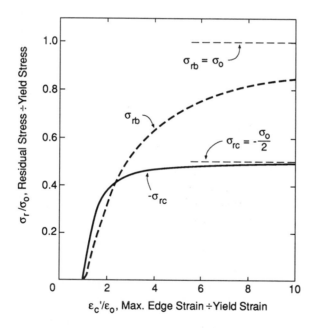

Figure 13.12   Residual stress at the edge of a rectangular beam, σ_{rc}, and at the elastic-plastic boundary, σ_{rb}, as functions of the maximum edge strain reached prior to unloading.

13.3.2 Analysis Extended to the Interior of the Beam

Now consider interior locations in the beam. The residual stress distribution consists of straight-line segments with slope changes at $y = \pm y_b$ as shown in Fig. 13.10(c). This arises from the fact that the residual stress distribution is the sum of the distributions for σ' and $\Delta\sigma$. Since the σ' distribution has slope changes at $y = \pm y_b$, and the $\Delta\sigma$ distribution has none, their sum σ_r has slope changes at $\pm y_b$ only. Comparing Figs. 13.10(a) and (c), it is seen that points in the beam that were yielded, $y > y_b$, follow a stress-strain path similar to either (d) or (e), depending on the sign of the residual stress. Points not yielded, $y < y_b$, deform only along the elastic line, but the stresses do not return to zero, as for (f).

The residual stress distribution can thus be completely described if its value at the elastic-plastic boundary, σ_{rb}, is determined in addition to σ_{rc}. Recall that the location of the elastic-plastic boundary y_b is related to the maximum strain reached by Eq. 13.14.

$$y_b = \frac{\sigma_o c}{E \varepsilon_c'} \tag{13.45}$$

Since this point has reached but not exceeded the yield stress, σ_{rb} is related to the corresponding residual strain by the elastic modulus.

$$\sigma_{rb} = E \varepsilon_{rb} \tag{13.46}$$

Plane sections remaining plane requires that the distribution of residual strain be linear, so that

$$\frac{\varepsilon_{rb}}{y_b} = \frac{\varepsilon_{rc}}{c} \tag{13.47}$$

Substituting y_b from Eq. 13.45 and ε_{rb} from Eq. 13.46 gives

$$\sigma_{rb} = \frac{\sigma_o \varepsilon_{rc}}{\varepsilon_c'} \tag{13.48}$$

Hence, σ_{rb} can be easily obtained from ε_{rc}, which is given by Eq. 13.43. Values of σ_{rb} are found to increase from zero at $\varepsilon_c'/\varepsilon_o = 1$ to σ_o at large values of $\varepsilon_c'/\varepsilon_o$. The trend is shown by the dashed line in Fig. 13.12.

To summarize, for a rectangular beam of an elastic, perfectly plastic material, deformed to a given value of strain, the equations given provide a complete description of the residual stress distribution. Values are needed for σ_{rc} and σ_{rb} from Eq. 13.42 and 13.48, and the residual stress distribution consists of straight line segments as shown in Fig. 13.10(c). The residual strain distribution is linear, with ε_{rc} being given by Eq. 13.43. Analogous residual stress distributions occur for other cross-sectional shapes and types of elasto-plastic stress-strain curve. Closed form analysis similar to that just described can be performed in some cases, and in other cases numerical analysis is required.

13.4 PLASTICITY OF CIRCULAR SHAFTS IN TORSION

Circular shafts loaded beyond the point of yielding in torsion, either solid or hollow, can be analyzed using procedures similar to those applied above for bending. The assumption that plane sections remain plane is again employed. (For noncircular sections, this assumption is violated, and more sophisticated analysis is needed that is not covered here.) Analysis of torsion of circular shafts requires stress-strain curves for a state of pure shear. These are not generally known but can be estimated from the more commonly available uniaxial stress-strain curves. We thus need to discuss estimation of stress-strain curves for shear before proceeding with analysis of circular shafts.

13.4.1 Stress-Strain Curves for Shear

Circular shafts in torsion are stressed at all points in pure planar shear. In this state of stress, the only nonzero component of stress is τ_{xy}, the x-y plane being taken to be parallel to the surface of the shaft, and the z-axis normal to the surface, as illustrated in Fig. 13.13(a). To estimate stress-strain curves for this case from uniaxial ones, the principal stresses and strains can be evaluated and used with equations based on deformation theory of plasticity as described in the previous chapter.

For such a pure shear τ_{xy}, one of the principal axes is the z-axis, and the other two lie in the x-y plane and are rotated $45°$ with respect to the x-y axes. The principal normal stresses are related to τ_{xy} by

$$\sigma_1 = -\sigma_2 = \tau_{xy}, \quad \sigma_3 = 0 \tag{13.49}$$

which was previously illustrated in Fig. 4.5. For an isotropic, homogeneous material subjected to such a τ_{xy}, the only nonzero component of strain is γ_{xy}. The strains on the principal axes are

$$\varepsilon_1 = -\varepsilon_2 = \frac{\gamma_{xy}}{2}, \quad \varepsilon_3 = 0 \tag{13.50}$$

These principal stresses, strains, and directions can be verified by applying Mohr's circle as described in Chapter 6.

Since pure shear is a special case of plane stress, the desired stress-strain curve for shear can be obtained from Eq. 12.39 used with $\lambda = \sigma_2/\sigma_1 = -1$. Substitution of this

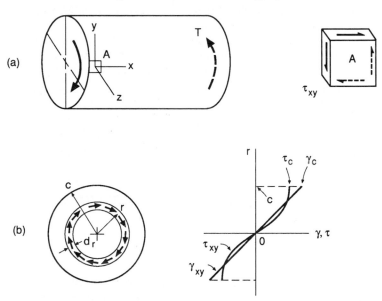

Figure 13.13 Circular shaft subjected to pure torsion. A state of pure shear stress occurs, with its magnitude varying with radius r as determined by a linear distribution of shear strain.

value, and also σ_1 and ε_1 from just above, gives

$$\gamma_{xy} = \frac{2\nu - 1}{2G(1 + \nu)}\tau_{xy} + \sqrt{3}f\left(\sqrt{3}\tau_{xy}\right) \tag{13.51}$$

where the substitution $E = 2G(1 + \nu)$ from Eq. 4.12 has also been employed. This general relationship can be used with any particular form of uniaxial stress-strain curve, $\bar{\varepsilon} = f(\bar{\sigma})$, to obtain the corresponding τ vs. γ curve.

For example, consider a uniaxial curve in the form of the Ramberg-Osgood equation.

$$\bar{\varepsilon} = f(\bar{\sigma}) = \frac{\bar{\sigma}}{E} + \left(\frac{\bar{\sigma}}{H}\right)^{\frac{1}{n}} \tag{13.52}$$

Use of this function in Eq 13.51 gives

$$\gamma_{xy} = \frac{\tau_{xy}}{G} + \sqrt{3}\left(\frac{\sqrt{3}\tau_{xy}}{H}\right)^{\frac{1}{n}} \tag{13.53}$$

It is useful to write this equation in a mathematical form similar to the uniaxial equation.

$$\gamma_{xy} = \frac{\tau_{xy}}{G} + \left(\frac{\tau_{xy}}{H_\tau}\right)^{\frac{1}{n}} \tag{13.54}$$

where n is the same as for the uniaxial case and the new constant H_τ is related to the constants from the uniaxial curve by

$$H_\tau = \frac{H}{3^{(n+1)/2}} \tag{13.55}$$

For an elastic, perfectly plastic stress-strain curve, no $\bar{\varepsilon} = f(\bar{\sigma})$ can be defined beyond yielding. This case can be handled by noting that the yield strength τ_o for pure shear can be related to the uniaxial value by applying Eq. 12.34 with $\lambda = -1$. At yielding, $\bar{\sigma} = \sigma_o$ and $\sigma_1 = \tau_{xy} = \tau_o$, so that

$$\tau_o = \frac{\sigma_o}{\sqrt{3}} \tag{13.56}$$

Hence, the relationship is

$$\tau_{xy} = G\gamma_{xy} \qquad \left(\tau_{xy} \le \tau_o\right) \qquad (a)$$
$$\tau_{xy} = \tau_o = \frac{\sigma_o}{\sqrt{3}} \qquad \left(\gamma_{xy} \ge \frac{\tau_o}{G}\right) \qquad (b) \tag{13.57}$$

For other discontinuous stress-strain curves with a distinct yield point, Eqs. 13.56 and 13.57(a) still apply, but Eq. 13.57(b) must be replaced by an appropriate expression derived from Eq. 13.51.

13.4.2 Analysis of Circular Shafts

Analysis of circular shafts loaded beyond yielding in torsion proceeds in a manner quite similar to that described above for bending. Consider the cross section of a shaft as shown in Fig. 13.13(b). Due to the radial symmetry, the shear stress $\tau_{xy} = \tau$ is constant for all points at a distance r from the shaft axis. The contribution to the torque of an annular element of area as shown is

$$dT = \text{(stress)}\text{(area)}\text{(distance)} = (\tau)(2\pi r \, dr)(r) \tag{13.58}$$

Integrating from $r = 0$ to the outer surface, $r = c$, gives the torque.

$$T = 2\pi \int_0^c \tau r^2 \, dr \tag{13.59}$$

For a hollow shaft with inner radius c_1 and outer radius c_2, this equation needs to be modified as follows:

$$T = 2\pi \int_{c_1}^{c_2} \tau r^2 \, dr \tag{13.60}$$

To evaluate either integral, a particular stress-strain curve for shear, $\gamma = f_\tau(\tau)$, is needed along with the assumption that plane sections remain plane during twisting. This requires a linear distribution of shear strain, $\gamma_{xy} = \gamma$.

$$\frac{\gamma}{r} = \frac{\gamma_c}{c} \tag{13.61}$$

where γ_c is the shear strain at $r = c$.

For a case of simple power hardening with no distinct yield point, we have the following stress-strain curve:

$$\gamma = \left(\frac{\tau}{H_3}\right)^{\frac{1}{n_3}} \tag{13.62}$$

Substituting this and Eq. 13.61 into Eq. 13.59, and evaluating the integral for a solid shaft, gives

$$T = \frac{2\pi c^3 \tau_c}{n_3 + 3} = \frac{2\pi c^3 H_3 \gamma_c^{n_3}}{n_3 + 3} \tag{13.63}$$

For a solid shaft and an elastic, perfectly plastic stress-strain curve, Eq. 13.57, the analysis that is needed parallels that leading to Eq. 13.18 for bending with a similar σ-ε curve. The result is

$$T = \frac{\pi c^3 \tau_c}{2} \qquad (\tau_c \leq \tau_o)$$

$$T = \frac{\pi c^3 \tau_o}{6}\left[4 - \left(\frac{\tau_o}{G\gamma_c}\right)^3\right] \qquad (\gamma_c \geq \tau_o/G) \tag{13.64}$$

As a final case, for a solid shaft and a Ramberg-Osgood type stress-strain curve, Eq. 13.54, analysis similar to that used for bending to obtain Eq. 13.27 gives

$$
T = 2\pi c^3 \tau_c \left[\frac{\dfrac{1}{4} + \dfrac{2n+1}{3n+1}\beta_\tau + \dfrac{n+2}{2n+2}\beta_\tau^2 + \dfrac{1}{n+3}\beta_\tau^3}{(1+\beta_\tau)^3} \right] \tag{13.65}
$$

where β_τ is the ratio of the plastic to elastic shear strain at the shaft surface, $r = c$, so that

$$
\beta_\tau = \frac{\gamma_{pc}}{\gamma_{ec}}, \quad \gamma_{pc} = \left(\frac{\tau_c}{H_\tau}\right)^{\frac{1}{n}}, \quad \gamma_{ec} = \frac{\tau_c}{G}, \quad \gamma_c = \gamma_{ec} + \gamma_{pc}
$$

Equation 13.65 thus constitutes a relationship between torque and surface shear stress that can be written explicitly, and also an implicit relationship between torque and surface shear strain.

Residual shear stresses for torsion behave in an analogous manner to those for bending and can be analyzed by a similar procedure.

13.5 NOTCHED MEMBERS

Notched engineering members are often subjected to loads in service that cause localized yielding. The resulting plastic strains are of special interest in estimating fatigue lives using the strain-based approach, which is the subject of the next chapter. Gross plastic deformation in notched members must also be avoided in engineering design by providing a sufficient safety factor against overload. This portion of the chapter provides engineering tools that address these needs. Before proceeding, the reader may wish to review the nomenclature for notched members as given in Section 9.2.2.

A good starting point is to consider the behavior of notched members over a wide range of applied loads as illustrated in Fig. 13.14. At low loads, the behavior is everywhere elastic and a simple linear relationship prevails. Localized plastic deformation begins when the stress at the notch exceeds the yield strength of the material. Plastic deformation then spreads over a region of increasing size for increasing load, until the entire cross section of the member has yielded. The behavior of load versus local notch strain is similar to the curve shown. Such a curve exhibits three regions corresponding to the three types of behavior: (a) no yielding, (b) local yielding, and (c) fully plastic yielding. Some plastic zones for local yielding at a notch in polycarbonate plastic are shown in Fig. 13.15.

13.5.1 Elastic Behavior and Initial Yielding

For elastic behavior, the notch stress σ can be estimated from the nominal stress S and the elastic stress concentration factor k_t.

$$
\sigma = k_t S \quad (\sigma \le \sigma_o) \tag{13.66}
$$

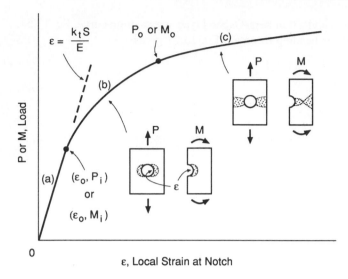

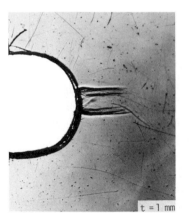

$$\varepsilon = \frac{k_t S}{E}$$

P_o or M_o

(c)

(b)

P or M, Load

(a)

(ε_o, P_i)
or
(ε_o, M_i)

0

ε, Local Strain at Notch

Figure 13.14   Load vs. local strain behavior of a notched member showing three regions of behavior: (a) no yielding, (b) local yielding, and (c) fully plastic yielding.

t = 1 mm   t = 2 mm   t = 5

Figure 13.15   Plastic zones at notches in polycarbonate plastic. The notches are 3 mm deep and have radii of $\rho = 2$ mm, and the thickness varies from 1 mm (left), and 2 mm (center), to 5 mm (right). Thickness affects the development of the plastic zone, as the state of stress is altered by different degrees of transverse constraint. (Photos courtesy of Prof. H. Nisitani, Kyushu University, Fukuoka, Japan.   Published in [Nisitani 85]; reprinted with permission from *Engineering Fracture Mechanics*, Pergamon Press, Oxford, UK.)

For axial or bending loads, this σ is the stress at the bottom of the notch in a direction parallel to S, specifically σ_{yn} as shown in Fig. 13.16. Values of k_t may be obtained from Figs. 10.1 and 10.2 and from various handbooks as noted in Chapter 10. Except where the notch radius ρ is small compared to the thickness t, the stress σ_{zn} in the thickness direction will be small compared to σ_{yn}. Also, $\sigma_{xn} = 0$ due to the free surface normal to the x direction, so that the state of stress is approximately uniaxial with $\sigma_{yn} = \sigma$. The

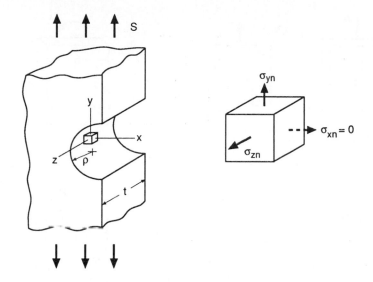

Figure 13.16   Coordinate system for the local stresses and strains at a notch.

corresponding strain is then simply

$$\varepsilon = \frac{k_t S}{E} \quad \left(\varepsilon \le \frac{\sigma_o}{E} \right) \tag{13.67}$$

Noting that S/E can be considered to be a nominal (average) strain, k_t is not only a stress concentration factor but is also a strain concentration factor. Equations 13.66 and 13.67 apply only until σ reaches the yield strength σ_o, beyond which they are not valid. For cases of shear loading, τ, γ, and G are similarly used with an appropriate k_t value, provided of course that the yield strength in shear is not exceeded.

Consider a plate with a central hole, with dimensions as defined in Fig. 13.17(a). Assuming that k_t is defined based on net section nominal stresses, we have

$$S = \frac{P}{2(b-a)t} \tag{13.68}$$

where the nomenclature a and b parallels that previously used for cracked members in Chapter 8, but is here applied for notches with blunt ends of some definite radius ρ. Yielding first occurs at the load denoted P_i, where $k_t S = \sigma_o$, for which Eq. 13.68 gives

$$P_i = \frac{2(b-a)t}{k_t} \sigma_o \tag{13.69}$$

which is called the *initial yielding load*.

For notched bending members, nominal stress is generally defined by applying the elastic bending formula to the cross section of depth $(b-a)$ remaining after removal of

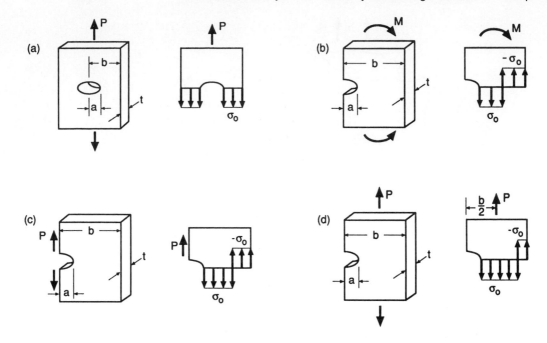

Figure 13.17 Geometries and stress distributions corresponding to fully plastic yielding for four cases of notched members.

the notch. For a rectangular bending member with a single edge notch as in Fig. 13.17(b), we thus have

$$S = \frac{6M}{(b-a)^2 t} \tag{13.70}$$

which can be obtained by substituting $c = (b-a)/2$ into Eq. 13.2. The *initial yielding moment* occurs when $k_t S = \sigma_o$, so that

$$M_i = \frac{(b-a)^2 t}{6k_t}\sigma_o \tag{13.71}$$

Similar equations based on the particular definition of nominal stress being used can be obtained for any other case.

13.5.2 Fully Plastic Yielding: Rough Estimates

Beyond the point of yielding, the local notch strains are larger than would be estimated from elastic analysis. (Compare the solid and dashed lines in Fig. 13.14.) Yielding still may be confined to a relatively small volume fraction of material, but a larger volume yields as loads increase. When yielding spreads to the entire cross-sectional area, the situation is described as *fully plastic yielding*. Beyond this point, small increases in load cause large increases in notch strain. The overall displacement, which up to this point has been nearly linear, also begins to increase rapidly with load. For an elastic, perfectly plastic material, no further increase in load is possible, and the load versus strain curve becomes flat.

Rough estimates of fully plastic yielding loads can be easily made. These are based on stress distributions for a perfectly plastic material as shown in Fig. 13.17. For axial loading with bilateral symmetry, the fully plastic load is approximately the load that causes the stress on the net area to reach the yield strength. For the particular example of a plate with a central hole or other central notch this gives

$$\sigma_o = \frac{P_o}{2(b-a)t}, \qquad P_o = 2(b-a)t\sigma_o \qquad (13.72)$$

where P_o is called *fully plastic load*. For a rectangular bending member with a single edge notch, the *fully plastic moment* of Eq. 13.20 can be applied to the cross section of depth $(b-a)$ remaining after removal of the notch. Substituting $c = (b-a)/2$ gives

$$M_o = \frac{(b-a)^2 t}{4}\sigma_o \qquad (13.73)$$

Analogous equations are easily developed for other cases of pure bending.

If there is combined bending and tension, as due to an eccentric load, unsymmetrical stress distributions are needed as in Figs. 13.17(c) and (d). The analysis given above of eccentric loading, Fig. 13.8 and Eq. 13.33, applies to such cases if the member involved has constant thickness. For example, for an edge-notched plate with loaded faces as in Fig. 13.17(c), substitute $y_e = (a+b)/2$ and $c = (b-a)/2$ into Eq. 13.33 to obtain the fully plastic load.

$$P_o = bt\sigma_o\left[-\alpha - 1 + \sqrt{2\left(1 + \alpha^2\right)}\right] \qquad (13.74)$$

where $\alpha = a/b$. Even if the load is centered, single edge notch creates a bending situation as in Fig. 13.17(d). In this case, the geometry corresponds to $y_e = a/2$ and $c = (b-a)/2$, so that the fully plastic load from Eq. 13.33 is

$$P_o = bt\sigma_o\left[-\alpha + \sqrt{2\alpha^2 - 2\alpha + 1}\right] \qquad (13.75)$$

where again $\alpha = a/b$.

Equations 13.72 through 13.75 are also useful for obtaining fully plastic loads for cracked members, where in this case the dimension a is interpreted as crack length. Note that all of these equations are included in Fig. 8.45 and were used in Chapters 8 and 11 for cracked bodies. Other configurations of loading and geometry can be similarly analyzed to estimate fully plastic loads for notched or cracked members. As already noted in Chapter 8, such estimates represent lower bounds due to the influences of strain hardening and constrained yielding.

13.5.3 Estimates of Notch Stress and Strain for Local Yielding

Few closed-form solutions exist for determining notch strains during plastic deformation. Numerical analysis, as by finite elements, can be used, but nonlinear elasto-plastic stress-strain relationships complicate such analysis and increase costs compared to linear-elastic analysis. Although nonlinear numerical analysis is sometimes necessary, various approximate methods for estimating notch stresses and strains have also been developed. Of these, *Neuber's rule* is the most widely used and will now be described.

Consider monotonic loading of a notched member having an elasto-plastic stress-strain curve as shown in Fig. 13.18(a). The maximum stress σ_{yn} and the corresponding strain ε_{yn} at the notch are of interest, where these will be denoted below simply as σ and ε. Once plastic deformation begins at the notch, the ratio of the notch stress to the nominal stress falls below the value k_t that applies for linear-elastic behavior. As already noted and illustrated in Fig. 13.14, strains show the opposite trend, exceeding values corresponding to elastic behavior. It therefore becomes necessary to define separate stress and strain concentration factors.

$$k_\sigma = \frac{\sigma}{S}, \quad k_\varepsilon = \frac{\varepsilon}{e} \tag{13.76}$$

where e is nominal strain, in particular the value from the material's stress-strain curve corresponding to S. The trends of these quantities with increasing notch strain are illustrated in Fig. 13.18(b).

Neuber's rule states simply that the geometric mean of the stress and strain concentration factors remains equal to k_t during plastic deformation.

$$\sqrt{k_\sigma k_\varepsilon} = k_t \tag{13.77}$$

For axial loading with bilateral symmetry, the approximately uniform stress distribution during fully plastic yielding causes both σ and S to have similar values. Hence, k_σ tends toward unity for large strains, so that the equation above suggests that k_ε is limited to the value k_t^2.

If fully plastic yielding does not occur, $e = S/E$ applies. This permits a useful equation for local yielding to be obtained from Neuber's rule by substituting Eq. 13.76 along with $e = S/E$ into Eq. 13.77. After simple manipulations we obtain

$$\sigma\varepsilon = \frac{(k_t S)^2}{E} \tag{13.78}$$

For a given material, geometry, and applied load, hence, for known E, k_t, and S, the product of stress and strain is thus a known constant. Since stress and strain are also related by the stress-strain curve of the material, a solution can be obtained for their values. Noting that $\sigma\varepsilon = constant$ is simply a hyperbola, a graphical solution is quite easy as illustrated in Fig. 13.18(a). If the local notch stress does not exceed the yield

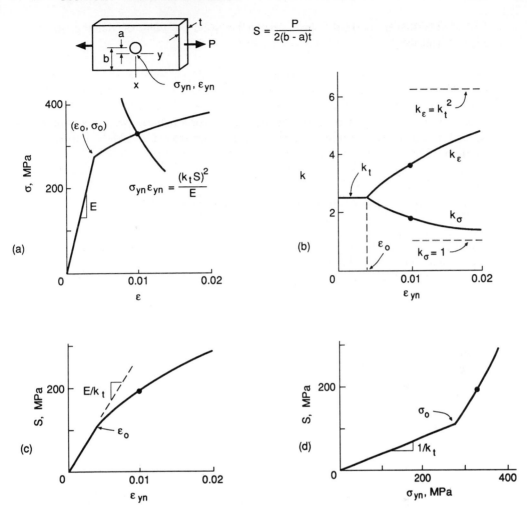

Figure 13.18 For a given notched member and stress-strain curve (a), Neuber's rule may be used to estimate local notch stresses and strains, σ and ε, corresponding to a particular value of nominal stress S. Stress and strain concentration factors vary as in (b), and S vs. ε and S vs. σ trends are shown in (c) and (d).

strength, Eq. 13.78 still gives the correct solution consistent with linear-elastic behavior. Use of Eq. 13.78 for a number of different values of S gives an S versus ε trend as in Fig. 13.18(c), and an S versus σ trend as in (d). Any points at a common S-value on these curves, such as the points marked by black circles, correspond to a hyperbola intersection as in (a), and also to particular values of k_σ and k_ε as in (b). Note that S is always proportional to the applied load, so that S can be thought of as simply a convenient way of representing the load. In contrast, both σ and ε are distinct variables and have nonlinear relationships with S.

For an elastic, perfectly plastic material beyond the point of yielding, strains are easily calculated by substituting $\sigma = \sigma_o$ into Eq. 13.78.

$$\varepsilon = \frac{(k_t S)^2}{\sigma_o E} \quad (\varepsilon \geq \sigma_o/E) \tag{13.79}$$

Useful closed-form equations may also be obtained for a material with power-hardening beyond yielding by substituting either stress or strain from Eq. 12.8(b) into Eq. 13.78 and solving for the other quantity.

$$\sigma = H_1 \left[\frac{(k_t S)^2}{E H_1} \right]^{\frac{n_1}{n_1+1}} , \quad \varepsilon = \left[\frac{(k_t S)^2}{E H_1} \right]^{\frac{1}{n_1+1}} \tag{13.80}$$

If a Ramberg-Osgood stress-strain curve is used, no closed form solution for stresses and strains is possible. Elimination of ε between Eq. 12.13 and 13.78 gives an equation involving stress σ that can be solved by trial and error or other numerical procedure.

$$(k_t S)^2 = \sigma^2 + \sigma E \left(\frac{\sigma}{H} \right)^{\frac{1}{n}} \tag{13.81}$$

Substitution of σ into Eq. 12.13 then gives ε.

13.5.4 Discussion

It is reasonable to use Neuber's rule in the form of Eq. 13.78 for all loads except those approaching fully plastic yielding. This is the case despite the fact that S values may exceed σ_o. Such a situation can arise, for example, in bending where fully plastic yielding does not occur until $S = 1.5\sigma_o$. (Substitute M_o from Eq. 13.73 into Eq. 13.70 to verify this.) Where nominal yielding does occur, a special version of Neuber's rule is sometimes used. See the paper by Seeger and Heuler (1980) in the References for details.

In the treatment above, plane stress was assumed, resulting in a uniaxial state of stress at the notch. However, if the notch radius ρ is small compared to the thickness t, the stress σ_{zn} of Fig. 13.16 will not be small. In fact, the situation near the notch for quite small ρ/t will approach plane strain, so that $\varepsilon_{zn} \approx 0$. A similar circumstance occurs in axisymmetric geometries, such as shafts, if ρ is small compared to the other dimensions of the shaft. Thus, it may be desirable to employ the plane strain assumption instead of plane stress.

Under plane strain, σ_{xn} is still zero due to the free surface, but now $\varepsilon_{zn} = 0$ applies. In this case, the equations of deformation plasticity theory, Eqs. 12.19 to 12.26, indicate that a tensile stress σ_{zn} develops as follows:

$$\sigma_{zn} = \tilde{\nu}\sigma_{yn} \tag{13.82}$$

where

$$\tilde{v} = \frac{v\bar{\sigma} + 0.5E\bar{\varepsilon}_p}{E\bar{\varepsilon}} \tag{13.83}$$

The quantity $\tilde{v}$ can be thought of as a *generalized Poisson's ratio*. Its value approaches the elastic Poissons ratio v if the plastic strains are small, increasing smoothly to 0.5 for large plastic strains. A stress-strain curve for plane strain for use with Neuber's rule may be obtained by picking a number of points on the uniaxial (same as effective) stress-strain curve, $\bar{\varepsilon} = f(\bar{\sigma})$. For each such point, calculate $\tilde{v}$ from Eq. 13.83. Since $\sigma_{xn} = 0$, we have plane stress in the y-z plane with $\lambda = \tilde{v}$. Substituting this λ into Eq. 12.39, or Eq. 12.42 for the Ramberg-Osgood form, thus gives a point $(\varepsilon_1, \sigma_1) = (\varepsilon_{yn}, \sigma_{yn})$ on the desired stress-strain curve. A number of such points provide a curve for use with Neuber's rule.

Neuber's rule may also be used for shear stresses and strains, as in torsionally loaded notched shafts. Equation 13.78 applies with σ and ε being replaced with τ and γ, and k_t and S being defined as appropriate for the particular case. For more complex situations, such as combined bending and torsion, Neuber's rule may still be used as described in the papers of Hoffmann and Seeger (1985) in the References.

It is important to keep in mind in all uses of Neuber's rule that it is an approximation. Strains so estimated generally tend to be reasonably accurate or somewhat larger than those from more accurate nonlinear numerical analysis, or from careful measurements. Where a numerical analysis is done based on linear-elastic material behavior, stresses at notches are not the correct values if they exceed the yield strength. However, these fictitious stresses can be treated as values of $k_t S$ and used with Neuber's rule to estimate the stress and strain as affected by yielding. Another approximate nonlinear numerical procedure that can be used in a manner similar to Neuber's rule is the *strain energy density method* as described by Glinka (1985).

13.5.5 Residual Stresses at Notches

If a notched member is loaded sufficiently for local yielding to occur, and then the load removed, residual stresses will remain. This is illustrated for an overload in tension in Fig. 13.19. As the nominal stress S is increased to (a) and then to (b), local yielding occurs at the notch. When the load is removed (c), a tensile residual strain and a compressive residual stress remain as shown. Stress distributions are also shown for the three situations (a), (b), and (c). If the overload is compressive, analogous behavior occurs with a tensile residual stress resulting.

This behavior is similar to that already discussed for bending. As before, the elastic recovery of the material upon unloading results in the most intensely deformed regions having a residual stress that is opposite in sign to the peak stress previously reached. Since equilibrium of forces requires that the integral sum of the stresses must be zero after the load is removed, some interior regions retain stresses of the same sign as the overload.

Consider the case of an elastic, perfectly plastic material, and assume that local yielding occurs due to an applied nominal stress S'. The local notch stress that occurs

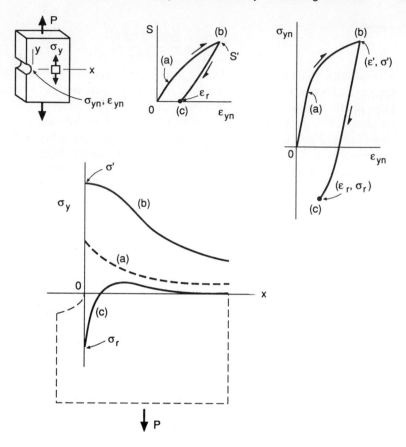

Figure 13.19   Residual stress and strain remaining after local yielding in a notched member. For loading to (a) there is no yielding, but at (b) there is, causing the stress distribution to flatten. After unloading (c), a tensile residual strain and a compressive residual stress remain at the notch surface.

at S' is in this case limited to σ_o, and the corresponding strain can be estimated from Neuber's rule using Eq. 13.79.

$$\sigma' = \sigma_o, \quad \varepsilon' = \frac{\left(k_t S'\right)^2}{\sigma_o E} \tag{13.84}$$

On unloading, there may or may not be yielding. Stress-strain paths and stress distributions for the two cases were previously illustrated in Fig. 10.27.

   If there is no yielding on unloading, the residuals are the elastic changes during unloading added to the values reached prior to unloading.

$$\sigma_r = \sigma' - k_t S', \quad \varepsilon_r = \varepsilon' - \frac{k_t S'}{E} \tag{13.85}$$

where the changes are negative in sign. Substituting for σ' and ε' from Eq. 13.84 gives

$$\sigma_r = \sigma_o - k_t S', \qquad \varepsilon_r = \frac{k_t S'}{E} \left[\frac{k_t S'}{\sigma_o} - 1 \right] \tag{13.86}$$

However, if $k_t S'$ is greater that $2\sigma_o$, the value of σ_r from above is less than $-\sigma_o$, which is impossible. This indicates that compressive yielding occurs and Eq. 13.86 does not apply.

   If compressive yielding does occur on unloading, the residual stress is limited to $-\sigma_o$. The change in strain $\Delta\varepsilon$ can be estimated from Neuber's rule using Eq. 13.78 by noting that σ changes by $\Delta\sigma = -2\sigma_o$ while S decreases from S' to zero.

$$\Delta\varepsilon = \frac{\left(k_t S'\right)^2}{\Delta\sigma\, E} = -\frac{\left(k_t S'\right)^2}{2\sigma_o E} \tag{13.87}$$

Adding the (negative) changes in stress and strain to the values σ' and ε' reached prior to unloading gives the residuals.

$$\sigma_r = -\sigma_o, \qquad \varepsilon_r = \frac{\left(k_t S'\right)^2}{2\sigma_o E} \qquad \left(k_t S' \geq 2\sigma_o\right) \tag{13.88}$$

Noting that $k_t S' > \sigma_o$ is required for initial yielding, the residual stresses at the notch for prior loading in tension are always compressive, and the strain tensile, for either case. This is evident from study of Eqs. 13.86 and 13.88. For a compressive overload, the same analysis applies except for appropriate sign changes.

   The methodology just applied for elastic, perfectly plastic behavior can be extended to other types of stress-strain curves, for which qualitatively similar effects occur. The stress and strain σ' and ε' at the overload S' must satisfy both the stress-strain curve and Neuber's rule.

$$\varepsilon' = f(\sigma'), \qquad \sigma'\varepsilon' = \frac{\left(k_t S'\right)^2}{E} \tag{13.89}$$

Assume that the unloading stress-strain behavior follows the factor-of-two expanded path as in a spring-slider rheological model as discussed in Chapter 12, specifically Eq. 12.53. Applying this along with Neuber's rule to the changes $\Delta\sigma$ and $\Delta\varepsilon$ during unloading then requires that two equations be satisfied as follows:

$$\Delta\varepsilon = 2f\left(\frac{\Delta\sigma}{2}\right), \qquad \Delta\sigma\,\Delta\varepsilon = \frac{(k_t \Delta S)^2}{E} \tag{13.90}$$

where $\Delta S = S'$ is the change in S during unloading. If both Eqs. 13.89 and 13.90 are solved, so that (ε', σ') and $(\Delta\varepsilon, \Delta\sigma)$ are known, the residuals can be obtained from

$$\sigma_r = \sigma' - \Delta\sigma, \qquad \varepsilon_r = \varepsilon' - \Delta\sigma \tag{13.91}$$

If a specific unloading stress-strain path (as in Fig. 12.13) is known for the material, this path can be used instead of the factor-of-two assumption.

Example 13.2

An axially loaded notched plate has a stress concentration factor of $k_t = 3$. It is made of a material having a Ramberg-Osgood stress-strain curve, Eq. 12.13, with constants of $E = 69$ GPa, $H = 690$ MPa, and $n = 0.15$.

(a) Estimate the local notch stress and strain, σ_{yn} and ε_{yn}, caused by a nominal stress of $S = 200$ MPa.

(b) Estimate the residual stress and strain remaining after removal of the load.

Solution (a) Both Neuber's rule, Eq. 13.78, and the stress-strain curve must be satisfied.

$$\varepsilon = f(\sigma) = \frac{\sigma}{E} + \left(\frac{\sigma}{H}\right)^{\frac{1}{n}} = \frac{\sigma}{69,000} + \left(\frac{\sigma}{690}\right)^{\frac{1}{0.15}}$$

$$\sigma\varepsilon = \frac{(k_t S)^2}{E} = \frac{(3 \times 200)^2}{69,000} = 5.217 \text{ MPa}$$

A graphical or numerical solution is needed to find σ and ε. For a graphical solution, the two equations above are plotted, and their intersection gives the desired values of σ and ε as shown below.

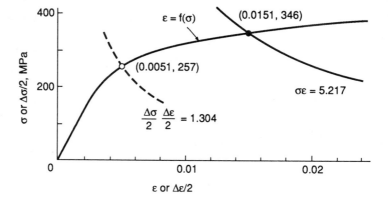

Figure E13.2

Accurate values from a numerical solution using Newton's method with Eq. 13.81 are

$$\sigma = 346.2 \text{ MPa}, \qquad \varepsilon = 0.01507 \qquad \qquad \textbf{Ans.}$$

(b) Since a specific unloading stress-strain path for the material is not known, we will assume that the factor-of-two expansion is followed. Applying both this and Neuber's rule to the unloading event with $\Delta S = S' = 200$ MPa gives two equations that must be satisfied.

$$\frac{\Delta\varepsilon}{2} = f\left(\frac{\Delta\sigma}{2}\right) = \frac{\Delta\sigma}{2(69,000)} + \left(\frac{\Delta\sigma}{2 \times 690}\right)^{\frac{1}{0.15}}$$

$$\Delta\sigma \, \Delta\varepsilon = \frac{(k_t \Delta S)^2}{E} = \frac{(3 \times 200)^2}{69,000} = 5.217 \text{ MPa}$$

Again, a graphical or numerical solution is needed, with the result being

$$\Delta\sigma = 513.1 \text{ MPa}, \quad \Delta\varepsilon = 0.01017$$

The residual values are thus

$$\sigma_r = \sigma' - \Delta\sigma = -166.9 \text{ MPa}, \quad \varepsilon_r = \varepsilon' - \Delta\varepsilon = 0.00490 \qquad \textbf{Ans.}$$

where the σ' and ε' values used are σ and ε from (a). The estimated stress-strain path is similar to the one in Fig. 13.19. This can be plotted using $\varepsilon = f(\sigma)$ and $\Delta\varepsilon = 2f(\Delta\sigma/2)$ if desired.

Discussion For a graphical solution to the determination of $\Delta\sigma$ and $\Delta\varepsilon$, the effort can be minimized by working with half-ranges. The first equation needed, $\Delta\varepsilon/2 = f(\Delta\sigma/2)$, is the same curve $\varepsilon = f(\sigma)$ as already plotted above, and the Neuber's rule equation needed is

$$\frac{\Delta\sigma}{2}\frac{\Delta\varepsilon}{2} = \frac{\left(k_t \dfrac{\Delta S}{2}\right)^2}{E} = \frac{5.217}{4} = 1.304$$

The intersection point is $(\Delta\varepsilon/2, \Delta\sigma/2)$ as shown by the dashed line above, and these values are of course doubled to obtain $\Delta\varepsilon$ and $\Delta\sigma$.

13.6 CYCLIC LOADING

Analysis of stresses and strains for cyclic loading is needed for dealing with such engineering situations as vibratory loading, earthquake loading, and fatigue. The preceding analysis of plastic deformation can be extended to cyclic loading by idealizing the stress-strain behavior of the material according to the spring and slider rheological models of the previous chapter. In particular, cycle-dependent hardening or softening and creep-relaxation behavior are assumed to be absent. Such an idealized material has identical cyclic and monotonic stress-strain curves, $\varepsilon = f(\sigma)$, and it has hysteresis loop curves that obey the factor-of-two assumption, $\Delta\varepsilon/2 = f(\Delta\sigma/2)$, which is Eq. 12.53. In applications, the stable cyclic stress-strain curve is used for $\varepsilon = f(\sigma)$. Both constant amplitude cyclic loading and irregular variation of load with time can be considered on this basis.

13.6.1 Bending

To develop a methodology for dealing with cyclic loading, first consider the example of cyclic loading of a beam due to loads that all lie in a plane of symmetry of the beam. For simplicity, assume that the cross section is symmetrical about the neutral axis. Let the moment vary cyclically between two values, M_{max} and M_{min}, as illustrated in Fig. 13.20. Assuming that acceleration effects are small, equilibrium of forces exists at any given time and requires that the integral of Eq. 13.6 must always be satisfied. Equation 13.6 therefore applies in the following form at times corresponding to M_{max} and M_{min}.

$$M_{max} = 2\int_0^c \sigma_{max} t y \, dy, \quad M_{min} = 2\int_0^c \sigma_{min} t y \, dy \qquad (13.92)$$

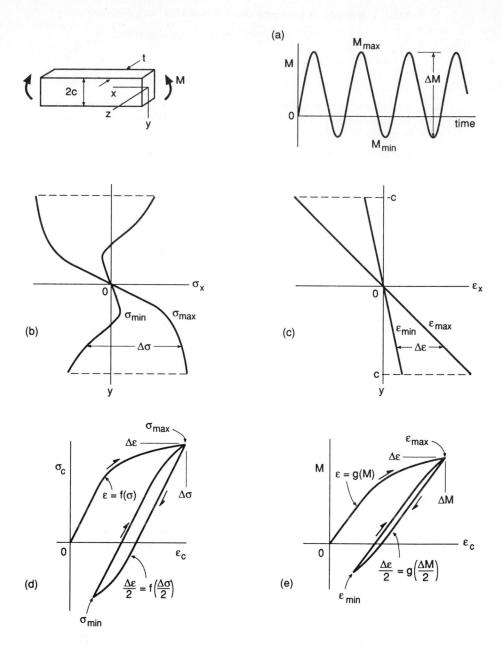

Figure 13.20 Beam subjected to a cyclic moment (a), which causes changes in the stress and strain distributions as in (b) and (c). The stress vs. strain behavior is illustrated in (d), and the moment vs. strain behavior in (e).

where σ_{max} is a function of y and constitutes the stress distribution that exists at a time corresponding to $M = M_{max}$, and similarly σ_{min} is the stress distribution at $M = M_{min}$. Such distributions are illustrated in Fig. 13.20(b).

Considering the range of moment ΔM, the properties of integrals permits the following simplification:

$$\Delta M = M_{max} - M_{min} = 2 \int_0^c \Delta \sigma \; t y \; dy \tag{13.93}$$

where $\Delta \sigma = \sigma_{max} - \sigma_{min}$ is the distribution of stress ranges with y. Since plane sections are expected to remain plane even during cyclic loading, the strain range at any distance y from the neutral axis is linearly related to that at the edge of the beam, as shown in Fig. 13.20(c).

Equation 13.93 states in effect that analysis of the changes in stress and strain beyond a point where the loading changes direction can be done in the same manner as analysis of monotonic loading. Actual values of stress and strain can be found by superimposing the changes on the values that existed when the direction changed.

$$\varepsilon_{min} = \varepsilon_{max} - \Delta\varepsilon, \qquad \sigma_{min} = \sigma_{max} - \Delta\sigma \tag{13.94}$$

where the above apply not only at the edge of the beam, but also to the entire distributions of these stresses and strains with y. The $(\varepsilon_{max}, \sigma_{max})$ values can be obtained from M_{max} by using the monotonic (same as cyclic) stress-strain curve in Eq. 13.6. For unloading from M_{max} to M_{min}, and for the subsequent reloading from M_{min} to M_{max}, the same $\Delta\sigma$ and $\Delta\varepsilon$ values occur as a result of Eq. 13.93 and $\Delta\varepsilon/2 = f(\Delta\sigma/2)$ applying for both events having the same ΔM. Hence, the stress and strain distributions when the moment reaches M_{max} the second time are estimated to be the same as those that previously existed. As a result of the stable behavior of the idealized material, subsequent cycles are exact repetitions of the first one.

To apply the above principles, it is convenient to divide both sides of Eq. 13.93 by two, so that both the moment and stress are expressed as cyclic amplitudes, $M_a = \Delta M/2$ and $\sigma_a = \Delta\sigma/2$.

$$M_a = 2 \int_0^c \sigma_a t y \; dy \tag{13.95}$$

Let the result of the integration of Eq. 13.6, applied for monotonic loading to M_{max}, be expressed as an explicit or implicit function giving the strain at the edge of the beam.

$$\varepsilon_{c\,max} = g\,(M_{max}) \tag{13.96}$$

Since the monotonic and cyclic stress-strain curves of the idealized material are the same, Eq. 13.95 gives the same functional dependence for the amplitudes of moment and edge strain as in Eq. 13.96 for monotonic loading, so that

$$\varepsilon_{ca} = g\,(M_a) \tag{13.97}$$

Stresses and strains due to cyclic loading can then be determined using Eqs. 13.96 and 13.97, along with plane sections remaining plane. In such analysis, note that the monotonic (same as cyclic) stress-strain curve $\varepsilon = f(\sigma)$ of the idealized material is being used for both the maximum and amplitude values.

$$\varepsilon_{\max} = f(\sigma_{\max}), \quad \varepsilon_a = f(\sigma_a) \tag{13.98}$$

For example, consider a rectangular beam made of an elastic, perfectly plastic material. The needed analytical result $\varepsilon = g(M)$ is provided by substituting $E\varepsilon_c = \sigma_c$ into Eq. 13.2 for straining below yielding, and for straining beyond yielding by solving Eq. 13.18 for ε_c, the strain at the edge of the beam.

$$\varepsilon_c = \frac{3M}{2tc^2E} \qquad (\varepsilon_c \le \sigma_o/E) \quad \text{(a)}$$

$$\tag{13.99}$$

$$\varepsilon_c = \frac{\sigma_o}{E}\sqrt{\frac{tc^2\sigma_o}{3\left(tc^2\sigma_o - M\right)}} \qquad (\varepsilon_c \ge \sigma_o/E) \quad \text{(b)}$$

Assuming that $M_{\max}$ is sufficient to cause yielding, the stress-strain curve and Eq. 13.99(b) give $\sigma_{\max}$ and $\varepsilon_{\max}$ at $y = c$.

$$\sigma_{c\,\max} = \sigma_o, \quad \varepsilon_{c\,\max} = \frac{\sigma_o}{E}\sqrt{\frac{M_o}{3\left(M_o - M_{\max}\right)}} \tag{13.100}$$

where the fully plastic moment $M_o = tc^2\sigma_o$ from Eq. 13.20 is introduced as a convenience. There may or may not be yielding during each half-cycle of loading. Considering either possibility and applying the stress-strain curve as $\varepsilon_a = f(\sigma_a)$ and Eq. 13.99 as $\varepsilon_{ca} = g(M_a)$ gives

$$\sigma_{ca} = E\varepsilon_{ca}, \quad \varepsilon_{ca} = \frac{3M_a}{2tc^2E} \qquad (\varepsilon_{ca} \le \sigma_o/E)$$

$$\tag{13.101}$$

$$\sigma_{ca} = \sigma_o, \quad \varepsilon_{ca} = \frac{\sigma_o}{E}\sqrt{\frac{M_o}{3\left(M_o - M_a\right)}} \qquad (\varepsilon_{ca} \ge \sigma_o/E)$$

The values of the stresses and strains at $M_{\min}$ are given by subtracting the changes from the maximum values, that is, by applying Eq. 13.94 and noting that $\Delta\varepsilon = 2\varepsilon_a$ and $\Delta\sigma = 2\sigma_a$.

13.6.2 Generalized Methodology for Other Cases

The procedure just described for analyzing cyclic bending uses the results of analysis performed in the same manner as for monotonic loading. Such a methodology is applicable for other geometries and modes of loading, relevant theoretical discussion and proofs from the viewpoint of plasticity theory being given in the papers by Mroz (1967, 1973). The following restrictions and comments apply: The same idealizations of the stress-strain behavior must be retained, namely, stable behavior always following the cyclic stress-strain curve and hysteresis loop curves obeying $\Delta\varepsilon/2 = f(\Delta\sigma/2)$. Also, if there are multiple applied loads, they must not result in nonproportional loading in regions of yielding, that is, the ratios of the principal stresses must remain at least approximately constant. States of stress other than uniaxial can be handled by applying deformation plasticity theory to the cyclic stress-strain curve in the same manner as done in Chapter 12 for monotonic curves.

Consider the strain at some location of interest, such as the edge strain for a beam in bending, but now also include other cases, such as the surface shear strain in a circular shaft, or the strain at the notch surface in a notched member. Assume that this strain, denoted ε, can be related explicitly or implicitly to the applied load by stress-strain analysis that considers plastic deformation, perhaps using some of the equations given earlier in this chapter.

$$\varepsilon = g(S) \tag{13.102}$$

where the applied load, such as an axial load, bending moment, torque, pressure, or a combination of these, is represented generically as a nominal stress S.

Cyclic loading with a maximum value S_{max} and an amplitude $S_a = \Delta S/2$ can be analyzed using this $\varepsilon = g(S)$. Analysis to obtain $\varepsilon = g(S)$ is done just as for monotonic loading using a stress-strain curve $\varepsilon = f(\sigma)$, which is already adjusted to correspond to the particular state of stress involved. The specific curve used is the same as the stable (half-life) cyclic stress-strain curve for the material and state of stress of interest, so that $\varepsilon_a = f(\sigma_a)$ is the cyclic stress-strain curve. The monotonic-loading analytical result $\varepsilon = g(S)$ can then be used for cyclic loading to obtain peak values and cyclic variations of stresses and strains. Assuming that the mean load during cycling is positive (tensile), $R > -1$, we have

$$\varepsilon_{max} = g\,(S_{max}) = f\,(\sigma_{max}) \qquad \text{(a)}$$

$$\varepsilon_a = g\,(S_a) = f\,(\sigma_a) \qquad \text{(b)} \tag{13.103}$$

$$\frac{\Delta\varepsilon}{2} = g\left(\frac{\Delta S}{2}\right) = f\left(\frac{\Delta\sigma}{2}\right) \qquad \text{(c)}$$

where (b) and (c) are equivalent but are both given to indicate that either amplitudes or ranges can be obtained, where all amplitudes are half the corresponding ranges. Values at S_{min} are obtained by subtracting the change from the values at S_{max}.

$$\varepsilon_{min} = \varepsilon_{max} - \Delta\varepsilon, \qquad \sigma_{min} = \sigma_{max} - \Delta\sigma \tag{13.104}$$

The procedure represented by Eqs. 13.103 and 13.104, and the corresponding stress-strain and load strain paths, are illustrated for the bending example by Fig. 13.20(d) and (e).

If the loading is completely reversed ($R = -1$), the amplitude and maximum values are identical, so that Eqs. 13.103(a) and (b) give the same result. If the cyclic loading extends farther into the negative (compressive) direction than into tension, $R < -1$, Eq. 13.103(a) needs to be replaced by

$$\varepsilon_{min} = -g\,(-S_{min}) = -f\,(-\sigma_{min}) \tag{13.105}$$

Ranges are then added to these minimum values to compute maximum values, Eq. 13.104 still applying.

Consider a notched member and approximate analysis using Neuber's rule. If the stress-strain curve obeys the Ramberg-Osgood form, $\varepsilon = g(S)$ is given implicitly by Neuber's rule and the stress-strain curve, Eqs. 13.78 and 12.13. Recall that solving these equations is equivalent to intersecting hyperbolas with the stress-strain curve as in Example 13.2. Hence, we use either the graphical approach or a numerical solution of Eq. 13.81, substituting $S = S_{max}$ and solving for $\sigma = \sigma_{max}$ and $\varepsilon = \varepsilon_{max}$. Also, we substitute $S = S_a$ and solve for $\sigma = \sigma_a$ and $\varepsilon = \varepsilon_a$. If a more exact analysis is desired, $\varepsilon = g(S)$ could be obtained by finite element analysis of the plastic deformation, which would be performed just as for monotonic loading except for use of the cyclic stress-strain curve.

Example 13.3

A notched plate made of the AISI 4340 steel of Table 12.1 has an elastic stress concentration factor of $k_t = 2.8$. The nominal stress is cycled between $S_{max} = 750$ and $S_{min} = 50$ MPa. Assume that the stress-strain behavior can be approximated using the stable cyclic stress-strain curve and that the behavior is similar to a spring and slider rheological model. Then estimate the stress-strain response.

Solution Use Neuber's rule and the cyclic stress-strain curve as in Eq. 13.103 to estimate both the maximums and amplitudes of the local notch stress and strain. For the initial monotonic response, assumed to follow the cyclic stress-strain curve, we have

$$\varepsilon_{max} = f(\sigma_{max}) = \frac{\sigma_{max}}{E} + \left(\frac{\sigma_{max}}{H'}\right)^{\frac{1}{n'}} = \frac{\sigma_{max}}{207{,}000} + \left(\frac{\sigma_{max}}{1655}\right)^{\frac{1}{0.131}}$$

$$\sigma_{max}\varepsilon_{max} = \frac{(k_t S_{max})^2}{E} = \frac{(2.8 \times 750)^2}{207{,}000} = 21.30$$

Solving as in Ex. 13.2 gives

$$\sigma_{max} = 972 \text{ MPa}, \qquad \varepsilon_{max} = 0.02192 \qquad \textbf{Ans.}$$

For the cyclic loading, it is convenient to work in terms of amplitudes using Neuber's rule and the same stress-strain curve, but evaluate these for

$$S_a = \frac{S_{max} - S_{min}}{2} = 350 \text{ MPa}$$

The needed equations are

$$\varepsilon_a = f(\sigma_a) = \frac{\sigma_a}{E} + \left(\frac{\sigma_a}{H'}\right)^{\frac{1}{n'}} = \frac{\sigma_a}{207{,}000} + \left(\frac{\sigma_a}{1655}\right)^{\frac{1}{0.131}}$$

$$\sigma_a\varepsilon_a = \frac{(k_t S_a)^2}{E} = \frac{(2.8 \times 350)^2}{207{,}000} = 4.640$$

Solving these gives

$$\sigma_a = 755 \text{ MPa}, \qquad \varepsilon_a = 0.00615$$

Finally, the minimum values for cyclic loading are

$$\sigma_{min} = \sigma_{max} - 2\sigma_a = -538 \text{ MPa}$$

$$\varepsilon_{min} = \varepsilon_{max} - 2\varepsilon_a = 0.00962 \qquad \textbf{Ans.}$$

where cycle-dependent relaxation and creep are assumed not to occur.

The estimated stress-strain response is qualitatively similar to the one shown for bending in Fig. 13.20(d). It can be plotted if desired by using $\varepsilon = f(\sigma)$ for the monotonic portion and $\Delta\varepsilon = 2f(\Delta\sigma/2)$ for both branches of the hysteresis loop. (Look ahead to Example 14.3 to see the actual plot.)

13.6.3 Application to Irregular Load vs. Time Histories

Estimates of stress-strain response for cyclic loading, as just described, can be extended to irregular variations of load with time. A stress-strain analysis done as for monotonic loading, but using the cyclic stress-strain curve, is of course still needed. This analysis can then be applied in the form $\Delta\varepsilon/2 = g(\Delta S/2)$ during irregular loading histories, specifically to all load excursions that correspond to stress-strain paths following the curve shape $\Delta\varepsilon/2 = f(\Delta\sigma/2)$.

Consider the example of Fig. 13.21. A notched member as in (a) is made of a material having a cyclic stress-strain curve, $\varepsilon_a = f(\sigma_a)$, as shown in (b). The load-strain curve, $\varepsilon_a = g(S_a)$, from Neuber's rule for this case is also shown. The load history of (c) is repeatedly applied, resulting in the load versus notch strain response of (d) and the local notch stress-strain response of (e). As a convenience, the load history has already been ordered to start at the largest absolute value of load. Recalling the behavior of the rheological model from Chapter 12, this results in all subsequent stress-strain paths following $\Delta\varepsilon/2 = f(\Delta\sigma/2)$, with origins at their respective hysteresis loop tips.

Analysis proceeds as follows: First, the stress and strain for point A are found from S_A by applying Eq. 13.103(a).

$$\varepsilon_A = g(S_A) = f(\sigma_A) \qquad (13.106)$$

Then Eq. 13.103(c) is applied to load ranges from the given history as follows:

$$\frac{\Delta\varepsilon_{AB}}{2} = g\left(\frac{\Delta S_{AB}}{2}\right) = f\left(\frac{\Delta\sigma_{AB}}{2}\right), \qquad \frac{\Delta\varepsilon_{BC}}{2} = g\left(\frac{\Delta S_{BC}}{2}\right) = f\left(\frac{\Delta\sigma_{BC}}{2}\right) \qquad (13.107)$$

and similarly for the following additional ranges: ΔS_{AD}, ΔS_{DE}, ΔS_{EF}, ΔS_{FG}, and ΔS_{EH}. Events $\Delta S_{CB'}$, $\Delta S_{GF'}$, $\Delta S_{HE'}$, and $\Delta S_{DA'}$ do not need to be analyzed as the results are the same as for the other branches of the corresponding hysteresis loops. The results of this analysis are sufficient to establish stress and strain values for all peaks and valleys in the load history, and also all of the intervening σ-ε paths. Overall ranges similar to ΔS_{AD} can be analyzed by ignoring the included minor load excursion, in this case event B-C-B'. This is the case because the *memory effect* causes the stress-strain path beyond B' to be the same as it would have been if event B-C-B' had not occurred.

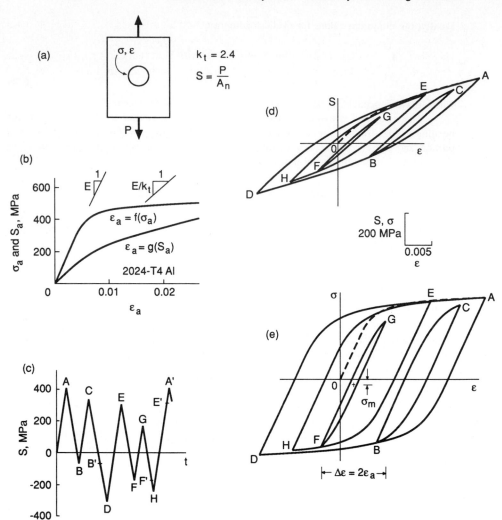

Figure 13.21 Analysis of a notched member subjected to an irregular load vs. time history. Notched member (a), having cyclic stress-strain and load-strain curves as in (b), is subjected to load history (c). The resulting load vs. notch strain response is shown in (d), and the local stress-strain response at the notch in (e). (Adapted from [Dowling 89]; copyright ©ASTM; reprinted with permission.)

However, the equations used do not apply to events such as ΔS_{CD}, since the stress-strain path involves portions of more than one hysteresis loop and so does not obey $\Delta\varepsilon/2 = f(\Delta\sigma/2)$.

The same logic can be applied to completely analyze load histories of any length, as long as the underlying assumptions of stable material behavior, and of at least approximately proportional loading, are satisfied. Beginning the analysis at the most extreme value of load is merely a convenience. Any starting point can be used if a more general logic based on spring and slider rheological models is employed.

Analysis of irregular load histories as just described is needed for making fatigue life estimates using the strain-based approach, which is considered in the next chapter.

13.6.4 Discussion

The methodology just described for cyclic loading is consistent with the analysis of residual stresses presented earlier. Consider starting from zero and applying a particular load, as described by a nominal stress S', and then returning to zero. Analysis of this event is identical to that for the first cycle of constant amplitude loading with $S_{max} = S'$ and $S_a = S'/2$, the resulting values of σ_{min} and ε_{min} being the residual stress and strain.

The procedure described clearly involves idealized materials behavior. Cycle-dependent hardening or softening is at least roughly included by the use of the stable cyclic stress-strain curve. However, the approach does not include cycle-dependent creep-relaxation or the details of the hardening or softening behavior. The degree of approximation involved with omitting these transient behaviors may in a few cases be a problem. The ease of application of the approach described nevertheless makes it a useful engineering tool, as in its use for the strain-based approach to fatigue in Chapter 14.

More serious limitations may be encountered if multiple applied loads cause non-proportional loading during plastic deformation. Time-dependent creep-relaxation behavior, as for metals at high temperature, is of course also beyond the scope of the method described. If such complexities need to be analyzed, a number of computer programs are available that incorporate sophisticated models of material behavior into analysis by finite elements. (See Dhalla and Gallagher (1982) for a guide to some of these.) However, caution is needed to assure that the stress-strain modeling is appropriate and specific to the material and situation of interest. Computer programs sometimes employ stress-strain models that are mathematically convenient but which poorly represent the behavior of real materials.

13.7 SUMMARY

For symmetrical bending of beams, and for torsion of solid or hollow circular shafts, stresses and strains can be readily analyzed for situations where plastic deformation occurs. Plane sections are assumed to remain plane, and an integral must be evaluated for the particular form of stress-strain curve of interest. Analysis of this type can be extended to combined axial and bending loads and to estimating residual stresses and strains. For members containing notches, stresses and strains at the notch are often estimated using Neuber's rule in the form of Eq. 13.78. Satisfying both Neuber's rule and the particular stress-strain curve of interest provides a solution. Also, this procedure may be extended to estimating residual stresses and strains at notches.

Plastic deformation in bending or torsion, and also localized plastic deformation at notches, causes strains to occur that exceed those from elastic analysis for the given loads. Large strains develop as yielding spreads over the entire cross section, which situation is called fully plastic yielding. Estimates of fully plastic loads may be made by assuming perfectly plastic behavior over the entire cross section. For notched members, this is done using the area remaining after subtraction of the notch.

Some specific analytical results presented are listed in Table 13.1, and additional cases are suggested as exercises.

Cyclic loading is readily analyzed if the behavior of the material is idealized to follow the stable (half-life) cyclic stress-strain curve. For this curve in the form $\varepsilon = f(\sigma)$, hysteresis loop curves can be approximated as obeying $\Delta\varepsilon/2 = f(\Delta\sigma/2)$. A useful expediency is to ignore the transient cycle-dependent stress-strain behavior, namely creep-relaxation and the progressive hardening or softening behavior.

On this basis, stress-strain analysis may be performed just as for monotonic loading except for the use of the stable cyclic stress-strain curve. For applied loading characterized by a nominal stress S that cycles between S_{max} and S_{min}, let the strain of interest be functionally related to S by an analytical result expressed in the form $\varepsilon = g(S)$. For cyclic loading that is completely reversed or which has a tensile (positive) mean level, $R \geq 1$, maximums and amplitudes (half-ranges) of stress and strain can be determined by applying this result as follows:

$$\varepsilon_{max} = g(S_{max}) = f(\sigma_{max}), \quad \varepsilon_a = g(S_a) = f(\sigma_a) \tag{13.108}$$

Minimums of stress and strain are obtained by subtracting ranges from maximum values. Similar analysis can also be applied for loading that is biased in the compressive direction, $R < -1$.

TABLE 13.1   CASES ANALYZED

Case	Equation No.
Bending of Rectangular Beams	
(a) Simple power-hardening material	13.11
(b) Elastic, perfectly plastic material	13.18
(c) Ramberg-Osgood material	13.27
(d) Bending plus axial loading: fully plastic yielding	13.33, 13.34
(e) Residual stress-strain for case (b)	13.42, 13.43
Torsion of Solid Circular Shafts	
(f) Simple power-hardening material	13.63
(g) Elastic, perfectly plastic material	13.64
(h) Ramberg-Osgood material	13.65
Axial and Bending Loading of Notched Members	
(i) Initial yielding	13.69, 13.71
(j) Fully plastic yielding	13.72-13.75
(k) Local σ-ε: elastic, perfectly plastic material	13.79
(l) Local σ-ε: elastic, power-hardening material	13.80
(m) Local σ-ε: Ramberg-Osgood material	13.81
(n) Residual stress-strain for case (k)	13.86, 13.88

Analysis of cyclic loading can be extended to irregular variation of load with time by applying $\Delta\varepsilon/2 = g(\Delta S/2)$ to all load excursions that correspond to stress-strain paths following $\Delta\varepsilon/2 = f(\Delta\sigma/2)$. The memory effect is invoked at points where stress-strain hysteresis loops close, after which S vs. ε and σ vs. ε return to the paths established prior to starting the loop.

NEW TERMS AND SYMBOLS

finite element method

fully plastic load or moment: P_o, M_o

fully plastic yielding

generalized Poisson's ratio, $\tilde{\nu}$

initial yielding

initial yield load or moment: P_i, M_i

local notch stress and strain: σ_{yn}, ε_{yn}, etc.

local yielding

Neuber's rule

residual stress and strain, σ_r and ε_r

stress and strain concentration factors, k_σ and k_ε

REFERENCES

CRANDALL, S. H. and N. C. DAHL, eds. 1959 *An Introduction to the Mechanics of Solids*, McGraw-Hill, New York.

DHALLA, A. K. and R. H. GALLAGHER. 1982 "Computational Methods for Nonlinear Structural Analyses," *Pressure Vessels and Piping: Design Technology—A Decade of Progress*, Am. Soc. of Mechanical Engineers, New York, pp. 185–201.

GLINKA, G. 1985 "Energy Density Approach to Calculation of Inelastic Stress-Strain Near Notches and Cracks," *Engineering Fracture Mechanics*, Vol. 22, No. 3, pp. 485–508. See also Vol. 22, No. 5, pp. 839–854.

HOFFMANN, M. and T. SEEGER. 1985 "A Generalized Method for Estimating Multiaxial Elastic-Plastic Notch Stresses and Strains: Part 1, Theory; and Part II, Applications and General Discussion," *Jnl. of Engineering Materials and Technology*, ASME, Vol. 107, Oct. 1985, pp. 250–260.

MROZ, Z. 1967 "On the Description of Anisotropic Workhardening," *Jnl. of the Mechanics and Physics of Solids*, Vol. 15, May 1967, pp. 163–175.

MROZ, Z. 1973 "Boundary-Value Problems in Cyclic Plasticity," *Second Int. Conf. on Structural Mechanics in Reactor Technology*, W. Berlin, Germany, Vol. 6B, Part L, Paper L7/6.

REDDY, J. N. 1993 *An Introduction to the Finite Element Method*, 2nd ed., McGraw-Hill, New York.

RICE, R. C., ed. 1988 *Fatigue Design Handbook*, 2nd ed., SAE Pub. No. AE-10, Society of Automotive Engineers, Warrendale, Pa.

SEEGER, T. and P. HEULER. 1980 "Generalized Application of Neuber's Rule," *Jnl. of Testing and Evaluation*, ASTM, Vol. 8, No. 4, pp. 199–204.

SHAMES, I. H. and F. A. COZZARELLI. 1992 *Elastic and Inelastic Stress Analysis*, Prentice Hall, Englewood Cliffs, NJ.

SMITH, J. O. and O. M. SIDEBOTTOM. 1965 *Inelastic Behavior of Load-Carrying Members*, John Wiley, New York.

TIMOSHENKO, S. 1941 *Strength of Materials, Part II, Advanced Theory and Problems*, D. Van Nostrand, New York. (Also reprinted by R. E. Krieger, Malabar, Fl., 1983.)

PROBLEMS AND QUESTIONS

Section 13.2

13.1 A beam has a symmetrical diamond-shaped cross section as shown below, and the material obeys a simple power-hardening stress-strain curve, Eq. 13.9.
   (a) For pure bending with the z-axis as the neutral axis, derive an equation relating bending moment M and the strain ε_c at $y = c$.
   (b) Modify the beam deflection analysis of Ex. 13.1 to obtain an equation for the mid-span deflection for this new cross section.

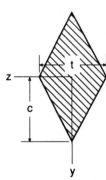

Figure P13.1

13.2 Proceed as in Prob. 13.1(a) and (b) except change the beam cross section to the one shown below.

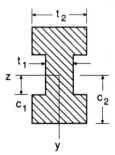

Figure P13.2

13.3 A rectangular beam of depth $2c$ and thickness t is made of a material that has a stress-strain curve of the form

$$\sigma = \sigma^* \sinh^{-1} \frac{\varepsilon}{\varepsilon^*}$$

where σ^* and ε^* are constants, with the initial tangent modulus to this stress-strain curve being $E_t = \sigma^*/\varepsilon^*$. Derive an equation relating bending moment M and edge strain ε_c for this case.

13.4 Verify Eq. 13.33 by directly applying equilibrium of forces and moments to the freebody diagram of Fig. 13.8(b).

13.5 A rectangular beam of depth $2c$ and thickness t is subjected to pure bending.
(a) Derive an equation relating moment M and edge strain ε_c for an elastic, linear-hardening stress-strain curve, Eq. 12.4.
(b) Does your solution reduce to the elastic case for incipient yielding at the beam edge? Does it reduce to Eq. 13.18 for $\delta = 0$?

13.6 A rectangular beam has depth $2c = 50$ and thickness $t = 25$ mm. It is made of a material having a Ramberg-Osgood stress-strain curve, Eq. 12.13, with $E = 69$ GPa, $H = 690$ MPa, and $n = 0.15$.
(a) Estimate and plot the curve relating moment M and edge strain ε_c out to $\varepsilon_c = 0.03$. (PC Problem)
(b) Does your result compare favorably with Fig. 13.7?

13.7 Aluminum alloy 7075-T651 has a monotonic stress-strain curve for uniaxial stress of the Ramberg-Osgood form, Eq. 12.13, with $E = 71$ GPa, $H = 586$ MPa, and $n = 0.0445$. Assume that a rectangular beam of this material, of depth $2c = 25$ and thickness $t = 12$ mm, is subjected to pure bending. (PC Problem)
(a) Make a plot of moment M versus edge strain ε_c, out to $\varepsilon_c = 0.04$, as predicted by Eq. 13.27.
(b) Approximate the material's stress-strain curve as an elastic, perfectly plastic one, with σ_o corresponding to the 0.2% offset yield strength. Then repeat (a) using this new σ-ε curve with Eq. 13.18, and comment on the comparison of the two results.

13.8 For a material with a simple power-hardening stress-strain curve, Eq. 13.9, derive an equation relating moment M and edge strain ε_c for a beam with a circular cross section under pure bending. (Suggestion: Consult the *definite integrals* section of a table of integrals, and note that a closed-form solution can be obtained in terms of the standard Gamma function, more specifically the Beta function.)

13.9 A simply supported beam has a rectangular cross section, and it is subjected to symmetrical pure bending due to a uniformly distributed load of w N/m. Derive an equation giving the maximum deflection in terms of w, beam dimensions, and material constants, where the material has a simple power-hardening stress-strain curve, Eq. 13.9.

13.10 Proceed as in Prob. 13.9, except change the beam to have one fixed end and a concentrated load P at the other end.

Section 13.3

13.11 Consider three identical rectangular beams, of depth $2c = 40$ and thickness $t = 20$ mm, loaded in symmetrical pure bending. All are made of an elastic, perfectly plastic material of yield strength $\sigma_o = 400$ MPa and elastic modulus $E = 200$ GPa. The first beam is loaded to an edge strain of $\varepsilon_c' = 0.002$ and then unloaded, the second to $\varepsilon_c' = 0.004$, and the third to $\varepsilon_c' = 0008$. For each of the three beams:
(a) Calculate the moment M' required to reach ε_c', and the residual stress and residual strain, σ_{rc} and ε_{rc}, remaining after unloading.

(b) Determine and plot the stress distribution, both at the maximum moment M' and after removal of M'.

(c) Comment on the differences among the three cases analyzed.

Section 13.4

13.12 Derive Eq. 13.64.

13.13 Derive Eq. 13.65, and show that it reduces to the correct elastic case for small plastic strains, and to Eq. 13.63 for large plastic strains.

13.14 For a simple power-hardening stress-strain curve, Eq. 13.62, and for a thick-walled tube with inner radius c_1 and other radius c_2:

(a) Derive an equation for torque T as a function of the maximum shear strain γ_c.

(b) Check your result to be sure that it reduces to Eq. 13.63 as c_1 approaches zero.

13.15 Aluminum alloy 2024-T351 has a uniaxial stress-strain curve for monotonic loading of Ramberg-Osgood form, Eq. 12.13, with $E = 73.1$ GPa, $H = 527$ MPa, and $n = 0.0663$, and also a value of Poisson's ratio of $v = 0.33$. (PC Problem)

(a) Estimate the shear stress-strain curve for this material.

(b) A solid round shaft of this material of radius $c = 9.53$ mm is loaded in torsion. Estimate and plot the curve relating torque T and surface shear strain out to $\gamma_c = 0.1$.

(c) For a gage length of $L = 152.4$ mm, corresponding torque versus angle of twist data are given below. Plot these points on the curve from (b) and comment on the comparison. (Hint: Angle of twist θ in radians is related to shear strain by $\theta = \gamma_c L / c$.)

T, N·m	θ, degrees
0	0
82.5	2
164	4
229	6
297	10
339	15
377	25
407	40
438	60
467	90

Section 13.5

13.16 The notched member of Fig. 13.18 has an elastic stress concentration factor of $k_t = 2.5$. Estimate the local notch stress and strain, σ_{yn} and ε_{yn}, and also k_σ and k_ε, for a nominal stress of $S = 192$ MPa. The material has an elastic, power-hardening stress-strain curve, Eq. 12.8, with $E = 69$ GPa, $H_1 = 834$ MPa, and $n_1 = 0.20$. Compare your results with the points marked with a black dot in Fig. 13.18.

13.17 For an elastic, linear-hardening material, Eq. 12.4, use Neuber's rule to derive equations for local notch stress and strain as functions of material constants, stress concentration factor k_t, and nominal stress S. Assume that yielding occurs at the notch. Does your result reduce to Eq. 13.79 for $\delta = 0$?

13.18 Aluminum alloy 7075-T651 has a monotonic stress-strain curve for uniaxial stress of the Ramberg-Osgood form, Eq. 12.13, with $E = 71$ GPa, $H = 586$ MPa, and $n = 0.0445$. A plate with a central round hole has dimensions, as defined in Fig. 10.1(a), of width $w = 76.2$, hole diameter $d = 19.05$, and thickness $t = 6.35$ mm. A strain gage was mounted in the hole to measure the strain ε_{yn}, with values being given below for various levels of axial load P, for monotonic loading of the plate. Estimate the P vs. ε_{yn} curve expected from Neuber's rule, plot this curve along with the test data, and comment on the comparison. (PC Problem)

P, Load, kN	ε_{yn}, Notch Strain
44.5	0.0042
66.7	0.0064
89.0	0.0087
111.2	0.0121
133.4	0.0174
155.7	0.0251
177.9	0.0366

13.19 Proceed as in Prob. 13.18, except approximate the stress-strain curve as an elastic, perfectly plastic one, with σ_o corresponding to the 0.2% offset yield strength.

13.20 A notched member has an elastic stress concentration factor of $k_t = 2.5$, and it is made of an elastic, perfectly plastic material having an elastic modulus of $E = 70$ GPa and a yield strength of $\sigma_o = 280$ MPa. Estimate residual stresses and strains for unloading after loading to nominal stresses of (a) $S = 100$, (b) $S = 175$, and (c) $S = 252$ MPa. In each case, plot the stress-strain path for loading and unloading.

13.21 For the situation of Prob. 13.18:
(a) Estimate σ_{yn} and ε_{yn} for $P = 111.2$ kN based on Neuber's rule.
(b) Then estimate the residual stress and strain at the notch that would be expected if the load were removed at this point.
(c) Plot the stress-strain path for (a) and (b). (PC Problem)
(d) Repeat (a), (b), and (c) for $P = 155.7$ kN.

Section 13.6

13.22 A notched plate of 2024-T351 Al is loaded in bending and has a stress concentration factor of $k_t = 2.5$. Estimate the local notch stress-strain response if the nominal stress is cycled between $S_{max} = 350$ and $S_{min} = -100$ MPa. The cyclic stress-strain curve is given by constants in Table 12.1.

13.23 Rectangular beams of RQC-100 steel have a depth of $2c = 6.35$ and a thickness of $t = 12.7$ mm. Five of these are subjected to completely reversed cyclic pure bending. Following cycle-dependent softening, the combinations of moment and edge strain for stable behavior in these tests are given below. Make a plot of M_a vs. ε_{ca} and compare these data with the curve from Eq. 13.27 applied to cyclic loading. The stable cyclic stress-strain curve of this steel in Ramberg-Osgood form is given by constants in Table 12.1. (PC Problem)

No.	M_a, Moment Amplitude, N·m	ε_{ca}, Edge Strain Amplitude
1	39.0	0.0022
2	46.4	0.0030
3	58.6	0.0050
4	68.9	0.0100
5	79.0	0.0180

Source: Data courtesy of R. W. Landgraf; see [Dowling 78].

14

Strain-Based
Approach to Fatigue

14.1 INTRODUCTION

The strain-based approach to fatigue considers the plastic deformation that may occur in localized regions where fatigue cracks begin, as at edges of beams and at stress raisers. Stresses and strains in such regions are analyzed and used as a basis for life estimates. Such a procedure permits detailed consideration of fatigue situations where local yielding is involved, which is often the case for ductile metals at relatively short lives. However, the approach also applies where there is little plasticity at long lives, so that it is a comprehensive approach that can be used in place of the stress-based approach.

The strain-based approach differs significantly from the stress-based approach, which is described in Chapters 9 and 10. Its features are highlighted in Fig. 14.1.

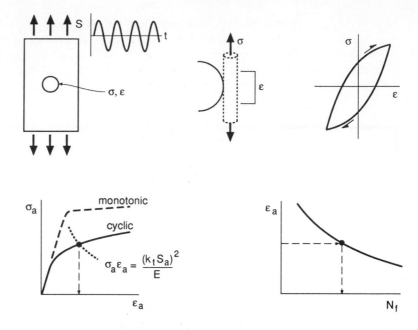

Figure 14.1 The strain-based approach to fatigue, in which local stress and strain, σ and ε, are estimated for the location where cracking is most likely. The effects of local yielding are included, and the material's cyclic stress-strain and strain-life curves from smooth axial test specimens are employed.

Recall that the stress-based approach emphasizes nominal (average) stresses rather than local stresses and strains, and that it employs elastic stress concentration factors and empirical modifications thereof. Employment of the *cyclic stress-strain curve* is a unique feature of the strain-based approach, as is the use of a *strain versus life curve*, instead of a nominal stress versus life (*S-N*) curve. A similarity between the two approaches is that neither includes specific analysis of crack growth as in the fracture mechanics approach of Chapter 11.

The strain-based approach was initially developed in the late 1950s and early 1960s in response to the need to analyze fatigue problems involving fairly short fatigue lives. The particular applications were nuclear reactors and jet engines, specifically cyclic loading associated with their operating cycles, especially cyclic thermal stresses. Subsequently, it became clear that the service loadings of many machines, vehicles, and structures include occasional severe events that can best be evaluated using a strain-based approach. One example is the loading on automotive suspension parts caused by potholes, high speed turns, or unusually rough roads. Another is the transient disturbance of electrical power systems, in some cases caused by lighting strikes, which can produce large mechanical vibrations in the turbines and generators of a power plant. Additional examples include loadings on aircraft due to gusts of wind in storms, and also loads due to combat maneuvers of fighter aircraft.

In this chapter, we will use the definitions and nomenclature given early in Chapter 9. Since the strain-based approach has some similarities to the stress-based approach, certain concepts employed will be related to those introduced in Chapters 9 and 10. Also, we will draw upon the information in Chapters 12 and 13 on plastic deformation. Of particular importance is the cyclic stress-strain curve, Eq. 12.58.

$$\varepsilon_a = \frac{\sigma_a}{E} + \left(\frac{\sigma_a}{H'}\right)^{\frac{1}{n'}} \tag{14.1}$$

Other material from Section 12.5 will be used, as will Neuber's rule from Section 13.5, and the general procedure for analyzing cyclic loads, Section 13.6. Therefore, while studying the present chapter, the reader should refer to these earlier chapters as needed for review.

14.2 STRAIN VERSUS LIFE CURVES

A strain versus life curve is a plot of strain amplitude versus cycles to failure. Such a curve is used by the strain-based approach in making life estimates in a manner analogous to the use of the S-N curve in the stress-based approach.

14.2.1 Strain-Life Tests and Equations

The test procedure generally used to obtain a strain-life curve is to apply completely reversed ($R = -1$) cycling between constant strain limits. Axial loading of an unnotched test specimen is most commonly employed as shown in Fig. 14.2. Strains are measured by an extensometer attached over a gage length within which the diameter is constant. The behavior of metal samples during such tests has already been discussed in Section 12.5 in connection with cyclic stress-strain curves. The same tests are simply continued until the specimen fails by fatigue, and results for several different strain amplitudes ε_a provide the desired curve. A schematic diagram and a curve fitted to actual data are given by the curves labeled *total* in Figs. 14.3 and 14.4, respectively. A log-log plot is usually used for strain-life curves.

Recall that the strain amplitude can be divided into elastic and plastic parts.

$$\varepsilon_a = \varepsilon_{ea} + \varepsilon_{pa} \tag{14.2}$$

where the elastic strain amplitude is related to the stress amplitude by $\varepsilon_{ea} = \sigma_a/E$. The plastic strain amplitude ε_{pa} is a measure of the width of the stress-strain hysteresis loop. (See Fig. 12.17.) It is also useful to plot ε_{ea} and ε_{pa} separately versus the number of cycles to failure N_f. As for the cyclic stress-strain curve, values from a hysteresis loop recorded near half of the fatigue life are usually used as being representative of

Figure 14.2 Test specimen, extensometer, and grips for strain-controlled fatigue testing. (Photo by G. K. McCauley, Virginia Tech.)

the approximately stable behavior after most cycle-dependent hardening or softening is complete. Thus, for a given test, the constant value ε_a and the stable values of ε_{ea} and ε_{pa} are plotted versus N_f, so that three plotted points are obtained on a strain-life plot as shown by the dashed line of Fig. 14.4.

If data from several tests are plotted, the elastic strains often give a straight line of shallow slope on a log-log plot, and the plastic strains a straight line of steeper slope. Equations can then be fitted to these lines.

$$\varepsilon_{ea} = \frac{\sigma_a}{E} = \frac{\sigma_f'}{E}\left(2N_f\right)^b, \quad \varepsilon_{pa} = \varepsilon_f'\left(2N_f\right)^c \quad \text{(a,b)} \qquad (14.3)$$

In the above, b and c are slopes on the log-log plot, assuming of course that the decades on the logarithmic scales in the two directions are equal. The intercept constants σ_f'/E

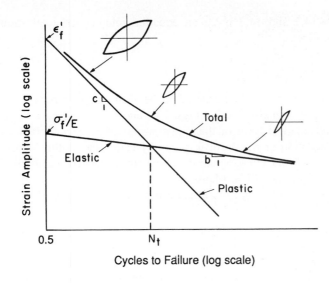

Figure 14.3 Elastic, plastic, and total strain vs. life curves. (Adapted from [Landgraf 70]; copyright ©ASTM; reprinted with permission.)

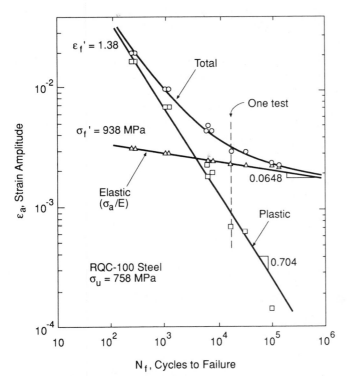

Figure 14.4 Strain vs. life curves for RQC-100 steel. For each of several tests, elastic, plastic, and total strain data points are plotted vs. life, and fitted lines are also shown. (From the author's data on the ASTM Committee E9 material.)

and ε'_f are by convention evaluated at $N_f = 0.5$, requiring use of the quantity $(2N_f)$ in the equations. The four constants needed are illustrated in Fig. 14.3.

Combining Equations 14.2 and 14.3 gives an equation relating the total strain amplitude ε_a and life.

$$\varepsilon_a = \frac{\sigma_f'}{E} \left(2N_f\right)^b + \varepsilon_f' \left(2N_f\right)^c \tag{14.4}$$

The quantities σ_f', b, ε_f', and c are considered to be material properties. This equation corresponds to the curves labeled *total* in Figs. 14.3 and 14.4. To obtain N_f for a given value of ε_a, the mathematical form of this equation requires either a graphical or numerical solution. An equation of this form is generally called the Coffin-Manson relationship, which name arises from the separate development of related equations in the late 1950s by both L. F. Coffin and S. S. Manson.

Note that Eq. 14.3(a) provides a stress-life relationship.

$$\sigma_a = \sigma_f' \left(2N_f\right)^b \tag{14.5}$$

Hence, if data over a wide range of lives are used to evaluate the strain-life constants, Eq. 14.4 includes a stress-life curve as its limiting case for small plastic strains. Equations 14.4 and 14.5 can thus be used up to quite long lives, at which point the slope b generally decreases, apparently approaching zero for materials with a distinct fatigue limit. (See Chapter 9.)

Strain-life data are available for a variety of engineering metals, as are values of the constants for the strain-life and cyclic stress-strain curves. Published collections of such information are referenced at the end of this chapter, and values of the constants for some representative metals are given in Table 14.1. Strain-life data are also available for some polymers, curves for two ductile polymers being shown in Fig. 14.5. Note that forms different than Eq. 14.4 would be needed to fit these data.

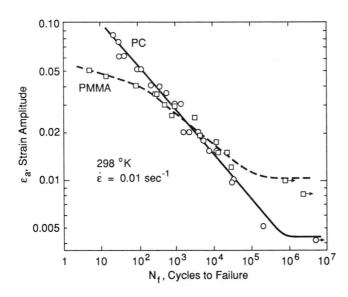

Figure 14.5 Strain-life curves for polycarbonate and polymethyl methacrylate. The shallow slope for PMMA at short lives is associated with crazing. (Data from [Beardmore 75].)

TABLE 14.1 CYCLIC STRESS-STRAIN AND STRAIN-LIFE CONSTANTS FOR SELECTED ENGINEERING METALS.[1]

Material	Source	Tensile Properties				Cyclic σ-ε Curve			Strain-Life Curve			
		σ_o	σ_u	$\tilde{\sigma}_{fB}$	% RA	E	H'	n'	σ_f'	b	ε_f'	c
(a) Steels												
SAE 1015 (normalized)	(8)	227 (33.0)	415 (60.2)	725 (105)	68	206,000 (29,900)	1058 (153)	0.24	976 (142)	−0.14	0.76	−0.59
Man-Ten[2] (hot rolled)	(7)	322 (46.7)	557 (80.8)	990 (144)	67	203,000 (29,500)	1096 (159)	0.187	1089 (158)	−0.115	0.912	−0.606
RQC-100 (roller Q & T)	(2)	683 (99.0)	758 (110)	1186 (172)	64	200,000 (29,000)	903 (131)	0.0905	938 (136)	−0.0648	1.38	−0.704
SAE 1045 (HR & norm.)	(6)	382 (55.4)	621 (90.1)	985 (143)	51	202,000 (29,400)	1258 (182)	0.208	948 (137)	−0.092	0.260	−0.445
SAE 4142 (As Q, 670 HB)	(1)	1619 (235)	2450 (355)	2580 (375)	6	200,000 (29,000)	2810 (407)	0.040	2550 (370)	−0.0778	0.0032	−0.436
SAE 4142 (Q & T, 560 HB)	(1)	1688 (245)	2240 (325)	2650 (385)	27	207,000 (30,000)	4140 (600)	0.126	3410 (494)	−0.121	0.0732	−0.805
SAE 4142 (Q & T, 450 HB)	(1)	1584 (230)	1757 (255)	1998 (290)	42	207,000 (30,000)	2080 (302)	0.093	1937 (281)	−0.0762	0.706	−0.869
SAE 4142 (Q & T, 380 HB)	(1)	1378 (200)	1413 (205)	1826 (265)	48	207,000 (30,000)	2210 (321)	0.133	2140 (311)	−0.0944	0.637	−0.761
AISI 4340[2] (Aircraft Qual.)	(3)	1103 (160)	1172 (170)	1634 (237)	56	207,000 (30,000)	1655 (240)	0.131	1758 (255)	−0.0977	2.12	−0.774
AISI 4340 (409 HB)	(1)	1371 (199)	1468 (213)	1557 (226)	38	200,000 (29,000)	1910 (277)	0.123	1879 (273)	−0.0859	0.640	−0.636
Ausformed H-11 (660 HB)	(1)	2030 (295)	2580 (375)	3170 (460)	33	207,000 (30,000)	3475 (504)	0.059	3810 (553)	−0.0928	0.0743	−0.7144
(b) Other Metals												
2024-T351 Al	(1)	379 (55.0)	455 (66.0)	558 (81.0)	25	73,100 (10,600)	662 (96.0)	0.070	927 (134)	−0.113	0.409	−0.713
2024-T4 Al[3] (Prestrained)	(4)	303 (44.0)	476 (69.0)	631 (91.5)	35	73,100 (10,600)	738 (107)	0.080	1294 (188)	−0.142	0.327	−0.645
7075-T6 Al	(5)	469 (68.0)	578 (84)	744 (108)	33	71,000 (10,300)	977 (142)	0.106	1466 (213)	−0.143	0.262	−0.619
Ti-6Al-4V (soln. tr. & age)	(1)	1185 (172)	1233 (179)	1717 (249)	41	117,000 (17,000)	1772 (257)	0.106	2030 (295)	−0.104	0.841	−0.688
Inconel X (Ni base, annl.)	(1)	703 (102)	1213 (176)	1309 (190)	20	214,000 (31,000)	1855 (269)	0.120	2255 (327)	−0.117	1.16	−0.749

Notes: [1]The tabulated values either have units of MPa (ksi), or they are dimensionless. [2]Test specimens prestrained except at short lives, also periodically overstrained at long lives. [3]For nonprestrained tests, use same constants except $\sigma_f' = 900(131)$ and $b = -0.102$.

Sources: Data in (1) [Conle 84]; (2) author's data on the ASTM Committee E9 material; (3) [Dowling 73]; (4) [Dowling 89] and [Topper 70]; (5) [Endo 69] and [Raske 72]; (6) [Leese 85]; (7) [Wetzel 77] pp. 41 and 66; (8) [SAE 89].

Example 14.1

Data are given below for completely reversed strain-controlled fatigue tests on RQC-100 steel. The elastic modulus is $E = 200$ GPa, and σ_a and ε_{pa} are measured near $N_f/2$. Determine constants for the strain-life curve.

ε_a	σ_a, MPa	ε_{pa}	N_f, cycles
0.0202	631	0.01695	227
0.0100	574	0.00705	1,030
0.0045	505	0.00193	6,450
0.0030	472	0.00064	22,250
0.0023	455	0.00010	110,000

Source: The author's data on the ASTM Committee E9 material.

Solution   Two graphical or least-squares fits are needed to obtain the constants σ_f', b, ε_f', and c, one for Eq. 14.5, and one for Eq. 14.3(b). Taking logarithms of both sides of Eq. 14.5 gives

$$\log \sigma_a = b \log \left(2N_f\right) + \log \sigma_f'$$

This has the form

$$y = mx + d$$

where

$$y = \log \sigma_a \qquad \text{(dependent variable)}$$

$$x = \log \left(2N_f\right) \qquad \text{(independent variable)}$$

$$m = b, \qquad d = \log \sigma_f'$$

Performing a least-squares fit using the σ_a and N_f data above gives

$$m = b = -0.05524 \qquad \qquad \text{Ans.}$$

$$d = 2.9406, \qquad \sigma_f' = 10^d = 872 \text{ MPa} \qquad \text{Ans.}$$

To obtain the remaining two constants, take logarithms of both sides of Eq. 14.3(b).

$$\log \varepsilon_{pa} = c \log \left(2N_f\right) + \log \varepsilon_f'$$

Similarly, we use

$$y = mx + d$$

where in this case

$$y = \log \varepsilon_{pa}, \qquad x = \log \left(2N_f\right)$$

$$m = c, \qquad d = \log \varepsilon_f'$$

Performing a second least-squares fit, this time using the ε_{pa} and N_f data, gives

$$m = c = -0.8177 \qquad \textbf{Ans.}$$

$$d = 0.5164, \qquad \varepsilon'_f = 10^d = 3.284 \qquad \textbf{Ans.}$$

All of these constants could also be evaluated graphically in an approximate manner from straight lines drawn on log-log plots of σ_a vs. N_f and ε_{pa} vs. N_f. The procedure would be similar to that in Ex. 12.1, except that intercepts σ'_f and ε'_f occur at $N_f = 0.5$ due to the form of the equations.

Discussion   The constants obtained differ somewhat from those in Table 14.1 due to the use here of an abbreviated set of data. If desired, the constants obtained can be employed to plot the two straight lines and a total-strain curve on log-log coordinates, and these compared with the given data, in a manner similar to Fig. 14.4. The specific equations to be plotted are

$$\varepsilon_{ea} = \frac{\sigma_a}{E} = \frac{\sigma'_f}{E}\left(2N_f\right)^b = \frac{872}{200,000}\left(2N_f\right)^{-0.0552}$$

$$\varepsilon_{pa} = \varepsilon'_f\left(2N_f\right)^c = 3.28\left(2N_f\right)^{-0.818}$$

$$\varepsilon_a = \varepsilon_{ea} + \varepsilon_{pa} = 0.00436\left(2N_f\right)^{-0.0552} + 3.28\left(2N_f\right)^{-0.818}$$

14.2.2 Comments on Strain-Life Equations and Curves

At long lives, the first (elastic strain) term of Eq. 14.4 is dominant, as the plastic strains are relatively small, and the curve approaches the elastic strain line. This corresponds to a thin hysteresis loop as shown in Fig. 14.3. Conversely, at short lives, the plastic strains are large compared to the elastic strains, the curve approaches the plastic strain line, and the hysteresis loops are fat. At intermediate lives, near the crossing point of the elastic and plastic strain lines, the two types of strain are of similar magnitude. The crossing point is identified as the *transition fatigue life*, N_t. An equation relating N_t to the other constants can be obtained by using the substitution $\varepsilon_{ea} = \varepsilon_{pa}$ to combine Eqs. 14.3(a) and (b), which gives

$$N_t = \frac{1}{2}\left(\frac{\sigma'_f}{\varepsilon'_f E}\right)^{\frac{1}{c-b}} \tag{14.6}$$

For a given material, the transition fatigue life N_t locates a boundary between fatigue behavior involving substantial plasticity and behavior involving little plasticity. The value of N_t is thus the most logical point for separating low-cycle and high-cycle fatigue. Special analysis of plasticity effects by the strain-based approach may be needed if lives around or less than N_t are of interest. Conversely, the S-N approach, which is based primarily on elastic analysis, may suffice at lives longer than N_t.

From Chapter 12, the plastic strain term of the cyclic stress-strain curve gives

$$\sigma_a = H'\varepsilon_{pa}^{n'} \tag{14.7}$$

If N_f is eliminated between Eqs. 14.3(b) and 14.5, and the result compared with this equation, the constants for the strain-life curve can be related to those for the cyclic strain-life curve.

$$n' = \frac{b}{c}, \qquad H' = \frac{\sigma_f'}{\left(\varepsilon_f'\right)^{b/c}} \tag{14.8}$$

Thus, of the six constants H', n', σ_f', b, ε_f', and c, only four are independent. However, it is common practice to make three separate fits of data using Eqs. 14.3(b), 14.5, and 14.7, so that the above relationships among the constants are satisfied only approximately. In other words, reported values for the six constants may not be exactly mutually consistent according to Eqs. 14.8.

In a few cases, there may be fairly large inconsistencies among the six constants. This arises in situations where the data do not fit the assumed mathematical forms very well. In particular, data points of σ_a vs. N_f or ε_{pa} vs. N_f may depart somewhat from log-log straight lines. In such cases, it should be assured that the strain-life constants σ_f', b, ε_f', and c used for Eq. 14.4 still give a reasonable representation of the total strain data, ε_a vs. N_f. Also, the H' and n' values used should be the ones actually fitted to the σ_a vs. ε_{pa} data, not values calculated from Eq. 14.8.

At very short lives, and especially so for ductile materials, the strain may be sufficiently large that true stresses and strains, as defined in Section 5.5, differ significantly from the more usual engineering values. In such cases, σ_a, ε_{pa}, and ε_a in the above equations should be replaced by true stress and strain values, $\tilde{\sigma}_a$, $\tilde{\varepsilon}_{pa}$, $\tilde{\varepsilon}_a$. If this is done, these equations often give a reasonable representation of fatigue data over a wide range that includes very short lives. Based on this, and if a tension test is interpreted as a fatigue test where failure occurs at $N_f = 0.5$ cycles, then the intercept constants σ_f' and ε_f' should be the same as the true fracture stress and strain from a tension test, $\tilde{\sigma}_f$ and $\tilde{\varepsilon}_f$. Although values of σ_f' and ε_f' are best obtained from fitting actual fatigue data, there is often reasonable agreement with $\tilde{\sigma}_f$ and $\tilde{\varepsilon}_f$. Note that the convenience of such a direct comparison explains why σ_f' and ε_f' are defined as intercepts at $N_f = 0.5$.

14.2.3 Trends for Engineering Metals

The large amount of data available for engineering metals permits some generalizations and trends to be stated concerning strain-life curves for this class of materials. Details follow.

As explained just above, the intercept constants for the strain-life curve are expected to be similar to the true fracture stress and strain from a tension test.

$$\sigma_f' \approx \tilde{\sigma}_f, \qquad \varepsilon_f' \approx \tilde{\varepsilon}_f \tag{14.9}$$

A ductile metal has a high value of $\tilde{\varepsilon}_f$ and a low value of $\tilde{\sigma}_f$. Hence, the plastic strain line, as in Fig. 14.3, tends to be high, and the elastic strain line low. This results in a steep strain-life curve as illustrated in Fig. 14.6. Also, the transition fatigue life N_t will be relatively long. The opposite trend generally occurs for a strong but relatively brittle metal, for which a high $\tilde{\sigma}_f$ and a low $\tilde{\varepsilon}_f$ correspond to a flatter strain-life curve and a

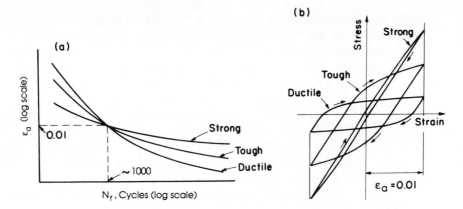

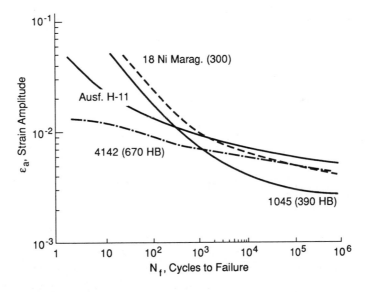

Figure 14.6 Trends in strain-life curves for strong, tough, and ductile metals. (Adapted from [Landgraf 70]; copyright ©ASTM; reprinted with permission.)

relatively low value of N_t. A *tough* material, which has intermediate values of both $\tilde{\sigma}_f$ and $\tilde{\varepsilon}_f$, tends to have a strain-life curve and N_t value between the two extremes. It is noteworthy that the strain-life curves for a wide variety of engineering metals tend to all pass near the strain $\varepsilon_a = 0.01$ for a life of $N_f = 1000$ cycles.

Strain-life curves for various steels that exhibit trends as just discussed are shown in Fig. 14.7. The variation of N_t with mechanical properties is illustrated by plotting its

Figure 14.7 Strain-life curves for four representative hardened steels. (Adapted from [Landgraf 68]; used with permission.)

value versus hardness for various steels in Fig. 14.8. Hardness of course varies inversely with ductility, so that N_t decreases as hardness is increased.

Some generalizations may also be made concerning the strain-life slope constants b and c. Values around $c = -0.6$ are common, and the relatively narrow range of $c = -0.5$ to -0.8 appears to include most engineering metals. A typical value of the elastic strain slope is $b = -0.085$. Relatively steep elastic slopes, around $b = -0.12$, are common for soft metals, such as annealed metals, and shallow slopes, nearer $b = -0.05$, are common for highly hardened metals. This trend contributes to the overall trend already noted of steep versus shallow strain-life curves for ductile versus brittle metals, respectively.

For steels with ultimate tensile strengths below about $\sigma_u = 1400$ MPa, recall from Chapter 10 that a fatigue limit occurs near 10^6 cycles at a stress amplitude around $\sigma_a = \sigma_u/2$. This establishes one point that must satisfy the stress-life relationship, Eq. 14.5. If the estimate $\sigma_f' \approx \tilde{\sigma}_f$ is also applied, Eq. 14.5 gives

$$b = -\frac{1}{6.3} \log \frac{2\tilde{\sigma}_f}{\sigma_u} \tag{14.10}$$

In other cases, where the fatigue limit (or long-life fatigue strength) at N_e cycles is given by $\sigma_a = m_e \sigma_u$, the estimate becomes

$$b = -\frac{\log \dfrac{\tilde{\sigma}_f}{m_e \sigma_u}}{\log (2N_e)} \tag{14.11}$$

where some approximate m_e and N_e values for various classes of engineering metals are given in Chapter 10. Higher ratios $\tilde{\sigma}_f/\sigma_u$ apply for more ductile metals, so that Eqs. 14.10 and 14.11 are consistent with the trends noted above for b.

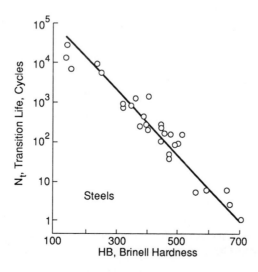

Figure 14.8 Transition fatigue life vs. hardness for a wide range of steels. (Adapted from [Landgraf 70]; copyright ©ASTM; reprinted with permission.)

14.2.4 Factors Affecting Strain-Life Curves; Surface Finish

If a hostile chemical environment or elevated temperature is present, smaller numbers of cycles to failure are expected, especially for lower frequencies where the environment has more time to act. At temperatures exceeding about one-half of the absolute melting temperature of a given material, nonlinear deformations due to time-dependent creep-relaxation behavior generally become significant. Strain-life and cyclic stress-strain curves then become dependent on test frequency. For example, such effects will occur at a sufficiently elevated temperature for any structural engineering metal, and they occur at room temperature for low-melting-temperature metals, such as lead and tin, and also for most polymers. Analysis of creep-relaxation and its effect on life, where this occurs in combination with cyclic loading, involves the problem area called *creep-fatigue interaction*. Some additional comments on this topic are given later in Chapter 15, and examples of recent research papers in the area can be found in Solomon et al. (1988).

In contrast to *S-N* curves at long lives, strain-life curves at relatively short lives are not highly sensitive to such factors as surface finish and residual stress. Residual stresses that are initially present are quickly removed by local yielding and cycle-dependent relaxation if cyclic plastic strains are present, and so these have only limited effect at lives around and below N_t. Surface finish is important in high-cycle fatigue because most of the life at the low stresses involved is spent initiating a crack. However, if significant plastic strains are present, a small crack (or crack-like damage) starts relatively early in the life, even if the surface is smooth. Most of the life is thus spent in growing small cracks into the material at some depth, where the surface finish cannot have an effect. The importance of crack growth effects at relatively short lives was previously illustrated by Fig. 9.17, and it is further illustrated by strain-life data in Fig. 14.9.

A reasonable method of modifying the strain-life curve to include the effect of surface finish is to change only the elastic slope b. Assume that all strain-life constants for smooth axial test specimens under zero mean stress are known, as well as the fatigue limit (or long-life fatigue strength) at N_e cycles, which is expressed in terms of the ultimate tensile strength as $\sigma_a = m_e \sigma_u$. The new value of b is then given by an expression similar to Eq. 14.11.

$$b = -\frac{\log \dfrac{\sigma_f'}{m_s m_e \sigma_u}}{\log (2N_e)} \tag{14.12}$$

where m_s is the surface finish factor from Chapter 10.

14.3 MEAN STRESS EFFECTS

Mean stress effects, as discussed in Chapters 9 and 10, need to be evaluated in applying the strain-based approach. In particular, the strain-life curve for completely reversed loading needs to be modified if a mean stress is present. It is useful to think of a family

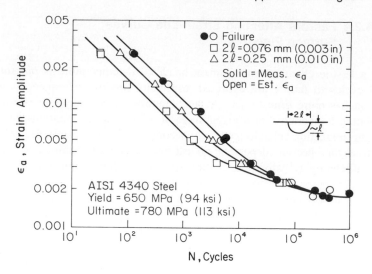

Figure 14.9 Strain-life curves for failure, and for two specific crack sizes, in an alloy steel. (Adapted from [Dowling 79a]; used with permission.)

of strain-life curves, where the particular one to be used depends on the mean stress. Test data illustrating this situation for an alloy steel are shown in Fig. 14.10.

14.3.1 Mean Stress Tests

For a cyclic strain test conducted with a nonzero mean strain, cycle-dependent relaxation of the mean stress is likely as described in Chapter 12 and illustrated by Fig. 12.25. If the plastic strain amplitude is not large, some of the mean stress is likely to remain. The life will then be affected by this mean stress. Note that mean strain by itself has little effect on fatigue life unless the value is quite large, that is, the effect is primarily due to whatever mean stress is present.

Alternatively, controlled stress tests can be run. In this case, there can be no relaxation of the mean stress, but cycle-dependent creep can occur. Large amounts of this type of deformation may accumulate and result in a failure similar to that from a tension test. However, the situation of primary interest here is plasticity in localized regions, in which large cyclic creep deformations are generally prevented by the stiffness of the surrounding elastic material. Hence, results from controlled stress tests are of present interest only if the failure is not dominated by cycle-dependent creep.

Regardless of the test procedure used, data can be obtained for evaluating the effect of mean stress on life. Discussion follows of various alternative methods of quantifying this mean stress effect.

14.3.2 Mean Stress Parameter of Morrow

The approach suggested by J. Morrow can be expressed as an equation giving the equivalent completely reversed stress amplitude, σ_{ar}, which is expected to produce the same

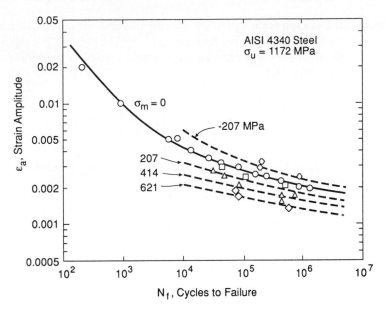

Figure 14.10 Mean stress effect on the strain-life curve of an alloy steel, with dashed curves from the mean stress parameter of Morrow. Most test specimens were overstrained prior to testing, and most with $N_f > 10^5$ cycles were also periodically overstrained. (Data from [Dowling 73].)

life as a given combination of amplitude σ_a and mean σ_m.

$$\sigma_{ar} = \frac{\sigma_a}{1 - \dfrac{\sigma_m}{\sigma_f'}} \qquad (14.13)$$

where the constant σ_f' is the same as for the stress-life curve. This expression is similar to Eq. 9.13, differing in that σ_f' replaces the ultimate tensile strength σ_u. Recall from Chapter 9 that the use of σ_u generally causes a conservative bias. Also, note that σ_f' is similar to the true fracture strength $\tilde{\sigma}_f$, and the latter is larger than σ_u except for brittle materials, where it is approximately the same. Compared to the use of σ_u, the above equation thus tends to be less conservative. In fact, the choice of σ_f' represents an attempt to obtain an accurate estimate, rather than a conservative one.

The stress-life equation given earlier can be generalized to include cases of nonzero mean stress by combining it with Eq. 14.13. To do so, σ_a in Eq. 14.5 needs to be replaced by σ_{ar}, as this equation applies only for zero mean stress.

$$\sigma_{ar} = \sigma_f' \left(2N_f\right)^b \qquad (14.14)$$

Hence, combining the above two equations and manipulating the result gives

$$\sigma_a = \left(\sigma_f' - \sigma_m\right) \left(2N_f\right)^b \qquad (14.15)$$

Note that this expression represents a family of stress-life curves, all parallel on a log-log plot, but with intercepts depending on σ_m.

The strain-life curve can also be generalized based on the Morrow parameter. First, we rearrange Eq. 14.15.

$$\sigma_a = \sigma_f' \left[\left(1 - \frac{\sigma_m}{\sigma_f'} \right)^{\frac{1}{b}} (2N_f) \right]^b \tag{14.16}$$

Comparing Eqs. 14.14 and 14.16, the effect of mean stress on life is such that N_f is replaced by the quantity N^*.

$$N^* = N_f \left(1 - \frac{\sigma_m}{\sigma_f'} \right)^{\frac{1}{b}} \tag{14.17}$$

This same modification must also apply to the strain-life curve, so that the desired family of strain-life curves is given by

$$\varepsilon_a = \frac{\sigma_f'}{E} (2N^*)^b + \varepsilon_f' (2N^*)^c \tag{14.18}$$

where the actual life, N_f, for a given combination of ε_a and σ_m is obtained from N^* using Eq. 14.17.

$$N_f = \frac{N^*}{\left(1 - \dfrac{\sigma_m}{\sigma_f'} \right)^{\frac{1}{b}}} \tag{14.19}$$

By substituting N^* into Eq. 14.18, a single equation for the family of strain-life curves can be obtained.

$$\varepsilon_a = \frac{\sigma_f'}{E} \left(1 - \frac{\sigma_m}{\sigma_f'} \right) (2N_f)^b + \varepsilon_f' \left(1 - \frac{\sigma_m}{\sigma_f'} \right)^{\frac{c}{b}} (2N_f)^c \tag{14.20}$$

This expression is similar to the original strain-life equation except that the intercept constants are in effect modified for any particular nonzero value of mean stress. It was used to plot the family of curves shown in Fig. 14.10.

Note that Eq. 14.18 is the same as the strain-life equation for $\sigma_m = 0$ except that N^* appears. The quantity N^* is thus the life calculated as if there were no mean stress effect, and the life N_f that is adjusted to include the effect of a mean stress can be obtained from this N^* by using Eq. 14.19.

These equations suggest a convenient graphical procedure for estimating life. First, plot the strain-life curve for zero mean stress. The life axis of course gives N_f directly for $\sigma_m = 0$. However, for nonzero mean stress, values read from the life axis are N^* values, which lead to the desired N_f values by the use of Eq. 14.19. The mean stress data of Fig. 14.10 are plotted on this basis in Fig. 14.11. The success of the Morrow parameter for this set of data is quite good, this being judged by the agreement of the data points for nonzero mean stress with the curve.

14.3.3 Modified Version of the Morrow Approach

The following modification of Eq. 14.20 is often used:

$$\varepsilon_a = \frac{\sigma'_f}{E} \left(1 - \frac{\sigma_m}{\sigma'_f} \right) (2N_f)^b + \varepsilon'_f (2N_f)^c \qquad (14.21)$$

The first (elastic strain) term is the same, but the mean stress dependence has been removed from the second (plastic strain) term. This has the effect of reducing the estimated effect of mean stress at relatively short lives. Members of the family of strain-life curves corresponding to Eq. 14.21 are compared with the alloy steel test data of Fig. 14.10 in Fig. 14.12.

Equation 14.21 does not lend itself to any graphical solution other than plotting several members of the family of curves and interpolating between them as necessary. However, this equation can be readily solved numerically by a computer program or subroutine written to handle the original strain-life equation. It is simply necessary to use a coefficient for the elastic term that is altered by the mean stress as indicated.

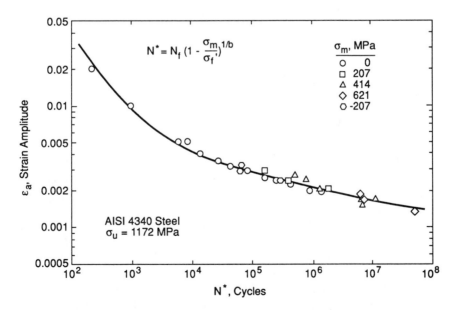

Figure 14.11   Mean stress data of Fig. 14.10 plotted vs. N^* according to the Morrow parameter.

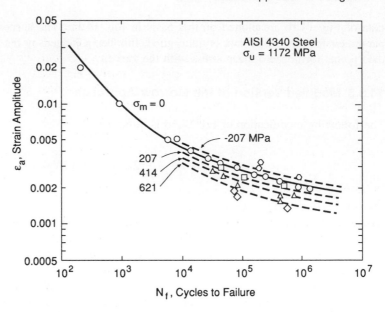

Figure 14.12 Family of strain-life curves given by the modified Morrow approach, and comparison with the data of Fig. 14.10.

14.3.4 Smith, Watson, and Topper (SWT) Parameter

This approach assumes that the life for any situation of mean stress depends on the product $\sigma_{\max}\varepsilon_a$, where by definition $\sigma_{\max} = \sigma_m + \sigma_a$.

$$\sigma_{\max}\varepsilon_a = h''\left(N_f\right) \tag{14.22}$$

Hence, the life is expected to be the same as for completely reversed ($\sigma_m = 0$) loading where this product has the same value. Let σ_{ar} and ε_{ar} be corresponding completely reversed stress and strain amplitudes, and note that for $\sigma_m = 0$ we have $\sigma_{\max} = \sigma_{ar}$. Eq. 14.22 thus becomes

$$\sigma_{\max}\varepsilon_a = \sigma_{ar}\varepsilon_{ar} \tag{14.23}$$

If the stress-life and strain-life curves for completely reversed loading are given by Eqs. 14.5 and 14.4, respectively, the function $h''(N_f)$ can be obtained by substituting these equations for σ_{ar} and ε_{ar}.

$$\sigma_{\max}\varepsilon_a = \sigma_f'\left(2N_f\right)^b\left[\frac{\sigma_f'}{E}\left(2N_f\right)^b + \varepsilon_f'\left(2N_f\right)^c\right] \tag{14.24}$$

which can be rearranged to obtain

$$\sigma_{\max}\varepsilon_a = \frac{\left(\sigma_f'\right)^2}{E}\left(2N_f\right)^{2b} + \sigma_f'\varepsilon_f'\left(2N_f\right)^{b+c} \tag{14.25}$$

A convenient graphical procedure is to make a plot of the quantity $\sigma_{max}\varepsilon_a$ versus N_f using Eqs. 14.25, which requires only the constants from $\sigma_m = 0$ test data. Then for any situation involving a nonzero mean stress, enter this plot with the value of the product $\sigma_{max}\varepsilon_a$ to obtain N_f. Such a plot for the alloy steel of Fig. 14.10 is shown in Fig. 14.13. The success of the SWT parameter for this particular case can be judged by the extent to which the data for nonzero mean stress agree with the curve.

Equation 14.25 can of course also be solved numerically. A computer program or subroutine written for handling the original strain-life equation can be used for this purpose by taking advantage of the similar mathematical forms of Eqs. 14.4 and 14.25.

14.3.5 Comments on Mean Stress Approaches

All three of the above approaches are in current use, and no consensus exists that any one of them is superior to the others. The unmodified Morrow approach seems to work reasonably well for steels, and in at least some cases gives better results than the SWT parameter. However, for some aluminum alloys, the true fracture strength $\tilde{\sigma}_f$ and the fitted stress-life constant σ_f' may differ significantly due to the stress-life data not fitting the form of Eq. 14.5 very well. In these cases, better agreement with test data is obtained by replacing the ratio σ_m/σ_f' in Eqs. 14.17 to 14.20 with $\sigma_m/\tilde{\sigma}_f$, while leaving σ_f' elsewhere in the equations.

One justification for using the modified Morrow approach is that the resulting reduced effect of σ_m at short lives may offset the modest bias in estimated life that arises

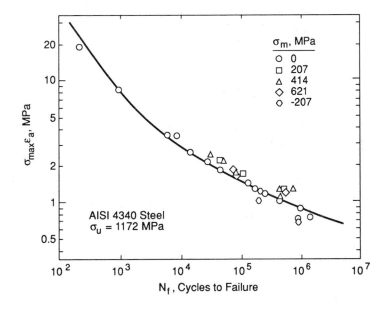

Figure 14.13   Plot of the Smith, Watson, and Topper parameter vs. life for the data of Fig. 14.10.

from neglecting the transient relaxation of mean stress in notched members. The SWT parameter appears to give good results for a wide range of materials and is a good choice for general use. However, it may give nonconservative estimates for compressive mean stresses. (Such a trend is apparent in Fig. 14.13.) Where refinement is desired, the choice among these parameters can only be based on fatigue test data involving mean stresses. Specific tests may need to be conducted on the material of interest, unless such data can be found in the literature for material that is at least similar.

Example 14.2

The RQC-100 steel of Table 14.1 is subjected to cycling with a strain amplitude of $\varepsilon_a = 0.004$ and a tensile mean stress fo $\sigma_m = 100$ MPa. How many cycles can be applied before fatigue cracking is expected?

First Solution   Three alternate methods of estimating life with a mean stress present are given, specifically Eqs. 14.18, 14.21, and 14.25. Let us proceed with the first of these, the Morrow parameter. Substituting the constants E, σ'_f, b, ε'_f and c from Table 14.1 for RQC-100 steel into Eq. 14.18, we have

$$\varepsilon_a = \frac{938}{200,000} (2N^*)^{-0.0648} + 1.38 (2N^*)^{-0.704}$$

Substituting the given $\varepsilon_a = 0.004$ and solving numerically for N^* gives

$$N^* = 8124 \text{ cycles}$$

(A similar but approximate value of N^* could be obtained by entering the total strain curve in Fig. 14.4 with $\varepsilon_a = 0.004$.) Since N^* does not include the effect of mean stress, its value must be substituted along with $\sigma_m = 100$ MPa into Eq. 14.19 to obtain an N_f value that does include this effect.

$$N_f = N^* \left(1 - \frac{\sigma_m}{\sigma'_f}\right)^{-\frac{1}{b}} = 8124 \left(1 - \frac{100}{938}\right)^{\frac{1}{0.0648}} = 1426 \text{ cycles} \qquad \textbf{Ans.}$$

Second Solution   To apply the modified Morrow approach, simply substitute the same material constants into Eq. 14.21.

$$\varepsilon_a = \frac{938}{200,000} \left(1 - \frac{\sigma_m}{938}\right) (2N_f)^{-0.0648} + 1.38 \left(N_f\right)^{-0.704}$$

Substituting $\sigma_m = 100$ MPa and simplifying gives

$$\varepsilon_a = \frac{838}{200,000} (2N_f)^{-0.0648} + 1.38 \left(2N_f\right)^{-0.704}$$

We then enter this equation with $\varepsilon_a = 0.004$ and solve numerically for N_f.

$$N_f = 6597 \text{ cycles} \qquad \textbf{Ans.}$$

Third Solution   For the SWT approach, Eq. 14.25, we need the product $\sigma_{max}\varepsilon_a$. Thus, first apply the cyclic stress-strain curve with constants from Table 14.1 to obtain σ_a.

$$\varepsilon_a = \frac{\sigma_a}{E} + \left(\frac{\sigma_a}{H'}\right)^{\frac{1}{n'}} = \frac{\sigma_a}{200,000} + \left(\frac{\sigma_a}{903}\right)^{\frac{1}{0.0905}}$$

Entering this equation with $\varepsilon_a = 0.004$ and solving numerically for σ_a gives

$$\sigma_a = 501 \text{ MPa}, \qquad \sigma_{max} = \sigma_m + \sigma_a = 100 + 501 = 601 \text{ MPa}$$

$$\sigma_{max}\varepsilon_a = 601(0.004) = 2.404 \text{ MPa}$$

Next we substitute material constants into Eq. 14.25.

$$\sigma_{max}\varepsilon_a = \frac{(938)^2}{200,000}\left(2N_f\right)^{2(-0.0648)} + (938)(1.38)(2N_f)^{-0.0648-0.704}$$

$$\sigma_{max}\varepsilon_a = 4.399(2N_f)^{-0.1296} + 1294\left(2N_f\right)^{-0.7688}$$

Entering this equation with $\sigma_{max}\varepsilon_a$ from above and solving numerically yields N_f.

$$N_f = 5091 \text{ cycles} \qquad\qquad \textbf{Ans.}$$

Discussion Numerical solutions for N_f as required above can all be done using a single computer program. Since the same mathematical form is encountered in each case, it is simply necessary to use appropriate values of the coefficients and exponents that apply to the two terms involving $(2N_f)$. A simple trial and error approach will suffice, or Newton's method could be employed. For the latter, a good initial guess can be obtained for Eq. 14.18, 14.21, or 14.25 by the following procedure: (1) Calculate N_f with the first term discarded. (2) Calculate N_f with the second term discarded. (3) Take the larger value from (1) or (2) as the initial guess.

14.4 MULTIAXIAL STRESS EFFECTS

Fatigue under multiaxial loading where plastic deformations occur is currently an area of active research. Reasonable estimates are possible for relatively simple situations, but considerable uncertainty exists as to the best procedure for complex nonproportional loadings, where the ratios of the principal stresses change, and where the principal axes may also rotate. Given this situation, the discussion that follows first considers some simple but limited approaches. Then an introductory discussion is given of possible approaches for more complex loadings. Before proceeding, the reader may wish to review Section 9.8, which considers multiaxial stresses for the stress-based approach, and also Section 12.3 on stress-strain relationships for three dimensions.

14.4.1 Effective Strain Approaches

Consider situations where all cyclic loadings have the same frequency and are either in-phase or $180°$ out-of-phase. It is then reasonable to define an effective strain amplitude that is proportional to the cyclic amplitude of the octahedral shear strain.

$$\bar{\varepsilon}_a = \frac{\bar{\sigma}_a}{E} + \bar{\varepsilon}_{pa} \tag{14.26}$$

The quantities $\bar{\sigma}_a$ and $\bar{\varepsilon}_{pa}$ are obtained from Eqs. 12.21 and 12.22 by substituting amplitudes of the principal stresses and plastic strains. A negative sign is employed for amplitude quantities that are 180° out-of-phase with amplitudes selected as positive.

The fatigue life for multiaxial loading is postulated to depend on the value of this effective strain. For uniaxial loading

$$\bar{\varepsilon}_a = \varepsilon_{1a} \quad (\sigma_2 = \sigma_3 = 0) \tag{14.27}$$

The quantity ε_{1a} is simply the uniaxial strain amplitude, which is related to life by Eq. 14.4, the strain-life equation from uniaxial test data. Hence

$$\bar{\varepsilon}_a = \frac{\sigma_f'}{E} \left(2N_f\right)^b + \varepsilon_f' \left(2N_f\right)^c \tag{14.28}$$

where the first and second terms correspond to elastic and plastic components of the effective strain, so that

$$\bar{\sigma}_a = \sigma_f' \left(2N_f\right)^b, \quad \bar{\varepsilon}_{pa} = \varepsilon_f' \left(2N_f\right)^c \quad \text{(a,b)} \tag{14.29}$$

Consider the special case of plane stress

$$\sigma_{2a} = \lambda\sigma_{1a}, \quad \sigma_{3a} = 0, \quad \varepsilon_{1a} = \varepsilon_{e1a} + \varepsilon_{p1a} \tag{14.30}$$

where the notation of Chapter 12 is used except for the added subscript a to indicate amplitude quantities. Noting that the (x, y, z) axes are the principal $(1, 2, 3)$ axes, Equations 12.19, 12.25, and 12.34 can be combined with Eqs. 14.28 to 14.30 to obtain an equation for the strain-life curve in terms of the first principal strain.

$$\varepsilon_{1a} = \frac{\dfrac{\sigma_f'}{E} \left(1 - \nu\lambda\right) \left(2N_f\right)^b + \varepsilon_f' \left(1 - 0.5\lambda\right) \left(2N_f\right)^c}{\sqrt{1 - \lambda + \lambda^2}} \tag{14.31}$$

This equation can be used along with the Ramberg-Osgood type stress-strain curve for biaxial loading, Eq. 12.42, where all stresses and strains are interpreted as amplitude quantities.

For the special state of plane stress which is pure shear, Eqs. 13.49 and 13.50 apply and $\lambda = -1$. The equation above then reduces to

$$\gamma_{xya} = \frac{\sigma_f'}{\sqrt{3}G} \left(2N_f\right)^b + \sqrt{3}\varepsilon_f' \left(2N_f\right)^c \tag{14.32}$$

where γ_{xya} is the shear strain amplitude and G is the shear modulus. The corresponding Ramberg-Osgood type of stress-strain curve has already been derived as Eq. 13.53.

14.4.2 Extensions of the Effective Strain Approach

Reasonable results are given by equations similar to those above for combined loadings that are in-phase or 180° out-of-phase. However, some systematic trends have been noticed that have led to attempted improvements. In particular, the cyclic amplitude of the hydrostatic stress appears to have an additional effect not accounted for by the octahedral shear strain. This quantity is the average of the amplitudes of the principal stresses.

$$\sigma_{ha} = \frac{\sigma_{1a} + \sigma_{2a} + \sigma_{3a}}{3} \tag{14.33}$$

Some notable special cases are:

(a) Pure planar shear stress, $\lambda = -1$

$$\sigma_{ha} = 0 \quad (\sigma_{1a} = -\sigma_{2a}; \ \sigma_{3a} = 0) \tag{14.34}$$

(b) Uniaxial stress, $\lambda = 0$

$$\sigma_{ha} = \frac{\sigma_{1a}}{3} \quad (\sigma_{2a} = \sigma_{3a} = 0) \tag{14.35}$$

(c) Equal biaxial stresses, $\lambda = 1$

$$\sigma_{ha} = \frac{2\sigma_{1a}}{3} \quad (\sigma_{1a} = \sigma_{2a}; \ \sigma_{3a} = 0) \tag{14.36}$$

The specific trend noticed is that for a given value of $\bar{\varepsilon}_a$ the life is shorter for larger values of σ_{ha}.

As an example of an approach including this effect, Mowbray (1980) suggests that the controlling plastic strain variable is the modified effective strain quantity given by

$$\hat{\varepsilon}_{pa} = \left[\frac{\bar{\varepsilon}_{pa}}{1 - \mu \left(\dfrac{\sigma_{ha}}{\bar{\sigma}_a} \right)} \right] \left[\frac{3 - \mu}{3} \right] \tag{14.37}$$

where μ is an empirical constant that appears to have a value around 0.7 for various engineering metals. The first factor in the product above gives the dependence on σ_{ha}, specifically on its ratio to the effective stress amplitude, $\bar{\sigma}_a$. The second factor merely scales the value so that $\hat{\varepsilon}_{pa} = \varepsilon_{p1a}$ for the uniaxial case, that is, so that $\hat{\varepsilon}_{pa}$ is an equivalent uniaxial strain. Thus, $\hat{\varepsilon}_{pa}$ replaces $\bar{\varepsilon}_{pa}$ in the plastic strain versus life relationship, Eq. 14.29(b). For plane stress, this leads to a modification of Eq. 14.31, wherein ε_f'

is replaced by a quantity $\hat{\varepsilon}_f$, which is

$$\hat{\varepsilon}_f = \left[\frac{3\sqrt{1-\lambda+\lambda^2} - \mu(1+\lambda)}{(3-\mu)\sqrt{1-\lambda+\lambda^2}}\right]\varepsilon_f' \tag{14.38}$$

The effective strain approach can also be extended to handle mean stress effects. A logical approach would be to assume that the controlling mean stress variable is the noncyclic component of the hydrostatic stress as previously applied for the stress-based approach in Section 9.8. Equation 14.31, or the modified version with $\hat{\varepsilon}_f$, can then be generalized to include this effect in a manner parallel to the derivations above leading to Eqs. 14.18, 14.21, or 14.25.

14.4.3 Critical Plane Approaches

Effective stress or strain approaches as described in Chapter 9 and just above are difficult to apply to loading situations that are complicated by nonproportional loading. For such loading, where the principal stress axes rotate during cycling, as in Fig. 9.40, trends occur that cannot be explained except by a *critical plane approach*. In such an approach, stresses and strains during cyclic loading are determined for various orientations (planes) in the material, and the stresses and strains acting on the most severely loaded plane are used to predict fatigue failure.

Cracks virtually always have irregular shapes due to growth through the grain structure of the material. Thus, growth due to a shear stress alone tends to be difficult due to mechanical interlocking and friction effects involving irregularities on the faces of cracks as shown in Fig. 14.14. Stresses and strains normal to the crack plane may have a major effect on the behavior, accelerating the growth if they tend to open the crack. This situation has led to a number of proposals for critical plane approaches. For example, Fatemi and Socie (1988) suggest a relationship similar to

$$\gamma_{ac}\left(1 + \frac{\alpha\sigma_{\text{max}c}}{\sigma_o'}\right) = \frac{\tau_f'}{G}(2N_f)^b + \gamma_f'(2N_f)^c \tag{14.39}$$

(a) 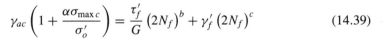 (b)

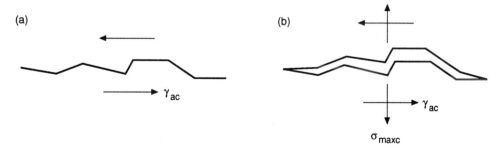

Figure 14.14   Crack under pure shear (a), where irregularities retard growth, compared to a situation (b), where a normal stress acts to open the crack, enhancing its growth. (Adapted from [Socie 87]; used with permission of ASME.)

where

γ_{ac} = largest amplitude of shear strain for any plane

$\sigma_{\max c}$ = peak tensile stress normal to the plane of γ_{ac}, occurring at any time during the

γ_{ac} cycle

α = additional empirical constant; $\alpha \approx 0.6$ to 1.0

σ'_o = cyclic yield strength

The quantities γ_{ac} and $\sigma_{\max c}$ are illustrated in Fig. 14.14. The constants τ'_f, b, γ'_f, and c give the strain-life curve from completely reversed tests in pure shear, specifically torsion tests on thin-walled tubes. If not known, these constants can be estimated from the ones from uniaxial loading using Eq. 14.32, or from a modified version of this based on Eq. 14.38. Implementation of this general type of an approach for complex variable amplitude loading requires considering the possibility of failure on a number of different planes.

An additional complexity is that there are two distinct modes of crack initiation and early growth, namely growth on planes of high shear stress (mode II) or growth in planes of high tensile stress (mode I). Shear cracking is most likely at high strains but may occur even at low strains for pure shear loading. Tensile cracking is most likely for equal biaxial stresses ($\lambda = 1$) but is also common for uniaxial loading. The occurrence of a given mode depends on the type of loading and the magnitude of the strain, and the details vary for different materials. Shear dominated cracking is addressed by Eq. 14.39 or other analogous relationship. A reasonable approach for tensile stress dominated cracking is to employ the Smith, Watson, and Topper parameter, Eq. 14.25. The quantity ε_a is interpreted as the largest amplitude of normal strain for any plane, and $\sigma_{\max}$ is the maximum normal stress on the same plane as ε_a, specifically the peak value during the ε_a cycle. The shortest life estimated from either the SWT parameter used in this way or Eq. 14.39 is the final life estimate. The necessity of making two calculations reflects the two possible modes of cracking.

No consensus currently exists as to the exact approach that should be used for complex situations of multiaxial loading. Various approaches are summarized and compared in the SAE publication edited by Leese and Socie (1989), and others are described in Kussmaul et al. (1991). For complex nonproportional loadings, determining the stresses and strains of interest for critical-plane fatigue life estimates for an engineering component will generally require the use of fairly sophisticated analysis based on incremental plasticity theory.

14.5 LIFE ESTIMATES FOR STRUCTURAL COMPONENTS

So far in this chapter we have considered only relationships between stresses and strains and life. To use these for given cyclic loadings on a structural component, such as a beam, a shaft, or a notched member, stresses and strains need to be first determined from applied loads. Thus, stress-strain analysis from Chapter 13 needs to be combined with the fatigue life relationships from the earlier portions of this chapter. Sections 13.5 and 13.6, which respectively consider notched members and cyclic loading, are especially needed.

We will use the same simplifying assumptions for the behavior of the material as in Chapters 12 and 13. The transient effects of cycle-dependent hardening or softening and creep-relaxation are thus not considered. In particular, the stress-strain behavior is idealized to always follow the stable cyclic stress-strain curve according to a multistage spring and slider rheological model. Such a material has identical monotonic and cyclic stress-strain curves. Where there are multiple applied loads, we will consider only cases with at least approximately proportional loading in regions of yielding, that is, cases where the ratios of the principal stresses remain at least approximately constant. The strain-based approach can be extended to handle more complex cases by employing a critical-plane approach as described above, but we will not pursue the topic that far here.

14.5.1 Constant Amplitude Loading

Based on the idealized behavior being assumed for the material, the cyclic and monotonic stress-strain curves are the same.

$$\varepsilon = f(\sigma), \quad \varepsilon_a = f(\sigma_a) \tag{14.40}$$

The specific function used is often the Ramberg-Osgood form, Eq. 14.1, and the constants for this curve are evaluated from stable behavior in cyclic strain tests. As in the rheological model, unloading and reloading during cycling is approximated as following stress-strain paths that are expanded with a scale factor of two relative to the above curve.

$$\frac{\Delta\varepsilon}{2} = f\left(\frac{\Delta\sigma}{2}\right) \tag{14.41}$$

Recall that origins for the $\Delta\sigma$ vs. $\Delta\varepsilon$ curves are the points where the direction of straining changes as illustrated in Fig. 12.14.

A stress-strain analysis of the component of interest is needed that is done just as for monotonic loading but using the cyclic stress-strain curve. Let this result be expressed as a function, which may be explicit or implicit, of a generic variable S that denotes load, moment, nominal stress, etc., as applicable in the particular case.

$$\varepsilon = g(S) \tag{14.42}$$

As discussed in Section 13.6, the material behavior assumed permits this analysis to be applied to cyclic loading. For constant amplitude loading which is biased in the tensile direction, the maximum stress and strain can be estimated from

$$\varepsilon_{\max} = g(S_{\max}) = f(\sigma_{\max}) \quad (R \geq -1) \tag{14.43}$$

Also, the amplitudes or ranges can be estimated from

$$\varepsilon_a = g\,(S_a) = f\,(\sigma_a)\,, \qquad \frac{\Delta\varepsilon}{2} = g\left(\frac{\Delta S}{2}\right) = f\left(\frac{\Delta\sigma}{2}\right) \qquad (14.44)$$

where these two equivalent equations are both given merely as a convenience. The application of Eqs. 14.43 and 14.44, which are the same as Eq. 13.103, is illustrated by Fig. 13.20. For loading that is biased in compression, $R < -1$, the details differ in a straightforward manner, and Eq. 13.105 is needed.

Once σ_{max}, ε_{max}, $\sigma_a = \Delta\sigma/2$, and $\varepsilon_a = \Delta\varepsilon/2$ are known, other quantities of interest follow easily. The local notch mean stress is of special interest for fatigue life prediction.

$$\sigma_m = \sigma_{max} - \sigma_a \qquad (14.45)$$

Substitution of the values of ε_a and σ_m then allow the fatigue life N_f to be estimated from a strain-life relation such as Eq. 14.18 or 14.21. Alternatively, σ_{max} and ε_a can be used with the life equation based on the SWT parameter, Eq. 14.25. Such a procedure can be applied for bending, torsional, or notched members using analytical results $\varepsilon = g(S)$ such as those given in Chapter 13.

As an example, consider analysis of a notched member using Neuber's rule, followed by a fatigue life estimate. If Eq. 14.1 is used for the cyclic stress-strain curve, the function $\varepsilon = g(S)$ is the implicit one obtained from using this σ-ε relation with Neuber's rule, Eq. 13.78. For cyclic loading with a nonzero mean level, Neuber's rule can be used twice with the cyclic (same as monotonic) stress-strain curve, once for S_{max} and once for S_a. Thus, two equations need to be solved to obtain σ_{max} and ε_{max}.

$$\varepsilon_{max} = \frac{\sigma_{max}}{E} + \left(\frac{\sigma_{max}}{H'}\right)^{\frac{1}{n'}}, \qquad \frac{(k_t S_{max})^2}{E} = \sigma_{max}\varepsilon_{max} \qquad (14.46)$$

Similarly for σ_a and ε_a, we need to solve

$$\varepsilon_a = \frac{\sigma_a}{E} + \left(\frac{\sigma_a}{H'}\right)^{\frac{1}{n'}}, \qquad \frac{(k_t S_a)^2}{E} = \sigma_a\varepsilon_a \qquad (14.47)$$

Such calculations have already been illustrated in Examples 13.2 and 13.3.

The fatigue notch factor k_f, as discussed in Chapter 10, is often employed in place of k_t in these equations. Use of the empirically based parameter k_f improves the accuracy of life prediction for sharp notches. However, see the discussion on crack growth effects near the end of this chapter for another viewpoint suggesting that k_t should be retained.

The steps necessary to perform a life estimate for constant amplitude loading of a notched member are summarized by Fig. 14.15. Step (1) is to assemble the needed input information related to the loading, geometry, and material involved. Then (2) the values of $(\varepsilon_{max}, \sigma_{max})$ and $(\varepsilon_a, \sigma_a)$ are obtained from Eqs. 14.46 and 14.47, which could be done using a graphical solution as illustrated. The estimated stress-strain response can then be plotted, and the local notch mean stress σ_m determined, which is shown as step (3). Finally (4) the life is obtained using ε_a and σ_m with Eq. 14.18 or 14.21, or ε_a and σ_{max} with Eq. 14.25.

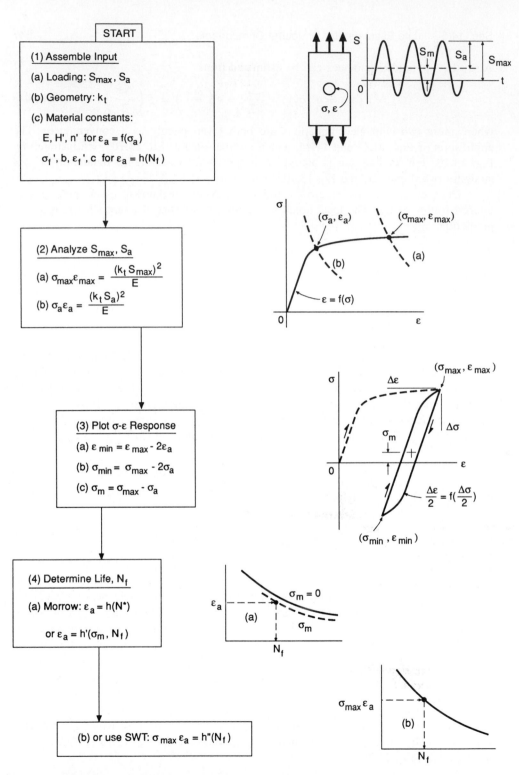

Figure 14.15 Steps required in strain-based life prediction for a notched member under constant amplitude loading.

Example 14.3

A notched plate made of the AISI 4340 (aircraft quality) steel of Table 14.1 has an elastic stress concentration factor of $k_t = 2.8$. If the nominal stress is cycled between $S_{max} = 750$ and $S_{min} = 50$ MPa, how many cycles can be applied before fatigue cracking is expected?

Solution   The identical situation has already been analyzed in Ex. 13.3 to estimate the local notch stresses and strains. In particular, the cyclic stress-strain curve and Neuber's rule were used as in Eqs. 14.46 and 14.47 to obtain both the maximums and amplitudes of stress and strain. The resulting values are

$$\sigma_{max} = 972 \text{ MPa}, \quad \varepsilon_{max} = 0.02192$$
$$\sigma_a = 755 \text{ MPa}, \quad \varepsilon_a = 0.00615$$

Note that the steps followed in Ex. 13.3 correspond to (1) and (2) in Fig. 14.15. Plotting the stress-strain response as in step (3) gives the result shown in Figure E14.3.

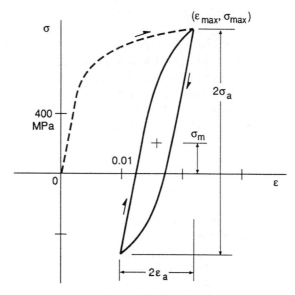

Figure E14.3

The mean stress during cycling is

$$\sigma_m = \sigma_{max} - \sigma_a = 972 - 755 = 217 \text{ MPa}$$

where cycle-dependent relaxation is assumed not to occur.

The life can now be estimated using σ_m and ε_a in either Eq. 14.18 or 14.21, or using σ_{max} and ε_a in Eq. 14.25, as already illustrated by Ex. 14.2. Pursuing the first of these, that is, using the Morrow parameter, we substitute material constants from Table 14.1 into Eq. 14.18.

$$\varepsilon_a = \frac{1758}{207,000}(2N^*)^{-0.0977} + 2.12\,(2N^*)^{-0.774}$$

Substituting ε_a from above and solving numerically for N^* gives

$$N^* = 3005 \text{ cycles}$$

The life N_f with the effect of σ_m included is given by Eq. 14.19.

$$N_f = N^*\left(1 - \frac{\sigma_m}{\sigma_f'}\right)^{-\frac{1}{b}} = 3005\left(1 - \frac{217}{1758}\right)^{\frac{1}{0.0977}} = 780 \text{ cycles} \qquad \textbf{Ans.}$$

14.5.2 Irregular Load vs. Time Histories

The methodology just described can be extended to irregular variations of load with time. This requires applying stress-strain analysis for irregular load versus time histories as described in Section 13.6 and illustrated by Fig. 13.21, which is repeated here as Fig. 14.16 for convenience. Once such an analysis is done, a fatigue life estimate may be readily made using the Palmgren-Miner rule from Chapter 9.

The strain amplitude and mean stress of each closed stress-strain hysteresis loop are first needed. Consider the example of Fig. 14.16. There are four closed loops as shown in (e), specifically corresponding to load excursions B-C-B', F-G-F', E-H-E', and A-D-A'. The strain amplitudes ε_a and mean stress σ_m for each is calculated from the (ε, σ) coordinates of the loop tips. For example, these quantities are labeled in (e) for loop F-G-F'. The life to failure N_f corresponding to each hysteresis loop can then be determined from its combination of strain amplitude and mean stress using a strain-life relationship, such as Eqs. 14.18 or 14.21. The SWT parameter may also be used with ε_a and $\sigma_{\max}$ in Eq. 14.25, the $\sigma_{\max}$ used being simply the highest stress in the loop, such as σ_G for loop F-G-F'. Having obtained the N_f value for each loop, the Palmgren-Miner rule can be applied. Each closed stress-strain hysteresis loop is considered to represent a cycle, and for repeating histories, the Palmgren-Miner rule is used in the form of Eq. 9.24.

A connection exists between the stable stress-strain behavior for irregular load histories and rainflow cycle counting, which is described in Section 9.9. First, assume that the load history starts at the peak or valley having the largest absolute value of load, reordering as in Fig. 9.46(a) and (b) being done if necessary. Each closed stress-strain hysteresis loop then corresponds to a cycle counted by the rainflow procedure. This arises because the cycle counting condition of Fig. 9.45 is satisfied at the same points in the load history where the memory effect acts and the stress-strain path returns to one previously established. (Such points are also the points where segments are skipped in

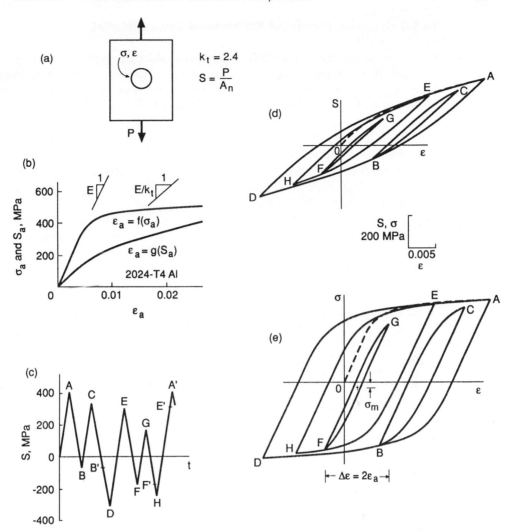

Figure 14.16 Analysis of a notched member subjected to an irregular load vs. time history. Notched member (a), having cyclic stress-strain and load-strain curves as in (b), is subjected to load history (c). The resulting load vs. notch strain response is shown in (d), and the local stress-strain response at the notch in (e). (Adapted from [Dowling 89]; copyright ©ASTM; reprinted with permission.)

the load history where the memory effect acts and the stress-strain path returns to one previously established. (Such points are also the points where segments are skipped in the rheological model as described in Section 12.4.) Thus, the use of closed stress-strain hysteresis loops to identify cycles is equivalent to rainflow cycle counting. Furthermore, rainflow cycle counting is now seen to possess a physical justification based on this correspondence with elasto-plastic stress-strain behavior.

14.5.3 Stepwise Procedure for Irregular Load Histories

It is useful to summarize the various steps required for stress-strain modeling and analysis, and then life prediction, for an irregular load history applied to an engineering component. The idealized material behavior described above and the logic and equations of Section 13.6 apply. Input information is assumed to exist as follows:

(a) Known history of load (S) versus time expressed as a list of peaks and valleys in order of occurrence.

(b) Material constants for the cyclic stress-strain and strain-life curves. The former, $\varepsilon_a = f(\sigma_a)$, is generally specified by Eq. 14.1 and constants E, H', and n', and the latter by Eq. 14.4 and constants σ'_f, b, ε'_f, and c.

(c) Analytical result in the form $\varepsilon = g(S)$ for the strain at the expected failure location in the component. This stress-strain analysis is done as for monotonic loading, using $\varepsilon = f(\sigma)$ as the stress-strain curve.

The stepwise procedure is as follows and is also illustrated in Fig. 14.17:

(1) Assume that the load (S) history repeats itself continuously and that it starts at the highest absolute value of S. Reorder the history as in Fig. 9.46(a) and (b) if necessary.

(2) Establish the stress and strain corresponding to the first (highest absolute value) peak or valley of S, called S_A. One of the following equations applies depending on the sign of S_A:

$$\varepsilon_A = g(S_A) = f(\sigma_A) \qquad (S_A > 0)$$
$$\varepsilon_A = -g(-S_A) = -f(-\sigma_A) \qquad (S_A < 0) \tag{14.48}$$

(3) Then proceed from each peak or valley of S to the next, while following the S-ε and σ-ε curve shapes given by

$$\frac{\Delta\varepsilon}{2} = g\left(\frac{\Delta S}{2}\right) = f\left(\frac{\Delta\sigma}{2}\right) \tag{14.49}$$

where ΔS, $\Delta\varepsilon$, and $\Delta\sigma$ are measured from new origins at each point corresponding to a peak or valley of S. Use the strain value at each peak or valley of S to establish the corresponding point on the σ-ε plot.

(4) However, at points such as X' in Fig. 9.45, where the rainflow cycle counting condition is met, and where the stress-strain memory effect also acts, take the following actions:

(a) Record the information that a cycle has been counted that ranges between the values $(S_X, \varepsilon_X, \sigma_X)$ and $(S_Y, \varepsilon_Y, \sigma_Y)$.

(b) Remove the event X-Y-X' from the load history to be used for further analysis. For the ΔS vs. $\Delta\varepsilon$ and $\Delta\sigma$ vs. $\Delta\varepsilon$ paths beyond the point X', return to the origin that was being used prior to reaching point X.

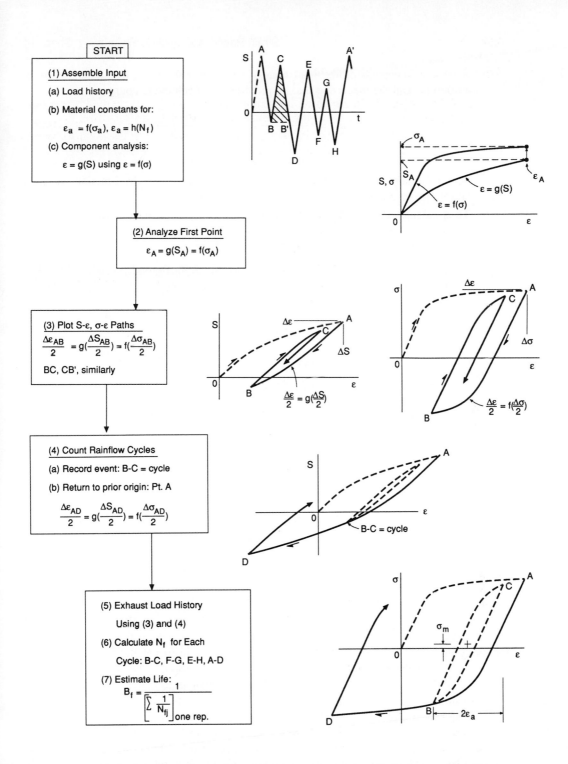

Figure 14.17 Life prediction for an irregular load vs. time history. The stepwise procedure given in the text is illustrated using the situation of the previous figure.

(5) After one repetition of the load history has been exhausted, calculate the strain amplitude and the mean and maximum stress for each cycle counted.

$$\varepsilon_{aj} = \frac{|\varepsilon_X - \varepsilon_Y|}{2}, \quad \sigma_{mj} = \frac{\sigma_X + \sigma_Y}{2} \tag{14.50}$$

$$\sigma_{\max j} = \text{MAX}(\sigma_X, \sigma_Y)$$

where the subscript j identifies the jth cycle counted, and the function MAX( ) simply picks the more tensile of the two stress values.

(6) Then calculate the life N_f corresponding to each cycle counted, using

$$\varepsilon_{aj} = h'(\sigma_{mj}, N_{fj}) \tag{14.51}$$

where the particular equation used is one of the strain-life relationships related to the Morrow parameter, Eq. 14.18 or 14.21. Alternatively, use the expression based on the SWT parameter, Eq. 14.25.

$$\sigma_{\max j} \varepsilon_{aj} = h''(N_{fj}) \quad (\sigma_{\max j} > 0)$$
$$N_{fj} = \infty \quad (\sigma_{\max j} \leq 0) \tag{14.52}$$

(7) Finally, estimate the number of repetitions to failure from the Palmgren-Miner rule in the form of Eq. 9.24.

$$B_f = \frac{1}{\left[\sum \dfrac{N_j}{N_{fj}}\right]_{\text{one rep.}}} \tag{14.53}$$

where the N_j are all unity if each cycle is treated individually as it is completed.

In most applications, the graphical procedure described above for modeling the S-ε and σ-ε paths would be accompanied by, or even replaced by, an equivalent program on a digital computer. Computer programs for accomplishing all of the above steps are described in Wetzel (1977), specifically in the papers therein by Landgraf et al. and by Brose, and also in Socie and Morrow (1980). The programming strategy used to follow the S-ε and σ-ε paths can employ the rules corresponding to the behavior of the rheological model as given in Section 12.4. These rules are applied not only for the σ-ε response, but also for the S-ε response, where the curve $\varepsilon = g(S)$ is used just as if it were a stress-strain curve. However, the use of discrete segments is not a necessity. One strategy that has been successfully employed is to make prior calculations of a number of points along the smooth curves, storing these in an array to be drawn upon while actually performing the S-ε and σ-ε modeling. Another is to calculate only the (ε, σ) coordinates of points corresponding to peaks and valleys of S.

The procedure described can be used for multiaxial loading. Adjustments as described in Chapter 12 and earlier in this chapter must of course be made to the cyclic stress-strain and strain-life relationships, and mean stress effects need to be handled appropriately. However, if multiple external loads cause significantly nonproportional loading in regions of yielding, more sophisticated analysis is needed using incremental plasticity theory, perhaps combined with finite element analysis on a digital computer. See Kussmaul et al. (1991), especially the paper therein by Hoffmann et al., for additional discussion.

Example 14.4

The nominal stress history shown in Figure E14.4a is repeatedly applied to a bending member made of SAE 1045 steel in the hot-rolled and normalized condition. If the member contains a notch with a stress concentration factor of $k_t = 3.0$, how many times can this history be applied before fatigue cracking is expected?

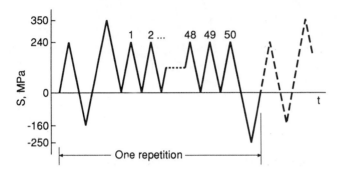

Figure E14.4a

Solution   We will apply Neuber's rule with the steps given in Section 14.5.3 and illustrated in Fig. 14.17. Before proceeding, the first peak and valley in the history are moved to the end, and the second peak is repeated at the end, so that the highest absolute value of S occurs first and last. This is shown in Fig. E14.4b along with rainflow cycle counting of the history according to the procedure of Section 9.9.

We next find the stress and strain for the first peak S_A from Eq. 14.48, specifically by applying Neuber's rule to analyze monotonic loading that follows the cyclic stress-strain curve, $\varepsilon = f(\sigma)$.

$$\varepsilon_A = \frac{\sigma_A}{E} + \left(\frac{\sigma_A}{H'}\right)^{\frac{1}{n'}}, \qquad \sigma_A \varepsilon_A = \frac{(k_t S_A)^2}{E}$$

Substituting $k_t = 3.0$, $S_A = 350$ MPa, and material constants from Table 14.1, allows these equations to be solved iteratively for stress and strain. The resulting values are

$$\sigma_A = 474.0 \text{ MPa}, \qquad \varepsilon_A = 1.1513 \times 10^{-2}$$

Further analysis of the load history involves applying Eq. 14.49, hence following stress-strain paths according to $\Delta\varepsilon/2 = f(\Delta\sigma/2)$. For loading beyond any point where a rainflow cycle is completed, the origin for the stress-strain path shifts back to a point prior to the beginning of the cycle in a manner consistent with the stress-strain memory effect. The resulting stress-strain response is shown in Fig. E14.4b.

Applying Eq. 14.49 to the cyclic stress-strain curve and to Neuber's rule, we have

$$\frac{\Delta\varepsilon_{XY}}{2} = \frac{\Delta\sigma_{XY}}{2E} + \left(\frac{\Delta\sigma_{XY}}{2H'}\right)^{\frac{1}{n'}}, \qquad \frac{\Delta\sigma_{XY}}{2}\frac{\Delta\varepsilon_{XY}}{2} = \frac{(k_t \Delta S_{XY})^2}{4E}$$

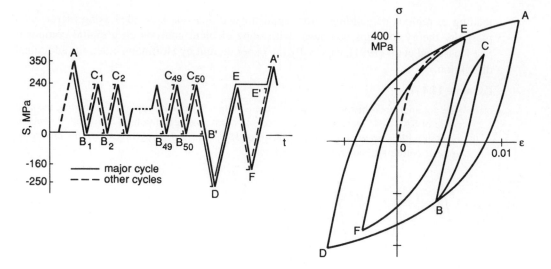

Figure E14.4b

where subscripts XY indicate ranges for any appropriate loading event X-Y. These equations can be entered with a value of ΔS_{XY} and solved iteratively for $\Delta \sigma_{XY}$ and $\Delta \varepsilon_{XY}$, with material constants and k_t being of course substituted first. For load range A-B_1, entering the above equations with $\Delta S_{AB} = 350$ MPa gives $\Delta \sigma_{AB}$ and $\Delta \varepsilon_{AB}$ as listed on the second line of Table E14.4(a). The stress and strain σ_B and ε_B at point B can then be computed by subtracting the ranges A-B from the values at point A, with the result being given in the third line of the table.

For load range B_1-C_1, the same equations are entered with $\Delta S_{BC} = 240$ MPa. The resulting stress and strain ranges, and the computation of σ_C and ε_C, are given in the fourth and fifth lines of the table. Load range C_1-B_2 produces the same stress and strain ranges, as do subsequent ranges B-C and C-B up to B', at which point 50 rainflow cycles B-C are complete. For loading beyond point B', the memory effect dictates that the stress-strain path must follow the one underway before these cycles were started. Hence, the origin shifts back to point A, and range A-D is evaluated, allowing σ_D and ε_D to be determined as also indicated in Table E14.4(a), lines 6 and 7. Similar analysis continues to the end of the load history. A cycle E-F is completed at point E', and the origin shifts back to point D for loading beyond E'. The major cycle A-D is finally completed at the end of the load history, point A'.

Stress and strain values are now available for each peak and valley in the load history. A life calculation based on the P-M rule can now proceed in a straightforward manner. If the Morrow parameter is used for the mean stress effect, the following calculations are needed for each of cycles B-C, E-F, and A-D:

$$\varepsilon_a = \frac{\Delta \varepsilon}{2}, \qquad \sigma_m = \frac{\sigma_{max} + \sigma_{min}}{2}$$

$$\varepsilon_a = \frac{\sigma'_f}{E} (2N^*)^b + \varepsilon'_f (2N^*)^c$$

$$N_f = N^* \left(1 - \frac{\sigma_m}{\sigma'_f} \right)^{-\frac{1}{b}}$$

A numerical solution is required to obtain N^*, and additional material constants from Table 14.1 are needed. The results are given in Table E14.4(b).

TABLE E14.4(a)

Line No.	Point or Range	S MPa	Comment	σ MPa	ε %	ΔS MPa	$\Delta \sigma$ MPa	$\Delta \varepsilon$ %
1	A	350	$\varepsilon = f(\sigma)$, 0 to A	474.0	1.1513	—	—	—
2	A-B_1	—	Start using $\Delta\varepsilon/2 = f(\Delta\sigma/2)$	—	—	350	701.5	0.7780
3	B_1	0	$\sigma_A - \Delta\sigma_{AB} = \sigma_B$	−227.5	0.3733	—	—	—
4	B_1-C_1, C_1-B_2 etc. to B'	—	50 cycles B-C	—	—	240	573.5	0.4475
5	C_1, C_2, etc.	240	$\sigma_B + \Delta\sigma_{BC} = \sigma_C$	346.0	0.8208	—	—	—
6	A-D	—	Origin at A	—	—	600	890.9	1.8004
7	D	−250	$\sigma_A - \Delta\sigma_{AD} = \sigma_D$	−416.9	−0.6491	—	—	—
8	D-E	—	—	—	—	490	818.1	1.3075
9	E	240	$\sigma_D + \Delta\sigma_{DE} = \sigma_E$	401.2	0.6584	—	—	—
10	E-F, F-E'	—	Cycle E-F	—	—	400	747.3	0.9539
11	F	−160	$\sigma_E - \Delta\sigma_{EF} = \sigma_F$	−346.1	−0.2955	—	—	—
12	D-A'	—	Origin at D; Cycle A-D	—	—	600	890.9	1.8004
13	A'	350	$\sigma_D + \Delta\sigma_{DA'} = \sigma_{A'}$	474.0	1.1513	—	—	—

TABLE E14.4(b)

Cycle	N_j	ε_a	σ_m, MPa	N^*	N_{fj}
B-C	50	2.237×10^{-3}	59.3	2.129×10^5	1.055×10^5
E-F	1	4.770×10^{-3}	27.6	1.207×10^4	8.753×10^3
A-D	1	9.002×10^{-3}	28.6	1.803×10^3	1.293×10^3

The quantities N_j are the numbers of cycles at each level for one repetition of the load history. Each ε_a is obtained from the corresponding $\Delta\varepsilon$ in Table E14.4(a), and each σ_m from the two applicable stress values in the same table. Hence, the estimated life is finally obtained from the P-M rule in the form of Eq. 14.53.

$$B_f = \cfrac{1}{\left[\sum \cfrac{N_j}{N_{fj}}\right]_{\text{one rep.}}} = \cfrac{1}{\cfrac{50}{1.055 \times 10^5} + \cfrac{1}{8.753 \times 10^3} + \cfrac{1}{1.293 \times 10^3}}$$

$$B_f = 734 \text{ repetitions} \qquad\qquad\qquad\qquad\qquad\qquad\qquad\qquad\qquad \textbf{Ans.}$$

Comment   This life estimate may have a conservative bias, as the estimated mean stresses for all cycles are tensile, and some relaxation of mean stress is likely to occur. This is especially true for the major cycle over the first several repetitions, and relaxation within each repetition during the B-C cycles may also occur in a manner similar to Fig. 12.26.

14.5.4 Simplified Procedure for Irregular Histories

A simplification of the procedure described is often employed that does not require a detailed knowledge of the loading sequence. It becomes necessary to have only a summary of the load history in the form of a matrix (as in Fig. 9.47) giving the numbers of rainflow-counted cycles at various combinations of load range and mean load. The procedure is illustrated in Fig. 14.18 for one of the cycles, specifically F-G, of the example of Fig. 14.16. Consistent with the assumption that only the cycle counting result for the load history is known, the values of S_F and S_G are considered to be known. However, the exact location of this cycle within the loading sequence is assumed to be unknown, and this affects the local notch mean stress for the cycle, and hence also its N_f value. The result of a similar situation occurring for all cycles except the largest one is that the analysis does not provide a single answer, but rather places generally narrow bounds on the calculated life.

Using F-G as a typical cycle, the key to making the simplified life estimate is to note that both load-strain loop F-G and stress-strain loop F-G must lie within the corresponding loop for the major (largest) cycle in the history, in this case cycle A-D. Since the loads S_F and S_G are known, this places limits on the mean strain of cycle F-G. In particular, in Fig. 14.18(a), load-strain loop F-G could be so far to the right that it is attached to the lower branch of loop A-D at P, or so far to the left that it is attached to the upper branch at Q.

These limits on the strains for loop F-G confine the corresponding stress-strain loop as shown in (b). Bounds σ_{mP} and σ_{mQ} can be placed on the mean stress of loop F-G by calculating the peak and valley stresses and strains for the two extreme possibilities. This is done by applying Eq. 14.48 to S_A, and then applying Eq. 14.49 for load ranges A-P, A-D, D-Q, and F-G. Knowing the bounds on the mean stress for cycle F-G then allows bounds to be placed on the corresponding number of cycles to failure, N_{fFG}.

The upper bound values on N_f are similarly obtained for all cycles in the history, and these are used with the P-M rule to obtain the upper bound on calculated life for the

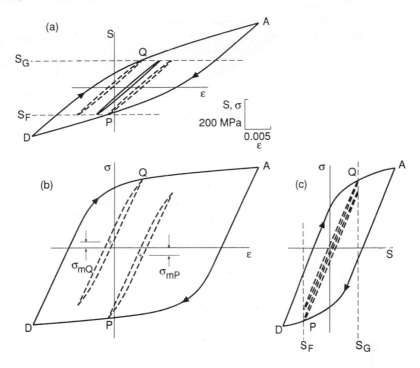

Figure 14.18 Simplified procedure that places bounds on the mean stress effect. For cycle F-G of Fig. 14.16, the mean stress must lie between the values σ_{mP} and σ_{mQ}. (Adapted from [Dowling 89]; copyright ©ASTM; reprinted with permission.)

irregular load history. The lower bound N_f values are similarly used to obtain the lower bound on the life. Such a procedure was applied to the transmission load history of Fig. 9.47 for the case of a notched member and material combination for which test data are available. Estimated bounds on the life are compared with this test data in Fig. 14.19. Reasonable agreement is obtained considering the degree of scatter in the data.

Note that summarizing the load history as its rainflow matrix results in a loss of detail relative to the original ordered list of peaks and valleys. A number of different sequences of peaks and valleys have this same matrix, and each potentially results in a different calculated life. This explains the situation of a bounded life calculation, the bounds being the extremes from all possible sequences that give the rainflow matrix used.

14.6 DISCUSSION

Additional discussion is useful, one purpose of which is to note some relationships and contrasts between the methodology presented in this chapter and that from earlier chapters describing other approaches to fatigue.

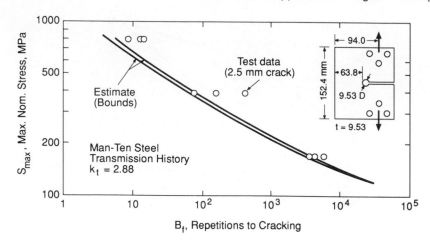

Figure 14.19 Maximum nominal stress vs. the number of repetitions to cracking, for repeated application of the SAE transmission history of Fig. 9.47 to the notched member and material indicated. (Adapted from [Dowling 87b]; used with permission; ©Society of Automotive Engineers. See [Wetzel 77] for data.)

14.6.1 Strain-Based Versus Stress-Based Approaches

The strain-based approach handles local plasticity effects in a more rational and detailed manner than does the stress-based approach of Chapters 9 and 10. Hence, it is generally the preferred approach for analyzing short fatigue lives, as judged by the transition fatigue life of the material, unless only rough estimates are needed. The strain-based approach can be used at long lives where elastic strains dominate, in which case it becomes equivalent to a stress-based approach. The stress-based approach itself thus need never be used. However, some of its features can be incorporated into the strain-based approach. The surface finish modification described above is one example of this.

In some cases, *S-N* curves may be available from testing of members very similar to the actual component of interest. Examples might be welded joints, built-up riveted beams, or vehicle axles, of a specific design and material. *S-N* curves from such members automatically include the effects of various complexities that are difficult to evaluate, such as the complex metallurgy and geometry of welds, fretting effects, or effects of surface hardening processes. Where such *S-N* curves are available, it may be advantageous to use these with a nominal-stress-based approach as described in Chapter 10, except with the actual *S-N* curve replacing an estimated one. Rainflow cycle counting and the Palmgren-Miner rule should be used for variable amplitude loading, along with a simple modified Goodman equation mean stress adjustment as in Eq. 10.21. Although the advantages of detailed analysis of local yielding effects by the strain-based approach are lost, the advantage gained by automatic inclusion of other complexities will sometimes outweigh this loss. The choice of an approach in such cases will be dictated by the details of the particular situation. It may even be useful to perform design or analysis using more than one approach and compare the results.

14.6.2 Mean Stresses and Plasticity Effects

A key feature of the strain-based approach is that fatigue life estimates are made based on local stresses and strains in the region where fatigue cracking is expected to start, as at the edge of a beam or in the bottom of a notch. Hence, even though one of the mean stress parameters used, Eq. 14.13, has the same form as those used for the stress-based approach, its manner of application is fundamentally different. The mean stress used is the one that occurs locally, and its value is obtained by specifically analyzing the local plastic deformation. However, note that the approach of Juvinall, as summarized in Chapter 10, does attempt to account for local yielding to an extent by effectively using Eq. 10.27.

An example of analysis by the strain-based approach involving mean stresses is provided by Figs. 14.20 and 14.21. The procedure described above for constant amplitude loading (Fig. 14.15) is applied to zero-to-maximum ($R = 0$) loading of notched plates of 2024-T351 Al. Stress-strain paths for four different load levels are shown in Fig. 14.20, and life calculations made from these are in reasonable agreement with test data as shown in Fig. 14.21. For the highest loads in this example, the mean stresses are near zero, as in Fig. 14.20(a). For decreasing load, the ratio $k_{fm} = \sigma_m/S_m$ of local to nominal mean stress increases. It becomes equal to k_t where the load is sufficiently low that no yielding occurs, as in (d). The overall trend is similar to that discussed in connection with the stress-based approach; see Fig. 10.14. Similarly, the ratio $k'_f = \sigma_a/S_a$ varies in a manner similar to Fig. 10.11.

Thus, recalling the discussion of trends and estimation procedures for S-N curves in Chapter 10, it is apparent that the strain-based approach provides a rational basis for specifically evaluating the effect of plastic deformation on S-N curves. The rough empirical estimates involving k_{fm} and k'_f and their trends with life, and with ductile versus brittle materials, are not necessary with the strain-based approach.

It is noteworthy that the procedure described above for estimating local mean stresses does not consider cycle-dependent creep-relaxation effects. Actual stress-strain responses at notches are expected to be similar to Fig. 14.22. Both creep and relaxation occur simultaneously, and these may interact with the cycle-dependent hardening or softening that is also occurring. More detailed modeling of the stress-strain response, as already discussed to an extent in Section 12.5, is a possibility. In particular, values of mean stress could be revised to reflect relaxation according to Eq. 12.60 or other analogous relationship. The improved accuracy of fatigue life prediction resulting from this may in a few cases justify the increased complexity of the analysis. However, an impediment to such more refined analysis is the necessity of obtaining additional material constants to describe the transient behavior. Note that the major effect of lowered mean stress due to local yielding is already included as a separate effect, as in Fig. 14.20.

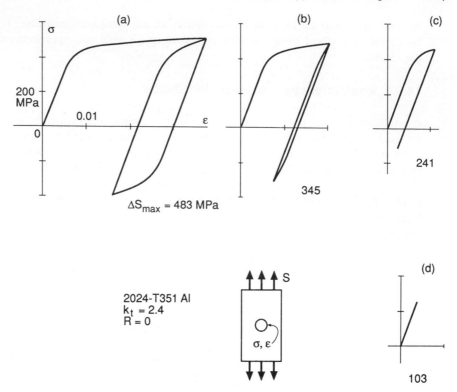

Figure 14.20 Estimated stress-strain responses for an aluminum alloy, for various levels of zero-to-maximum loading on plates with central round holes. (Adapted from [Dowling 87b]; used with permission; ©Society of Automotive Engineers.)

14.6.3 Sequence Effects Related to Mean Stresses

The rational handling of mean stress effects by the strain-based approach is especially important for irregular load versus time histories. Consider the two load histories of Fig. 14.23. They differ only in that the initial severe loading cycle has two different sequences. No difference in these two situations would be predicted by the stress-based approach, as the mean nominal stress for the subsequent lower level cycles is zero in both cases.

Stress-strain responses for the two load histories as shown were estimated from the procedure of Fig. 14.17 as described above, using the σ-ε and S-ε curves for the 2024-T4 Al notched member from Fig. 14.16. Mean stresses are present for the lower level cycles, and these differ in sign as a result of the sequence of the initial severe cycle. Large differences in the fatigue life can occur in such situations, and these are predicted with reasonable accuracy by the strain-based approach. It is of course of special concern that the stress-based approach, as usually applied, does not predict the detrimental effect

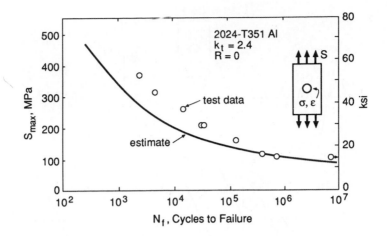

Figure 14.21 Life estimates for the situation of Fig. 14.20 using the SWT parameter, and also corresponding test data. (Data from [Wetzel 68].)

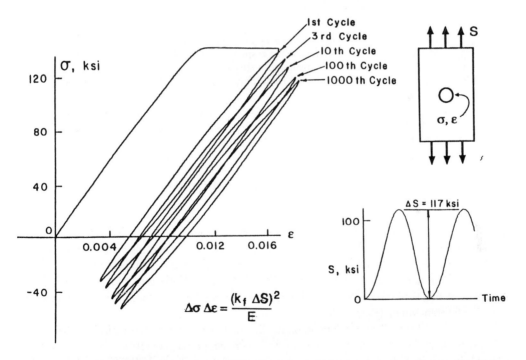

Figure 14.22 Simulation of the stress-strain behavior at a notch, for zero-to-maximum loading of Ti-811 with $k_f = 1.75$. A smooth specimen was subjected to cyclic loading, with the direction of straining being reversed whenever Neuber's rule using k_f was satisfied. (Adapted from [Stadnick 72]; copyright ©ASTM; reprinted with permission.)

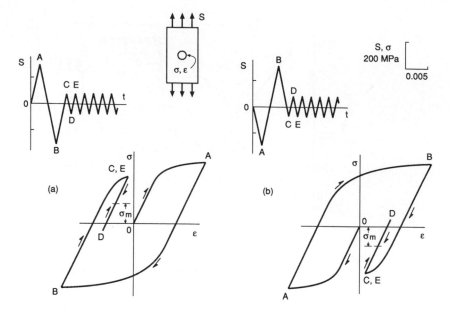

Figure 14.23 Two load histories applied to a notched member ($k_t = 2.4$), and the estimated notch stress-strain responses for 2024-T4 Al. The high-low overload in (a) produces a tensile mean stress, and the low-high overload in (b) produces the opposite. (Adapted from [Dowling 82b]; used with permission of ASME.)

of history (a). The approach in Juvinall's books that is related to Eq. 10.27, where an elastic, perfectly plastic material is assumed, could be extended to rationally analyze irregular load histories, but no such extension is described there.

14.6.4 Sequence Effects Related to Physical Damage to the Material

Sequence effects on life also occur which are associated with physical damage to the material caused by occasional severe cycles. Small cracks, and also the preceding slip-band damage, etc., usually occur at shorter fractions of the total life at higher strain levels. This is evident in the strain-life curves of Fig. 14.9, and also in the stress-life curves of Fig. 9.17. Such behavior creates a situation where a difficulty exists in applying the Palmgren-Miner (P-M) rule.

The P-M rule does not depend on the linearity with cycles of any physical parameter characterizing fatigue damage. The physical damage, perhaps crack size or crack density, can increase in a nonlinear manner with life fraction as illustrated in Fig. 14.24(a). Let the life fraction for N_j cycles, applied at a particular stress level S_j, be termed U_j.

$$U_j = \frac{N_j}{N_{fj}} \tag{14.54}$$

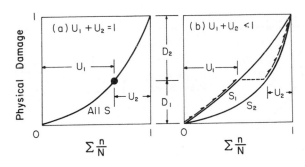

Figure 14.24 Physical damage vs. life fraction, where the relationship is (a) unique and (b) nonunique. (From [Dowling 87b]; used with permission; ©Society of Automotive Engineers.)

If the stress level is changed from S_1 to S_2 during the life, the P-M rule requires that

$$U_1 + U_2 = 1 \tag{14.55}$$

This is obeyed for Fig. 14.24(a), but not for (b). In the latter case, the damage curves differ for the two different stress levels, with the result that the summation of life fractions at failure is not unity. If the damage proceeds at a higher rate for S_1 than for S_2, the summation is less than unity, which is the particular case illustrated.

This is precisely the situation that occurs if a relatively small number of cycles at a high stress level are followed by, or mixed with, cycling at a lower stress level. Summations of life fractions less than unity can thus occur as a result of the same degree of physical damage occurring at shorter fractions of the life at high stress levels. A reasonable engineering approach to this problem is to obtain strain-life or S-N curves by applying a few high-level cycles prior to each test, so as to pre-damage the material. A life fraction of around 0.01 or 0.02 appears to generally be sufficient to lower the strain-life curve so that nonconservative life estimates are avoided. An example of such a strain-life curve for prestrained material is shown in Fig. 14.25.

In steels that have a distinct fatigue limit, effects of this sort extend to stresses below the fatigue limit. Periodic overstrains have an especially severe effect and appear to essentially eliminate the fatigue limit as shown in Fig. 14.26. Therefore, for irregular loading where some stress levels substantially exceed the fatigue limit, infinite life should not be assumed for cycles below this level. Limited data as in Fig. 14.26 suggest that extrapolation below the fatigue limit of the strain-life curve obtained at shorter lives could be used to deal with this situation. On a stress-life plot, this corresponds to linear extrapolation on log-log coordinates as already employed in Ex. 10.5.

14.6.5 Crack Growth Effects

In the usual manner of obtaining strain-life curves, the N_f values correspond to failure or to substantial cracking in small (typically 5 to 10 mm diameter) axial test specimens. It is generally observed that life predictions made on this basis correspond to an *engineering size crack* that is easily visible with the naked eye, hence of size on the order of 1 to 5 mm. The existence of such a crack is often considered to constitute failure of the component. However, this rather loose definition of failure is not always sufficient. For

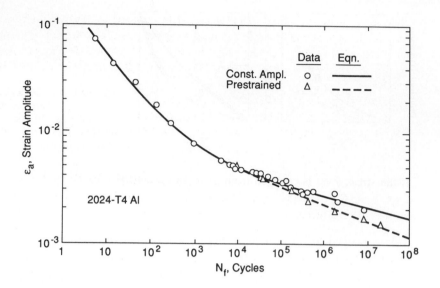

Figure 14.25 Effect of initial overstrain (10 cycles at $\varepsilon_a = 0.02$) on the strain-life curve of an aluminum alloy. (Adapted from [Dowling 89] as based on data from [Topper 70]; copyright ©ASTM; reprinted with permission.)

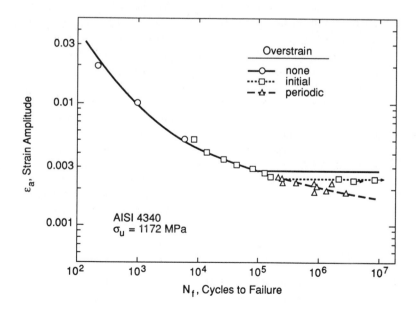

Figure 14.26 Effects of both initial and periodic overstrains on the strain-life curve for an alloy steel. The fatigue limit for the no overstrain case is estimated from test data on similar material. (Data from [Dowling 73].)

example, it may be desired to predict the life required to develop a crack of a definite size, so that the remaining life required to grow this crack to failure can be estimated using fracture mechanics as described in Chapter 11. Strain-life curves corresponding to specific small crack sizes as in Fig. 14.9 are needed if this is to be done in a rigorous manner.

On this basis, the strain-based approach is said to provide the *crack initiation life* N_i to a crack size a_i, and the fracture mechanics approach the life N_{if} to grow the crack from a_i to the failure size a_f. Hence, the total life to failure, N_f, is

$$N_i + N_{if} = N_f \tag{14.56}$$

If a_i is chosen to be too short, then the behavior of the crack may be affected by local notch plasticity or other complexities affecting small cracks, so that the use of linear-elastic fracture mechanics is compromised. On the other hand, a_i cannot be too long, as the notch surface strains used in the strain-based approach do not apply at too great a distance from the notch. A reasonable choice as an initiation size is the crack length l' defined by Eq. 8.21. The length l' varies with geometry and is generally around 0.1ρ to 0.2ρ, where ρ is the notch tip radius. However, this degree of rigor is not always needed. Additional discussion is provided in Dowling (1979) and Socie, Dowling, and Kurath (1984).

From Section 10.2 recall that the principal reason why an empirical fatigue notch factor k_f is needed appears to be related to the growth of cracks near the notch. Based on limited test data, the initiation of small cracks at notches appears to be governed by k_t, not by k_f. Hence, if the life is separated into crack initiation and growth phases as described above, it is reasonable to use k_t rather than k_f with the strain-based approach to estimate N_i. For sharp notches, this procedure will predict early initiation of cracks which are then expected from fracture mechanics to grow slowly or not at all. Such *nonpropagating cracks* are indeed commonly observed.

14.7 SUMMARY

The strain-based approach employs estimates for the stresses and strains that occur at locations where fatigue cracking is likely to start, such as at edges of beams and in notches. The behavior of the material is characterized using the stable cyclic stress-strain curve and the strain-life curve from uniaxial loading.

$$
\begin{aligned}
\varepsilon_a &= \frac{\sigma_a}{E} + \left(\frac{\sigma_a}{H'}\right)^{\frac{1}{n'}} \\
\varepsilon_a &= \frac{\sigma_f'}{E}\left(2N_f\right)^b + \varepsilon_f'\left(2N_f\right)^c
\end{aligned}
\tag{14.57}
$$

Mean stress also affects the fatigue life, so that the strain-life curve often needs to be generalized to include this effect according to one of Eqs. 14.18, 14.21, or 14.25.

Strain-life and cyclic stress-strain curves vary for different engineering metals and processing histories in a manner that can generally be correlated with other mechanical properties. For example, ductile metals usually have good resistance to fatigue at high strains corresponding to short lives, but poor resistance at long lives, and vice versa for highly strengthened metals. Elevated temperature and hostile chemical environments affect strain-life curves, and time-dependent creep-relaxation behavior may complicate analysis at high temperatures.

If multiaxial loading occurs, then the cyclic stress-strain and strain-life curves need to be used in more general form. For biaxial loading that is proportional or approximately so, the relationships needed are Eqs. 12.42 and 14.31, with the latter perhaps being modified using Eq. 14.38. For nonproportional loading, the more advanced incremental plasticity theory is needed for relating stresses and strains, and a critical-plane approach is appropriate for estimating fatigue life.

To apply the strain-based approach to an engineering component, such as a beam or a notched member, an analysis relating applied load and strain at the expected failure location is needed. In uncomplicated cases, proportional loading is assumed, as is stable stress-strain behavior without cycle or time-dependent creep-relaxation. A stress-strain analysis is needed that is done just as for monotonic loading, except that the cyclic stress-strain curve is used. For example, Neuber's rule can be used for notched members with a stress-strain curve $\varepsilon = f(\sigma)$, where $\varepsilon_a = f(\sigma_a)$ is the cyclic stress-strain curve. The result of this analysis is given by a load-strain curve $\varepsilon = g(S)$, where S quantifies the applied load.

A starting point for analyzing cyclic loading may be established by applying monotonic loading to the highest absolute value of S. Where this S is positive, we use

$$\varepsilon_{\max} = g(S_{\max}) = f(\sigma_{\max}) \tag{14.58}$$

Load-strain and stress-strain paths for cyclic loading are then estimated by

$$\frac{\Delta\varepsilon}{2} = g\left(\frac{\Delta S}{2}\right) = f\left(\frac{\Delta\sigma}{2}\right) \tag{14.59}$$

where new origins are established for the ΔS vs. $\Delta\varepsilon$ and $\Delta\sigma$ vs. $\Delta\varepsilon$ curves at points where the direction of loading changes. It is sometimes convenient to apply this equation in terms of amplitude quantities, $\varepsilon_a = \Delta\varepsilon/2$, etc., so that

$$\varepsilon_a = g(S_a) = f(\sigma_a) \tag{14.60}$$

After the stress-strain response is determined, mean stresses are easily obtained from

$$\sigma_m = \sigma_{\max} - \sigma_a \tag{14.61}$$

A life estimate may then be made by substituting ε_a and σ_m into an appropriate strain-life curve, or the Smith, Watson, and Topper parameter $\sigma_{\max}\varepsilon_a$ can be employed. For constant amplitude loading, the entire procedure is summarized by Fig. 14.15.

If the load varies in an irregular manner with time, Eqs. 14.58 to 14.60 still apply, and these may be used to estimate the local stress-strain response to a given load history. The memory effect is needed and causes both the load-strain and stress-strain paths to form closed loops and then to return to the path established before starting the loop. Either the rainflow cycle counting method of Chapter 9 or the multistage spring and slider rheological model of Chapter 12 can be used to locate points where the memory effect acts. Since closed stress-strain hysteresis loops are also rainflow cycles, fatigue life estimates using the Palmgren-Miner rule follow readily once the local stress-strain response is determined. A detailed procedure and example are given in Section 14.5, and the procedure is summarized by Fig. 14.17.

Use of the local notch mean stress, rather than the nominal (average) mean stress, provides a rational basis for analyzing the effects of plastic deformation, including certain effects of loading sequence for irregular load versus time histories. Additional sequence effects are caused by physical damage to the material during occasional severe loading cycles. These can be handled by using strain-life curves for test specimens that have been subjected to plastic deformation before testing or periodically during testing. The strain-based approach can also be used in combination with crack growth life estimates by fracture mechanics to obtain total fatigue lives for crack initiation plus growth.

NEW TERMS AND SYMBOLS

crack initiation life, N_i
critical plane approach
effective strain approach
engineering size crack
modified Morrow approach
Morrow parameter
sequence effect

Smith, Watson, and Topper (SWT)
parameter, $\sigma_{max}\varepsilon_a$
strain-based approach
strain-life (Coffin-Manson) curve:
σ_f', b, ε_f', c
transition fatigue life, N_t

REFERENCES

(a) General References

ASTM. 1990 *Annual Book of ASTM Standards*, Vol. 03.01, Am. Soc. for Testing and Materials, Philadelphia, Pa. See No. E606, "Constant Amplitude Low-Cycle Fatigue Testing."

BANNANTINE, J. A., J. J. COMER, and J. L. HANDROCK. 1990 *Fundamentals of Metal Fatigue Analysis*, Prentice Hall, Englewood Cliffs, N.J.

DOWLING, N. E. 1979 "Notched Member Fatigue Life Predictions Combining Crack Initiation and Propagation," *Fatigue of Engineering Materials and Structures*, Vol. 2, No. 2, pp. 129–138.

FATEMI, A. and D. F. SOCIE. 1988 "A Critical Plane Approach to Multiaxial Fatigue Damage Including Out-of-Phase Loading," *Fatigue and Fracture of Engineering Materials and Structures*, Vol. 11, No. 3, pp. 149–165.

KUSSMAUL, K. F., D. L. MCDIARMID, and D. F. SOCIE, eds. 1991 *Fatigue Under Biaxial and Multiaxial Loading*, Mechanical Engineering Publications Ltd., London, UK.

LANDGRAF, R. W. 1970 "The Resistance of Metals to Cyclic Deformation," *Achievement of High Fatigue Resistance in Metals and Alloys*, ASTM STP 467, Am. Soc. for Testing and Materials, Philadelphia, Pa., pp. 3–36.

LEESE, G. E. and D. SOCIE, eds. 1989 *Multiaxial Fatigue: Analysis and Experiments*, SAE Pub. No. AE-14, Soc. of Automotive Engineers, Warrendale, Pa.

MOWBRAY, D. F. 1980 "A Hydrostatic Stress Sensitive Relationship for Fatigue Under Biaxial Stress Conditions," *Journal of Testing and Evaluation*, ASTM, Vol. 8, No. 1, Jan. 1980, pp. 3–8.

RICE, R. C., ed. 1988 *Fatigue Design Handbook*, 2nd ed., SAE Pub. No. AE-10, Soc. of Automotive Engineers, Warrendale, Pa.

SOCIE, D. F., N. E. DOWLING, and P. KURATH. 1984 "Fatigue Life Estimation of Notched Members," *Fracture Mechanics: Fifteenth Symposium*, ASTM STP 833, Am. Soc. for Testing and Materials, Philadelphia, Pa., pp,. 284–299.

SOCIE, D. F. and J. MORROW. 1980 "Review of Contemporary Approaches to Fatigue Damage Analysis," *Risk and Failure Analysis for Improved Performance and Reliability*, J. J. Burke and V. Weiss, eds., Plenum Pub. Corp., New York, pp. 141–194.

SOLOMON, H. D., et al., eds. 1988 *Low Cycle Fatigue*, ASTM STP 942, Am. Soc. for Testing and Materials, Philadelphia, Pa.

WETZEL, R. M., ed. 1977 *Fatigue Under Complex Loading: Analyses and Experiments*, SAE Pub. No. AE-6, Soc. of Automotive Engineers, Warrendale, Pa.

(b) Sources for Material Properties

BOLLER, C. and T. SEEGER. 1987 *Materials Data for Cyclic Loading*, 5 vols., Elsevier Science Pubs., Amsterdam.

SAE. 1989 "Technical Report on Low Cycle Fatigue Properties of Wrought Materials," SAE J1099, Information Report, Soc. of Automotive Engineers, Warrendale, Pa. See also L. E. Tucker, R. W. Landgraf, and W. R. Brose, "Proposed Technical Report on Fatigue Properties for the SAE Handbook," SAE Paper No. 740279, Automotive Engineering Congress, Detroit, Mich., 1974.

NRIM. 1978 *NRIM Fatigue Data Sheets*, National Research Institute for Metals, Tokoyo, Japan; starting in 1978 and continuing.

PROBLEMS AND QUESTIONS

Section 14.2

14.1 Using the two straight lines plotted for RQC-100 steel in Fig. 14.4:

    (a) Estimate your own values of the constants σ_f', b, ε_f', and c for the strain-life curve. The elastic modulus is $E = 200$ GPa. Do your values agree reasonably well with those given?

    (b) Using your results from (a), estimate the constants H' and n' for the cyclic stress-strain curve. Compare your result to the values in Table 14.1, and comment on possible reasons for any differences.

14.2 Consider the strain-life curves in Fig. 14.7 for SAE 4142 (670 HB) steel, and for ausformed H-11 steel, and also the information given in Table 14.1. How do the strain-life curves differ? Are the differences what might be expected from the tensile properties?

14.3 Plot strain-life curves for the four heat treatments of SAE 4142 steel in Table 14.1 all together on one graph. Comment on any trends you observe and how these correlate with the tensile properties and hardness. (Suggestion: Include the transition fatigue life N_t in your discussion.) (PC Problem)

14.4 How do the strain-life curves for the two ductile polymers in Fig. 14.5 differ from those typical of engineering metals? Consider both qualitative trends, such as the shape of the curves, and also quantitative trends, such as the life corresponding to $\varepsilon_a = 0.01$.

14.5 For hot-rolled and normalized SAE 1045 steel with an elastic modulus of $E = 202$ GPa, some fatigue data from completely reversed tests are listed below. The tabulated values of σ_a and ε_{pa} were measured near $N_f/2$. (PC Problem)

(a) Determine constants for the strain-life curve. Compare the results with those listed in Table 14.1, which were obtained from a larger set of data.

(b) Plot both the data and the fitted curves for elastic, plastic, and total strain. Do your constants provide a good representation of the data?

(c) Proceed as in (a) and (b) for the cyclic stress-strain curve.

ε_a	σ_a, MPa	ε_{pa}	N_f, cycles
0.0200	524	0.01741	257
0.0100	459	0.00774	1,494
0.0060	410	0.00398	6,749
0.0040	352	0.00227	19,090
0.0030	315	0.00144	36,930
0.0020	270	0.00067	321,500
0.0015	241	0.00031	2,451,000

Source: Data in [Leese 85].

14.6 An ASTM A470-7 (Ni-Cr-Mo-V) steel used for large shafts in turbine-generator units has tensile properties of yield $\sigma_o = 741$ MPa, ultimate $\sigma_u = 858$ MPa, true fracture strength $\tilde{\sigma}_{fB} = 1386$ MPa, and 71% reduction in area, and the elastic modulus is $E = 207$ GPa. Fatigue data from controlled-strain tests are given below, with the tabulated values of σ_a and ε_{pa} being measured near $N_f/2$. Proceed as in Prob. 14.5, except omit the comparison with Table 14.1. (PC Problem)

ε_a	σ_a, MPa	ε_{pa}	N_f, cycles
1.98×10^{-2}	711	1.61×10^{-2}	206
1.19×10^{-2}	634	8.57×10^{-3}	475
7.97×10^{-3}	594	4.87×10^{-3}	1,250
5.44×10^{-3}	551	2.50×10^{-3}	3,700
4.12×10^{-3}	539	1.26×10^{-3}	6,700
3.42×10^{-3}	531	7.3×10^{-4}	15,000
2.89×10^{-3}	527	2.6×10^{-4}	54,000
2.32×10^{-3}	481	–	202,500
2.17×10^{-3}	452	–	618,000
2.04×10^{-3}	427	–	1,165,000

Source: Data in [Dowling 82a].

14.7 Based on the constants in Table 14.1:

    (a) Plot strain-life curves for steels SAE 1015 and 4142 (450 HB).

    (b) Noting that these curves are from polished test specimens, modify the curves for both materials to include the effect of a typical machined surface finish. How do the effects of surface finish differ for the two steels? (Suggestion: Use information from Chapter 10 or from any mechanical engineering design book.)

14.8 For all materials in Table 14.1:

    (a) Plot σ_f' versus the $\tilde{\sigma}_{fB}$ on linear-linear coordinates. How good is the correlation, and what trends are evident?

    (b) Plot ε_f' versus the true fracture strain $\tilde{\varepsilon}_f$ on log-log coordinates, and answer the same questions. (Hint: Calculate $\tilde{\varepsilon}_f$ from % RA.)

14.9 Use the strain-life constants given for each material in Table 14.1 to calculate the strain amplitude ε_a corresponding to $N_f = 1000$ cycles. Then prepare a histogram of the frequency of occurrence of various ε_a values, and comment on the result obtained. (PC Problem)

Section 14.3

14.10 Estimate the life for RQC-100 steel with $\varepsilon_a = 0.004$ as in Ex. 14.2, except change the mean stress to the compressive value of $\sigma_m = -100$ MPa. (PC Problem)

    (a) Using the Morrow parameter.

    (b) Using the modified Morrow approach.

    (c) Using the SWT parameter.

14.11 For aluminum alloy 2024-T351, apply the method indicated to estimate the life corresponding to the following combinations of strain amplitude and mean stress: (PC Problem)

    (a) $\varepsilon_a = 0.008$, $\sigma_m = +70$ MPa,   Morrow parameter

    (b) $\varepsilon_a = 0.008$, $\sigma_m = -70$ MPa,   Morrow parameter

    (c) $\varepsilon_a = 0.008$, $\sigma_m = +70$ MPa,   SWT parameter

    (d) $\varepsilon_a = 0.008$, $\sigma_m = -70$ MPa,   SWT parameter

14.12 Compare Figs. 14.10 and 14.12. Does either the Morrow approach or its modified version appear to be superior for this material? Are there any specific trends of disagreement with the data in either case, and if so, what are these trends?

14.13 Compare Figs. 14.11 and 14.13. Does either the Morrow approach or the SWT approach appear to be superior for this material? Are there any specific trends of disagreement with the data in either case, and if so, what are these trends?

14.14 Consider the normalized amplitude-mean diagram for 7075-T6 Al in Fig. 9.38, and also the constants for this material in Table 14.1.

    (a) Does the Morrow parameter, Eq. 14.13, give good results for this material? Is the agreement improved if σ_f' is replaced by $\tilde{\sigma}_f$? Is the agreement for either σ_f' or $\tilde{\sigma}_f$ better than that for the similar expression using σ_u, Eq. 9.13?

    (b) Based on (a), write Eq. 14.20 in the form that you would use for this material, making any modifications necessary.

14.15 Some strain-life data at nonzero mean stresses are given below for prestrained 2024-T4 Al.

    (a) Using these values, plot data points of the SWT parameter, $\sigma_{max}\varepsilon_a$, versus N_f on log-log coordinates. Also plot the curve expected from the constants in Table 14.1, and comment on the success of the SWT parameter for this material. (PC Problem)

(b) Using the modified Morrow approach, Eq. 14.21, plot the members of the family of strain-life curves that correspond to mean stresses of $\sigma_m = 0$, 72, 142, and 290 MPa. Then plot the test data and comment on the agreement with the curves.

ε_a	σ_a, MPa	σ_m, MPa	N_f, cycles
0.00345	245	71.7	37,800
0.00232	165	71.7	244,600
0.00172	122	71.7	760,000
0.00410	291	142	11,000
0.00303	215	141	30,000
0.00254	181	144	58,500
0.00198	143	143	158,000
0.00148	105	142	437,100
0.00121	85.5	144	820,000
0.00250	178	292	27,700
0.00149	106	292	200,000
0.00109	76.5	287	747,000

Source: Data in [Topper 70].

14.16 Using the data for prestrained 2024-T4 Al from Prob. 14.15: (PC Problem)

(a) Plot ε_a versus N^* from Eq. 14.17 on log-log coordinates. Also plot the strain-life curve from the material constants, and comment on the success of the Morrow parameter for this data.

(b) If the correlation of the data on the ε_a vs. N^* plot from (a) is not very good, try replotting the data with N^* calculated using the true fracture strength $\tilde{\sigma}_f$ in place of σ_f' in Eq. 14.17. Is the agreement improved?

Section 14.4

14.17 Some shear strain vs. life data from torsion of thin-walled tubes are given below for hot-rolled and normalized SAE 1045 steel. The shear strain amplitude γ_a was constant in each test, and the τ_a and γ_{pa} values were measured near $N_f/2$. Some properties of this steel are listed in Table 14.1, and the shear modulus is $G = 79.1$ GPa.

γ_a	τ_a, MPa	γ_{pa}	N_f, cycles
0.0250	267	0.0217	502
0.0150	234	0.0120	1,372
0.0082	197	0.0057	6,998
0.0050	165	0.0029	33,840
0.0040	158	0.0020	70,020
0.0030	148	0.0012	546,000

Source: Data in [Leese 85].

(a) Plot these data on log-log coordinates, compare them with the curve predicted by Eq. 14.32, and comment on the success of this equation.

(b) Add the γ_a vs. N_f equation to your plot that is obtained by using the hydrostatic stress adjustment of Mowbray. Does this improve the agreement with the data?

14.18 For the hot-rolled and normalized SAE 1045 steel of Table 14.1, plot a family of estimated strain-life curves, ε_{1a} vs. N_f, for biaxial loadings specified by $\lambda = -1.0, -0.5, 0, 0.5,$ and 1.0. How is the curve for $\lambda = -1$ related to the γ_a vs. N_f curve for pure shear? (PC Problem)

14.19 Derive an equation analogous to Eq. 14.31 for ε_{1a} vs. N_f for the case of axisymmetric loading.

$$\sigma_{2a} = \sigma_{3a} = \lambda' \sigma_{1a}$$

Also, for the hot-rolled and normalized SAE 1045 steel of Table 14.1, plot the particular curve that applies for $\lambda' = 0.5$, along with the one for uniaxial loading, and comment on the comparison.

Section 14.5.1
(All PC Problems)

14.20 For RQC-100 steel, a plate with a round hole has dimensions, as defined in Fig. 10.1(a), of $w = 150, d = 12,$ and $t = 2.5$ mm. Estimate the life to fatigue cracking for the following constant amplitude cyclic loadings:

(a) $\Delta P = 120$ kN at $R = -1$.

(b) $\Delta P = 165$ kN at $R = -1$.

(c) $\Delta P = 120$ kN at $R = 0$.

(d) $\Delta P = 165$ kN at $R = 0$.

14.21 For the same component and material as in Prob. 14.20, estimate the life for constant amplitude cyclic loading between $P_{max} = 130$ and $P_{min} = 35$ kN.

14.22 A bar made from a flat plate of RQC-100 steel has dimensions, as defined in Fig. 10.1(d), of $w_2 = 60, w_1 = 50,$ fillet radius $\rho = 1.25,$ and thickness $t = 6.5$ mm. It is subjected to constant amplitude cyclic bending in the plane of the plate between a maximum moment of $M_{max} = 1200$ N·m and an oppositely directed minimum moment of $M_{min} = -100$ N·m. Estimate the life to fatigue cracking.

14.23 Plates with round holes of 2024-T351 Al are axially loaded and have an elastic stress concentration factor from Fig. 10.1(a) of $k_t = 2.4$. Make estimates of the life to fatigue cracking for the following constant amplitude loadings:

(a) $\Delta S = 275$ MPa at $R = -1$.

(b) $\Delta S = 410$ MPa at $R = -1$.

(c) $\Delta S = 240$ MPa at $R = 0$.

(d) $\Delta S = 340$ MPa at $R = 0$.

Your results for (c) and (d) should be in reasonable agreement with the stress-strain responses of Fig. 14.20 and the S_{max} versus N_f curve of Fig. 14.21.

14.24 In selecting materials for use in notched components, a useful basis of comparison is to plot the quantity $k_t S_a$ based on Neuber's rule versus cycles to failure.

$$k_t S_a = \sqrt{\sigma_a \varepsilon_a E}$$

The right hand side can be calculated from the stress-life and strain-life equations, and the left hand side is a measure of the fatigue strength of a notched member of given k_t for completely reversed loading.

(a) Compare the four heat treatments of SAE 4142 steel from Table 14.1 on this basis, and comment on the comparison.

(b) Why is this approach better than simply comparing strain-life curves?

14.25 Rectangular beams of RQC-100 steel have a depth of $2c = 6.35$ mm and a thickness of $t = 12.7$ mm. Five of these were subjected to completely reversed cyclic pure bending under controlled edge strains, resulting in fatigue lives as listed below, where M_a is the stable moment measured near $N_f/2$.

(a) Use the strain-based approach to estimate moment amplitudes corresponding to several fatigue lives over the range $N_f = 10^2$ to 10^6 cycles. Then plot a curve of M_a versus N_f. Also plot the test data and comment on the comparison with the curve.

(b) Also plot the M_a vs. N_f curve using one of the S-N curve estimates from Chapter 10, and comment on the success of this estimate compared to (a).

(c) The strains for these tests are given in Prob. 13.23. Does the strain-life curve for bending agree with that for axial loading?

No.	M_a, Moment Amplitude, N·m	N_f, cycles
1	39.0	240,000
2	46.4	32,990
3	58.6	4,882
4	68.9	1,335
5	79.0	438

Source: Data courtesy of R. W. Landgraf; see [Dowling 78].

Sections 14.5.2 to 14.5.4
(All PC Problems)

14.26 AISI 4340 steel with $\sigma_u = 1468$ MPa is made into a notched shaft that is loaded in bending and has a stress concentration factor of $k_t = 3.0$. This shaft is a part in a helicopter, and the nominal bending stress history for each flight in simplified form is shown below. Estimate the number of flights necessary to develop a crack in this part if $S_{a2} = 70$ MPa for: (a) $S_{max} = 275$ MPa, and (b) $S_{max} = 480$ MPa. Make a qualitative sketch of the local stress-strain response to guide your solution.

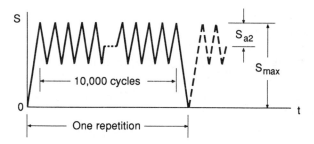

Figure P14.26

14.27 A notched tension member of RQC-100 steel has a stress concentration factor of $k_t = 2.5$. How many repetitions of the nominal stress history shown below can be applied before fatigue cracking is expected? Make a qualitative sketch of the local stress-strain response to guide your solution.

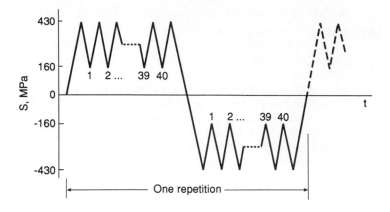

Figure P14.27

14.28 The load histories of Fig. 14.23(a) and (b) are applied to AISI 4340 steel ($\sigma_u = 1468$ MPa) made into the form of axially loaded plates with a round hole having $k_t = 2.5$. The completely reversed overload cycles are applied at a nominal stress amplitude of $S_{a1} = 700$ MPa, and they occur at intervals of 2×10^5 cycles of the lower level loading, which is applied at $S_{a2} = 205$ MPa.

  (a) Estimate the life for the two different types of overload and comment on any effect of the overload sequence.

  (b) Modify your analysis to include an estimate of relaxation of the local mean stress and the effect of this on life. Is the effect significant? (Suggestion: See Eqs. 12.60 and 12.61.)

14.29 The nominal stress history shown below is applied to AISI 4340 steel ($\sigma_u = 1468$ MPa) made into the form of an axially loaded plate with a round hole having $k_t = 2.5$. The value of S_{max} is 620 MPa and $N_2 = 3000$ cycles.

  (a) Estimate the number of repetitions required to cause fatigue cracking. Draw a qualitative sketch of the local stress-strain response to guide your solution.

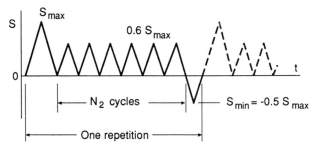

Figure P14.29

(b) Estimate the fatigue limit for this steel for a finely ground surface finish. Does this value affect your analysis in (a)?

14.30 Hot-rolled and normalized SAE 1045 steel is used to make a notched bending member having an elastic stress concentration factor of $k_t = 3.0$. The following sequence of nominal stress peaks and valleys is repeatedly applied: $-300, +100, -170, +100, -170, +300, -100, +300$ MPa, and then back to -300 MPa to start over. Estimate the number of repetitions of this sequence required to cause fatigue cracking.

14.31 For the vehicle transmission load history of Fig. 9.47 applied to the notched test specimen of Man-Ten steel in Fig. 14.19, make life estimates for nominal stress levels of $S_{max} = 175$, 400, and 800 MPa.

(a) Assume that there are no effects of mean stress.

(b) Include mean stress effects by determining the average of upper and lower bounds on life.

Section 14.6

14.32 In Figs. 14.19 and 14.21, the lives for notched members are somewhat longer than calculated from strain-life curves. Comment on the details of this trend and suggest one or more possible reasons for it.

14.33 Consider the situation of Fig. 14.21, and note that fatigue crack growth constants and a K_c value for similar material are given in Table 11.1. The actual test data for $R = 0$ loading and other details are given in Prob. 10.18. (PC Problem)

(a) Estimate numbers of cycles N_{if} required for crack growth for both $S_{max} = 240$ and 340 MPa. (Suggestion: To obtain an initial crack length, collapse the hole of diameter d to form a crack of tip-to-tip length $2a = d$. Then analyze crack growth in the resulting center-cracked plate as an approximation to the actual case.)

(b) Add the crack growth lives N_{if} from (a) to crack initiation lives N_i from Prob. 14.23(c) and (d), where N_i is taken to be the life from strain-based analysis. Is agreement with the test data significantly improved by including crack growth in the analysis?

15

Time-Dependent Behavior: Creep and Damping

15.1 INTRODUCTION

Elastic and plastic strains are commonly idealized as appearing instantly upon the application of stress. Further deformation that occurs gradually with time is called *creep strain*. Creep is often important in engineering design, as in applications involving high temperature, such as steam turbines in power plants, jet and rocket engines, and nuclear reactors. Some more mundane examples of creep are the failure of light-bulb filaments, the gradual loosening of plastic eyeglass frames, slow deformation leading to rupture of plastic pipe, and the movement of ice in glaciers.

For metals and crystalline ceramics, creep deformation is sufficiently large to be of importance in a given material only above a temperature that is generally in the range of 30 to 60% of its absolute melting temperature. Large creep strains can occur in polymers and glasses above the particular material's glass transition temperature T_g, as discussed in Chapters 2 and 3. Thus, polymers that are in a leathery or rubbery state are susceptible to creep, and this is often the case even at room temperature. Concrete creeps at room temperature, but the process becomes slower with time, so that only small additional strains occur after the first year or so.

Selection of an appropriate material is likely to be a critical factor in a creep-sensitive design. Engineering metals used in high temperature service generally contain alloying elements such as chromium, nickel, and cobalt, with the percentages of these expensive materials increasing with temperature resistance. The large differences that exist in temperature resistance among various classes of engineering metals are illustrated by short-time creep data in Figure 15.1. New tough ceramic materials offer opportunities for greater temperature resistance than even the best metal alloys. Conversely, polymers are severely limited in their temperature resistance, but some are of course more resistant than others.

Other environmental effects, such as oxidation and environmental cracking, are also more likely to cause difficulty as chemical activity increases with temperature. For example, the resistance to oxidation in air and creep-related failure are compared for various classes of heat-resisting metals in Figs. 15.2. Although such other environmental factors are not treated in this chapter, they nevertheless also need to be considered in design for high-temperature service. Another complexity that often occurs in engineering situations is cyclic loading combined with time-dependent deformation. A harmful creep-fatigue interaction that accelerates the fatigue process can then occur. In other situations, the dissipation of energy resulting from time-dependent deformation during cycle loading, called *materials damping*, may be beneficial in limiting mechanical vibrations.

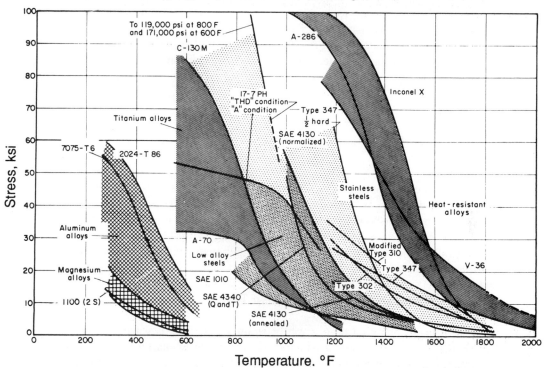

Figure 15.1   Stress vs. temperature for 3% total deformation in 10 min for various engineering metals. (From [van Echo 58]; used with permission.)

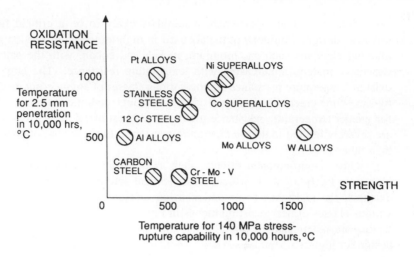

Figure 15.2 Relative creep and oxidation resistance of various classes of engineering metals. (Adapted from [Sims 78]; used with permission.)

The engineering methods that have been developed for analyzing and predicting creep behavior provide tools that can be used in design to avoid failures due to creep. One concern is excessive deformation. Another is *creep rupture*, which is a separation (fracture) of the material that can occur as a result of the creep process. In what follows, we will first consider creep testing and physical mechanisms, which topics provide needed introductory information prior to emphasizing engineering methods in the remainder of the chapter.

15.2 CREEP TESTING

The most common method of creep testing is simply to apply a constant axial load, either in tension or compression, to a bar or cylinder of the material of interest. Since the load is to be held constant for long periods of time, simple dead weights and a lever system may be used as shown in Fig. 15.3. The creep strain is measured with time, and the time at rupture is recorded if this occurs during the test. Tests on a given material are generally done at various stresses and temperatures, and test durations can range from less than one minute to several years.

15.2.1 Behavior Observed in Creep Tests

The behavior observed on a graph of strain versus time is usually similar to Fig. 15.4. There is an initial nearly instantaneous occurrence of elastic and perhaps also plastic strain, followed by the gradual accumulation of creep strain. The strain rate, $\dot{\varepsilon} = d\varepsilon/dt$, hence the slope of the ε vs. t plot, is at first relatively high. However, $\dot{\varepsilon}$ decreases and often becomes approximately constant, at which point the *primary* or *transient* stage of

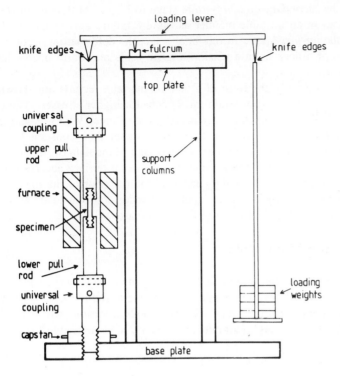

Figure 15.3   Schematic of a creep testing machine. (From [Evans 85] p. 39; used with permission.)

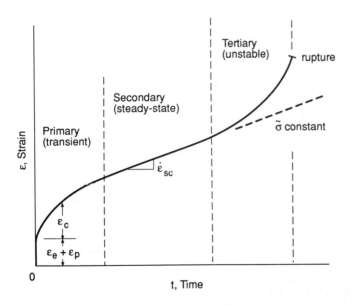

Figure 15.4   Strain vs. time behavior during creep under constant load, hence constant engineering stress, and the three stages of creep.

creep is said to end, and the *secondary* or *steady-state* stage to begin. At the end of the secondary stage, $\dot{\varepsilon}$ increases in an unstable manner as rupture failure approaches, with this portion being called the *tertiary* stage. In this final stage, the deformation becomes localized by the formation of a neck as in a tension test, or voids may form inside the material, or both may occur.

Some creep data for a metal in the form of strain versus time records are shown in Fig. 15.5. As might be expected, higher strain rates occur for higher stresses. These data are from the work of Andrade, a notable early investigator of creep who was active around 1910, and were obtained under constant true stress, rather than constant load, using the apparatus shown in Fig. 15.6. No tertiary stage occurs in the data of Fig. 15.5, probably as a result of the constant true stress. Note that the usual tertiary acceleration of creep is in at least some cases due not to any change in the behavior of the material itself, but rather due to the decreasing cross-sectional area under constant load, which causes the true stress to increase, hence also causing $\dot{\varepsilon}$ to increase.

15.2.2 Representing Creep Test Results

The results of a single creep test can be summarized by giving the following four quantities: stress σ, temperature T, steady-state creep rate $\dot{\varepsilon}_{sc}$, and time to rupture t_r. A variety of alternatives exist for presenting the data from a series of tests at various stresses and temperatures. For example, stress versus strain rate, σ versus $\dot{\varepsilon}_{sc}$, can be plotted for each

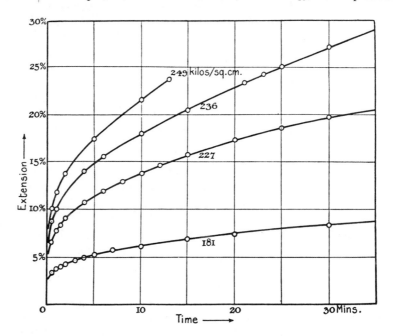

Figure 15.5 Creep curves for lead at 17°C from the early work of Andrade, where 1 kg/cm² = 0.0981 MPa. (From [Andrade 14]; used with permission.)

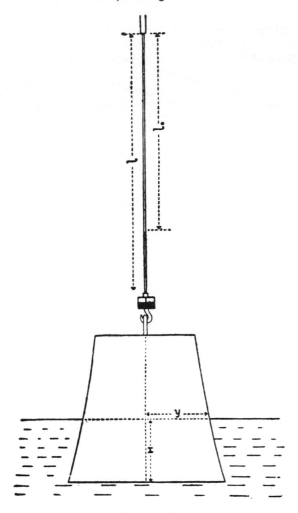

Figure 15.6 Apparatus used by Andrade for creep testing under constant true stress. The mass M is shaped according to $y = \frac{1}{l_o+x}\sqrt{\frac{Ml_o}{\rho\pi}}$ and is suspended in a liquid of mass density ρ, so that the force decreases with strain to maintain constant true stress during uniform elongation. (From [Andrade 10]; used with permission.)

of several temperatures as illustrated in Fig. 15.7. Another useful presentation is to plot stress versus life, σ versus t_r, for various temperatures as illustrated by Fig. 15.8. As seen in these examples, logarithmic scales are often used.

A σ versus t_r plot is analogous to an S-N curve for fatigue, except that the life is a rupture time rather than a number of cycles. Stress-life data are also often represented by plotting σ vs. T with lines for different times to failure t_r. Some curves plotted on this basis for three tempers of an alloy steel are shown in Fig. 15.9. A plot of this type is of course simply an alternative means of presenting the same information as a plot similar to Fig. 15.8. In addition to the rupture life, the life to reach a particular value of strain may be of interest. Stress-life plots for three values of strain, and also for rupture, all for a single temperature, are shown for a polymer in Fig. 15.10.

Stress-strain curves for various constant values of time, called *isochronous stress-strain curves*, are often needed. These are constructed from strain versus time data

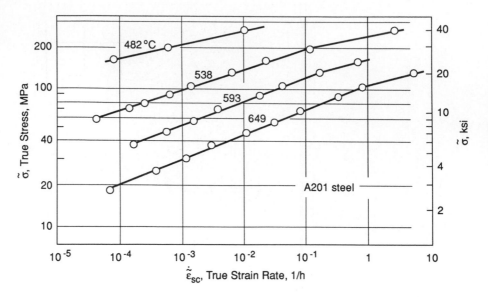

Figure 15.7 Stress versus steady-state strain rate at various temperatures for a carbon steel used for pressure vessels. (Adapted from [Randall 57]; copyright ©ASTM; reprinted with permission.)

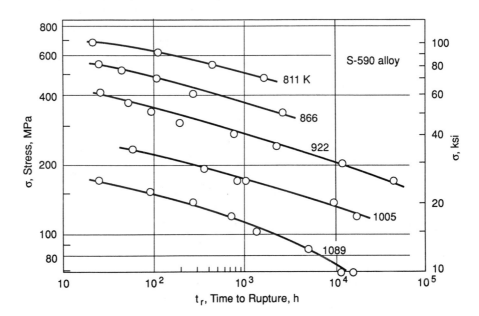

Figure 15.8 Stress vs. rupture life curves for S-590, an iron-based heat-resisting alloy. (Data from [Goldhoff 59a].)

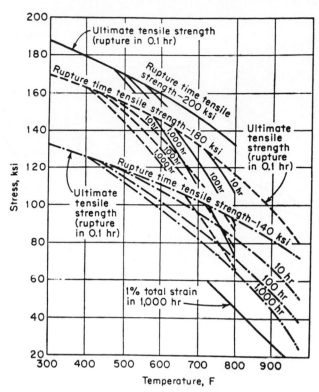

Figure 15.9 Creep rupture behavior for three tempers of AISI 4340 steel, plotted as stress vs. temperature curves for various lives. (From [Horger 65] p. 16; used with permission of ASME.)

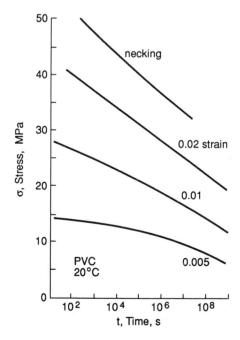

Figure 15.10 Stress-life curves for unplasticized polyvinyl chloride for three values of strain, and also for the onset of rupture by necking. (Adapted from [EVC 89]; used with permission.)

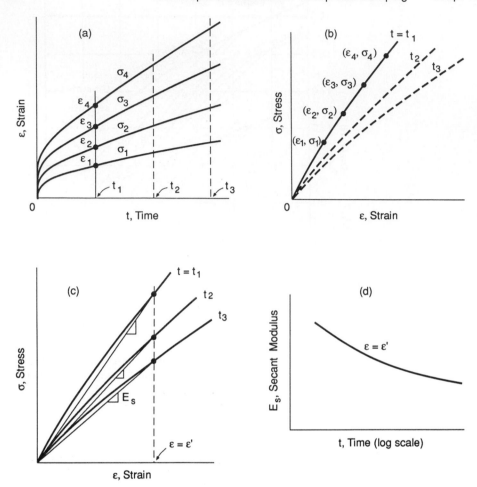

Figure 15.11 Construction of isochronous stress-strain curves and secant modulus curves. Strain-time data (a) for several stress levels can be used to obtain isochronous σ-ε curves as in (b). Secant moduli E_s can then be obtained from the isochronous σ-ε curves as shown in (c), and E_s is often plotted vs. the corresponding time as in (d).

for several stress levels as shown in Fig. 15.11(a) and (b). Strains corresponding to a particular time, such as $t = t_1$, are obtained as shown in (a). These are then plotted versus the corresponding stress values as in (b), forming the isochronous stress-strain curve for time t_1. Similar curves can be constructed for other values of time, such as t_2 and t_3, so that a family of stress-strain curves is obtained. For polymers, it is a common practice to use isochronous stress-strain curves to determine secant moduli, E_s, corresponding to specific values of strain, as shown for strain $\varepsilon = \varepsilon'$ in (c). Such E_s values are then plotted versus time to characterize the behavior of the material as in (d). Secant modulus

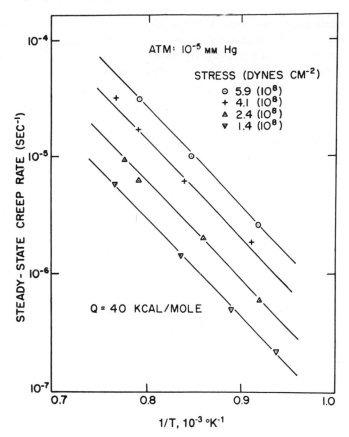

Figure 15.12   Steady-state creep rate vs. reciprocal of absolute temperature for single crystals of titanium oxide ceramic, TiO_2, also called *rutile*, tested under compression in a vacuum, where 1 dyne/cm^2 = 10^{-7} MPa. (Adapted from [Hirthe 63]; reprinted by permission of the American Ceramic Society.)

versus time curves for more than one value of strain may be obtained and plotted as a family of curves.

Of particular significance are plots of $\dot{\varepsilon}_{sc}$ versus $1/T$, the reciprocal of absolute temperature, where a log scale is used for $\dot{\varepsilon}_{sc}$, and a linear scale for $1/T$. Such a plot with lines for various stresses is shown for a ceramic in Fig. 15.12. Also useful is a similar plot with t_f replacing $\dot{\varepsilon}_{sc}$. Slopes on these plots give apparent *activation energies*, a subject that is discussed below.

15.3 PHYSICAL MECHANISMS OF CREEP

The physical mechanisms causing creep differ markedly for different classes of materials. In addition, even for a given material, different mechanisms act at various combinations of stress and temperature. Motions of atoms, vacancies, dislocations, or molecules within a solid material occur in a time-dependent manner, and they occur more rapidly at higher temperatures. Such motions are important in explaining creep behavior and fall within the broad category of behavior called *diffusion*.

15.3.1 Viscous Creep

The viscosity of a liquid, as employed in fluid mechanics, is the ratio of the applied shear stress to the resulting rate of shear strain.

$$\eta_\tau = \frac{\tau}{\dot{\gamma}} \tag{15.1}$$

where η_τ is the shear viscosity. Similarly, a tensile viscosity can be defined.

$$\eta = \frac{\sigma}{\dot{\varepsilon}} \tag{15.2}$$

For an ideal viscous substance that is incompressible, $\eta = 3\eta_\tau$. Also, a constant value of η corresponds to the behavior of an ideal dashpot with force vs. displacement behavior of the form $P = c\dot{x}$, where the constant c is analogous to η. (See Section 4.2).

Some nominally solid materials behave in a manner similar to a liquid with a very high viscosity. This is generally the case for amorphous solids such as silica glass and some polymers. An amorphous (glassy) solid is formed if a liquid is very viscous just before freezing, as this results in the molecules being unable to arrange themselves into an orderly crystalline array. Also, materials that are normally crystalline can be made in an amorphous form by very rapid cooling, so that there is insufficient time for crystallization.

In response to a stress applied to an amorphous solid, molecules or groups of molecules move relative to one another in a time-dependent manner, resulting in creep deformation. Such molecular motions constitute a diffusion process that is enhanced if the temperature is increased. This occurs because a temperature increase is intimately related to an increase of the average oscillations of atoms about their equilibrium positions. Greater oscillations result in more frequent stress-driven molecular rearrangements that contribute to creep deformation.

Such a situation is a case of *thermal activation*. From the physics involved, the rate of a thermally activated process is expected to be governed by an equation of the following form, which is called the *Arrhenius equation*.

$$\dot{\varepsilon} = Ae^{\frac{-Q}{RT}} \tag{15.3}$$

where the rate involved here is the strain rate $\dot{\varepsilon}$, and Q is a special physical constant called the *activation energy*. It is a measure of the energy barrier that must be overcome for molecular motion to occur. For Q in units of calories/mole and absolute temperature T in kelvins (K), the quantity $R \approx 2$ is the universal gas constant in units of cal/(K·mole).

The value of Q for viscous flow is ideally a constant for a given material, but may change if the physical mechanism is altered due to a sufficient shift in temperature or stress.

Equation 15.3 in effect specifies the variation with temperature of the viscosity η of Eq. 15.2. A combined equation can thus be written that gives the effects of both stress and temperature.

$$\dot{\varepsilon} = A_1 \sigma e^{\frac{-Q}{RT}} \qquad (15.4)$$

where the new coefficient A_1 depends mainly on the material, but as for Q it may change if the physical mechanism is altered due to a sufficient shift in temperature or stress. The proportionality of strain rate and stress seen above is a key feature of creep due to viscous flow. The form of the temperature dependence is similar for all thermally activated rate processes, including some other creep mechanisms to be discussed below.

15.3.2 Creep in Polymers

At temperatures below the glass transition temperature T_g of a given polymer, creep effects are relatively small. Above T_g, creep effects rapidly become significant. As T_g for common polymers is often in the range -100 to $+200°C$, this temperature may be exceeded around or even below room temperature. For a primarily crystalline thermoplastic such as polyethylene, viscous flow occurs at temperatures substantially above T_g, especially upon approaching the melting temperature. Note that the secondary (hydrogen and van der Waals) bonds that hold the carbon-based molecular chains to one another below T_g are weak above T_g. Thus, creep can occur by the molecular chains sliding past one another in a viscous manner. The process is made easier in linear polymers if the molecular chains are shorter, and it is enhanced by the absence of obstacles to sliding, such as bulky side groups on the molecules, or cross-linking between the chain-like molecules. The stress and temperature dependence of this type of viscous flow at least roughly obey Eq. 15.4.

However, the behavior is more complicated at intermediate temperatures, only modestly above T_g, where the behavior is leathery or rubbery. Here, sliding of the molecular chains is more difficult, and they are more easily entangled with one another, particularly if the chains are long. Such entanglements give the material an increasing resistance as deformation proceeds, so that the rate of creep decreases as deformation progresses. Other obstacles to sliding, such as bulky side groups or cross-linking, have a similar effect, so that these also tend to limit creep if they are present.

Such obstacles to sliding of the molecules also give the material a memory of its before-deformation shape. In particular, after removal of the applied stress, the stretched and distorted chain segments between entanglement or cross-link points act somewhat like springs that tend to cause the prior creep deformation to disappear (recover) with time. This self-limiting creep and recovery behavior is a departure from simple viscous behavior that is similar to that of the transient creep rheological model of Fig. 4.14(c). In a real polymer, elastic deformation is of course also present, and some nonrecoverable viscous deformation occurs in addition to the recoverable portion. The simplest rheological model that has behavior even roughly similar to that of a polymer is thus one that combines elastic, steady-state creep, and transient creep elements as shown in Fig. 15.13.

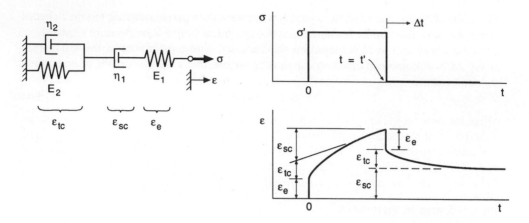

Figure 15.13 Creep and recovery behavior in a rheological model that combines elastic, steady-state creep, and transient creep elements. The elastic strain ε_e in spring E_1 is recovered immediately upon unloading, whereas the transient creep strain ε_{tc} in parallel combination (E_2, η_2) is recovered slowly with time. The steady-state creep strain ε_{sc} in dashpot η_1 is never recovered.

15.3.3 Diffusional Flow in Crystalline Materials

Crystalline materials commonly used in engineering include metals and their alloys and the engineering ceramics. Some ceramics contain a crystalline phase in combination with an glassy phase, such as porcelain and fired clay brick. As a result of their similarities in structure, crystalline materials (or phases) have roughly similar physical mechanisms for creep deformation.

The mechanisms most frequently encountered can be separated into two broad classes, which are termed *diffusional flow* and *dislocation creep*. Diffusional flow can occur at low stress but requires relatively high temperature. This mechanism involves the movement of vacancies (holes) in the crystal lattice, as illustrated by Fig. 15.14.

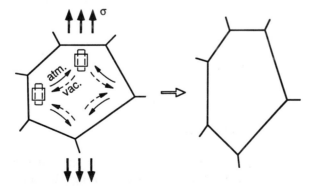

Figure 15.14 Mechanism of creep due to diffusional flow of vacancies between grain boundaries in a crystalline material.

The boundaries between crystal grains are disorderly regions from which vacancies can move into the crystal lattice. Although the anisotropy of the crystal grains and boundaries causes complex variations of stress and strain, grain boundaries that are oriented normal to an applied tensile stress are generally subjected to tensile strains normal to their surfaces, and those parallel to compressive strains. Vacancies are less easily accommodated near boundaries under tensile strain, as there is an increased likelihood that an atom or ion from elsewhere will move in to fill the space, and conversely more easily accommodated near boundaries that are under compressive strain. The result is that there is a movement of vacancies as shown by the dashed lines in Fig. 15.14. This is equivalent to an opposite movement of atoms or ions as shown by the solid lines. Continued transfer of material by this means causes the grain to deform, contributing to a macroscopic strain, with the deformation being time dependent as a result of the diffusion process requiring a finite time to occur.

If the vacancies move through the crystal lattice, the behavior is called *Nabarro-Herring creep*. Also, the resulting strain rate is approximately proportional to the stress and inversely proportional to the square of the average grain diameter.

$$\dot{\varepsilon} \propto \frac{\sigma}{d^2} \tag{15.5}$$

However, if the vacancies instead move along grain boundaries, the behavior is termed *Coble creep*. The dependence on stress is similar, but the dependence on grain size is altered.

$$\dot{\varepsilon} \propto \frac{\sigma}{d^3} \tag{15.6}$$

This proportionality of strain rate and stress indicates that both types of diffusional flow are essentially viscous processes. In addition, as any diffusion process is expected to be thermally activated, Eq. 15.3 applies, and the stress and temperature dependence are thus similar to the form of Eq. 15.4. However, more detailed analysis and also comparison with experimental data indicate that an inverse temperature dependence also appears in the coefficient. The resulting relationships are thus

$$\dot{\varepsilon} = \frac{A_2 \sigma}{d^2 T} e^{\frac{-Q_v}{RT}} \quad (N\text{-}H \text{ creep})$$

$$\dot{\varepsilon} = \frac{A_2' \sigma}{d^3 T} e^{\frac{-Q_b}{RT}} \quad (\text{Coble creep}) \tag{15.7}$$

where A_2 and A_2' are new material constants. The activation energies Q_v and Q_b for creep are similar to those for diffusion, Q_v for self-diffusion of the material in its own crystal lattice, and Q_b for diffusion along grain boundaries.

15.3.4 Dislocation Creep in Crystalline Materials

Dislocation creep involves the more drastic motion of dislocations, which are line defects, rather than only of vacancies, which are point defects. Consequently, high stresses are required, but the effect can occur at intermediate temperatures where diffusional flow is small. The mechanisms are complex and not fully understood, but *dislocation climb* is thought to be important.

Consider Fig. 15.15. Due to the effect of an applied stress, an edge dislocation moves along a crystal lattice plane by the stepwise slip process described in Chapter 2, which is also called *glide*. On encountering an obstacle, such as a precipitate particle or an immobile entanglement of other dislocations, further deformation requires that the dislocation move to another lattice plane. Such a motion is termed *climb* and requires a rearrangement of atoms, again by vacancy diffusion. The cumulative effect of a large number of such climb events is to permit more glide, hence more macroscopic deformation, than could otherwise occur. The deformation is time-dependent because the climb process is time-dependent. As grain boundaries are not a major factor, the strain rate is not strongly affected by grain size, but the resistance to the climb process is such that there is a strong stress dependence.

$$\dot{\varepsilon} \propto \sigma^m \tag{15.8}$$

where m varies with material and test conditions but is typically on the order of 5.

Since a diffusion process is involved, the temperature dependence is still governed by behavior similar to Eq. 15.3. The additional (but relatively weak) inverse dependence on temperature as in Eq. 15.7 occurs here also, but grain size is no longer a primary variable. Hence, the equation has the form

$$\dot{\varepsilon} = \frac{A_3 \sigma^m}{T} e^{\frac{-Q}{RT}} \tag{15.9}$$

Although the details of the mechanism may vary with material, or even with temperature for a given material, the process nevertheless generally appears to be diffusion controlled. This is supported by the fact that Q for high-temperature creep generally agrees with that for self-diffusion of the material in its own crystal lattice.

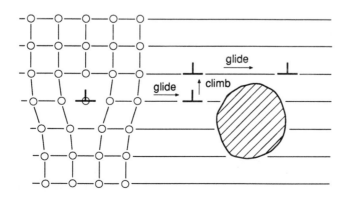

Figure 15.15 Climb of an edge dislocation, permitting continued glide past an obstacle, and enabling deformation to proceed.

15.3.5 Evaluation of Activation Energies

Activation energies for creep can be estimated from slopes of lines on plots of $\log \dot{\varepsilon}$ versus $1/T$ for constant stress, such as the lines in Fig. 15.12. Assume that the simple form of Eq. 15.3 applies and take logarithms to the base ten of both sides.

$$\log \dot{\varepsilon} = \log A - \frac{Q}{RT} \log e = \log A - 0.217 Q \left(\frac{1}{T} \right) \tag{15.10}$$

where for the second form $\log_{10} e$ is evaluated and the substitution $R \approx 2$ cal/(K·mole) is made, so that Q is in units of cal/mole and T in kelvins (K). The units for the logarithmic scale used for $\dot{\varepsilon}$ and the linear scale used for $1/T$ will not generally be the same, so that raw slope values may need to be adjusted before equating them to $0.217Q$.

If Eq. 15.3 is obeyed, the activation energy Q will be the same for all values of stress. Hence, data of $\log \dot{\varepsilon}$ versus $1/T$ for various stresses would form parallel straight lines. This is seen to be at least approximately the case in Fig. 15.12. An analogous procedure may be applied to find activation energies where the behavior obeys Eq. 15.7 or 15.9. In particular, a plot of the quantity $\log(\dot{\varepsilon}T)$ versus $1/T$ produces straight lines all having the same slope proportional to Q, but with intercepts depending on stress σ. Alternatively, if the constant m is known, a plot of $\log(\dot{\varepsilon}T/\sigma^m)$ versus $1/T$ can be made. This produces a single line of slope proportional to Q for all stresses.

Care is needed in employing values of activation energy Q, as large shifts in temperature or stress may change the mechanism sufficiently to alter Q. This is demonstrated by some data for high-purity aluminum shown in Fig. 15.16. In this case, Q for creep at relatively high temperature is found to be independent of both stress and strain, and it is constant and close to the self-diffusion value. However, below about half the absolute melting temperature, Q decreases and becomes variable, as a result of other mechanisms occurring that are not controlled by the same activation energy.

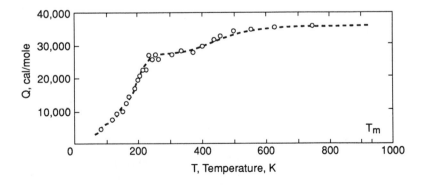

Figure 15.16 Activation energy for creep of pure aluminum as a function of the absolute temperature. (Adapted from [Sherby 57]; reprinted with permission from *Acta Metallurgica*, Pergamon Press, Oxford, UK.)

15.3.6 Deformation Mechanism Maps

For a given material, a *deformation mechanism map* can be drawn that shows what deformation mechanism is dominant for any given combination of stress and temperature. An example for a pure metal is provided by Fig. 15.17. In addition to the creep mechanisms that occur in this crystalline material, a region is shown where elastic deformation is dominant. Above the yield stress, plastic deformation by dislocation glide is the dominant type of deformation. A theoretical limit on the strength is also shown, this corresponding to the theoretical shear strength, $\tau_b \approx G/10$, that can cause shear of crystal planes even if no dislocation motion occurs. (See Section 2.4.)

Details of the map of course differ with the material and its processing. For example, in crystalline materials, diffusional flow is more likely to be an important factor if the grain size is small. This results from the inverse dependence of strain rate on grain size as in Eq. 15.7. Hence, where this mechanism is dominant, large grain size is beneficial. Dislocation creep is relatively less important in crystalline ceramics than in metals, as the crystal lattice in ceramics is inherently more resistant to dislocation motion.

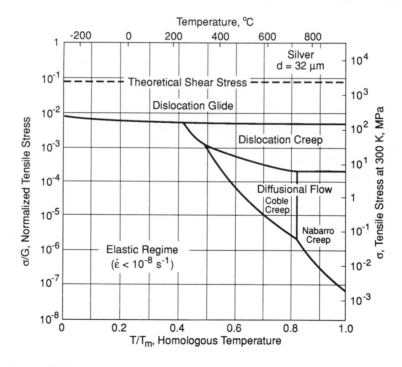

Figure 15.17 Deformation mechanism map of stress vs. temperature for pure silver with a grain size of 32 μm. On the left scale, tensile stresses are normalized to the shear modulus, which decreases with increasing temperature. Homologous temperature T/T_m is the ratio of temperature to melting temperature, both absolute. (Adapted from [Ashby 72]; reprinted with permission from *Acta Metallurgica*, Pergamon Press, Oxford, UK.)

Detailed mechanism maps are given in Frost and Ashby (1982) for a number of metals and ceramics. The more detailed maps include lines of constant strain rate, making them especially useful. Constants for equations similar to those above are also given for each region of the maps. However, many of the materials represented are relatively pure metals and compounds, and only a limited number are materials used for engineering purposes. Until more comprehensive information of this type becomes available, it will be necessary to rely more directly on empirical representation of test data, or on *time-temperature parameters*, as discussed shortly below.

15.3.7 Creep in Concrete

Although concrete can be described in general terms as a crystalline ceramic, its complex structure results in distinctive creep mechanisms and behavior. Concrete contains cement paste that has been chemically combined with water in the hydration reaction that hardens the concrete. There is also unhydrated paste present that is slowly converted by the hydration reaction as time passes. Additional unreacted water is present in pores, between micro-layers of the hydrated cement, and chemically adsorbed (weakly attached by secondary chemical bonds) to the hydrated paste. Finally, there is aggregate (sand and stone) in a gradation of sizes.

Creep in concrete occurs mostly in the cement paste, with the aggregate acting to limit the deformation. Thus, elastic strains build up in the aggregate and oppose the creep in the paste, causing the creep rate to decrease. Data illustrating this trend are shown in Fig. 15.18. The detailed mechanisms of creep in the cement paste are complex and not completely understood. One possibility is that an applied stress squeezes the unreacted water in some voids and causes it to move by viscous flow to another location, permitting a time-dependent distortion of the microstructure. Another is that adsorbed water lubricates layers and particles of hydrated cement and allows them to slide relative to one another. Microcracking at interfaces between aggregate and paste also progresses with time and contributes to creep. In addition, at high stresses, mechanisms similar to those in more ordinary crystalline ceramics may be significant.

It is likely that a combination of some or all of these and perhaps other mechanisms act. Since creep occurs primarily in the cement paste, the elastic deformations that build up in the aggregate reverse the creep strain when the load is removed. This recovery

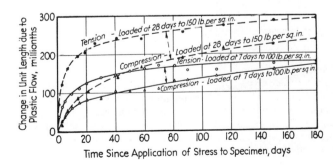

Figure 15.18 Creep strain in μm/m for concrete in tension and compression for two different times of curing. (From [Davis 37]; copyright © ASTM; reprinted with permission.)

behavior, and also the decreasing creep rate under load, is similar to the behavior of the transient creep rheological model of Fig. 4.14(c). However, the creep strain due to microcracking is not recovered on unloading, and general elastic deformation also occurs. Thus, the behavior is roughly similar to the rheological model of Fig. 15.13. More complex models with additional stages, and special nonlinear springs and dashpots, are used in actual applications.

15.4 TIME-TEMPERATURE PARAMETERS AND LIFE ESTIMATES

Creep deformation can proceed to the point of rupture of the material by the development of cracking, crazing, or other damage, which results from the intense strain. For example, in crystalline materials, voids may appear along grain boundaries or at other points of localized stress concentration, such as precipitate particles, by a process called *creep cavitation*. An example is shown in Fig. 15.19. The enlarging and joining of grain boundary or other voids then causes cracks, and can progress to the point of fracture, called creep rupture. However, if the temperature is sufficiently high in a ductile and relatively pure metal, the process of *dynamic recrystallization* can occur, in which these voids are essentially repaired as they try to form. Large deformations are then possible, and failure eventually occurs by necking. Creep rupture of ductile polymers is generally preceded by large uniform or necking deformations.

In engineering design where creep must be considered, there must be neither excessive deformation nor rupture within the desired service life, which is likely to be lengthy, perhaps 20 years or more. However, test-time limitations result in creep data generally being available only out to 1,000 hours (42 days), or sometimes 10,000 h (14 months), but seldom to 100,000 h (11 years). For estimating the behavior at low strain rates and long times, one possible approach is to estimate creep strains for the service temperature of interest by extrapolating the appropriate σ vs. $\dot{\varepsilon}$ curve, such as one of those in Fig. 15.7, back to low $\dot{\varepsilon}$ values. Similarly, rupture lives, or lives to a particular

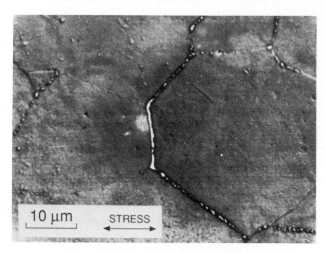

10 μm   STRESS

Figure 15.19 Grain boundary cavitation and cracking due to creep, in a tantalum alloy (T-111), tested under creep-fatigue interaction with temperature variation between 200 and 1150°C. (Adapted from [Sheffler 72].)

strain value, could be estimated by extrapolating stress-life plots, such as Figs. 15.8 or 15.10, to long lives. However, such extrapolations do not work very well, as the slopes of fitted lines on log-linear or log-log plots of σ vs. $\dot{\varepsilon}$ or σ vs. t_r may not be constant, or they may be constant over only limited ranges of these variables. Abrupt slope changes may occur due to a shift in the creep mechanism, so that extrapolation is not valid. In other words, one cannot extrapolate across the boundaries for various creep mechanisms on a deformation mechanism map.

A more successful approach is to use data from relatively short time tests, but at temperatures above the service temperature of interest, to estimate the behavior for the longer time at the service temperature. Under these circumstances, a common physical mechanism for tests and service is more likely than for extrapolation at a constant temperature. Such an approach involves the use of a *time-temperature parameter*. We will consider three approaches of this type, namely the Sherby-Dorn parameter, the Larson-Miller parameter, and the Manson-Haferd parameter.

15.4.1 Sherby-Dorn (*S-D*) Parameter

The Arrhenius rate equation is the basis of the Sherby-Dorn (*S-D*) time-temperature parameter. A key assumption is that the activation energy for creep is constant. First, write Eq. 15.3 in differential form and note that the coefficient is a function of stress, $A = A(\sigma)$.

$$d\varepsilon = A(\sigma)\, e^{\frac{-Q}{RT}}\, dt \tag{15.11}$$

Then integrate both sides of the equation, and discard the constant of integration so that only the steady-state creep strain appears.

$$\varepsilon_{sc} = A(\sigma)\, t e^{\frac{-Q}{RT}} \tag{15.12}$$

This equation suggests that creep strains for a given stress form a unique curve if plotted versus the quantity

$$\theta = t e^{\frac{-Q}{RT}} \tag{15.13}$$

which is termed the *temperature-compensated time*. Some supporting test data for aluminum alloys are shown in Fig. 15.20.

To formally define the *S-D* parameter, we use logic as follows: The creep strain at rupture is observed to be fairly constant for a given value of temperature-compensated time to rupture, θ_r, as for the data for any one material in Fig. 15.20. Hence, θ_r depends only on stress. Now take logarithms to the base ten of both sides of Eq. 15.13, noting that at $\theta = \theta_r$ the time is $t = t_r$, the rupture time.

$$P_{SD} = \log \theta_r = \log t_r - 0.217 Q \left(\frac{1}{T}\right) \tag{15.14}$$

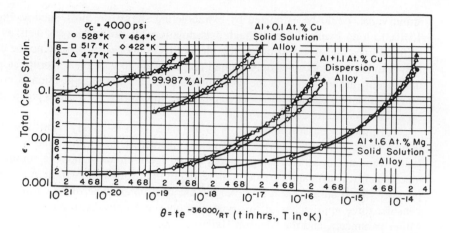

Figure 15.20 Creep strain vs. temperature-compensated time for aluminum and dilute alloys tested at $\sigma = 27.6$ MPa at various temperatures. (From [Orr 54]; used with permission.)

where P_{SD} is the S-D parameter, and $\log_{10} e$ and $R \approx 2$ cal/(K·mole) are evaluated. Units of hours are employed for t_r.

Values of the activation energy Q for use with the S-D parameter can be obtained by plotting creep-rupture data on coordinate axes of $\log t_r$ versus $1/T$ as illustrated in Fig. 15.21. A family of parallel straight lines is expected, one for each value of stress. These lines all have slopes given by $0.217Q$, and each intercept at $1/T = 0$ can be interpreted as the P_{SD} value for that stress. Hence, to calculate values of both Q and P_{SD}, the minimum information needed is two points on such a straight line for constant stress. Some typical values of Q are 90,000 cal/mole for various steels and

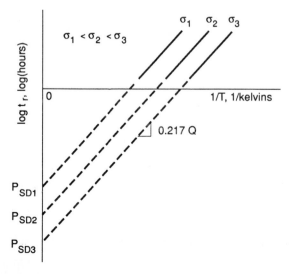

Figure 15.21 Graphical interpretation for the Sherby-Dorn parameter, with constant slope proportional to the activation energy Q.

stainless steels, and 36,000 cal/mole for pure aluminum and dilute alloys. These and a few additional values for specific engineering metals are listed in Table 15.1.

Once Q is known, stress-life data can be employed to make a plot of P_{SD} versus stress as shown for a heat-resisting iron-base alloy in Fig. 15.22. The data for all stresses

TABLE 15.1 SOME VALUES OF CONSTANTS FOR TIME-TEMPERATURE PARAMETERS[1]

Material	Sherby-Dorn	Larson-Miller	Manson-Haferd	
	Q, cal/mole	C	T_a, K	log t_a
Various steels and stainless steels[2,3]	$\approx$ 90,000	$\approx$ 20	—	—
Pure aluminum and dilute alloys[2]	$\approx$ 36,000	—	—	—
S-590 alloy (Fe base)[4]	85,000[5]	17	172	20
A-286 stainless steel[4]	91,000	20	367	16
Nimonic 80A (Ni base)[4]	91,000	18	311	16
1Cr-1Mo-0.25V steel[4]	110,000	22	311	18

Note: [1]Values given are for temperatures T in kelvins (K) and rupture times t_r in hours.

Sources: [2]Values in [Orr 54].   [3]Value in [Larson 52].   [4]Values in [Goldhoff 59a] and [Goldhoff 59b].   [5]Revised value from [Conway 69].

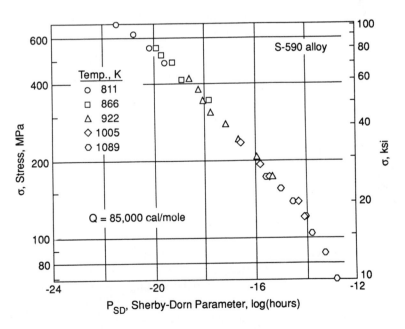

Figure 15.22 Correlation using the Sherby-Dorn parameter of creep-rupture data for S-590 alloy. (Data from [Goldhoff 59a].)

and temperatures should fall together along a single curve, with the success of such correlation of the data being a measure of the success of the parameter for any particular set of data. Using such a plot and Eq. 15.14, rupture times t_r can be determined for particular values of stress and temperature. The test data used to obtain the P_{SD} versus σ plot generally involve shorter rupture times than the service lives of interest. Hence, test data at relatively short t_r and high temperature are being used to predict the behavior at longer t_r and lower temperature.

Service lives in creep situations may be limited by excess deformation rather than by rupture. It is then useful to identify a particular value of creep strain, such as 1% or 2%, that is considered to represent failure. The S-D parameter can be used in this situation also, the rupture life t_r in Eq. 15.14 being simply replaced by t_f, the time to reach the strain of interest. The values of P_{SD} used of course need to be obtained from data for t_f rather than t_r.

Example 15.1

An engineering component made of an alloy steel is subject to creep under simple tension at a stress of 150 MPa. What is the highest temperature that can be permitted if the component must function for 40 days, and if a safety factor of 10 on life is required? The same material was subjected to creep in a test at 150 MPa at 530°C, in which it ruptured in 260 hours.

Solution   Since the stress is the same in the test as in service, a value of the Sherby-Dorn parameter of Eq. 15.14 can be calculated from the test data and used to make life estimates for the component in service. This value is

$$P_{SD} = \log t_r - \frac{0.217Q}{T} = \log (260 \text{ h}) - \frac{0.217(90,000)\text{K}}{(530 + 273)\text{K}} = -21.91.$$

In calculating P_{SD}, the units of $0.217Q$ are kelvins (K) if Q is in cal/mole, so that the test temperature of 530°C is converted to units of K when it is substituted. Lacking more specific information, the approximate value of Q from the first line of Table 15.1 was employed.

For a service life of 40 days and a safety factor of 10 on life, the component must be able to survive for 400 days, that is, $t_r = 9600$ hours. Equation 15.14 can be solved for temperature T, and this t_r substituted along with P_{SD} from above.

$$T = \frac{0.217Q}{\log t_r - P_{SD}} = \frac{0.217(90,000)}{\log (9600 \text{ h}) - (-21.91)} = 754 \text{ K} = 481°C \qquad \textbf{Ans.}$$

15.4.2 Larson-Miller (*L-M*) Parameter

The time-temperature parameter of Larson and Miller is an analogous approach to that of Sherby and Dorn, but different assumptions and therefore different equations are used. The L-M parameter can also be derived starting from Eq. 15.13 by substituting $\theta = \theta_r$ and $t = t_r$ for the rupture time, and similarly, taking logarithms to the base ten of both sides. However, we then rearrange the result to obtain

$$P_{LM} = 0.217Q = T (\log t_r + C) \qquad (15.15)$$

where

$$C = -\log \theta_r$$

Units of kelvins (K) for T and hours for t_r will be used here. However, in much of the literature related to the $L\text{-}M$ parameter, the temperature is in degrees Fahrenheit, here denoted T_F. The parameter is then given by

$$P'_{LM} = (T_F + 460)(\log t_r + C) \tag{15.16}$$

so that

$$P'_{LM} = 1.8 P_{LM} \tag{15.17}$$

The units for C are unaffected as t_r is in units of hours in all cases.

   The value of C can be interpreted as an extrapolated intercept on a plot of $\log t_r$ versus $1/T$ as shown in Fig. 15.23. A family of straight lines is expected for various stress values, with all of these lines having a common intercept at $1/T = 0$ of $\log t_r = -C$. The slopes are the values of P_{LM} corresponding to each stress. Hence, to calculate values of both C and P_{LM}, the minimum information needed is two points on such a straight line for constant stress. Note that this approach has the same theoretical basis as the $S\text{-}D$ parameter, differing in that the activation energy is assumed not to be constant but to vary with stress. Also, θ_r is assumed not to vary with stress but instead to be a material constant as given by C.

   Once the constant C is known, values of P_{LM} from stress-life data can be plotted against stress, as shown in Fig. 15.24. Such a plot is used in a manner similar to a P_{SD} vs. σ curve to estimate rupture times. Values of the constant C for rupture of various steels and other structural engineering metals are often near 20, so that this value may be used as an estimate if more specific information is not available. (See Table 15.1 for some values.) The $L\text{-}M$ parameter may also be used to estimate times t_f corresponding

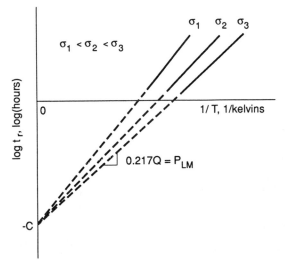

Figure 15.23 Graphical interpretation for the Larson-Miller parameter, with $-C$ being the common intercept of lines of varying slope.

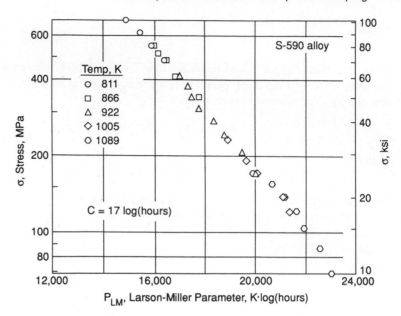

Figure 15.24 Correlation using the Larson-Miller parameter of creep-rupture data for S-590 alloy. (Data from [Goldhoff 59a].)

to a limiting creep strain prior to rupture, provided of course that the needed t_f data are available. Another use of the L-M parameter, or of any other time-temperature parameter, is in comparing and ranking materials. Such a comparison for several materials is shown in Fig. 15.25. Higher curves indicate more resistant materials.

Example 15.2

Reconsider Ex. 15.1 using the Larson-Miller parameter.

Solution The logic is the same as before, except that now P_{LM} from Eq. 15.15 is employed with the constant $C = 20$ from the first line of Table 15.1. We first calculate P_{LM} from the test.

$$P_{LM} = T \, (\log t_r + C) = [(530 + 273) \, \text{K}] \left[\log (260 \, \text{h}) + 20\right] = 18{,}000$$

Solving Eq. 15.15 for temperature T and substituting $t_r = 9600$ h along with this P_{LM} value gives

$$T = \frac{P_{LM}}{\log t_r + C} = \frac{18{,}000}{\log 9600 + 20} = 751 \, \text{K} = 478°\text{C} \qquad \textbf{Ans.}$$

15.4.3 Manson-Haferd (M-H) Parameter

This parameter is based on the empirical observation that plots of $\log t_r$ vs. T, rather than vs. $1/T$, often form straight lines for constant stress. Further, a family of such straight lines for various stresses is assumed to converge at a point $(T_a, \log t_a)$ as shown

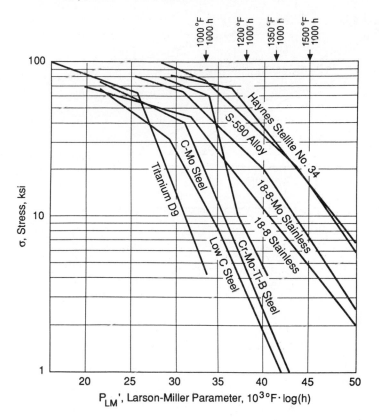

Figure 15.25 Comparison of several metals using the Larson-Miller parameter. (Adapted from [Larson 52]; used with permission of ASME.)

in Fig. 15.26. The values of both T_a and $\log t_a$ are considered to be material constants. For any given stress, the M-H parameter is given by the inverse of the slope of the line.

$$P_{MH} = \frac{T - T_a}{\log t_r - \log t_a} \tag{15.18}$$

Some values of the constants T_a and $\log t_a$ are given in Table 15.1. The additional empirical constant compared to the other parameters is noted to require data at two stress levels rather than one. Once T_a and $\log t_a$ are known for a given material, this parameter can be employed in the same manner as the others. An example plot of P_{MH} vs. σ is shown in Fig. 15.27.

Equation 15.18 can be used with temperature in kelvins (K) and t_r in hours. However, as for the L-M parameter, much of the literature related to the M-H parameter employs Fahrenheit temperatures while still using hours for t_r. For these units

$$P'_{MH} = \frac{T_F - T'_a}{\log t_r - \log t_a} \tag{15.19}$$

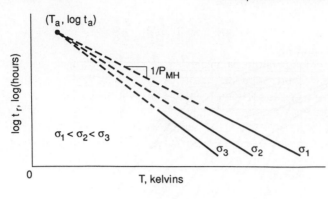

Figure 15.26 Plot used to obtain the constants T_a and $\log t_a$ for the Manson-Haferd parameter.

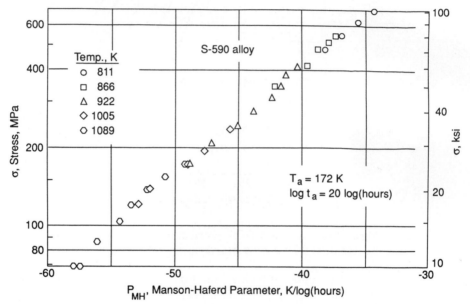

Figure 15.27 Correlation using the Manson-Haferd parameter of creep-rupture data for S-590 alloy. (Data from [Goldhoff 59a].)

with the value of $\log t_a$ being the same and

$$P'_{MH} = 1.8 P_{MH}, \qquad T'_a = 1.8 T_a - 460 \qquad (15.20)$$

15.4.4 Discussion

The three time-temperature parameters discussed seem to differ considerably from one another, and none are consistent with the creep rate equations that are currently the most widely accepted, namely Eqs. 15.7 and 15.9, due to the added (but weak) dependence on the inverse of temperature. Nevertheless, they seem to often give reasonable results.

For example, note that all have a similar and reasonable ability to correlate the same set of data in Figs. 15.22, 15.24, and 15.27.

The success of any time-temperature parameter approach depends upon a similar physical mechanism of creep occurring in the relatively short-time, high-temperature tests as in the typically longer-time, lower-temperature service situation. Thus, the test data should involve times as close to the service application as possible, so that the extrapolation involved is not so extreme that an entirely new mechanism is encountered. A good general rule is that the test data should extend to about 10% of the desired service life, such as tests out to 17,500 hours = 2 years for a 20-year service life. Caution is advised where this must be compromised.

Of the three parameters discussed and others that have been developed, no one has emerged as being clearly superior. See Conway (1969) and Goldhoff (1974) for additional discussion of time-temperature parameters.

15.5 CREEP FAILURE UNDER VARYING STRESS

If stresses change infrequently, stress-life curves and time-temperature parameters can still be employed to make life estimates. However, if stress changes occur so often that the cyclic loading begins to cause fatigue damage, a more complex situation exists that requires special analysis.

15.5.1 Creep Rupture under Step Loading

Rough estimates of life to creep rupture can be made by applying a *time-fraction rule* to stress-life plots in the same manner that the Palmgren-Miner rule is used with cyclic lives for fatigue.

$$\sum \frac{\Delta t_i}{t_{ri}} = 1$$

$$B_f \left(\sum \frac{\Delta t_i}{t_{ri}} \right)_{\text{one rep.}} = 1$$

(15.21)

Life is in this case expressed in terms of time, and stress versus rupture life curves similar to Fig. 15.8 are used. The quantity Δt_i is the time spent at stress level σ_i, and t_{ri} is the corresponding rupture life. For the second version, the summation is done for one repetition of a loading sequence that occurs a number of times, and B_f is the number of repetitions to failure. If the temperature also changes, then the appropriate stress-life curve is used for each temperature step. A similar procedure can be employed where failure is considered to be the accumulation of a particular amount of creep strain. Stress-life curves for the particular strain value, as from Fig. 15.10, are of course used.

Failure lives for use with these equations also can be obtained from time-temperature parameters. Each stress level σ_i is used to obtain a parameter value P_i, which is then solved for t_{ri} employing the appropriate temperature. Temperature variations are thus easily included.

15.5.2 Creep-Fatigue Interaction

Practical applications at high temperature often involve both creep and fatigue, and these phenomena may act together in a synergistic manner. For example, various components of aircraft jet engines experience periods of both fluctuating and steady stress, due to the complex situation of thermal stresses caused by large temperature variations combined with cyclic loading, as the aircraft flies at constant speed, changes speed, lands and shuts down the engines, etc. High-temperature components in nuclear reactors and various pressure vessels are also subjected to combined creep and fatigue.

One simple approach is to sum the life fractions due to both creep and fatigue, thus combining the Palmgren-Miner rule, Eq. 9.23, and the time fraction rule of Eq. 15.21.

$$\sum \frac{\Delta t_i}{t_{ri}} + \sum \frac{\Delta N_i}{N_{fi}} = 1 \tag{15.22}$$

However, this approach is very rough because the physical processes of creep and fatigue are distinct, so that a simple addition of effects cannot be expected to be accurate. In particular, in engineering metals, creep damage may involve grain boundary cracking, whereas the damage due to the fatigue portion of the loading may be concentrated in slip bands within the crystal grains.

Where creep and fatigue interact, the frequency of cycling is important, as slow frequencies give creep more time to contribute to the damage. One approach developed with such effects in mind is the *frequency-modified fatigue* approach of L. F. Coffin. The cyclic stress-strain and strain-life relationships are generalized so that the various material constants become functions of temperature and frequency.

Even the details of the time variation of stress and strain can be important. This is illustrated by some high-temperature test data on an engineering metal in Fig. 15.28. When compared on the basis of equal inelastic (creep plus plastic) strain ranges, the different waveforms can have very different cyclic lives. This results from the complexities of the physical process of damage in the material. For example, the loading with creep in compression only, called PC, is sometimes (but not always) the most severe. This can occur where an oxidizing environment, perhaps only air, causes an oxide surface layer to form during the compressive creep loading. Subsequent rapid loading into tension cracks this oxide, leading to early cracking of the metal beneath. Data of the type in Fig. 15.28 form the basis of the *strain-range partitioning* approach developed by S. S. Manson and co-workers. Four types of test are run, namely tests involving mainly plastic deformation and little creep (PP), creep mainly in tension (CP), creep mainly in compression (PC), and creep in both tension and compression (CC). Life predictions are then made for engineering components on the basis of the life fractions spent in each of the four types of loading.

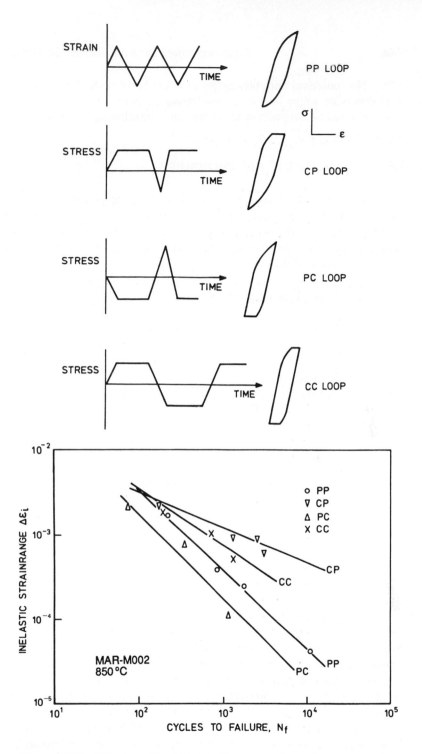

Figure 15.28 Effect on life of intermittent creep in tension, compression, or both, during cyclic loading of the cast Ni-base alloy MAR-M002 at 850°C. Stress-strain hysteresis loops have the shapes shown. (Adapted from [Antunes 78]; used with permission; first published by AGARD/NATO.)

No consensus currently exists as to the best approach to creep-fatigue interaction, and this is an active area of current research. Some recent research and review papers can be found in Solomon et al. (1988) and Amzallag et al. (1982).

15.6 STRESS-STRAIN-TIME RELATIONSHIPS

To analyze the stresses and strains in engineering components subject to creep, it is necessary to have stress-strain relationships that include the time dependency. Relationships suggested by simple linear rheological models are often useful for polymers, but more complex behavior also occurs in these materials, and especially in metals, that requires special consideration.

15.6.1 Linear Viscoelasticity

Strain-time equations for a constant applied stress are given for various rheological models in Fig. 15.29. The equations for (a), (b), and (c) are developed in Chapter 4. (See Eqs. 4.38, 4.45, and 4.46.) The relationship for (c) is the same as for (b) except that the elastic displacement of spring E_1 is added. Since (d) is simply the series combination of (a) and (b), the strains from these two simply add to give the relationship for (d).

$$\varepsilon = \frac{\sigma}{E_1} + \frac{\sigma t}{\eta_1} + \frac{\sigma}{E_2}\left(1 - e^{\frac{-E_2 t}{\eta_2}}\right) \tag{15.23}$$

The three terms in this equation correspond respectively to the instantaneous elastic strain in spring E_1, the steady-state creep strain in dashpot η_1, and the transient creep strain in the (E_2, η_2) parallel combination.

$$\varepsilon_e = \frac{\sigma}{E_1}, \quad \varepsilon_{sc} = \frac{\sigma t}{\eta_1}, \quad \varepsilon_{tc} = \frac{\sigma}{E_2}\left(1 - e^{\frac{-E_2 t}{\eta_2}}\right) \tag{15.24}$$

Strain rates from differentiation with respect to time are also of interest.

$$\dot{\varepsilon}_e = 0, \quad \dot{\varepsilon}_{sc} = \frac{\sigma}{\eta_1}, \quad \dot{\varepsilon}_{tc} = \frac{\sigma}{\eta_2}e^{\frac{-E_2 t}{\eta_2}} \tag{15.25}$$

Hence, the steady-state creep strain has a constant rate, as it should, and the transient creep strain proceeds at a decreasing rate, with the limiting value $\varepsilon_{tc} = \sigma/E_2$ being approached for large t. For a given stress, such a trend of strain with time is similar to that observed in real materials, except that no tertiary creep stage occurs in the model. (See Figs. 15.4 and 15.5.)

From examining the equations above and Fig. 15.29, it is apparent, for any of these models, that for a given time t there is a simple proportionality between stress and strain, and also between stress and strain rate. A similar proportionality applies for any model constructed from combinations of linear springs and dashpots, so that such models are said to exhibit *linear viscoelasticity*. As a result of this situation, the stress-strain relationship for any given value of time, that is, the *isochronous stress-strain curve*, is a straight line. This is illustrated in Fig. 15.30 for the four models of Fig. 15.29.

For the steady-state creep model of Fig. 15.30(a), the dashpot does not deform except after a finite time, but its strain is unlimited for long loading times. Consequently, the stress-strain curve has a slope E_1 for small t, but zero for large t. In contrast, the transient creep model (*b*) cannot deform instantaneously, and for long times its displacement is limited by the spring. Hence, the response is rigid for small t, and has a slope E_2 for large t. For the transient-creep-plus-elastic model (*c*), only spring E_1 deforms for small t, but for large t the dashpot has no effect and the stiffness is that corresponding to E_1 and E_2 in series.

$$E_e = \frac{E_1 E_2}{E_1 + E_2} \tag{15.26}$$

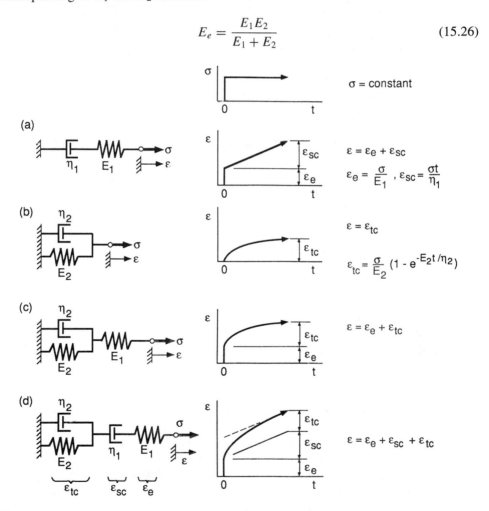

Figure 15.29   Strain vs. time behavior for four viscoelastic models.

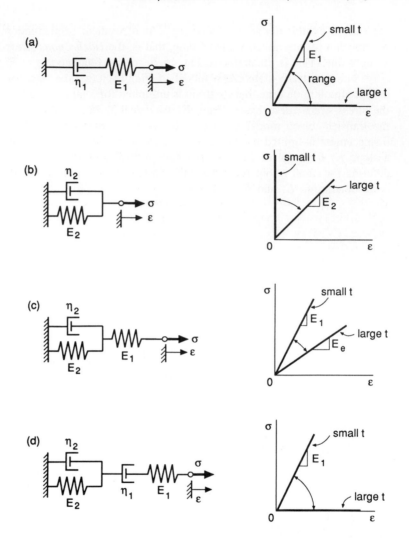

Figure 15.30 Limits on the linear isochronous stress-strain curves for four linear viscoelastic models.

Finally, for (d), the slope is E_1 for short times, but zero at long times due to the deformation of dashpot η_1 not being constrained.

For design applications with polymers, it is common practice to assume that the behavior follows linear viscoelasticity, with the time-dependent elastic modulus being taken as the secant modulus E_s to mildly nonlinear isochronous stress-strain curves. Manufacturers of polymers usually make this information available in the form of E_s versus time curves as in Fig. 15.11(d). Since the exact composition and processing of even nominally identical polymers varies considerably, generic data have limited usefulness. Specific data from manufacturers or from new laboratory tests are preferable.

15.6.2 Nonlinear Creep Equations

From the discussion above on mechanisms, it is apparent that linear viscoelastic behavior, where strain rate is proportional to stress, $\dot{\varepsilon} \propto \sigma$, sometimes occurs. This is the case for ideal viscous flow, Eq. 15.4, and where low-stress creep in crystalline materials is dominated by diffusional flow of the Nabarro-Herring or Coble types, Eq. 15.7. However, for dislocation creep in crystalline materials, strain rates are not proportional to stress, there being instead a strong power dependence according to Eq. 15.9. Polymers and concrete may behave in a linear viscoelastic manner at low stresses, but not generally at high stresses. Isochronous stress-strain curves for a polymer and a concrete are shown in Figs. 15.31 and 15.32. These materials are clearly nonlinear at the relatively high stresses involved.

Thus, more general stress-strain-time relationships than those provided by linear viscoelastic models are often needed. This situation has led to a wide variety of nonlinear relationships that employ expressions involving powers of σ, such as

$$\varepsilon = \varepsilon_i + B\sigma^m t + D\sigma^\alpha \left(1 - e^{-\beta t}\right) \tag{15.27}$$

The quantities B, m, D, α, and β are empirical constants from creep data for a given material and temperature. The instantaneous strain ε_i can include both elastic and plastic

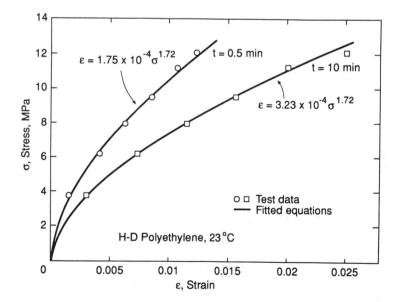

Figure 15.31   Two isochronous stress-strain curves for high-density polyethylene in tension.

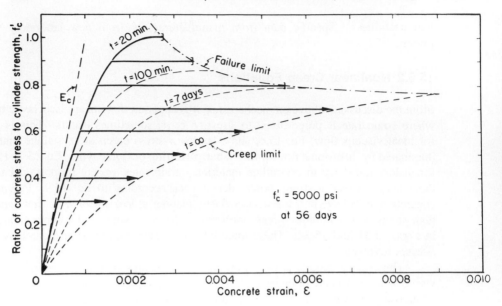

Figure 15.32 Isochronous stress strain curves (dashed lines) for a concrete tested in compression after curing 56 days, where f'_c is the ultimate strength in compression. As the strain rates decrease with time and appear to approach zero, a creep limit curve can be drawn. (From [Rusch 60]; used with permission.)

parts as in Eq. 12.13.

$$\varepsilon_i = \varepsilon_e + \varepsilon_p = \frac{\sigma}{E} + \left(\frac{\sigma}{H}\right)^{\frac{1}{n}} \tag{15.28}$$

In a manner similar to Eq. 15.23, the second term of Eq. 15.27 is the steady-state (secondary) creep strain, and the third is the transient (primary) creep strain.

In Marin (1962), a useful form of Eq. 15.27 is employed that is the special case where $1/n = \alpha = m$.

$$\varepsilon = \frac{\sigma}{E} + \left[B_1 + B_2 t + B_3\left(1 - e^{-\beta t}\right)\right]\sigma^m \tag{15.29}$$

This equation can be expressed in the same mathematical form as the elasto-plastic relationship of Eq. 12.13.

$$\varepsilon = \frac{\sigma}{E} + \left(\frac{\sigma}{H_c}\right)^{\frac{1}{n_c}} \tag{15.30}$$

where

$$n_c = \frac{1}{m}, \quad H_c = \left[B_1 + B_2 t + B_3\left(1 - e^{-\beta t}\right)\right]^{-\frac{1}{m}}$$

This can be considered to be an extension of Eq. 12.13 that includes the additional variable of time t.

At times sufficiently large for the transient straining to be essentially complete, Eq. 15.27 gives a simplified expression for steady-state creep.

$$\varepsilon = \varepsilon_i + B\sigma^m t + D\sigma^\alpha, \qquad \dot{\varepsilon}_{sc} = B\sigma^m \qquad \text{(a,b)} \qquad (15.31)$$

where the third term of (a) is the limiting value of transient creep strain. The temperature dependence of the steady-state creep rate can be estimated from an activation energy as described earlier. For example, for power-law creep as in Eq. 15.9, comparison with Eq. 15.31(b) gives the constant B in a temperature-dependent form.

$$B = \frac{A_3}{T} e^{\frac{-Q}{RT}} \qquad (15.32)$$

Numerous other equations giving creep strains have been employed, such as

$$\varepsilon = \varepsilon_i + D_3\sigma^\delta t^\phi, \qquad \varepsilon = \frac{\sigma}{E} + D_3\sigma^\delta t^\phi \qquad \text{(a,b)} \qquad (15.33)$$

where the exponent ϕ is in the range zero to unity, and a power-type stress dependence is seen to be included. In the second form, the instantaneous strain ε_i is assumed to consist of elastic strain only. Values of constants for this equation are given for some engineering metals at specific temperatures in Table 15.2. Other relationships used feature a time function that is logarithmic or hyperbolic.

TABLE 15.2   SOME CONSTANTS FOR EQ. 15.33(b)

Material	Temperature	E	D_3	δ	ϕ
	°C	MPa (ksi)	for MPa, hours (for ksi, hours)		
SAE 1035 steel[1]	524	161,000 (23,300)	1.58×10^{-11} (4.78×10^{-8})	4.15	0.40
Copper alloy 360[1]	371	85,500 (12,400)	4.26×10^{-9} (1.06×10^{-5})	4.05	0.87
Pure nickel[2]	700	150,000 (21,700)	2.42×10^{-6} (3.02×10^{-4})	2.50	0.28
7075-T6 Al[2]	316	36,500 (5,300)	1.35×10^{-13} (1.00×10^{-7})	7.00	0.33
Cr-Mo-V steel[2]	538	152,000 (22,000)	1.15×10^{-9} (1.07×10^{-7})	2.35	0.34

Notes: [1]Constants from [Chu 70] based on 1-hour creep tests.   [2]From [Lubahn 61] pp. 159, 255, and 574, based on creep data extending to 300, 18, and 10^4 hours, respectively.

$$\varepsilon = \varepsilon_i + D_4 \log(1 + \beta_4 t) \tag{15.34}$$

$$\varepsilon = \varepsilon_i + \frac{D_5 t}{1 + \beta_5 t} \tag{15.35}$$

These latter two relationships, and modifications and extensions of them, are used for transient creep, as is Eq. 15.27 with $B = 0$. They are especially useful for materials where the behavior is dominated by transient creep, such as concrete. Note that Eq. 15.35 approaches a limiting value for infinite time, as does Eq. 15.27 with $B = 0$. All other expressions above give unlimited strains for large t.

Constants for nonlinear stress-strain-time equations as just described are not generally available in comprehensive tabular or similar form for various materials. This is due to the lack of a consensus as to which of the many equations to use, and also due to the constants changing not only with material but also with stress and temperature. Changes in constants must of course be expected if the stress or temperature changes sufficiently that a different physical mechanism of creep occurs. Recall that the dominant creep mechanism depends on the region of stress and temperature involved as shown on a deformation mechanism map, such as Fig. 15.17. Hence, it is generally necessary in engineering applications to employ creep data for the material and conditions of interest, either from the literature or from new laboratory tests, to obtain the needed constants.

An exception is that steady-state creep constants for a wide range of stresses and temperatures are given in Frost and Ashby (1982) for representative metals and ceramics, some of which are engineering materials. Also, creep rates for the relatively pure metals and ceramics included can generally be used as worst case estimates for similar engineering materials. This opportunity arises because commercial materials virtually always have a more disordered microstructure, and thus slower creep rates, than the nearest pure material of similar grain size.

15.7 CREEP DEFORMATION UNDER VARYING STRESS

In the discussion so far, we have considered situations where the applied stress and temperature are held constant. However, many engineering applications involve stresses or temperatures or both that vary, including situations where creep and fatigue occur simultaneously. A brief introduction to creep under varying stress will be given here. More detail can be found in the books on creep by Penny and Marriott (1971), on plastics by Crawford (1987), and on concrete by Neville et al. (1983).

15.7.1 Recovery of Creep Strain

Recovery is the time-dependent disappearance of creep strain after the removal of some or all of the applied stress. Behavior after removal of stress was previously illustrated for simple rheological models in Fig. 4.14. There is no recovery of creep strain for simple steady-state creep models, but all creep strain is recovered in the spring and slider parallel combination of the transient creep model. For the combined rheological model of Fig. 15.13, the elastic strain ε_e in spring E_1 is recovered instantly upon unloading, and the transient creep strain ε_{tc} in the parallel combination is recovered slowly with time. The steady-state creep strain ε_{sc} in the dashpot remains.

To explore this behavior of the ideal model in detail, let a constant stress σ' be maintained for a time t'. From Fig. 15.29(b), the strain in the transient creep element at t' is

$$\varepsilon_2' = \varepsilon_{tc} = \frac{\sigma'}{E_2} \left(1 - e^{\frac{-E_2 t'}{\eta_2}} \right) \tag{15.36}$$

After removal of the stress, it is easily shown (see Prob 4.24) that this strain decreases and approaches zero at infinite time according to

$$\varepsilon_2 = \varepsilon_2' e^{\frac{-E_2 \Delta t}{\eta_2}} \tag{15.37}$$

where $\Delta t = t - t'$ is the time elapsed since removal of the stress. Now consider recovery for the entire model. The elastic strain in spring E_1 is recovered instantly, but the strain in the steady-state creep element, dashpot η_1, remains unchanged.

$$\varepsilon_1 = \varepsilon_{sc} = \frac{\sigma' t'}{\eta_1} \tag{15.38}$$

The strain-time response of the entire model after removal of σ' can now be obtained by combining ε_1 and ε_2, with ε_2' from Eq. 15.36 also being substituted.

$$\varepsilon = \frac{\sigma' t'}{\eta_1} + \frac{\sigma'}{E_2} \left(1 - e^{\frac{-E_2 t'}{\eta_2}} \right) e^{\frac{-E_2 \Delta t}{\eta_2}} \tag{15.39}$$

This equation corresponds to the decreasing strain curve after unloading at $t = t'$ in Fig. 15.13.

Analogous but more complex behavior occurs in real materials. In particular, the governing equations are not generally linear with stress, and strains initially classified as transient may not be recovered fully. In polymers and concrete, considerable portions of the creep strain are often recovered. However, for metals or for cases of simple viscous flow in polymers or glass, there is generally only a relatively small amount of recovery. In modeling real materials, additional transient creep elements in series or special nonlinear springs and dashpots are sometimes used.

15.7.2 Stress Relaxation

Consider a material and temperature combination where creep occurs under constant applied stress, but employ a new type of test in which the specimen is quickly loaded to some given strain and then held. The stress will decrease with time, and this loss of stress is called *relaxation*. Equations for the stress variation with time in two simple rheological models are given in Fig. 15.33. [The equation for (a) was previously derived in Ex. 4.4, and that for (b) is requested as Prob. 4.25.] In a relaxation test, the stress

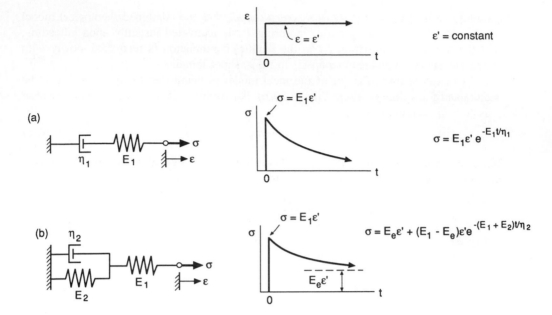

Figure 15.33   Relaxation under constant strain for two linear viscoelastic models.

appears to approach a stable value after a long time, which value may be only a little below the initial stress, or it may be much lower, depending on the material, temperature, and strain level involved. What essentially occurs during relaxation is that some of the elastic strain that appears on initial rapid loading is slowly replaced by creep strain, with the total of the two being constant according to the constraint of the test.

Real materials of course behave in a more complex manner than do simple linear viscoelastic models. For example, consider stress relaxation during constant strain for a material where the plastic strain and the transient creep strain are small, so that the behavior is dominated by elastic strain and steady-state creep strain. Assume that the rate of creep strain is related to stress by a power relationship as in Eq. 15.9, and further assume that this applies even during the decreasing stress situation of relaxation.

$$\dot{\varepsilon}_c = B\sigma^m \qquad (15.40)$$

where the constants B and m apply for a given material and temperature. The total strain is held constant at the value ε', so that

$$\varepsilon_e + \varepsilon_c = \varepsilon', \qquad \dot{\varepsilon}_e + \dot{\varepsilon}_c = 0 \qquad (15.41)$$

where $\varepsilon_e = \sigma/E$ is the elastic strain, and the second equation is obtained by differentiating the first with respect to time.

Since $\dot{\varepsilon}_e = \dot{\sigma}/E$, the equations above combine to give

$$\frac{1}{E}\frac{d\sigma}{dt} + B\sigma^m = 0 \qquad (15.42)$$

This differential equation is easily solved by integration.

$$\int_0^t dt = -\frac{1}{BE}\int_{\sigma_i}^\sigma \frac{d\sigma}{\sigma^m} \tag{15.43}$$

Since only elastic strain occurs on the rapid initial loading, the initial stress at the beginning of relaxation, $t = 0$, is

$$\sigma_i = E\varepsilon' \tag{15.44}$$

After performing the integration and then some manipulation, equations as follows are obtained for the variation of stress with time:

$$\sigma = \frac{\sigma_i}{\left[tBE(m-1)\sigma_i^{m-1}+1\right]^{\frac{1}{m-1}}} \qquad (m \neq 1)$$

$$\sigma = \sigma_i e^{-BEt} \qquad (m = 1) \tag{15.45}$$

This analysis in effect generalizes the case of Fig. 15.33(a) so that the dashpot has a nonlinear response according to

$$\eta_1 = \frac{\sigma}{\dot\varepsilon} = \frac{1}{B\sigma^{m-1}} \tag{15.46}$$

The special case of $m = 1$ is seen to be equivalent to the original linear model with $\eta_1 = 1/B$.

   Equation 15.45 with m as appropriate can be used to approximate the behavior of crystalline materials in the power-law (dislocation creep) and viscous (diffusional flow) regions, respectively, of the deformation mechanism map. However, caution is advised, as it needs to be remembered that transient creep strain effects are neglected in doing so. Relaxation behavior in real materials is also estimated using other mathematical expressions, with these sometimes being based on rheological models of varying degrees of complexity.

15.7.3 Step Loading of Linear Viscoelastic Models

Consider a series of loading steps as shown in Fig. 15.34(a). An increase in stress causes additional instantaneous strain (elastic plus plastic), additional transient creep strain, and increased steady-state strain rates. A decrease in stress causes instantaneous loss of some of the elastic strain and a decrease in strain rates, and perhaps even some recovery of transient creep strain if the resulting stress is low. A number of approaches exist for predicting strain versus time behavior under these conditions.

   First, consider the behavior of any linear viscoelastic rheological model, that is, any combination of standard linear springs and dashpots. Any such model can be shown to obey the *Boltzmann superposition principle*. This principle states simply that the creep

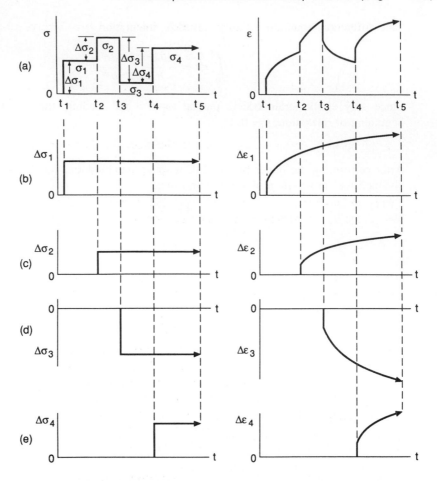

Figure 15.34 The Boltzmann superposition principle for linear viscoelastic models and materials. An applied stress history and the resulting strain response is shown in (a), and the stress changes and resulting strains that are superimposed to obtain this response are shown in (b–e).

strain at any time is the sum of the strains due to each change in stress, $\Delta\sigma$, that has occurred, where each $\Delta\sigma$ is considered to act continuously as an applied stress starting from the time it occurs to any later time. An application is illustrated in Fig. 15.34. Stress σ_1 is applied at time t_1, but at time t_2 the stress changes to σ_2. After this change, the creep strain due to $\Delta\sigma_1 = \sigma_1$ continues to accumulate, but an additional creep strain must be added due to the additional stress $\Delta\sigma_2 = \sigma_2 - \sigma_1$. The amount of this additional creep strain is the same as that due to a stress equal to $\Delta\sigma_2$ applied by itself starting at time t_2. Similarly, following t_3, the creep strains due to $\Delta\sigma_1$ and $\Delta\sigma_2$ continue, but the strain for $\Delta\sigma_3$ must now be added. In this particular example, the $\Delta\sigma_3 = \sigma_3 - \sigma_2$ to be added is negative, so that subtraction occurs.

In general, the stress change to reach the ith load step from the previous level is

$$\Delta \sigma_i = \sigma_i - \sigma_{i-1} \qquad (15.47)$$

where the stress during the ith step is the sum of all changes that have occurred so far.

$$\sigma_i = \sum \Delta \sigma_i \qquad (15.48)$$

Let the linear strain-time relationship for the model be represented by

$$\varepsilon = \sigma \ f(t) \qquad (15.49)$$

where various examples of $f(t)$ are available from Fig. 15.29. The strain at any later time t due to $\Delta \sigma_i$ acting as an applied stress starting at time t_i is thus

$$\Delta \varepsilon_i = \Delta \sigma_i \ f(t - t_i) \qquad (15.50)$$

where $(t - t_i)$ is the time since the stress change $\Delta \sigma_i$. According to the superposition principle, the total strain is the sum of all of the strains $\Delta \varepsilon_i$ due to each $\Delta \sigma_i$ that has occurred.

$$\varepsilon = \sum \Delta \varepsilon_i = \sum \Delta \sigma_i \ f(t - t_i) \qquad (15.51)$$

This summation is equivalent to the graphical procedure of simply adding the strains due to each $\Delta \sigma$ at any desired time as illustrated by Fig. 15.34(b-e).

The behavior of real materials is often too complex to be represented by linear viscoelasticity. Hence, nonlinear extensions of this approach are sometimes used, as are other approaches. For materials such as metals where recovery behavior is relatively unimportant, it may be reasonable to use certain approaches that neglect recovery behavior entirely, such as *time hardening* and *strain hardening*, as discussed next.

15.7.4 Time-Hardening and Strain-Hardening Rules

These methods are illustrated in Fig. 15.35 for the stress history shown in (a). Estimated strains are shown in (b) for the two approaches. For time hardening, creep (not total) strain versus time curves are first plotted for all stress levels involved as shown in (c). Whenever the stress changes, the deformation is assumed to proceed according to the curve for the new stress, starting at the point on this curve corresponding to the actual value of time. For example, after the stress changes to σ_2 at time t_2, the strain-time curve for σ_2 is used to estimate the behavior until the stress again changes. These segments of strain-time response for each stress level are then combined to estimate the overall strain response as shown in (b). Changes in the elastic and plastic strain may also need to be included. In this example, instantaneous changes in elastic strain are shown in (b) along with the segments of creep strain versus time response from (c).

Assume that the creep strain vs. time curves are given by

$$\varepsilon_c = f_2(\sigma, t) \qquad (15.52)$$

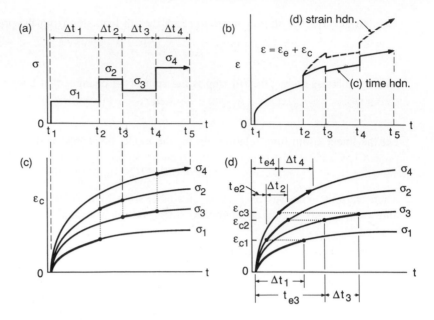

Figure 15.35 A step stress history (a) and estimated strain responses (b), based on time hardening (c), and strain hardening (d).

The change in creep strain during the ith stress level is thus obtained from the curve for σ_i, and it is the difference between the creep strains corresponding to times t_i and t_{i+1}.

$$\Delta\varepsilon_{ci} = f_2\left(\sigma_i, t_{i+1}\right) - f_2\left(\sigma_i, t_i\right) \tag{15.53}$$

The accumulated creep strain at any time t is thus the summation of the creep strains increments for all σ_i that have occurred.

$$\varepsilon_c = \sum \Delta\varepsilon_{ci} \tag{15.54}$$

For example, consider creep according to Eq. 15.33(b). Summing the elastic strain due to the current value of stress and the accumulated creep strain from Eq. 15.54, the total strain is

$$\varepsilon = \varepsilon_e + \varepsilon_c = \frac{\sigma}{E} + D_3 \sum \sigma_i^\delta \left(t_{i+1}^\phi - t_i^\phi\right) \tag{15.55}$$

Strain hardening is similar to time hardening except that the deformation after a stress change is assumed to start at a point on the new creep strain versus time curve that corresponds to the actual value of creep strain, rather than of time, as illustrated in Fig. 15.35(d). At time t_2, the stress changes to σ_2, and the curve for σ_2 is used for a time period $\Delta t_2 = t_3 - t_2$. The starting point on the σ_2 curve corresponds to the creep strain ε_{c1} reached at the end to the σ_1 step. Hence, it does not correspond to the real time t_2 but rather to a fictitious time, t_{e2}. Similarly, at time t_3 the curve for σ_3 is used for the time period $\Delta t_3 = t_4 - t_3$ starting at the fictitious time t_{e3} corresponding to the creep strain ε_{c2} at the end of the previous step. For each step, such as the ith one, the t_{ei} value must be calculated from the creep strain reached at the end of the previous step by solving Eq. 15.52 for t_{ei}.

$$\varepsilon_{c(i-1)} = f_2(\sigma_i, t_{ei}) \tag{15.56}$$

The creep strain occurring during the ith step can then be computed.

$$\Delta\varepsilon_{ci} = f_2(\sigma_i, t_{ei} + \Delta t_i) - f_2(\sigma_i, t_{ei}) \tag{15.57}$$

The accumulated creep strain is then given by summing these $\Delta\varepsilon_{ci}$. For example, for creep according to Eq. 15.33(b), the above equations give

$$t_{ei} = \left(\frac{\varepsilon_{c(i-1)}}{D_3\sigma_i^\delta}\right)^{\frac{1}{\phi}}, \quad \varepsilon = \frac{\sigma}{E} + D_3 \sum \sigma_i^\delta \left[(t_{ei} + \Delta t_i)^\phi - t_{ei}^\phi\right] \tag{15.58}$$

Strain hardening and time hardening give different results except for creep strain curves that are linear, that is, for steady-state creep where $\dot{\varepsilon}_c$ for a given stress does not vary with time. Although neither procedure can be regarded as anything other than a rough approximation, strain hardening does appear to be more accurate for engineering metals than time hardening. Such a trend might be expected as it is more logical to assume that the effect of prior deformation is more closely related to the amount of strain that has occurred rather than simply to the amount of time elapsed. If step changes in temperature occur instead of, or in addition to, stress changes, then time hardening and strain hardening can still be used by introducing strain-time curves for more than one temperature. One approach is to employ temperature compensated time, θ from Eq. 15.13, in place of the actual time t in the equations given above.

15.8 CREEP UNDER MULTIAXIAL STRESS

Multiaxial stresses of course often occur in engineering components, but creep data and constants are generally based on uniaxial tests. Thus, special methodology is needed for generalizing uniaxial data to handle multiaxial situations.

15.8.1 Creep Deformation under Multiaxial Stress

Consider an ideal linear viscous fluid that is incompressible, called a Newtonian fluid. Incompressibility requires that the volumetric strain be zero, hence also that the volumetric strain rate be zero. Hence, Eq. 4.18 gives

$$\dot{\varepsilon}_x + \dot{\varepsilon}_y + \dot{\varepsilon}_z = 0 \tag{15.59}$$

Now consider a uniaxial stress σ_x and the resulting strain rate.

$$\dot{\varepsilon}_x = \frac{\sigma_x}{\eta} \quad (\sigma_y = \sigma_z = 0) \tag{15.60}$$

where η is the tensile viscosity of Eq. 15.2. The strain rates in the other two directions are the same and thus can be obtained by invoking Eq. 15.59.

$$\dot{\varepsilon}_y = \dot{\varepsilon}_z = -\frac{\dot{\varepsilon}_x}{2} = -\frac{1}{2}\left(\frac{\sigma_x}{\eta}\right) \quad (\sigma_y = \sigma_z = 0) \tag{15.61}$$

If stresses also occur in the other directions, these produce additional strains in a similar manner, so that strain rates are given by

$$\dot{\varepsilon}_x = \frac{1}{\eta}\left[\sigma_x - 0.5\left(\sigma_y + \sigma_z\right)\right] \quad (a)$$

$$\dot{\varepsilon}_y = \frac{1}{\eta}\left[\sigma_y - 0.5\left(\sigma_x + \sigma_z\right)\right] \quad (b) \tag{15.62}$$

$$\dot{\varepsilon}_z = \frac{1}{\eta}\left[\sigma_z - 0.5\left(\sigma_x + \sigma_y\right)\right] \quad (c)$$

Shear stresses and strains may also be present, for which the shear viscosity η_τ of Eq. 15.1 applies. Pursuing logic similar to that leading to Eq. 4.12 gives $\eta_\tau = \eta/3$, so that the only independent constant is η, and the equations for shear strain rates can be written as

$$\dot{\gamma}_{xy} = \frac{3}{\eta}\tau_{xy}, \quad \dot{\gamma}_{yz} = \frac{3}{\eta}\tau_{yz}, \quad \dot{\gamma}_{zx} = \frac{3}{\eta}\tau_{zx} \tag{15.63}$$

These relationships are analogous to Hooke's Law, Eqs. 4.10 and 4.11, except that they involve strain rates. Poisson's ratio η is replaced by 0.5, also E by η, and G by $\eta/3$.

The equations above can be extended to cases where the stress versus strain rate relationship is nonlinear by interpreting η as a secant modulus on a stress versus strain rate plot.

$$\eta = \frac{\bar{\sigma}}{\bar{\dot{\varepsilon}}} \tag{15.64}$$

where $\bar{\sigma}$ and $\bar{\dot{\varepsilon}}$ are the *effective stress* and the *effective strain rate*, which in terms of principal stresses and corresponding strain rates are

$$\bar{\sigma} = \frac{1}{\sqrt{2}}\sqrt{(\sigma_1 - \sigma_2)^2 + (\sigma_2 - \sigma_3)^2 + (\sigma_3 - \sigma_1)^2} \quad (a)$$

$$\tag{15.65}$$

$$\bar{\dot{\varepsilon}} = \frac{\sqrt{2}}{3}\sqrt{(\dot{\varepsilon}_1 - \dot{\varepsilon}_2)^2 + (\dot{\varepsilon}_2 - \dot{\varepsilon}_3)^2 + (\dot{\varepsilon}_3 - \dot{\varepsilon}_1)^2} \quad (b)$$

Note that $\bar{\sigma}$ is the same as Eq. 12.21 and $\bar{\dot{\varepsilon}}$ is analogous to Eq. 12.22. We now have a situation analogous to deformation plasticity theory, as described in Section 12.3, except for strain rates replacing strains. In a similar manner, the effective stress versus strain rate relationship is the same as the uniaxial one.

$$\bar{\dot{\varepsilon}} = g\left(\bar{\sigma}\right) \tag{15.66}$$

For example, for creep obeying Eq. 15.29, the effective strain rate is

$$\dot{\bar{\varepsilon}} = \left(B_2 + B_3 \beta e^{-\beta t} \right) \bar{\sigma}^m = B_4 \bar{\sigma}^m \tag{15.67}$$

where the coefficient B_4 is the indicated function of time. Equations 15.62 and 15.63 are in this case employed by making the substitution

$$\frac{1}{\eta} = \frac{\dot{\bar{\varepsilon}}}{\bar{\sigma}} = B_4 \bar{\sigma}^{m-1} \tag{15.68}$$

This permits strain rates to be calculated for any desired state of stress and for any time, t.

Example 15.3

For a given material and temperature, the uniaxial creep behavior follows Eq. 15.33(b). A thin-walled tubular pressure vessel of radius r and wall thickness b has closed ends and is made of this material. Develop an equation for the relative change in radius, $\Delta r / r$, as a function of time t and a constant pressure p in the vessel.

Solution   We first generalize the uniaxial stress-strain curve to an effective stress-strain curve.

$$\bar{\varepsilon} = \frac{\bar{\sigma}}{E} + D_3 \bar{\sigma}^\delta t^\phi, \quad \dot{\bar{\varepsilon}} = D_3 \phi \bar{\sigma}^\delta t^{\phi-1}$$

The viscosity is thus both stress and time dependent.

$$\frac{1}{\eta} = \frac{\dot{\bar{\varepsilon}}}{\bar{\sigma}} = D_3 \phi \bar{\sigma}^{\delta-1} t^{\phi-1}$$

This relationship can now be used with Eq. 15.62 to solve our particular problem, which involves stresses and strains as follows:

$$\sigma_1 = \frac{pr}{b}, \quad \sigma_2 = \frac{pr}{2b}, \quad \sigma_3 \approx 0$$

$$\varepsilon_1 = \frac{\Delta(2\pi r)}{2\pi r} = \frac{\Delta r}{r}$$

where σ_1 and ε_1 are in the hoop direction and σ_2 is in the longitudinal direction, and where these stresses are noted to be principal stresses.

Letting the (x, y, z) directions be the principal $(1, 2, 3)$ directions, Eq. 15.62(a) gives

$$\dot{\varepsilon}_1 = \frac{1}{\eta} \left[\sigma_1 - 0.5 \left(\sigma_2 + \sigma_3 \right) \right]$$

$$\dot{\varepsilon}_1 = D_3 \phi \bar{\sigma}^{\delta-1} t^{\phi-1} \left[\frac{pr}{b} - 0.5 \left(\frac{pr}{2b} \right) \right]$$

Also, $\bar{\sigma}$ is given by Eq. 15.65(a)

$$\bar{\sigma} = \frac{1}{\sqrt{2}} \sqrt{ \left(\frac{pr}{b} - \frac{pr}{2b} \right)^2 + \left(\frac{pr}{2b} \right)^2 + \left(-\frac{pr}{b} \right)^2 } = \frac{\sqrt{3} pr}{2b}$$

Substituting this $\bar{\sigma}$ into the expression for $\dot{\varepsilon}_1$ and simplifying gives

$$\dot{\varepsilon}_1 = \frac{\sqrt{3}}{2} D_3 \phi t^{\phi-1} \left(\frac{\sqrt{3}pr}{2b} \right)^{\delta}$$

Since the pressure p is constant, we can integrate with respect to time to obtain the creep strain.

$$\varepsilon_{c1} = \int_0^t \dot{\varepsilon}_1 \, dt = \frac{\sqrt{3}}{2} D_3 t^{\phi} \left(\frac{\sqrt{3}pr}{2b} \right)^{\delta}$$

We now need the elastic strain from Eq. 4.10.

$$\varepsilon_{e1} = \frac{1}{E} [\sigma_1 - \nu (\sigma_2 + \sigma_3)] = \left(1 - \frac{\nu}{2} \right) \left(\frac{pr}{bE} \right)$$

Adding the elastic and creep strains finally gives the desired result.

$$\frac{\Delta r}{r} = \varepsilon_1 = \varepsilon_{e1} + \varepsilon_{c1}$$

$$\frac{\Delta r}{r} = \left(1 - \frac{\nu}{2} \right) \left(\frac{pr}{bE} \right) + \frac{\sqrt{3}}{2} D_3 t^{\phi} \left(\frac{\sqrt{3}pr}{2b} \right)^{\delta} \qquad \textbf{Ans.}$$

Discussion   Another approach would be to integrate Eqs. 15.62 to 15.64 with respect to time, assuming that all stresses are constant. This gives equations for the creep strains that are of the same form as the equations of deformation plasticity theory, Eqs. 12.25 and 12.26, differing in that E_p is replaced according to

$$\frac{1}{E_p} = \int_0^t \frac{1}{\eta} \, dt$$

The solution of the corresponding elasto-plastic problem of Example 12.3 can thus be used. It is first necessary to note that our stress-strain relationship is of the Ramberg-Osgood from as in Eq. 15.30, where

$$n_c = \frac{1}{\delta}, \qquad H_c = \frac{1}{\left(D_3 t^{\phi} \right)^{\frac{1}{\delta}}}$$

Hence, substituting n_c and H_c into the result of Ex. 12.3, while taking care to use b for thickness and t for time, gives the answer to the present problem, which is found to be the same as above.

15.8.2 Creep Failure under Multiaxial Stress

For multiaxial loading, it is logical to use stress-life curves or time-temperature parameters by simply replacing the uniaxial stress with the effective stress $\bar{\sigma}$ of Eq. 15.65. This is true both for predicting rupture times t_r and times t_f to limiting strain values. However, caution is advised in that only limited multiaxial creep data are available to verify such an approach.

Creep rupture under multiaxial loading is also affected to an extent by the maximum principal stress σ_1 . One approach is to use the following effective stress for creep rupture.

$$\bar{\sigma}_c = \alpha \sigma_1 + (1 - \alpha)\bar{\sigma} \qquad (15.69)$$

where $\bar{\sigma}$ is from Eq. 15.65, and α is a material constant with a value between zero and unity that must be evaluated from multiaxial testing. The quantity $\bar{\sigma}_c$ is then considered to be equivalent to a uniaxial stress. If a value of α is unavailable, another possibility is to take the worst possible case for α between zero and unity, which corresponds to

$$\bar{\sigma}_c = \text{MAX}\,(\sigma_1, \bar{\sigma}) \qquad (15.70)$$

For additional discussion of these equations, see the paper by Leckie and Hayhurst in Pomeroy (1978).

15.9 COMPONENT STRESS-STRAIN ANALYSIS

Stress-strain analysis of engineering components where time-dependent deformation occurs can be performed for simple cases using isochronous stress-strain curves. If these curves are approximately linear, corresponding to linear viscoelastic behavior, only linear-elastic stress analysis is needed. For nonlinear isochronous σ-ε curves, analysis is done in the same manner as for elasto-plastic stress-strain curves that are not time dependent. Hence, various analytical results from Chapter 13 can be adapted to creep situations.

For complex geometry and loading where there is time-dependent deformation, it is becoming increasingly common to use computerized numerical analysis, as by finite elements. This subject will not be included here, but introductory discussions are given in Penny and Marriott (1971), Kraus (1980), and Dhalla and Gallagher (1982).

15.9.1 Linear Viscoelastic Behavior

Assume that it is reasonable to represent the behavior of a given material using a rheological model built up of combinations of linear springs and dashpots. As already explained, the isochronous stress-strain curves are then all straight lines, which can be represented by

$$\varepsilon = \sigma\, f(t) \qquad (15.71)$$

where the particular $f(t)$ that applies depends on the model. A time-dependent modulus is often used.

$$E(t) = \frac{1}{f(t)} = \frac{\sigma}{\varepsilon} \qquad (15.72)$$

where $E(t)$ is simply the slope of the isochronous stress-strain curve. Since the stress-strain relationship is linear, component stress-strain analysis for any particular time t can be done using a stress analysis based on linear-elastic behavior. The analysis is unaffected except that the elastic modulus varies with time.

The situation for a rectangular beam under pure bending is illustrated in Fig. 15.36. For all values of time, the stress distribution is linear and unchanging according to the bending formula from linear-elastic analysis.

$$\sigma = \frac{My}{I_z} = \frac{3My}{2bc^3} \tag{15.73}$$

The strain for any position y in the beam at any time t is then obtained by combining this with Eq. 15.71 or 15.72

$$\varepsilon = \frac{3My}{2bc^3} f(t) = \frac{3My}{2bc^3} \frac{1}{E(t)} \tag{15.74}$$

Analysis of linear-elastic behavior can be similarly applied to other situations, such as more complex bending problems, shafts in torsion, pressure vessels, geometries containing stress raisers, etc. The stress distributions from linear-elastic analysis apply in all cases. To determine strains where nonuniaxial stress states occur, it is necessary to apply both Hooke's Law, Eq. 4.10, and the equations for multiaxial viscous flow, Eq. 15.62.

If the isochronous stress-strain curves are nonlinear, but not grossly so, it may still be reasonable to employ linear-elastic analysis as just described. Values of $E(t)$ are replaced by values of the secant modulus, $E_s(t)$ as defined in Fig. 15.11. This represents an approximation that may be useful for engineering purposes. For example, such an approach is used for plastics in designing components against excessive deflection in long-term use. Curves of $E_s(t)$ vs. time, as in Fig. 15.11(d), are thus needed to characterize the material.

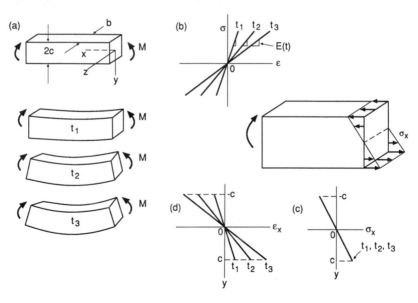

Figure 15.36   Behavior under constant applied moment of a beam (a) made of a linear viscoelastic material (b). The stress distribution is linear and constant with time (c), while the strain maintains a linear distribution as it increases (d).

15.9.2 Nonlinear Isochronous Stress-Strain Curves

If the isochronous stress-strain curves are markedly nonlinear, then analysis similar to that described in Chapter 13 for plastic deformation is needed. The isochronous stress-strain curve for any particular time is used just as if it were an elasto-plastic stress-strain curve. This is illustrated in Fig. 15.37 for a rectangular beam under static pure bending. The stress distribution must resist a bending moment that does not change with time. However, as time passes, the shape of the stress distribution must adjust as a result of a changing shape of the isochronous stress-strain curve. A linear strain distribution is a reasonable assumption for all values of time, but the magnitude of the strain increases.

For example, assume for the beam material in Fig. 15.37 that all strains except the steady-state creep strains are small and that these are given by

$$\varepsilon = B\sigma^m t \tag{15.75}$$

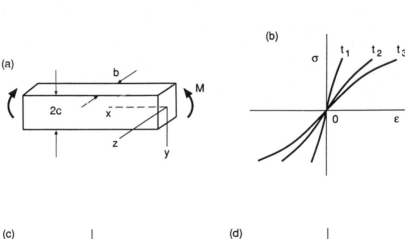

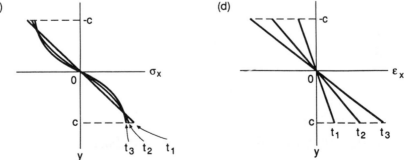

Figure 15.37   Behavior under constant moment of a beam (a) made of a material with nonlinear isochronous stress-strain curves (b). The stress distribution may change its shape with time (c), while the strain distributions remain linear as the strain increases (d).

For any particular time t, this equation represents simple power hardening as in Eq. 13.9, with constants as follows:

$$n_2 = \frac{1}{m}, \quad H_2 = \frac{1}{(Bt)^{\frac{1}{m}}} \tag{15.76}$$

The analytical result of Eq. 13.11 can now be employed by making these substitutions. Denoting the beam thickness as b to avoid confusion with time t, this gives

$$M = \frac{2mbc^2\sigma_c}{1 + 2m} = \frac{2mbc^2}{1 + 2m}\left(\frac{\varepsilon_c}{Bt}\right)^{\frac{1}{m}} \tag{15.77}$$

where σ_c, ε_c are the stress and strain at the edge of the beam. The relationship between strain and moment depends on time as expected. In this particular case, the stress distribution is nonlinear but does not adjust with time unless m differs for the various isochronous curves.

Other analytical results from Chapter 13 can be applied to creep problems in a similar manner if the isochronous stress-strain curves are represented by equations of the same mathematical form as those used in each case. Recall that various analyses for bending, torsion, and notched members are given in Chapter 13, or suggested in the exercises there, for stress-strain curves of several forms. The power-law form of many of the stress-strain-time equations for creep makes the simple power-hardening and Ramberg-Osgood forms particularly convenient for use.

Consider isochronous stress-strain curves that have the form of the Ramberg-Osgood equation

$$\varepsilon = \varepsilon_e + \varepsilon_c = \frac{\sigma}{E} + \left(\frac{\sigma}{H_c}\right)^{\frac{1}{n_c}} \tag{15.78}$$

where the second term is creep strain, but may also include plastic strain. The constant n_c is a value appropriate for creep, and H_c varies with time to provide a family of curves. One example has already been given as Eq. 15.29. As another example, if Eq. 15.33(b) applies, we have

$$n_c = \frac{1}{\delta}, \quad H_c = \frac{1}{\left(D_3 t^\phi\right)^{\frac{1}{\delta}}} \tag{15.79}$$

Equation 15.78 can thus be used as a basis for straightforwardly adapting a number of the analytical results of Chapter 13 to creep.

15.10 ENERGY DISSIPATION (DAMPING) IN MATERIALS

Materials subjected to cyclic loading absorb energy, some of which may be stored as potential energy within the structure of the material, but most of which is dissipated as heat to the surroundings. Such energy dissipation may be small, and even difficult to measure, but it is nevertheless always present. Otherwise, vibrations (say in a tuning fork of the material) would never decay, and the physically impossible situation of a perpetual motion machine would exist. Energy dissipation in materials, termed *damping* or *internal friction*, is caused by a wide range of physical mechanisms, depending on the material, temperature, and frequency of cyclic loading involved. Any physical mechanism that causes creep can cause damping, but other mechanisms that act at low stresses are not associated with macroscopic creep effects. The small strains associated with such low-stress damping phenomena are called *anelastic strains*, and low-stress damping itself is called *anelastic damping*. Damping also occurs as a result of plastic deformation.

Damping behavior of materials is of practical importance as the degree of damping affects the behavior under vibratory loading. In particular, higher damping results in lower stresses under forced vibration near resonance, and also in more rapid decay of free vibration. Damping behavior may thus affect the choice of materials in vibration-sensitive applications, such as turbine blades.

15.10.1 Damping Behavior of Rheological Models

The transient-creep-plus-elastic rheological model exhibits behavior similar to low stress (anelastic) damping as summarized in Fig. 15.38. A sinusoidal stress is assumed to be applied.

$$\sigma = \sigma_a \sin \omega t \tag{15.80}$$

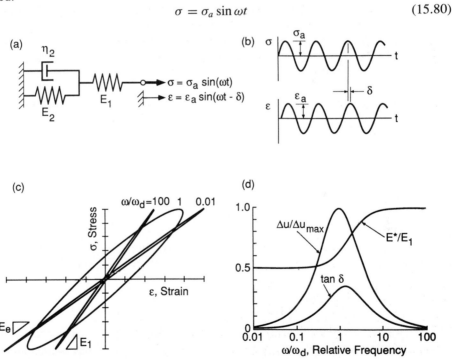

Figure 15.38   Behavior of a transient-creep-plus-elastic rheological model under a sinusoidal stress. The curves shown correspond to $E_1/E_2 = 1$.

where σ_a is the stress amplitude and ω the angular frequency. The strain response of the model at any frequency is sinusoidal, but there is a phase shift, as specified by a phase angle δ, relative to the sine wave of the stress.

$$\varepsilon = \varepsilon_a \sin(\omega t - \delta) \tag{15.81}$$

The two equations above are the parametric equations for an ellipse. Hence, the stress-strain response forms an elliptical hysteresis loop as shown in (c).

The area inside this loop is the energy absorbed in each cycle per unit volume of material, which is call the *unit damping energy*, Δu. Evaluating the loop area gives an equation for Δu.

$$\Delta u = \pi \sigma_a \varepsilon_a \sin \delta \tag{15.82}$$

Both δ and Δu exhibit maxima when plotted versus the angular frequency ω. For this particular model, the Δu maximum occurs at the frequency $\omega_d = E_2/\eta_2$. The quantity ($\varepsilon_a \sin \delta$) is the strain from Eq. 15.81 when σ is zero. Hence, it is the half-width of the elliptical hysteresis loop and is a measure of the nonlinearity in strain. This quantity, sometimes called the *remnant displacement*, is roughly analogous to the plastic strain amplitude, $\varepsilon_{pa} = \Delta \varepsilon_p/2$, of Fig. 12.17. Since ($\varepsilon_a \sin \delta$) is proportional to Δu, it exhibits the same type of frequency dependence as Δu.

At frequencies high compared to ω_d, the dashpot is essentially rigid. Deformation is thus prevented in spring E_2 and occurs only in E_1, so that the response is essentially linear with stiffness E_1. Conversely, at frequencies low compared to ω_d, there is sufficient time for free movement of the dashpot, so that it has little effect. The elliptical loop again reduces to a straight line, but in this case with the lower stiffness E_e corresponding to E_1 and E_2 in series. (See Eq. 15.26.) The stiffness or elastic modulus, called E^*, thus varies with frequency, increasing from E_e at low frequency to E_1 at high frequency. The transition between the two occurs in the neighborhood of the energy dissipation peak.

For any rheological model composed of linear springs and dashpots, that is, for any linear viscoelastic model, elliptical hysteresis loops are formed and the strain response is proportional to the applied stress. Hence, the strain amplitude ε_a is proportional to the stress amplitude σ_a. Also, the phase shift δ does not depend on σ_a, but only on ω. Hence, the elliptical hysteresis loops for various σ_a at a given frequency all have the same proportions and differ only in size. From Eq. 15.82, this results in the unit damping energy being proportional to the square of the stress amplitude.

$$\Delta u = J \sigma_a^2 \tag{15.83}$$

where J is a constant for a given set of model constants and frequency. More complex rheological models that are useful for damping include series combinations of several transient creep elements with differing constants. Such a model exhibits several peaks on a plot of δ or Δu vs. ω, one for each transient creep element. Also, elements having more complex nonlinear behavior are sometimes used.

15.10.2 Definitions of Variables Describing Damping

A number of different definitions are in use for describing the damping behavior of any model or material. These are summarized in Table 15.3. The *unit damping energy* Δu, also often denoted D, has already been defined, as have the *phase angle* δ and the *remnant displacement*. The *loss coefficient* Q^{-1} is defined as

$$Q^{-1} = \tan \delta \tag{15.84}$$

where δ is the phase angle (in radians) as already discussed. The symbol η is also often used for Q^{-1}, but we will avoid that nomenclature here due to our use of the symbol η for the quite different variable of viscosity, as in dashpot constants. The variable Q itself is called the *quality factor*.

The ratio of the stress amplitude to the strain amplitude is called the *dynamic modulus* or the *absolute modulus*.

$$E^* = \frac{\sigma_a}{\varepsilon_a} \tag{15.85}$$

TABLE 15.3   DEFINITIONS FOR MATERIALS DAMPING

Symbol	Name	Definition[1]	Small damping approximation[2]
δ	Phase angle (radians)	Phase shift of ε relative to σ	(Same)
$\varepsilon_a \sin \delta$	Remnant displacement	ε at $\sigma = 0$	$\varepsilon_a \delta$
E^*	Dynamic or absolute modulus	$E^* = \sigma_a/\varepsilon_a$	(Same)
Δu or D	Unit damping energy	Area in σ-ε loop: $\Delta u = \pi \sigma_a \varepsilon_a \sin \delta$	$\Delta u = \dfrac{\pi \sigma_a^2 \delta}{E^*}$
u_e	Elastic strain energy	$u_e = \dfrac{\sigma_a \varepsilon_a \cos \delta}{2}$	$u_e = \dfrac{\sigma_a^2}{2E^*}$
Q^{-1} or η	Loss coefficient	$Q^{-1} = \tan \delta = \dfrac{\Delta u}{2\pi u_e}$	$Q^{-1} = \delta$
Q	Quality factor	$Q = \dfrac{2\pi u_e}{\Delta u}$	$Q = 1/\delta$
E'	Storage modulus	$E' = E^* \cos \delta$	$E' = E^*$
Δ_t	Log decrement	$\Delta_t = \dfrac{\Delta u}{2u_e} = \pi Q^{-1}$	$\Delta_t = \pi \delta$

Notes: [1]The equations involving trigonometric functions apply for elliptical hysteresis loops only.   [2]Applicable for small damping for any loop shape.

which is illustrated in Fig. 15.39. Another frequently used measure of stiffness is the *storage modulus*.

$$E' = \frac{\sigma'}{\varepsilon_a} = \frac{\sigma_a \cos \delta}{\varepsilon_a} = E^* \cos \delta \tag{15.86}$$

This is the slope of a line from the origin to the maximum strain point on the elliptical hysteresis loop, which occurs at the stress $\sigma = \sigma_a \cos \delta$ as shown. The storage modulus is conventionally used to define the elastic strain energy at the peak strain, as also shown in Fig. 15.39, so that this quantity is

$$u_e = \frac{\sigma' \varepsilon_a}{2} = \frac{\sigma_a \varepsilon_a \cos \delta}{2} \tag{15.87}$$

From this and Eqs. 15.82 and 15.84, the loss coefficient is related to the energies Δu and u_e by

$$Q^{-1} = \frac{\Delta u}{2\pi u_e} \tag{15.88}$$

One additional definition that is often used is the *log decrement*, $\Delta_t = \pi Q^{-1}$. If the damping is small, such as $Q^{-1} = 0.1$ or less, then some of the equations involved in these definitions simplify as shown in the last column of Table 15.3.

15.10.3 Low Stress Mechanisms in Metals

At low stresses in engineering metals, a variety of damping mechanisms occur, each of which behave in a manner similar to the linear viscoelastic model discussed above. This results in a number of different peaks in energy dissipation occurring as frequency or temperature is changed as illustrated in Fig. 15.40.

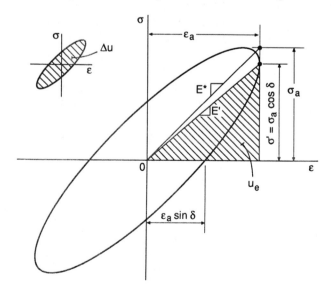

Figure 15.39 Definitions for an elliptical hysteresis loop.

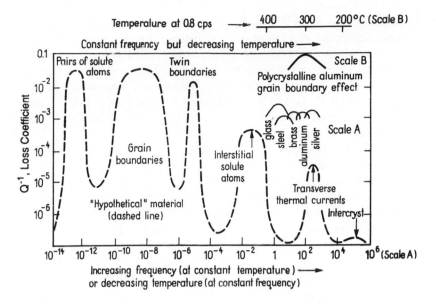

Figure 15.40 Damping peaks for a hypothetical and several real metals, and the associated microstructural mechanisms. (Adapted from [Lazan 68] p. 39; used with permission.)

An example of such a mechanism is the *Snoek effect*. This involves interstitial solute atoms in a body-centered-cubic (BCC) metal, such as carbon or nitrogen in iron, as illustrated in Fig. 15.41. Such interstitial atoms are small compared to iron atoms, so that they can occupy the normally unoccupied positions in the middle of the cube edges of the BCC iron crystal structure, somewhat distorting the structure in doing so. If a tensile stress is applied as shown, interstitial atoms along cube edges that are approximately normal to the applied stress are further squeezed by the Poisson contraction that occurs. They tend to jump to cube edges that are more parallel to the applied stress, where the tensile strain provides added space to accommodate them.

If the frequency of loading is very high, the interstitials have insufficient time to move, and so the effect does not occur. Conversely, if the frequency is very low, the

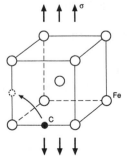

Figure 15.41 Mechanism of the Snoek effect, involving motion of interstitial atoms in a BCC metal structure.

interstitials can move freely. In both cases, the strain in the material is in phase with the stress, but for the lower frequency the material deforms more and thus has a slightly lower elastic modulus. However, at intermediate cyclic loading frequencies corresponding to the time required for a jump to occur, the strain response is slower than the applied stress, a phase lag occurs, and the dissipation of energy is at a maximum. The applied stress can then be said to be in resonance with the jumping of the interstitials. Since the jumping process is thermally activated, a peak in energy dissipation can also be observed by varying the temperature at a constant frequency.

Another example is the *thermoelastic effect*, also called thermal current damping. As Poisson's ratio is usually somewhat less than the value of $\nu = 0.5$ corresponding to the volume being constant, the volume of a stressed solid increases during the tensile portion of cyclic loading. If the loading is so rapid that heat exchange with the surroundings does not have time to occur, the increased volume results in a temperature decrease, hence also in a thermal contraction. Conversely, for rapid loading under compression, there is decreased volume, increased temperature, and thermal expansion. Hence, rapid cyclic loading is accompanied by thermal strains that are in the opposite direction of the mechanical strains, resulting in an apparent stiffening of the material. The elastic modulus that occurs is referred to as the *adiabatic* (meaning constant heat) value. But under slow cycling, where heat exchange with the surroundings can occur freely, there are no thermal strains opposing the mechanical strains. The elastic modulus then takes on the lower *isothermal* value. At intermediate frequencies that correspond to the time required for heat flow, the strain response exhibits a phase lag and there is a relative maximum in energy dissipation.

15.10.4 Additional Mechanisms and Trends

Other damping mechanisms that occur involve various time-dependent movements of impurity atoms or vacancies (point defects), movements of dislocations (line defects), and sliding of grain boundaries. In ferromagnetic materials, an effect called *magnetoelastic damping* is important, in which energy dissipation results from rotations in the directions of the microscopic magnetic domains that occur in such a material. This effect is unusual in that it is independent of frequency and is more strongly dependent on stress amplitude than expected from the ideal model discussed above. In particular, Δu is proportional to the cube of stress, rather than the square as in Eq. 15.83. In addition, the effect may saturate and become independent of stress above a critical level, and it may be sensitive to mean stress.

The gross dislocation motions that result in slip of crystal planes, and hence in plastic deformation, can cause large amounts of energy dissipation called *plastic strain damping*. Such effects are inactive at low stresses, but for engineering metals at temperatures where creep effects are small, they become the dominant mechanism of damping at high stresses. The behavior is dramatically different from that of the linear viscoelastic model, being insensitive to frequency and very sensitive to stress. Hysteresis loops due to plastic deformation are not elliptical; rather they are pointed as in Fig. 12.17. Based on the approximation for the shape of hysteresis loop curves discussed in Chapter 12, plastic strain damping can be calculated from the area inside the stress-strain hysteresis loop. For a cyclic stress-strain curve of Ramberg-Osgood form, Eq. 12.59 applies, giving

$$\Delta u = 4 \left(\frac{1 - n'}{1 + n'} \right) \sigma_a \varepsilon_{pa} = \frac{4 \left(1 - n' \right) \sigma_a^{\frac{n'+1}{n'}}}{(1 + n') (H')^{\frac{1}{n'}}} \tag{15.89}$$

where Eq. 12.11 is used to obtain the second form, and H' and n' are the constants for the cyclic stress-strain curve. For multiaxial loading, adding the contributions to energy dissipation in the three principal directions gives this equation in more general form.

$$\Delta u = 4 \left(\frac{1 - n'}{1 + n'} \right) \bar{\sigma}_a \bar{\varepsilon}_{pa} \tag{15.90}$$

Let Eq. 15.83 be generalized to

$$\Delta u = J \sigma_a^d \tag{15.91}$$

Noting that $n' = 1/7$ is typical, the exponent on stress from Eq. 15.89 is

$$d = \frac{n' + 1}{n'} \approx 8 \tag{15.92}$$

This indicates a very strong dependence on stress in contrast to $d = 2$ for ideal anelastic damping.

Polymers and elastomers are likely to exhibit behavior similar to ideal linear viscoelasticity ($d = 2$) if the stress levels are not excessively high. Some of the mechanisms that operate to cause peaks in energy dissipation involve chain molecule motions, such as rotations, translations, or coiling and uncoiling, of interior or end segments of chains. The segments involved can be long or short, and motions of side groups can also cause damping. In general, larger moving entities cause damping peaks at lower frequencies for a given temperature, or at higher temperatures for a given frequency.

At stresses where metals are used for engineering purposes, the damping is generally quite low, involving loss coefficients around $Q^{-1} = 0.01$ or less. In contrast, the values for polymers at service stresses may be much larger, such as $Q^{-1} = 0.1$ or more. Correspondingly large phase angles and variations in the elastic modulus (E^* or E') then also occur. An example is shown in Fig. 15.42. Such large damping may be disadvantageous if excessive heat is generated, but it is often beneficial in quickly damping any vibrations that develop. High damping is of course often intentionally employed to mitigate sound or vibration by the use of polymers and especially elastomers.

Some trend curves for damping energy in engineering metals at room temperature are suggested by Lazan (1968) and are shown in Fig. 15.43. For nonmagnetic metals, the stress exponent for the middle of the data trend is a little larger than the ideal viscoelastic value of $d = 2$ up to about 80% of the fatigue limit σ_e, where the slope of about $d = 8$ begins due to plastic strain damping. Ferromagnetic metals have higher damping at

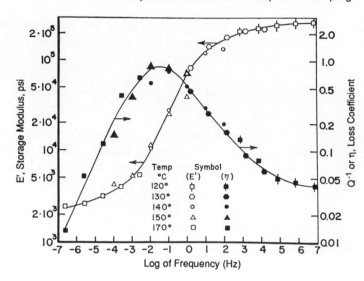

Figure 15.42 Stiffness and damping of an aromatic epoxy at various temperatures as a function of frequency. (Adapted from [Kaelble 65]; copyright © ASTM; reprinted with permission.)

low stresses and a value of $d \approx 3$, followed by a region of $d \approx 1$, which prevails until $d \approx 8$ begins around the fatigue limit. For nonmagnetic metals, the following continuous relationship is also suggested by Lazan as approximating the line segments shown:

$$\Delta u = 7 \left(\frac{\sigma_a}{\sigma_e} \right)^{2.3} + 41 \left(\frac{\sigma_a}{\sigma_e} \right)^{8} \quad \frac{\text{kJ}}{\text{m}^3}$$

$$\Delta u = \left(\frac{\sigma_a}{\sigma_e} \right)^{2.3} + 6 \left(\frac{\sigma_a}{\sigma_e} \right)^{8} \quad \frac{\text{in·lb}}{\text{in}^3}$$

(15.93)

The scatterband in Fig. 15.43 results not only from different behavior for different materials, but also from various frequencies being represented in the data available. Shifts in frequency or temperature may cause significant changes in behavior, so that these curves should be used as a rough guide only.

15.10.5 Damping in Engineering Components

The discussion so far of damping has considered only the behavior of uniformly stressed samples of material, rather than engineering components, such as beams, shafts, etc., that contain nonuniform distributions of stress. For ideal viscoelastic behavior, $d = 2$, the loss coefficient Q^{-1} for a component is the same as for the material. Otherwise, Q^{-1} depends on the geometry and so is not the same as for a uniformly stressed material.

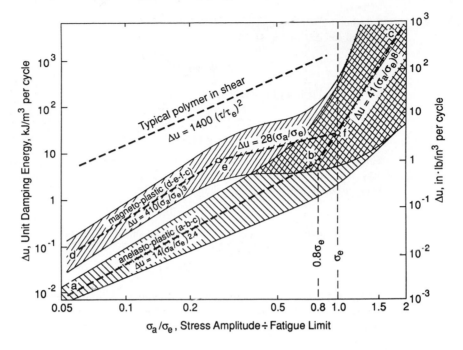

Figure 15.43 Trends for damping in ferromagnetic and nonferromagnetic metals, and also for a typical viscoelastic polymer. For the two classes of metals, ranges of behavior and idealized relations are given, with the equations being for Δu in kJ/m³. (Adapted from [Lazan 68] p. 139; used with permission.)

The total damping energy ΔU for a component can be obtained by integrating over its volume. A second integration over the volume yields the elastic energy for the component, and the ratio of these is the loss factor for the component.

$$\Delta U = \int \Delta u \, dV, \quad U_e = \int u_e \, dV, \quad Q_v^{-1} = \frac{\Delta U}{2\pi U_e} \tag{15.94}$$

Such analysis is treated in detail in Lazan (1968) for cases where elastic stress analysis is appropriate. If large plastic deformations occur, the methods of Chapter 13 for beams and shafts can be extended to calculate damping energies with Δu given by Eq. 15.89. Numerical analysis, as by finite elements, may be employed for complex geometries and loadings.

Example 15.4

Consider a rectangular beam of depth $2c$, thickness b, and length L, that is subjected to a cyclic pure bending moment M_a about zero mean. The strains are sufficiently large that a nonlinear stress distribution occurs, and the material follows a cyclic stress-strain curve of the Ramberg-Osgood form, Eq. 12.58. Obtain an equation for the total energy ΔU dissipated in each cycle of loading as a function of the stresses and strains at the beam edge.

Solution   The damping energy in the member is obtained by combining Eq. 15.91 with the integral for ΔU of Eq. 15.94.

$$\Delta U = \int \Delta u \, dV = \int J\sigma_a^d \, dV$$

where J and d from Eq. 15.89 apply.

$$J = \frac{4\left(1 - n'\right)}{\left(1 + n'\right)\left(H'\right)^{\frac{1}{n'}}}, \qquad d = \frac{n' + 1}{n'}$$

Since we have pure bending, the stress amplitude varies with distance y from the neutral axis, but not with position x along the beam length. (The coordinate axes of Fig. 13.3 are being used.) Hence, a suitable volume element is

$$dV = Lb \, dy$$

Symmetry permits the integral to be evaluated on only one side of the beam, so that it becomes

$$\Delta U = 2JLb \int_0^c \sigma_a^d \, dy$$

This integral can be evaluated by a procedure similar to that leading to the bending analysis of Eq. 13.27. We first make the reasonable physical assumption that plane sections remain plane even during the cyclic plastic deformation. Thus, Eq. 13.22 applies to the strain amplitudes.

$$y = \frac{c}{\varepsilon_{ca}}\varepsilon_a, \qquad dy = \frac{c}{\varepsilon_{ca}}d\varepsilon_a$$

where ε_a is strain amplitude, and ε_{ca} is the particular value at the beam edge, $y = c$. The cyclic stress-strain curve (Eq. 12.58) and its differential (as in Eq. 13.24) are also needed.

$$\varepsilon_a = \frac{\sigma_a}{E} + \left(\frac{\sigma_a}{H'}\right)^{\frac{1}{n'}}, \qquad d\varepsilon_a = \left[\frac{1}{E} + \frac{1}{n'\sigma_a}\left(\frac{\sigma_a}{H'}\right)^{\frac{1}{n'}}\right]d\sigma_a$$

Substituting $d\varepsilon_a$ into the expression for dy, and then substituting the result into the integral for ΔU, gives a form with stress as the only variable.

$$\Delta U = \frac{2JLbc}{\varepsilon_{ca}} \int_0^{\sigma_{ca}} \sigma_a^{\frac{n'+1}{n'}}\left[\frac{1}{E} + \frac{1}{n'\sigma_a}\left(\frac{\sigma_a}{H'}\right)^{\frac{1}{n'}}\right]d\sigma_a$$

where σ_{ca} is the stress amplitude at $y = c$, and the expression involving n' has been substituted for d.

The integral is now readily evaluated. After doing so, substituting for J, and performing some manipulation, we obtain

$$\Delta U = \frac{8Lbc}{H^{\frac{1}{n'}}}\left(\frac{1 - n'}{1 + n'}\right)\sigma_{ca}^{\frac{n'+1}{n'}}\left[\frac{\dfrac{n'}{2n'+1} + \dfrac{1}{2+n'}\beta}{1 + \beta}\right] \qquad \textbf{Ans.}$$

where

$$\beta = \frac{\varepsilon_{pca}}{\varepsilon_{eca}}, \qquad \varepsilon_{pca} = \left(\frac{\sigma_{ca}}{H'}\right)^{\frac{1}{n'}}, \qquad \varepsilon_{eca} = \frac{\sigma_{ca}}{E}, \qquad \varepsilon_{ca} = \varepsilon_{eca} + \varepsilon_{pca}$$

in which ε_{pca} and ε_{eca} are plastic and elastic strain amplitudes, respectively, at $y = c$.

15.11 SUMMARY

15.11.1 Creep: Introductory Aspects

Creep tests of materials are most commonly performed by applying various levels of constant stress to uniaxial specimens. Data can be obtained on strains, strain rates, and rupture lives. In addition to strain vs. time plots, the data may be used to construct stress vs. life plots, where life is the time to rupture or to a particular value of strain. Stress-strain curves corresponding to given times, called isochronous stress-strain curves, are also useful. Creep test results may be employed as a basis for materials selection, and in analysis of trial component designs, to assure that an adequate safety factor exists against excessive deformation or creep rupture.

The physical mechanisms associated with creep deformation vary widely with the material and with the combination of stress and temperature involved. Behavior similar to simple viscous flow, where stress and strain rate are proportional, may occur. This is the case for glasses, for polymers at temperatures significantly above T_g, and for crystalline materials (metals and ceramics) at high temperature but low stress. For the latter, the mechanism of viscous flow is often diffusional flow involving movement of vacancies through the crystal lattice or along grain boundaries. At relatively high stresses in crystalline materials, the dominant creep mechanisms involve movement of dislocations, which process is highly sensitive to stress, so that strain rates are proportional to stress raised to a power on the order of five.

Creep mechanisms in polymers around and above T_g, but not very close to the melting temperature, involve various relatively complex motions and interactions of the long chain-like molecules. Effects such as entanglement of the chains can cause increased resistance as deformation proceeds, and also a tendency for much of the creep deformation to be recovered if the load is removed. Creep in concrete involves distinctive mechanisms associated with time-dependent deformation in the cement paste and movement of water in pores. Elastic strains in the aggregate oppose and limit the creep strain, and also cause a strong recovery behavior to occur.

The temperature dependence of creep strain rates is often similar to that predicted by thermal activation according to the Arrhenius equation.

$$\dot{\varepsilon} = Ae^{-\frac{Q}{RT}} \tag{15.95}$$

The activation energy Q varies with the mechanism. The coefficient A also depends on the mechanism and includes stress and sometimes grain size dependence of a type that varies with the mechanism, and also an additional (but weak) inverse dependence on temperature.

15.11.2 Creep: Engineering Analysis

For making life estimates, time-temperature parameters are generally used rather than more direct extrapolation of stress-life curves. For example, creep-rupture lives for various combinations of stress and temperature can be used with a material constant C to calculate values of the Larson-Miller Parameter.

$$P_{LM} = T \left(\log t_r + C\right) \tag{15.96}$$

A plot of P_{LM} versus stress can then be used to estimate times to rupture t_r for situations not represented in the original data. Extrapolation beyond that data by up to a factor of ten in life is reasonable as long as a new creep mechanism is not encountered. Other time-temperature parameters used include the Sherby-Dorn parameter, Eq. 15.14, and the Manson-Haferd parameter, Eq. 15.18. Where stress levels vary, a time-fraction rule, used similarly to the Palmgren-Miner rule for fatigue, provides rough life estimates. If creep is combined with cyclic loading, life fractions for creep may be combined with those for fatigue from the Palmgren-Miner rule to roughly estimate the combined effect. However, complex interactions can occur and more detailed analysis may be needed.

A variety of stress-strain-time relationships have been proposed that can be used for making engineering estimates of behavior. For steady-state creep, a power law dependence of strain rate on stress is often applicable.

$$\dot{\varepsilon}_{sc} = B\sigma^m \tag{15.97}$$

Additional terms, or an altered mathematical form, are used to describe transient creep, for which $\dot{\varepsilon}$ varies with time. Situations of varying stress are also of interest, including recovery of creep strain after removal of stress, relaxation of stress under constant strain, and step loading. Various alternatives exist for handling step loading. Where linear viscoelastic behavior is a reasonable approximation, a superposition procedure can be used as in Fig. 15.34. Time hardening or strain hardening (Fig. 15.35) are often employed for step loading of metals, with the latter appearing generally to be superior.

For multiaxial stresses, creep strains can be estimated using equations similar to those of plasticity theory but applied to strain rates.

$$\dot{\varepsilon}_x = \frac{1}{\eta} \left[\sigma_x - 0.5 \left(\sigma_y + \sigma_z\right)\right], \text{ etc.} \tag{15.98}$$

The tensile viscosity η appears, and since creep strains cause little volume change, 0.5 replaces Poisson's ratio. For the general case of linear viscoelastic behavior, η varies with time but not with stress. However, η may be treated as a stress-dependent variable to handle cases where nonlinear isochronous stress-strain curves occur. Rupture lives for multiaxial states of stress can be estimated by assuming that the effective stress $\bar{\sigma}$ has the same effect as a numerically equal uniaxial stress. A secondary dependence on the maximum principal stress can also be added according to Eq. 15.69.

For engineering components, ordinary linear-elastic stress analysis can be applied if the material behavior can be approximated as following linear viscoelasticity. Strains at various times are then obtained from the time-dependent elastic modulus $E(t)$ of the material. Where the isochronous (constant time) stress-strain curves are nonlinear, analysis is done just as for an elasto-plastic stress-strain curve. However, since such an analysis applies for only a single value of time, several repetitions of the analysis for various values of time may be needed. Isochronous stress-strain curves sometimes have the Ramberg-Osgood form.

$$\varepsilon = \frac{\sigma}{E} + \left(\frac{\sigma}{H_c}\right)^{\frac{1}{n_c}} \qquad (15.99)$$

where n_c is the appropriate value for creep and H_c is a function of time, so that a family of isochronous stress-strain curves is represented. Elasto-plastic stress-strain analysis for this form can then be used in a straightforward manner for creep.

15.11.3 Materials Damping

Energy dissipation during cyclic loading, called *damping*, can occur as a result of any of the physical mechanisms that cause creep deformation. However, other mechanisms that do not cause significant creep strain are also important in damping. Examples include various low stress (anelastic) damping mechanisms in metals, such as the Snoek effect and thermoelastic coupling, and also plastic deformation at high stresses. In general, the damping energy per cycle, per unit volume, varies with stress amplitude according to

$$\Delta u = J\sigma_a^d \qquad (15.100)$$

where the exponent d depends on the material and the mechanism that is dominant for a given stress, temperature, and frequency.

The damping behavior of metals at low stresses, and also of polymers in general, is often similar to the behavior of linear viscoelastic models that contain transient creep elements. As a result, peaks in energy dissipation occur at certain frequencies for constant temperature, and the energy dissipated in each cycle is approximately proportional to the square of stress, $d \approx 2$. In many steels, ferromagnetic damping provides a relatively high level of energy dissipation at low stresses that may be beneficial in limiting vibrations in engineering components. This type of damping is not sensitive to frequency and differs significantly from ideal viscoelastic damping in other respects as well. At stresses around and above the fatigue limit in metals, the energy dissipation is insensitive to frequency and is dominated by plastic deformation. The energy dissipated in each cycle is then highly sensitive to stress, typically being proportional to the eighth power of stress, $d \approx 8$.

NEW TERMS AND SYMBOLS

(a) Creep

activation energy, Q	creep rupture
Boltzmann superposition principle	deformation mechanism map
Coble creep	diffusion
creep cavitation	diffusional flow
creep exponent, m	dislocation creep
creep-fatigue interaction	effective strain rate, $\bar{\dot{\varepsilon}}$
creep recovery	isochronous stress-strain curve

Larson-Miller constant, C

linear viscoelasticity

Manson-Haferd constants: T_a, $\log t_a$

Nabarro-Herring creep

primary (transient) creep

secondary (steady-state) creep

strain-hardening rule

stress relaxation

temperature compensated time, θ

tertiary creep

thickness, b

time-dependent elastic modulus: $E(t)$, $E_s(t)$

time-fraction rule

time-hardening rule

time-temperature parameters:

   Sherby-Dorn, P_{SD}

   Larson-Miller, P_{LM}

   Manson-Haferd, P_{MH}

time to rupture, t_r

universal gas constant, R

viscosity: η_τ, η

(b) Damping

anelastic damping

component energies: ΔU, U_e

damping exponent, d

dynamic modulus, E^*

elastic strain energy, u_e

loss coefficient: $Q^{-1} = \tan \delta$

magnetoelastic damping

materials damping (internal friction)

phase shift, δ

plastic strain damping

remnant displacement

Snoek effect

thermoelastic effect

unit damping energy, Δu

REFERENCES

(a) General References

AMZALLAG, C., B. N. LEIS, and P. RABBE, eds. 1982 *Low Cycle Fatigue and Life Prediction*, ASTM STP 770, Am. Soc. for Testing and Materials, Philadelphia, Pa.

ASTM. 1990 *Annual Book of ASTM Standards*, Am. Soc. for Testing and Materials, Philadelphia, Pa. See Vol. 03.01: E139, "Conducting Creep, Creep-Rupture, and Stress-Rupture Tests of Metallic Materials"; E328, "Stress Relaxation Tests for Materials and Structures." Vol. 04.02: C512, "Creep of Concrete in Compression." Vol. 08.02: D2990, "Tensile, Compressive, and Flexural Creep and Creep-Rupture of Plastics"; D2991, "Testing Stress-Relaxation of Plastics."

CONWAY, J. B. 1969 *Stress-Rupture Parameters: Origin, Calculation, and Use*, Gordon and Breach, New York.

CRAWFORD, R. J. 1987 *Plastics Engineering*, 2nd ed., Pergamon Press, Oxford, UK.

DHALLA, A. K. and R. H. GALLAGHER. 1982 "Computational Methods for Nonlinear Structural Analyses," *Pressure Vessels and Piping: Design Technology—A Decade of Progress*, Am. Soc. of Mechanical Engineers, New York, pp. 185–201.

EVANS, R. W. and B. WILSHIRE. 1985 *Creep of Metals and Alloys*, The Institute of Metals, London, UK.

FROST, H. J. and M. F. ASHBY. 1982 *Deformation Mechanism Maps: The Plasticity and Creep of Metals and Ceramics*, Pergamon Press, Oxford, UK.

GOLDHOFF, R. M. 1974 "Towards the Standardization of Time-Temperature Parameter Usage in Elevated Temperature Data Analysis," *Jnl. of Testing and Evaluation*, ASTM, Vol. 2, No. 5, pp. 387–424.

KRAUS, H. 1980 *Creep Analysis*, John Wiley, New York.

LAZAN, B. J. 1968 *Damping of Materials and Members in Structural Mechanics*, Pergamon Press, Oxford, UK.

LEVY, S. and J. H. DUBOIS. 1984 *Plastic Product Design Engineering Handbook*, 2nd ed., Chapman and Hall, New York.

MARIN, J. 1962 *Mechanical Behavior of Engineering Materials*, Prentice Hall, Englewood Cliffs, N.J.

NEVILLE, A. M., W. H. DILGER, and J. J. BROOKS. 1983 *Creep of Plain and Structural Concrete*, Construction Press, New York.

PENNY, R. K. and D. L. MARRIOTT. 1971 *Design for Creep*, McGraw-Hill Book Co., New York.

POMEROY, C. D., ed. 1978 *Creep of Engineering Materials*, Mechanical Engineering Publications, Ltd., London, UK.

POWELL, P. C. 1983 *Engineering with Polymers*, Chapman and Hall, London, UK.

SHAMES, I. H. and F. A. COZZARELLI. 1992 *Elastic and Inelastic Stress Analysis*, Prentice Hall, Englewood Cliffs, N.J.

SOLOMON, H. D. et al., eds. 1988 *Low Cycle Fatigue*, ASTM STP 942, Am. Soc. for Testing and Materials, Philadelphia, Pa.

(b) Sources for Material Properties

ASHBY, M. F., C. GANDHI, and D. M. R. TAPLIN. 1979 "Fracture Mechanisms Maps and Their Construction for F.C.C. Metals and Alloys," *Acta Metallurgica*, Vol. 27, p. 699–729.

BRADLEY, E. F., ed. 1979 *Source Book on Materials for Elevated-Temperature Applications*, Am. Soc. for Metals, Metals Park, Oh.

CLAUS, F. J. 1969 *Engineer's Guide to High Temperature Materials*, Addison-Wesley, Reading, Ma.

CONWAY, J. B. and P. N. FLAGELLA. 1971 *Creep-Rupture Data for the Refractory Metals to High Temperatures*, Gordon and Breach, New York.

HALLOWELL, J. B., ed. 1989 *Structural Alloys Handbook*, 3 vols., Metals and Ceramics Information Center, Battelle Columbus Labs, Columbus, Oh.

INCO. 1972 *High Temperature High Strength Nickel Base Alloys*, 3rd ed., International Nickel Co., New York.

ROTHMAN, M. F., ed. 1988 *High-Temperature Property Data: Ferrous Alloys*, ASM International, Metals Park, Oh.

PROBLEMS AND QUESTIONS

Section 15.2

15.1 Based on Fig. 15.7, and considering only strains rates below 0.1 h^{-1}, obtain approximate constants B and m for an equation of the form

$$\dot{\varepsilon}_{sc} = B\sigma^m$$

Do so for each of the four temperatures. Is the value of m constant? Are the strain rates in this data highly sensitive to stress? (Suggestion: Use the straight lines on the graph, and read stresses on each line at two contrasting values of strain rate, extrapolating where needed.)

15.2 A 40% tin, 60% lead alloy solder wire of diameter 3.15 mm is subjected to creep by hanging weights from lengths of the wire. Length changes measured over a 254 mm gage length after various elapsed times are given below for three different weights.

(a) Plot the family of strain vs. time curves that results.

(b) Characterize the behavior. Is it dominated by either transient or steady-state creep, or do significant amounts of both occur?

(c) Determine steady-state creep rates, $\dot{\varepsilon}_{sc}$, for each value of weight and plot these on log-log coordinates versus the corresponding stresses. Does a straight line provide a reasonable fit to the data? If so, find values of constants B and m for the relationship $\dot{\varepsilon}_{sc} = B\sigma^m$.

	Length change, mm		
Time, min	4.54 kg	6.80 kg	9.07 kg
0	0	0	0
0.25	0.28	0.46	0.69
0.5	0.36	0.66	0.94
1	0.48	0.91	1.45
2	0.71	1.40	2.36
4	1.09	2.24	4.09
6	1.47	3.00	5.72
8	1.83	3.38	7.26
12	2.54	4.90	10.41
16	3.23	6.38	13.64
20	3.91	7.82	16.74

15.3 From the data in Fig. 15.8, construct an approximate plot with curves of stress versus temperature for various times to failure. In particular, plot three curves, one corresponding to each of the three rupture times 10^2, 10^3, and 10^4 hours.

15.4 Using the information in Fig. 15.9 for AISI 4340 steel with $\sigma_u = 140$ ksi, plot approximate stress versus rupture time curves for temperatures of 600, 800, and 950°F. Employ a logarithmic scale for the rupture time.

15.5 Plot approximate isochronous stress-strain curves for lead at 17°C based on the data in Fig. 15.5, doing so for times of $t = 3$, 10, and 30 min. Are the curves linear?

Section 15.3

15.6 Consider metals, polymers, and concrete. Which of these classes of materials generally exhibit strong recovery of creep strain after unloading, and which do not? Briefly explain in terms of the physical mechanisms of creep why the strains are generally recovered, or why they are not recovered, for each class of materials.

15.7 Compare the diffusional flow and dislocation creep mechanisms of crystalline materials. In particular, considering the effects of grain size, stress, and temperature, how do the trends in behavior differ?

15.8 For AISI 304 stainless steel with a grain size of 200 μm, constants for Eq. 15.9 are as follows for stress σ in units of MPa, temperature T in kelvins (K), and strain rates $\dot{\varepsilon}$ in s^{-1}. (Constants based on [Frost 82] p. 62.)

$$A_3 = 1.1 \times 10^{-4} \quad (T = 900 \text{ K})$$

$$A_3 = 4.6 \times 10^{-4} \quad (T = 1200 \text{ K})$$

$$A_3 = 2.0 \times 10^{-3} \quad (T = 1450 \text{ K})$$

$$m = 7.5, \qquad Q/R = 33{,}700 \text{ K}$$

Make a log-log plot of stress σ versus strain rate $\dot{\varepsilon}$, showing lines for the three different temperatures. Consider stresses in the range 50 to 500 MPa for 900 K, 15 to 150 MPa for 1200 K, and 6 to 60 MPa for 1450 K.

15.9 Using the constants from the previous problem for 304 SS at 1200 K, plot isochronous stress-strain curves for times of 1 minute, 1 hour, and 1 week. Consider stresses such that the strains extend to about $\varepsilon = 0.02$ in each case. Assume that only elastic strains need to be added to creep strains from Eq. 15.9, and that the elastic modulus at 1200 K is $E = 120{,}000$ MPa. (PC Problem)

15.10 Approximately confirm the value of activation energy Q shown on Fig. 15.12.

15.11 Consider the deformation mechanism map for pure silver of Fig. 15.17. Approximately draw or trace this map and also draw additional maps that show qualitatively how the map changes due to:
(a) Decreased yield strength due to less cold work in processing of the material.
(b) Increased yield strength due to more cold work.
(c) Increased grain size.
(d) Decreased grain size.

Section 15.4

15.12 In a creep test at 640°C, a stainless steel specimen loaded in tension at a stress of 80 MPa failed by creep rupture after 1000 hours. This same material is loaded in tension, also at 80 MPa, in an engineering application where the temperature is 585°C. Estimate the permissible actual service life in days if a safety factor of 20 on life is required.

15.13 For the situation of the previous problem, what is the highest permissible temperature for an actual service life of 90 days at 80 MPa?

15.14 Alloy S-590 is subjected in a service application to a uniaxial tension stress of 140 MPa at 650°C.
(a) Using Fig. 15.22, estimate the creep rupture life based on the Sherby-Dorn time-temperature parameter.
(b) If a safety factor of two on stress is required, what actual service life can be allowed?
(c) Based on your answers for (a) and (b), what is the safety factor on life that corresponds to the factor of two on stress?

15.15 Proceed as in the previous problem, except use the Manson-Haferd parameter and Fig. 15.27.

15.16 For aluminum alloyed with 1.6% Mg, the relationship between stress and the Sherby-Dorn parameter, as determined from uniaxial creep-rupture data, is approximately

$$\sigma = -11.3P_{SD} - 124 \quad (25 \leq \sigma \leq 85 \text{ MPa})$$

where P_{SD} is based on units of hours and kelvins (K) as described in the text, and σ is in MPa. (Equation based on data in [Orr 54].)
(a) Estimate the rupture life for the following four situations:
(1) $\sigma = 50$ MPa, $T = 200°C$
(2) $\sigma = 50$ MPa, $T = 150°C$
(3) $\sigma = 25$ MPa, $T = 200°C$
(4) $\sigma = 25$ MPa, $T = 150°C$
(b) Comment on the trends in rupture life with stress and temperature. Is the life very sensitive to temperature and stress?

15.17 The aluminum alloy of the previous problem is used in an application where the applied stress is 28 MPa. The part must not fail by creep rupture before 10,000 hours, and a safety factor of 1.5 on stress is required.
(a) What is the highest operating temperature that is permissible?
(b) What is the safety factor on life that is achieved by using the factor of 1.5 on stress?
(c) What is the safety margin in temperature corresponding to the factor of 1.5 on stress?
(d) If an operating temperature of 155°C is later found to be necessary, what is the highest stress that can be permitted?

15.18 Using the Larson-Miller plot for alloy S-590 of Fig. 15.24:
(a) Fit a straight line that gives a reasonable representation of the data, and obtain the corresponding equation relating stress and P_{LM}.
(b) Then use your result to estimate and plot stress-life curves for creep rupture for temperatures of 500, 650, and 800°C. A log-log plot is suggested, and the plot should cover rupture lives between 100 and 100,000 hours for each temperature.

15.19 Creep-rupture data are given in Table P15.19 for the Cr-Mo-V alloy steel in Table 15.1.
(a) Calculate values of the Sherby-Dorn parameter and plot these versus stress. (PC Problem)
(b) Proceed as in (a), except use the Larson-Miller parameter.
(c) Proceed as in (a), except use the Manson-Haferd parameter.
(d) Which of the three parameters provide reasonable correlations of the data? Is any one parameter noticeably better in this respect than the others?

Section 15.5

15.20 A stainless steel has creep rupture behavior at two stress levels as shown in Table P15.20. This same material is used in an engineering component subject to creep. For 21 hours of each day, the stress is 210 MPa and the temperature is 480°C. For the remaining 3 hours of each day, the stress is 62 MPa and the temperature is 700°C. Estimate the number of days to rupture.

15.21 For 10 hours during each week of service in a particular application, the aluminum alloy of Prob. 15.16 is subjected to a stress of 40 MPa and a temperature of 150°C. However,

TABLE P15.19

Temperature °C	Stress MPa	Rupture time hours
482	621	37
482	565	975
482	538	3,581
482	483	9,878
538	552	7
538	469	213
538	414	1,493
538	338	5,108
538	262	10,447
593	417	18
593	345	167
593	276	615
593	200	2,200
593	152	6,637
649	276	19
649	207	102
649	172	125
649	138	331
732	138	3.7
732	103	8.9
732	69	31.8

Source: Data in [Goldhoff 59b].

TABLE P15.20

Stress, MPa	Temperature, °C	Rupture time, hours
210	540	1000
62	730	1000

for one additional hour in each week, the stress is 25 MPa and the temperature 200°C. If a safety factor of ten on life is required, what is the useful service life?

Section 15.6

15.22 Using the constants in Table 15.2 for Cr-Mo-V steel at 538°C: (PC Problem)
   (a) Plot isochronous stress-strain curves for times of 1, 10, 10^2, 10^3, and 10^4 hours. Do the curves differ greatly? Comment on any trends that are apparent from these curves.
   (b) Plot strain-time curves out to 1000 hours for stresses of 75, 150, and 200 MPa. Comment on the trends from these curves also.

15.23 Strain versus time data obtained by hanging weights from strips of high-density polyethylene are plotted below.
  (a) Obtain isochronous stress-strain data for times of $t = 30$, 200, and 600 seconds, and plot these on log-log coordinates.
  (b) If reasonably straight lines are obtained, find constants C and δ for relationships of the form

$$\varepsilon = C\sigma^\delta$$

  for each value of t. If the lines are nearly parallel, use the same value of the slope δ for all of them, but different values of C.
  (c) Now attempt to find values of D_3 and ϕ for Eq. 15.33 so that this equation (with ε_i neglected) represents the stress-strain-time behavior. (PC Problem)

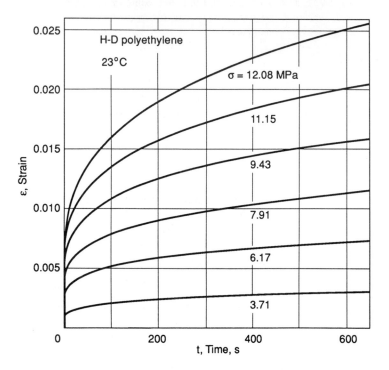

Figure P15.23

Section 15.7

15.24 A material behaves according to Eq. 15.33(b), and the creep rate from this equation is assumed to apply even during stress relaxation.
  (a) For a strain ε' that is quickly applied and held constant, develop an equation for the stress vs. time behavior during relaxation.
  (b) Does your result reduce to Eq. 15.45 for the special case of $\phi = 1$?

15.25 A bolt used at 630°C is made of 304 stainless steel with a grain size of $d = 50~\mu m$. The bolt is tightened to a preload stress of 30 MPa. Loss of this preload may occur due to creep dominated by diffusional flow along grain boundaries, that is, Coble creep according to Eq. 15.7. Constants apply as follows: (Constants based on [Frost 82] p. 62.)

$$Q_b/R = 20,100~\text{K}$$

$$A_2' = 7.73 \times 10^{-12}~\frac{\text{K} \cdot \text{m}^5}{\text{MN} \cdot \text{s}}$$

$$E = 151,000~\text{MPa (at 630°C)}$$

(a) After what time period is half the bolt preload lost due to creep?

(b) What grain size is needed to avoid losing more than half the preload in one year?

(c) What reduced temperature with the original grain size would avoid loss of half the preload in one year? The value of E increases about 100 MPa for each degree Celsius decrease in temperature. (PC Problem)

15.26 For the rheological model of Fig. 15.33(b), verify the equation given for the stress vs. time response during relaxation under a constant strain ε'.

15.27 Verify Eq. 15.39 using the Boltzman superposition principle.

15.28 For creep according to Eq. 15.9, show that time hardening and strain hardening give the same result.

Section 15.8

15.29 For a given material and temperature, the uniaxial creep behavior follows Eq. 15.33(b). Develop a corresponding equation $\gamma = f(\tau)$ for creep in pure planar shear, where the constants are expressed in terms of D_3, δ, and ϕ from uniaxial test data.

15.30 Proceed as in Example 15.3 except change the pressure vessel to a thin-walled spherical one of radius r and wall thickness b.

15.31 A stainless steel pressure vessel is a cylindrical tube with closed ends, wall thickness 1 cm, and diameter at mid-thickness 20 cm. It is loaded with an internal pressure of 5 MPa and an axial load in tension of 100 kN. The operating temperature is 900 K, and the material has constants as given in Prob. 15.8.

(a) What is the rate of the effective (uniaxial equivalent) strain $\bar{\dot{\varepsilon}}$?

(b) What are the percentage increases in the vessel length and diameter in 10 years?

(c) Does the design seem to be reasonable from the standpoint of creep deformation?

15.32 A thin-walled tube with closed ends has a diameter at mid-thickness of 125 mm and a wall thickness of 3 mm, and it is loaded with with an internal pressure of 1.7 MPa. The material is the aluminum alloy of Prob. 15.16, and the temperature is 150°C. Estimate the life to creep rupture.

Section 15.9

15.33 A rectangular beam of depth 2c and thickness b is subjected to pure bending due to a moment M that is held constant with time. Develop an equation giving the maximum strain as a function of M and time t for creep behavior according to (a) $\varepsilon_c = B\sigma^m t$, and (b)

$\varepsilon_c = D_3 \sigma^\delta t^\phi$. In both cases, assume that the instantaneous elastic and plastic strains are small.

15.34 A solid circular shaft of radius c is subjected to a torque T that is held constant with time. Develop equations giving the surface shear strain as a function of torque T and time t for materials that have uniaxial creep behavior according to (a) and (b) of Prob. 15.33. Assume that the instantaneous (elastic and plastic) strains are small.

15.35 A high-density polyethylene material is made into a cantilever beam of depth $2c = 10$ mm and thickness $b = 6.2$ mm, and a 1.2 kg weight is applied to the end of the beam. What strain is expected at a point on the edge of the beam ($y = c$) which is 90 mm from the weight after (a) 30 seconds and (b) 10 minutes? You may use the isochronous stress-strain curves of Fig. 15.31.

15.36 A beam of 7075-T6 Al has a rectangular cross section and is used at a temperature of 316°C. The material has stress-strain constants as given in Table 15.2. A moment of 2.7 kN·m is applied, a safety factor of 1.5 on load is required, and the creep strain cannot exceed 1% in 100 hours. Select a beam size such that the depth $2c$ is twice the thickness b, that is, find $b = c$.
(a) Assuming that the elastic deformation is negligible.
(b) Including elastic deformation effects. (PC Problem)

Section 15.10

15.37 For the ideal viscoelastic model of Fig. 15.38, qualitatively sketch the σ-ε responses for three different values of σ_a, where $\sigma_{a1} < \sigma_{a2} < \sigma_{a3}$, all of which are applied at the same frequency. How do Δu and u_e vary with σ_a? How do δ and Q^{-1} vary with σ_a?

15.38 Use Eq. 12.59 to verify Eq. 15.89.

15.39 Consider a rectangular beam of depth $2c$, thickness b, and length L that is subjected to a cyclic moment of amplitude M_a about zero mean. Assume that the energy dissipation of the material is given by Eq. 15.83, and that the damping is small, so that the distribution of stress over the beam depth is approximately linear, and also so that the last column of Table 15.3 applies.
(a) Obtain an equation giving the energy ΔU dissipated in each cycle of loading.
(b) Evaluate the loss coefficient Q^{-1}. How is the value related to that for uniaxial loading of the material?
(c) Use your result from (a) to obtain ΔU for a cantilever beam with one end fixed and a cyclic load P_a at the free end.

15.40 Proceed as in (a) and (b) of the previous problem, but consider the more general situation of small damping where the exponent $d \neq 2$, as in Eq. 15.91.

15.41 Consider completely reversed cyclic loading at a strain amplitude ε_a of an elastic, perfectly plastic material having yield strength σ_o and elastic modulus E.
(a) Write an expression for the unit damping energy Δu as a function of ε_a.
(b) Use (a) to obtain the energy ΔU dissipated by cyclic pure bending of a rectangular beam of depth $2c$, thickness b, and length L. Express the result as a function of the strain amplitude ε_{ca} at the beam edge, where ε_{ca} exceeds σ_o/E.

15.42 A material loaded in pure shear has a Ramberg-Osgood type cyclic stress-strain curve of the form of Eq. 13.54.

(a) Show that the unit damping energy can be approximated by

$$\Delta u = 4 \left(\frac{1 - n'}{1 + n'} \right) \tau_a \gamma_{pa}$$

where τ_a and γ_{pa} are the amplitudes of shear stress and plastic shear strain, respectively.

(b) Consider a completely reversed cyclic torque T_a applied to a shaft of radius c made of such a material. Develop an equation giving the energy ΔU dissipated in each cycle of loading as an implicit function of the shear strain amplitude γ_{ca} at the shaft surface, where γ_{ca} is sufficiently large to cause a significantly nonlinear stress distribution.

Bibliography

[Andrade 10]   E. M. da C. ANDRADE. 1910 "On the Viscous Flow in Metals and Allied Phenomena," *Proc. Royal Soc., Series A*, London, UK, Vol. 84, pp. 1–12.

[Andrade 14]   E. M. da C. ANDRADE. 1914 "The Flow in Metals under Large Constant Stresses," *Proc. Royal Soc., Series A*, London, UK, Vol. 90, pp. 329–342.

[Antunes 78]   V. T. A. ANTUNES and P. HANCOCK. 1978 "Strainrange Partitioning of MAR-M002 over the Termperature Range 750°C–1040°C," *Characterization of Low Cycle High Temperature Fatigue by the Strainrange Partitioning Method*, AGARD-CP-243, North Atlantic Treaty Organization, Advisory Group for Aerospace Research and Development, Neuilly-sur-Seine, France, pp. 5–1 to 5–9.

[Ashby 72]   M. F. ASHBY. 1972 "A First Report on Deformation Mechanism Maps," *Acta Metallurgica*, Vol. 20, pp. 887–897.

[Ashby 86]   M. F. ASHBY and D. R. H. JONES. 1986 *Engineering Materials 2: An Introduction to Microstructures, Processing and Design*, Pergamon Press, Oxford, UK.

[ASM 87]   ASM. 1987 *Engineered Materials Handbook, Vol. 1: Composites*, ASM International, Metals Park, Oh.

[ASM 88]   ASM. 1988 *Engineered Materials Handbook, Vol. 2: Engineering Plastics*, ASM International, Metals Park, Oh.

[ASTM 90a]   ASTM. 1990 *Annual Book of ASTM Standards*, Am. Soc. for Testing and Materials, Philadelphia, Pa.

[ASTM 90b]   ASTM. 1990 *Annual Book of ASTM Standards*, Am. Soc. for Testing and Materials, Philadelphia, Pa. See No. E466, "Conducting Constant Amplitude Axial Fatigue Tests of Metallic Materials," Vol. 03,01; and No. E1049, "Cycle Counting in Fatigue Analysis," Vol. 03.01; and also No. D671,"Flexural Fatigue of Plastics by Constant Amplitude of Force," Vol. 08.01.

752

[ASTM 91]   ASTM. 1991 *Annual Book of ASTM Standards*, Am. Soc. for Testing and Materials, Philadelphia, Pa. See No. E399, "Standard Test Method for Plane-Strain Fracture Toughness of Metallic Materials," Vol. 03.01; and Nos. E561, E813, and E1290 in Vol. 03.01; and also No. D5045 in Vol. 08.03.

[Barsom 71]   J. M. BARSOM. 1971 "Fatigue-Crack Propagation in Steels of Various Yield Strengths," *Jnl. of Engineering for Industry, Trans. ASME, Series B*, Vol. 93, No. 4, Nov. 1971, pp. 1190–1196.

[Barsom 75]   J. M. BARSOM. 1975 "Development of the AASHTO Fracture-Toughness Requirements for Bridge Steels," *Engineering Fracture Mechanics*, Vol. 7, No. 3, Sept. 1975, pp. 605–618.

[Barsom 87]   J. M. BARSOM and S. T. ROLFE. 1987 *Fracture and Fatigue Control in Structures*, 2nd ed., Prentice Hall, Englewood Cliffs, NJ.

[Bates 69]   R. C. BATES and W. G. CLARK, JR. 1969 "Fractography and Fracture Mechanics," *Trans. of the Am. Soc. for Metals*, Vol. 62, pp. 380–389.

[Beardmore 75]   P. BEARDMORE and S. RABINOWITZ. 1975 "Fatigue Deformation of Polymers," *Treatise on Materials Science and Technology, Vol. 6: Plastic Deformation of Materials*, R. J. Arsenault, ed., Academic Press, New York, pp. 267–331.

[Boller 56]   K. H. BOLLER. 1956 "Fatigue Properties of Various Glass-Fiber- Reinforced Plastic Laminates," WADC TR 55-389, Wright Air Dev. Ctr., Wright Patterson Air Force Base, Oh.

[Boswell 59]   C. C. BOSWELL, JR. and R. A. WAGNER. 1959 "Fatigue in Rotary-Wing Aircraft," *Metal Fatigue*, G. Sines and J. L. Waisman, eds., McGraw-Hill, New York, pp. 355–375.

[Boyer 85]   H. E. Boyer and T. L. Gall, eds. 1985 *Metals Handbook: Desk Edition*, American Society for Metals, Metals Park, Oh.

[Bridgman 44]   P. W. BRIDGMAN. 1944 "The Stress Distribution at the Neck of a Tension Specimen," *Trans. of the Am. Soc. for Metals*, Vol. 32, pp. 553–574.

[Bridgman 52]   P. W. BRIDGMAN. 1952 *Studies in Large Plastic Flow and Fracture*, McGraw-Hill, New York.

[Brockenbrough 81] R. L. BROCKENBROUGH and B. G. JOHNSTON. 1981 *USS Steel Design Manual*, ADUSS 27-3400-04, United States Steel Corp., Monroeville, Pa.

[Broek 86]   D. BROEK. 1986 *Elementary Engineering Fracture Mechanics*, 4th ed., Kluwer Academic Pubs., Dordrecht, The Netherlands.

[Budinski 92]   K. G. BUDINSKI. 1992 *Engineering Materials: Properties and Selection*, 4th ed., Prentice Hall, Englewood Cliffs, NJ.

[Bullens 48]   D. K. BULLENS. 1948 *Steel and Its Heat Treatment*, Vol. 1, John Wiley, New York.

[Bush 74]   S. H. BUSH. 1974 "Structural Materials for Nuclear Power Plants," *Jnl. of Testing and Evaluation*, ASTM, Vol. 2, No. 6, Nov. 1974, pp. 435–462.

[Buxbaum 73]   O. BUXBAUM. 1973 "Methods of Stress Measurement Analysis for Fatigue Life Evaluation," *Fatigue Life Prediction for Aircraft Structures and Materials*, AGARD-LS-62, North Atlantic Treaty Organization, Advisory Group for Aerospace Research and Development, Neuilly-sur-Seine, France, pp. 2–1 to 2–19.

[Campbell 82]   J. E. CAMPBELL, W. W. GERBERICH and J. H. UNDERWOOD, eds. 1982 *Application of Fracture Mechanics for Selection of Metallic Structural Materials*, Am. Soc. for Metals, Metals Park, Oh.

[Carlson 59]   R. G. CARLSON. 1959 *Fatigue Studies of Inconel*, Contract No., BMI-1335, UC-25 Metallurgy and Ceramics, Battelle Memorial Institute, Columbus, Oh., June 26, 1959.

[Chang 78]   J. B. CHANG, et al. 1978 "Improved Methods for Predicting Spectrum Loading Effects: Phase 1 Report, Vol. 1," AFFDL-TR-79-3036, Air Force Flight Dynamics Laboratory, Wright-Patterson AFB, Oh.

[Chu 70]   S. C. CHU and O. M. SIDEBOTTOM. 1970 "Creep of Metal Torsion-Tension Members Subjected to Nonproportionate Load Changes," *Experimental Mechanics*, June 1970.

[Clark 70]   W. G. CLARK, JR. and E. T. WESSEL. 1970 "Application of Fracture Mechanics Technology to Medium Strength Steels," *Review of Developments in Plane Strain Fracture Toughness Testing*,

W. F. Brown, Jr. ed., ASTM STP 463, American Society for Testing and Materials, Philadelphia, Pa., pp. 160–190.

[Clark 71]   W. G. CLARK, JR. 1971 "Fracture Mechanics in Fatigue," *Experimental Mechanics*, Vol. 11, No. 9, Sept. 1971, pp. 421–428.

[Clark 76]   W. G. CLARK, JR. and J. D. LANDES. 1976 "An Evaluation of Rising Load K_{Iscc} Testing," *Stress Corrosion—New Approaches*, ASTM STP 610, Am. Soc. for Testing and Materials, Philadelphia, Pa., pp. 108–127.

[Coffin 50]   L. F. COFFIN. 1950 "The Flow and Fracture of a Brittle Material," *Jnl. of Applied Mechanics*, Vol. 17 (*Trans. ASME*, Vol. 72), Sept. 1950, pp. 233–248.

[Collins 81]   J. A. COLLINS. 1981 *Failure of Materials in Mechanical Design*, John Wiley, New York.

[Conle 84]   F. A. CONLE, R. W. LANDGRAF and F. D. RICHARDS. 1984 *Materials Data Book: Monotonic and Cyclic Properties of Engineering Materials*, Ford Motor Co., Scientific Research Staff, Dearborn, Mich.

[Conway 69]   J. B. CONWAY. 1969 *Stress-Rupture Parameters: Origin, Calculation, and Use*, Gordon and Breach, New York.

[Cottell 56]   G. A. COTTELL. 1956 "Lessons to be Learnt from Failures in Service," *Proc. of the Int. Conf. on Fatigue of Metals*, London, Sept. 1956, and New York, Nov. 1956, The Institution of Mechanical Engineers, London, UK, pp. 563–569.

[Crandall 59]   S. H. CRANDALL and N. C. DAHL, eds. 1959 *An Introduction to the Mechanics of Solids*, McGraw-Hill, New York.

[Creyke 82]   W. E. C. CREYKE, I. E. J. SAINSBURY and R. MORRELL. 1982 *Design with Non-Ductile Materials*, Applied Science, London, UK.

[Dauskardt 90] R. H. DAUSKARDT, D. B. MARSHALL and R. O. RITCHIE. 1990 "Cyclic Fatigue Crack Propagation in Magnesia-Partially-Stabilized Zirconia," *Jnl. of the Am. Ceramic Society*, Vol. 73, No. 4, pp. 893–903.

[Davis 37]   R. E. DAVIS, H. E. DAVIS and E. H. BROWN. 1937 "Plastic Flow and Volume Changes in Concrete," *Proc. of the Am. Soc. for Testing and Materials*, Vol. 37, Part II, pp. 317–331.

[Davis 43]   E. A. DAVIS. 1943 "Increase of Stress with Permanent Strain and Stress-Strain Relations in the Plastic State for Copper under Combined Stresses," *Jnl. of Applied Mechanics, Trans. ASME*, Vol. 65, Dec. 1943, pp. A187–A196.

[Davis 45]   E. A. DAVIS. 1945 "Yielding and Fracture of Medium-Carbon Steels under Combined Stresses," *Jnl. Applied Mechanics, Trans. ASME*, Vol. 67, Mar. 1945, pp. A13–A24.

[Dennis 86]   K. R. DENNIS. 1986 "Fatigue Crack Growth of Gun Tube Steel under Spectrum Loading," MS Thesis, Engineering Science and Mechanics Dept., Virginia Polytechnic Institute and State University, Blacksburg, Va., May 1986.

[Dowling 72]   N. E. DOWLING. 1972 "Fatigue Failure Predictions for Complicated Stress-Strain Histories," *Jnl. of Materials*, ASTM, Vol. 7, No. 1, Mar. 1972, pp. 71–87.

[Dowling 73]   N. E. DOWLING. 1973 "Fatigue Life and Inelastic Strain Response under Complex Histories for an Alloy Steel," *Journal of Testing and Evaluation*, ASTM, Vol. 1, No. 4, Jul. 1973, pp. 271–287.

[Dowling 77]   N. E. DOWLING. 1977 "Fatigue-Crack Growth Rate Testing at High Stress Intensities," *Flaw Growth and Fracture*, ASTM STP 631, Am. Soc. for Testing and Materials, Philadelphia, Pa., pp. 139–158.

[Dowling 78]   N. E. DOWLING. 1978 "Stress-Strain Analysis of Cyclic Plastic Bending and Torsion," *Jnl. of Engineering Materials and Technology*, ASME, Vol. 100, Apr. 1978, pp. 157–163.

[Dowling 79a]   N. E. DOWLING. 1979 "Notched Member Fatigue Life Predictions Combining Crack Initiation and Propagation," *Fatigue of Engineering Materials and Structures*, Vol. 2, No. 2, pp. 129–138.

[Dowling 79b]   N. E. DOWLING and W. K. WILSON. 1979 "Analysis of Notch Strain for Cyclic Loading," *Fifth Int. Conf. on Structural Mechanics in Reactor Technology*, Vol. L, Paper L13/4, North-Holland Pub. Co., Amsterdam.

[Dowling 82a]   N. E. DOWLING. 1982 "Torsional Fatigue Life of Power Plant Equipment Rotating Shafts," Report No. DOE/RA29353-1 (DE8300913), U.S. Dept. of Energy, Office of Electrical Energy Systems, Washington, DC, Sept. 1982.

[Dowling 82b]   N. E. DOWLING. 1982 "Fatigue Failure Predictions for Complex Load Versus Time Histories," Section 7.4 of *Pressure Vessels and Piping: Design Technology—1982—A Decade of Progress*, S. Y. Zamrik and D. Dietrich, eds., Book No. G00213, Am. Soc. of Mechanical Engineers, New York. Also pub. in *Journal of Engineering Materials and Technology*, ASME, Vol. 105, Jul. 1983, pp. 206-214, with Erratum, Oct. 1983, p. 321.

[Dowling 83]   N. E. DOWLING. 1983 "Growth of Short Fatigue Cracks in an Alloy Steel," Paper No. 83-PVP-94, ASME 4th National Congress on Pressure Vessel and Piping Technology, June 19-24, 1983, Portland, Or.

[Dowling 87a]   N. E. DOWLING. 1987 "J-Integral Estimates for Cracks in Infinite Bodies," *Engineering Fracture Mechanics*, Vol. 26, No. 3, pp. 333–348.

[Dowling 87b]   N. E. Dowling. 1987 "A Review of Fatigue Life Prediction Methods," *Durability by Design*, SAE Pub. No. SP-730, Soc. of Automotive Engineers, Warrendale, Pa., Paper No. 871966.

[Dowling 89]   N. E. DOWLING and A. K. KHOSROVANEH. 1989 "Simplified Analysis of Helicopter Fatigue Loading Spectra," J. M. Potter and R. T. Watanabe, eds., *Development of Fatigue Loading Spectra*, ASTM STP 1006, Am. Soc. for Testing and Materials, Philadelphia, Pa., pp. 150–171.

[Endo 69]   T. ENDO and J. MORROW. 1969 "Cyclic Stress-Strain and Fatigue Behavior of Representative Aircraft Metals," *Journal of Materials*, ASTM, Vol. 4, No. 1, Mar. 1969, pp. 159–175.

[Evans 85]   R. W. EVANS and B. WILSHIRE. 1985 *Creep of Metals and Alloys*, The Institute of Metals, London, UK.

[EVC 89]   EVC. 1989 "Engineering Design with Unplasticized Polyvinyl Chloride," European Vinyls Corp. (UK) Ltd., Cheshire, UK.

[Farag 89]   M. M. FARAG. 1989 *Selection of Materials and Manufacturing Processes for Engineering Design*, Prentice Hall International (UK) Ltd., Hertfordshire, UK.

[Felbeck 84]   D. K. FELBECK and A. G. ATKINS. 1984 *Strength and Fracture of Engineering Solids*, Prentice Hall, Englewood Cliffs, NJ.

[Floreen 79]   S. FLOREEN and R. H. KANE. 1979 "Effects of Environment on High-Temperature Fatigue Crack Growth in a Superalloy," *Metallurgical Transactions*, Vol. 10A, Nov. 1979, pp. 1745–1751.

[Forrest 62]   P. G. FORREST. 1962 *Fatigue of Metals*, Pergamon Press, Oxford, UK, and Addison-Wesley, Reading, Ma.

[French 50]   R. S. FRENCH and W. R. HIBBARD. 1950 "Effect of Solute Atoms on Tensile Deformation of Copper," *Jnl. of Metals; Trans. AIME*, Vol. 188, pp. 53–58.

[French 56]   H. J. French. 1956 "Some Aspects of Hardenable Alloy Steels," *Jnl. of Metals; Trans. AIME*, Vol. 206, pp. 770–782.

[Frost 59]   N. E. FROST. 1959 "A Relation Between the Critical Alternating Propagation Stress and Crack Length for Mild Steel," *Proc. of the Institution of Mechanical Engineers*, London, UK, Vol. 173, No. 35, pp. 811–827.

[Frost 82]   H. J. FROST and M. F. ASHBY. 1982 *Deformation Mechanism Maps: The Plasticity and Creep of Metals and Ceramics*, Pergamon Press, Oxford, UK.

[Geil 65]   P. H. GEIL. 1965 "Polymer Morphology," *Chemical and Engineering News*, Vol. 43, No. 33, Aug. 16, 1965, pp. 72–84.

[Gerberich 79]   W. W. GERBERICH and N. R. MOODY. 1979 "A Review of Fatigue Fracture Topology Effects on Threshold and Growth Mechanisms," *Fatigue Mechanisms*, J. T. Fong, ed., ASTM STP 675, Am. Soc. for Testing and Materials, Philadelphia, Pa., pp. 292–341.

[Gohn 64]   G. R. GOHN. 1964 "Fatigue in Electronic and Magnetic Materials," *Fatigue—An Interdisciplinary Approach*, J. J. Burke et al., eds., Proc. of the 10th Sagamore Army Materials Research Conf., Syracuse University Press, Syracuse, N.Y., pp. 287–315.

[Goldhoff 59a]   R. M. GOLDHOFF. 1959 "Which Method for Extrapolating Stress-Rupture Data?" *Materials in Design Engineering*, Vol. 49, No. 4, pp. 93–97.

[Goldhoff 59b]   R. M. GOLDHOFF. 1959 "Comparison of Parameter Methods for Extrapolating High Temperature Data," *Trans. of the Am. Soc. of Mechanical Engineers*, Vol. 81, pp. 629–644.

[Grassi 49]   R. C. GRASSI and I. CORNET. 1949 "Fracture of Gray Cast Iron Tubes under Biaxial Stresses," *Jnl. of Applied Mechanics*, Vol. 16 (*Trans. ASME*, Vol. 71), June 1949, pp. 178–182.

[Griggs 36]   D. T. GRIGGS. 1936 "Deformation of Rocks Under High Confining Pressures," *Jnl. of Geology*, Vol. 44, pp. 541–577.

[Grover 66]   H. J. GROVER. 1966 *Fatigue of Aircraft Structures*, NAVAIR 01-1A-13, Naval Air Systems Command, Department of the Navy, Washington, DC.

[Hartmann 59]   E. C. HARTMANN and F. M. HOWELL. 1959 "Laboratory Fatigue Testing of Materials," *Metal Fatigue*, G. Sines and J. L. Waisman, eds., McGraw-Hill, New York, pp. 89–111.

[Hayden 65]   H. W. HAYDEN, W. G. MOFFATT and J. WULFF. 1965 *The Structure and Properties of Materials, Vol. III: Mechanical Behavior*, John Wiley, New York.

[Hertzberg 75]   R. W. HERTZBERG, J. A. MANSON and M. SKIBO. 1975 "Frequency Sensitivity of Fatigue Processes in Polymeric Solids," *Polymer Engineering and Science*, Vol. 15, No. 4, Apr. 1975, pp. 252–260.

[Heywood 62]   R. B. HEYWOOD. 1962 *Designing Against Fatigue of Metals*, Reinhold, New York.

[Herring 89]   S. D. HERRING. 1989 *From the Titanic to the Challenger: An Annotated Bibliography on Technological Failures of the Twentieth Century*, Garland Publishing, Inc., New York.

[Hill 83]   R. HILL. 1983 *The Mathematical Theory of Plasticity*, Oxford University Press, London, UK.

[Hirthe 63]   W. M. HIRTHE and J. O. BRITTAIN. 1963 "High Temperature Steady-State Creep in Rutile," *Jnl. of the Am. Ceramic Soc.*, Vol. 46, No. 9, pp. 411–417.

[Hock 26]   L. HOCK and S. BOSTROM. 1926 "Beitrage zur Thermodynamic des Joule-Effektes am Rohkautschuk," *Kautschuk*, June 1926, pp. 130–136.

[Horger 65]   O. J. HORGER, ed. 1965 *ASME Handbook: Metals Engineering Design*, 2nd ed., McGraw-Hill, New York.

[Howell 55]   F. M. HOWELL and J. L. MILLER. 1955 "Axial Stress Fatigue Strengths of Several Structural Aluminum Alloys," Proc. of the Am. Soc. for Testing and Materials, Vol. 55, pp. 955–968.

[Hudson 69]   C. M. HUDSON. 1969 "Effect of Stress Ratio on Fatigue Crack Growth in 7075-T6 and 2024-T3 Aluminum Alloy Specimens," NASA TN D-5390, National Aeronautics and Space Administration, Langley Research Center, Hampton, Va.

[Hunter 54]   M. S. HUNTER and W. G. FRICKE, JR. 1954 "Metallographic Aspects of Fatigue Behavior of Aluminum," *Proc. of th Am. Soc. for Testing and Materials*, Vol. 54, pp. 717–732.

[Hunter 56]   M. S. HUNTER and W. G. FRICKE, Jr. 1956 "Fatigue Crack Propagation in Aluminum Alloys," *Proc. of the Am. Soc. for Testing and Materials*, Vol. 56, pp. 1038–1046.

[Imhof 73]   E. J. IMHOF and J. M. BARSOM. 1973 "Fatigue and Corrosion-Fatigue Crack Growth of 4340 Steel at Various Yield Strengths," *Progress in Flaw Growth and Fracture Toughness Testing*, ASTM STP 536, Am. Soc. for Testing and Materials, Philadelphia, Pa., pp. 182–205.

[Jaeger 69]   J. C. JAEGER. 1969 *Elasticity, Fracture, and Flow with Engineering and Geological Applications*, 3rd ed., Methuen, London, UK.

[Jones 70]   M. H. JONES and W. F. BROWN, JR. 1970 "The Influence of Crack Length and Thickness in Plane Strain Fracture Toughness Tests," *Review of Developments in Plane Strain Fracture Toughness Testing*, W. F. Brown, Jr., ed., ASTM STP 463, Am. Soc. for Testing and Materials, Philadelphia, Pa., pp. 160–190.

[Juran 88]   R. JURAN, et al., eds. 1988 *Modern Plastics Encyclopedia*, Published annually by McGraw-Hill, New York.

[Juvinall 67]   R. C. JUVINALL. 1967 *Stress, Strain and Strength*, McGraw-Hill, New York.

[Kaelble 65]   D. H. KAELBLE. 1965 "Micromechanisms and Phenomenology of Damping in Polymers," *Internal Friction, Damping, and Cyclic Plasticity*, ASTM STP 378, Am. Soc. for Testing and Materials, Philadelphia, Pa.

[Kelly 86]   A. KELLY and N. H. MACMILLAN. 1986 *Strong Solids*, 3rd. ed., Clarendon Press, Oxford, UK.

[Kelly 89]   A. Kelly, ed. 1989 *Concise Encyclopedia of Composite Materials*, Pergamon Press, Oxford, UK.

[Kim 81]   K. KIM and A. MUBEEN. 1981 "Relationship Between Differential Stress Intensity Factor and Crack Growth Rate in Cyclic Tension in Westerly Granite," *Fracture Mechanics Methods for Ceramics, Rocks, and Concrete*, ASTM STP 745, Am. Soc. for Testing and Materials, Philadelphia, Pa., pp. 157–168.

[Klesnil 80]   M. KLESNIL and P. LUKAS. 1980 *Fatigue of Materials*, Czechoslovak Academy of Sciences, Prague, and Elsevier Scientific, Amsterdam.

[Knott 79]   J. F. KNOTT. 1979 "An Introduction to Fracture Mechanics, Part III," *The Welder*, Vol. 41, No. 202, pp. 6–9.

[Kuhn 52]   P. KUHN and H. F. HARDRATH. 1952 "An Engineering Method for Estimating Notch Size Effect in Fatigue Tests on Steel," NACA TN 2805, National Advisory Committee for Aeronautics, Washington, DC.

[Landgraf 68]   R. W. LANDGRAF. 1968 "Cyclic Deformation and Fatigue Behavior of Hardened Steels," T & AM Report No. 320, Dept. of Theoretical and Applied Mechanics, University of Illinois, Urbana, Ill.

[Landgraf 69]   R. W. LANDGRAF, J. MORROW and T. ENDO. 1969 "Determination of the Cyclic Stress-Strain Curve," *Journal of Materials*, ASTM, Vol. 4, No. 1, Mar. 1969, pp. 176–188.

[Landgraf 70]   R. W. LANDGRAF. 1970 "The Resistance of Metals to Cyclic Deformation," *Achievement of High Fatigue Resistance in Metals and Alloys*, ASTM STP 467, Am. Soc. for Testing and Materials, Philadelphia, Pa., pp. 3–36.

[Landgraf 80]   R. W. LANDGRAF and F. A. CONLE. 1980 "The Development and Use of Mechanical Properties in Industry," *Use of Computers in Managing Material Property Data, MPC-14*, J. A. Graham, ed., Am. Soc. of Mechanical Engineers, New York, pp. 1–12.

[Landgraf 88]   R. W. LANDGRAF and R. A. CHERNENKOFF. 1988 "Residual Stress Effects on Fatigue of Surface Processed Steels," *Analytical and Experimental Methods for Residual Stress Effects in Fatigue*, ASTM STP 1004, R. L. Champoux et al., eds., Am. Soc. for Testing and Materials, Philadelphia, Pa., pp. 1–12.

[Larson 52]   F. R. LARSON and J. MILLER. 1952 "A Time-Temperature Relationship for Rupture and Creep Stresses," *Trans. of the Am. Soc. of Mechanical Engineers*, Vol. 74, pp. 765–771.

[Lazan 68]   B. J. LAZAN. 1968 *Damping of Materials and Members in Structural Mechanics*, Pergamon Press, Oxford, UK.

[Leese 85]   G. E. LEESE and J. MORROW. 1985 "Low Cycle Fatigue Properties of a 1045 Steel in Torsion," *Multiaxial Fatigue*, K. J. Miller and M. W. Brown, eds., ASTM STP 853, Am. Soc. for Testing and Materials, Philadelphia, Pa., pp. 482–496.

[Lessells 40]   J. M. LESSELLS and C. W. MACGREGOR. 1940 "Combined Stress Experiments on a Nickel-Chrome-Molybdenum Steel," *Jnl. of the Franklin Institute*, Vol. 230, Aug. 1940, pp. 163–181.

[Logsdon 76]   W. A. LOGSDON. 1976 "Elastic Plastic (J_{Ic}) Fracture Toughness Values: Their Experimental Determination and Comparison with Conventional Linear Elastic (K_{Ic}) Fracture Toughness Values for Five Materials," *Mechanics of Crack Growth*, ASTM STP 590, Am. Soc. for Testing and Materials, Philadelphia, Pa., pp. 43–60.

[Lubahn 61]   J. D. LUBAHN and R. P. FELGAR. 1961 *Plasticity and Creep of Metals*, John Wiley, New York.

[Luken 87]   R. C. LUKEN, JR. 1987 "Fracture Behavior of CPM 10V," MS Thesis, Dept. of Materials Engineering, Virginia Polytechnic Institute and State University, Blacksburg, Va.

[Lyle 74]   A. K. LYLE. 1974 "Glass Composition Design and Development," *Handbook of Glass Manufacture*, Vol. 1, F. V. Tooley, ed., Books for Industry, Inc., New York, pp. 3–17.

[MacGregor 52]   C. W. MacGregor and N. Grossman. 1952 "Effects of Cyclic Loading on Mechanical Behavior of 24S-T4 and 75S-T6 Aluminum Alloys and SAE 4130 Steel," NACA TN 2812, National Advisory Committee for Aeronautics, Washington, DC.

[Madeyski 78]   A. Madeyski and L. Albertin. 1978 "Fractographic Methods of Evaluation of the Cyclic Stress Amplitude in Fatigue Failure Analysis," *Fractography in Failure Analysis*, B. M. Strauss and W. H. Cullen, Jr., eds., ASTM STP 645, Am. Soc. for Testing and Materials, Philadelphia, Pa., pp. 73–83.

[Marandet 77]   B. Marandet and G. Sanz. 1977 "Evaluation of the Toughness of Thick Medium-Strength Steels …" *Flaw Growth and Fracture*, ASTM STP 631, Am. Soc. for Testing and Materials, Philadelphia, Pa., pp. 72–95.

[Marin 40]   J. Marin and R. L. Stanley. 1940 "Failure of Aluminum Subjected to Combined Stresses," *Welding Journal; Welding Research Supplement*, Feb. 1940, pp. 74s–80s.

[Mattson 59]   R. L. Mattson and J. G. Roberts. 1959 "Effect of Residual Stresses Induced by Strain Peening upon Fatigue Strength," G. M. Rassweiler and W. L. Grube, eds., *Internal Stresses and Fatigue in Metals*, Elsevier, Amsterdam, pp. 337–360.

[MILHDBK 83]   MILHDBK. 1983 *Military Standardization Handbook: Metallic Materials and Elements for Aerospace Vehicle Structures*, MIL-HDBK-5D, 2 Vols., U.S. Dept. of Defense and Federal Aviation Administration, Naval Publications and Forms Ctr., Philadelphia, Pa.

[Munz 81]   D. Munz, G. Himsolt and J. Eschweiler. 1981 "Effect of Stable Crack Growth on Fracture Toughness Determination for Hot-Pressed Silicon Nitride at Elevated Temperatures," *Fracture Mechanics for Ceramics, Rocks, and Concrete*, S. W. Freiman and E. R. Fuller, Jr., eds., ASTM STP 745, Am. Soc. for Testing and Materials, Philadelphia, Pa., pp. 69–84.

[Muvdi 57]   B. B. Muvdi, G. Sachs and E. P. Klier. 1957 "Axial Load Fatigue Properties of High Strength Steels," *Proc. Am. Soc. for Testing and Materials*, Vol. 57, pp. 655–666.

[Nadai 41]   A. Nadai and M. J. Manjoine. 1941 "High-Speed Tension Tests at Elevated Temperatures," *Jnl. of Applied Mechanics, Trans. ASME*, Vol. 8, No. 2, June 1941, pp. A-77 to A-91.

[Naghdi 58]   P. M. Naghdi, F. Essenburg and W. Koff. 1958 "An Experimental Study of Initial and Subsequent Yield Surfaces in Plasticity," *Jnl. of Applied Mechanics*, ASME, Vol. 25, June 1958, pp. 201–213.

[Newman 86]   J. C. Newman and I. S. Raju. 1986 "Stress-Intensity Factor Equations for Cracks in Three-Dimensional Finite Bodies Subjected to Tension and Bending Loads," *Computational Methods in the Mechanics of Fracture*, S. N. Atluri, ed., Elsevier North Holland, New York.

[Nisitani 81]   H. Nisitani and K. Takao. 1981 "Significance of Initiation, Propagation, and Closure of Microcracks in High Cycle Fatigue of Ductile Metals," *Engineering Fracture Mechanics*, Vol. 15, No. 3-4, pp. 445–456.

[Nisitani 84]   H. Nisitani and N. Noda. 1984 "Stress Concentration of a Cylindrical Bar with a V-Shaped Circumferential Groove under Torsion, Tension, or Bending," *Engineering Fracture Mechanics*, Vol. 20, No. 5/6, pp. 743–766.

[Nisitani 85]   H. Nisitani and H. Hyakutake. 1985 "Condition for Determining the Static Yield and Fracture of a Polycarbonate Plate Specimen with Notches," *Engineering Fracture Mechanics*, Vol. 22, No. 3, pp. 359–368.

[Novak 69]   S. R. Novak and S. T. Rolfe. 1969 "Modified WOL Specimen for K_{Iscc} Environmental Testing," *Journal of Materials*, ASTM, Vol. 4, No. 3, Sept. 1969, pp. 701–728.

[NRC 89]   NRC. 1989 *Materials Science and Engineering for the 1990's: Maintaining Competitiveness in the Age of Materials*, Report of the National Research Council, Committee on Materials Science and Engineering; National Academy Press, Washington, DC.

[NTSB 89]   NTSB. 1989 "Aircraft Accident Report—Aloha Airlines Flight 243, Boeing 737-200, N73711, Near Maui, Hawaii, April 28, 1988," Report No. NTSB/AAR-89/03, National Transportation Safety Board, Washington, DC.

[Orange 67]   T. W. Orange. 1967 "Fracture Toughness of Wide 2014-T6 Aluminum Sheet at $-320°$F," NASA TN D-4017, National Aeronautics and Space Administration, Washington, DC.

[Orr 54]	R. L. ORR, O. D. SHERBY and J. E. DORN. 1954 "Correlations of Rupture Data for Metals at Elevated Temperature," *Trans. of the Am. Soc. for Metals*, Vol. 46, pp. 113–128.
[Paris 64]	P. C. PARIS. 1964 "The Fracture Mechanics Approach to Fatigue," *Fatigue—An Interdisciplinary Approach*, J. J. BURKE, et al., eds., Proc. of the 10th Sagamore Army Materials Research Conf., Syracuse University Press, Syracuse, N.Y., pp. 107-127.
[Paris 72]	P. C. PARIS, et al. 1972 "Extensive Study of Low Fatigue Crack Growth Rates in A533 and A508 Steels," *Stress Analysis and Growth of Cracks, Proceedings of the 1971 National Symposium on Fracture Mechanics, Part I*, ASTM STP 513, Am. Soc. for Testing and Materials, Philadelphia, Pa., pp. 141–176.
[Pearson 86]	H. S. PEARSON and R. G. DOOMAN. 1986 "Fracture Analysis of Propane Tank Explosion," *Case Histories Involving Fatigue and Fracture Mechanics*, C. M. Hudson and T. P. Rich, eds., ASTM STP 918, Am. Soc. for Testing and Materials, Philadelphia, Pa., pp. 65–77.
[Peterson 59a]	R. E. PETERSON. 1959 "Analytic Approach to Stress Concentration Effect in Fatigue of Aircraft Materials," *Proc. of the Symp. on Fatigue of Aircraft Structures*, WADC TR-59-507, Wright Air Development Ctr., Oh., pp. 273–299.
[Peterson 59b]	R. E. PETERSON. 1959 "Notch-Sensitivity," *Metal Fatigue*, G. Sines and J. L. Waisman, eds., McGraw-Hill, New York, pp. 293–306.
[Peterson 74]	R. E. PETERSON. 1974 *Stress Concentration Factors*, John Wiley, New York.
[Raghava 73]	R. RAGHAVA, R. M. CADDELL and G. S. Y. YEH. 1973 "The Macroscopic Yield Behavior of Polymers," *Jnl. of Materials Science*, Chapman & Hall Ltd., London, Vol. 8, pp. 225–232.
[Rahka 86]	K. RAHKA. 1986 "Use of Material Characterization to Complement Fracture Mechanics in the Analysis of Two Pressure Vessels for Further Service in a Hydrogenating High-Temperature Process," *Case Histories Involving Fatigue and Fracture Mechanics*, C. M. Hudson and T. P. Rich, eds., ASTM STP 918, Am. Soc. for Testing and Materials, Philadelphia, Pa., pp. 3–30.
[Randall 57]	P. N. RANDALL. 1957 "Constant-Stress Creep Rupture Tests of a Killed Carbon Steel," *Proc. of the Am. Soc. for Testing and Materials*, Vol. 57, pp. 854–876.
[Raske 72]	D. T. RASKE. 1972 "Section and Notch Size Effects in Fatigue," PhD Thesis, Dept. of Theoretical and Applied Mechanics, University of Illinois, Urbana, Ill.
[Reed 83]	R. P. REED, J. H. SMITH and B. W. CHRIST. 1983 "The Economic Effects of Fracture in the United States: Part 1," Special Pub. No. 647-1, U. S. Dept. of Commerce, National Bureau of Standards, U.S. Govt. Printing Office, Washington, DC.
[Richards 61]	C. W. RICHARDS. 1961 *Engineering Materials Science*, Wadsworth, Belmont, Cal.
[Richards 70]	F. D. Richards and R. M. Wetzel. 1970 "Mechanical Testing of Materials Using an Analog Computer," Tech. Report No. SR 70-126, Scientific Research Staff, Ford Motor Co., Dearborn, Mich., Sept. 1970. See also *Materials Research and Standards*, ASTM, Vol. 11, No. 2, pp. 19–22, 51.
[Riddell 74]	M. N. RIDDELL. 1974 "A Guide to Better Testing of Plastics," *Plastics Engineering*, Vol. 30, No. 4, Apr. 1974, pp. 71–78.
[Ritchie 77]	R. O. RITCHIE. 1977 "Near-Threshold Fatigue Crack Propagation in Ultra-High Strength Steel," *Jnl. of Engineering Materials and Technology*, ASME, Vol. 99, Jul. 1977, pp. 195–204.
[Rosenberger 68]	P. C. ROSENBERGER. 1968 "Fatigue Behavior of Smooth and Notched Specimens of Man-Ten Steel," M.S. Thesis, Dept. of Theoretical and Applied Mechanics, University of Illinois, Urbana, Ill.
[Rusch 60]	H. RUSCH. 1960 "Researches Toward a General Flexural Theory for Structural Concrete," *Proc. of the Am. Concrete Institute*, Vol. 57, pp. 1–26.
[SAE 89]	SAE. 1989 "Technical Report on Low Cycle Fatigue Properties of Wrought Materials," SAE J1099, Information Report, Society of Automotive Engineers, Warrendale, Pa. See also L. E. Tucker, R. W. Landgraf and W. R. Brose, "Proposed Technical Report on Fatigue Properties for the SAE Handbook," SAE Paper No. 740279, Automotive Engineering Congress, Detroit, Mich., 1974.
[Schijve 62]	J. SCHIJVE and D. BROEK. 1962 "Crack Propagation Tests Based on a Gust Spectrum With Variable Amplitude Loading," *Aircraft Engineering*, Vol. 34, pp. 314–316.

[Schwartz 84]   M. M. SCHWARTZ. 1984 *Composite Materials Handbook*, McGraw-Hill, New York.

[Sheffler 72]   K. D. SHEFFLER and G. S. DOBLE. 1972 "Influence of Creep Damage on the Low Cycle Thermal-Mechanical Fatigue Behavior of Two Tantalum-Base Alloys," Report No. NASA CR 121001, TRW ER-7592, National Aeronautics and Space Administration, Lewis Research Ctr., Cleveland, Oh.

[Sherby 57]   O. D. SHERBY, J. L. LYTTON and J. E. DORN. 1957 "Activation Energies for Creep of High-Purity Aluminum," *Acta Metallurgica*, Vol. 5, pp. 219–227.

[Sims 78]   C. T. SIMS. 1957 "High-Temperature Alloys in High-Technology Systems," *High Temperature Alloys for Gas Turbines*, C. Coutsouradis et al., eds., Applied Science Pubs., London, UK, pp. 13–65.

[Sinclair 52]   G. M. SINCLAIR and W. J. CRAIG. 1952 "Influence of Grain Size on Work Hardening and Fatigue Characteristics of Alpha Brass," *Trans. of the Am. Soc. for Metals*, Vol. 44, pp. 929–948.

[Sinclair 53]   G. M. SINCLAIR and T. J. DOLAN. 1953 "Effect of Stress Amplitude on Statistical Variability in Fatigue Life of 75S-T6 Aluminum Alloy," *Trans. of the Am. Soc. of Mechanical Engineers*, Jul. 1953, pp. 867–872.

[Sines 59]   G. SINES. 1959 "Behavior of Metals under Complex Static and Alternating Stresses," *Metal Fatigue*, G. Sines and J. L. Waisman, eds., McGraw-Hill, New York, pp. 145–169.

[Sines 75]   G. SINES and M. ADAMS. 1975 "The Use of Brittle Materials as Compressive Structural Elements," Paper No. 75-DE-23, *Am. Soc. of Mechanical Engineers*.

[Socie 87]   D. SOCIE. 1987 "Multiaxial Fatigue Damage Models," *Jnl. of Engineering Materials and Technology*, ASME, Vol. 109, Oct. 1987, pp. 293–298.

[Srawley 76]   J. E. SRAWLEY. 1976 "Wide Range Stress Intensity Factor Expressions for ASTM E399 Standard Fracture Toughness Specimens," *Int. Jnl. of Fracture Mechanics*, Vol. 12, Jun. 1976, pp. 475–476.

[Stadnick 72]   S. J. STADNICK and J. MORROW. 1972 "Techniques for Smooth Specimen Simulation of the Fatigue Behavior of Notched Members," *Testing for Prediction of Material Performance in Structures and Components*, ASTM STP 515, Am. Soc. for Testing and Materials, Philadelphia, Pa., pp. 229–252.

[Steigerwald 70]   E. A. STEIGERWALD. 1970 "Crack Toughness Measurements of High-Strength Steels," *Review of Developments in Plane Strain Fracture Toughness Testing*, W. F. Brown, Jr., ed., ASTM STP 463, Am. Soc. for Testing and Materials, Philadelphia, Pa., pp. 102–123.

[Stonesifer 76]   F. R. STONESIFER. 1976 "Effects of Grain Size and Temperature on Sub-Critical Crack Growth in A533 Steel," NRL Memorandum Report 3400, Naval Research Laboratory, Washington, DC.

[Stubbington 61]   C. A. STUBBINGTON and P. J. E. FORSYTH. 1961 "Some Corrosion-Fatigue Observations on a High-Purity Aluminum-Zinc-Magnesium Alloy and Commercial D. T. D. 683 Alloy," *Jnl. of the Inst. of Metals*, London, UK, Vol. 90, 1961–62, pp. 347–354.

[Swift 71]   T. SWIFT. 1971 "Development of the Fail-safe Design Features of the DC-10," *Damage Tolerance in Aircraft Structures*, ASTM STP 486, Am. Soc. for Testing and Materials, Philadelphia, Pa., pp. 164–214.

[Tada 85]   H. TADA, P. C. PARIS and G. R. IRWIN. 1985 *The Stress Analysis of Cracks Handbook*, 2nd. ed., Paris Productions Inc., 226 Woodbourne Dr., St. Louis, Mo. 63105.

[Thomas 87]   D. W. THOMAS. 1987 "Vehicle Modeling and Service Loads Analysis," *Durability by Design*, SAE Pub. No. SP-730, Soc. of Automotive Engineers, Warrendale, Pa., Paper No. 871940.

[Timoshenko 83]   S. P. TIMOSHENKO. 1983 *History of Strength of Materials*, McGraw-Hill, New York.

[Tobler 78]   R. L. TOBLER and R. P. REED. 1978 "Fatigue Crack Growth Resistance of Structural Alloys at Cryogenic Temperatures," K. D. Timmerhaus, et al., eds., *Advances in Cryogenic Engineering*, Vol. 24, Plenum Press, New York, pp. 82–90.

[Tobolsky 65]   A. V. TOBOLSKY. 1965 "Some Viewpoints on Polymer Physics," *Structure and Properties of Polymers*, Jnl. of Polymers Sciences, Part C, Polymer Symposia, No. 9, John Wiley, New York, pp. 157–191.

[Topper 70]   T. H. Topper and B. I. Sandor. 1970 "Effects of Mean Stress and Prestrain on Fatigue Damage Summation," *Effects of Environment and Complex Load History on Fatigue Life*, ASTM STP 462, Am. Soc. for Testing and Materials, Philadelphia, Pa., pp. 93–104.

[van Echo 58]   J. A. van Echo. 1958 "Short-Time Creep of Structural Sheet Metals," *Short-Time High Temperature Testing*, A. H. Levy, ed., Am. Soc. for Metals, Metals Park, Oh., pp. 58–91.

[Van Vlack 89]   L. H. Van Vlack. 1989 *Elements of Materials Science and Engineering*, 6th. ed., Addison-Wesley, Reading, Ma.

[Venkateswaran 88] A. Venkateswaran, K. Y. Donaldson, and D. P. H. Hasselman. 1988 "Role of Intergranular Damage-Induced Decrease in Young's Modulus in the Nonlinear Deformation and Fracture of an Alumina at Elevated Temperatures," *Jnl. of the Am. Ceramic Society*, Vol. 71, No. 7, Jul. 1988, pp. 565–576.

[Voss 88]   H. Voss and J. Karger-Kocsis. 1988 "Fatigue Crack Propagation in Glass-Fibre and Glass-Sphere Filled PBT Composites," *Int. Jnl. of Fatigue*, Vol. 10, No. 1, Jan. 1988, pp. 3–11.

[Waisman 59]   J. L. Waisman. 1959 "Factors Affecting Fatigue Strength," *Metal Fatigue*, G. Sines and J. L. Waisman, eds., McGraw-Hill, New York, pp. 7–35.

[Warren 38]   B. E. Warren and J. Biscoe. 1938 "Fourier Analysis of X-ray Patterns of Soda-silica Glass," *Jnl. of the Am. Ceramic Soc.*, Vol. 21, pp. 259–265.

[Wei 65]   R. P. Wei. 1965 "Fracture Toughness Testing in Alloy Development," *Fracture Toughness Testing and its Applications*, ASTM STP 381, Am. Soc. for Testing and Materials, Philadelphia, Pa., pp. 279–289.

[Wetzel 68]   R. M. Wetzel. 1968 "Smooth Specimen Simulation of Fatigue Behavior of Notches," *Journal of Materials*, ASTM, Vol. 3, No. 3, Sept. 1968, pp. 646–657.

[Wetzel 77]   R. M. Wetzel, ed. 1977 *Fatigue Under Complex Loading: Analyses and Experiments*, SAE Pub. No. AE-6, Soc. of Automotive Engineers, Warrendale, Pa.

[Whyte 75]   R. R. Whyte, ed. 1975 *Engineering Progress Through Trouble*, The Institution of Mechanical Engineers, London, UK.

[Wiederhorn 77]   S. M. Wiederhorn. 1977 "Dependence of Lifetime Predictions on the Form of the Crack Propagation Equation," *Fracture 1977, Vol. 3; Proc. of the 4th Int. Conf. on Fracture*, D. M. R. Taplin, ed., University of Waterloo Press, Waterloo, Ontario, Canada, pp. 893–901.

[Williams 87]   J. G. Williams. 1987 "Fracture Mechanics of Polymers and Adhesives," *Fracture of Non-Metallic Materials*, K. P. Herrmann and L. H. Larsson, eds., Kluwer Academic Pubs., Dordrecht, The Netherlands.

[Wood 70]   H. A. Wood et al., eds. 1970 "Proceedings of the Air Force Conference on Fatigue and Fracture of Aircraft Structures and Materials," AFFDL TR 70-144, Air Force Flight Dynamics Laboratory, Wright-Patterson AFB, Oh.

[Zachariasen 32]   W. H. Zachariasen. 1932 "The Atomic Arrangement in Glass," *Jnl. of the Am. Chemical Soc.*, Vol. 54, pp. 3841–3851.

[Zwikker 54]   C. Zwikker. 1954 *Physical Properties of Solid Materials*, Pergamon Press, London, UK.

Index

FREQUENTLY USED SYMBOLS

Roman Letters

a	Crack length	m	Fatigue limit reduction factor; fatigue crack growth exponent; creep exponent
A	Cross-sectional area; creep coefficients		
b	Maximum possible crack length; stress-life exponent; thickness where t is time	M	Bending moment
		n	Strain hardening exponent
B	Bulk modulus; Bridgman correction factor; number of repetitions of a cyclic loading sequence	N	Number of cycles
		p	Pressure
		P	Force (load)
c	Half-depth of beam; radius of shaft; plastic strain versus life exponent; dashpot constant	q	Notch sensitivity
		Q	Activation energy
C	Fatigue crack growth coefficient	Q^{-1}	Loss coefficient in damping
d	Diameter; damping exponent	r	Radius
e	Base of natural logarithms ($e = 2.718\ldots$)	R	Stress ratio for cyclic loading (minimum ÷ maximum); universal gas constant
E	Modulus of elasticity		
F	Finite width factor of fracture mechanics	S	Nominal (average) stress
G	Shear modulus; strain energy release rate	t	Time; thickness
h	Height; half-height of cracked bodies	T	Temperature; torque
H	Strength coefficient for stress-strain curves	u	Energy per unit volume
I	Area moment of inertia about an in-plane axis	U	Energy
J	Polar moment of inertia; J-integral; damping coefficient	v	Displacement
		V	Volume
k	Stress concentration factor; spring constant	w	Width
K	Stress intensity factor of fracture mechanics	x, y, z	Spatial coordinates
L	Length	X	Safety factor

Greek Letters

α	Coefficient of thermal expansion; angle; relative crack length ($\alpha = a/b$); Peterson constant	θ	Angle
		λ	Stress biaxiality ratio σ_2/σ_1
		ν	Poisson's ratio
β	Neuber constant	ρ	Notch-tip radius
γ	Shear strain; Walker exponent	σ	Normal stress at a point, or in a uniformly stressed member
δ	Slope reduction factor; crack-tip opening displacement; phase angle		
		τ	Shear stress
ε	Normal strain	ω	Angular velocity
η	Tensile viscosity		